ESSENTIALS OF

Biochemistry

ESSENTIALS OF
Biochemistry

Dr. U. Satyanarayana

M.Sc., Ph.D., F.I.C., F.A.C.B., F.A.C.B.I

Professor of Biochemistry & Director (Research)
Dr. Pinnamaneni Siddhartha Institute of Medical Sciences
(Dr. NTR University of Health Sciences)
Chinaoutpalli, Gannavaram (Mdl)
Krishna (Dist), A.P., India

Dr. U. Chakrapani

M.B.B.S., M.S., D.N.B.

ELSEVIER

Co-published with

Since 1960 BOOKS AND ALLIED

BOOKS & ALLIED Pvt. Ltd.

Essentials of Biochemistry, 3e
Satyanarayana and Chakrapani

ELSEVIER

RELX India Pvt. Ltd.

Published by RELX India Pvt. Ltd., under special arrangement with Books & Allied (P) Ltd.

First Published : 2002
Reprinted : 2003, 2004, 2005
Second Revised Edition (multicolour) : 2007
Reprinted : 2008, 2009, 2010
Revised Reprint : 2012
Reprinted : 2013, 2014, 2015, 2016, 2017, 2018
Third Edition : 2019
Third Revised & Updated Edition: 2021

ISBN: 978-81-312-6497-3

e-Book ISBN: 978-81-312-6498-0

Medical knowledge is constantly changing. As new information becomes available, changes in treatment, procedures, equipment and the use of drugs become necessary. The author, editors, contributors and the publisher have, as far as it is possible, taken care to ensure that the information given in this text is accurate and up-to-date. However, readers are strongly advised to confirm that the information, especially with regard to drug dose/usage, complies with current legislation and standards of practice. *Please consult full prescribing information before issuing prescriptions for any product mentioned in this publication.*

This edition of **Essentials of Biochemistry**, **3e** by Dr. U. Satyanarayana and Dr. U. Chakrapani is co-published by an arrangement between RELX India Private Limited and Books & Allied (P) Ltd.

RELX India Pvt. Ltd.

Registered Office: 818, 8th Floor Indraprakash Building, 21, Barakhamba Road, New Delhi-110001.

Corporate Office: 14th Floor, Building No. 10B, DLF Cyber City, Phase II, Gurgaon-122002, Haryana, India.

BOOKS & ALLIED (P) Ltd.

Registered Office: No. 1-E(1) 'Shubham Plaza', 83/1 Beliaghata Main Road, Kolkata-700010, West Bengal, India.

Authors sponsored and supported by ***Uppala Author-Publisher Interlinks***, Vijayawada-520008

Typeset at Books & Allied (P) Ltd., Kolkata-700010

Printed and bound in India at Rajkamal Electric Press, Kundli, Haryana.

PREFACE

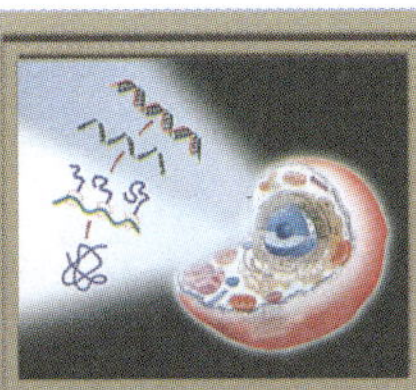

Preface to the Third Edition (Revised and updated)

This book "Essentials of Biochemistry" is now nineteen years old. The response to the previous editions is simply overwhelming. It gives me a great pleasure to present the Third edition (Revised and updated) of "Essentials of Biochemistry" with several unique/novel features, some high-lights are listed below

- A thorough revision and updating of each chapter with most recent advances
- An improved orientation and treatment of human biochemistry in health and disease with inputs from M.D. (Biochemistry) and M.D. (General Medicine) professors.
- An innovative approach to give medical concepts/clinical correlates, spread throughout each chapter (this certainly helps the students to understand the importance of biochemistry with special reference to health and disease)
- A wide variety of 40 case studies with relevant clinical and biochemical profiles (along with diagnosis and discussion)
- The third edition has been designed with a 3R motto – read, remember and reproduce

AN INVITATION TO READERS

I sincerely invite the readers (particularly students), experts on Biochemistry subject to feel free and write to me (Email ID : uppalasatya@yahoo.com) expressing frank opinions, critical comments and constructive suggestions. And this will help me to further improve the book in subsequent revisions.

Dr. U. SATYANARAYANA

Preface to the First Edition

Biochemistry is perhaps the most unique subject, dealing with the chemistry of life in any form, be it human, animal, plant or microorganism. This has made biochemistry a common core subject of study in several disciplines—medical, dental, pharmacy, physiotherapy, homeopathy, ayurveda, medical laboratory technology, nursing, veterinary, agriculture, biotechnology, biomedical engineering, home science and other branches of life sciences (botany, zoology, microbiology, genetics). Although the subject of biochemistry is rather vast, only basic concepts and principles (along with their applications) need to be learnt by the students of these disciplines.

This book 'Essentials of Biochemistry' is a redistillation (while the earlier book 'Biochemistry' represents distillation) of my knowledge encapsulating the essence of the subject in a nutshell. It will satisfy the curricular requirements of the beginners, learning biochemistry in various disciplines.

Many students learning biochemistry are afraid of complicated structures and reactions. This book will be a boon for such students, since most structures are deleted without compromising on the subject information, including the most recent advances.

This book, written in a simple language with colour illustrations, is designed to create interest and enthusiasm among the students to learn the subject to the maximum. The summary and biomedical/clinical concepts will help to develop clear concepts in biochemistry. The self-assessment exercises (essays, short notes, MCQs) will equip the students to boldly face the examinations.

Dr. U. SATYANARAYANA

ACKNOWLEDGEMENTS

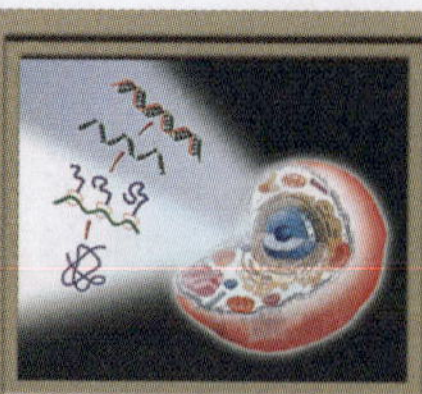

Acknowledgements to the Third Edition (Revised and updated)

I am grateful to (late) Mr Arunabha Sen (former Managing Director), Ms Ruba Sen, Managing Director and Mr Depayan Sen, Director, Books and Allied Pvt. Ltd. for their support and encouragement to bring out the third (revised and updated) edition of the book. I am indebted to Mr Shyamal Bhattacharya for his excellent page making and graphics work in the book. I am grateful to Mr Abhijit Ghosal for his help in cover design.

I thank Dr (Ms) U. B. Vijaya Lakshmi, MD, Professor of Biochemistry and Dr (Ms) P. Chudamani, MD, Assistant Professor of General Medicine, NRI Medical College, Guntur, AP., who have contributed to improve the book with regard to medical/clinically related Biochemistry. I acknowledge the contributions of Dr E. Suresh, MD, Assistant Professor of Biochemistry, Dr. PSIMS, Chinaoutpalli.

I express my gratitude to the officials of Elsevier (RELX India Pvt. Ltd.) for their constant support and encouragement – Mr Shanker Kaul, Managing Director; Mr Vidur Renjen, Director Education; Mr Sandeep Singh, Finance Manager; Ms Nimisha Goswami, Director Content Strategy – India/SEA; Ms Debyani Roy, Regional Manager IPR; Ms Shabina Nasim, Head of Content Project Management – APAC; Mr Arvind Kaul, Content Strategist; Mr Shubham Dixit, Content Project Manager; and Mr. Gopalakrishnan Venkatraman, Senior Graphic Artist.

I thank my family members (wife – Krishna Kumari, my younger son – Amrut Pani and my daughters-in-law – Hima Bindu and Oohasri; grand children – Maahe and Dev Sai) for their constant support. I am grateful to Uppala Author-Publisher Interlinks, Vijayawada for sponsoring and supporting me to bring out this edition.

Dr. U. SATYANARAYANA

Acknowledgements to the First Edition

I owe a deep debt of gratitute to my parents, the late Sri U. Venkata Subbaiah, and Smt. Vajramma for cultivating in me the habit of early rising. This book was written in the early morning hours (between 2 a.m.–6 a.m.), when the world around was fast asleep!

My thanks are due to my sons Dr. Chakrapani, MBBS and Amrutpani who made significant contribution in the preparation of this book. I also thank my wife, Krishna Kumari for her constant cooperation and support. I am indebted to Dr. C. Nageswara Rao and Dr. C. Sudha (Pinnamaneni Polyclinic, Vijayawada) for their encouragement in my academic pursuits.

I express my gratitude to Mr. Arunabha Sen, Director, Books & Allied (P) Ltd. Kolkata, for his wholehearted cooperation in bringing out this book. I thank Mr. Shyamal Bhattacharya for his wonderful page-making and graphic work. I am indebted to Mr. Abhijit Ghosal for the cover design of this book.

I express my gratefulness to Uppala Author—Publisher Interlinks, Vijayawada, for sponsoring and supporting me to write this book.

Dr. U. SATYANARAYANA

CONTENTS

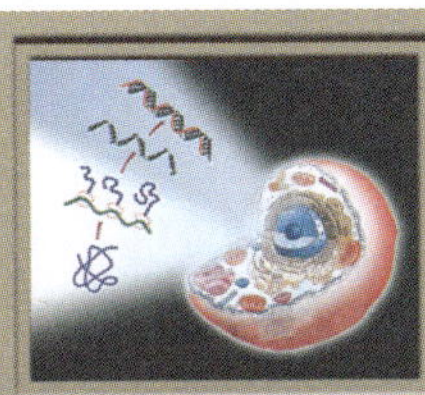

BIOMOLECULES AND THE CELL

1

The living matter is composed of mainly six elements—***carbon***, ***hydrogen***, ***oxygen***, ***nitrogen***, ***phosphorus*** and ***sulfur***. These elements together constitute about 90% of the dry weight of the human body. Several other functionally important elements are also found in the cells. These include Ca, K, Na, Cl, Mg, Fe, Cu, Co, I, Zn, F, Mo and Se.

Carbon—a unique element of life

Carbon is the most predominant and versatile element of life. It possesses a unique property to form infinite number of compounds. This is attributed to the ability of carbon to form stable covalent bonds and C—C chains of unlimited length. It is estimated that about ***90% of compounds*** found in living system invariably ***contain carbon***.

Chemical molecules of life

Life is composed of lifeless chemical molecules. A single cell of the bacterium, *Escherichia coli* contains about 6,000 different organic compounds. It is believed that man may contain about 100,000 different types of molecules, although only a few of them have been characterized.

Complex biomolecules

The organic compounds such as ***amino acids***, ***nucleotides*** and ***monosaccharides*** serve as the ***monomeric units*** or building blocks of complex biomolecules—***proteins***, ***nucleic acids*** (DNA and RNA) and ***polysaccharides***, respectively. The important biomolecules (macromolecules) with their respective building blocks and major functions are given in ***Table 1.1***. As regards ***lipids***, it may be noted that they are ***not biopolymers*** in a strict sense, but majority of them contain fatty acids.

Structural hierarchy of an organism

The ***macromolecules*** (proteins, lipids, nucleic acids and polysaccharides) form supramolecular assemblies (e.g. membranes) which in turn organize into organelles, cells, tissues, organs and finally the whole organism.

TABLE 1.1 The major complex biomolecules of cells

Biomolecule	*Building block (repeating unit)*	*Major functions*
1. Protein	Amino acids	Fundamental basis of structure and function of cell (static and dynamic functions).
2. Deoxyribonucleic acid (DNA)	Deoxyribonucleotides	Repository of hereditary information.
3. Ribonucleic acid (RNA)	Ribonucleotides	Essentially required for protein biosynthesis.
4. Polysaccharide (glycogen)	Monosaccharides (glucose)	Storage form of energy to meet short term demands.
5. Lipid	Fatty acids, glycerol	Storage form of energy to meet long term demands; structural components of membranes.

Table 1.2 Chemical composition of a normal man (weight 65 kg)

Constituent	Percent (%)	Weight (kg)
Water	61.6	40
Protein	17.0	11
Lipid	13.8	9
Carbohydrate	1.5	1
Minerals	6.1	4

Chemical composition of man

The chemical composition of a normal man, weighing 65 kg, is given in ***Table 1.2***. Water is the solvent of life and contributes to more than 60% of the weight. This is followed by protein (mostly in muscle) and lipid (mostly in adipose tissue). The carbohydrate content is rather low which is in the form of glycogen.

THE CELL

The cell is the structural and functional unit of life. It may be also regarded as the ***basic unit of biological activity.***

The concept of cell originated from the contributions of Schleiden and Schwann (1838). However, it was only after 1940, the complexities of cell structure were exposed.

Prokaryotic and eukaryotic cells

The cells of the living kingdom may be divided into two categories

1. **Prokaryotes** (*Greek* : pro – before; karyon – nucleus) ***lack a well defined nucleus*** and possess relatively simple structure. These include the various bacteria.

2. **Eukaryotes** (*Greek* : eu – true; karyon – nucleus) possess a ***well defined nucleus*** and are more complex in their structure and function. The higher organisms (animals and plants) are composed of eukaryotic cells.

A comparison of the characteristics between prokaryotes and eukaryotes is listed in ***Table 1.3***.

EUKARYOTIC CELL

The human body is composed of about 10^{14} cells. There are about 250 types of specialized cells in the human body e.g. erythrocytes, nerve cells, β-cells of pancreas. An eukaryotic cell is generally 10 to 100 μm in diameter. A diagrammatic representation of a typical rat liver cell is depicted in ***Fig.1.1***. The plant cell differs from an animal cell by possessing a rigid cell wall (mostly composed of cellulose) and ***chloroplasts***. The latter are the sites of ***photosynthesis***.

The cell consists of well defined subcellular organelles, enveloped by a plasma membrane. By differential centrifugation of tissue homogenate, it is possible to isolate each cellular organelle in a relatively pure form (***Refer Chapter 41***). The distribution of major enzymes and metabolic pathways in different cellular organelles is given in the chapter on enzymes (***Refer Table 6.4***). The subcellular organelles are briefly described in the following pages.

Nucleus

Nucleus is the ***largest cellular organelle***, surrounded by a double membrane nuclear envelope. The outer membrane is continuous with the membranes of endoplasmic reticulum. At certain intervals, the two nuclear membranes have nuclear pores with a diameter of about 90 nm. These pores permit the free passage of the products synthesized in the nucleus into the surrounding cytoplasm.

Table 1.3 Comparison between prokaryotic and eukaryotic cells

Characteristic	Prokaryotic cell	Eukaryotic cell
1. Size	Small (generally 1-10 mm)	Large (generally 10-100 mm)
2. Cell membrane	Cell is enveloped by a rigid cell wall	Cell is enveloped by a flexible plasma membrane
3. Sub-cellular organelles	Absent	Distinct organelles are found (e.g. mitochondria, nucleus, lysosomes)
4. Nucleus	Not well defined; DNA is found as nucleoid, histones are absent	Nucleus is well defined, surrounded by a membrane; DNA is associated with histones
5. Cell division	Usually fission and no mitosis	Mitosis

Medical Concepts/ Clinical Correlates

- **Accumulation of lipofuscin, a pigment rich in lipids and proteins, in the cell has been implicated in ageing process.**
- **Leakage of lysosomal enzymes into the cell degrades several functional macro-molecules and this may lead to certain disorders (e.g. arthritis).**

Nucleus contains ***DNA***, the repository of genetic information. Eukaryotic DNA is associated with basic proteins (histones) in the ratio of 1 : 1, to form ***nucleosomes***. An assembly of nucleosomes constitutes ***chromatin*** fibres of chromosomes (*Greek*: chroma – colour; soma – body). Thus, a single human chromosome is composed of about a million nucleosomes. The number of chromosomes is a characteristic feature of the species. Humans have 46 chromosomes, compactly packed in the nucleus.

The nucleus of the eukaryotic cell contains a dense body known as ***nucleolus***. It is rich in RNA.

The ground material of the nucleus is often referred to as ***nucleoplasm***. It is rich in enzymes such as DNA polymerases and RNA polymerases.

Mitochondria

The mitochondria (*Greek*: mitos – thread; chondros – granule) are the centres for the cellular respiration and energy metabolism. They are regarded as the ***power houses of the cell*** with variable size and shape. Mitochondria are rod-like or filamentous bodies, usually with dimensions of 1.0-3 μm. About 2,000 mitochondria, occupying about $^1/_5$th of the total cell volume, are present in a typical cell.

The mitochondria are composed of a double membrane system (***Fig.11.4***). The outer membrane is smooth and completely envelops the organelle. The inner membrane is folded to form ***cristae*** (*Latin* – crests) which occupy a larger surface area. The internal chamber of mitochondria is referred to as ***matrix*** or ***mitosol.***

The components of ***electron transport chain*** and ***oxidative phosphorylation*** (flavoprotein, cytochromes b, c_1, c, a and a_3 and coupling factors) are buried in the ***inner mitochondrial membrane***. The matrix contains several enzymes concerned with the energy metabolism of carbohydrates, lipids and amino acids (e.g., citric acid cycle, β-oxidation). ATP, the energy currency, generated in mitochondria is exported to all parts of the cell to provide energy for cellular work.

The mitochondrial matrix contains a circular double stranded DNA (***mtDNA***), RNA and ribosomes. It is estimated that about 10% of the mitochondrial proteins are produced in the mitochondria.

The structure and functions of mitochondria closely resemble prokaryotic cells. It is hypothesized that mitochondria have evolved from aerobic bacteria. Further, it is believed that during evolution, the aerobic bacteria developed a symbiotic relationship with primordial anaerobic eukaryotic cells that ultimately led to the arrival of aerobic eukaryotes.

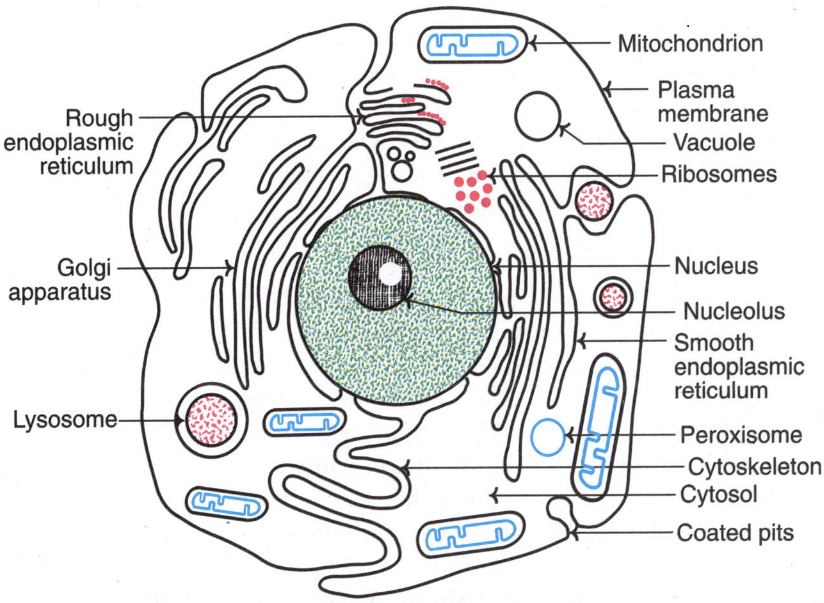

Fig.1.1 : Diagrammatic representation of a rat liver cell.

Endoplasmic reticulum

The network of membrane enclosed spaces that extends throughout the cytoplasm constitutes endoplasmic reticulum (ER). Some of these thread-like structures extend from the nuclear pores to the plasma membrane.

A large portion of the ER is studded with ribosomes to give a granular appearance which is referred to as ***rough endoplasmic reticulum***. Ribosomes are the factories of protein biosynthesis.

The smooth endoplasmic reticulum does not contain ribosomes. It is involved in the synthesis of lipids (triacylglycerols, phospholipids, sterols) and metabolism of drugs, besides supplying Ca^{2+} for the cellular functions.

Golgi apparatus

Eukaryotic cells contain a unique cluster of membrane vesicles known as ***dictyosomes*** which, in turn, constitute Golgi apparatus (or Golgi complex). The newly synthesized proteins are handed over to the Golgi apparatus which catalyse the addition of carbohydrates, lipids or sulfate moieties to the proteins. These chemical modifications are necessary for the transport of proteins across the plasma membrane.

Lysosomes

Lysosomes are spherical vesicles enveloped by a single membrane. Lysosomes are regarded as the digestive tract of the cell, since they are actively involved in digestion of cellular substances—namely proteins, lipids, carbohydrates and nucleic acids. Lysosomal enzymes are categorized as ***hydrolases***. These include the following enzymes (with substrate in brackets) — α-glucosidase (glycogen), cathepsins (proteins), lipases (lipids), ribonucleases (RNA).

The pH of the lysosomal matrix is more acidic (pH < 5) than the cytosol (pH~7) and this facilitates the degradation of different compounds. The lysosomal enzymes are responsible for maintaining the cellular compounds in a dynamic state, by their degradation and recycling. The degraded products leave the lysosomes, usually by diffusion, for reutilization by the cell. Sometimes, however, certain residual products, rich in lipids and proteins, collectively known as ***lipofuscin*** accumulate in the cell. Lipofuscin is the ***age pigment*** or wear and tear pigment which has been implicated in ageing process.

The digestive enzymes of cellular compounds are confined to the lysosomes in the best interest of the cell. Escape of these enzymes into cytosol will destroy the functional macromolecules of the cell and result in many complications. The occurrence of several diseases (e.g. arthritis, muscle diseases, allergic disorders) has been partly attributed to the release of lysosomal enzymes.

Peroxisomes

Peroxisomes, also known as ***microbodies***, are single membrane cellular organelles. They are spherical or oval in shape and contain the enzyme ***catalase***. Catalase protects the cell from the toxic effects of H_2O_2 by converting it to H_2O and O_2. Peroxisomes are also involved in the oxidation of long chain fatty acids ($> C_{18}$) and synthesis of plasmalogens and glycolipids. Plants contain ***glyoxysomes***, a specialized type of peroxisomes, involved in the glyoxylate pathway.

Cytosol and cytoskeleton

The ***cellular matrix*** is collectively referred to as cytosol. Cytosol is basically a compartment containing several enzymes, metabolites and salts in an aqueous gel like medium. More recent studies however, indicate that the cytoplasm actually contains a complex network of protein filaments, spread throughout, that constitutes cytoskeleton. The cytoplasmic filaments are of three types–microtubules, actin filaments and intermediate filaments. The filaments which are polymers of proteins are responsible for the structure, shape and organization of the cell.

INTEGRATION OF CELLULAR FUNCTIONS

The various intracellular processes and biochemical reactions are tightly controlled and integrated. Division of a cell into two daughter cells is good example of the orderly occurrence of an integrated series of cellular reactions.

Apoptosis is the ***programmed cell death*** or ***cell suicide***. This occurs when the cell has fulfilled its biological functions. Apoptosis may be regarded as a ***natural cell death*** and it differs from the cell death caused by injury due to radiation, anoxia etc. Programmed cell death is a highly regulated process.

CARBOHYDRATES

2

Carbohydrates are the most abundant organic molecules in nature. They are primarily composed of the elements ***carbon, hydrogen and oxygen***. The name carbohydrate literally means 'hydrates of carbon'.

Carbohydrates may be defined as polyhydroxyaldehydes or ketones or compounds which produce them on hydrolysis. The term 'sugar' is applied to carbohydrates soluble in water, and sweet to taste.

Functions of carbohydrates

Carbohydrates participate in a wide range of functions

1. They are the most abundant dietary source of energy (4 Cal/g) for all organisms.

2. Carbohydrates are ***precursors for*** many organic compounds (***fats, amino acids***).

3. Carbohydrates (as glycoproteins and glycolipids) participate in the structure of cell membrane (***glycocalyx***) and cellular functions such as cell growth, adhesion and fertilization.

4. Carbohydrates also serve as the storage form of energy (***glycogen***) to meet the immediate energy demands of the body.

The term ***glycome*** represents the entire ***spectrum of carbohydrates of an organism***, present in either free or complex form of molecules. ***Glycomics*** is the study of glycomes that includes physiological, pathological, genetic and various other aspects. Broadly, ***glycobiology*** is the comprehensive ***study of carbohydrates in health and disease***.

CLASSIFICATION OF CARBOHYDRATES

Carbohydrates are often referred to as saccharides (*Greek* : sakcharon–sugar). They are broadly classified into 3 groups—***monosaccharides, oligosaccharides and polysaccharides***. This categorization is based on the number of sugar units. Mono- and oligosaccharides are sweet to taste, crystalline in character and soluble in water, hence they are commonly known as ***sugars***.

Monosaccharides

Monosaccharides (*Greek* : mono-one) are the simplest group of carbohydrates and are often referred to as simple sugars. They have the general formula $C_n(H_2O)_n$, and they cannot be further hydrolysed.

Aldoses : When the functional group in monosaccharides is an ***aldehyde*** $\left(\overset{\displaystyle H}{\underset{|}{}}\!\!-\overset{|}{C}=O\right)$, they are known as aldoses e.g. glyceraldehyde, glucose.

Ketoses : When the functional group is a ***keto*** or ***carbonyl*** $\left(-\overset{|}{C}=O\right)$ group, they are referred to as ketoses e.g. dihydroxyacetone, fructose.

Based on the number of carbon atoms, the monosaccharides are regarded as ***trioses (3C), tetroses (4C), pentoses (5C), hexoses (6C) and heptoses (7C)***. These terms along with functional groups are used while naming monosaccharides. For instance, ***glucose is a aldohexose while fructose is a ketohexose***.

The common monosaccharides and disaccharides of biological importance are given in the ***Table 2.1***.

TABLE 2.1 Monosaccharides and disaccharides of biological importance

Monosaccharides	*Occurrence*	*Biochemical importance*
Trioses		
Glyceraldehyde	Found in cells as phosphate	Glyceraldehyde 3-phosphate is an intermediate in glycolysis
Dihydroxyacetone	Found in cells as phosphate	Its 1-phosphate is an intermediate in glycolysis
Tetroses		
D-Erythrose	Widespread	Its 4-phosphate is an intermediate in HMP pathway of carbohydrate metabolism
Pentoses		
D-Ribose	Widespread as a constituent of RNA and nucleotides	For the structure of RNA and nucleotide coenzymes (ATP, NAD^+, $NADP^+$)
D-Deoxyribose	As a constituent of DNA	For the structure of DNA
Hexoses		
D-Glucose	As a constituent of polysaccharides (starch, glycogen, cellulose) and disaccharides (lactose, sucrose)	The 'sugar fuel' of life; excreted in urine in diabetes. Structural unit of cellulose in plants
D-Galactose	As a constituent of lactose (milk sugar)	Converted to glucose, failure leads to galactosemia
D-Fructose	Fruits and honey, as a constituent of sucrose and inulin	Its phosphates are intermediates of glycolysis
Disaccharides	*Occurrence*	*Biochemical importance*
Sucrose	As a constituent of cane sugar and beet sugar, pineapple	Most commonly used table sugar supplying calories
Lactose	Milk sugar	Carbohydrate source to breast fed infants.
Maltose	Occurs in germinating seeds	Intermediate in the digestion of starch

Oligosaccharides

Oligosaccharides (*Greek* : oligo-few) contain 2-10 monosaccharide molecules which are liberated on hydrolysis. Based on the number of monosaccharide units present, the oligosaccharides are subdivided to ***disaccharides, trisaccharides*** etc.

Polysaccharides

Polysaccharides (*Greek* : poly-many) are polymers of monosaccharide units with high molecular weight (up to a million). They are usually tasteless (non-sugars) and form colloids with water. Polysaccharides are of two types—***homopolysaccharides*** and ***heteropolysaccharides***.

MONOSACCHARIDES—STRUCTURAL ASPECTS

Stereoisomerism is an important character of monosaccharides. Stereoisomers are the compounds that have the same structural formulae but differ in their spatial configuration.

A ***carbon*** is said to be ***asymmetric when it is attached to four different atoms or groups***. The number of asymmetric carbon atoms (n) determines the possible isomers of a given compound which is equal to 2^n. Glucose contains 4 asymmetric carbons and thus has 16 isomers.

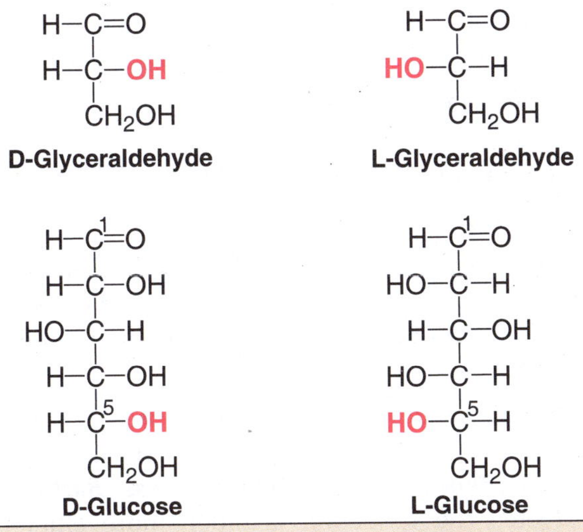

Fig.2.1 : *D- and L-forms of glucose compared with D- and L-glyceraldehydes (the reference carbohydrate).*

Glyceraldehyde—the reference carbohydrate

Glyceraldehyde (triose) is the simplest monosaccharide with one asymmetric carbon atom. It exists as two stereoisomers and has been chosen as the ***reference carbohydrate*** to represent the structure of all other carbohydrates.

D- and L-isomers

The D- and L-isomers are mirror images of each other. The spatial orientation of —H and —OH groups on the carbon atom (C_5 for glucose) that is adjacent to the terminal primary alcohol carbon determines whether the sugar is D- or L-isomer. If the —OH group is on the right side, the sugar is of D-series, and if on the left side, it belongs to L-series. The structures of D- and L-glucose based on the reference monosaccharide, D- and L-glyceraldehyde (glycerose) are depicted in ***Fig.2.1***.

It may be noted that the ***naturally occurring monosaccharides*** in the mammalian tissues are mostly of ***D-configuration***. The enzyme machinery of cells is specific to metabolise D-series of monosaccharides.

Optical activity of sugars

Optical activity is a characteristic feature of compounds with ***asymmetric carbon*** atom. When a beam of polarized light is passed through a solution of an optical isomer, it will be rotated either to the right or left. The terms ***dextrorotatory*** **(d+)** and ***levorotatory*** **(l–)** are used to compounds that respectively rotate the plane of polarized light to the right or to the left.

Epimers

If two monosaccharides ***differ*** from each other in their ***configuration around a single specific carbon*** (other than anomeric) atom, they are referred to as *epimers* to each other (***Fig.2.2***). For instance, ***glucose and galactose are epimers*** with regard to carbon 4 (C_4-epimers). That is, they differ in the arrangement of —OH group at C_4. Glucose and mannose are epimers with regard to carbon 2 (C_2-epimers).

STRUCTURE OF GLUCOSE

For a better understanding of glucose structure, let us consider the formation of hemiacetals and hemiketals, respectively produced when an aldehyde or a ketone reacts with alcohol.

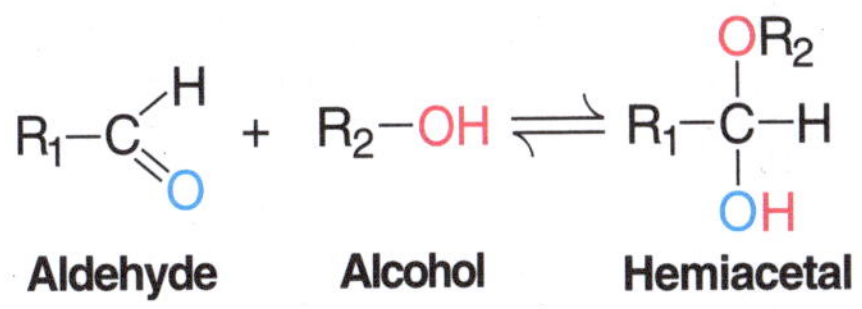

The hydroxyl group of monosaccharides can react with its own aldehyde or keto functional group to form hemiacetal and hemiketal. Thus, the aldehyde group of glucose at C_1 reacts with alcohol group at C_5 to form two types of cyclic hemiacetals namely ***α and β***, as depicted in ***Fig.2.3***.

Anomers—mutarotation

The ***α and β cyclic forms of D-glucose*** are known as anomers. They differ from each other in the configuration only around C_1 known as ***anomeric carbon (hemiacetal carbon)***. In case of α anomer, the —OH group held by anomeric carbon is on the opposite side of the group —CH_2OH of sugar ring. The reverse is true for β-anomer. The anomers differ in certain physical and chemical properties.

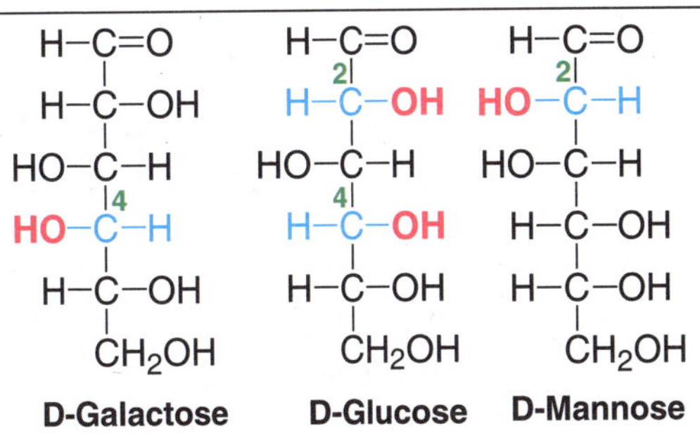

Fig.2.2 : *Structures of epimers (glucose and galactose are C_4-epimers while glucose and mannose are C_2-epimers).*

Medical Concepts/ Clinical Correlates

- **Dextrose (glucose in solution in dextrorotatory form) is frequently used in medical practice.**
- **Fructose is abundantly found in the semen which is utilized by the sperms for energy.**
- **Several diseases are associated with carbohydrates e.g., diabetes mellitus, glycogen storage diseases, galactosemia, lactose intolerance.**
- **Fermentation of sucrose (of chocolates and candies), resulting in organic acids, is mainly responsible for dental caries (destruction of teeth).**

Mutarotation : The α and β anomers of glucose have different optical rotations. The specific optical rotation of a freshly prepared glucose (α anomer) solution in water is +112.2° which gradually changes and attains an equilibrium with a constant value of +52.7°. In the presence of alkali, the decrease in optical rotation is rapid. The optical rotation of β-glucose is +18.7°. ***Mutarotation is defined as the change in the specific optical rotation representing the interconversion of α and β forms of D-glucose to an equilibrium mixture.*** Mutarotation depicted in ***Fig. 2.3***, is summarized in the next column.

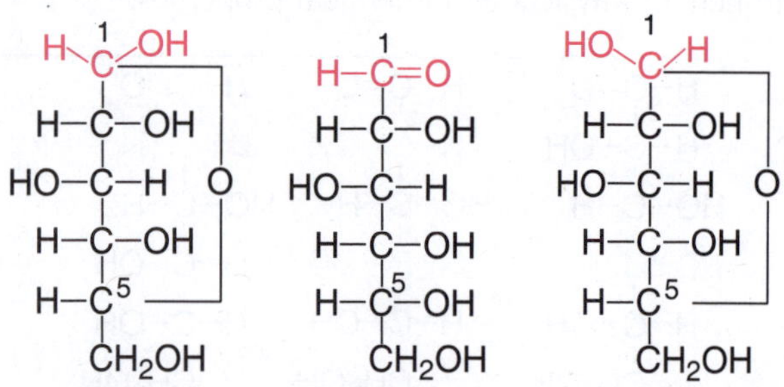

Fig.2.3 : *Mutarotation of glucose representing α and β anomers (Fischer projections).*

$$\begin{array}{ccccc} \alpha\text{-D-Glucose} & \rightleftharpoons & \text{Equilibrium mixture} & \rightleftharpoons & \beta\text{-D-Glucose} \\ +112.2^\circ & & +52.7^\circ & & +18.7^\circ \end{array}$$

(Specific optical rotation $[\alpha]^{20}_{D}$)

REACTIONS OF MONOSACCHARIDES

Reducing properties

The ***sugars*** are classified as ***reducing or non-reducing***. The reducing property is attributed to the ***free aldehyde*** or ***keto group*** of anomeric carbon.

In the laboratory, many tests are employed to identify the reducing action of sugars. These include ***Benedict's test***, ***Fehling's test***, ***Barfoed's test*** etc. The reduction is much more efficient in the alkaline medium than in the acid medium.

The ***enediol forms*** or sugars reduce cupric ions (Cu^{2+}) of copper sulphate to cuprous ions (Cu^{+}), which form a yellow precipitate of cuprous hydroxide or a red precipitate of cuprous oxide (***Refer Appendix A*** for reactions tests of carbohydrates in the laboratory).

Oxidation

Depending on the oxidizing agent used, the ***terminal aldehyde*** (***or keto***) or the terminal alcohol or both the groups may be oxidized. For instance, consider glucose :

1. Oxidation of aldehyde group (CHO → COOH) results in the formation of gluconic acid.

2. Oxidation of terminal alcohol group (CH_2OH → COOH) leads to the production of glucuronic acid.

Reduction

When treated with reducing agents such as sodium amalgam, the ***aldehyde or keto group*** of monosaccharide is reduced to corresponding alcohol.

The important monosaccharides and their corresponding alcohols are shown below.

D-Glucose → D-Sorbitol
D-Galactose → D-Dulcitol
D-Ribose → D-Ribitol

Sorbitol and ***dulcitol*** when accumulate in tissues in large amounts cause strong osmotic effects leading to swelling of ***cells***, and certain pathological conditions. e.g. ***cataract***, peripheral neuropathy, nephropathy. ***Mannitol is useful to reduce intracranial tension by forced diuresis*** (increased urine output).

Dehydration

When treated with concentrated sulfuric acid, monosaccharides undergo dehydration with an elimination of 3 water molecules. Thus, hexoses give hydroxymethyl furfural while pentoses give furfural on dehydration. These furfurals can condense with phenolic compounds (α-naphthol) to form coloured products. This is the chemical basis of the popular ***Molisch test***. In case of oligo- and polysaccharides, they are first hydrolysed to monosaccharides by acid, and this is followed by dehydration.

Osazone formation

Phenylhydrazine in acetic acid, when boiled with reducing sugars, forms osazones. The ***first two carbons*** (***C_1 and C_2***) are ***involved*** in osazone formation. The sugars that differ in their configuration on these two carbons give the same type of osazones, since the difference is masked by binding with phenylhydrazine. Thus ***glucose***, ***fructose*** and mannose give the same type (***needle-shaped***) of osazones.

Reducing disaccharides also give osazones—maltose sunflower-shaped and lactose powder-puff shaped.

GLYCOSIDES

Glycosides are formed when the hemiacetal or hemiketal hydroxyl group (of anomeric carbon) of a carbohydrate reacts with a hydroxyl group of another carbohydrate or a non-carbohydrate (e.g. methyl alcohol, phenol, glycerol). The bond so formed is known as ***glycosidic bond*** and the ***non-carbohydrate*** moiety (when present) is referred to as ***aglycone***.

DERIVATIVES OF MONOSACCHARIDES

There are several derivatives of monosaccharides, some of which are physiologically important

1. **Amino sugars :** When one or more hydroxyl groups of the monosaccharides are replaced by amino groups, the products formed are amino sugars e.g. D-glucosamine, D-galactosamine. They are present as constituents of heteropolysaccharides.

N-Acetylneuraminic acid (NANA) is a derivative of N-acetylmannose and pyruvic acid. It is an important constituent of glycoproteins and glycolipids. The term ***sialic acid*** is used to include NANA and its other derivatives.

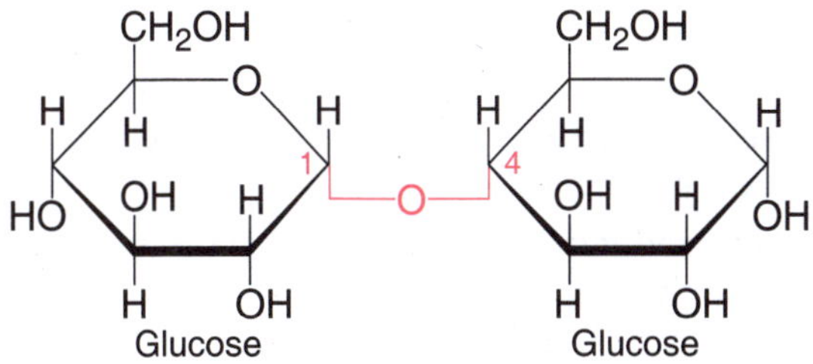

Maltose

(α-D-glucosyl ($1 \rightarrow 4$) α-D-glucose)

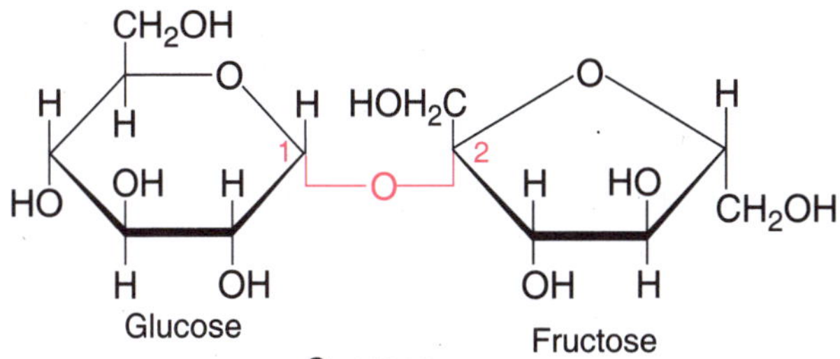

Sucrose

(α-D-glucosyl ($1 \rightarrow 2$) β-D-fructose)

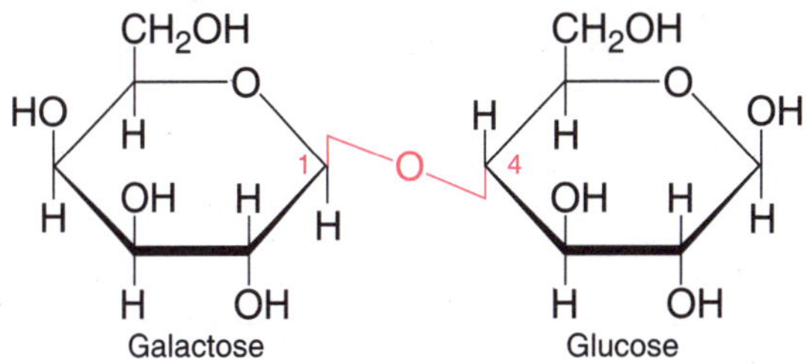

Lactose

(β-D-galactosyl ($1 \rightarrow 4$) β-D-glucose)

Fig.2.4 : Structures of disaccharides— maltose, sucrose and lactose.

2. **Deoxysugars :** These are the sugars that contain ***one oxygen less*** than that present in the parent molecule. The groups —CHOH and $—CH_2OH$ become $—CH_2$ and $—CH_3$ due to the absence of oxygen. ***D-2-Deoxyribose*** is the most important deoxysugar since it is a structural constituent of DNA (in contrast to D-ribose in RNA).

3. **L-Ascorbic acid** (vitamin C) **:** This is a ***water-soluble vitamin***, the ***structure of*** which closely resembles that of a ***monosaccharide***.

DISACCHARIDES

Among the oligosaccharides, disaccharides are the most common (***Fig.2.4***). As is evident from the name, a disaccharide consists of two monosaccharide units (similar or dissimilar) held together by a ***glycosidic bond***. They are crystalline, water soluble and sweet to taste. The disaccharides are of two types

1. **Reducing** disaccharides with ***free aldehyde*** or ***keto group*** e.g. maltose, lactose.

2. **Non-reducing** disaccharides with ***no free aldehyde or keto group*** e.g. sucrose, trehalose.

Maltose

Maltose is composed of ***two α-D-glucose units*** held together by α (1 → 4) glycosidic bond. The free aldehyde group present on C_1 of second glucose answers the reducing reactions, besides the osazone formation (sunflower-shaped). Maltose can be hydrolysed by dilute acid or the enzyme maltase to liberate two molecules of α-D-glucose.

Sucrose

Sucrose (***cane sugar***) is the sugar of commerce, mostly produced by sugar cane and sugar beets. Sucrose is made up of ***α-D-glucose and β-D-fructose***. The two monosaccharides are held together by a glycosidic bond ($\alpha_1 \rightarrow \beta_2$), between C_1 of α-glucose and C_2 of β-fructose. The reducing groups of glucose and fructose are involved in glycosidic bond, hence sucrose is a ***non-reducing sugar***, and it ***cannot form osazones***.

Inversion of sucrose

Sucrose, as such is dextrorotatory (+66.5°). But, when hydrolysed, sucrose becomes levorotatory (–28.2°). The process of change in optical rotation from dextrorotatory (+) to levorotatory (–) is referred to as inversion. The hydrolysed mixture of sucrose, containing glucose and fructose, is known as ***invert sugar***.

Lactose

Lactose is more commonly known as ***milk sugar*** since it is the disaccharide found in milk. Lactose is composed of ***β-D-galactose and β-D-glucose*** held together by β (1 → 4) glycosidic bond. The anomeric carbon of C_1 glucose is free, hence lactose exhibits reducing properties and forms osazones (powder-puff or hedgehog shape).

Lactose of milk is the most important carbohydrate in the nutrition of young mammals. It is hydrolysed by the intestinal enzyme lactase to glucose and galactose.

Sucralose–an artificial sweetener

Sucralose (***trichlorosucrose***) is about ***500 times sweeter than sucrose***. It is not digested, hence no calories are supplied. Further, it is acceptable in terms of taste, safety and stability.

Sucralose is used in baking, soft drinks and products that require longer shelf-life. It does not promote dental caries, and is ***safe for consumption by diabetics and non-diabetics***. Sucralose is one of the ***most used artificial sweeteners***. The other common sugar substitutes include ***saccharin*** and ***cyclamate***.

POLYSACCHARIDES

Polysaccharides (or simply ***glycans***) consist of ***repeat units of monosaccharides or their derivatives,*** held together by glycosidic bonds. They are primarily concerned with two important functions-structural, and storage of energy.

Polysaccharides are of two types

1. **Homopolysaccharides** which on hydrolysis yield only a single type of monosaccharide. They are named based on the nature of the monosaccharide unit. Thus, ***glucans*** are polymers of glucose whereas ***fructosans*** are polymers of fructose.

2. **Heteropolysaccharides** on hydrolysis yield a mixture of a few monosaccharides or their derivatives.

HOMOPOLYSACCHARIDES

Starch

Starch is the ***carbohydrate reserve of plants*** which is the most important dietary source for higher animals, including man. High content of starch is found in cereals, roots, tubers, vegetables etc. Starch is a homopolymer composed of D-glucose units held by α-glycosidic bonds. It is known as ***glucosan or glucan***.

Starch consists of two polysaccharide components-water soluble ***amylose*** (15-20%) and a water insoluble ***amylopectin*** (80-85%). Chemically, amylose is a long unbranched chain with 200—1,000 D-glucose units held by α (1 → 4) glycosidic linkages. Amylopectin, on the other hand, is a branched chain with α (1 → 6) glycosidic bonds at the branching points and α (1 → 4) linkages everywhere else. Amylopectin molecule containing a few thousand glucose units looks like a branched tree (20–30 glucose units per branch).

Inulin

Inulin is a polymer of fructose i.e., ***fructosan.*** It occurs in dahlia bulbs, garlic, onion etc. It is a low molecular weight (around 5,000) polysaccharide

Medical Concepts/ Clinical Correlates

- **Accumulation of sorbitol and dulcitol in the tissues may cause certain pathological conditions e.g. cataract, nephropathy, peripheral neuropathy.**
- **The non-digestible carbohydrates such as cellulose, play a significant role in human nutrition. These include decreasing the intestinal absorption of glucose and cholesterol, and increasing bulk of feces to avoid constipation.**
- **The mucopolysaccharide hyaluronic acid serves as a lubricant and shock absorbant in joints.**
- **The enzyme hyaluronidase of semen degrades the gel (contains hyaluronic acid) around the ovum. This allows effective penetration of sperm into the ovum.**
- **The mucopolysaccharide heparin is an anticoagulant (prevents blood clotting).**
- **Streptomycin is a glycoside employed in the treatment of tuberculosis.**

easily soluble in water. Inulin is not utilized by the body. It is used for assessing kidney function through ***measurement of glomerular filtration rate (GFR).***

Glycogen

Glycogen is the ***carbohydrate reserve in animals***, hence often referred to as ***animal starch***. It is present in high concentration in liver, followed by muscle, brain etc. Glycogen is also found in plants that do not possess chlorophyll (e.g. yeast, fungi).

The structure of glycogen is similar to that of amylopectin with more number of branches. Glucose is the repeating unit in glycogen joined together by α (1 $\rightarrow$ 4) glycosidic bonds, and α (1 $\rightarrow$ 6) glycosidic bonds at branching points.

Cellulose

Cellulose occurs ***exclusively in plants*** and it is the most abundant organic substance in plant kingdom. It is a predominant constituent of plant cell wall. Cellulose is totally absent in animal body.

Cellulose is composed of ***β-D-glucose*** units linked by **β *(1 → 4) glycosidic bonds***. Cellulose cannot be digested by mammals—including man—due to lack of the enzyme that cleaves β-glycosidic bonds (α amylase breaks α bonds only). Certain ruminants and herbivorous animals contain microorganisms in the gut which produce enzymes that can cleave β-glycosidic bonds. Hydrolysis of cellulose yields a disaccharide ***cellobiose,*** followed by ***β-D-glucose***.

Cellulose, though not digested, has great importance in human nutrition. It is a major constituent of ***fiber***, the ***non-digestable carbohydrate***. The functions of dietary fiber include decreasing the absorption of glucose and cholesterol from the intestine, besides increasing the bulk of feces.

HETEROPOLYSACCHARIDES

When the polysaccharides are ***composed of different types of sugars or their derivatives***, they are referred to as heteropolysaccharides or ***hetero-glycans***.

MUCOPOLYSACCHARIDES

Mucopolysaccharids are heteroglycans made up of repeating units of sugar derivatives, namely ***amino sugars and uronic acids***. These are more commonly known as ***glycosaminoglycans (GAG)***. Acetylated amino groups, besides sulfate and carboxyl groups are generally present in GAG structure.

Some of the mucopolysaccharides are found in combination with proteins to form ***mucoproteins*** or ***mucoids*** or ***proteoglycans***. Mucoproteins may contain up to 95% carbohydrate and 5% protein.

Mucopolysaccharides are essential components of tissue structure. The extracellular spaces of tissues (particularly connective tissue-cartilage, skin, blood vessels, tendons) consist of collagen and elastin fibers embedded in a matrix or ground substance. The ground substance is predominantly composed of GAG.

The important mucopolysaccharides include hyaluronic acid, chondroitin 4-sulfate, heparin, dermatan sulfate and keratan sulfate.

Hyaluronic acid

Hyaluronic acid is an important GAG found in the ground substance of synovial fluid of joints and vitreous humor of eyes. It is also present as a ground substance in connective tissues and forms a gel around the ovum. Hyaluronic acid serves as a ***lubricant and shock absorbant in joints***.

Hyaluronic acid is composed of alternate units of ***D-glucuronic acid*** and ***N-acetyl D-glucosamine***. These two molecules form disaccharide units held together by β (1 → 3) glycosidic bond. Hyaluronic acid contains about 250–25,000 disaccharide units (held by β 1 → 4 bonds) with a molecular weight up to 4 million.

Chondroitin sulfate

Chondroitin 4-sulfate (*Greek* : chondros-cartilage) is a major constituent of various mammalian tissues (bone, cartilage, tendons, heart, valves, skin, cornea etc.). Structurally, it is comparable with hyaluronic acid. Chondroitin 4-sulfate consists of repeating disaccharide units composed of D-glucuronic acid and N-acetyl D-galactosamine 4-sulfate.

Heparin

Heparin is ***an anticoagulant***, (prevents blood clotting) that occurs in blood, lung, liver, kidney, spleen etc. Heparin helps in the release of the enzyme lipoprotein lipase which causes clearing the turbidity of lipemic plasma.

Heparin is composed of alternating units of ***N-sulfo D-glucosamine 6-sulfate and glucuronate 2-sulfate***.

Dermatan sulfate

The name dermatan sulfate is derived from the fact that this compound mostly ***occurs in the skin***. It is structurally related to chondroitin 4-sulfate.

GLYCOPROTEINS

Several ***proteins*** are covalently ***bound to carbohydrates*** which are referred to as glycoproteins. The carbohydrate content of glycoprotein varies from 1% to 90% by weight. Sometimes the term ***mucoprotein*** is used for glycoprotein with carbohydrate concentration more than 4%. Glycoproteins are very widely distributed in the cells and perform variety of functions. These include their role as enzymes, hormones, transport proteins, structural proteins and receptors.

Blood group substances

The blood group ***antigens*** (of erythrocyte membrane) contain carbohydrates as glycoproteins or glycolipids. N-Acetylgalactosamine, galactose, fucose, sialic acid etc. are found in the blood group substances. The carbohydrate content also plays a determinant role in blood grouping.

SUMMARY

1. *Carbohydrates are the polyhydroxyaldehydes or ketones, or compounds which produce them on hydrolysis. The term sugar is applied to carbohydrates soluble in water and sweet to taste.*
2. *Carbohydrates are broadly classified into 3 groups–monosaccharides, oligosaccharides and polysaccharides.*
3. *If two monosaccharides differ in their structure around a single carbon atom, they are known as epimers. Glucose and galactose are C_4-epimers.*
4. *D-Glucose is the most predominant naturally occurring aldose/monosaccharide. Glucose exists as α and β anomers with different optical rotations. The interconversion of α and β anomeric forms with change in the optical rotation is known as mutarotation.*
5. *Among the oligosaccharides, disaccharides are the most common. These include the reducing disaccharides namely lactose (milk sugar) and maltose (malt sugar) and the non-reducing sucrose (cane sugar).*
6. *Polysaccharides are the polymers of monosaccharides or their derivatives, held together by glycosidic bonds. Homopolysaccharides are composed of a single monosaccharide (e.g., starch, glycogen, cellulose, inulin). Heteropolysaccharides contain a mixture of few monosaccharides or their derivatives (e.g., mucopolysaccharides).*

LIPIDS

3

Lipids (*Greek* : lipos–fat) are of great importance to the body as the chief concentrated storage form of energy, besides their role in cellular structure and various other biochemical functions. As such, lipids are a ***heterogeneous group*** of compounds.

Lipids may be regarded as organic substances relatively insoluble in water, soluble in organic solvents (alcohol, ether etc.), actually or potentially related to fatty acids and utilized by the living cells.

Unlike the polysaccharides, proteins and nucleic acids, lipids are not polymers. They are mostly small molecules.

Lipidomics is a recent term used for the comprehensive ***study of lipids*** that includes chemistry, functions, metabolism, pathology and various other aspects.

Classification of lipids

Lipids are broadly classified into simple, complex, derived and miscellaneous lipids, which are further subdivided as follows

1. **Simple lipids :** Esters of fatty acids with alcohols. These are mainly of two types

   (a) **Fats and oils** (triacylglycerols) **:** These are ***esters of fatty acids with glycerol***. The difference between fat and oil is only physical. Thus, oil is a liquid while fat is a solid at room temperature.

   (b) **Waxes :** Esters of fatty acids (usually long chain) with alcohols other than glycerol. Cetyl alcohol is most commonly found in waxes.

2. **Complex (or compound) lipids :** Esters of fatty acids with alcohols ***containing additional groups*** such as ***phosphate***, ***nitrogenous base***, carbohydrate, protein etc. They are further divided :

   (a) **Phospholipids :** Lipids ***containing phosphoric acid*** and frequently a nitrogenous base. This is in addition to alcohol and fatty acids.

      (i) **Glycerophospholipids :** These phospholipids contain glycerol as the alcohol e.g., ***lecithin***, ***cephalin***.

      (ii) **Sphingophospholipids :** Sphingosine is the alcohol in this group of phospholipids e.g., ***sphingomyelin***.

   (b) **Glycolipids :** These lipids contain a fatty acid, ***carbohydrate*** and nitrogenous base. The alcohol is sphingosine, hence they are also called as glycosphingolipids. Glycerol and phosphate are absent e.g., ***cerebrosides***, gangliosides.

   (c) **Lipoproteins :** Macromolecular complexes of lipids with proteins.

   (d) **Other complex lipids :** Sulfolipids, aminolipids and lipopolysaccharides are among the other complex lipids.

3. **Derived lipids :** These are the derivatives obtained on the hydrolysis of group I and group 2 lipids which possess the characteristics of lipids. These include glycerol and other alcohols, fatty acids, mono- and diacylglycerols, lipid soluble vitamins, steroid hormones, hydrocarbons and ketone bodies.

4. **Miscellaneous lipids :** These include a large number of compounds possessing the characteristics of lipids e.g., carotenoids, squalene, hydrocarbons such as pentacosane (in bees wax), terpenes etc.

NEUTRAL LIPIDS : The lipids which are uncharged are referred to as neutral lipids. These are mono-, di-, and triacylglycerols, cholesterol and cholesteryl esters.

Functions of lipids

Lipids perform several important functions

1. They are the ***concentrated fuel reserve*** of the body (triacylglycerols).

2. Lipids are the constituents of ***membrane structure*** and regulate the membrane permeability (phospholipids and cholesterol).

3. They serve as a source of ***fat soluble vitamins*** (A, D, E and K).

4. Lipids are important as cellular ***metabolic regulators*** (steroid hormones and prostaglandins).

FATTY ACIDS

Fatty acids are ***carboxylic acids with hydrocarbon side chain***. They are the simplest form of lipids.

Occurrence

Fatty acids mainly occur in the esterified form as major constituents of various lipids. They are also present as free (unesterified) fatty acids.

Even and odd carbon fatty acids

Most of the fatty acids that occur in ***natural lipids*** are of ***even carbons*** (usually 14C–20C). This is due to the fact that biosynthesis of fatty acids mainly occurs with the sequential addition of 2 carbon units. ***Palmitic acid (16C)*** and stearic acid (18C) are the most common. Among the ***odd chain*** fatty acids, ***propionic acid (3C)*** and valeric acid (5C) are well known.

Saturated and unsaturated fatty acids

Saturated fatty acids ***do not contain double bonds***, while unsaturated fatty acids contain one or more double bonds (***Table 3.1***). Both saturated and unsaturated fatty acids almost equally occur in the natural lipids. Fatty acids with one double bond are known as monounsaturated and those with 2 or more double bonds are collectively known as ***polyunsaturated fatty acids (PUFA)***.

Shorthand representation of fatty acids

Instead of writing the full structures, biochemists employ shorthand notations (by numbers) to represent fatty acids. The general rule is that the total number of carbon atoms are written first, followed by the number of double bonds and finally the (first carbon) position of double bonds, starting from the carboxyl end. Thus, saturated fatty acid, palmitic acid is written as 16 : 0, oleic acid as 18 : 1; 9, arachidonic acid as 20 : 4; 5, 8, 11, 14.

ESSENTIAL FATTY ACIDS

The fatty acids that cannot be synthesized by the body and, therefore, ***should be supplied in the diet*** are known as essential fatty acids (EFA). Chemically, they are polyunsaturated fatty acids, namely ***linoleic acid*** (18 : 2; 9, 12) and ***linolenic acid*** (18 : 3; 9, 12, 15). ***Arachidonic acid*** (20 : 4; 5, 8, 11, 14) becomes essential, if its precursor linoleic acid is not provided in the diet in sufficient amounts. The structures of EFA are given in the ***Table 3.1***.

Functions of EFA : Essential fatty acids are required for the ***membrane structure and function***, transport of cholesterol, formation of lipoproteins, ***prevention of fatty liver*** etc. They are also needed for the synthesis of another important group of compounds, namely ***eicosanoids*** (***Chapter 25***).

Deficiency of EFA : The deficiency of EFA results in ***phrynoderma*** or ***toad skin***, characterized by the presence of horny eruptions on the posterior and lateral parts of limbs, on the back and buttocks, loss of hair and poor wound healing.

TRANS FATTY ACIDS

Trans fatty acids (**TFA**) possess double bonds and are ***formed during hydrogenation*** (saturation) of natural oils. TFA are ***widely used in food industry*** (fast foods and ready to eat packaged foods) to ***increase the shelf life of various foods***.

TFA are known to be highly ***detrimental to health***, as their consumption is associated with ***increased risk*** of various diseases including diabetes, and ***cardiovascular diseases***. (***Refer chapter 23***)

TRIACYLGLYCEROLS

Triacylglycerols (formerly triglycerides) are the ***esters of glycerol with fatty acids***. The fats and oils that are widely distributed in both plants and animals are chemically triacylglycerols. They are insoluble in water and non-polar in character and commonly known as ***neutral fats***.

Fats as stored fuel : Triacylglycerols are the most abundant group of lipids that primarily function as fuel reserves of animals. The fat reserve of normal humans (men 20%, women 25% by weight) is sufficient to meet the body caloric requirements for 2–3 months.

Structures of acylglycerols : Monoacylglycerols, diacylglycerols and triacylglycerols, respectively consisting of one, two and three molecules of fatty acids esterified to a molecule of glycerol, are known (***Fig.3.1***). Among these, triacylglycerols are the most important biochemically.

Triacylglycerols of plants have higher content of unsaturated fatty acids compared to that of animals.

PROPERTIES OF TRIACYLGLYCEROLS

A few important properties of triacylglycerols, which have biochemical relevance, are discussed below.

1. **Hydrolysis :** Triacylglycerols undergo stepwise enzymatic hydrolysis to finally liberate free fatty acids and glycerol. The process of hydrolysis, catalysed by ***lipases*** is important for digestion of fat in the gastrointestinal tract, and fat mobilization from the adipose tissues.

TABLE 3.1 Selected examples of biochemically important fatty acids

Common name	Systematic name	Abbreviation*	Structure
I. Saturated fatty acids			
Acetic acid	Ethanoic acid	2 : 0	CH_3COOH
Propionic acid	n-Propanoic acid	3 : 0	CH_3CH_2COOH
Butyric acid	n-Butanoic acid	4 : 0	$CH_3(CH_2)_2COOH$
Valeric acid	n-Pentanoic acid	5 : 0	$CH_3(CH_2)_3COOH$
Caproic acid	n-Hexanoic acid	6 : 0	$CH_3(CH_2)_4COOH$
Caprylic acid	n-Octanoic acid	8 : 0	$CH_3(CH_2)_6COOH$
Capric acid	n-Decanoic acid	10 : 0	$CH_3(CH_2)_8COOH$
Lauric acid	n-Dodecanoic acid	12 : 0	$CH_3(CH_2)_{10}COOH$
Myristic acid	n-Tetradecanoic acid	14 : 0	$CH_3(CH_2)_{12}COOH$
Palmitic acid	n-Hexadecanoic acid	16 : 0	$CH_3(CH_2)_{14}COOH$
Stearic acid	n-Octadecanoic acid	18 : 0	$CH_3(CH_2)_{16}COOH$
Arachidic acid	n-Cicosanoic acid	20 : 0	$CH_3(CH_2)_{18}COOH$
Behenic acid	n-Docosanoic acid	22 : 0	$CH_3(CH_2)_{20}COOH$
Lignoceric acid	n-Tetracosanoic acid	24 : 0	$CH_3(CH_2)_{22}COOH$
II. Unsaturated fatty acids			
Palmitoleic acid	*cis*-9-Hexadecenoic acid	16 : 1; 9	$CH_3(CH_2)_5CH = CH(CH_2)_7COOH$
Oleic acid	*cis*-9-Octadecenoic acid	18 : 1; 9	$CH_3(CH_2)_7CH = CH(CH_2)_7COOH$
Linoleic acid**	*cis*, *cis*-9,12-Octadeca-dienoic acid	18 : 2; 9, 12	$CH_3(CH_2)_4CH = CHCH_2CH = CH(CH_2)_7COOH$
Linolenic acid**	All *cis*–9,12,15-Octa-decatrienoic acid	18 : 3; 9, 12, 15	$CH_3CH_2CH = CHCH_2CH = CHCH_2CH = CH(CH_2)_7COOH$
Arachidonic acid	All *cis*–5,8,11,14-Eicosatetraenoic acid	20 : 4; 5, 8, 11, 14	$CH_3(CH_2)_4CH = CHCH_2CH = CHCH_2CH = CHCH_2CH = CH(CH_2)_3COOH$

* *Total number of carbon atoms, followed by the number of double bonds and the first carbon position of the double bond(s).*

** *Essential fatty acids.*

Medical Concepts/ Clinical Correlates

- Lipids are important to the body as constituents of membranes, source of fat soluble (A, D, E and K) vitamins and metabolic regulators (steroid hormones and prosta-glandins).
- The unsaturated fatty acids–linoleic and linolenic acid—are essential to humans, the deficiency of which causes phryno-derma or toad skin.
- Trans fatty acids, widely used in food industry, are detrimental to health-linked with increased risk of diabetes, heart diseases etc.

2. **Saponification :** The hydrolysis of triacylglycerols by alkali to produce glycerol and soaps is known as saponification.

$$\text{Triacylglycerol} + 3\ \text{NaOH} \longrightarrow \text{Glycerol} + 3\ \text{R-COONa (soaps)}$$

3. **Rancidity :** Rancidity is the term used to represent the ***deterioration of fats and oils*** resulting in an unpleasant taste. Fats containing unsaturated fatty acids are more susceptible to rancidity.

Antioxidants : The ***substances which can prevent the occurrence of oxidative rancidity*** are known as antioxidants. Trace amounts of antioxidants such as tocopherols (vitamin E), hydroquinone, gallic acid and α-naphthol are added to the commercial preparations of fats and oils to prevent rancidity. Propyl gallate, butylated hydroxyanisole (BHA) and butylated hydroxytoluene (BHT) are the antioxidants used in food preservation.

4. **Lipid peroxidation *in vivo*:** In the living cells, lipids undergo oxidation to ***produce peroxides and free radicals which can damage the tissue***. The free radicals are believed to cause inflammatory diseases, ***ageing***, ***cancer***, ***atherosclerosis*** etc. It is fortunate that the cells possess antioxidants such as vitamin E, urate and superoxide dismutase to prevent *in vivo* lipid peroxidation (***Chapter 34***).

PHOSPHOLIPIDS

These are complex or ***compound lipids*** containing phosphoric acid, in addition to fatty acids, nitrogenous base and alcohol (***Fig.3.2***). There are two classes of phospholipids

1. Glycerophospholipids (or phosphoglycerides) that contain glycerol as the alcohol.
2. Sphingophospholipids (or sphingomyelins) that contain sphingosine as the alcohol.

Glycerophospholipids

Glycerophospholipids are the major lipids that occur in biological membranes. They consist of glycerol 3-phosphate esterified at its C_1 and C_2 with fatty acids. Usually, C_1 contains a saturated fatty acid while C_2 contains an unsaturated fatty acid.

1. **Phosphatidic acid :** This is the simplest phospholipid. It does not occur in good concentration in the tissues. Basically, phosphatidic acid is an ***intermediate in the synthesis of triacylglycerols and phospholipids***.

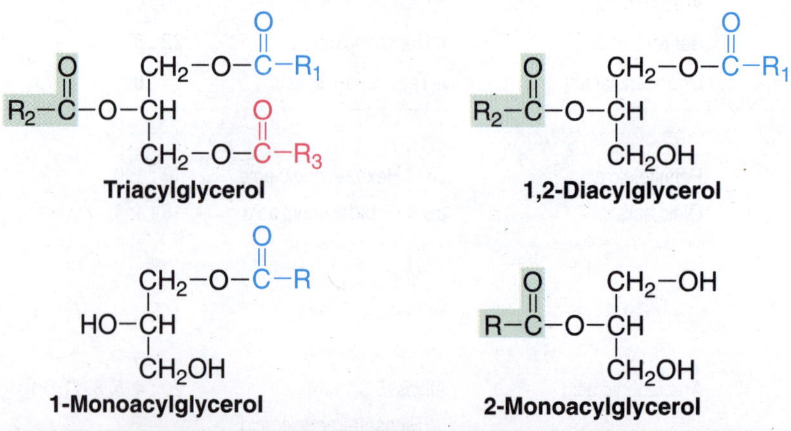

Fig.3.1 : *General structures of acylglycerols (For palmitoyl R = $C_{15}H_{31}$; for stearoyl R = $C_{17}H_{35}$; For linoceoyl R = $C_{17}H_{31}$)*

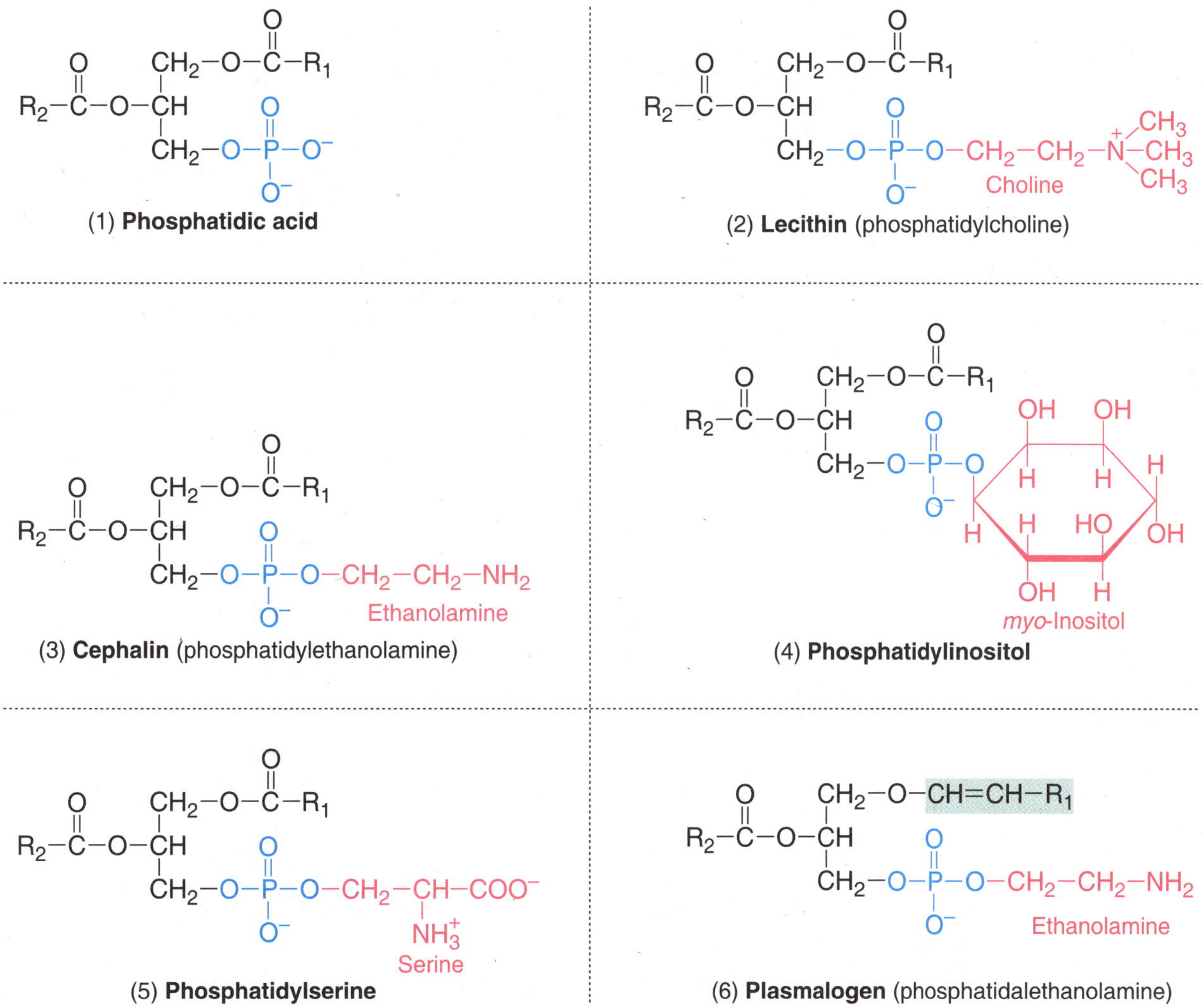

Fig.3.2 : *Structures of phospholipids.*

2. **Lecithins (*phosphatidylcholine*) :** These are the most abundant group of phospholipids in the cell membranes. Chemically, lecithin (*Greek* : lecithos—egg yolk) is a phosphatidic acid with choline as the base.

(a) **Dipalmitoyl lecithin** is an important phosphatidylcholine found ***in lungs***. It is a ***surface active agent*** (***lung surfactant***) and prevents the adherence of inner surface of the lungs due to surface tension. ***Respiratory distress syndrome*** in infants is a disorder characterized by the absence of dipalmitoyl lecithin.

(b) **Lysolecithin** is formed by removal of the fatty acid either at C_1 or C_2 of lecithin.

3. **Cephalins (*phosphatidylethanolamine*) :** ***Ethanolamine*** is the ***nitrogenous base*** present in cephalins. Thus, lecithin and cephalin differ with regard to the base.

4. **Phosphatidylinositol :** The steroisomer ***myo-inositol*** is attached to phosphatidic acid to give phosphatidylinositol(PI). This is an important component of cell membranes. The action of certain hormones (e.g. oxytocin, vasopressin) is mediated through PI.

5. **Phosphatidylserine :** The amino acid serine is present in this group of glycerophospholipids. ***Phosphatidylthreonine*** is also found in certain tissues.

6. **Plasmalogens :** When a fatty acid is attached by an ***ether*** linkage at C_1 of glycerol in the

Medical Concepts/ Clinical Correlates

- The phospholipid – dipalmitoyl lecithin prevents the adherence of inner surface of the lungs, the absence of which is associated with respiratory distress syndrome in infants.
- Lecithin-sphingomyelin (L/S) ratio of amniotic fluid is an indicator of fetal maturity. L/S ratio above 2 indicates full lung maturity.
- Phospholipids are important for the synthesis and transport of lipoproteins and reverse transport of cholesterol.

glycerophospholipids, the resultant compound is plasmalogen. Phosphati***dal***ethanolamine is the most important which is similar in structure to phospha-ti***dyl***ethanolamine but for the ether linkage (in place of ester). An unsaturated fatty acid occurs at C_1. Brain and muscle contain a good concentration (about 10% of phospholipids) of plasmalogens.

7. **Cardiolipin :** It is so named as it was first isolated from heart muscle. Structurally, a cardiolipin consists of two molecules of phosphatidic acid held by an additional glycerol through phosphate groups. It is an important component of inner mitochondrial membrane and ***essential for mitochondrial function***. Decreased cardiolipin levels may result in mitochondrial dysfunction, aging, hypothyroidism, cardioskeletal myopathy (***Barth syndrome***).

Sphingomyelins

Sphingosine is an amino alcohol present in sphingomyelins (sphingophospholipids). They do not contain glycerol at all. Sphingosine is attached by an amide linkage to a fatty acid to produce ***ceramide***. Sphingomyelins are important constituents of myelin and are found in good quantity in brain and nervous tissues.

Functions of phospholipids

Phospholipids constitute an important group of compound lipids that perform a wide variety of functions

1. In association with proteins, phospholipids form the structural ***components of membranes*** and regulate membrane permeability.

2. Phospholipids (lecithin, cephalin and cardiolipin) in the mitochondria are responsible for ***maintaining the conformation of electron transport chain*** components and thus cellular respiration.

3. Phospholipids are essential for the synthesis of different lipoproteins and thus participate in the ***transport of lipids***.

4. Accumulation of fat in liver (fatty liver) can be prevented by phospholipids, hence they are regarded as ***lipotropic factors***.

5. Arachidonic acid, an unsaturated fatty acid liberated from phospholipids serves as a ***precursor for*** the synthesis of ***eicosanoids*** (prostaglandins, prostacyclins, thrombaxanes etc.).

6. Phospholipids (phosphatidylinositol) are involved in ***signal transmission*** across membranes.

GLYCOLIPIDS

Glycolipids (***glycosphingolipids***) are important constituents of cell membrane and nervous tissues (particularly the brain). ***Cerebrosides*** are the simplest form of glycolipids. They contain a ceramide (sphingosine attached to a fatty acid) and one or more sugars. ***Galactocerebroside*** (galacto-sylceramide) and glucocerebroside are the most important glycolipids.

Gangliosides : These are predominantly found in ganglions and are the most complex form of glycosphingolipids. They are the derivatives of cerebrosides and contain one or more molecules of N-acetylneuraminic acid (NANA), the most important sialic acid.

LIPOPROTEINS

Lipoproteins are molecular complexes of lipids with proteins. They are the transport vehicles for lipids in the circulation. There are five types of lipoproteins, namely ***chylomicrons, very low density***

lipoproteins (VLDL), low density lipoproteins (LDL), high density lipoproteins (HDL) and free fatty acid-albumin complexes. Their structure, separation, metabolism and diseases are discussed together (***Chapter 14***).

STEROIDS

Steroids are the compounds containing a cyclic steroid nucleus (or ring) namely ***cyclopentanoperhydrophenanthrene (CPPP)***. It consists of a phenanthrene nucleus (rings A, B and C) to which a cyclopentane ring (D) is attached.

The structure and numbering of CPPP are shown in ***Fig.3.3***. The steroid nucleus represents saturated carbons, unless specifically shown as double bonds. The methyl side chains (19 and 18) attached to carbons 10 and 13 are shown as single bonds. At carbon 17, steroids usually contain a side chain.

There are several steroids in the biological system. These include ***cholesterol, bile acids, vitamin D, sex hormones*** and ***adrenocortical hormones***. If the steroid contains one or more hydroxyl groups it is commonly known as ***sterol*** (means solid alcohol).

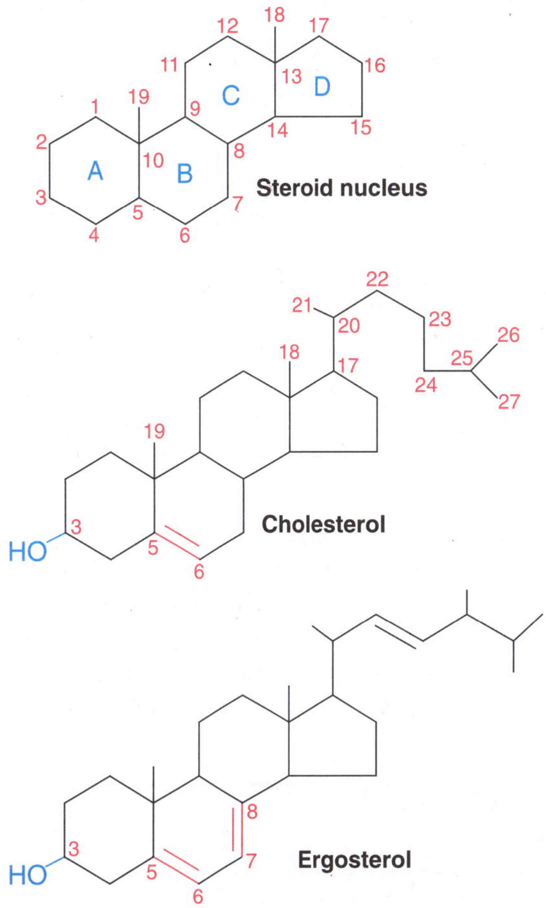

Fig.3.3 : Structures of steroids (A, B, C—Perhydrophenanthrene; D-Cyclopentane).

CHOLESTEROL

Cholesterol is ***exclusively found in animals*** and is the most abundant animal sterol. It is widely distributed in all cells and is a major component of cell membranes and lipoproteins. Cholesterol (*Greek* : chole—bile) was first isolated from bile. Cholesterol literally means 'solid alcohol from bile'.

Structure and occurrence

The structure of cholesterol (***$C_{27}H_{46}O$***) is depicted in ***Fig.3.3***. It has ***one hydroxyl*** group at C_3 and ***a double bond*** between C_5 and C_6. An 8 carbon aliphatic side chain is attached to C_{17}. Cholesterol contains a total of 5 methyl groups.

Cholesterol is found in association with fatty acids to form cholesteryl esters (esterification occurs at the OH group of C_3).

Properties and reactions : Cholesterol is an yellowish crystalline solid. The crystals, under the microscope, show a notched ( ) appearance. Cholesterol is insoluble in water and soluble in organic solvents such as chloroform, benzene, ether etc.

Several reactions given by cholesterol are useful for its qualitative identification and quantitative estimation. These include ***Salkowski test***, ***Liebermann-Burchard reaction*** and ***Zak's test***.

Functions of cholesterol : Cholesterol functions as an ***insulating cover*** for the transmission of electrical impulses in the nervous tissue. Cholesterol performs several other biochemical functions which include its role in ***membrane structure and function***, in the synthesis of bile acids, ***hormones*** (sex and cortical) and ***vitamin D*** (for details, ***refer Chapters 7*** and ***19***).

Ergosterol

Ergosterol occurs ***in plants***. It is also found as a structural constituent of membranes in yeast and fungi. Ergosterol is an important precursor for vitamin D. When exposed to light, the ring B of ergosterol opens and it is converted to ergocalciferol, a compound containing vitamin D activity.

AMPHIPATHIC LIPIDS

As per definition, lipids are insoluble (hydrophobic) in water. This is primarily due to the predominant presence of hydrocarbon groups. However, some of the lipids possess polar or hydrophilic groups which tend to be soluble in water. Molecules which contain ***both hydrophobic and hydrophilic groups*** are known as ***amphipathic*** (*Greek* : amphi-both, pathos—passion).

Examples of amphipathic lipids : Among the lipids, ***fatty acids***, ***phospholipids***, sphingolipids, bile salts and ***cholesterol*** (to some extent) are amphipathic in nature.

Orientation amphipathic lipids : When the amphipathic lipids are mixed in water (aqueous phase), the polar groups (heads) orient themselves towards aqueous phase while the non-polar (tails) orient towards the opposite directions. This leads to the formation of ***micelles***. Micelle formation, facilitated by bile salts is very ***important for lipid digestion and absorption*** (***Chapter 8***).

Membrane bilayers

In case of biological membranes, a bilayer of lipids is formed orienting the polar heads to the outer aqueous phase on either side and the nonpolar tails into the interior. The formation of a lipid bilayer is the basis of membrane structure.

Liposomes : They are produced when amphipathic lipids in aqueous medium are subjected to ***sonification***. They have intermittent aqueous phases in the lipid bilayer. Liposomes in combination with tissue specific antigens are used as carriers of drugs to target tissues.

Emulsions : These are formed when nonpolar lipids (e.g. triacylglycerols) are mixed with water. The particles are larger in size and stabilized by emulsifying agents (usually amphipathic lipids), such as bile salts and phospholipids.

SUMMARY

1. *Lipids are the organic substances relatively insoluble in water, soluble in organic solvents (alcohol, ether), actually or potentially related to fatty acids and are utilized by the body.*
2. *Lipids are classified into simple (fats and oils), complex (phospholipids, glycolipids), derived (fatty acids, steriod hormones) and miscellaneous (carotenoids).*
3. *Fatty acids are the major constituents of various lipids. The polyunsaturated fatty acids (PUFA) namely linoleic acid and linolenic acid are the essential fatty acids that need to be supplied in the diet.*
4. *Triacylglycerols (simply fats) are the esters of glycerol with fatty acids. They are found in adipose tissue and primarily function as fuel reserve of animals.*
5. *Phospholipids are complex lipids containing phosphoric acid. Glycerophospholipids contain glycerol as the alcohol and these include lecithin, cephalin, phosphatidylinositol, plasmalogen and cardiolipin.*
6. *Steroids contain the ring cyclopentanoperhydrophenanthrene. The steroids of biological importance include cholesterol, bile acids, vitamin D, sex hormones and cortical hormones. A steroid containing one or more hydroxyl groups is known as sterol.*

4 PROTEINS AND AMINO ACIDS

Proteins are the ***most abundant organic molecules of the living system***. They occur in every part of the cell and constitute about 50% of cellular dry weight. Proteins form the fundamental basis of structure and function of life.

Functions of proteins

Proteins perform a great variety of specialized and essential functions in the living cells. These functions may be broadly grouped as ***static*** (***structural***) and ***dynamic***.

Structural functions : Certain proteins perform '***brick and mortar***' roles and are primarily responsible for structure and strength of body. These include ***collagen*** and ***elastin*** found in bone matrix, vascular system and other organs and ***α-keratin*** present in epidermal tissues.

Dynamic functions : The dynamic functions of proteins are more diversified in nature. These include proteins acting as ***enzymes, hormones, blood clotting factors, immunoglobulins***, storage proteins, besides their function in genetic control, muscle contraction, respiration etc. Proteins performing dynamic functions are appropriately regarded as '***the working horses***' of cell.

Proteins are polymers of amino acids

Proteins on complete hydrolysis (with concentrated HCl) yield L-α-amino acids. Therefore, proteins are the polymers of ***L-α-amino acids.***

A typical protein is born at translation (***protein biosynthesis***), and matures through post-translational process (Refer ***Chapter 25***). A protein performs specific and specialized functions, ages and dies (gets degraded) to release component amino acids.

STANDARD AMINO ACIDS

As many as 300 amino acids occur in nature—of these, ***only 20***—known as standard amino acids are repeatedly ***found in the structure of proteins***, isolated from different forms of life—animal, plant and microbial. This is because of the universal nature of the genetic code available for the incorporation of only 20 amino acids when the proteins are synthesized in the cells. The process in turn is controlled by DNA, the genetic material of the cell. After the synthesis of proteins, some of the incorporated amino acids undergo modifications to form their derivatives.

AMINO ACIDS

Amino acids are a group of organic compounds containing two ***functional groups—amino and carboxyl***. The amino group (***—NH_2***) is basic while the carboxyl group (***—COOH***) is acidic in nature.

General structure of amino acids

The amino acids are termed as α-amino acids, if both the carboxyl and amino groups are attached to the same carbon atom, as depicted below

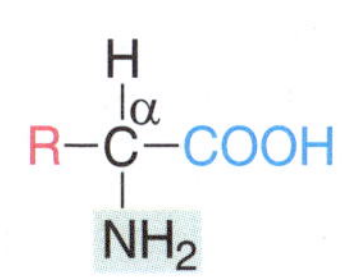

General structure

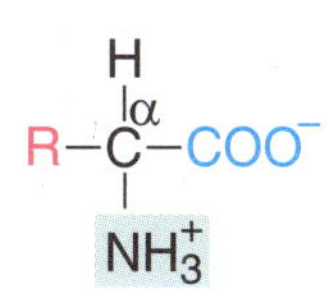

Exists as ion

The α-carbon atom binds to a side chain represented by R which is different for each of the 20 amino acids found in proteins. The amino acids mostly exist in the ionized form in the biological system (shown in previous page).

Optical isomers of amino acids

If a carbon atom is attached to four different groups, it is ***asymmetric*** and therefore exhibits ***optical isomerism***. The amino acids (except glycine) possess four distinct groups (R, H, COO^-, NH_3^+) held by α-carbon. Thus all the amino acids (except glycine where R = H) have optical isomers.

The structures of ***L- and D-amino acids*** are written based on the configuration of ***L- and D-glyceraldehyde*** as shown in ***Fig.4.1***. The proteins are composed of L-α-amino acids

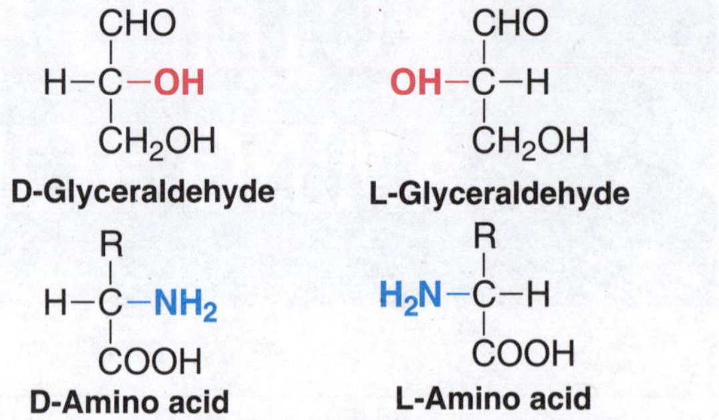

Fig.4.1 : *D- and L-forms of amino acid based on the structure of glyceraldehyde.*

Classification of amino acids

There are different ways of classifying the amino acids based on the structure and chemical nature, nutritional requirement, metabolic fate etc.

A. Amino acid classification based on the structure : A ***comprehensive*** classification of amino acids is based on their structure and chemical nature. Each amino acid is assigned a 3 letter or 1 letter symbol. These symbols are commonly used to represent the amino acids in protein structure. The 20 amino acids found in proteins are divided into seven distinct groups.

In ***Table 4.1***, the different groups of amino acids, their symbols and structures are given.

B. Nutritional classification of amino acids : The twenty amino acids (***Table 4.1***) are required for the synthesis of variety of proteins, besides other biological functions. However, all these 20 amino acids need not be taken in the diet. Based on the nutritional requirements, amino acids are grouped into two classes—essential and non-essential.

1. **Essential or indispensable amino acids :** The amino acids which ***cannot be synthesized by the body*** and, therefore, need to be supplied through the diet are called essential amino acids. They are required for proper growth and maintenance of the individual. The ten amino acids listed below are essential for humans (and also rats) :

   Arginine, **V**aline, **H**istidine, **I**soleucine, **L**eucine, **L**ysine, **M**ethionine, **P**henylalanine, **T**hreonine, **T**ryptophan.

   [The code ***A.V. HILL, MP., T. T.*** (first letter of each amino acid) may be memorized to recall essential amino acids. Other useful codes are H. VITTAL, LMP and MATTVILPhLy.]

   Of the ten listed above, two amino acids namely **a**rginine and **h**istidine can be partly synthesized by adult humans, hence these are considered as ***semi–essential amino acids*** (remember **Ah**, to recall). Thus, 8 amino acids are absolutely essential while 2 are semi-essential.

2. **Non-essential or dispensable amino acids :** The body can synthesize about 10 amino acids to meet the biological needs, hence they need not be consumed in the diet. These are—glycine, alanine, serine, cysteine, aspartate, asparagine, glutamate, glutamine, tyrosine and proline.

3. **Conditionally essential or indispensable amino acids :** The synthesis of some non-essential amino acids is limited, although it is adequate to meet the body's requirement in a healthy state. In certain (patho-physiological) conditions like surgery, wound, sepsis, thermal injury etc., the ***non-essential amino acids become*** conditionally ***essential*** (indispensable). ***Arginine, cysteine, glycine, glutamine, proline*** and ***tyrosine*** are considered as conditionally essential amino acids, and ***need to be supplemented in the diet during pathophysiological conditions***.

C. Amino acid classification based on their metabolic fate : The carbon skeleton of amino acids can serve as a precursor for the synthesis of glucose (***glycogenic***) or fat (***ketogenic***) or both. From metabolic view point, amino acids are divided into three groups (Refer ***Table 15.4***).

TABLE 4.1 Structural classification of L-α-amino acids found in proteins

Name	*Symbol* *3 letters*	 *1 letter*	*Structure*	*Special group present*
I. Amino acids with aliphatic side chains				
1. Glycine	Gly	G	$H-CH(NH_3^+)-COO^-$	
2. Alanine	Ala	A	$CH_3-CH(NH_3^+)-COO^-$	
3. Valine	Val	V	$(H_3C)_2CH-CH(NH_3^+)-COO^-$	Branched chain
4. Leucine	Leu	L	$(H_3C)_2CH-CH_2-CH(NH_3^+)-COO^-$	Branched chain
5. Isoleucine	Ile	I	$CH_3-CH_2-CH(CH_3)-CH(NH_3^+)-COO^-$	Branched chain
II. Amino acids containing hydroxyl (—OH) groups				
6. Serine	Ser	S	$CH_2(OH)-CH(NH_3^+)-COO^-$	Hydroxyl
7. Threonine	Thr	T	$H_3C-CH(OH)-CH(NH_3^+)-COO^-$	Hydroxyl
Tyrosine	Tyr	Y	See under aromatic	Hydroxyl

Table 4.1 contd. next page

Name	Symbol: 3 letters	Symbol: 1 letter	Structure	Special group present
III. Sulfur containing amino acids				
8. Cysteine	Cys	C	$HS-CH_2-CH(NH_3^+)-COO^-$	Sulfhydryl
Cystine	—	—	$^-OOC-CH(NH_3^+)-CH_2-S-S-CH_2-CH(NH_3^+)-COO^-$	Disulfide
9. Methionine	Met	M	$CH_3-S-CH_2-CH_2-CH(NH_3^+)-COO^-$	Thioether
IV. Acidic amino acids and their amides				
10. Aspartic acid	Asp	D	$^-OOC-\overset{\beta}{CH_2}-\overset{\alpha}{CH}(NH_3^+)-COO^-$	β-Carboxyl
11. Asparagine	Asn	N	$H_2N-C(=O)-CH_2-CH(NH_3^+)-COO^-$	Amide
12. Glutamic acid	Glu	E	$^-OOC-\overset{\gamma}{CH_2}-\overset{\beta}{CH_2}-\overset{\alpha}{CH}(NH_3^+)-COO^-$	γ-Carboxyl
13. Glutamine	Gln	Q	$H_2N-C(=O)-CH_2-CH_2-CH(NH_3^+)-COO^-$	Amide
V. Basic amino acids				
14. Lysine	Lys	K	$NH_3^+-\overset{\varepsilon}{CH_2}-\overset{\delta}{CH_2}-\overset{\gamma}{CH_2}-\overset{\beta}{CH_2}-\overset{\alpha}{CH}(NH_3^+)-COO^-$	ε-Amino
15. Arginine	Arg	R	$H_2N-C(=NH_2^+)-NH-CH_2-CH_2-CH_2-CH(NH_3^+)-COO^-$	Guanidino
16. Histidine	His	H	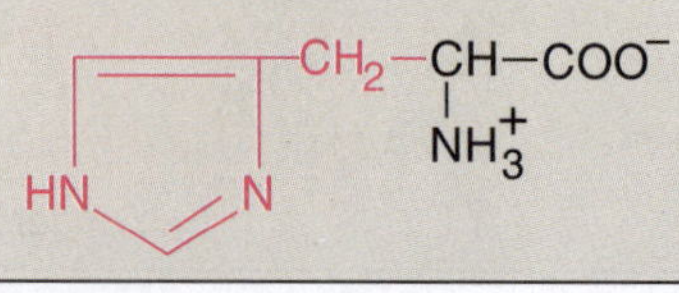	Imidazole

Table 4.1 contd. next page

Name	Symbol		Structure	Special group present
	3 letters	1 letter		
VI. Aromatic amino acids				
17. Phenylalanine	Phe	F	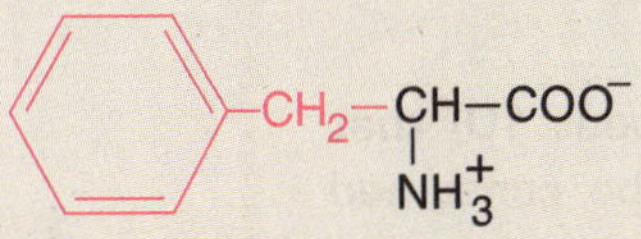	Benzene or phenyl
18. Tyrosine	Tyr	Y	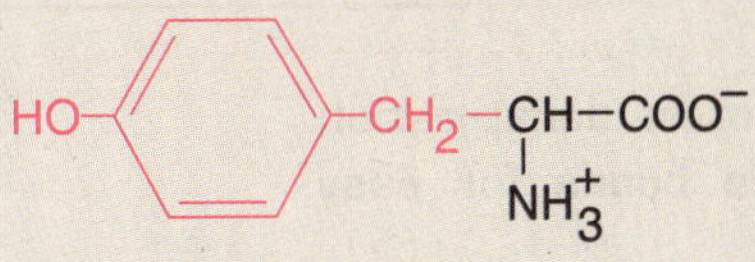	Phenol
19. Tryptophan	Trp	W	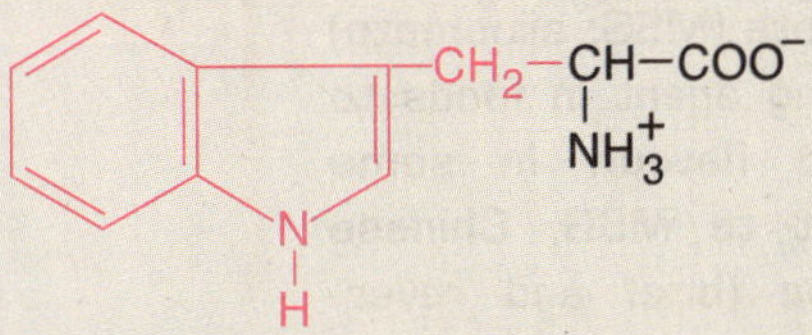	Indole
VII. Imino acid				
20. Proline	Pro	P	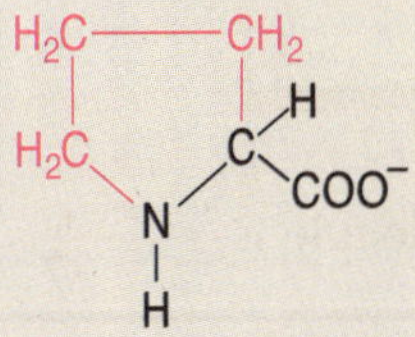 or 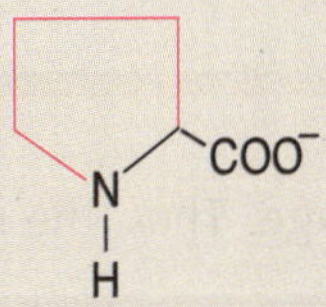	Pyrrolidine

(*Note* : *R group is shown in red*)

Selenocysteine—the 21st amino acid

Besides the 20 amino acids commonly occurring in proteins, another amino acid namely selenocysteine is also found in certain enzymes/ proteins e.g. ***glutathione peroxidase***, ***thioredoxin reductase***. Selenocysteine contains the trace element selenium in place of sulfur atom of cysteine.

Properties of amino acids

A. Physical properties

1. **Solubility :** Most of the amino acids are usually soluble in water and insoluble in organic solvents.

2. **Melting points :** Amino acids generally melt at higher temperatures, often above 200°C.

3. **Taste :** Amino acids may be sweet (Gly, Ala, Val), tasteless (Leu) or bitter (Arg, Ile). ***Monosodium glutamate*** (MSG; ***ajinmoto***) is employed as a flavoring agent in food industry to increase taste and ***flavor***. In some individuals intolerant to MSG, ***Chinese restaurant syndrome*** (brief and reversible flu-like symptoms) is observed.

4. **Optical properties :** All the amino acids ***except glycine*** possess optical isomers due to the presence of asymmetric α-carbon atoms.

5. **Amino acids as ampholytes :** Amino acids contain both acidic (—COOH) and basic (—NH_2) groups. They can ***donate or accept a proton***, hence amino acids are regarded as ampholytes.

Zwitterion or dipolar ion : The name *zwitter* is derived from the German word which means *hybrid*. Zwitter ion (or dipolar ion) is a hybrid molecule containing ***positive and negative ionic groups***.

The amino acids rarely exist in a neutral form with free carboxylic (—COOH) and free amino (—NH_2) groups. In strongly ***acidic pH*** (low pH), the amino acid is positively charged (***cation***) while in strongly ***alkaline pH*** (high pH), it is negatively charged (***anion***). Each amino acid has a characteristic pH (e.g. leucine, pH 6.0) at which it carries both positive and negative charges, and exists as a zwitterion (***Fig.4.2***).

Isoelectric pH (symbol ***pI***) is defined as the pH at which a ***molecule exists as a zwitterion*** or dipolar

Medical Concepts/ Clinical Correlates

- **Half of the amino acids (about 10) that occur in proteins have to be consumed by humans in the diet, hence they are essential.**
- **Cooking results in protein denaturation exposing more peptide bonds for easy digestion.**
- **Monosodium glutamate (MSG; ajinomoto) is used as a flavoring agent in foods to increase taste and flavour. In some individuals intolerant to MSG, Chinese restaurant syndrome (brief and reversible flu-like symptoms) is observed.**

ion and carries no net charge. Thus, the molecule is electrically neutral.

B. Chemical properties

The general reactions of amino acids are mostly due to the presence of two functional groups namely carboxyl (—COOH) group and amino (—NH_2) group.

Reactions due to —COOH group

1. Amino acids form salts (—COONa) with bases, and esters (—COOR′) with alcohols.
2. **Decarboxylation :** Amino acids undergo decarboxylation to produce corresponding amines.

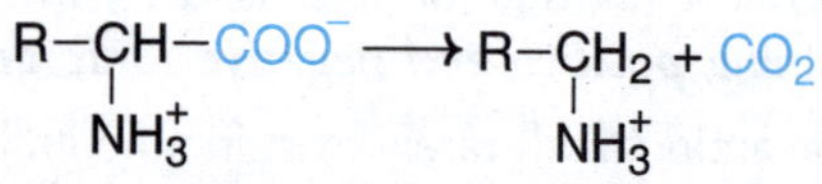

   This reaction assumes significance in the living cells due to the formation of many biologically important amines. These include histamine, tyramine and γ-amino butyric acid (GABA) from the amino acids histidine, tyrosine and glutamate, respectively.
3. **Reaction with ammonia :** The carboxyl group of dicarboxylic amino acids reacts with NH_3 to form amide

   Aspartic acid + NH_3 ⟶ Asparagine

   Glutamic acid + NH_3 ⟶ Glutamine

Reactions due to —NH_2 group

4. The amino groups behave as bases and combine with acids (e.g. HCl) to form salts (—$NH_3^+Cl^-$).
5. **Reaction with ninhydrin :** The α-amino acids react with ninhydrin to form a purple, blue or pink colour complex (Ruhemann's purple).

   Amino acid + Ninhydrin ⟶ Keto acid + NH_3 + CO_2 + Hydrindantin

   Hydrindantin + NH_3 + Ninhydrin ⟶ ***Ruhemann's purple***

   Ninhydrin reaction is effectively used for the quantitative determination of amino acids and proteins. (***Note*** **:** Proline and hydroxyproline give yellow colour with ninhydrin).
6. **Colour reactions of amino acids :** Amino acids can be identified by specific colour reactions (***Refer Appendix B***).
7. **Transamination :** Transfer of an amino group from an amino acid to a keto acid to form

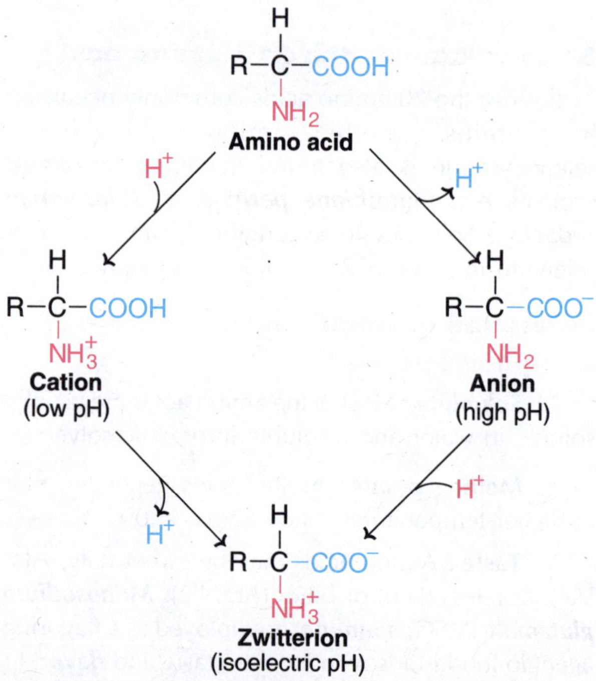

***Fig.4.2 :** Existence of an amino acid as cation, anion and zwitterion.*

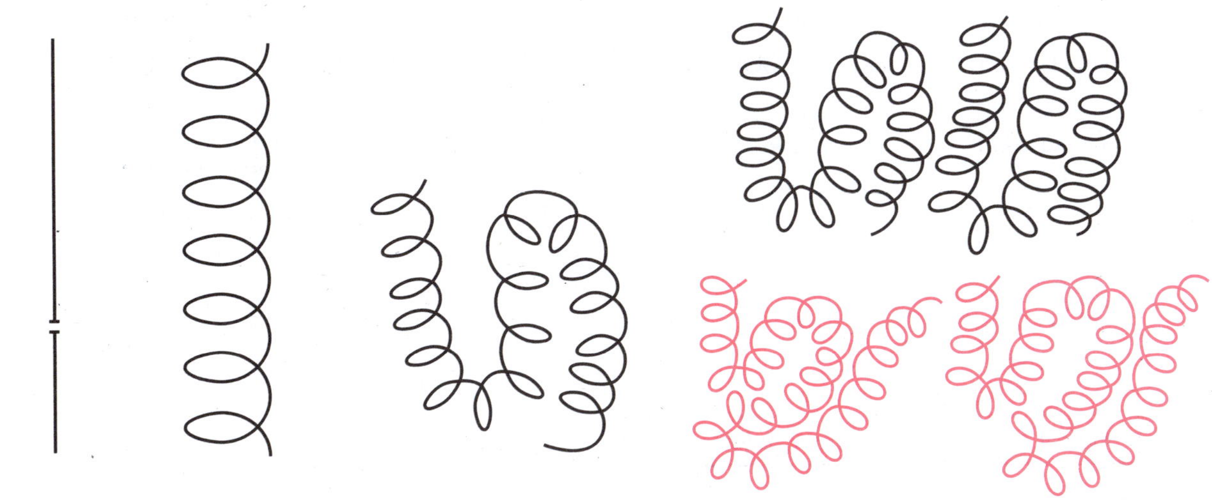

Fig.4.3 : Diagrammatic representation of protein structure (Note : The four subunits of two types in quaternary structure).

a new amino acid is a very important reaction in amino acid metabolism (details given in ***Chapter 15***).

8. **Oxidative deamination :** The amino acids undergo oxidative deamination to liberate free ammonia (***Refer Chapter 15***).

STRUCTURE OF PROTEINS

Proteins are the polymers of L-α-amino acids. The structure of proteins is rather complex which can be divided into 4 levels of organization (***Fig.4.3***) :

1. **Primary structure :** The ***linear sequence of amino acids*** forming the backbone of proteins (polypeptides).

2. **Secondary structure :** The ***spatial arrangement of protein by*** twisting of the polypeptide chain.

3. **Tertiary structure :** The ***three dimensional structure*** of a functional protein.

4. **Quaternary structure :** Some of the proteins are composed of two or more polypeptide chains referred to as subunits. The ***spacial arrangement of these subunits*** is known as quaternary structure.

The term ***protein*** is generally used for a polypeptide containing ***more than 50 amino acids.*** In recent years, however, some authors have been using '***polypeptide***' even if the number of amino acids is a few hundreds. They prefer to use protein to an assembly of polypeptide chains with quaternary structure.

PRIMARY STRUCTURE OF PROTEIN

Each protein has a unique sequence of amino acids which is determined by the genes contained in DNA. The primary structure of a protein is largely responsible for its function.

Peptide bond

The amino acids are held together in a protein by covalent peptide bonds or linkages. These bonds are rather strong and serve as the cementing material between the individual amino acids (considered as bricks).

Formation of a peptide bond : When the ***amino group*** of an amino acid combines with the ***carboxyl group*** of another amino acid, a peptide bond is formed (***Fig.4.4***). Note that a ***dipeptide*** will have ***two amino acids*** and one peptide (not two) bond. Peptides containing more than 10 amino acids (decapeptide) are referred to as ***polypeptides***.

Determination of primary structure

The primary structure comprises the identification of constituent amino acids with regard to their quality, quantity and sequence in a protein structure.

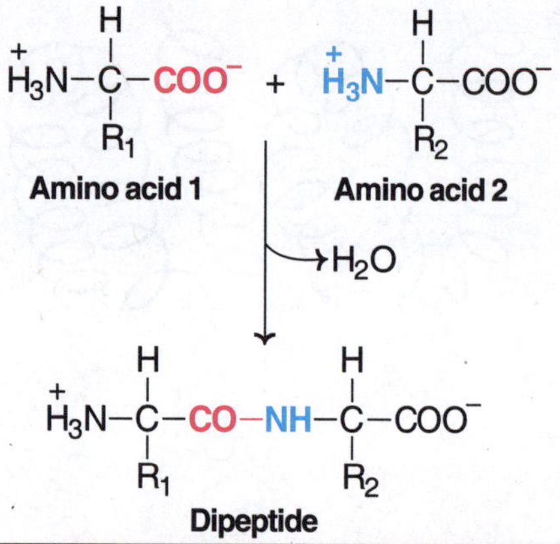

Fig.4.4 : *Formation of a peptide bond.*

A ***pure sample of a protein or a polypeptide is essential*** for the determination of primary structure which involves 3 stages :

1. Determination of amino acid composition.

2. Degradation of protein or polypeptide into smaller fragments.

3. Determination of the amino acid sequence.

SECONDARY STRUCTURE OF PROTEIN

The ***conformation of polypeptide chain by twisting or folding*** is referred to as secondary structure. The amino acids are located close to each other in their sequence. Two types of secondary structures, **α-helix** and **β-sheet**, are mainly identified.

α-Helix

α-Helix is the ***most common*** spiral structure of protein. It has a rigid arrangement of polypeptide chain. α-Helical structure was proposed by Pauling and Corey (1951) which is regarded as one of the milestones in the biochemistry research. The salient features of α-helix (***Fig.4.5***) are given below

1. The α-helix is a ***tightly packed*** coiled structure with amino acid side chains extending outward from the central axis.

2. The α-helix is ***stabilized by*** extensive ***hydrogen bonding***. It is formed between H atom attached to peptide N, and O atom attached to peptide C. The ***hydrogen bonds*** are individually weak but collectively, they are strong enough to stabilize the helix.

3. All the ***peptide bonds*** except the first and last in a polypeptide chain participate in ***hydrogen bonding***.

4. Each turn of α-helix contains ***3.6 amino acids*** and travels a ***distance of 0.54 nm***. The spacing of each amino acid is 0.15 nm.

5. α-Helix is a ***stable conformation*** formed spontaneously with the lowest energy.

6. The right handed α-helix is more stable than left handed helix (a right handed helix turns in the

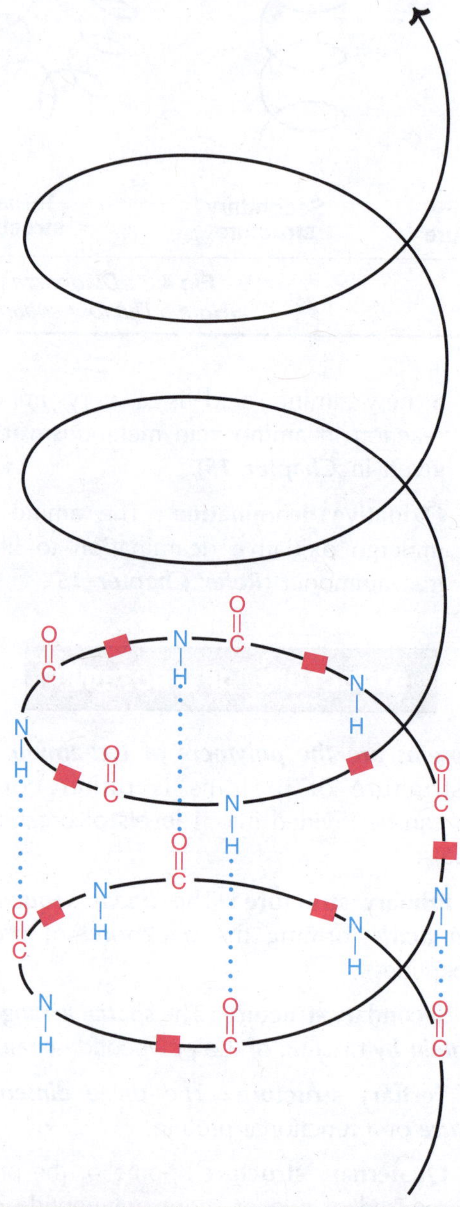

Fig.4.5 : *Diagrammatic representation of secondary structure of protein—a right handed α-helix*

(-Indicate $-\overset{H}{\overset{|}{C}}-R$ *groups of amino acids; dotted blue lines are hydrogen bonds; Note that only a few hydrogen bonds shown for clarity).*

direction that the fingers of right hand curl when its thumb points in the direction the helix rises).

β-Pleated sheet

This is the second type of structure (hence β after α) proposed by Pauling and Corey. β-Pleated sheets (or simply β-sheets) are ***composed of two or more segments of fully extended peptide chains***. In the β-sheets, the hydrogen bonds are formed between the neighbouring segments of polypeptide chain(s).

TERTIARY STRUCTURE OF PROTEIN

The ***three-dimensional arrangement of protein structure*** is referred to as tertiary structure. It is a compact structure with hydrophobic side chains held interior while the hydrophilic groups are on the surface of the protein molecule. This type of arrangement ensures stability of the molecule.

Bonds of tertiary structure : Besides the hydrogen bonds, ***disulfide bonds*** (—S—S), ***ionic interactions*** (electrostatic bonds) and ***hydrophobic interactions*** also contribute to the tertiary structure of proteins.

Domains : The term domain is used to represent the ***basic units of protein structure (tertiary) and functions***. A polypeptide with 200 amino acids normally consists of two or more domains.

QUATERNARY STRUCTURE OF PROTEIN

A great majority of the proteins are composed of single polypeptide chains. Some of the proteins, however, consist of two or more polypeptides which may be identical or unrelated. Such proteins are termed as ***oligomers*** and possess quaternary structure. The individual polypeptide chains are known as ***monomers, protomers*** or ***subunits***. A ***dimer*** consits of ***two*** polypeptides while a ***tetramer*** has ***four***.

Bonds in quaternary structure : The monomeric subunits are held together by non-convalent bonds namely ***hydrogen bonds***, ***hydrophobic interactions and ionic bonds***.

Importance of oligomeric proteins : These proteins play a significant role in the ***regulation of metabolism*** and cellular function.

Examples of oligomeric proteins : *Hemoglobin*, aspartate transcarbamoylase, ***lactate dehydrogenase***.

Example of protein structure

Structure of human insulin : Insulin consists of two polypeptide chains A and B (***Fig.4.6***). The A chain has glycine at the N-terminal end and asparagine at the C-terminal end. The B chain has phenylalanine and alanine at the N- and C-terminal ends, respectively. Originally, insulin is synthesized as a single polypeptide ***preproinsulin*** which undergoes proteolytic processing to give ***proinsulin*** and finally ***insulin***.

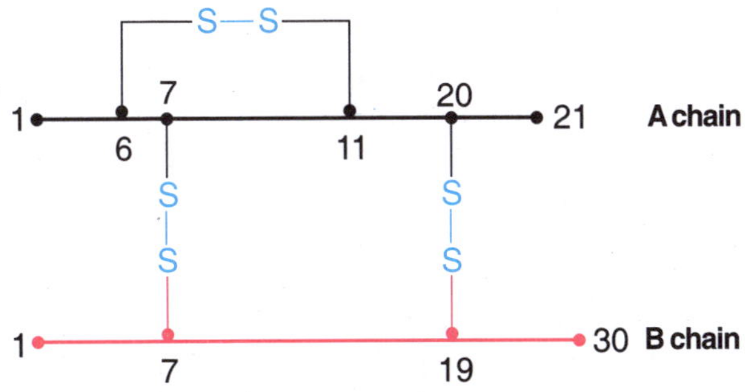

***Fig.4.6 :** Diagrammatic representation of human insulin structure*
***(Note :** A and B polypeptide chains are held by two disulfide linkages).*

PROPERTIES OF PROTEINS

1. **Solubility :** Proteins form ***colloidal solutions*** instead of true solutions in water. This is due to huge size of protein molecules.

2. **Molecular weight :** The proteins vary in their molecular weights, which in turn, is dependent on the number of amino acid residues. ***Each amino acid*** on an average contributes to a ***molecular weight of about 110***. A few proteins with their molecular weights are listed below :

   Insulin – 5,700; Myoglobin – 17,000;

   Hemoglobin – 64,450; Serum albumin – 69,000

3. **Shape :** There is a wide variation in the protein shape. It may be ***globular*** (insulin), oval (albumin) ***fibrous*** or elongated (fibrinogen).

4. **Isoelectric pH :** At isoelectric pH (pI), the proteins exist as ***zwitterions*** or ***dipolar ions***. They are electrically neutral (do not migrate in the electric field) with minimum solubility, maximum precipitability and least buffering capacity. The isoelectric pH(pI) for some proteins are given here

   Casein – 4.6; Human albumin – 4.7

   Hemoglobin – 6.7

In acidic medium (lower than pI), the proteins are present as cations while in alkaline medium (above pI), they are found as anions. Both the cationic and

Medical Concepts/ Clinical Correlates

- **Collagen is the most abundant protein in mammals. It is rich in glycine, and contains hydroxyproline and hydroxylysine.**
- **Several biologically important peptides are known in the living organisms. These include glutathione for the maintenance of RBC structure and integrity; oxytocin that causes uterus contraction; vasopressin that stimulates retention of water by kidneys; enkephalins that inhibit the sense of pain in the brain.**
- **Homocysteine has been implicated as a risk factor in the onset of coronary heart diseases.**
- **Heat coagulation test is most commonly employed to detect the presence of albumin in urine.**

anionic forms of proteins move in the electric field. This property is utilized in the ***separation of protein by electrophoresis***.

DENATURATION

The phenomenon of ***disorganization of native protein structure*** is known as denaturation. Denaturation results in the loss of secondary, tertiary and quaternary structure of proteins.

Agents of denaturation

Physical agents : ***Heat***, violent shaking, X-rays, ***UV radiation***.

Chemical agents : ***Acids, alkalies, organic solvents*** (***ether***, alcohol), salts of heavy metals (Pb, Hg), urea, salicylate.

Characteristics of denaturation

1. The native helical structure of protein is lost (***Fig.4.7***).

2. The ***primary structure*** of a protein with peptide linkages remains ***intact*** i.e., peptide bonds are not hydrolysed.

3. The protein ***loses*** its ***biological activity***.

4. Denatured protein becomes insoluble in the solvent in which it was originally soluble.

5. Denatured protein is ***more easily digested***. This is due to increased exposure of peptide bonds to enzymes. Cooking causes protein denaturation and, therefore, cooked food (protein) is more easily digested.

6. Denaturation is ***usually irreversible***. For instance, omelette can be prepared from an egg (protein-albumin) but the reversal is not possible.

Coagulation : The term 'coagulum' refers to a semi-solid viscous precipitate of protein. Irreversible denaturation results in coagulation. Coagulation is optimum and requires lowest temperature at isoelectric pH. Albumins and globulins are coagulable proteins. ***Heat coagulation test is commonly used to detect the presence of albumin in urine***.

CLASSIFICATION OF PROTEINS

Proteins are classified in several ways. Three major types of classifying proteins ***based on their function, chemical nature and solubility properties and nutritional importance*** are discussed here.

A. Functional classification of proteins

Based on the function they perform, proteins are classified in the following groups (with examples)

1. **Structural proteins :** ***Keratin*** of hair and nails, collagen of bone.

2. **Enzymes or catalytic proteins :** ***Hexokinase***, pepsin.

3. **Transport proteins :** ***Hemoglobin***, serum albumin.

4. **Hormonal proteins :** ***Insulin***, growth hormone.

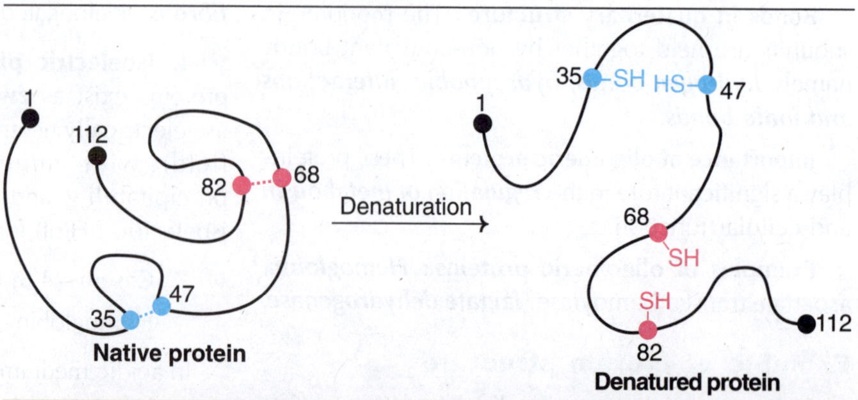

Fig.4.7 : Denaturation of a protein.

TABLE 4.2 Summary of classification of proteins

- **PROTEINS**
  - Simple
    - Globular proteins
      - Albumins
      - Globulins
      - Glutelins
      - Prolamines
      - Histones
      - Globins
      - Protamines
    - Scleroproteins
      - Collagens
      - Elastins
      - Keratins
  - Conjugated
    - Nucleoproteins
    - Glycoproteins Mucoproteins
    - Lipoproteins
    - Phosphoproteins
    - Chromoproteins
    - Metalloproteins
  - Derived
    - Primary
      - Coagulated proteins
      - Proteans
      - Metaproteins
    - Secondary
      - Proteoses
      - Peptones
      - Polypeptides
      - Peptides

5. **Contractile proteins :** Actin, myosin.
6. **Storage proteins :** Ovalbumin, glutelin.
7. **Genetic proteins :** Nucleoproteins.
8. **Defense proteins :** Snake venoms, Immunoglobulins.
9. **Receptor proteins** for hormones, viruses.

B. Protein classification based on chemical nature and solubility

This is a more comprehensive and popular classification of proteins. It is based on the amino acid composition, structure, shape and solubility properties. Proteins are broadly classified into 3 major groups (***Table 4.2***).

1. **Simple proteins :** They are composed of only amino acid residues.

2. **Conjugated proteins :** Besides the amino acids, these proteins contain a non-protein moiety known as ***prosthetic group*** or conjugating group.

3. **Derived proteins :** These are the denatured or degraded products of simple and conjugated proteins.

The above three classes are further sub-divided into different groups. The summary of protein classification is given in the ***Table 4.2***.

Among the ***simple proteins, globular proteins*** are spherical in shape, soluble in water or other solvents and digestable e.g., albumin, globulin. ***Scleroproteins*** (fibrous proteins) are fiber like in shape, insoluble in water and resistant to digestion e.g., collagen, keratin.

The ***conjugated proteins*** may contain ***prosthetic groups*** such as ***nucleic acid***, ***carbohydrate***, ***lipid***, metal etc. The primary derived proteins are produced by agents such as heat, acids, alkalies etc., while the secondary derived proteins are hydrolytic products of proteins.

C. Nutritional classification of proteins

From the nutritional point of view, proteins are classified into 3 categories

1. **Complete proteins :** These proteins ***have all the ten essential amino acids*** in the required proportion by the human body to promote good growth. e.g. ***egg albumin***, ***milk casein***.

2. **Partially incomplete proteins :** These proteins are partially lacking one or more essential amino acids and hence can promote moderate growth. e.g. ***wheat*** and rice proteins (limiting Lys, Thr).

3. **Incomplete proteins :** These proteins completely lack one or more essential amino acids. Hence they do not promote growth at all e.g. ***gelatin*** (***lacks Trp***), zein (lacks Trp, Lys).

BIOLOGICALLY IMPORTANT PEPTIDES

Generally, the term 'peptide' is applied when the number of constituent amino acids is less than 10. Some examples of biologically active peptides and their functions are described here.

1. **Glutathione :** It is a ***tripeptide*** composed of 3 amino acids. Chemically, glutathione is ***γ-glutamylcysteinylglycine***. It is widely distributed in nature and exists in reduced or oxidized states.

$$\underset{\text{Reduced}}{2G{-}SH} \rightleftharpoons \underset{\text{Oxidized}}{G{-}S{-}S{-}G}$$

Functions : The ***reversible oxidation-reduction of glutathione*** is important for many of its ***biological functions***. In a steady state, the cells generally maintain a ratio of about 100/1 of GSH to G—S—S—G.

(a) Glutathione serves as a coenzyme for certain enzymes e.g. prostaglandin PGE_2 synthetase, glyoxylase.

(b) It ***prevents the oxidation of sulfhydryl (—SH) groups*** of several proteins to disulfide (—S—S—) groups. This is essential for the protein function, including that of enzymes.

(c) Glutathione (reduced) performs specialized functions in erythrocytes

(i) It ***maintains RBC*** membrane ***structure*** and integrity.

(ii) It protects hemoglobin from getting oxidized by agents such as H_2O_2.

(d) Toxic amounts of peroxides and free radicals produced in the cells are scavanged by glutathione peroxidase (a selenium containing enzyme).

$$2\,GSH + H_2O_2 \xrightarrow{\text{Peroxidase}} G-S-S-G + 2\,H_2O$$

2. **Oxytocin :** It is a hormone secreted by posterior pituitary gland and contains 9 amino acids (nonapeptide). Oxytocin causes contraction of uterus.

3. **Vasopressin** (*antidiuretic hormone, ADH*) **:** ADH is also a nonapeptide produced by posterior pituitary gland. It stimulates kidneys to retain water and thus increases the blood pressure.

4. **Bradykinin and kallidin :** They are nona- and decapeptides, respectively. Both of them act as powerful vasodilators. They are produced from plasma proteins by snake venom enzymes.

5. **Gastrointestinal hormones :** Gastrin, secretin etc. are the gastrointestinal peptides which serve as hormones.

SUMMARY

1. *Proteins are nitrogen containing, most abundant organic macromolecules widely distributed in animals and plants. They perform structural and dynamic functions in the organisms.*
2. *Proteins are polymers composed of L-α-amino acids. They are 20 in number and classified into different groups based on their structure, chemical nature, nutritional requirement and metabolic fate.*
3. *Amino acids possess two functional groups namely carboxyl (—COOH) and amino (—NH_2). In the physiological system, they exist as dipolar ions commonly referred to as zwitterions.*
4. *The structure of protein is divided into four levels of organization. The primary structure represents the linear sequence of amino acids. The twisting and spatial arrangement of polypeptide chain is the secondary structure. Tertiary structure constitutes the three dimensional structure of a functional protein. The assembly of similar or dissimilar polypeptide subunits comprises quaternary structure.*
5. *Proteins are classified into three major groups. Simple proteins contain only amino acid residues (e.g. albumin). Conjugated proteins contain a non-protein moiety known as prosthetic group, besides the amino acids (e.g. glycoproteins). Derived proteins are obtained by degradation of simple or conjugated proteins.*
6. *In addition to proteins, several peptides perform biologically important functions. These include glutathione, oxytocin and vasopressin.*

NUCLEIC ACIDS AND NUCLEOTIDES

5

There are two types of nucleic acids, namely ***deoxyribonucleic acid*** (***DNA***) and ***ribonucleic acid*** (***RNA***). Primarily, nucleic acids serve as repositories and transmitters of genetic information.

Functions of nucleic acids

DNA is the chemical basis of heredity and may be regarded as the ***reserve bank of genetic information***. DNA is exclusively responsible for maintaining the identity of different species of organisms over millions of years. Further, every aspect of cellular function is under the control of DNA. The ***DNA*** is organized into ***genes***, the fundamental units of ***genetic information***. The genes control the protein synthesis through the mediation of RNA, as shown below

DNA ⟶ RNA ⟶ Protein

Components of nucleic acids

Nucleic acids are the polymers of nucleotides (***polynucleotides***) held by 3′ and 5′ phosphate bridges. In other words, nucleic acids are built up by the monomeric units—nucleotides (It may be recalled that protein is a polymer of amino acids).

NUCLEOTIDES

Nucleotides are composed of a ***nitrogenous base*** (also called as ***nucleobase***), a ***pentose sugar*** and a ***phosphate***. Nucleotides perform a wide variety of functions in the living cells.

BIOLOGICALLY IMPORTANT NUCLEOTIDES AND THEIR FUNCTIONS

1. ***Purine and pyrimidines nucleotides***, besides being structural components of DNA and RNA, are involved in a wide variety of ***metabolic functions***—energy metabolism, protein synthesis, control of enzyme activities etc.

2. Nucleotides are structural components of some ***coenzymes of B-complex vitamins*** e.g. FAD, NAD^+.

3. Sugar derivative of nucleotides namely ***UDP-glucose*** participates in the synthesis of glycogen.

4. ***CDP-acylglycerol*** (a lipid derivative) is an intermediate in ***lipid biosynthesis***.

5. ***cAMP*** and ***cGMP*** are ***second messengers*** for some hormonal action.

6. GTP and GDP are involved in a cascade of events of ***signal transduction*** pathways.

7. ***Synthetic analogs*** of nucleotides are employed in the ***treatment of cancer***, ***AIDS*** and suppression of immune response during organ transplantation (details given later).

STRUCTURE OF NUCLEOTIDES

As already stated, the nucleotide essentially consists of ***base***, ***sugar*** and ***phosphate***. The term nucleoside refers to base + sugar. Thus, nucleotide is nucleoside + phosphate.

Purines and pyrimidines

The nitrogenous bases found in nucleotides (and, therefore, nucleic acids) are ***aromatic heterocyclic compounds***. The bases are of two types—purines and pyrimidines. Their general structures are depicted in ***Fig.5.1***. Purines are

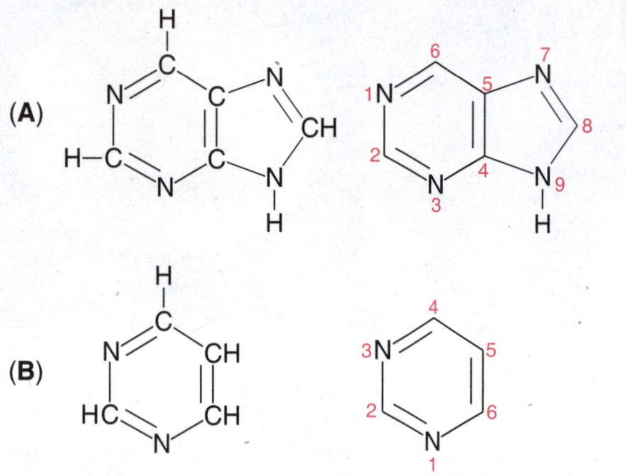

Fig.5.1 : *General structure of nitrogen bases (**A**) Purine (**B**) Pyrimidine (The positions are numbered according to the international system).*

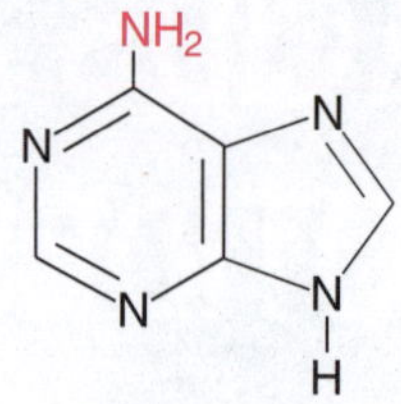

Adenine (A)
(6-aminopurine)

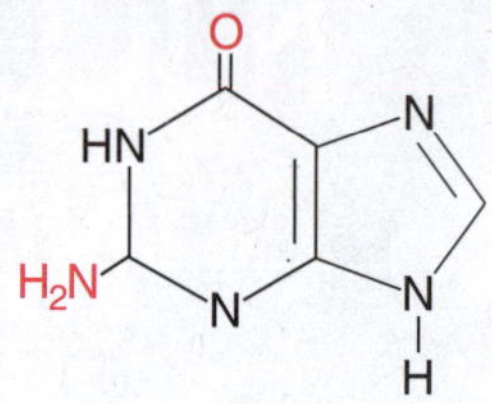

Guanine (G)
(2-amino 6-oxypurine)

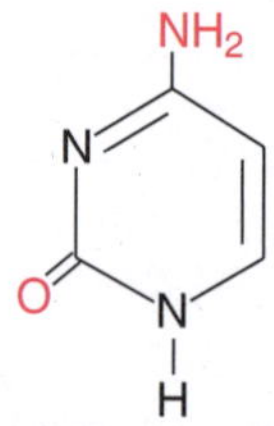

Cytosine (C)
(2-oxy 4-aminopyrimidine)

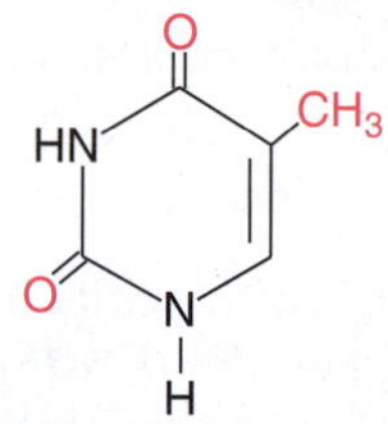

Thymine (T)
(2,4-dioxy-5 methylpyrimidine)

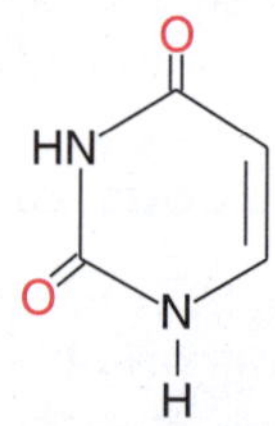

Uracil (U)
(2,4-dioxypyrimidine)

Fig.5.2 : *Structures of major purines (A, G) and pyrimidines (C, T, U) found in nucleic acids.*

numbered in the anticlockwise direction while pyrimidines are numbered in the clockwise direction. And this is an internationally accepted system to represent the structure of bases.

Major bases in nucleic acids

The structures of major purines and pyrimidines found in nucleic acids are shown in ***Fig.5.2***. DNA and RNA contain the same purines namely adenine (A) and guanine (G). Further, the pyrimidine cytosine (C) is found in both DNA and RNA. However, the nucleic acids differ with respect to the second pyrimidine base. ***DNA contains thymine (T) whereas RNA contains uracil (U)***. As is observed in the ***Fig.5.2***, thymine and uracil differ in structure by the presence (in T) or absence (in U) of a methyl group.

Sugars of nucleic acids

The five carbon monosaccharides (***pentoses***) are found in the nucleic acid structure. ***RNA*** contains ***D-ribose*** while ***DNA*** contains ***D-deoxyribose***. Ribose and deoxyribose differ in structure at C_2. Deoxyribose has one oxygen less at C_2 compared to ribose (***Fig.5.3***).

Nomenclature of nucleotides

The addition of a pentose sugar to base produces a nucleoside. If the sugar is ribose, ribonucleosides are formed. Adenosine, guanosine, cytidine and uridine are the ribonucleosides of A, G, C and U respectively. If the sugar is a deoxyribose, ***deoxyribonucleosides*** are produced.

The term mononucleotide is used when a single phosphate moiety is added to a nucleoside. Thus adenosine monophosphate (AMP) contains adenine + ribose + phosphate.

The principal bases, their respective nucleosides and nucleotides found in the structure of nucleic acids are given in ***Table 5.1***. Note that the ***prefix 'd'*** is used ***to indicate*** if the sugar is ***deoxyribose*** (e.g. dAMP).

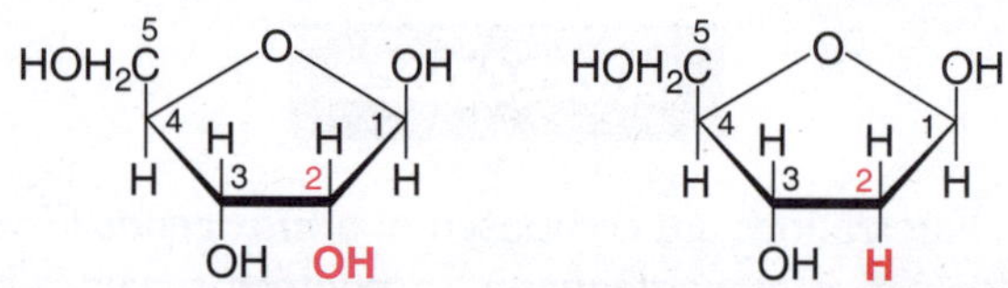

D-Ribose **D-2-Deoxyribose**

Fig.5.3 : *Structures of sugars present in nucleic acids (ribose in found in RNA and deoxyribose in DNA; Note the structural difference at C_2).*

Medical Concepts/ Clinical Correlates

- DNA is organized into genes, the fundamental units of genetic information. Genes control protein biosynthesis through the mediation of RNA.
- Nucleic acids are the polymers of nucleotides. Certain nucleotides serve as B-complex vitamin coenzymes (FAD, NAD^+, CoA), carriers of high energy intermediates (UDP-glucose, S-adenosylmethionine) and second messengers of hormonal action (cAMP, cGMP).
- Uric acid is a purine, and the end product of purine metabolism, that has been implicated in the disorder gout.
- Certain purine bases from plants such as caffeine (of coffee), theophylline (of tea) and theobromine (of cocoa) are of pharmacological interest.

The binding of nucleotide components

The atoms in the purine ring are numbered as 1 to 9 and for pyrimidine as 1 to 6 (See ***Fig.5.1***). The carbons of sugars are represented with an associated prime (′) for differentiation. Thus the pentose carbons are 1′ to 5′.

The pentoses are bound to nitrogenous bases by ***β-N-glycosidic bonds***. The N^9 of a purine ring binds with $C_{1(1')}$ of a pentose sugar to form a covalent bond in the purine nucleoside. In case of pyrimidine nucleosides, the glycosidic linkage is between N^1 of a pyrimidine and $C_{1'}$ of a pentose.

The hydroxyl groups of adenosine are esterified with phosphates to produce 5′- or 3′-monophosphates. 5′-Hydroxyl is the most commonly esterified, hence 5′ is usually omitted while writing nucleotide names. Thus AMP represents adenosine 5′-monophosphate. However, for adenosine 3′-monophosphate, the abbreviation 3′-AMP is used.

The structures of two selected nucleotides namely AMP and TMP are depicted in ***Fig.5.4***.

Nucleoside di- and triphosphates

Nucleoside monophosphates possess only one phosphate moiety (AMP, TMP). The addition of second or third phosphates to the nucleoside results in nucleoside diphosphate (e.g. ADP) or triphosphate (e.g. ATP), respectively.

STRUCTURE OF DNA

DNA is a ***polymer of deoxyribonucleotides*** (or simply ***deoxynucleotides***). It is composed of monomeric units namely deoxyadenylate (***dAMP***), deoxyguanylate (***dGMP***), deoxycytidylate (***dCMP***) and

TABLE 5.1 Principal bases, nucleosides and nucleotides

Base	*Ribonucleoside*	*Ribonucleotide (5′-monophosphate)*	*Abbreviation*
Adenine (A)	Adenosine	Adenosine 5′-monophosphate or adenylate	AMP
Guanine (G)	Guanosine	Guanosine 5′-monophosphate or guanylate	GMP
Cytosine (C)	Cytidine	Cytidine 5′-monophosphate or cytidylate	CMP
Uracil (U)	Uridine	Uridine 5′-monophosphate or uridylate	UMP
Base	*Deoxyribonucleoside*	*Deoxyribonucleotide (5′-monophosphate)*	*Abbreviation*
Adenine (A)	Deoxyadenosine	Deoxyadenosine 5′-monophosphate or deoxyadenylate	dAMP
Guanine (G)	Deoxyguanosine	Deoxyguanosine 5′-monophosphate or deoxyguanylate	dGMP
Cytosine (C)	Deoxycytidine	Deoxycytidine 5′-monophosphate or deoxycytidylate	dCMP
Thymine (T)	Deoxythymidine	Deoxythymidine 5′-monophosphate or deoxythymidylate	dTMP

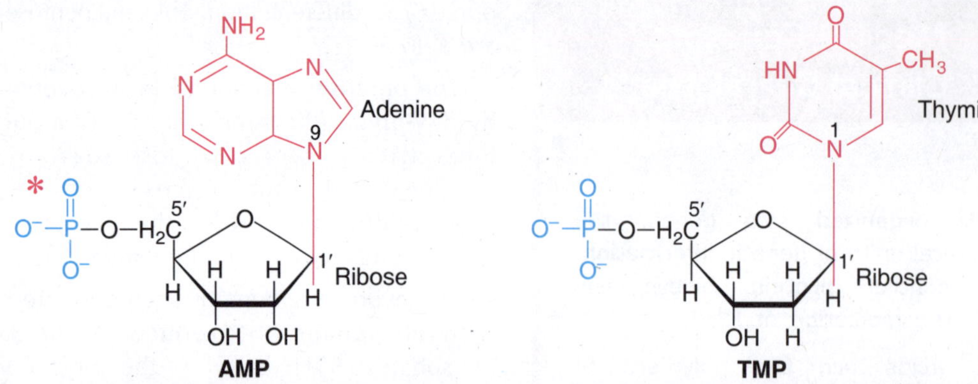

Fig.5.4 : *The structures of adenosine 5′-monophosphate (AMP) and thymidine 5′-monophosphate (TMP) (*-Addition of second or third phosphate gives adenosine diphosphate (ADP) and adenosine triphosphate (ATP) respectively).*

deoxythymidylate (***dTMP***) (It may be noted here that some authors prefer to use TMP for deoxythymidylate, since it is found only in DNA).

Chargaff's rule of DNA composition

Erwin Chargaff in late 1940s quantitatively analysed the DNA hydrolysates from different species. He observed that in all the species he studied DNA had equal numbers of adenine and thymine residues (A = T) and equal numbers of guanine and cytosine residues (G = C). This is known as Chargaff's rule of ***molar equivalence between the purines and pyrimidines in DNA*** structure.

Single stranded DNA, and RNAs which are usually single stranded, do not obey Chargaff's rule. However, double stranded RNA which is the genetic material in certain viruses satisfies Chargaff'ss rule.

DNA DOUBLE HELIX

The ***double helical*** structure of ***DNA*** was proposed by ***James Watson and Francis Crick*** in 1953 (Nobel Prize, 1962). The elucidation of DNA structure is considered as a ***milestone in*** the era of ***modern biology***. The structure of DNA double helix is comparable to a twisted ladder. The salient features of ***Watson-Crick model of DNA*** (now known as B-DNA) are given below (***Fig.5.5***).

1. The DNA is a right handed double helix. It consists of ***two polydeoxyribonucleotide chains*** (strands) twisted around each other on a common axis.

2. The two strands are ***antiparallel***, i.e., one strand runs in the 5′ to 3′ direction while the other in 3′ to 5′ direction. This is comparable to two parallel adjacent roads carrying traffic in opposite direction.

3. The width (or diameter) of a double helix is 20 A° (2 nm).

4. Each turn (pitch) of the helix is 34 A° (3.4 nm) with 10 pairs of nucleotides, each pair placed at a distance of about 3.4 A°.

5. The two strands are held together by ***hydrogen bonds*** formed by complementary base pairs. The A-T pair has 2 hydrogen bonds while G–C pair has 3 hydrogen bonds. The G–C is stronger by about 50% than A = T.

6. The complementary base pairing in ***DNA helix proves Chargaff's rule***. The content of adenine equals to that of thymine (***A*** = ***T***), and guanine equals to that of cytosine (***G*** = ***C***).

OTHER TYPES OF DNA STRUCTURE

It is now recognized that besides double helical structure, DNA also exists in certain unusual structures. It is believed that such structures are important for molecular recognition of DNA by proteins and enzymes. This is in fact needed for the DNA to discharge its functions in an appropriate manner. Some selected ***unusual structures of DNA*** are ***bent DNA***, ***triple-stranded DNA*** and ***four-standed DNA***.

Organization of DNA in the cell

In prokaryotic cells (e.g. bacteria) with no distinct nucleus, the DNA is rather loosely packed. In contrast, the eukaryotic DNA is found in association with

Medical Concepts/ Clinical Correlates

- Synthetic analogs of bases (5-fluorouracil, 6-mercaptopurine, 6-azauridine) are used to inhibit the growth of cancer cells.
- Certain antitumor drugs (e.g. cisplatin) can produce bent DNA structure and damage it.
- MicroRNAs and silencing RNAs, capable of inhibiting gene expression, are potential agents for therapeutic purposes.

basic proteins namely ***histones*** to form ***nucleosomes***. The nucleosomes in turn coil to produce chromatin in which form DNA is present in the chromosomes.

STRUCTURE OF RNA

RNA is a ***polymer of ribonucleotides*** held together by 3′,5′-phosphodiester bridges. Although RNA has certain similarities with DNA structure, they have several specific differences

1. **Pentose :** The sugar in ***RNA is ribose*** in contrast to deoxyribose in DNA.

2. **Pyrimidine :** RNA ***contains*** the pyrimidine ***uracil*** in place of thymine (in DNA).

3. **Single strand :** RNA is usually a single stranded polynucleotide. However, this strand may fold at certain places to give a double stranded structure, if complementary base pairs are in close proximity.

4. **Chargaff's rule—not obeyed :** Due to the single stranded nature, there is no specific relation between purine and pyrimidine contents. Thus the guanine content is not equal to cytosine (as is the case in DNA).

5. **Orcinol colour reaction :** RNAs can be histologically identified by orcinol colour reaction due to the presence of ribose.

TYPES OF RNA

The three major types of RNAs with their respective cellular composition are given below

1. ***Messenger RNA (mRNA)*** : *5–10%*
2. ***Transfer RNA (tRNA)*** : *10–20%*
3. ***Ribosomal RNA (rRNA)*** : *50–80%*

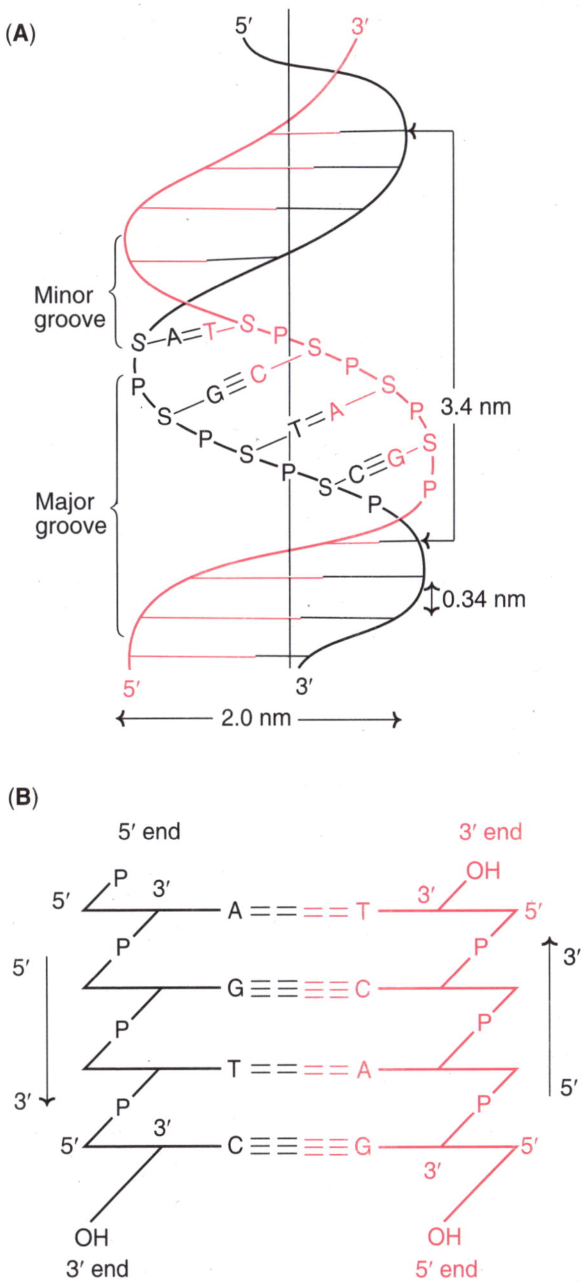

Fig.5.5 : (A) Watson-Crick model of DNA helix (B) Complementary base pairing in DNA helix..

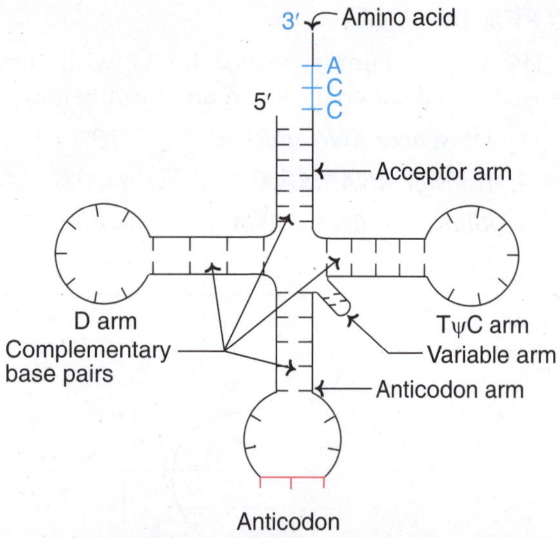

Fig.5.6 : Structure of transfer RNA.

Besides the three RNAs referred above, other RNAs are also present in the cells. These include ***heterogeneous nuclear RNA*** (***hnRNA***), small nuclear RNA (***snRNA***), small nucleolar RNA (***snoRNA***) and small cytoplasmic RNA (***scRNA***).

The RNAs are synthesized from DNA, and are primarily involved in the process of protein biosynthesis (***Chapter 25***). The RNAs vary in their structure and function.

Messenger RNA (mRNA)

The mRNA is synthesized in the nucleus (in eukaryotes) as ***heterogeneous nuclear RNA*** (***hnRNA***). hnRNA on processing liberates the functional mRNA which enter the cytoplasm to participate in protein synthesis. mRNA has high molecular weight with short half-life.

Transfer RNA (tRNA)

Transfer RNA (***soluble RNA***) molecule contains 71-80 nucleotides (mostly 75) with a molecular weight of about 25,000. There are at least 20 species of tRNAs corresponding to 20 amino acids present in protein structure.

The structure of tRNA depicted in ***Fig.5.6*** resembles that of a clover leaf. tRNA contains mainly four arms, each arm contains a base paired stem.

1. **The acceptor arm :** This arm is capped with a sequence CCA (5′ to 3′). The amino acid is attached to the acceptor arm.

2. **The anticodon arm :** This arm, with the three specific nucleotide bases (anticodon), is responsible for the recognition of triplet codon of mRNA. The codon and anticodon are complementary to each other.

3. **The D arm :** It is so named due to the presence of dihydrouridine.

4. **The TΨC arm :** This arm contains a sequence of T, pseudouridine (represented by psi, Ψ) and C.

5. **The variable arm :** This arm is the most variable in tRNA.

Ribosomal RNA (rRNA)

The ribosomes are the factories of protein synthesis. They are composed of two major nucleoprotein complexes–60S subunit and 40S

Table 5.3 Major Cellular RNAs and their function(s)

Type of RNA	*Abbreviation*	*Function(s)*
Messenger RNA	mRNA	Transfers genetic information from genes to ribosomes to synthesize proteins.
Heterogeneous nuclear RNA	hnRNA	Serves as a precursor for mRNA.
Transfer RNA	tRNA	Transfers amino acid to mRNA for protein biosynthesis.
Ribosomal RNA	rRNA	Provides structural framework for ribosomes.
Small nuclear RNA	snRNA	Involved in mRNA processing.
Small cytoplasmic RNA	scRNA	Involved in the selection of proteins for export.
Micro-RNA and silencing RNA	miRNA and siRNA	Both inhibit gene expression.

subunit. The 60S subunit contains 28S rRNA, 5S rRNA and 5.8S rRNA while the 40S subunit contains 18S rRNA. The function of rRNAs in ribosomes is not clearly known. It is believed that they play a significant role in the binding of mRNA to ribosomes and protein synthesis.

Other RNAs

The various RNAs and their respective functions are summarised in ***Table 5.2***.

CATALYTIC RNAs—RIBOZYMES

In certain instances, the RNA component of a ribonucleoprotein (RNA in association with protein) is catalytically active. Such RNAs are termed as ribozymes.

Ribonuclease P (RNase P) is a ribozyme containing protein and RNA component. It cleaves tRNA precursors to generate mature tRNA molecules.

RNA molecules are known to adapt tertiary structure just like proteins (i.e. enzymes). The specific conformation of RNA may be responsible for its function as biocatalyst. It is believed that ***ribozymes*** (RNAs) were functioning as ***catalysts before the occurrence of protein enzymes***, during the course of evolution.

SUMMARY

1. *DNA is the chemical basis of heredity organized into genes, the basic units of genetic information.*
2. *RNAs (mRNA, tRNA and rRNA) are produced by DNA which in turn carry out protein synthesis.*
3. *Nucleic acids are the polymers of nucleotides (polynucleotides) held by 3′ and 5′ phosphodiester bridges. A nucleotide essentially consists of base + sugar (nucleoside) and phosphate.*
4. *Besides being the constituents of nucleic acid structure, nucleotides perform a wide variety of cellular functions (e.g. energy carriers, metabolic regulators, second messengers etc.)*
5. *Both DNA and RNA contain the purines-adenine (A) and guanine (G) and the pyrimidine-cytosine (C). The second pyrimidine is thymine (T) in DNA while it is uracil (U) in RNA.*
6. *The structure of DNA is a double helix (Watson-Crick model) composed of two antiparallel strands of polydeoxynucleotides twisted around each other. The strands are held together by 2 or 3 hydrogen bonds formed between the bases i.e. A = T; G ... C. DNA structure satisfies Chargaff's rule that the content of A is equal to T and that of G equal to C.*

ENZYMES

6

Enzymes are biocatalysts—the catalysts of life. A ***catalyst*** is defined as a ***substance that increases the velocity*** or rate ***of a chemical reaction*** without itself undergoing any change in the overall process.

Enzymes may be defined as biocatalysts synthesized by living cells. They are protein in nature, colloidal and thermolabile in character, and specific in their action.

In recent years, certain ***non-protein enzymes*** (chemically ***RNA***) have also been identified.

NOMENCLATURE AND CLASSIFICATION

In the early days, the enzymes were given ***names*** by their discoverers ***in an arbitrary manner***. For example, the names pepsin, trypsin and chymotrypsin convey no information about the function of the enzyme or the nature of the substrate on which they act.

Enzymes are sometimes considered under two broad categories : (a) **Intracellular enzymes**—They are functional within cells where they are synthesized. (b) **Extracellular enzymes**—These enzymes are active outside the cell; all the digestive enzymes belong to this group.

The International Union of Biochemistry (***IUB***) appointed an ***Enzyme Commission*** in 1961. This committee made a thorough study of the existing enzymes and devised some basic principles for the classification and nomenclature of enzymes. Since 1964, the ***IUB system of enzyme classification*** has been in force. Enzymes are divided into ***six major classes*** (in that order). Each class on its own represents the general type of reaction brought about by the enzymes of that class.

1. **Oxidoreductases :** Enzymes involved in ***oxidation-reduction*** reactions.

2. **Transferases :** Enzymes that catalyse the ***transfer of functional groups***.

3. **Hydrolases :** Enzymes that bring about ***hydrolysis*** of various compounds.

4. **Lyases :** Enzymes specialised in the ***addition or removal of water***, ammonia, CO_2 etc.

5. **Isomerases :** Enzymes involved in all the ***isomerization*** reactions.

6. **Ligases :** Enzymes catalysing the ***synthetic reactions*** (*Greek* : ligate—to bind) where two molecules are joined together and ATP is used.

[The word ***OTHLIL*** (first letter in each class) may be memorised to remember the six classes of enzymes in the correct order].

Each class in turn is subdivided into many sub-classes which are further divided. A four digit ***E***nzyme ***C***ommission (***E.C.***) number is assigned to each enzyme representing the class (first digit), sub-class (second digit), sub-sub class (third digit) and the individual enzyme (fourth digit).

In the ***Table 6.1***, selected examples for the six classes of enzymes are given.

Although the IUB names for the enzymes are specific and unambiguous, they have not been accepted for general use as they are complex and cumbersome to remember. Therefore, the ***trivial names, along with the E.C. numbers*** as and when needed, are ***commonly used and widely accepted***.

CHEMICAL NATURE OF ENZYMES

All the enzymes are invariably proteins. In recent years, however, a few RNA molecules have been shown to function as enzymes. Each enzyme has its own tertiary structure and specific conformation which is very essential for its catalytic activity. The functional unit of the enzyme is known as ***holoenzyme*** which is often made up of ***apoenzyme*** (the protein part) and a ***coenzyme*** (non-protein organic part).

FACTORS AFFECTING ENZYME ACTIVITY

The contact between the enzyme and substrate is the most essential pre-requisite for enzyme activity. The important factors that influence the velocity of the enzyme reaction are discussed hereunder

1. Concentration of enzyme

As the concentration of the ***enzyme is increased, the velocity of the reaction proportionately increases*** (***Fig.6.1***). In fact, this property of enzyme is made use in determining the serum enzymes for the diagnosis of diseases.

TABLE 6.1 Classification of enzymes

*Enzyme class with examples**	*Reaction catalysed*
1. **Oxidoreductases** Alcohol dehydrogenase (alcohol : NAD^+ oxidoreductase E.C. 1.1.1.1.), cytochrome oxidase, L- and D-amino acid oxidases	Oxidation ⟶ Reduction $AH_2 + B \longrightarrow A + BH_2$
2. **Transferases** Hexokinase (ATP : D-hexose 6-phosphotransferase, E.C. 2.7.1.1.), transaminases, transmethylases, phosphorylase	Group transfer $A-X+B \longrightarrow A+B-X$
3. **Hydrolases** Lipase (triacylglycerol acyl hydrolase E.C. 3.1.1.3), choline esterase, acid and alkaline phosphatases, pepsin, urease	Hydrolysis $A-B+H_2O \longrightarrow AH+BOH$
4. **Lyases** Aldolase (fructose-bisphosphate aldolase, E.C. 4.1.2.13), fumarase, histidase	Addition ⟶ Elimination $A-B+X-Y \longrightarrow AX-BY$
5. **Isomerases** Triose phosphate isomerase (D-glyceraldehyde 3-phosphate ketoisomerase, E.C. 5.3.1.1), retinol isomerase, phosphohexose isomerase	Interconversion of isomers $A \longrightarrow A'$
6. **Ligases** Glutamine synthetase (L-glutamate ammonia ligase, E.C. 6.3.1.2), acetyl CoA carboxylase, succinate thiokinase	Condensation (usually dependent on ATP) $A+B \xrightarrow[\text{ATP} \quad \text{ADP + Pi}]{} A-B$

* *For one enzyme in each class, systematic name along with E.C. number is given in the brackets.*

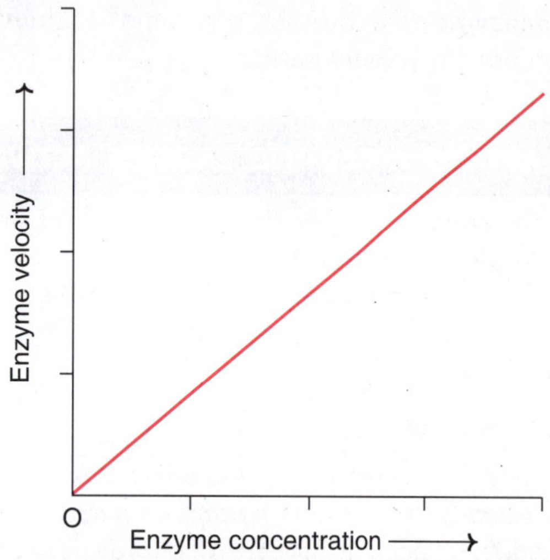

Fig.6.1 : Effect of enzyme concentration on enzyme velociy.

2. Concentration of substrate

Increase in the substrate concentration gradually ***increases the velocity of enzyme reaction*** within the limited range of substrate levels. A rectangular hyperbola is obtained when velocity is plotted against the substrate concentration (***Fig.6.2***). Three distinct phases of the reaction are observed in the graph.

Enzyme kinetics and K_m value : The enzyme (E) and substrate (S) combine with each other to form an unstable enzyme-substrate complex (ES) for the formation of product (P).

$$E + S \underset{k_2}{\overset{k_1}{\rightleftharpoons}} ES \xrightarrow{k_3} E + P$$

Here k_1, k_2 and k_3 represent the velocity constants for the respective reactions, as indicated by arrows.

K_m, the ***Michaelis-Menten constant*** (or ***Briggs and Haldane constant***), is given by the formula

$$K_m = \frac{k_2 + k_3}{k_1}$$

The following equation is obtained after suitable algebraic manipulation.

$$v = \frac{V_{max}\,[S]}{K_m + [S]} \quad \text{equation (1)}$$

where v = Measured velocity,
V_{max} = Maximum velocity,
S = Substrate concentration,
K_m = Michaelis – Menten constant.

K_m* or the *Michaelis-Menten constant is defined as the ***substrate concentration*** (expressed in moles/lit) ***to produce half-maximum velocity*** in an enzyme catalysed reaction. It indicates that half of the enzyme molecules (i.e. 50%) are bound with the substrate molecules when the substrate concentration equals the K_m value.

K_m value is a constant and a characteristic feature of a given enzyme (comparable to a thumb impression or signature). It is a representative for measuring the strength of ES complex. A ***low K_m value indicates a strong affinity between enzyme and substrate***, whereas a high K_m value reflects a weak affinity between them. For majority of enzymes, the K_m values are in the range of 10^{-5} to 10^{-2} moles. It may, however,

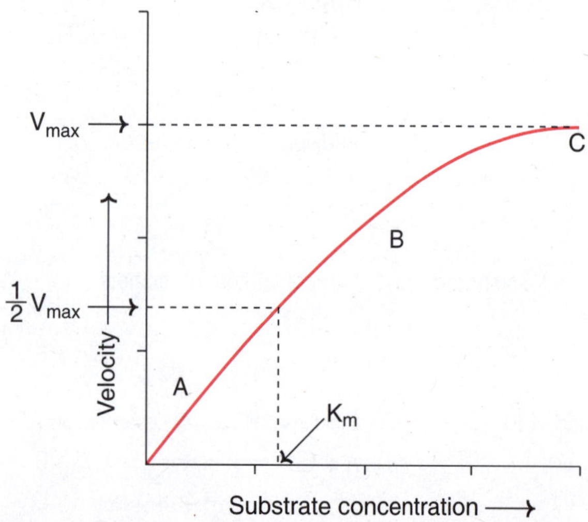

Fig.6.2 : Effect of substrate concentration on enzyme velocity (A-linear; B-curve; C-almost unchanged).

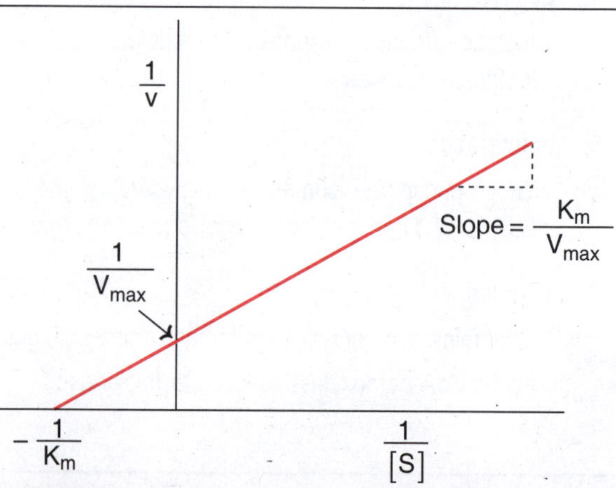

Fig.6.3 : Lineweaver-Burk double reciprocal plot.

be noted that K_m is not dependent on the concentration of enzyme.

Lineweaver-Burk double reciprocal plot : For the determination of K_m value, the substrate saturation curve (***Fig.6.2***) is not very accurate since V_{max} is approached asymptotically. By taking the reciprocals of the equation (1), a straight line graphic representation is obtained.

The Lineweaver-Burk plot is shown in ***Fig.6.3***. It is much easier to calculate the K_m from the intercept on x-axis which is $-(1/K_m)$. Further, the double reciprocal plot is ***useful in understanding the effect of various inhibitions*** (discussed later).

3. Effect of temperature

Velocity of an enzyme reaction ***increases with increase in temperature up*** to a maximum and then declines. A bell-shaped curve is usually observed (***Fig.6.4***).

The optimum temperature for most of the enzymes is between 40°C–45°C. However, a few enzymes (e.g. venom phosphokinases, muscle adenylate kinase) are active even at 100°C.

In general, when the enzymes are exposed to a temperature above 50°C, ***denaturation*** leading to derangement in the native (tertiary) structure of the protein and active site are seen. Majority of the enzymes become inactive at higher temperature (above 70°C).

4. Effect of pH

Increase in the hydrogen ion concentration (pH) considerably influences the enzyme activity and a

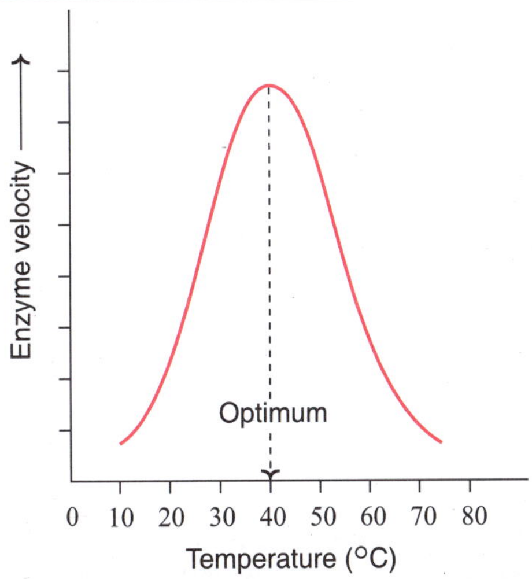

Fig.6.4 : Effect of temperature on enzyme velocity.

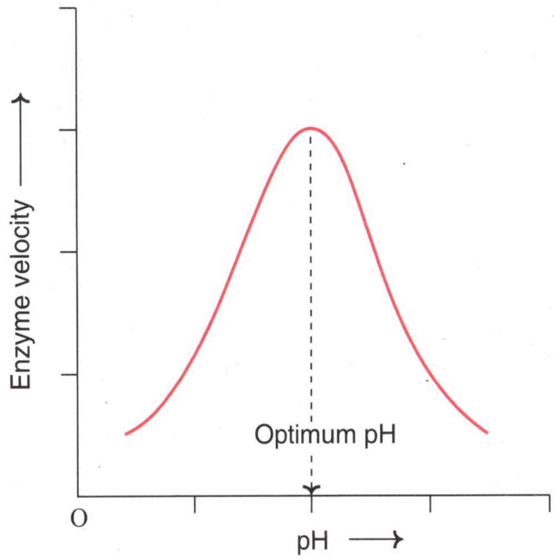

Fig.6.5 : Effect of pH on enzyme velocity.

bell-shaped curve is normally obtained (***Fig.6.5***). Each enzyme has an optimum pH at which the velocity is maximum.

Most of the enzymes of higher organisms show ***optimum activity around neutral pH*** (***6-8***). There are, however, many exceptions like pepsin (1-2), acid phosphatase (4-5) and alkaline phosphatase (10-11) for optimum pH.

5. Effect of product concentration

The accumulation of reaction products generally decreases the enzyme velocity. For certain enzymes, the products combine with the active site of enzyme and form a loose complex and, thus, inhibit the enzyme activity. In the living system, this type of inhibition is generally prevented by a quick removal of products formed.

6. Effect of activators

Some of the enzymes require certain inorganic ***metallic cations*** like Mg^{2+}, Mn^{2+}, Zn^{2+}, Ca^{2+}, Co^{2+}, Cu^{2+}, Na^{+}, K^{+} etc. for their optimum activity. Rarely, anions are also needed for enzyme activity e.g. chloride ion (Cl^-) for amylase.

Two categories of enzymes requiring metals for their activity are distinguished

(a) **Metal-activated enzymes :** The metal is not tightly held by the enzyme and can be exchanged easily with other ions e.g. ***ATPase*** (***Mg^{2+}*** and Ca^{2+}) Enolase (Mg^{2+}).

Medical Concepts/ Clinical Correlates

- The existence of life is unimaginable without the presence of enzymes—the biocatalysts.
- Competitive inhibitors of certain enzymes are of great biological significance. Allopurinol, employed in the treatment of gout, inhibits xanthine oxidase to reduce the formation of uric acid. The other competitive inhibitors include aminopterin used in the treatment of cancers, sulfanilamide as antibacterial agent and dicumarol as an anticoagulant.
- Penicillin antibiotics irreversibly inhibit serine containing enzymes of bacterial cell wall synthesis.

(b) **Metalloenzymes :** These enzymes hold the metals rather tightly which are not readily exchanged. e.g. alcohol dehydrogenase, carbonic anhydrase, alkaline phosphatase, carboxypeptidase and aldolase contain zinc.

Phenol oxidase (copper);

Pyruvate oxidase (manganese);

Xanthine oxidase (molybdenum);

Cytochrome oxidase (iron and copper).

ACTIVE SITE

Enzymes are big in size compared to substrates which are relatively smaller. Evidently, a small portion of the huge enzyme molecule is directly involved in the substrate binding and catalysis (***Fig.6.6***).

The active site (or active centre) of an enzyme is defined as the small region at which the substrate(s) binds and participates in the catalysis.

Salient features of active site

1. The existence of active site is ***due to the tertiary*** structure of protein resulting in ***three dimensional native conformation***.

2. The active site is made up of amino acids (known as ***catalytic residues***) which are far from each other in the linear sequence of amino acids (primary structure of protein). For instance, the enzyme ***lysozyme*** has 129 amino acids. The ***active site*** is formed by the contribution of amino acid ***residues*** numbered ***35, 52, 62, 63 and 101***.

3. The active site is not rigid in structure and shape. It is rather ***flexible*** to promote the specific substrate binding.

4. Generally, the active site possesses a ***substrate binding site*** and a ***catalytic site.*** The latter is for the catalysis of the specific reaction.

5. Of the 20 amino acids that could be present in enzyme structure, only some of them are ***repeatedly found at the active sites*** of various enzymes. These amino acids are ***serine***, ***aspartate***, ***histidine***, cysteine, lysine, arginine, glutamate, tyrosine etc. Among these amino acids, ***serine*** is the most frequently found at the active sites.

6. The substrate[S] binds the enzyme (E) at the active site to form enzyme-substrate complex (ES). The product (P) is released after the catalysis and the enzyme is available for reuse.

$$E + S \rightleftharpoons ES \longrightarrow E + P$$

ENZYME INHIBITION

Enzyme inhibitor is defined as a substance which binds with the enzyme and brings about a ***decrease in catalytic activity*** of that ***enzyme***. The inhibitor

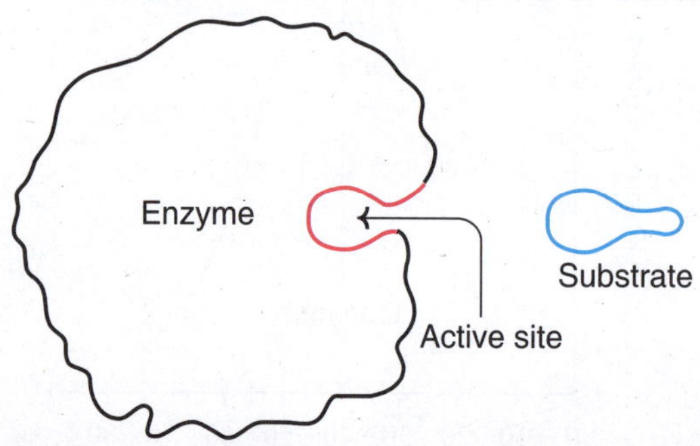

Fig.6.6 : *A diagrammatic representation of an enzyme with active site.*

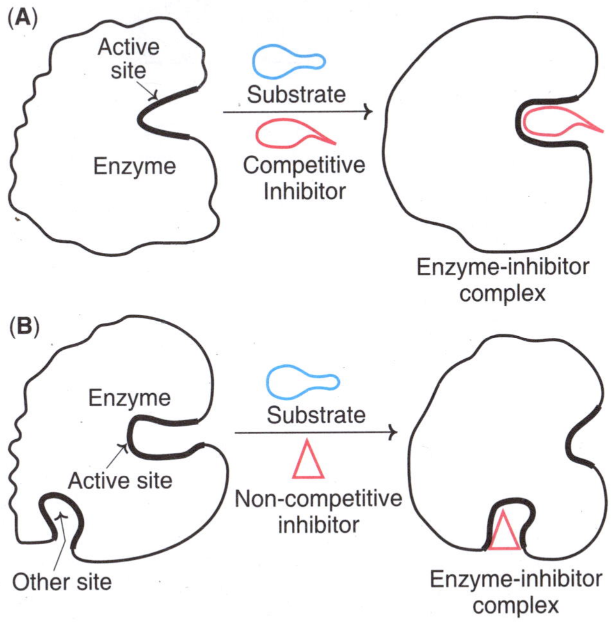

***Fig.6.7 :** A diagrammatic representation of an (**A**) Competitive and (**B**) Non-competitive inhibition.*

may be organic or inorganic in nature. There are three broad categories of enzyme inhibition

1. Reversible inhibition.
2. Irreversible inhibition.
3. Allosteric inhibition.

1. Reversible inhibition

The inhibitor binds non-covalently with enzyme and the enzyme inhibition can be reversed if the inhibitor is removed. The reversible inhibition is further sub-divided into

I. Competitive inhibition (***Fig.6.7A***)

II. Non-competitive inhibition (***Fig.6.7B***)

I. **Competitive inhibition :** The inhibitor (I) which closely resembles the real substrate (S) is regarded as a ***substrate analogue***. The ***inhibitor competes with substrate*** and binds at the active site of the enzyme but does not undergo any catalysis. As long as the competitive inhibitor holds the active site, the enzyme is not available for the substrate to bind.

The relative concentration of the substrate and inhibitor and their respective affinity with the enzyme determines the degree of competitive inhibition. The inhibition could be overcome by a high substrate concentration. In competitive inhibition, the ***K_m value increases*** whereas ***V_{max}*** remains ***unchanged***.

The enzyme ***succinate dehydrogenase*** (SDH) is a classical example of competitive inhibition with succinic acid as its substrate. ***Malonic acid has structural similarity with succinic acid*** and compete with the substrate for binding at the active site of SDH.

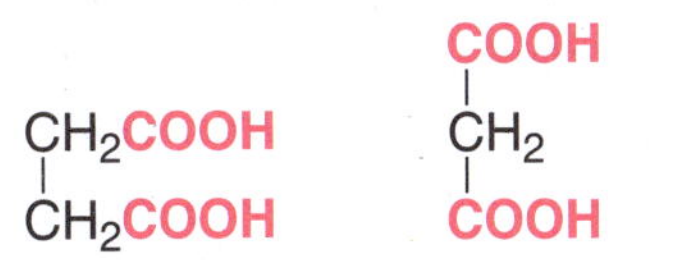

Succinic acid **Malonic acid**

Methanol is ***toxic*** to the body when it is converted to formaldehyde by the enzyme alcohol dehydrogenaes (ADH). Ethanol can compete with methanol for ADH. Thus, ***ethanol can be used in the treatment of methanol poisoning***.

Some more examples of the enzymes with substrates and competitive inhibitors of biological significance are given in ***Table 6.2***.

Antimetabolites : These are the chemical ***compounds that block the metabolic reactions*** by their inhibitory action on enzymes. Antimetabolites are usually structural analogues of substrates, and thus are competitive inhibitors (***Table 6.2***). They are in use for the treatment of cancer, gout etc. The term ***antivitamins*** is used for the antimetabolites which block the biochemical actions of vitamins causing deficiencies. e.g., sulphanilamide, dicumarol.

II. **Non-competitive inhibition :** The inhibitor binds at a site other than the active site on the enzyme surface (***Fig.6.7B***). This binding impairs the enzyme function. The ***inhibitor has no structural resemblance with the substrate***. However, there usually exists a strong affinity for the inhibitor to bind at the second site. In fact, the inhibitor does not interfere with the enzyme-substrate binding. But the catalysis is prevented, possibly due to a distortion in the enzyme conformation.

The non-competitive inhibitor generally binds with the enzyme as well as the ES complex.

For non-competitive inhibition, the ***K_m value is unchanged*** while ***V_{max} is lowered***.

Heavy metal ions (Ag^+, Pb^{2+}, Hg^{2+} etc.) can non-competitively inhibit the enzymes by binding with cysteinyl sulfhydryl groups.

2. Irreversible inhibition

The inhibitors ***bind covalently with the enzymes and inactivate them***, which is irreversible. These inhibitors are usually toxic substances which may be present naturally or man-made.

The ***penicillin antibiotics*** act as irreversible inhibitors of serine – containing enzymes, and ***block the bacterial cell wall synthesis***. ***Cyanide*** inhibits ***cytochrome oxidase*** (binds to iron atom) of electron transport chain. ***Fluoride inhibits enolase*** (by removing manganese), and thus ***glycolysis***.

Diisopropyl fluorophosphate (DFP) is a ***nerve gas*** developed by the Germans during Second World War. DFP irreversibly binds with enzymes containing serine at the active site, e.g. ***serine proteases***, ***acetylcholine esterase***.

Suicide inhibition : In this type of irreversible inhibition, the original inhibitor is converted to a more potent form by the same enzyme that ought to be inhibited e.g., ***allopurinol***, an inhibitor of xanthine oxidase, gets converted ***to alloxanthine***, a more effective inhibitor of the enzyme (Refer ***Chapter 17***).

3. Allosteric inhibition

The details of this type of inhibition are given under allosteric regulation as a part of the regulation of enzyme activity in the living system.

ENZYME SPECIFICITY

Enzymes are highly specific in their action when compared with the chemical catalysts. The occurrence of thousands of enzymes in the biological system might be due to the specific nature of enzymes. ***Three types*** of enzyme specificity are well-recognised

1. Stereospecificity, 2. Reaction specificity,
3. Substrate specificity.

Specificity is a characteristic property of the active site.

1. **Stereospecificity or optical specificity :** Stereoisomers are the compounds which have the same molecular formulae, but differ in their structural configuration.

The ***enzymes act only on one isomer*** and, therefore, exhibit stereoisomerism.

e.g. L-amino acid oxidase and D-amino acid oxidase act on L- and D-amino acids respectively.

Hexokinase acts on D-hexoses;
Glucokinase on D-glucose;
Amylase acts on α-glycosidic linkages;
Cellulase cleaves β-glycosidic bonds.

Stereospecificity is explained by considering three distinct regions of substrate molecule specifically binding with three complementary regions on the

TABLE 6.2 Selected examples of enzymes with their respective substrates and competitive inhibitors

Enzyme	*Substrate(s)*	*Inhibitor(s)*	*Significance of inhibitor(s)*
Xanthine oxidase	Hypoxanthine xanthine	Allopurinol	Used in the control of gout to reduce excess production of uric acid from hypoxanthine.
Monoamine oxidase	Catecholamines (epinephrine, norepinephrine)	Ephedrine, amphetamine	Useful for elevating catecholamine levels.
Dihydrofolate reductase	Dihydrofolic acid	Aminopterin, amethopterin, methotrexate	Employed in the treatment of leukemia and other cancers.
Acetylcholine esterase	Acetylcholine	Succinyl choline	Used in surgery for muscle relaxation, in anaesthetised patients.
Dihydropteroate synthase	Para aminobenzoic acid (PABA)	Sulfonilamide	Prevents bacterial synthesis of folic acid.
Vitamin K epoxide reductase	Vitamin K	Dicumarol	Acts as an anticoagulant.
HMG CoA reductase	HMG CoA	Lovastatin, compactin	Inhibit cholesterol biosynthesis

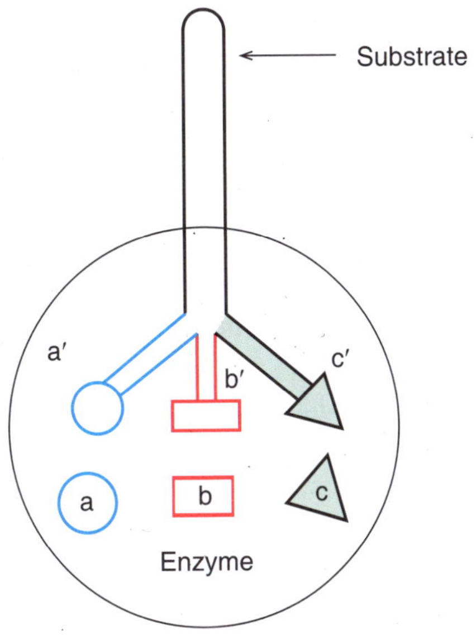

Fig.6.8 : Diagrammatic representation of stereo-specificity (a′, b′, c′)—three point attachment of substrate to the enzyme (a, b, c).

surface of the enzyme (***Fig.6.8***). The class of enzymes belonging to ***isomerases do not exhibit stereo-specificity***, since they are specialized in the interconversion of isomers.

2. **Reaction specificity :** The ***same substrate*** can undergo ***different types of reactions***, each catalysed by a separate enzyme and this is referred to as reaction specificity. An amino acid can undergo transamination, oxidative deamination, decarboxylation, racemization etc. The enzymes however, are different for each of these reactions.

3. **Substrate specificity :** The substrate specificity varies from enzyme to enzyme. It may be either absolute, relative or broad.

(a) **Absolute substrate specificity :** Certain enzymes ***act only on one substrate*** e.g. ***glucokinase*** acts on glucose to give glucose 6-phosphate, urease cleaves urea to ammonia and carbon dioxide.

(b) **Relative substrate specificity :** Some enzymes ***act on structurally related substances***. This, in turn, may be dependent on the specific group or a bond present. The action of trypsin and chymotrypsin is a good example for ***group specificity*** (***Refer Fig.8.4***). Trypsin hydrolyses peptide linkage involving arginine or lysine. Chymotrypsin cleaves peptide bonds attached to aromatic amino acids (phenylalanine, tyrosine and tryptophan). Examples of ***bond specificity***-glycosidases acting on glycosidic bonds of carbohydrates, lipases cleaving ester bonds of lipids etc.

(c) **Broad specificity :** Some enzymes ***act on closely related substrates*** which is commonly known as broad substrate specificity, e.g. ***hexokinase*** acts on glucose, fructose, mannose and glucosamine and not on galactose.

COENZYMES

Many enzymes require certain ***non-protein*** small additional factors, collectively referred to as cofactors for catalysis. The cofactors may be ***organic or inorganic*** in nature.

The non-protein, organic, low molecular weight and dialysable substance associated with enzyme function is known as coenzyme.

The functional enzyme is referred to as ***holoenzyme*** which is made up of a protein part ***(apoenzyme)*** and a non-protein part ***(coenzyme)***. The term ***activator*** is referred to the inorganic cofactor (like Ca^{2+}, Mg^{2+}, Mn^{2+} etc.) necessary to enhance enzyme activity.

Coenzymes are second substrates : Coenzymes are often regarded as the second substrates or ***co-substrates***, since they have affinity with the enzyme comparable with that of the substrates. Coenzymes ***undergo alterations*** during the enzymatic reactions, which are ***later regenerated***. This is in contrast to the substrate which is converted to the product.

Coenzymes participate in various reactions involving ***transfer of atoms or groups*** like hydrogen, aldehyde, keto, amino, acyl, methyl, carbon dioxide etc. Coenzymes play a decisive role in enzyme function.

Coenzymes from B-complex vitamins : Most of the coenzymes are the derivatives of water soluble B-complex vitamins. In fact, the ***biochemical functions of B-complex vitamins are exerted through their respective coenzymes***. The chapter on vitamins gives the details of structure and function of the coenzymes (***Chapter 7***). In ***Table 6.3***, a summary of the vitamin related coenzymes with their functions is given.

Non-vitamin coenzymes : Not all coenzymes are vitamin derivatives. There are some other organic substances, which have no relation with vitamins but function as coenzymes. They may be considered as non-vitamin coenzymes e.g. ***ATP, CDP, UDP*** etc.

Nucleotide coenzymes : Some of the coenzymes possess nitrogenous base, sugar and phosphate. Such coenzymes are, therefore, regarded as nucleotides e.g. NAD^+, $NADP^+$, FMN, FAD, coenzyme A, UDP etc.

Coenzymes do not decide enzyme specificity : A particular coenzyme may participate in catalytic reactions along with different enzymes. For instance, NAD^+ acts as a coenzyme for lactate dehydrogenase and alcohol dehydrogenase. In both the enzymatic reactions, NAD^+ is involved in hydrogen transfer. The ***specificity of the enzyme is mostly dependent on the apoenzyme and not on the coenzyme***.

MECHANISM OF ENZYME ACTION

Catalysis is the prime function of enzymes. The nature of catalysis taking place in the biological system is similar to that of non-biological catalysis. For any chemical reaction to occur, the reactants have to be in an activated state or transition state.

Enzymes lower activation energy : The energy required by the reactants to undergo the reaction is known as ***activation energy.*** The reactants when heated attain the activation energy. The catalyst (or the enzyme in the biological system) reduces the activation energy and this causes the reaction to proceed at a lower temperature. Enzymes do not alter the equilibrium constants, they only ***enhance the velocity of the reaction***.

The role of catalyst or enzyme is comparable with a tunnel made in a mountain to reduce the barrier as illustrated in ***Fig.6.9***. The enzyme lowers energy barrier of reactants, thereby making the reaction go faster. The ***enzymes reduce the activation energy of the reactants*** in such a way that all the biological systems occur at body temperature (below 40°C).

Enzyme-substrate complex formation

The prime requisite for enzyme catalysis is that the substrate (***S***) must combine with the enzyme (***E***) at the active site to form enzyme-substrate complex (***ES***) which ultimately results in the product formation (***P***).

$$E + S \rightleftharpoons ES \longrightarrow E + P$$

A few theories have been put forth to explain mechanism of enzyme-substrate complex formation.

TABLE 6.3 Coenzymes of B-complex vitamins*

Coenzyme (abbreviation)	Derived from vitamin	Atom or group transferred	Dependent enzyme (example)
Thiamine pyrophosphate (TPP)	Thiamine	Aldehyde or keto	Transketolase
Flavin mononucleotide (FMN)	Riboflavin	Hydrogen and electron	L - Amino acid oxidase
Flavin adenine dinucleotide (FAD)	Riboflavin	"	D - Amino acid oxidase
Nicotinamide adenine dinucleotide(NAD^+)	Niacin	"	Lactate dehydrogenase
Nicotinamide adenine dinucleotide phosphate ($NADP^+$)	"	"	Glucose 6-phosphate dehydrogenase
Lipoic acid	Lipoic acid	"	Pyruvate dehydrogenase complex
Pyridoxal phosphate (PLP)	Pyridoxine	Amino or keto	Alanine transaminase
Coenzyme A (CoA)	Pantothenic acid	Acyl	Thiokinase
Tetrahydrofolate (FH_4)	Folic acid	One carbon (formyl, methenyl etc.)	Formyl transferase
Biotin coenzyme	Biotin	CO_2	Pyruvate carboxylase
Methylcobalamin; Deoxyadenosyl cobalamin	Cobalamin	Methyl/isomerisation	Methylmalonyl CoA mutase

Details for each coenzyme are given in the chapter 7 on vitamins

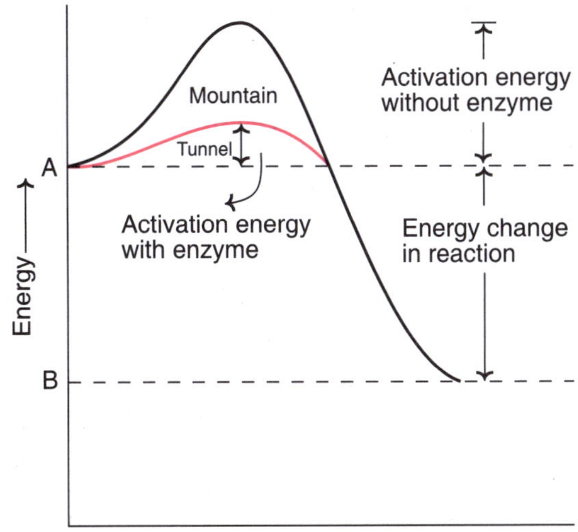

Fig.6.9 : *Effect of enzyme on activation energy of a reaction (A is the substrate and B is the product. Enzyme decreases activation energy).*

Lock and key model or Fischer's template theory

According to this model, the structure or conformation of the ***enzyme is rigid***. The substrate fits to the binding site (now active site) just as a key fits into the proper lock. Thus the ***active site*** of an enzyme is ***a rigid*** and pre-shaped template where only a specific substrate can bind (***Fig.6.10A***). This model does not give any scope for the flexible nature of enzymes, hence the model totally ***fails to explain*** many facts of ***enzymatic reactions***, the most important being the effect of allosteric modulators.

Induced fit theory or Koshland's model

Koshland, in 1958, proposed a more acceptable and realistic model for enzyme-substrate complex formation. As per this model, the ***active site is not rigid and pre-shaped***. The essential features of the substrate binding site are present at the nascent active site. The interaction of the ***substrate*** with the enzyme ***induces*** a fit or a ***conformation change in the enzyme***, resulting in the formation of a strong substrate binding site. Further, due to induced fit, the appropriate amino acids of the enzyme are repositioned to form the active site and bring about the catalysis (***Fig.6.10B***).

Induced fit model has ***sufficient experimental evidence*** from the X-ray diffraction studies. Koshland's model also explains the action of allosteric modulators and competitive inhibition on enzymes.

Substrate strain theory

In this model, the substrate is strained due to the induced conformation change in the enzyme (***Fig.6.10C***). It is also possible that when a substrate binds to the preformed active site, the ***enzyme induces a strain to the substrate***. The strained substrate leads to the formation of product. The concept of substrate strain explains the role of enzyme in increasing the rate of reaction.

In fact, a ***combination of the induced fit model with the substrate strain*** is considered to be ***operative*** in the enzymatic action.

MECHANISM OF ENZYME CATALYSIS

The formation of an enzyme-substrate complex (ES) is very crucial for the catalysis to occur. It is estimated that an enzyme catalysed reaction proceeds 10^6 to 10^{12} times faster than a non- catalysed reaction. It is worthwhile to briefly understand the ways and means through which the catalytic process takes place leading to the product formation. The enhancement in the rate of the reaction is mainly due to four processes :

1. Acid-base catalysis;
2. Substrate strain;
3. Covalent catalysis;
4. Entropy effects.

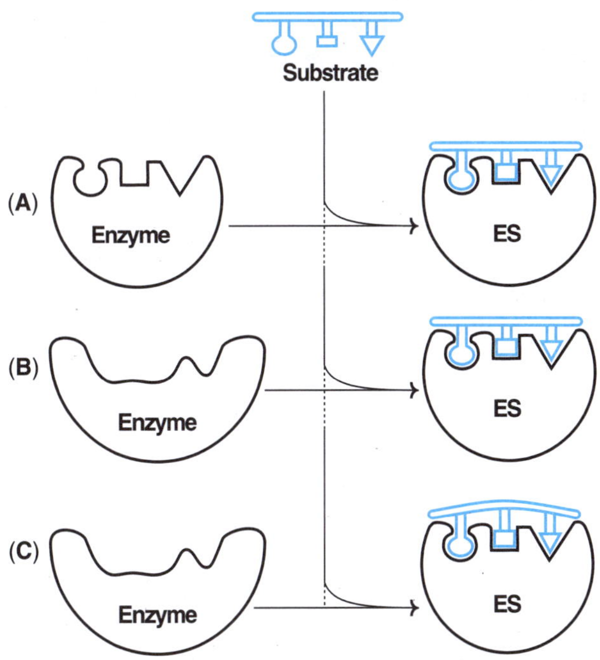

Fig.6.10 : *Mechanism of enzyme-substrate (ES) complex formation* ***(A)*** *Lock and key model;* ***(B)*** *Induced fit theory* ***(C)*** *Substrate strain theory.*

REGULATION OF ENZYME ACTIVITY IN THE LIVING SYSTEM

The various factors that influence enzyme activity in the laboratory (*in vitro*) have been discussed. Some of these factors like temperature and pH are quite constant in the living cell. In biological system, regulation of enzyme activities occurs at different stages in one or more of the following ways to achieve cellular economy.

1. Allosteric inhibition
2. Activation of latent enzymes
3. Compartmentation of metabolic pathways
4. Control of enzyme synthesis
5. Enzyme degradation
6. Isoenzymes

1. Allosteric regulation or allosteric inhibition

Some of the enzymes possess additional sites, known as allosteric sites (*Greek* : allo—other), besides the active site. Such enzymes are known as allosteric enzymes. The ***allosteric sites*** are unique places on the enzyme molecule.

Allosteric effectors : Certain substances referred to as allosteric modulators (effectors or modifiers) ***bind at the allosteric site and regulate the enzyme activity***. The enzyme activity is increased when a positive (+) allosteric effector binds at the allosteric site known as activator site. On the other hand, a negative (–) allosteric effector binds at the allosteric site called inhibitor site and inhibits the enzyme activity. e.g., for the enzyme phosphofructokinase (of glycolysis), ATP acts as an allosteric inhibitor, while ADP and AMP are allosteric activators.

Feedback regulation : The process of ***inhibiting the first step by the final product***, in a series of enzyme catalysed reactions of a metabolic pathway is referred to as feedback regulation. Look at the series of reactions given below

$$A \xrightarrow{e_1} B \xrightarrow{e_2} C \xrightarrow{e_3} D \xrightarrow{e_4} E$$

A is the initial substrate, B, C, and D are the intermediates and E is the end product, in a pathway catalysed by four different enzymes (e_1, e_2, e_3, e_4). The very first step ($A \rightarrow B$ by the enzyme e_1) is the most effective for regulating the pathway, by the final end product E. This type of control is often called ***negative feedback regulation*** since increased levels of end product will result in its (e_1) decreased synthesis. This is a real ***cellular economy to save the cell from the wasteful expenditure*** of synthesizing a compound which is already available within the cell.

Feedback inhibition or ***end product inhibition*** is a specialised type of allosteric inhibition necessary to control metabolic pathways for efficient cellular function.

Aspartate transcarbamoylase (ATCase) is a good example of an allosteric enzyme inhibited by a feedback mechanism. ATCase catalyses the very first reaction in pyrimidine biosynthesis.

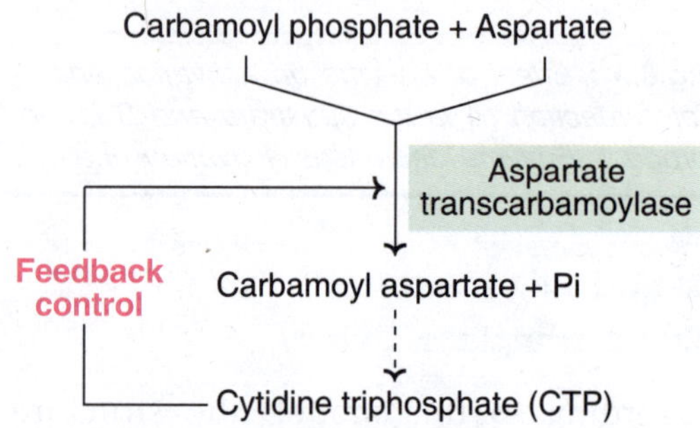

Carbamoyl phosphate undergoes a sequence of reactions for synthesis of the end product, CTP. When CTP accumulates, it allosterically inhibits the enzyme aspartate transcarbamoylase by a feedback mechanism.

2. Activation of latent enzymes

Latent enzymes, as such, are inactive. Some enzymes are synthesized as ***proenzymes*** or ***zymogens*** which undergo irreversible covalent activation by the breakdown of one or more peptide bonds. For instance, proenzymes—namely chymotrypsinogen, pepsinogen and plasminogen, are respectively—converted to the active enzymes chymotrypsin, pepsin and plasmin.

3. Compartmentation

There are certain substances in the body (e.g., fatty acids, glycogen) which are synthesized and also degraded. There is no point for simultaneous occurrence of both the pathways. Generally, the ***synthetic*** (anabolic) and ***breakdown*** (catabolic) ***pathways*** are ***operative*** in ***different cellular organelles*** to achieve maximum economy. For instance, enzymes for fatty acid synthesis are found in the cytosol whereas enzymes for fatty acid oxidation are present in the mitochondria.

TABLE 6.4 Distribution of certain enzymes and metabolic pathways in cellular organelles

Organelle	*Enzyme/metabolic pathway*
Cytoplasm	Aminotransferases; peptidases; glycolysis; hexose monophosphate shunt; fatty acid synthesis; purine and pyrimidine catabolism.
Mitochondria	Fatty acid oxidation; amino acid oxidation; Krebs cycle; urea synthesis; electron transport chain and oxidative phosphorylation.
Nucleus	Biosynthesis of DNA and RNA.
Endoplasmic reticulum (microsomes)	Protein biosynthesis; triacylglycerol and phospholipid synthesis; steroid synthesis and reduction; cytochrome P_{450}; esterase.
Lysosomes	Lysozyme; phosphatases; phospholipases; hydrolases; proteases; lipases; nucleases.
Golgi apparatus	Glucose 6-phosphatase; 5′-nucleotidase; glucosyl-and galactosyl-transferases.
Peroxisomes	Catalase; urate oxidase; D-amino acid oxidase; long chain fatty acid oxidation.

Mitochondrion is regarded as the power house of the cell where most of the energy producing pathways are located. The intracellular location of certain enzymes and metabolic pathways is given in ***Table 6.4***.

4. Control of enzyme synthesis

Most of the enzymes, particularly the rate limiting ones, are present in very low concentration. Nevertheless, the amount of the enzyme directly controls the velocity of the reaction, catalysed by that enzyme. ***Many rate limiting enzymes have short half-lives***. This helps in the efficient regulation of the enzyme levels.

There are two types of enzymes—(a) ***Constitutive enzymes*** (house-keeping enzymes)—the levels of which are not controlled and remain fairly constant. (b) ***Adaptive enzymes***—their concentrations increase or decrease as per body needs and are well-regulated. The synthesis of enzymes (proteins) is regulated by the genes (***Chapter 26***).

Induction and repression : The term induction is used to represent increased synthesis of enzyme while repression indicates its decreased synthesis. Induction or repression which ultimately determines the enzyme concentration at the gene level through the mediation of hormones or other substances.

5. Enzyme degradation

Enzymes are not immortal, since it will create a series of problems. There is a lot of variability in the half-lives of individual enzymes. For some, it is in days while for others in hours or in minutes, e.g. LDH_4—5 to 6 days; LDH_1—8 to 12 hours; amylase—3 to 5 hours.

In general, the key and ***regulatory enzymes are most rapidly degraded***. Though not always true, an enzyme with long half-life is usually sluggish in its catalytic activity.

6. Isoenzymes

Multiple forms of the same ***enzyme*** will also help in the regulation of enzyme activity, Many of the isoenzymes are tissue-specific. Although isoenzymes of a given enzyme catalyse the same reaction, they differ in K_m, V_{max} or both. e.g. isoenzymes of LDH and CPK.

UNITS OF ENZYME ACTIVITY

Enzymes are never expressed in terms of their concentration (as mg or µg etc.), but are expressed only as activities. Various methods have been introduced for the estimation of enzyme activities (particularly for the plasma enzymes). In fact, the activities have been expressed in many ways, like ***King-Armstrong units***, ***Somogyi units***, ***Reitman-Frankel*** units, spectrophotometric units etc.

Katal

In order to ***maintain uniformity*** in the expression of enzyme activities (as units) worldover, the ***Enzyme Commission of IUB*** has suggested radical changes. A new unit—namely katal (abbreviated as kat)—was introduced. ***One kat denotes the conversion of one***

mole substrate per second (mol/sec). Activity may also be expressed as millikatals (mkat), microkatals (μkat) and so on.

International Units (IU)

Some workers prefer to use standard units or SI units (System International). One SI unit or International Unit (IU) is defined as the amount of enzyme activity that catalyses the ***conversion of one micromol of substrate per minute***. SI units and katal are interconvertible.

$$1 \text{ IU} = 16.67 \text{ nkat}$$

Laboratory use of enzyme units

In the clinical laboratories, however, the units—namely katal or SI units—are yet to find a place. ***Many investigators still use the old units*** like King-Armstrong units, Somogyi units etc., while expressing the enzyme activities.

PLASMA/SERUM ENZYMES

Enzymes in the circulation are divided into two groups – plasma functional and plasma non-functional.

1. Plasma specific or plasma functional enzymes

Certain enzymes are normally present in the plasma and they have specific functions to perform. Generally, these enzyme activities are higher in plasma than in the tissues. They are mostly synthesized in the liver and enter the circulation e.g. lipoprotein lipase, plasmin, thrombin, choline esterase, ceruloplasmin etc.

2. Non-plasma specific or plasma non-functional enzymes

These enzymes are either totally absent or present at a low concentration in plasma compared to their levels found in the tissues. The digestive enzymes of the gastrointestinal tract (e.g. amylase, pepsin, trypsin, lipase etc.) present in the plasma are known as ***secretory enzymes.*** All the other plasma enzymes associated with metabolism of the cell are collectively referred to as ***constitutive enzymes*** (e.g. lactate dehydrogenase, transaminases, acid and alkaline phosphatases, creatine phosphokinase).

DIAGNOSTIC IMPORTANCE OF ENZYMES

Estimation of enzyme activities in biological fluids (particularly plasma/serum) is of great clinical importance. Particularly non-plasma specific enzymes are very important for the diagnosis and prognosis of several diseases.

The normal serum level of an enzyme indicates the balance between its synthesis and release in the routine cell turnover. The raised enzyme levels could be due to cellular damage, increased rate of cell turnover, proliferation of cells, increased synthesis of enzymes etc. Serum enzymes are conveniently used as ***markers*** to detect the cellular damage which ultimately helps ***in the diagnosis of diseases***.

(**Note :** Ther term ***biomarker*** refers to any ***laboratory analyte*** (***enzyme, protein, antigen,*** antibody, metabolite etc.) ***that is useful for the diagnosis/prognosis of any disease***. Biomarker is a vague term, and less frequently used by biochemists.)

A summary of the important enzymes useful for the diagnosis of specific diseases is given in ***Table 6.5***.

ISOENZYMES

The ***multiple forms of an enzyme*** catalysing the same reaction are ***isoenzymes*** or ***isozymes.*** They, however, differ in their physical and chemical properties which include the structure, electrophoretic and immunological properties, K_m and V_{max} values, pH optimum, relative susceptibility to inhibitors and degree of denaturation.

Isoenzymes of lactate dehydrogenase (LDH)

Among the isoenzymes, LDH had been the most thoroughly investigated.

LDH, whose systematic name is L-lactate-NAD^+ oxidoreductase (E.C. 1.1.1.27), catalyses the interconversion of lactate and pyruvate as shown below

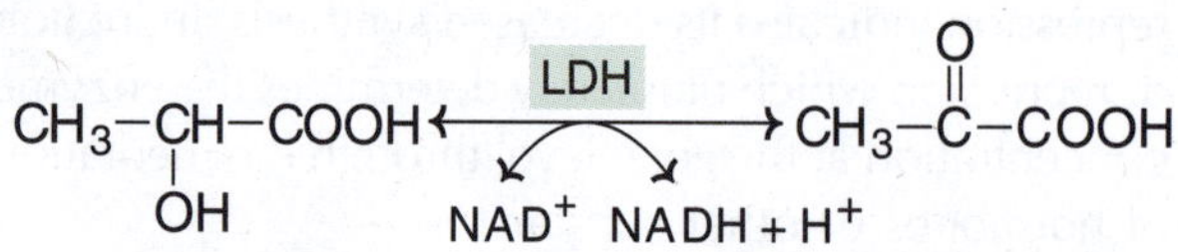

Lactic acid **Pyruvic acid**

LDH has five distinct isoenzymes LDH_1, LDH_2, LDH_3, LDH_4 and LDH_5. They can be separated by electrophoresis (cellulose or starch gel or agarose gel). LDH_1 has more positive charge and fastest in electrophoretic mobility while LDH_5 is the slowest.

Medical Concepts/ Clinical Correlates

- **Feedback (or end product) inhibition is a specialized form of allosteric inhibition that controls several metabolic pathways e.g. CTP inhibits aspartate trans-carbamoylase; Cholesterol inhibits HMG CoA reductase. The end product inhibition is utmost important to cellular economy since a compound is synthesized only when required.**
- **A few RNA molecules (ribozymes) function as non-protein enzymes. It is believed that ribozymes were functioning as biocatalysts before the occurrence of protein enzymes during evolution.**
- **Elevation in serum enzyme activities is of great importance for the diagnosis of several diseases. e.g. amylase–acute pancreatitis; ALT–hepatitis; alkaline phos-phatase–obstructive jaundice; CPK-MI;**

Structure of LDH isoenzymes : LDH is an oligomeric (tetrameric) enzyme made up of four polypeptide subunits. Two types of subunits namely M (for muscle) and H (for heart) are produced by different genes. M—subunit is basic while H subunit is acidic. The isoenzymes contain either one or both the subunits giving LDH_1 (H_4) to LDH_5 (M_4).

Diagnostic importance of LDH : Isoenzymes of LDH have immense value in the diagnosis of heart and liver related disorders (***Fig.6.11***). In healthy individuals, the activity of LDH_2 is higher than that of LDH_1 in serum. In the case of ***myocardial infarction, LDH_1*** is much greater than LDH_2 and this happens within 12 to 24 hours after infarction. Increased activity of ***LDH_5*** in serum is an indicator of ***liver diseases.*** LDH activity in the RBC is 80-100 times more than that in the serum. Hence for estimation of LDH or its isoenzymes, serum should be totally free from hemolysis or else false positive results will be obtained.

Isoenzymes of creatine phosphokinase

Creatine kinase (CK) or ***creatine phosphokinase*** (***CPK***) catalyses the interconversion of phospho-creatine (or creatine phosphate) to creatine.

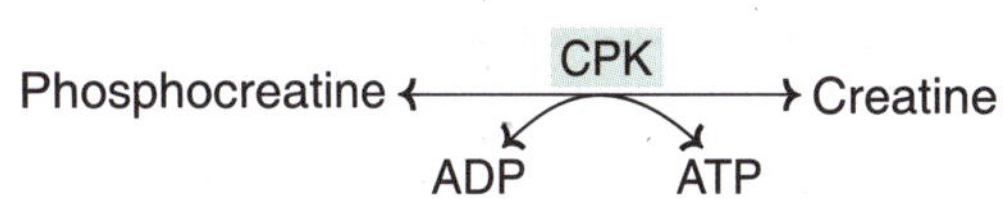

CPK exists as ***three isoenzymes***. Each isoenzyme is a dimer composed of two subunits—M (muscle) or B (brain) or both.

Table 6.5 Important enzymes in the diagnosis of diseases

Serum enzyme (elevated)	*Disease (most important)*
Amylase	Acute pancreatitis
Serum glutamate pyruvate transaminase (SGPT)	Liver diseases (hepatitis)
Serum glutamate oxaloacetate transaminase (SGOT)	Heart attacks (myocardial infarction)
Alkaline phosphatase	Rickets, obstructive jaundice
Acid phosphatase	Cancer of prostate gland
Lactate dehydrogenase (LDH)	Heart attacks, liver diseases
Creatine phosphokinase (CPK)	Myocardial infarction (early marker)
Aldolase	Muscular dystrophy
5′-Nucleotidase	Hepatitis
γ-Glutamyl transpeptidase (GGT)	Alcoholism

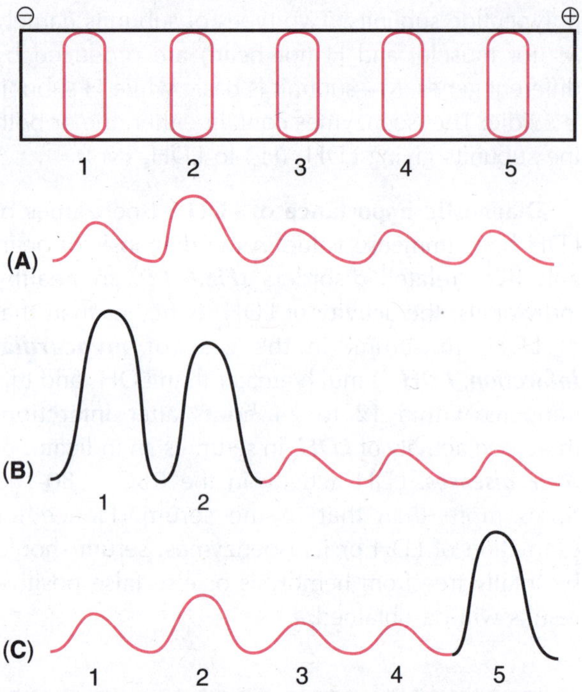

Fig.6.11 : *Electrophoresis of lactate dehydrogenase with relative proportions of isoenzymes* **(A)** *Normal serum* **(B)** *Serum from a patient of myocardial infarction (LDH_1 and LDH_2") **(C)** Serum from a patient of liver disease (LDH_5").*

Isoenzyme	*Subunit*	*Tissue of origin*
CPK_1	BB	Brain
CPK_2	MB	Heart
CPK_3	MM	Skeletal muscle

In healthy individuals, the isoenzyme CPK_2 (MB) is almost undetectable in serum with less than 2% of total CPK. After the myocardial infarction (MI), within the first 6-18 hours, CPK_2 increases in the serum to as high as 20% (against 2% normal). CPK_2 isoenzyme is not elevated in skeletal muscle disorders. Therefore, estimation of ***CPK_2 (MB) is the earliest reliable indication of myocardial infarction.***

Isoenzymes of alkaline phosphatase

As many as ***six isoenzymes*** of alkaline phosphatase have been identified. The enzyme is a monomer, the isoenzymes are due to the difference in the carbohydrate content (sialic acid residues). The most important ALP isoenzymes are α_1-ALP, α_2-heat labile ALP, α_2-heat stable ALP, pre-β ALP, γ-ALP etc.

Increase in α_2-heat labile ALP suggests hepatitis whereas pre β-ALP indicates bone diseases.

Isoenzymes of alcohol dehydrogenase

Alcohol dehydrogenase (ADH) has two heterodimer isoenzymes. Among the white Americans and Europeans, '**$\alpha\beta_1$**' isoenzyme is predominant whereas in Japanese and Chinese (Orientals) '**$\alpha\beta_2$**' is mostly present. The isomer $\alpha\beta_2$ more rapidly converts alcohol to acetaldehyde.

Accumulation of acetaldehyde is associated with tachycardia (increase in heart rate) and facial flushing among Orientals which is not commonly seen in whites. It is believed that ***Japanese and Chinese have increased sensitivity to alcohol due to the $\alpha\beta_2$-isoenzyme of ADH.***

ENZYME PATTERN IN DISEASES

For the right diagnosis of a particular disease, it is always better to estimate a few (three or more) serum enzymes, instead of a single enzyme. Examples of enzyme patterns in important diseases are given here.

Enzymes in myocardial infarction

The enzymes—namely creatine phosphokinase (***CPK***), aspartate transaminase (***AST***) and lactate dehydrogenase (***LDH***)—are important in the diagnosis of myocardial infarction (MI). The elevation of these enzymes in serum in relation to hours/days of MI is given in the ***Fig.6.12***.

Creatine phosphokinase (precisely isoenzyme ***MB***) is the first enzyme to be released into circulation within ***6-18 hours after the infarction***. Therefore, CPK estimation is highly useful for the early diagnosis of MI. This enzyme reaches a peak value within 24-30 hours and returns to normal level by the 2nd or 3rd day.

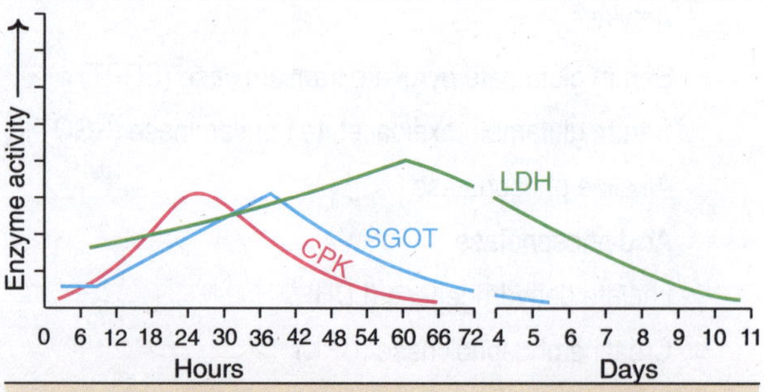

Fig.6.12 : *Enzyme pattern in myocardial infarction (CPK-Creatine phosphokinase; SGOT-Serum glutamate oxaloacetate transaminase; LDH-Lactate dehydrogenase).*

Aspartate transaminase (AST or SGOT) rises sharply after CPK, and reaches a peak within 48 hours of the myocardial infarction. AST takes 4-5 days to return to normal level.

Lactate dehydrogenase (LDH_1) generally rises from the second day after infarction, attains a peak by the 3rd or 4th day and takes about 10-15 days to reach normal level. Thus, LDH is the last enzyme to rise and also the last enzyme to return to normal level in MI.

[***Note :*** Certain non-enzymatic biomarkers (e.g. ***troponin I, myoglobin, brain natriuretic peptide***) are in recent use for the diagnosis of MI. The concentration of troponin I shots up in serum within 4 hours of MI and remains elevated for about 2 weeks.]

Enzymes in liver diseases

The following enzymes—when elevated in serum—are useful for the diagnosis of liver dysfunction due to viral hepatitis (jaundice), toxic hepatitis, cirrhosis and hepatic necrosis

1. ***Alanine transaminase***;
2. Aspartate transaminase;
3. Lactate dehydrogenase.

The enzymes that markedly increase in intrahepatic and extrahepatic cholestasis are

1. Alkaline phosphatase, 2. 5′-Nucleotidase

Serum ***γ-glutamyl transpeptidase*** is useful for the diagnosis of ***alcoholic liver diseases***.

Enzymes in muscle diseases

In the muscular dystrophies, probably due to the increased leakage of enzymes from the damaged cells, serum levels of certain muscle enzymes are increased. These include creatine phosphokinase, aldolase and aspartate transaminase. Of these, CPK is the most reliable indicator of muscular diseases, followed by aldolase.

Enzymes in cancers

Increase in the serum acid phosphatase (tartarate labile) is specific for the detection of prostatic carcinoma.

[***Note : Prostate specific antigen*** (PSA; Mol Wt. 32kD), though not an enzyme, is a more reliable marker for the detection of prostate cancer. Normal serum concentration of PSA is 1-4 ng/ml.]

A non-specific increase in certain enzymes like LDH, alkaline phosphatase and transaminase may be associated with malignancy in any part of the body.

β-Glucuronidase estimation in urine is useful in detecting the cancers of urinary bladder, pancreas etc.

SUMMARY

1. *Enzymes are the protein biocatalysts synthesized by the living cells. They are classified into six major classes—oxidoreductases, transferases, hydrolases, lyases, isomerases and ligases.*
2. *Factors like concentration of enzyme, substrate, temperature, pH etc. influence enzyme activity. The substrate concentration to produce half-maximal velocity is known as Michaelis constant (K_m value).*
4. *Enzyme activities are inhibited by reversible (competitive, non-competitive and un-competitive), irreversible and allosteric manner.*
5. *Many enzymes require the presence of non-protein substances called cofactors (coenzymes) for their action. Most of the coenzymes are derivatives of B-complex vitamins (e.g. NAD^+, FAD, TPP etc.)*
5. *Estimation of serum enzymes is of great help in the diagnosis of several diseases. Serum amylase and lipase are increased in acute pancreatitis; alanine transaminase in hepatitis; aspartate transaminase, lactate dehydrogenase (LDH) and creatine phosphokinase (CPK) in myocardial infarction; alkaline phosphatase in rickets and hyperparathyroidism; acid phosphatase in prostatic carcinoma; g-glutamyl transpeptidase in alcoholism.*
6. *Isoenzymes are the multiple forms of an enzyme catalysing the same reaction which however, differ in their physical and chemical properties. LDH has five isoenzymes while CPK has three. LDH_1 and CPK_2 are very important in the diagnosis of myocardial infarction.*

THE READER MUST REFER APPENDIX D (Page Nos : 422 - 431) FOR RELATED CLINICAL CASE STUDIES (Nos : 18, 19, 20, 21, 22, 23, 24)

VITAMINS

7

It is difficult to define vitamins precisely. ***Vitamins may be regarded as organic compounds required in the diet in small amounts to perform specific biological functions for normal maintenance of optimum growth and health of the organism.***

The bacterium ***E.coli*** does not require any vitamin, as it ***can synthesize all of them***. It is believed that during the course of evolution, the ability to synthesize vitamins was lost. Hence, the higher organisms have to obtain them from diet. The vitamins are required in small amounts, since their degradation is relatively slow.

Classification of vitamins

There are around 15 vitamins, essential for humans. They are classified as ***fat soluble*** (A, D, E and K) and ***water soluble*** (C and B-group) vitamins as shown in the ***Table 7.1***. The B-complex vitamins may be sub-divided into ***energy-releasing*** (B_1, B_2, B_6, biotin etc.) and ***hematopoietic*** (folic acid and B_{12}). Most of the ***water soluble vitamins exert the functions through their respective coenzymes*** while only one fat soluble vitamin (K) has been identified to function as a coenzyme.

Synthesis of vitamins by intestinal bacteria

Vitamins, as per the definition, are not synthesized in the body. However, the bacteria of the gut can produce some of the vitamins, required by man and animals.

As far as humans are concerned, it is believed that the normal ***bacterial synthesis*** and absorption of ***vitamin K*** and ***biotin may be sufficient*** to meet the body requirements. For other B-complex vitamins, the synthesis and absorption are relatively less. Administration of anitibiotics often kills the vitamin synthesizing bacteria present in the gut, hence additional consumption of vitamins is recommended.

Fat soluble vitamins—general

The four vitamins, namely vitamin A, D, E, and K are known as fat ***or lipid soluble***. Their availability in the diet, absorption and transport are ***associated with fat***. They are soluble in fats and oils and also the fat solvents (alcohol, acetone etc.). Fat soluble vitamins can be stored in liver and adipose tissue.

Water soluble vitamins—general

The water soluble vitamins are a ***heterogenous group of compounds*** since they differ chemically from each other. The only common character shared by them is their solubility in water. Most of these vitamins are readily excreted in urine and they are ***not toxic to the body***. Water soluble vitamins must be continuously supplied in the diet. ***Generally, vitamin deficiencies are multiple rather than individual with overlapping symptoms.***

The water souble vitamins form ***coenzymes*** (Refer ***Table 6.3***) that ***participate*** in a variety of ***biochemical reactions***. Most of these reactions are related to either energy generation or hematopoiesis.

Vitamers

The term vitamers represents the chemically similar substances that ***possess qualitatively similar vitamin activity***. e.g. retinol, retinal and retinoic acid are vitamers of ***vitamin A***; pyridoxine, pyridoxal and pyridoxamine are vitamers of vitamin B_6.

TABLE 7.1 Classification of vitamins

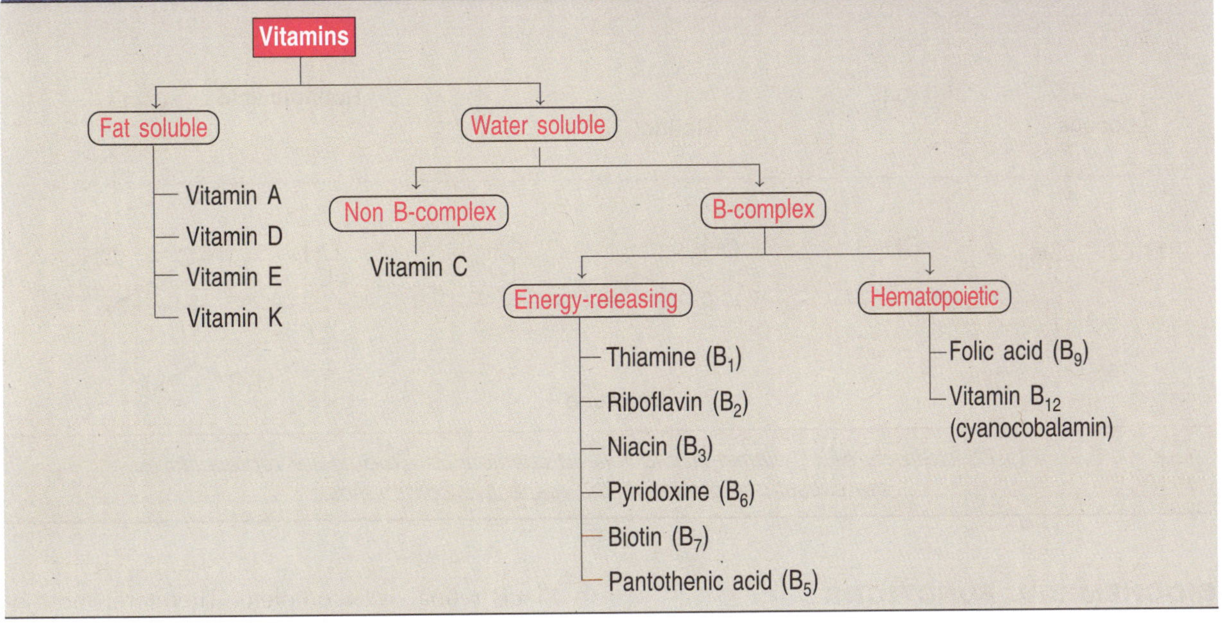

INDIVIDUAL VITAMINS

In the following pages, the individual members of the fat soluble and water soluble vitamins are discussed with regard to the chemistry, biochemical functions, recommended dietary/daily allowances (RDA), dietary sources, deficiency manifestations etc.

VITAMIN A

The fat soluble vitamin A, as such is present only in foods of animal origin. However, its ***provitamins carotenes*** are found in plants. Retinol, retinal and retinoic acid are regarded as ***vitamers*** of vitamin A.

Chemistry

In the recent years, the term vitamin A is collectively used to represent many structurally related and biologically active molecules (***Fig.7.1***).

1. **Retinol (vitamin A alcohol) :** It is a primary alcohol containing ***β-ionone ring***. The side chain has ***two isoprenoid units***.

2. **Retinal (vitamin A aldehyde) :** This is an aldehyde form ***obtained*** by the ***oxidation of retinol***.

3. **Retinoic acid (vitamin A acid) :** This is produced by the oxidation of retinal.

4. **β-Carotene (provitamin A) :** This is found in ***plant foods***. It is cleaved in the intestine to produce two moles of retinal. In humans, this conversion is inefficient, hence β-carotene possesses about ***one-sixth vitamin A activity*** compared to that of retinol.

Absorption, transport and mobilization

Dietary retinyl esters are hydrolysed by pancreatic or intestinal brush border hydrolases in the intestine, releasing retinol and free fatty acids. Carotenes are hydrolysed by ***β-carotene 15-15′–dioxygenase*** of intestinal cells to release 2 moles of retinal which is reduced to retinol. In the intestinal mucosal cells, retinol is reesterified to long chain fatty acids, incorporated into chylomicrons and transferred to the lymph. The retinol esters of chylomicrons are taken up by the liver and stored (***Fig.7.2***).

Retinol is transported in the circulation by the ***plasma retinol binding protein***. Many cells of target tissues contain a ***cellular retinol-binding protein*** that carries retinol to the nucleus and binds to the chromatin (DNA). It is here that retinol exerts its function in a manner ***analogous to*** that of a ***steroid hormone***.

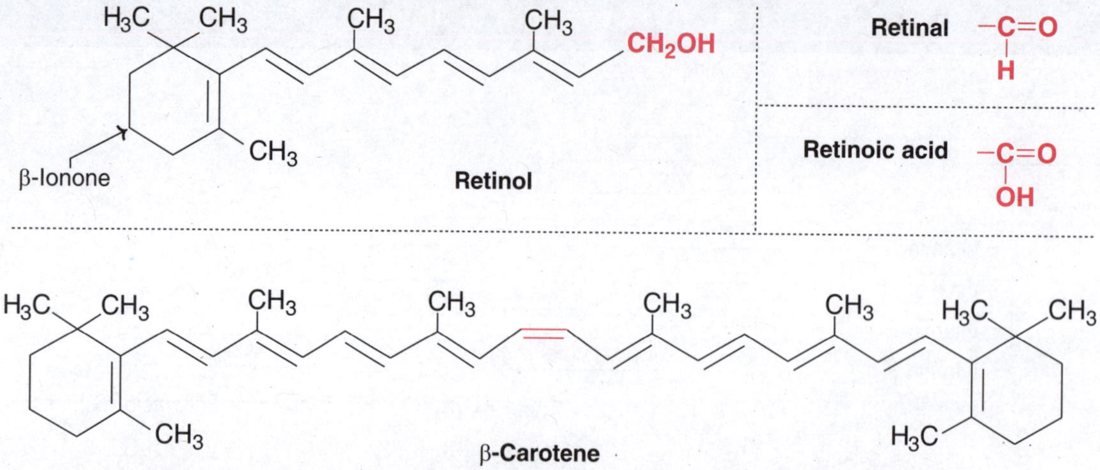

Fig.7.1 : *Structures of vitamin A and related compounds (Red colour represents the substituent groups in the respective compounds).*

BIOCHEMICAL FUNCTIONS

Vitamin A is necessary for a variety of functions such as ***vision***, ***proper growth*** and ***differentiation***, ***reproduction*** and ***maintenance of epithelial cells***.

Vitamin A and vision : The biochemical function of vitamin A in the process of vision was first elucidated by George Wald (Nobel Prize 1968). The events occur in a cyclic process known as ***Rhodopsin cycle*** or ***Wald's visual cycle*** (***Fig.7.3***).

Rods and cones

The retina of the eye possesses two types of cells—rods and cones. The human eye has about 10 million rods and 5 million cones. ***Rods*** are involved in ***dim light vision*** whereas cones are responsible for bright light and colour vision.

Wald's visual cycle

Rhodopsin is a conjugated protein present in rods. It contains ***11-cis retinal*** and the protein ***opsin***.

The primary event in visual cycle, on ***exposure to light***, is the isomerization of 11-*cis*-retinal to all-*trans* retinal. This leads to a ***conformational change in opsin*** which is responsible for the generation of nerve impulse. The all-*trans*-retinal is immediately isomerized by retinal isomerase (of retinal epithelium) to 11-*cis*-retinal. This combines with opsin to regenerate rhodopsin and complete the visual cycle (***Fig.7.3***). However, the conversion of all *trans*-retinal to 11-*cis* retinal is incomplete. Therefore, most of the all-*trans*-retinal is transported to the liver and converted to all-*trans* retinol by alcohol dehydrogenase. The all-*trans*-retinol undergoes isomerization to 11-*cis* retinol which is then oxidized to 11-*cis* retinal to participate in the visual cycle.

Other biochemical functions of vitamin A

1. ***Retinol*** and ***retinoic acid*** function almost like ***steroid hormones***. They regulate the protein synthesis and thus are involved in the cell growth and differentiation.

2. Vitamin A is essential to ***maintain healthy epithelial tissue***.

3. ***Carotenoids*** (most important β-carotene) function as ***antioxidants*** and reduce the risk of cancers initiated by free radicals and strong oxidants. β-Carotene is found to be beneficial to prevent heart attacks. This is also attributed to the antioxidant property.

Recommended dietary allowance (RDA)

The requirement of vitamin A for adults in the form of retinol (or retinol equivalents) is around 600 μg/day (or 4800 μg of β-carotene). It may be noted that one International Unit (IU) equals to 0.3 mg of retinol. The requirement increases in pregnant women, and lactating mothers.

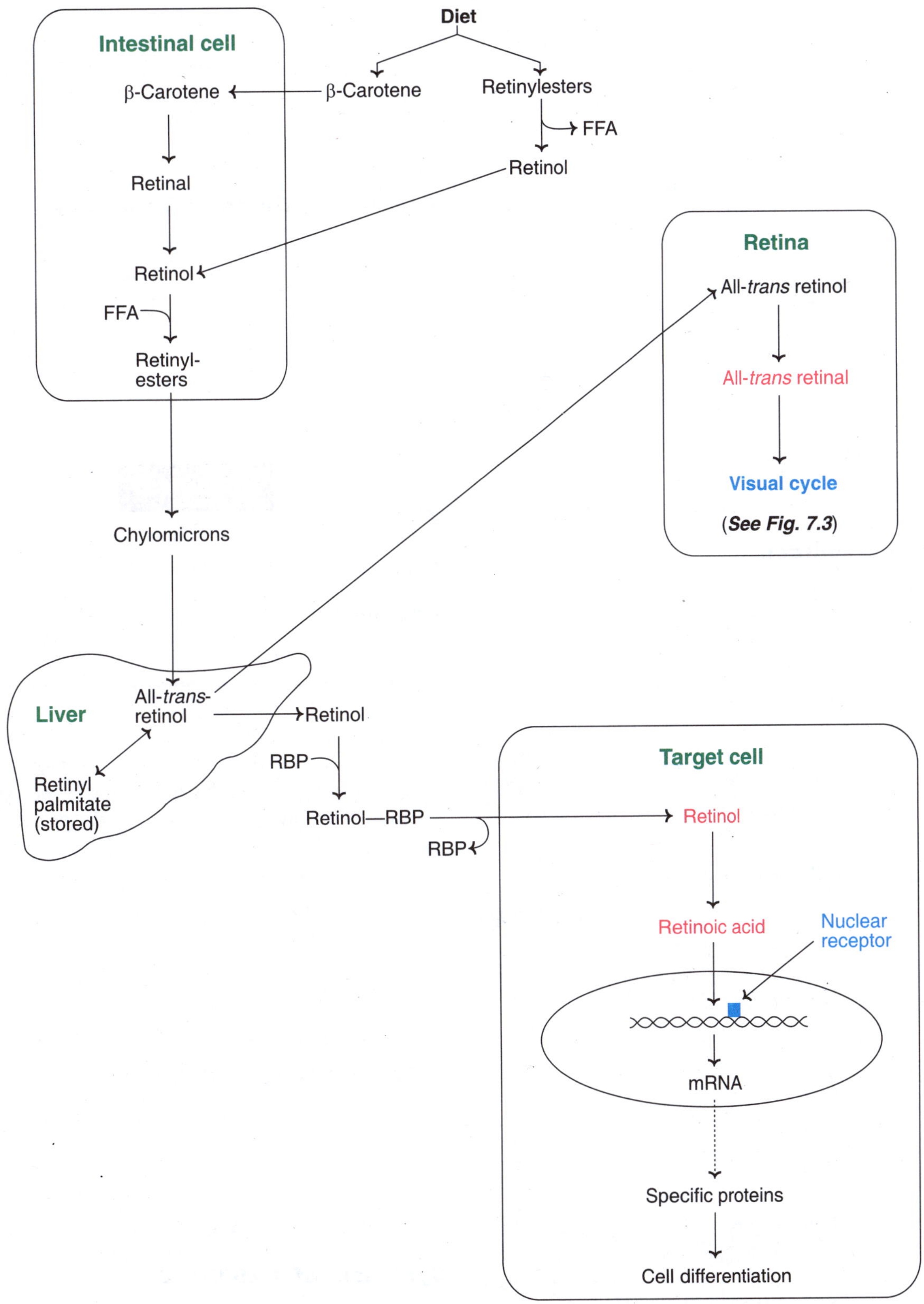

Fig.7.2 : *Summary of vitamin A absorption, transport and biochemical functions (FFA-Free fatty acid; RBP-Retinol binding protein).*

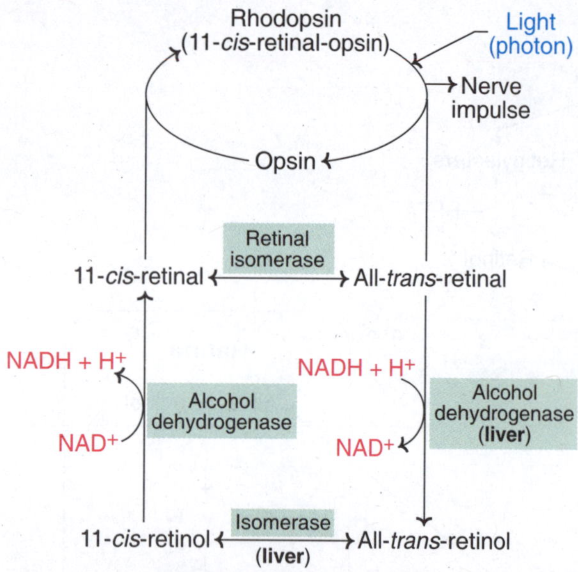

Fig.7.3 : Wald's visual cycle.

Dietary sources

Animal sources contain (preformed) vitamin A. The best sources are ***liver***, ***kidney***, egg yolk, milk, cheese and butter. ***Fish*** (cod or shark) ***liver oils*** are very rich in vitamin A.

Vegetable sources contain the provitamin A–***carotenes***. Yellow and dark green vegetables and fruits are good sources of carotenes e.g. carrots, spinach, amaranthus, pumpkins, mango, papaya etc.

Vitamin A deficiency

Deficiency manifestations of the eyes' : ***Night blindness*** (***nyctalopia***) is one of the earliest symptoms of vitamin A deficiency. The individuals have ***difficulty to see in dim light***.

Severe deficiency of vitamin A leads to ***xerophthalmia***. This is characterized by dryness in conjunctiva and cornea and keratinization of epithelial cells. In certain areas of conjunctiva, white triangular plaques known as ***Bitot's spots*** are seen.

If xerophthalmia persists for a long time, corneal ulceration and degeneration occur. This results in the destruction of cornea, a condition referred to as ***keratomalacia***, ***causing total blindness***. Vitamin A deficiency blindness is most common in children of the developing countries.

Hypervitaminosis A

Excessive consumption of vitamin A leads to toxicity. The symptoms of hypervitaminosis A include ***dermatitis*** (drying and redness of skin), ***enlargement of liver***, skeletal decalcification, tenderness of long bones, loss of weight, irritability, loss of hair, joint pains etc.

Beneficial effects of β-carotene

Increased consumption of β-carotene is associated with ***decreased*** incidence of ***heart attacks***, skin and lung cancers. This is attributed to the ***antioxidant role of β-carotene*** which is independent of its role as a precursor of vitamin A. Ingestion of high doses of β-carotene for long periods are not toxic like vitamin A.

VITAMIN D

Vitamin D is a fat soluble vitamin. It resembles ***sterols in structure and functions like a hormone***.

Chemistry

Cholecalciferol (vitamin D_3) is found in animals.Ergocalciferol (vitamin D_2) is formed from ergosterol and is present in plants. Both the sterols are similar in structure except that ergocalciferol has an additional methyl group and a double bond. ***Ergocalciferol*** and ***cholecalciferol*** are sources for vitamin D activity and are referred to as ***provitamins***.

During the course of cholesterol biosynthesis (***Chapter 14***), ***7-dehydrocholesterol*** is formed as an intermediate. On exposure to sunlight, 7-dehydrocholesterol is converted to cholecalciferol in the skin (dermis and epidermis). Vitamin D is regarded as a ***sun-shine vitamin***.

METABOLISM AND BIOCHEMICAL FUNCTIONS

Vitamins D_2 and D_3 as such are not biologically active. They are metabolized identically in the body and converted to active forms of vitamin D. The metabolism and biochemical functions of vitamin D are depicted in ***Fig.7.5***.

Synthesis of 1,25-DHCC

Cholecalciferol is first hydroxylated at 25th position to ***25-hydroxycholecalciferol*** (***25-OH*** D_3) by a specific hydroxylase present in liver. 25-OH D_3

is the major storage and circulatory form of vitamin D. Kidney possesses a specific enzyme, ***25-hydroxycholecalciferol 1-hydroxylase*** which hydroxylates 25-hydroxycholecalciferol at position 1 to produce ***1,25-dihydroxycholecalciferol (1,25-DHCC)***. 1,25-DHCC contains 3 hydroxy groups (1,3 and 25 carbon) hence referred to as ***calcitriol***. The synthesis of calcitriol is depicted in ***Figs.7.4*** and ***7.5***.

Biochemical functions

Calcitriol (1,25-DHCC) is the biologically active form of vitamin D. It ***regulates the plasma levels of calcium and phosphate***. Calcitriol acts at 3 different levels (intestine, kidney and bone) to maintain plasma calcium (normal 9-11 mg/dl).

1. **Action of calcitriol on the intestine :** Calcitriol ***increases the intestinal absorption of calcium and phosphate***. In the intestinal cells, calcitriol binds with a cytosolic receptor to form a calcitriol-receptor complex. This complex then approaches the nucleus and interacts with a specific DNA leading to the synthesis of a specific ***calcium binding protein***. This protein increases the calcium uptake by the intestine.

2. **Action of calcitriol on the bone :** In the osteoblasts of bone, calcitriol stimulates calcium uptake for deposition as calcium phosphate. Thus calcitriol is essential for ***bone formation***.

3. **Action of calcitriol on the kidney :** Calcitriol is also involved in ***minimizing the excretion of calcium and phosphate*** through the kidney, by decreasing their excretion and enhancing reabsorption.

Vitamin D is a hormone and not a vitamin—justification

Calcitriol (1,25-DHCC) is now considered as an important ***calciotropic hormone*** while cholecalciferol is the prohormone. The following characteristic features of vitamin D (comparable with hormone) justify its status as a hormone.

1. Vitamin D_3 (cholecalciferol) is ***synthesized in the skin*** by ultra-violet rays of sunlight.

2. The biologically active form of vitamin D, calcitriol is produced in the kidney.

3. Calcitriol has ***target organs***–intestine, bone and kidney, where it specifically acts.

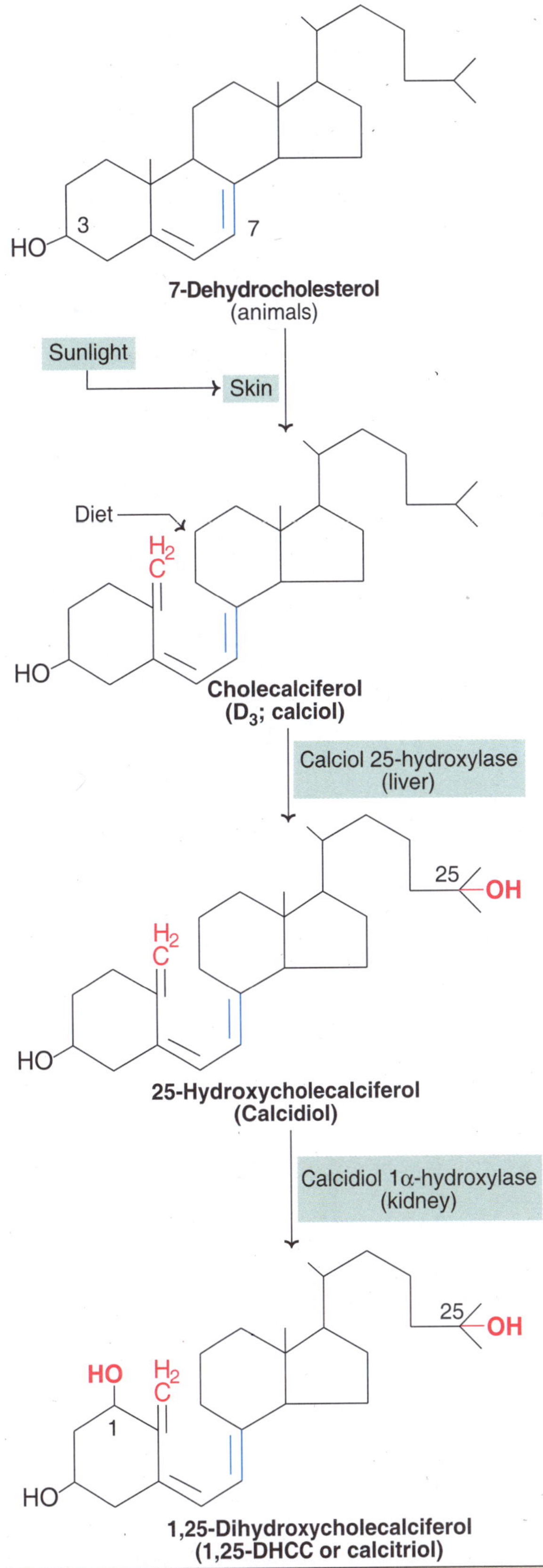

Fig.7.4 : *Biosynthesis of active form of vitamin D-calcitriol (1,25-DHCC).*

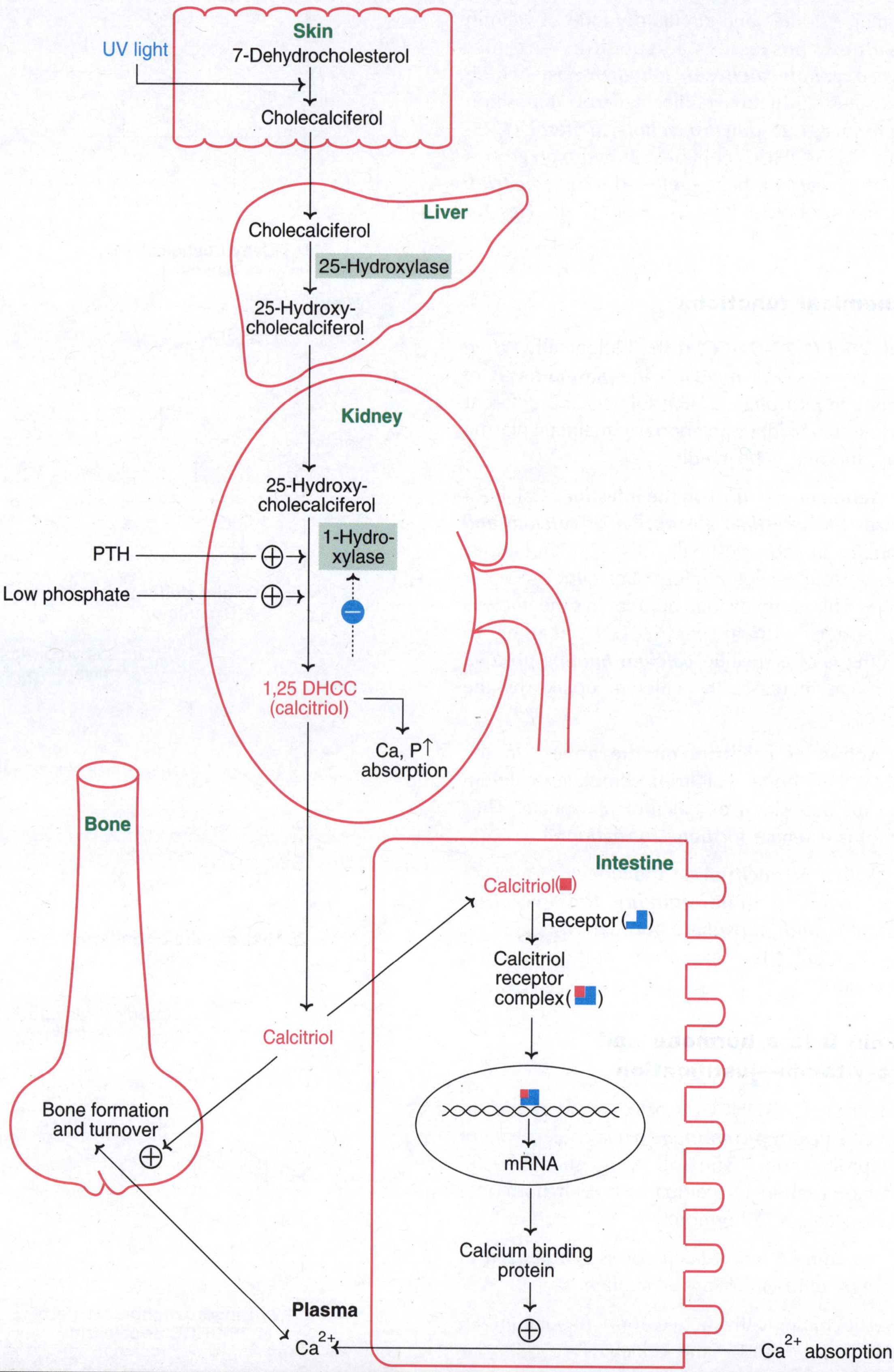

***Fig.7.5 :** Metabolism and biochemical functions of vitamin D (1,25-DHCC-1,25-Dihydroxycholecalciferol, also called as calcitriol is the active form of vitamin D; PTH-Parathyroid hormone).*

Medical Concepts/ Clinical Correlates

- It is believed that during the course of evolution, the ability to synthesize vitamins was lost by the higher organisms, hence they should be supplied through the diet.
- For humans, the normal intestinal bacterial synthesis of vitamin K and biotin is almost sufficient to meet the body's requirements.
- Administration of antibiotics often destroys the vitamin synthesizing bacteria in the gut, hence additional supplementation of vitamins is recommended during antibiotic therapy.
- Vitamin A deficiency causes night blindness; vitamin D deficiency rickets (in children) or osteomalacia (in adults); vitamin E deficiency minor neurological symptoms; vitamin K deficiency bleeding.

4. Calcitriol ***action is similar to steroid hormones***. It binds to a receptor in the cytosol and the complex acts on DNA to stimulate the synthesis of calcium binding protein.

5. Calcitriol synthesis is self-regulated by a ***feedback mechanism*** i.e., calcitriol decreases its own synthesis.

Recommended dietary allowance (RDA)

The daily requirement of vitamin D is ***400 International Units or 10 µg of cholecalciferol***. In countries with good sunlight (like India), the RDA for vitamin D is 200 IU (or 5 µg cholecalciferol).

Higher consumption of vitamin D may be beneficial

There is some recent evidence suggesting that ***higher vitamin D status in the body is protective against prediabetes, metabolic syndrome, various cancers etc***. It appears that the desirable levels of vitamin D are higher than the RDA. Perhaps, increased sunlight exposure may be capable of meeting body's vitamin D requirement. However, this is associated with increased risk of skin cancers.

Dietary sources

Good sources of vitamin D include ***fatty fish, fish liver oils, egg yolk*** etc. Milk is not a good source of vitamin D. In a natural way, ***exposure of skin to sunlight*** synthesizes vitamin D.

Deficiency symptoms

Deficiency of vitamin D leads to demineralization of bone. The result is ***rickets in children and osteomalacia in adults***. The word rickets is derived from an old *English* word wrickken, meaning to twist. Osteomalacia is derived from Greek (osteon-bone; malakia-softness). Vitamin D is often called as antirachitic vitamin.

Rickets in children is characterized by bone deformities due to incomplete mineralization, resulting in soft and pliable bones and delay in teeth formation. In rickets, the plasma level of calcitriol is decreased and ***alkaline phosphatase activity is elevated***.

In case of ***osteomalacia (adult rickets)*** demineralization of the bones occurs (bones become softer), increasing their susceptibility to fractures.

Renal rickets (renal osteodystrophy)

This is seen in patients with chronic renal failure. Renal rickets is mainly due to ***decreased synthesis of calcitriol in kidney***. It can be treated by administration of calcitriol.

Hypervitaminosis D

Vitamin D is stored mostly in liver and slowly metabolised. Among the vitamins, vitamin D is the ***most toxic in overdoses*** (10-100 times RDA). Toxic effects of hypervitaminosis D include demineralization of bones (resorption) and increased calcium absorption from the intestine, leading to elevated calcium in plasma (hypercalcemia). Hypervitaminosis D may lead to formation of stones in kidneys (renal calculi). High consumption of vitamin D is associated with loss of appetite, nausea, increased thirst, loss of weight etc.

VITAMIN E

Vitamin ***E (tocopherol)*** is a naturally occurring antioxidant. It is essential for normal reproduction in many animals, hence known as ***anti-sterility vitamin***.

Chemistry

Vitamin E is the name given to a group of tocopherols and tocotrienols. About eight tocopherols have been identified–α, β, γ, δ etc. Among these, ***α-tocopherol*** is the most active. The tocopherols are derivatives of ***6-hydroxy chromane (tocol) ring*** with isoprenoid (3 units) side chain. The antioxidant property is due to the chromane ring.

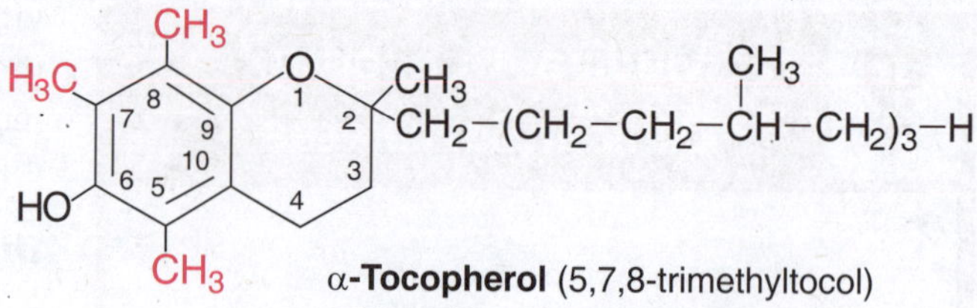

α-**Tocopherol** (5,7,8-trimethyltocol)
β-Tocopherol (5,8-dimethyltocol)
γ-Tocopherol (7,8-dimethyltocol)

*Fig. 7.6 : Structure of α-tocopherol (**Note** : The tocopherols differ in the substitution of methyl groups, represented in red).*

Biochemical functions

Most of the functions of vitamin E are related to its ***antioxidant*** property. It prevents the non-enzymatic oxidations of various cell components (e.g. unsaturated fatty acids) by molecular oxygen and free radicals such as superoxide (O_2^-) and hydrogen peroxide (H_2O_2). The element ***selenium helps*** in these functions.

Vitamin E is lipophilic in character and is found in association with lipoproteins, fat deposits and cellular membranes. ***It protects the polyunsaturated fatty acids (PUFA) from peroxidation reactions.*** Vitamin E acts as a scavenger and gets itself oxidized (to quinone form) by free radicals (R) and spares PUFA, as shown below

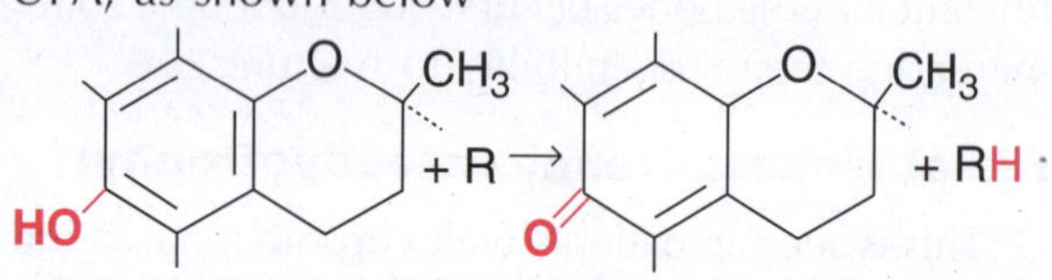

The biochemical functions of vitamin E, related either directly or indirectly to its antioxidant property, are given hereunder

1. Vitamin E is essential for the ***membrane structure and integrity*** of the cell, hence it is regarded as a ***membrane antioxidant***.

2. It prevents the peroxidation of polyunsaturated fatty acids in various tissues and membranes. It protects RBC from hemolysis by oxidizing agents (e.g. H_2O_2).

3. It is closely associated with ***reproductive functions and prevents sterility***. Vitamin E preserves and maintains germinal epithelium of gonads for proper reproductive function.

4. Vitamin E ***protects liver*** from being damaged by toxic compounds such as carbon tetrachloride.

5. It works in association with vitamins A, C, and β-carotene, to delay the onset of cataract.

6. Vitamin E has been recommended for the ***prevention of chronic diseases*** such as ***cancer and heart diseases***. Clinical trials in this regard are rather disappointing, hence it is no more recommended. However, some clinicians continue to use it particularly in subjects susceptible to heart attacks. It is believed that vitamin E prevents the oxidation of LDL. (**Note :** The oxidized LDL have been implicated to promote heart diseases.)

Vitamin E and selenium

The element selenium is found in the enzyme ***glutathione peroxidase*** that destroys free radicals. Thus, Se is also involved in antioxidant functions like vitamin E, and both of them ***act synergistically***. To a certain extent, Se can spare the dietary requirement of vitamin E, and vice versa.

Recommended dietary allowance (RDA)

A daily consumption of about ***10 mg (15 IU)*** of α-tocopherol for man and 8 mg (12 IU) for woman is recommended. One mg of α-tocopherol is equal to 1.5 IU. Vitamin E supplemented diet is advised for pregnant and lactating women.

Dietary sources

Many vegetable oils are rich sources of vitamin E. ***Wheat germ oil***, cotton seed oil, peanut oil, corn oil and sunflower oil are the good sources of this vitamin. It is also present in meat, milk, butter and eggs.

Deficiency symptoms

The symptoms of vitamin E deficiency vary from one animal species to another. In ***many animals***, the deficiency is associated with ***sterility***, degenerative changes in muscle, megaloblastic anaemia and changes in central nervous system.

Toxicity of vitamin E

Among the fat soluble vitamins (A, D, E, K), vitamin E is the least toxic.

VITAMIN K

Vitamin K is the only fat soluble vitamin with a ***specific coenzyme function***. It is required for the production of ***blood clotting factors***, essential for coagulation (in *German*–**K**oagulation; hence the name K for this vitamin).

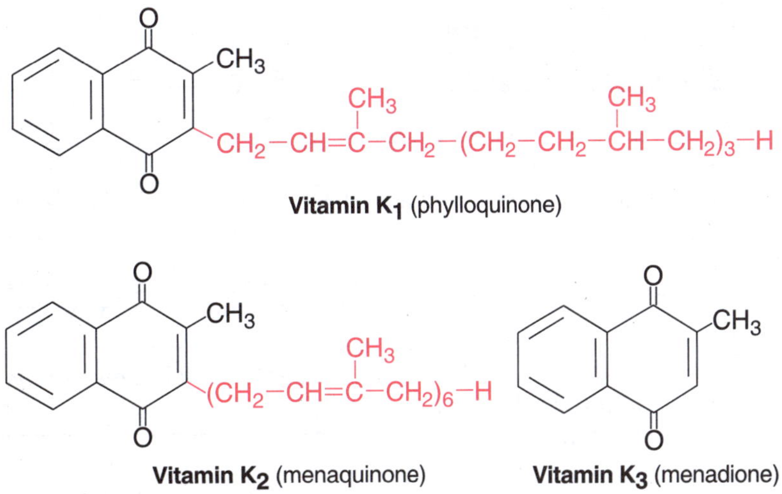

Fig.7.7 : Structures of vitamin K.

Chemistry

Vitamin K exists in different forms. Vitamin K_1 (phylloquinone) is present in plants. Vitamin K_2 (menaquinone) is produced by the intestinal bacteria and also found in animals. Vitamin K_3 (menadione) is a synthetic form.

All the three vitamins (K_1, K_2, K_3) are ***naphthoquinone derivatives (Fig.7.7)***. Isoprenoid side chain is present in vitamins K_1 and K_2. All the three are stable to heat. Their activity is, however, lost by oxidizing agents, irradiation, strong acids and alkalies.

Biochemical functions

The functions of vitamin K are concerned with blood clotting process. It brings about the ***post-translational*** (after protein biosynthesis in the cell) ***modification of certain blood clotting factors***. The clotting factors **II** (prothrombin), **VII**, **IX** and **X** are synthesized as inactive precursors (zymogens) in the liver. Vitamin K acts as a coenzyme for the ***carboxylation of glutamic acid*** (to form γ-***carboxyglutamate***) residues present in the proteins and this reaction is catalysed by a carboxylase (microsomal).

Recommended dietary allowance (RDA)

Strictly speaking, there is no RDA for vitamin K, since it can be ***adequately synthesized in the gut***. It is however, recommended that half of the body requirement is provided in the diet, while the other half is met from the bacterial synthesis. Accordingly, the ***suggested RDA*** for an adult is ***70-100 μg/day***.

Dietary sources

Cabbage, ***cauliflower***, tomatoes, alfa alfa, spinach and other green vegetables are good sources. It is also present in egg yolk, meat, liver, cheese and dairy products.

Deficiency symptoms

The deficiency of vitamin K is ***uncommon***, since it is present in the diet in sufficient quantity and/or is adequately synthesized by the intestinal bacteria.

Deficiency of vitamin K leads to the lack of active prothrombin in the circulation. The result is that blood coagulation is adversely affected. The individual bleeds profusely even for minor injuries. ***The blood clotting time is increased***.

Hypervitaminosis K

Administration of large doses of vitamin K produces hemolytic anaemia and jaundice, particularly in infants. The toxic effect is due to increased breakdown of RBC.

Antagonists of vitamin K

The compounds—namely ***heparin***, ***bishydroxy-coumarin***—act as anticoagulants and are antagonists to vitamin K. The salicylates and dicumarol are also antagonists to vitamin K. ***Dicumarol*** is structurally related to vitamin K and acts as a ***competitive inhibitor*** in the synthesis of active prothrombin.

VITAMIN C (ASCORBIC ACID)

Vitamin C is a water soluble ***versatile vitamin***. It plays an important role in human health and disease.

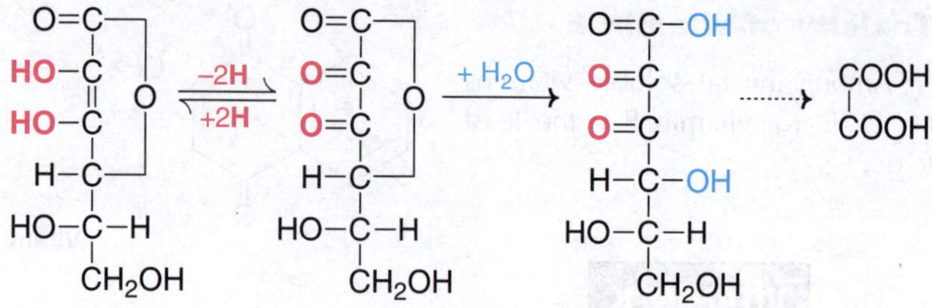

L-Ascorbic acid (reduced form) | **Dehydro L-ascorbic acid** (oxidized form) | **Diketo L-gulonic acid** | **Oxalic acid**

Fig.7.8 : *Structures of vitamin C (ascorbic acid) and its related compounds.*

Chemistry

Ascorbic acid is a hexose (6 carbon) derivative and closely ***resembles monosaccharides in structure*** (***Fig.7.8***). The acidic property of vitamin C is due to the enolic hydroxyl groups. It is a strong reducing agent. L-Ascorbic acid undergoes oxidation to form dehydroascorbic acid and this reaction is reversible. On hydration, dehydroascorbic acid is irreversibly converted to 2,3-diketogulonic acid which is inactive. Oxidation of ascorbic acid is rapid in the presence of copper. Hence vitamin C becomes inactive if the foods are prepared in copper vessels.

Biosynthesis and metabolism

Many animals can synthesize ascorbic acid from glucose via ***uronic acid pathway*** (***Chapter 13***). However, ***man,*** other primates, guinea pigs and bats cannot synthesize ascorbic acid due to the ***deficiency of*** a single enzyme namely ***L-gulonolactone oxidase***.

Biochemical functions

Most of the functions of vitamin C are related to its property to undergo reversible oxidation–reduction i.e., interconversion of ascorbic acid and dehydroascorbic acid.

1. **Collagen formation :** Vitamin C plays the role of a coenzyme in hydroxylation of proline and lysine while protocollagen is converted to collagen (i.e. ***post-translational modification***). The hydroxylation reaction is catalysed by ***lysyl hydroxylase*** (for lysine) and ***prolyl hydroxylase*** (for proline). This reaction is dependent on vitamin C.

Hydroxyproline and hydroxylysine are essential for the collagen cross-linking and the strength of the fiber. In this way, vitamin C is necessary for maintenance of normal connective tissue and wound healing process.

2. **Bone formation :** Bone tissues possess an ***organic matrix, collagen*** and the inorganic calcium, phosphate etc. Vitamin C is required for bone formation.

3. **Iron and hemoglobin metabolism :** Ascorbic acid ***enhances iron absorption by keeping it in the ferrous form***. This is due to the reducing property of vitamin C.

4. **Tryptophan metabolism :** Vitamin C is essential for the ***hydroxylation of tryptophan*** (enzyme-hydroxylase) to hydroxytryptophan in the synthesis of serotonin.

5. **Tyrosine metabolism :** Ascorbic acid is required for the oxidation of p-hydroxy phenylpyruvate (enzyme hydroxylase) to homogentisic acid in tyrosine metabolism. Further, ascorbic acid is also needed by dopamine β-hydroxylase, involved in the ***synthesis of catecholamines*** (norepinephrine and epinephrine).

6. **Folic acid metabolism :** The active form of the vitamin folic acid is tetrahydrofolate (FH_4). Vitamin C is ***needed for the formation of FH_4*** (enzyme-folic acid reductase).

7. **Synthesis of corticosteroid hormones :** Adrenal gland possesses high levels of ascorbic acid, particularly in periods of stress. It is believed that vitamin C is necessary for the ***hydroxylation reactions*** in the synthesis of corticosteroid hormones.

8. **Sparing action of other vitamins :** Ascorbic acid is a ***strong antioxidant***. It spares vitamin A, vitamin E, and some B-complex vitamins from oxidation.

9. **Immunological function :** Vitamin C enhances the ***synthesis of immunoglobulins*** (antibodies) and increases the phagocytic action of leucocytes.

10. **Preventive action on cataract :** Vitamin C reduces the risk of cataract formation.

11. **Preventive action on chronic diseases :** As an antioxidant, vitamin C reduces the risk of cancer, cataract, coronary heart diseases.

Recommended dietary allowance (RDA)

About ***40-50 mg*** vitamin C intake per day will meet the adult requirement. Additional intakes (20-40% increase) are recommended for women during pregnancy and lactation.

Dietary sources

Citrus fruits, gooseberry (***amla***), guava, green vegetables (cabbage, spinach), tomatoes, potatoes (particularly skin) are rich in ascorbic acid. High content of vitamin C is found in adrenal gland and gonads. ***Milk is a poor source of ascorbic acid.***

Deficiency symptoms

The deficiency of ascorbic acid results in ***scurvy***. This disease is characterized by spongy and ***sore gums, loose teeth***, anemia, swollen joints, decreased immunocompetence, delayed wound healing, haemorrhage, osteoporosis etc.

THIAMINE (VITAMIN B_1)

Thiamine (***anti-beri beri*** or antineuritic vitamin) is water soluble. It has a specific coenzyme, ***thiamine pyrophosphate (TPP)*** which is mostly associated with carbohydrate metabolism.

Chemistry

Thiamine contains a pyrimidine ring and a thiazole ring held by a methylene bridge (***Fig.7.9***). Thiamine is the only natural compound with thiazole ring.

The alcohol (OH) group of thiamine is esterfied with phosphate (2 moles) to form the ***coenzyme, thiamine pyrophosphate*** (***TPP or cocarboxylase***). The pyrophosphate moiety is donated by ATP and the reaction is catalysed by the enzyme thiamine pyrophosphate transferase.

Biochemical functions

The coenzyme, thiamine pyrophosphate or cocarboxylase is intimately connected with the energy releasing reactions in the carbohydrate metabolism (***Refer Chapter 13*** also).

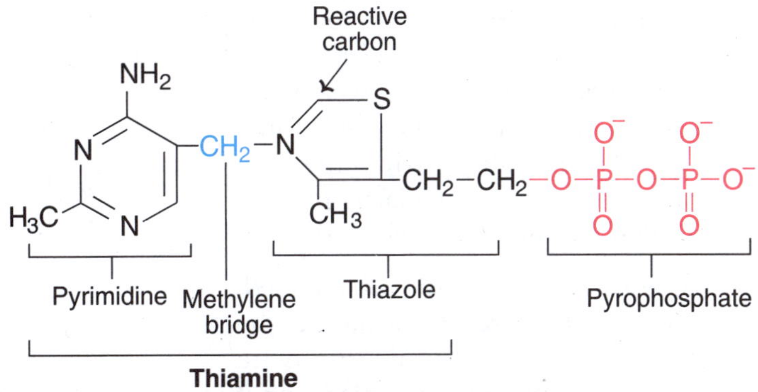

Fig.7.9 : *Structures of thiamine and thiamine pyrophosphate (TPP).*

1. The enzyme ***pyruvate dehydrogenase*** catalyses (oxidative decarboxylation) the irreversible conversion of pyruvate to acetyl CoA. This reaction is dependent on TPP, besides the other coenzymes (***Refer Fig.13.4***).

2. **α-Ketoglutarate dehydrogenase** is an enzyme of the ***citric acid cycle***. This enzyme is comparable with pyruvate dehydrogenase and requires TPP.

3. **Transketolase** is dependent on TPP. This is an enzyme of the hexose monophosphate shunt.

4. **The branched chain α-keto acid dehydrogenase (decarboxylase)** catalyses the oxidative decarboxylation of branched chain amino acids (valine, leucine and isoleucine) to the respective keto acids. This enzyme also requires TPP.

5. TPP plays an important role in the ***transmission of nerve impulse***. It is believed that TPP is required for acetylcholine synthesis and the ion translocation of neural tissue.

Recommended dietary allowance (RDA)

The daily requirement of thiamine depends on the intake of carbohydrate. A dietary supply of ***1-1.5 mg/day*** is recommended for adults (about 0.5 mg/1,000 Cals of energy).

Dietary sources

Cereals, pulses, oil seeds, nuts and yeast are good sources. Polishing of rice removes about 80% of thiamine. Vitamin B_1 is also present in animal foods like pork, liver, heart, kidney, milk etc.

Deficiency symptoms

The deficiency of vitamin B_1 results in a condition called ***beri-beri*** [*Sinhalese* : I cannot (said twice)]. Beri-beri is mostly seen in populations ***consuming***

exclusively polished rice as staple food. The early symptoms of thiamine deficiency are loss of appetite (anorexia), weakness, constipation, nausea, mental depression, peripheral neuropathy, irritability etc. Numbness in the legs complaints of ***'pins and needles sensations'*** are reported.

Carbohydrate metabolism is impaired. Accumulation of ***pyruvate*** occurs in the tissues which is harmful. Pyruvate concentration in ***plasma is elevated***, and it is also ***excreted in urine***.

Thiamine antagonists

Pyrithiamine and ***oxythiamine*** are the two important antimetabolites of thiamine.

RIBOFLAVIN (VITAMIN B_2)

Riboflavin through its coenzymes takes part in a variety of cellular ***oxidation–reduction reactions***.

Chemistry

Riboflavin contains ***6,7-dimethyl isoalloxazine*** (a heterocyclic 3 ring structure) attached to ***D-ribitol*** by a nitrogen atom. Ribitol is an open chain form of sugar ribose with the aldehyde group (CHO) reduced to alcohol (CH_2OH).

Coenzymes of riboflavin

Flavin mononucleotide (FMN) and flavin adenine dinucleotide (FAD) are the two coenzyme forms of riboflavin (***Fig.7.10***). The ribitol (5 carbon) is linked to a phosphate in FMN. FAD is formed from FMN by the transfer of an AMP moiety from ATP.

Biochemical functions

The flavin coenzymes (mostly FAD and to a lesser extent FMN) participate in many ***redox reactions*** responsible for energy production. The functional unit of both the coenzymes is isoalloxazine ring which serves as an acceptor of two hydrogen atoms (with electrons). FMN or FAD undergo identical reversible reactions accepting two hydrogen atoms forming $FMNH_2$ or $FADH_2$ (***Fig.7.11***).

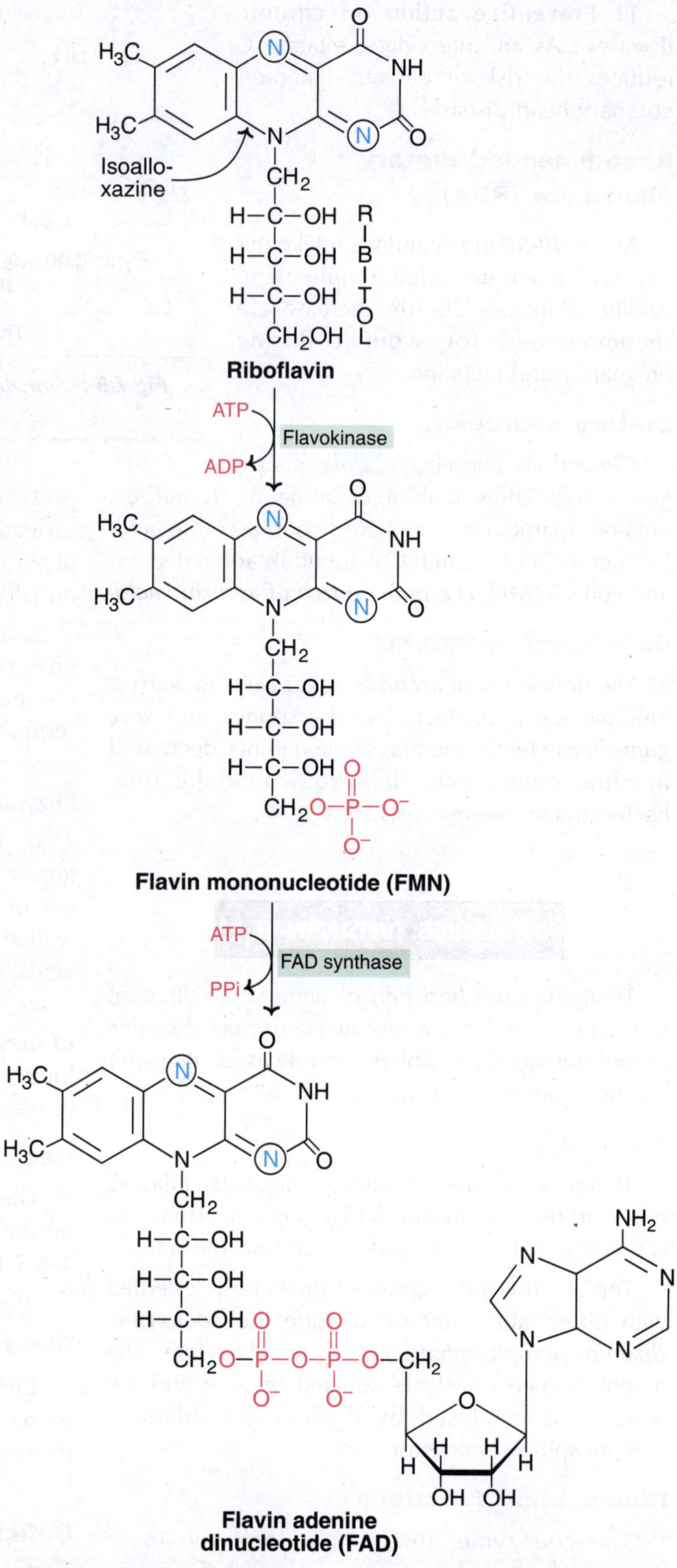

Fig.7.10 : Structures and biosynthesis of flavin mononucleotide (FMN) and flavin adenine dinucleotide (FAD).

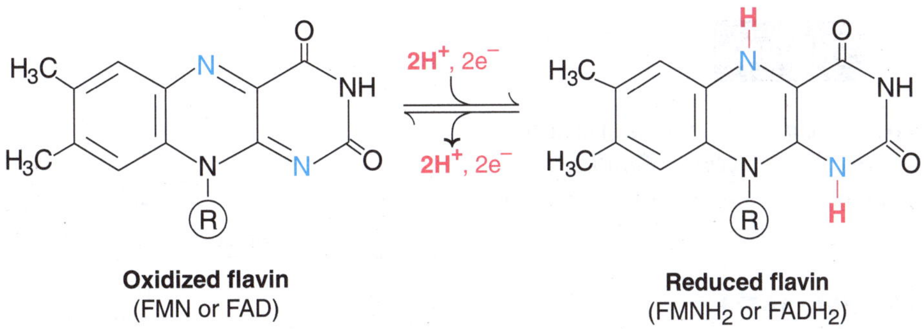

Fig.7.11 : *Participation of FMN or FAD in oxidation-reduction reactions (R-represents the rest of the structure of FMN or FAD as depicted in Fig. 7.10).*

Enzymes that use flavin coenzymes (FMN or FAD) are called ***flavoproteins***. Many flavoproteins contain metal atoms (iron, molybdenum etc.) which are known as ***metalloflavoproteins***.

The coenzymes, FAD and FMN are associated with several enzymes involved in ***carbohydrate***, ***lipid***, ***protein*** and purine metabolisms, besides the electron transport chain. A few examples are listed in ***Table 7.2***.

Recommended dietary allowance (RDA)

The daily requirement of riboflavin for an adult is ***1.2-1.7 mg***. Higher intakes (by 0.2-0.5 mg/day) are advised for pregnant and lactating women.

Dietary sources

Milk and milk products, ***meat***, eggs, liver, kidney are rich sources. Cereals, fruits, vegetables and fish are moderate sources.

Deficiency symptoms

Riboflavin deficiency symptoms include ***cheilosis*** (fissures at the corner of the mouth), ***glossitis*** (tongue smooth and purplish) and ***dermatitis***. Assay of the enzyme ***glutathione reductase*** in erythrocytes will be useful in assessing the riboflavin deficiency.

Antimetabolite : ***Galactoflavin*** is an antimetabolite of riboflavin.

TABLE 7.2 Selected examples of FAD and FMN dependent enzymes along with their respective reactions

Enzyme	*Reaction*
FAD dependent	
I. *Carbohydrate metabolism*	
(a) Pyruvate dehydrogenase complex	Pyruvate ⟶ Acetyl CoA
(b) α-Ketoglutarate dehydrogenase complex	α-Ketoglutarate ⟶ Succinyl CoA
(c) Succinate dehydrogenase	Succinate ⟶ Fumarate
II. *Lipid metabolism*	
(d) Acyl CoA dehydrogenase	Acyl CoA ⟶ α, β-Unsaturated acyl CoA
III. *Protein metabolism*	
(e) Glycine oxidase	Glycine ⟶ Glyoxylate + NH_3
(f) D-Amino acid oxidase	D-Amino acid ⟶ α-Keto acid + NH_3
IV. *Purine metabolism*	
(g) Xanthine oxidase	Xanthine ⟶ Uric acid
FMN dependent	
L-Amino acid oxidase	L-Amino acid ⟶ α-Keto acid + NH_3

NIACIN

Niacin or ***nicotinic acid*** (vitamin B_3) is also known as ***pellagra preventive*** (P.P.) ***factor of Goldberg***. The coenzymes of niacin (NAD^+ and $NADP^+$) can be synthesized by the essential amino acid, tryptophan.

Chemistry and synthesis of coenzymes

Niacin is a ***pyridine derivative***. Structurally, it is ***pyridine 3-carboxylic acid***. The amide form of niacin is known as niacinamide or nicotinamide.

Dietary nicotinamide, niacin and tryptophan (an essential amino acid) contribute to the synthesis of the coenzymes–***nicotinamide adenine dinucleotide (NAD^+) and nicotinamide adenine dinucleotide phosphate ($NADP^+$)***. The pathway for the biosynthesis of NAD^+ and $NADP^+$ is depicted in ***Fig.7.12***. Nicotinamide is deaminated in the body to niacin. Niacin then undergoes a series of reactions to produce NAD^+ and $NADP^+$. Tryptophan produces quinolinate which then forms nicotinate mononucleotide and, ultimately, NAD^+ and $NADP^+$. ***Sixty milligrams of tryptophan is equivalent to 1 mg of niacin for the synthesis of niacin coenzymes***. Nicotinamide, liberated on the degradation of NAD^+ and $NADP^+$ is mostly excreted in urine as ***N-methylnicotinamide***.

Biochemical functions

The coenzymes NAD^+ and $NADP^+$ are involved in a variety of ***oxidation-reduction reactions.*** They accept hydride ion (hydrogen atom and one electron : H^-) and undergo reduction in the pyridine ring. This results in the neutralization of positive charges. The nitrogen atom and the fourth carbon atom of nicotinamide ring participate in the reaction. While one atom of hydrogen (as hydride ion) from the substrate (AH_2) is accepted by the coenzyme, the other hydrogen ion (H^+) is released into the surrounding medium (***Fig.7.13***).

This reaction is reversed when NADH is oxidized to NAD^+. $NADP^+$ also functions like NAD^+ in the oxidation-reduction reactions.

A large number of enzymes (about 40) belonging to the class ***oxidoreductases*** are dependent on NAD^+ or $NADP^+$. Selected examples of enzymes and the reactions they catalyse are given in ***Table 7.3***.

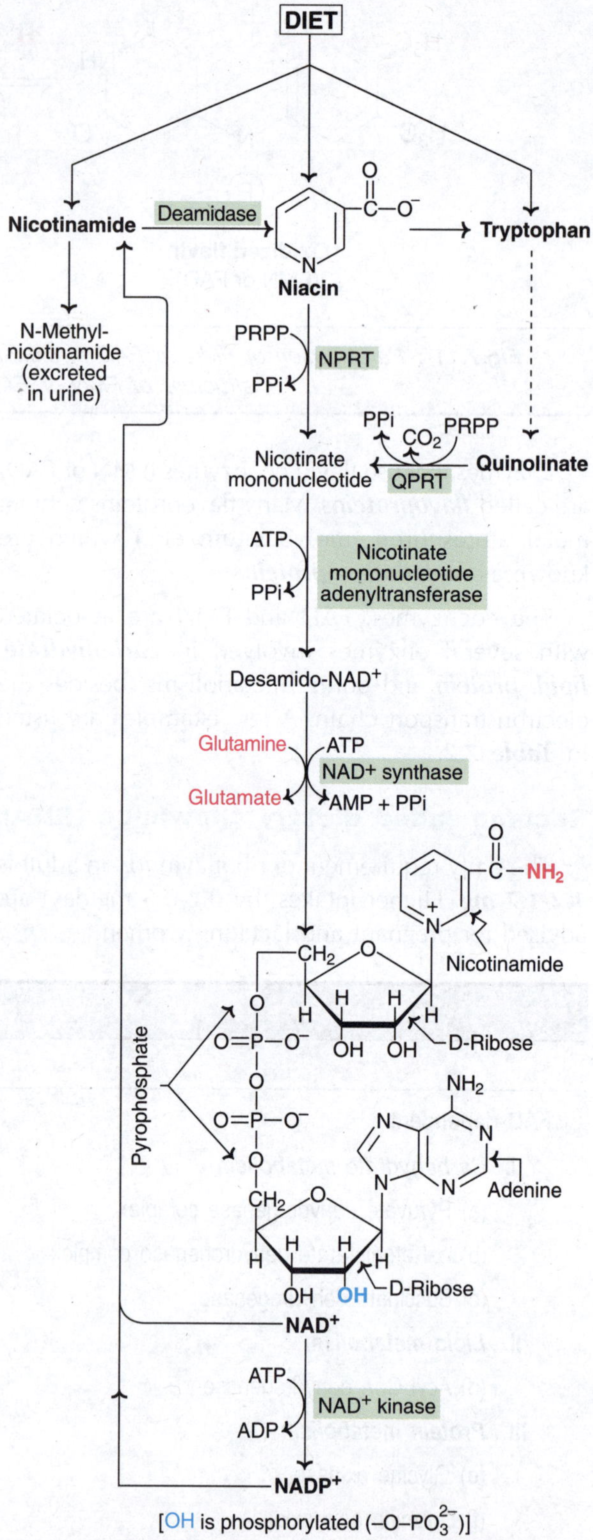

Fig.7.12 : Outline of the biosynthesis of nicotinamide nucleotides, NAD^+ and $NADP^+$ (NPRT-Nicotinate phosphoribosyltransferase; QPRT-Quinolinate phosphoribosyltransferase; PPi-Pyrophosphate; PRPP-Phosphoribosyl pyrophosphate).

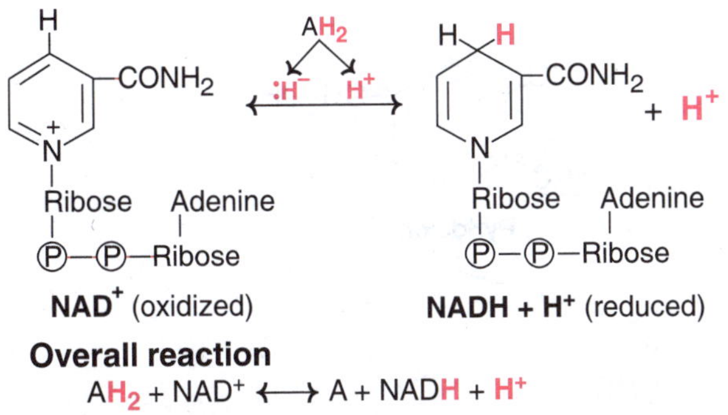

Fig.7.13 : *Mechanism of oxidation and reduction of nicotinamide coenzyme—NAD^+ (Note : Similar mechanism operates for $NADP^+$ also).*

Recommended dietary allowance (RDA)

The daily requirement of niacin for an adult is ***15-20 mg*** and for children, around 10-15 mg. Very often, the term ***niacin equivalents*** (*NE*) is used while expressing its RDA. ***One NE = 1 mg niacin or 60 mg of tryptophan***.

Dietary sources

The rich natural sources of niacin include ***liver***, ***yeast***, whole grains, cereals, pulses like beans and peanuts. Milk, fish, eggs and vegetables are moderate sources. The essential amino acid tryptophan can serve as a precursor for the synthesis of nicotinamide coenzymes. On an average, 1 g of a good quality protein containing about 60 mg of tryptophan is equivalent to 1 mg of niacin ***(conversion ratio 60 : 1)*** for the synthesis of nicotinamide co-enzymes.

Deficiency symptoms

Niacin deficiency results in a condition called ***pellagra*** (*Italian*: rough skin). The disease pellagra involves skin, gastrointestinal tract and central nervous system. The symptoms of pellagra are commonly referred to as ***three Ds***. The disease also progresses

TABLE 7.3 Selected examples of NAD^+ and $NADP^+$ dependent enzymes along with their respective reactions

Enzyme	*Reaction*
NAD^+ dependent	
I. *Carbohydrate metabolism*	
(a) Lactate dehydrogenase	Pyruvate ⟶ Lactate
(b) Pyruvate dehydrogenase complex	Pyruvate ⟶ Acetyl CoA
II. *Lipid metabolism*	
(c) β-Hydroxy acyl CoA dehydrogenase	β-Hydroxy acyl CoA ⟶ β-Keto acyl CoA
(d) Alcohol dehydrogenase	Alcohol ⟶ Acetaldehyde
III. *Protein metabolism*	
(e) Branched chain α-keto acid dehydrogenase	α-Keto acids of branched chain amino acids (Leu, Ile, Val) ⟶ Corresponding acyl CoA thioesters
NAD^+ or $NADP^+$ dependent	
(a) Glutamate dehydrogenase	Glutamate ⟶ α-Ketoglutarate + NH_3
(b) Isocitrate dehydrogenase	Isocitrate ⟶ Oxalosuccinate
$NADP^+$ dependent	
(a) Glucose 6-phosphate dehydrogenase	Glucose 6-phosphate ⟶ 6-Phosphogluconolactone
(b) Malic enzyme	Malate ⟶ Pyruvate
NADPH dependent	
(a) HMG CoA reductase	HMG CoA ⟶ Mevalonate
(b) Phenylalanine hydroxylase	Phenylalanine ⟶ Tyrosine
(c) Dihydrofolate reductase	Folic acid ⟶ Tetrahydrofolic acid.

in that order ***dermatitis***, ***diarrhea***, ***dementia***, and if not treated may rarely lead to ***death*** (4th D).

PYRIDOXINE (VITAMIN B_6)

Vitamin B_6 is used to collectively represent the three compounds namely ***pyridoxine***, ***pyridoxal and pyridoxamine*** (the ***vitamers*** of B_6).

Chemistry

Vitamin B_6 compounds are ***pyridine derivatives***. They differ from each other in the structure of a functional group attached to 4th carbon in the pyridine ring. Pyridoxine is a primary alcohol, pyridoxal is an aldehyde form while pyridoxamine is an amine (***Fig.7.14***).

Synthesis of coenzyme

The active form of vitamin B_6 is the coenzyme ***pyridoxal phosphate (PLP)***. PLP can be synthesized from the three compounds pyridoxine, pyridoxal and pyridoxamine. B_6 is excreted in urine as 4-pyridoxic acid. The different forms of B_6 and their interrelationship is depicted in ***Fig.7.14***.

Biochemical functions

Pyridoxal phosphate (PLP), the coenzyme of vitamin B_6 is found attached to the ε-amino group of lysine in the enzyme. PLP is closely associated with the metabolism of amino acids. The synthesis of certain specialized products such as serotonin, histamine, niacin coenzymes from the amino acids is dependent on pyridoxine. Pyridoxal phosphate participates in reactions like ***transamination***, ***decarboxylation***, ***deamination***, ***transsulfuration condensation*** etc.

1. **Transamination :** Pyridoxal phosphate is involved in the transamination reaction (by transaminase) ***converting amino acids to keto acids*** (***Refer Fig.15.3***).

2. **Decarboxylation :** Some of the α-amino acids undergo decarboxylation to form the respective amines. This is carried out by a group of enzymes called ***decarboxylases*** which are dependent on PLP. Many biogenic amines with important functions are synthesized by PLP decarboxylation.

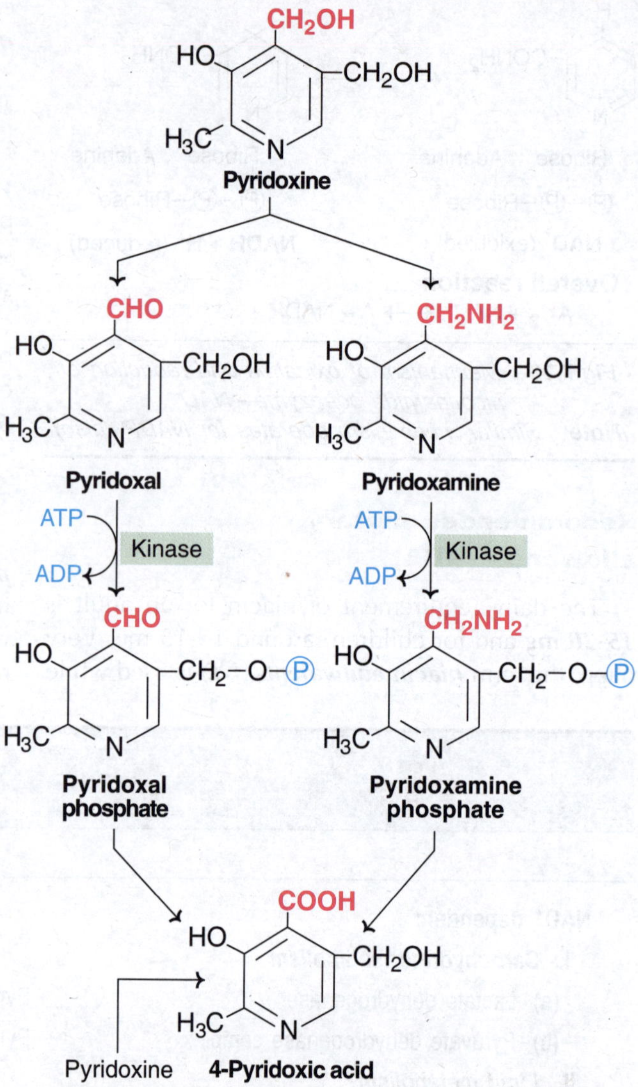

Fig.7.14 : Pyridoxine, its derivatives and coenzyme.

(a) **Serotonin** (5-hydroxytryptamine, 5 HT), produced from tryptophan is important in nerve impulse transmission (neurotransmitter). It regulates sleep, behaviour, blood pressure etc.

Tryptophan ⟶ 5-Hydroxytryptophan $\xrightarrow{PLP}$ (CO_2) 5-Hydroxytryptamine

(b) The synthesis of **catecholamines** (dopamine, norepinephrine and epinephrine) from tyrosine requires PLP. Catecholamines are involved in metabolic and nervous regulation.

Tyrosine ⟶ DOPA $\xrightarrow{PLP}$ (CO_2) Dopamine ⟶ Norepinephrine ⟶ Epinephrine

> **Medical Concepts/ Clinical Correlates**
>
> - Vitamin C deficiency causes scurvy. The manifestations of scurvy are related to the impairment in the synthesis of collagen.
> - Megadoses of vitamin C are used in common cold, wound healing etc.
> - β-Carotene, vitamin E and ascorbic acid serve as antioxidants and reduce the risk of heart attacks and cancers.
> - Long term use of isoniazid for the treatment of tuberculosis causes vitamin B_6 deficiency.
> - Folic acid supplementation lowers elevated plasma homocysteine levels, and reduces the risk associated with atherosclerosis and thrombosis.

3. Pyridoxal phosphate is required for the synthesis of ***δ-amino levulinic acid***, the precursor for heme synthesis.

$$\text{Glycine} + \text{Succinyl CoA} \xrightarrow[\text{PLP}]{\text{ALA Synthase}} \delta\text{-Amino-levulinic acid (ALA)} \dashrightarrow \text{Heme}$$

4. PLP plays an important role in the metabolism of sulfur containing amino acids. ***Trans-sulfuration*** (transfer of sulfur) from homocysteine to serine occurs in the synthesis of cysteine.

5. **Deamination** of hydroxyl group containing amino acids requires PLP.

$$\text{Serine} \xrightarrow[\text{PLP}]{\text{Dehydratase}} \text{Pyruvate} + NH_3$$

$$\text{Threonine} \xrightarrow[\text{PLP}]{\text{Dehydratase}} \alpha\text{-Ketobutyrate} + NH_3$$

6. **Serine** is synthesized from glycine by a PLP dependent enzyme hydroxymethyltransferase.

Recommended dietary allowance (RDA)

The requirement of pyridoxine for an adult is ***1.5-2.0 mg/day***. During pregnancy, lactation and old age, an intake of 2.5 mg/day is recommended.

Dietary sources

Animal sources such as ***egg yolk, fish, milk,*** meat are rich in B_6. Wheat, corn, cabbage, roots and tubers are good vegetable sources.

Deficiency symptoms

Pyridoxine deficiency is associated with ***neurological symptoms*** such as depression, irritability, nervousness and mental confusion. Convulsions and peripheral neuropathy are observed in severe deficiency. These symptoms are related to the decreased synthesis of biogenic amines (serotonin, GABA, norepinephrine and epinephrine). Demyelination of neurons is also observed in B_6 deficiency.

In pyridoxine deficiency, the synthesis of niacin coenzymes (NAD^+ and $NADP^+$) from tryptophan is impaired. ***Xanthurenic acid***, produced in high quantities is ***excreted in urine***, which serves as a reliable index for ***B_6 deficiency***.

Pyridoxine antagonists

Isoniazid, deoxypyridoxine and methoxy pyridoxine are the antagonists of vitamin B_6.

BIOTIN

Biotin (formerly known as ***anti-egg white injury factor, vitamin B_7 or vitamin H)*** is a sulfur containing B-complex vitamin. It directly participates as a coenzyme in the carboxylation reactions.

Williams and his associates (1940) showed that the egg white injury in rats and chicks was due to the presence of an anti-vitamin in egg white. The ***egg-white injury factor*** was identified as a glycoprotein–***avidin***, and ***biotin*** was called as ***anti-egg white injury factor***.

Chemistry

Biotin is a heterocyclic sulfur containing monocarboxylic acid. The structure is formed by fusion of ***imidazole*** and ***thiophene rings*** with a valeric acid side chain (***Fig.7.15***). Biotin is covalently bound

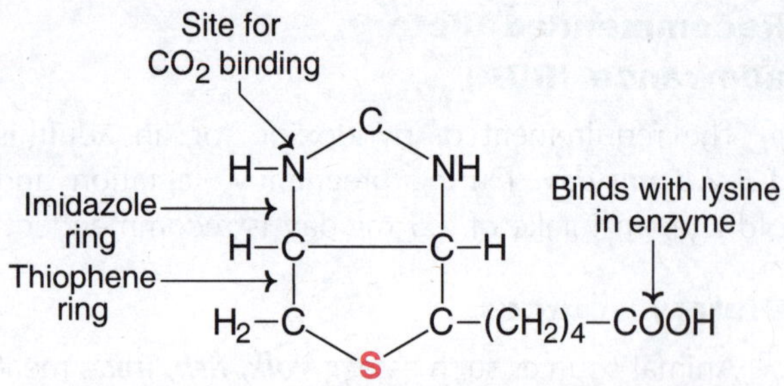

Fig.7.15 : Structure of biotin with binding sites.

to ε-amino group of lysine to form ***biocytin*** in the enzymes.

Biochemical functions

Biotin serves as a carrier of CO_2 in ***carboxylation reactions***. The reaction catalysed by ***pyruvate carboxylase***, converting pyruvate to oxaloacetate has been investigated in detail (***Fig.7.16***).

As a coenzyme, biotin is involved in various metabolic reactions.

1. **Gluconeogenesis and citric acid cycle :** The ***conversion of pyruvate to oxaloacetate*** by biotin dependent ***pyruvate carboxylase*** is essential for the synthesis of glucose from many non-carbohydrate sources. Oxaloacetate so formed is also required for the continuous operation of citric acid cycle.

2. **Fatty acid synthesis :** Acetyl CoA is the starting material for the synthesis of fatty acids. The very first step in fatty acid synthesis is a carboxylation reaction.

$$\text{Acetyl CoA} \xrightarrow[\text{Acetyl CoA carboxylase}]{\text{Biotin}} \text{Malonyl CoA}$$

Recommended dietary allowance (RDA)

A daily intake of about 20-30 µg is recommended for adults. In fact, biotin is normally synthesized by the intestinal bacteria. However, to what extent the synthesized biotin contributes to the body requirements is not clearly known.

Dietary sources

Biotin is widely distributed in both animal and plant foods. The rich sources are ***liver***, ***kidney***, egg yolk, milk, tomatoes, grains etc.

Deficiency symptoms

The symptoms of biotin deficiency include ***anemia, loss of appetite, nausea***, dermatitis, glossitis etc.

Biotin ***deficiency is uncommon***, since it is well distributed in foods and also supplied by the intestinal bacteria.

Antagonists

Desthiobiotin, biotin sulphonic acid are biotin antagonists.

PANTOTHENIC ACID

Pantothenic acid (*Greek* **:** pantos–everywhere) or vitamin B_5, formerly known as chick ***anti-dermatitis factor*** (or filtrate factor) is widely distributed in nature. It's metabolic role as ***coenzyme A*** is also widespread.

Chemistry and synthesis of coenzyme A

Pantothenic acid consists of two components, ***pantoic acid*** and ***β-alanine***, held together by a peptide linkage. Synthesis of coenzyme A from pantothenate occurs in a series of reactions (***Fig.7.17***). Pantothenate is first phosphorylated to which cysteine is added. Decarboxylation, followed by addition of AMP moiety

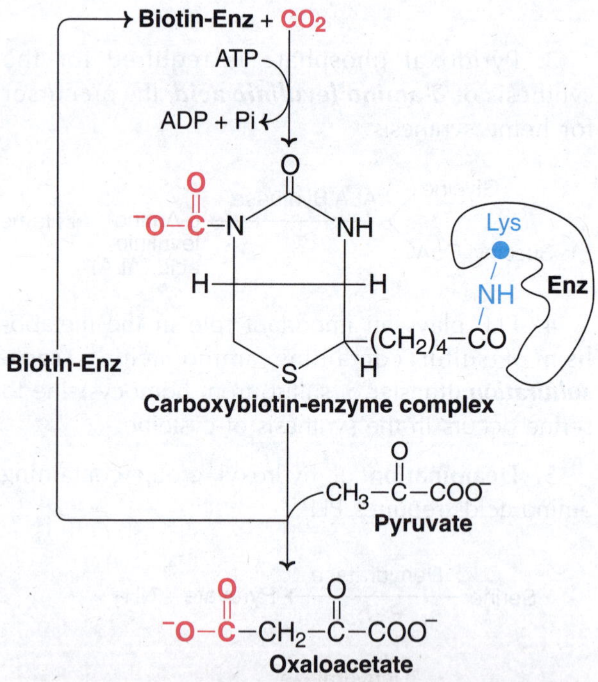

Fig.7.16 : Role of biotin in the carboxylation reaction, catalysed by the enzyme pyruvate carboxylase (Enz-Enzyme).

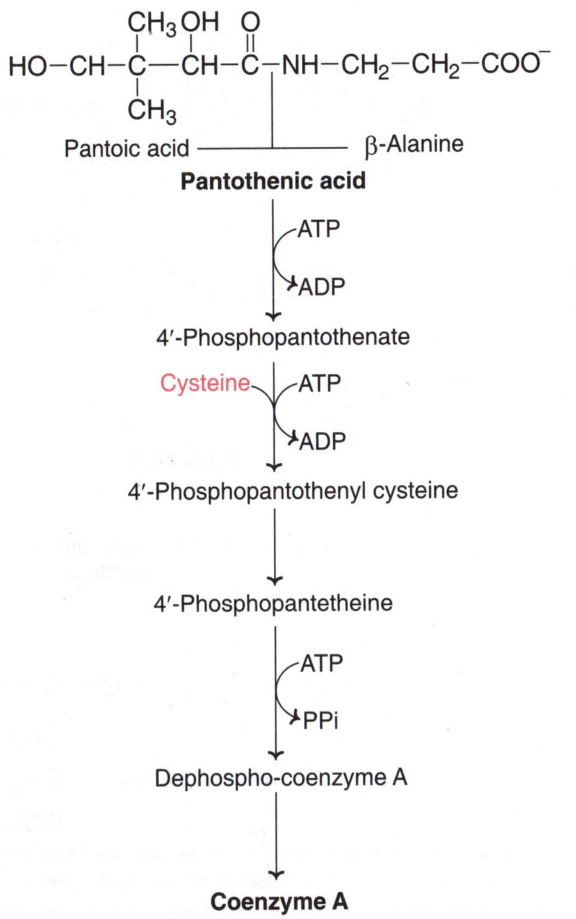

Fig.7.17 : *Summary of the synthesis of coenzyme A from pantothenic acid (**Note :** Pantothenic acid consists of pantoic acid and β-alanine).*

and a phosphate (each from ATP) results in coenzyme A. The structure of coenzyme A consists of pantothenic acid joined to β-mercaptoethanolamine (thioethanolamine) on one end. On the other side, pantothenic acid is held by a phosphate bridge to adenylic acid. The adenylic acid is made up of adenine, and a phosphate linked to carbon-3 of ribose.

Biochemical functions

The functions of pantothenic acid are exerted through ***coenzyme A or CoA (A for acetylation).*** Coenzyme A is a central molecule involved in all the metabolisms (carbohydrate, lipid and protein). It plays a unique role in integrating various metabolic pathways. More than 70 enzymes that depend on coenzyme A are known.

Coenzyme A serves as a ***carrier of activated acetyl or acyl groups*** (as thiol esters). This is comparable with ATP which is a carrier of activated phosphoryl groups.

A few examples of enzymes involved the participation of coenzyme A are given below.

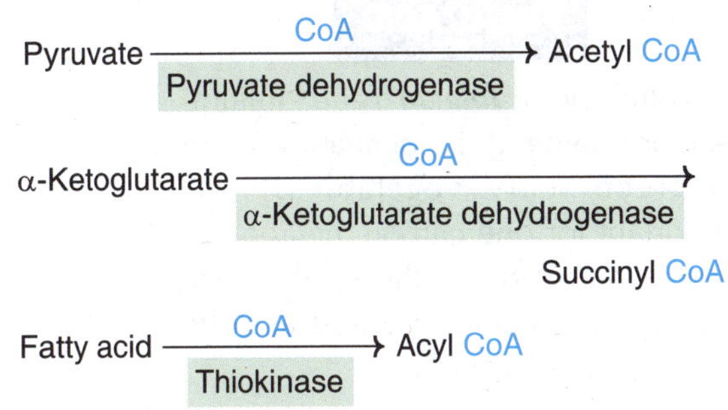

Coenzyme A may be regarded as a ***coenzyme of metabolic integration***, since acetyl CoA is a central molecule for a wide variety of biochemical reactions, as illustrated in ***Fig.7.18***.

Recommended dietary allowance (RDA)

The requirement of pantothenic acid for humans is not clearly known. A daily intake of about ***5-10 mg*** is advised for adults.

Dietary sources

Pantothenic acid is one of the ***most widely distributed*** vitamins found in plants and animals. The rich sources are egg, liver, meat, yeast, milk etc.

Deficiency symptoms

Pantothenic acid deficiency manifestations have not been reported in humans. Dr. Gopalan, a world

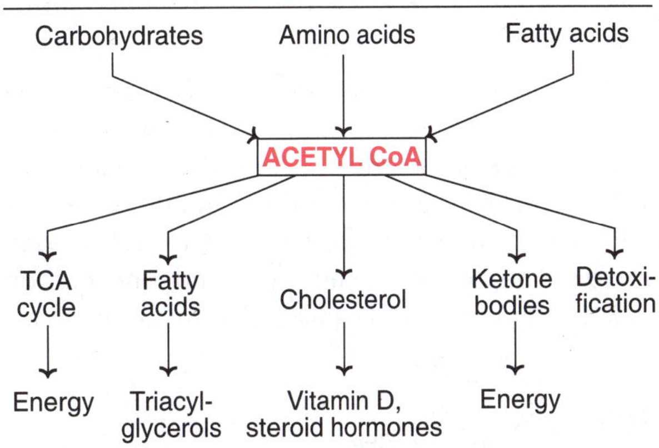

Fig.7.18 : *An overview of formation and utilization of acetyl CoA.*

renowned nutritionist from India, linked the ***burning feet syndrome*** (pain and numbness in the toes, sleeplessness, fatigue etc.) with pantothenic acid deficiency.

FOLIC ACID

Folic acid or ***folacin*** (*Latin* : folium-leaf) or Vitamin B_9 is abundantly found in green leafy vegetables. It is important for one carbon metabolism and is required for the synthesis of certain amino acids, purines and the pyrimidine—thymine.

Chemistry

Folic acid consists of three components– pteridine ring, p-amino benzoic acid (PABA) and glutamic acid (1 to 7 residues). Folic acid mostly has one glutamic acid residue and is known as ***pteroyl-glutamic acid (PGA).***

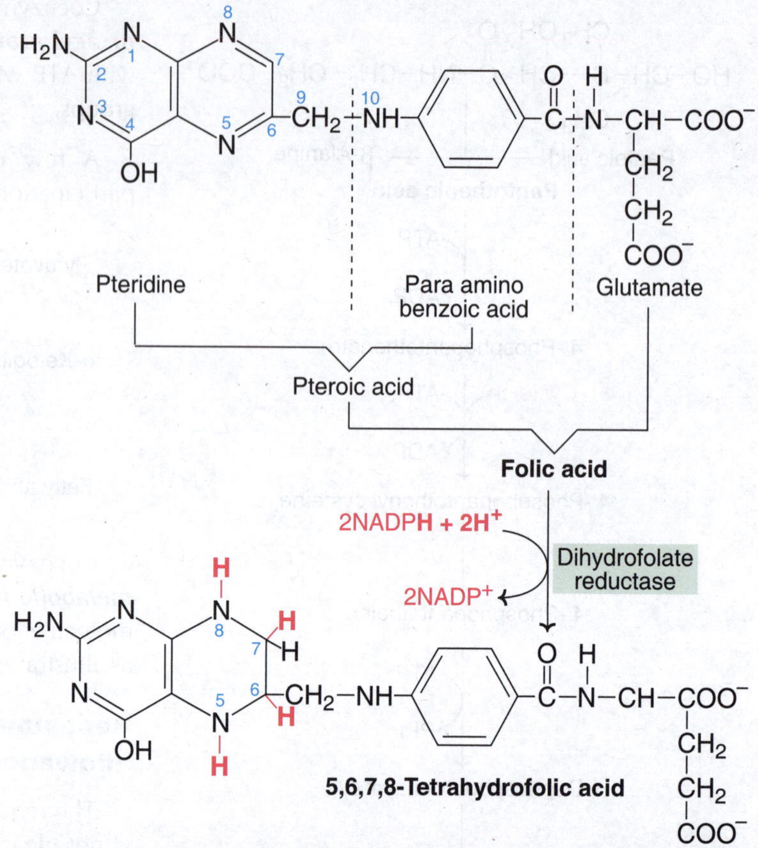

***Fig.7.19** : Conversion of folic acid to tetrahydrofolic acid (THF).*

The active form of folic acid is ***tetrahydrofolate (THF or FH_4).*** It is synthesized from folic acid by the enzyme ***dihydrofolate reductase***. The reducing equivalents are provided by 2 moles of NADPH. The hydrogen atoms are present at positions 5, 6, 7 and 8 of THF (***Fig.7.19***).

Biochemical functions

Tetrahydrofolate (THF or FH_4), the coenzyme of folic acid is actively involved in the ***one carbon metabolism***.

The one carbon units bind with THF at position N^5 or N^{10} or on both N^5 and N^{10} of pteroyl structure. The attachment of formyl (—CHO) at position 5 of THF gives N^5-formyl tetrahydrofolate which is commonly known as ***folinic acid or citrovorum factor***. The other commonly found one carbon moieties and their binding with THF are given below.

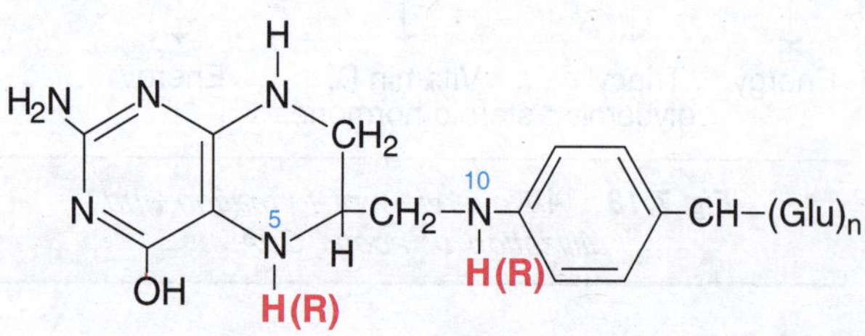

THF-1 carbon derivative	R group (one carbon unit)
N^5-Formyl THF	—CHO
N^{10}-Formyl THF	—CHO
N^5-Formimino THF	—CH=NH
N^5, N^{10}-Methenyl THF	≡CH
N^5, N^{10}-Methylene THF	$=CH_2$
N^5 – Methyl THF	$—CH_3$

The essential functions of THF in one carbon metabolism are summarized in ***Fig.7.20.***

The interrelationship between the various 1-carbon THF derivatives along with their involvement in the synthesis of different compounds is given in ***Fig.15.23*** (***Chapter 15***). Many important compounds are synthesized in one carbon metabolism.

1. ***Purines*** (carbon 2, 8) which are incorporated into DNA and RNA.

2. ***Pyrimidine*** nucleotide–deoxythymidylic acid (dTMP), involved in the synthesis of DNA.

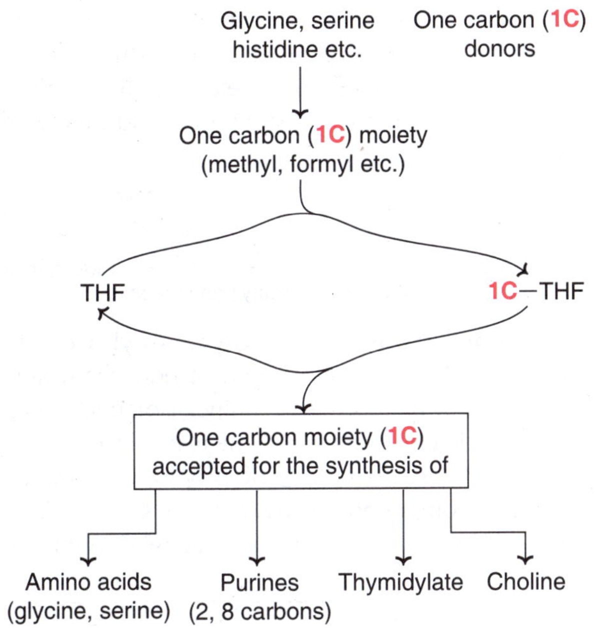

Fig.7.20 : An overview of one carbon metabolism (THF-Tetrahydrofolate).

3. ***Glycine***, serine, ethanolamine and choline are produced.

4. ***N-Formylmethionine***, the initiator of protein biosynthesis is formed.

Recommended dietary allowance (RDA)

The daily requirement of folic acid is around ***200 µg***. In the women, higher intakes are recommended during pregnancy (400 µg/day) and lactation (300 µg/day).

Dietary sources

Folic acid is widely distributed in nature. The rich sources are ***green leafy vegetables***, whole grains, cereals, liver, kidney, yeast and eggs. Milk is rather a poor source of folic acid.

Deficiency symptoms

Folic acid deficiency is probably the ***most common vitamin deficiency***, observed primarily ***in the pregnant women***, in both developed (including USA) and developing countries (including India).

The ***macrocytic anemia*** (abnormally large RBC) associated with megaloblastic changes in bone marrow is a characteristic feature of folate deficiency.

Folic acid deficiency in pregnant women may cause neural defects in the fetus. Hence, high doses of folic acid are recommended during pregnancy to prevent birth defects.

Formiminoglutamate (FIGLU), formed in histidine metabolism ***accumulates, and is excreted in urine***. Histidine load test utilizing the excretion of FIGLU in urine is used to assess folic acid deficiency.

Folic acid antagonists

Aminopterin and amethopterin (also called as methotrexate) are structural analogues of folic acid. They competitively ***inhibit dihydrofolate reductase*** and block the formation of THF. The biosynthesis of purines, thymine nucleotides, and hence DNA is impaired. This results in the blockage of cell proliferation. ***Aminopterin and methotrexate are successfully used in the treatment of many cancers, including leukemia***.

COBALAMIN (VITAMIN B_{12})

Vitamin B_{12} is also known as ***anti-pernicious anemia vitamin***. It is a unique vitamin, synthesized by only microorganisms and not by animals and plants. It was the last vitamin to be discovered.

Chemistry

The empirical formula of vitamin B_{12} (**cyanocobalamin**) is $C_{63}H_{90}N_{14}O_{14}PCo$. The structure of vitamin B_{12} consists of a corrin ring with a central ***cobalt*** atom (***Fig.7.21***). The corrin ring is almost similar to the tetrapyrrole ring structure found in other porphyrin compounds e.g. heme (with Fe) and chlorophyll (with Mg).

There are two coenzyme forms of vitamin B_{12}.

(a) **Methylcobalamin** in which cyanide is replaced by methyl group.

(b) **5′-Deoxyadenosyl cobalamin**, cyanide is replaced by 5′ deoxyadenosine forming an unusal carbon cobalt bond (***Fig.7.22***).

Absorption, transport and storage

The vitamin B_{12} is present in the diet in a bound form to proteins. B_{12} is liberated by the enzymes (acid hydrolases) in the stomach. The dietary source of B_{12} is known as ***extrinsic factor of Castle***. The

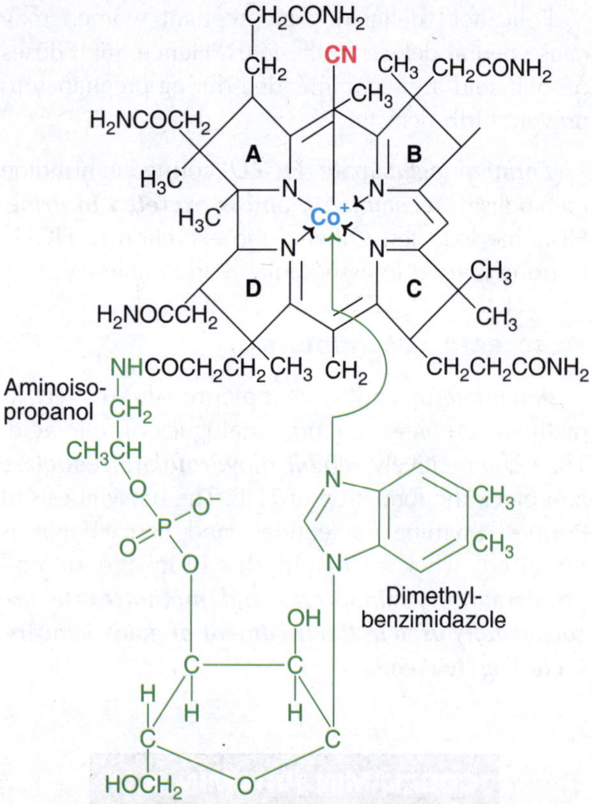

Fig.7.21 : Structure of vitamin B_{12} (cyanocobalamin).

stomach secretes a special protein called ***intrinsic factor (IF)***.

The cobalamin–IF complex travels through the gut. The complex binds to specific receptors on the surface of the mucosal cells of the ileum. The binding of the complex and entry of B_{12} into the mucosal cells is mediated by Ca^{2+} ions. In the mucosal cells, B_{12} is converted to methylcobalamin. It is then transported in the circulation in a bound form to proteins namely ***transcobalamins*** (TC-I, TC-II). Methylcobalamin which is in excess is taken up by the liver, converted to deoxyadenosyl B_{12} and stored in this form. It is believed that liver can store about 4-5 mg, an amount sufficient to meet the body requirements of B_{12} for 4-6 years.

Biochemical functions

About ten enzymes requiring vitamin B_{12} have been identified. Most of them are found in bacteria (glutamate mutase, ribonucleotide reductase etc.). There are only ***two reactions in mammals*** that are dependent on vitamin B_{12}.

1. **Synthesis of methionine from homocysteine :** Vitamin B_{12}, as methylcobalamin is used in this reaction. This is an important reaction involving N^5-methyl tetrahydrofolate from which tetrahydrofolate is liberated. This metabolic step signifies the interrelation between vitamin B_{12} and folic acid (for details, see later)

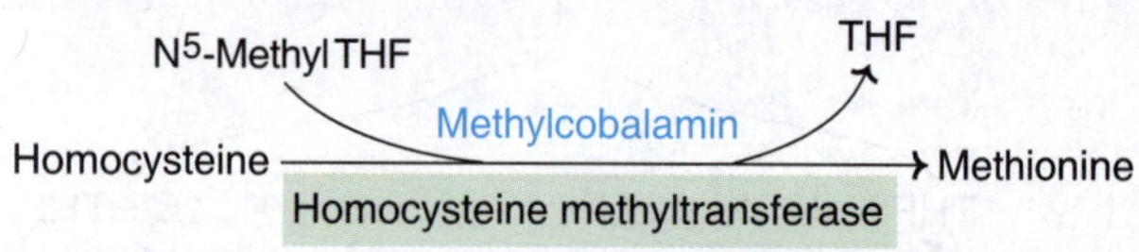

2. **Isomerization of methymalonyl CoA to succinyl CoA :** The degradation of odd chain fatty acids, certain amino acids (valine, isoleucine etc.) and pyrimidines (thymine and uracil) produce directly or through the mediation of propionyl CoA, an important compound methylmalonyl CoA. This is converted by the enzyme methylmalonyl CoA mutase

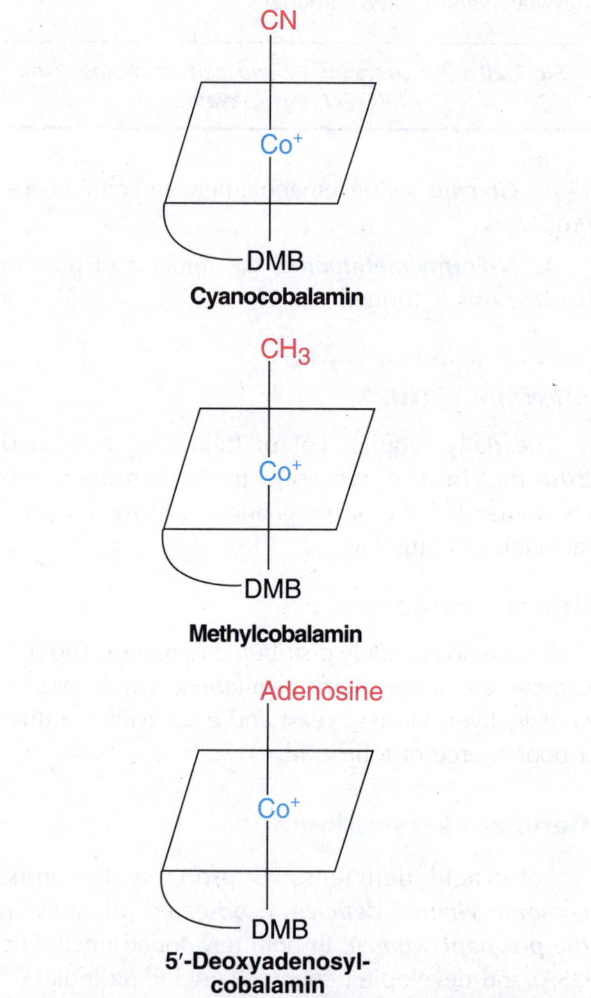

***Fig.7.22 :** Coenzyme derivatives of vitamin B_{12} (**Note :** Corrin ring represented diagrammatically is identical in all; DMB-Dimethylbenzimidazole).*

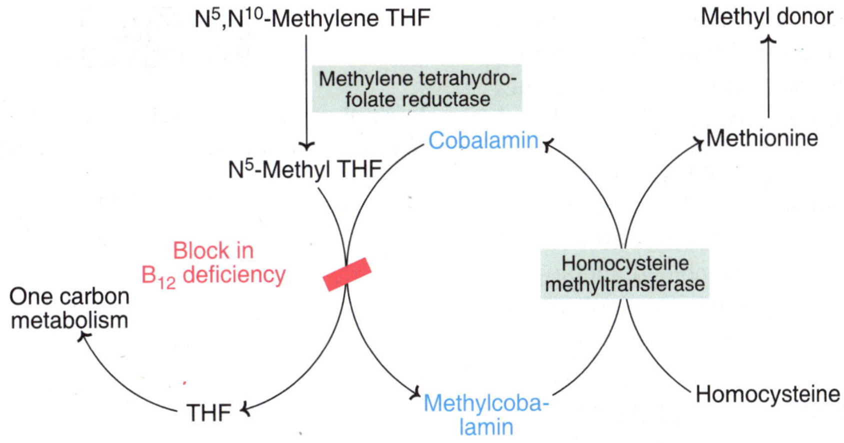

Fig. 7.23 : Interrelationship between folic acid and vitamin B_{12}.

to succinyl CoA in the presence of B_{12} coenzyme, deoxyadenosyl cobalamin. This reaction involves hydrogen transfer and intramolecular rearrangement. In ***B_{12} deficiency***, methylmalonyl CoA accumulates and is ***excreted in urine*** as ***methylmalonic acid***.

Recommended dietary allowance (RDA)

A daily intake of about ***1-2 μg*** of vitamin B_{12} is adequate to meet the adult requirements. For children, ***0.5-1.5 μg/day*** is recommended. During pregnancy and lactation, the requirement is ***3 μg/day***.

Dietary sources

Foods of ***animal origin*** are the only sources for vitamin B_{12}. The rich sources are ***liver, kidney, milk,*** curd, eggs, fish, pork and chicken. Curd is a better source than milk, due to the synthesis of B_{12} by *Lactobacillus*.

Vitamin B_{12} is synthesized only by micro-organisms (anaerobic bacteria). Plants cannot synthesize, hence B_{12} is never found in plant foods. Animals obtain B_{12} either by eating foods, derived from other animals or from the intestinal bacterial synthesis. Small quantities of vitamin ***B_{12}, synthesized by bacteria on the surface of fruits***, may partially meet the requirements.

Deficiency symptoms

The most important disease associated with vitamin B_{12} deficiency is ***pernicious anemia***. It is characterized by low hemoglobin levels, decreased number of erythrocytes and neurological manifestations.

The excretion of ***methylmalonic acid*** (elevated) in urine and estimation of serum B_{12} level are used to assess B_{12} deficiency.

Treatment

Vitamin B_{12} is administered in therapeutic doses (100-1000μg) intramuscularly. Folic acid administration can also reverse hematological abnormalities, observed in B_{12} deficiency. However, the neurological symptoms persist. Therefore, a combined supplementation of B_{12} and folate is employed to treat the patients with megaloblastic anemias.

INTERRELATION BETWEEN FOLIC ACID AND VITAMIN B_{12}—FOLATE TRAP OR METHYL TRAP HYPOTHESIS

The deficiency of either folate or vitamin B_{12} results in a similar type of anemia. This suggests a probable biochemical interrelation between these two vitamins. There is only one metabolic reaction known, common to folate and vitamin B_{12} (***Fig.7.23***).

In vitamin B_{12} deficiency, increased folate levels are observed in plasma. The activity of the enzyme ***homocysteine methyltransferase (methionine synthase) is low in B_{12} deficiency***. As a result, the only major pathway for the conversion of N^5-methyl THF to tetrahydrofolate is blocked and body THF pool is reduced. Essentially, almost the entire body folate becomes trapped as N^5-methyl THF. This is known as ***folate trap or methyl trap***. In this manner, B_{12} deficiency results in decreased folate coenzymes that leads to reduced nucleotide and DNA synthesis.

Although the tissue folate levels are adequate or high, there is a ***functional folate deficiency*** due to the lack of THF pool. The outcome is the development of ***megaloblastic anemia***. A combined therapy of vitamin B_{12} and folic acid is generally employed to treat the patients with megaloblastic anemia.

VITAMIN LIKE COMPOUNDS

Besides the vitamins described above, there are many other compounds present in foods as accessory factors. Although not essential in the diet, they perform many important functions in the body.

CHOLINE

Choline is ***trimethylhydroxy ethanolamine***.

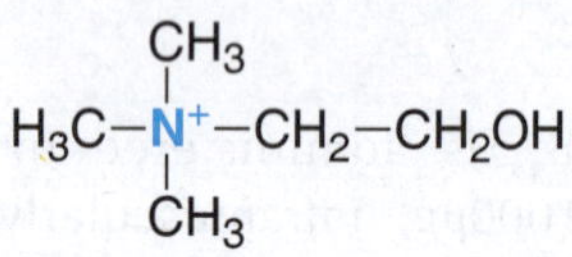

Best and Huntsman (1934) found that choline deficiency in rats produced fatty liver and established its nutritional importance. However, in a strict sense, choline is not a vitamin, since it is adequately synthesized in the body (from serine). It is also available from many dietary sources (e.g. milk, eggs, liver, cereals etc.).

Biochemical functions

1. Choline, as a component of ***phospholipids*** (lecithins) is involved in membrane structure and lipid transport.

2. Choline prevents the accumulation of fat in liver (as ***lipotropic factor***). It promotes the synthesis of phospholipids and lipoproteins and the disposal of ***triacylglycerols*** from liver.

3. Due to the presence of three methyl groups (one carbon fragments), choline is actively involved in ***one carbon metabolism***.

4. Choline is a precursor for the synthesis of acetylcholine which is required for transmission of nerve impulse.

INOSITOL

Inositol is ***hexahydroxy-cyclohexane***. It is also known as *myo*-inositol or *meso*-inositol.

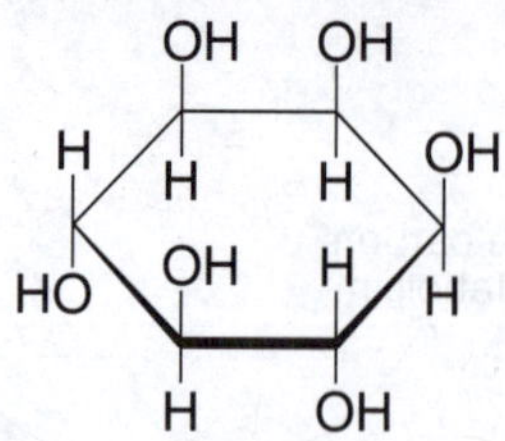

Biochemical functions

1. Inositol is required for the synthesis of ***phosphatidylinositol*** (lipositol) which is a constituent of cell membrane.

2. It acts as ***lipotropic factor*** (along with choline) and prevents the accumulation of fat in liver.

3. For some hormones, inositol acts as a ***second messenger*** at the membrane level for the release of Ca^{2+} ions.

4. Inositol concentration in the heart muscle is high, the significance of which however, is not known.

5. Phytin is hexaphosphate of inositol found is plants. It prevents the absorption of iron and calcium from the intestine.

LIPOIC ACID

Lipoic acid (thioctic acid) is a sulfur containing fatty acid (6, 8-dithiooctanoic acid). It exists in an oxidized and reduced form. Lipoic acid is fat as well as water soluble.

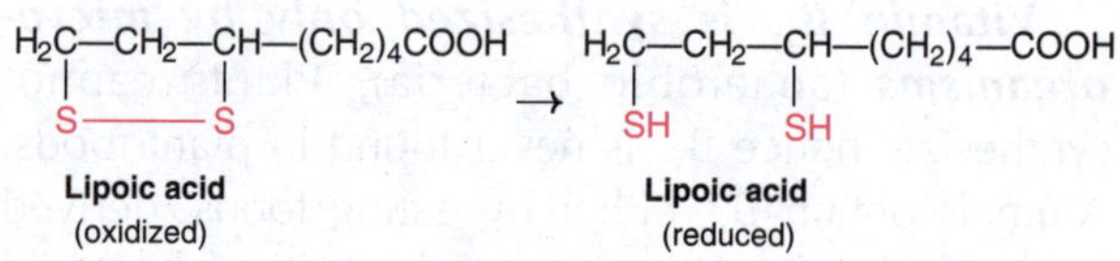

Biochemical functions

Lipoic acid is involved in the ***decarboxylation reactions*** along with other vitamins (thiamine, niacin, riboflavin and pantothenic acid). The conversion of pyruvate to acetyl CoA (by pyruvate dehydrogenase) and α-ketoglutarate to succinyl CoA (by α-ketoglutarate dehydrogenase) require lipoic acid.

Therapeutic uses of lipoic acid

In recent years, administration of high doses (100–600 mg/day) of lipoic acid (or dihydrolipoic acid) is gaining importance. Being fat and water soluble, it can comfortably reach various tissues. The therapeutic applications of lipoic acid are related to its antioxidant property (regarded as ***universal antioxidant***), some of them are listed

- Reduces the free radicals in brain that otherwise contribute to Alzheimer's disease and multiple sclerosis.
- Lipoic acid stimulates production of glutathione (GSH), besides helping in the recycling of vitamins E and C.
- Reduces insulin resistance, and brings down plasma low density lipoproteins.
- May be useful in the prevention of stroke and myocardial infarction.

PARA AMINOBENZOIC ACID

Para aminobenzoic acid (PABA) is a structural constituent of folic acid. PABA may be regarded as a ***vitamin in another vitamin*** (folic acid)

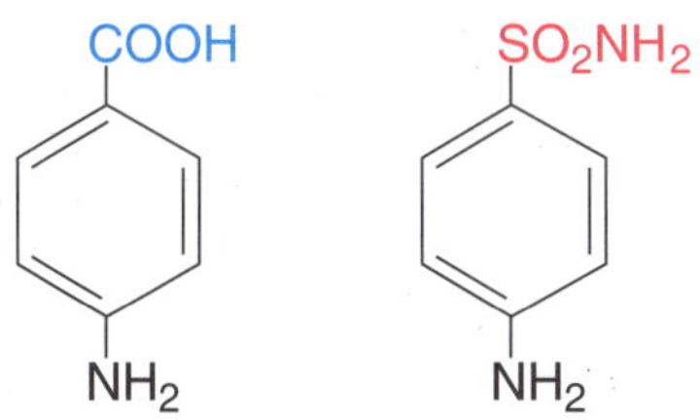

PABA **Sulfonilamide**

The specific functions of PABA in humans, except that it is a component of folic acid, have not been identified.

PABA is synthesized by the bacteria and is essential for their growth. The sulfa drug sulfanilamide (p-amino benzene sulfanilamide) is a structural analogue of PABA. Sulfanilamide competes with PABA and acts as a bacteriostatic agent. Ingestion of large doses of PABA will compete with the action of drugs and therefore should be avoided during sulfanilamide therapy (trade name—sulfonamides).

TABLE 7.4 Selected list of antivitamins/vitamin antagonists along with corresponding vitamins

Antivitamin/ vitamin antgonists	*Vitamin*
Dicumarol Warfarin	Vitamin K
Thiaminase Pyrithiamine Oxythiamine	Thiamine
Galactoflavin	Riboflavin
Isoniazid Deoxypyridoxine	Pyridoxine
Avidin Desthiobiotin	Biotin
Aminopterin Methotrexate Trimethoprim	Folic acid
Sulfanilamide	Para-aminobenzoic acid

BIOFLAVONOIDS

Szent-Gyorgi and his associates (1936) observed that flavonoids, isolated from lemon peel (known as citrin) were responsible for maintenance of normal capillary permeability. The term ***vitamin P*** (P for permeability) was used to this group of substances. However, they are commonly known as bioflavonoids.

Bioflavonoids act as antioxidants and protect ascorbic acid from being destroyed. It is suggested that this antioxidant property may be responsible for maintenance of capillary permeability. Bioflavonoids have been used to correct the vascular abnormality in humans.

ANTIVITAMINS

Antivitamins are ***antagonistic to*** (oppose and block) the ***action of vitamins***. They usually have structural similarities with vitamins. Administration of antivitamins causes vitamin deficiencies. The common antivitamins are discussed as antagonists for each vitamin.

SUMMARY

1. *Vitamins are accessory food factors required in the diet. They are classified as fat soluble (A, D, E and K) and water soluble (B-complex and C).*
2. *Vitamin A is involved in vision, proper growth, differentiation and maintenance of epithelial cells. Its deficiency results in night blindness.*
3. *The active form of vitamin D is calcitriol which functions like a steroid hormone and regulates plasma levels of calcium and phosphate. Vitamin D deficiency leads to rickets in children and osteomalacia in adults.*
4. *Vitamin E is a natural antioxidant necessary for normal reproduction in many animals.*
5. *Vitamin K has a specific coenzyme function.. It catalyses the carboxylation of glutamic acid residues in blood clotting factors (II, VII, IX and X) and converts them to active form.*
6. *Vitamin C (ascorbic acid) is involved in the hydroxylation of proline and lysine in the formation of collagen. Scurvy is caused by ascorbic acid deficiency. Therapeutic use of megadoses of vitamin C, to cure everything from common cold to cancer, has become controversial.*
7. *Thiamine (B_1), as a cocarboxylase (TPP) is involved in energy releasing reactions. The coenzymes of riboflavin (FAD and FMN) and niacin (NAD^+ and $NADP^+$) take part in a variety of oxidation-reduction reactions connected with energy generation.*
8. *Pyridoxal phosphate (PLP), the coenzyme of vitamin B_6, is mostly associated with amino acid metabolism. PLP participates in transamination, decarboxylation, deamination and condensation reactions.*
9. *Biotin (anti-egg white injury factor) participates as a coenzyme in carboxylation reactions of gluconeogenesis, fatty acid synthesis etc.*
10. *Coenzyme A (of pantothenic acid) is involved in the metabolism of carbohydrates, lipids and amino acids and their integration.*
11. *Tetrahydrofolate (THF), the coenzyme of folic acid participates in the transfer of one carbon units (formyl, methyl etc.) in amino acid and nucleotide metabolism. Megaloblastic anemia is caused by folic acid deficiency.*
12. *Vitamin B_{12} has two coenzymes, deoxyadenosylcobalamin and methylcobalamin. B_{12} deficiency results in pernicious anemia.*

THE READER MUST REFER APPENDIX D (Page Nos : 422 - 431) FOR RELATED CLINICAL CASE STUDIES (Nos : 3, 4, 5)

DIGESTION AND ABSOROPTION

8

Food is the basic and essential requirement of man for his very existence. The food we eat consists of carbohydrates, proteins, lipids, vitamins and minerals. The bulk of the food ingested is mostly in a complex macromolecular form which cannot, as such, be absorbed by the body.

Digestion is a process involving the hydrolysis of large and complex organic molecules of foodstuffs into smaller and preferably water-soluble molecules which can be easily absorbed by the gastrointestinal tract for utilization by the organism. Digestion of macromolecules also promotes the absorption of fat soluble vitamins and certain minerals.

Cooking of the food, and mastication (in the mouth) significantly improve the digestibility of foodstuffs by the enzymes.

Gastrointestinal tract

Digestion as well as absorption are complicated processes that occur in the gastrointestinal tract (***GIT***) involving many organs. The diagrammatic representation of GIT is depicted in ***Fig.8.1***, and the essential organs with their respective major functions are given in ***Table 8.1***. The digestive organs possess a large reserve capacity. For instance, pancreas secretes enzymes 5-10 fold higher than required for digestion of foods normally ingested.

The digestion and absorption of individual foods, namely carbohydrates, proteins, lipids and nucleic acids, is described here. The gastrointestinal hormones are discussed under hormones (***Chapter 20***).

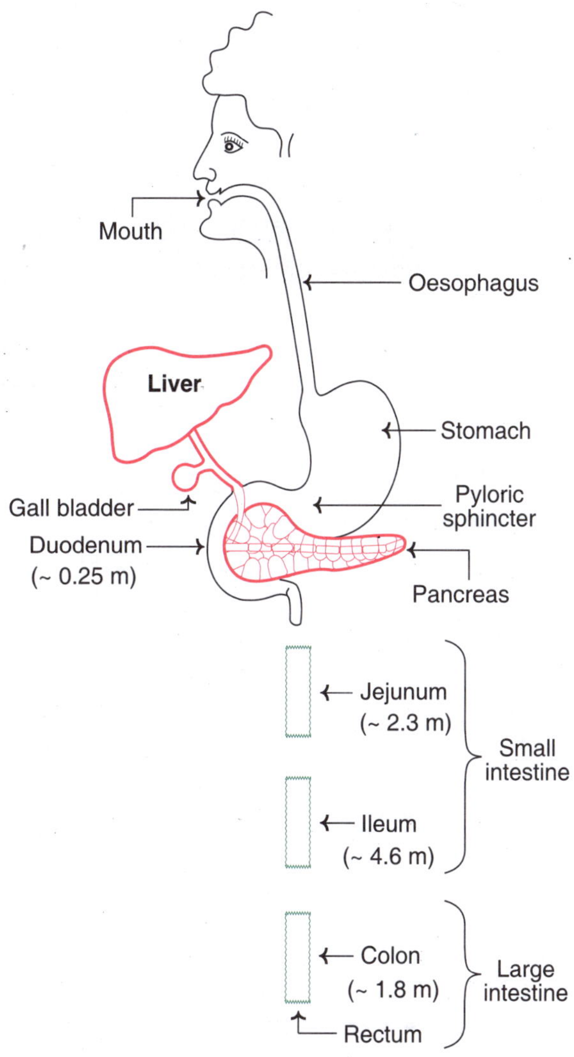

Fig.8.1 : *Diagrammatic representation of gastrointestinal tract.*

TABLE 8.1
Organs of gastrointestinal tract with their major functions in digestion and absorption

Organ	*Major function(s)*
Mouth	Production of saliva containing α-amylase; partial digestion of polysaccharides
Stomach	Elaboration of gastric juice with HCl and proteases; partial digestion of proteins
Pancreas	Release of $NaHCO_3$ and many enzymes required for intestinal digestion
Liver	Synthesis of bile acids
Gall bladder	Storage of bile
Small intestine	Final digestion of foodstuffs; absorption of digested products
Large intestine	Mostly absorption of electrolytes; bacterial utilization of certain non-digested and/or unabsorbed foods

CARBOHYDRATES

The principal dietary carbohydrates are ***polysaccharides (starch, glycogen), disaccharides (lactose, sucrose)*** and, to a minor extent, monosaccharides (glucose, fructose).

Digestion

The digestion of carbohydrates occurs briefly in mouth and largely in the intestine (***Fig.8.2***). The polysaccharides get hydrated during heating which is essential for their efficient digestion. The hydrolysis of glycosidic bonds is carried out by a group of enzymes called ***glycosidases***.

Digestion in the mouth : Carbohydrates are the only nutrients for which the digestion begins in the mouth to a significant extent. During the process of mastication, ***salivary α-amylase (ptyalin)*** acts on starch randomly and cleaves α-1,4 glycosidic bonds.

Carbohydrates not digested in the stomach : The enzyme salivary amylase is inactivated by high acidity (low pH) in the stomach. Consequently, the ongoing degradation of starch is stopped.

Digestion in the small intestine : The acidic dietary contents of the stomach, on reaching small intestine, are neutralized by bicarbonate produced by pancreas. The ***pancreatic α-amylase*** acts on starch and continues the digestion process. Amylase specifically ***acts on α-1,4 glycosidic bonds*** and not on α-1,6 bonds. The resultant products are disaccharides (maltose, isomaltose) and oligo-saccharides.

The final digestion of di- and oligosaccharides to monosaccharides primarily occurs at the mucosal lining of the upper jejunum. This is carried out by ***oligosaccharidases and disaccharidases.*** The enzyme ***sucrase*** is capable of hydrolysing a large quantity of table sugar (sucrose). In contrast, ***lactase (β-galactosidase)*** is the rate-limiting, and, consequently, the utilization of milk sugar (lactose) is limited in humans.

Absorption of monosaccharides

The principal monosaccharides produced by the digestion of carbohydrates are glucose, fructose and

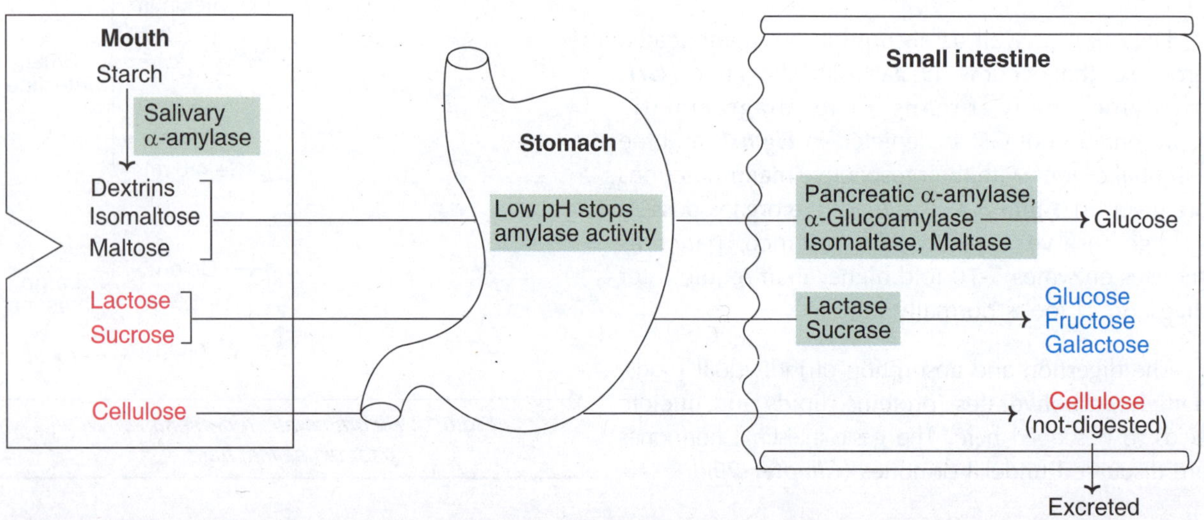

Fig.8.2 : Overview of digestion of carbohydrates.

galactose. Of these, glucose accounts for nearly 80% of the total monosaccharides. The absorption of sugars mostly takes place in the ***duodenum*** and ***upper jejunum*** of small intestine.

It is observed that hexoses are more rapidly absorbed than pentoses. Further, among the monosaccharides, ***galactose is most efficiently absorbed***, followed by glucose and fructose. Insulin has no effect on the absorption of sugars.

Mechanism of absorption

Different sugars possess different mechanisms for their absorption. Glucose is transported into the intestinal mucosal cells by a carrier mediated and energy requiring process (***Fig.8.3***).

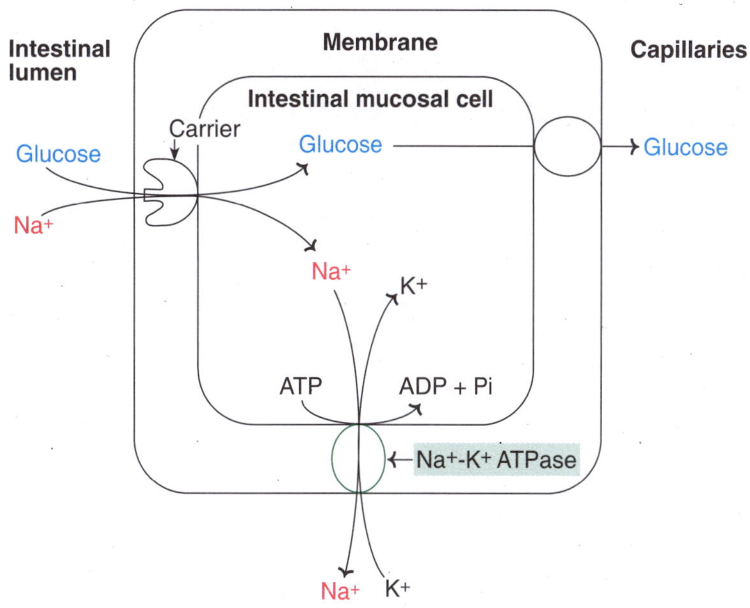

Fig.8.3 : *Transport of glucose across intestinal epithelium (**Note :** Transport amino acids also occurs by a similar mechanism; replace glucose in figure by amino acid).*

Glucose and ***Na***$^+$ share the same transport system (symport). The concentration of Na^+ is higher in the intestinal lumen compared to mucosal cells. Na^+, therefore, moves into the cells along its concentration gradient and simultaneously glucose is transported into the intestinal cells. This is mediated by the same carrier system. Thus, Na^+ diffuses into the cell and it drags glucose along with it. The intestinal Na^+ gradient is the immediate energy source for glucose transport. This energy is indirectly supplied by ATP since the reentry of Na^+ (against the concentration gradient) into the intestinal lumen is an energy-requiring active process. The enzyme ***Na^+-K^+ ATPase*** is involved in the transport of Na^+ in exchange of K^+ against the concentration gradient. (***Refer Chapter 33***)

Oral rehydration therapy (ORT) : ORT is the most common ***treatment of diarrhea***. The oral rehydration fluid contains glucose and sodium. Intestinal absorption of sodium is facilitated by the presence of glucose.

Non-digestible carbohydrates

The plant foods are rich in fibrous material which cannot be digested either by the human enzymes or intestinal bacteria. The ***fibers*** are chemically ***complex carbohydrates*** which include ***cellulose***, hemicellulose, pectins, lignin and gums. Fiber in nutrition is of special importance which is described under nutrition (***Chapter 23***).

Abnormalities of carbohydrate digestion

In general, humans possess an efficient system of carbohydrate digestion and absorption. Since only the monosaccharides are absorbed, any defect in the activities of ***disaccharidases*** results in the passage of undigested disaccharides into the large intestine. The disaccharides draw water from the intestinal mucosa by osmosis and cause osmotic diarrhea. Further, bacterial action of these ***undigested carbohydrates leads to flatulence*** (increased intestinal motility, cramps and irritation).

Lactose intolerance

Defect in the enzyme ***lactase*** (β-galactosidase) is the ***most common disaccharidase deficiency in humans***. It is estimated that more than half of the world's adult population is affected by lactose intolerance.

The treatment of lactose intolerance is quite simple. Elimination of lactose from the diet (severe restriction of milk and dairy products) will solve the problem.

Continued consumption of lactose by lactose intolerant individuals causes typical symptoms of ***flatulence***.

Medical Concepts/ Clinical Correlates

- **Cooking of food significantly improves the digestibility by enzymes.**
- **Lactose intolerance due to a defect in the enzyme lactase is very common. The treatment advocated is restriction of lactose milk and milk products in the diet.**
- **Flatulence, occurring after ingestion of certain non-digestible oligosaccharides, is characterized by increased intestinal motility, cramps and irritation.**
- **Direct intestinal absorption of proteins and polypeptides is observed in the infants, immediately after birth. This is important for the transfer of maternal immunoglobulins (via breast-feeding) to the offspring.**
- **In some adults, macromolecular (protein) absorption by intestine is responsible for antibody formation, often causing food allergy.**

The problem of flatulence

Flatulence is characterized by ***increased intestinal motility, cramps and irritation***. This occurs after ingestion of certain carbohydrates and is explained as follows.

Certain disaccharides (e.g. lactose in lactose intolerance) and oligosaccharides (e.g. raffinose) are not diagested by humans. They can be degraded by the bacteria present in ileum (lower part of small intestine) to liberate monosaccharides. The latter can be metabolized by the bacteria.

As the monosaccharides are utilized by the intestinal bacteria, gases such as ***hydrogen, methane and carbon dioxide***—besides lactate and short chain fatty acids—are released. These compounds cause flatulence.

The occurrence of flatulence after the ingestion of ***leguminous seeds*** (bengal gram, redgram, beans, peas, soya bean) is very common. They contain several ***nondigestible oligosaccharides*** by human intestinal enzymes. These compounds are degraded and utilised by intestinal bacteria causing flatulence. ***Raffinose*** containing galactose, glucose and fructose is a predominant oligosaccharide found in leguminous seeds.

PROTEINS

The proteins subjected to digestion and absorption are obtained from two sources—dietary and endogenous.

Proteins are degraded by a class of enzymes—namely ***hydrolases***—which specifically cleave the peptide bonds, hence known as peptidases. They are divided into two groups

1. **Endopeptidases (proteases)** which attack the internal peptide bonds and release peptide fragments, e.g. pepsin, trypsin.
2. **Exopeptidases** which act on the peptide bonds of terminal amino acids. Exopeptidases are subdivided into ***carboxypeptidases*** (act on C-terminal amino acid) and ***aminopeptidases*** (act on N-terminal amino acid).

I. Digestion of proteins by gastric secretion

Protein digestion begins in the stomach. ***Gastric juice*** produced by stomach contains hydrochloric acid and a protease proenzyme namely pepsinogen.

Hydrochloric acid : The pH of the stomach is < 2 due to the presence of HCl, ***secreted by parietal (oxyntic) cells of gastric gland***. This acid performs two important functions-***denaturation of proteins*** and killing of certain microorganisms. The denatured proteins are more susceptible to proteases for digestion.

Pepsin : Pepsin (*Greek* : pepsis—digestion) is produced by the serous cells of the stomach as pepsinogen, the inactive zymogen or proenzyme. Pepsinogen is converted to active pepsin either by autocatalysis, brought about by other pepsin molecules or by gastric HCl (pH < 2).

Pepsin digestion of proteins results in peptides and a few amino acids which act as stimulants for the release of the hormone cholecystokinin from the duodenum.

Rennin : This enzyme, also called chymosin, is found in the ***stomach of infants and children***. Rennin

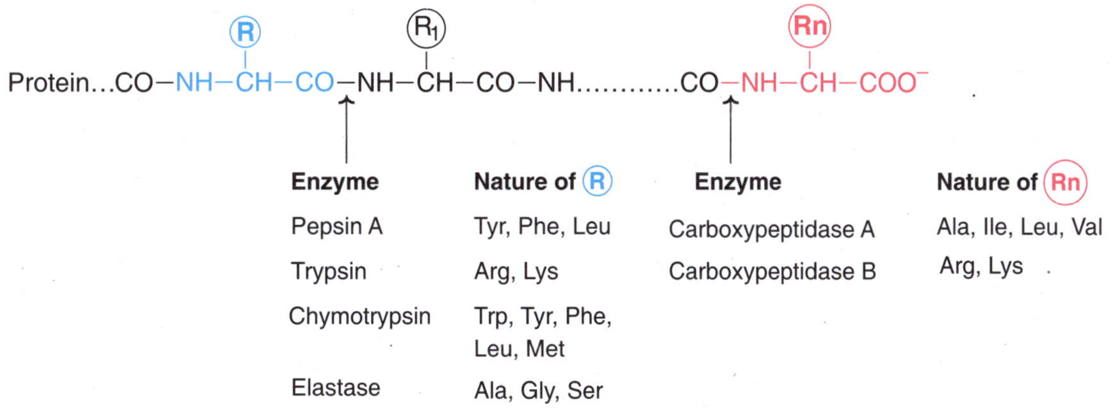

***Fig.8.4** : Digestion of proteins—Specificity of enzyme cleavage of peptide bonds.*

is involved in the curdling of milk. It converts milk protein casein to calcium paracaseinate which can be effectively digested by pepsin. Rennin is ***absent in adults***.

II. Digestion of proteins by pancreatic proteases

The proteases of pancreatic juice are secreted as zymogens (proenzymes) and then converted to active forms. These processes are initiated by the release of two polypeptide hormones, namely ***cholecystokinin*** and ***secretin*** from the intestine.

Specificity of pancreatic proteases : Trypsin, chymotrypsin and elastase are endopeptidases active at neutral pH. Gastric HCl is neutralized by pancreatic $NaHCO_3$ in the intestine and this creates favourable pH for the action of proteases.

The substrate specificity of pancreatic proteases is depicted in ***Fig.8.4***. For instance, trypsin cleaves the peptide bonds, the carbonyl (—CO—) group of which is contributed by arginine or lysine.

Action of carboxypeptidases : The pancreatic carboxypeptidases (A and B) are metalloenzymes that are dependent on Zn^{2+} for their catalytic activity, hence they are sometimes called ***Zn-proteases***.

The combined action of pancreatic proteases results in the formation of free amino acids and small peptides (2-8 amino acids).

III. Digestion of proteins by small intestinal enzymes

The luminal surface of intestinal epithelial cells contains ***aminopeptidases*** and ***dipeptidases***. Aminopeptidase is a non-specific exopeptidase which repeatedly cleaves N-terminal amino acids one by one to produce free amino acids and smaller peptides (***Fig.8.5***). The dipeptidases act on different dipeptides to liberate amino acids.

Absorption of amino acids and dipeptides

The free amino acids, dipeptides and to some extent tripeptides are absorbed by intestinal epithelial cells.

The small intestine possesses an efficient system to absorb free amino acids. L-Amino acids are more rapidly absorbed than D-amino acids. The transport of L-amino acids occurs by an active process (against a concentration gradient), in contrast to D-amino acids which takes place by a simple diffusion.

Mechanism of amino acid absorption

Amino acids are primarily absorbed by ***a similar*** mechanism, as described for the transport of ***D-glucose***. It is basically a ***Na^+-dependent active process*** linked with the transport of Na^+. As the Na^+ diffuses along the concentration gradient, the amino acid also enters the intestinal cell. Both Na^+ and amino acids share a common carrier and are transported together. The energy is supplied indirectly by ATP (for details, ***see absorption of monosaccharides*** and ***Fig.8.3***).

Abnormalities of protein digestion and amino acid absorption

Any defect in the pancreatic secretion impairs protein and fat digestion. This causes the loss of undigested protein in the feces along with the abnormal appearance of lipids. Deficiency of pancreatic secretion may be due to pancreatitis (see later), cystic fibrosis or surgical removal of pancreas.

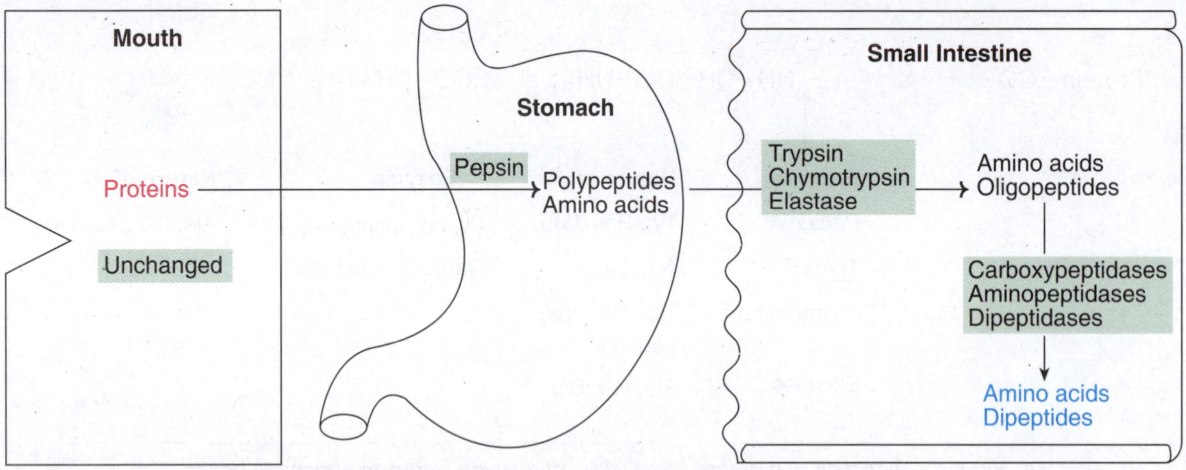

Fig.8.5 : Overview of digestion of proteins.

Hartnup's disease (neutral amino aciduria)

Hartnup is the name of the family in whom this disease was first discovered. It is characterized by the ***inability*** of intestinal and renal epithelial ***cells to absorb neutral amino acids***. Tryptophan absorption is most severely affected with a result that typical symptoms of ***pellagra*** are observed in the patients of Hartnup's disease. This is related to the ***impairment in the conversion of tryptophan to NAD$^+$ and NADP$^+$, the coenzymes of niacin***.

LIPIDS

Lipids are insoluble or sparingly soluble in aqueous solution. The digestive enzymes, however, are present in aqueous medium. This poses certain problems for the digestion and absorption of lipids. Fortunately, the digestive tract possesses specialized machinery to

1. Increase the surface area of lipids for digestion;
2. Solubilize the digested products for absorption.

Minor digestion of lipids in the stomach

In case of infants, the milk fat (with short chain fatty acids) can be hydrolysed by gastric lipase to some extent. This is because the stomach pH of infants is close to neutrality, ideal for gastric lipase action.

Emulsification of lipids in the small intestine

Emulsification is the phenomenon of ***dispersion of lipids into smaller droplets due to reduction in the surface tension***. This is accompanied by increase in the surface area of lipid droplets. Emulsification is ***essential*** for effective ***digestion*** of lipids, since the enzymes can act only on the surface of lipid droplets. More correctly, lipases act at the interfacial area between the aqueous and lipid phase.

The process of emulsification occurs by three complementary mechanisms

1. ***Detergent action*** of bile salts;
2. ***Surfactant action*** of degraded lipids;
3. ***Mechanical mixing*** due to peristalsis.

1. **Bile salts :** Bile salts are the most effective ***biological emulsifying agents***. They interact with lipid particles and the aqueous duodenal contents and convert them into smaller particles (emulsified droplets). Further, bile salts stabilize the smaller particles by preventing them from coalescing.

2. **Surfactant action of degraded lipids :** The initial digestive products of lipids (catalysed by lipase) namely free fatty acids, monoacylglycerols promote emulsification. These compounds along with phospholipids are known as surfactants.

3. Besides the action of bile salts and surfactants, the ***mechanical mixing due to peristalsis*** also helps in the emulsification of lipids. The smaller lipid emulsion droplets are good substrates for digestion.

Digestion of lipids by pancreatic enzymes

The pancreatic enzymes are primarily responsible for the degradation of dietary triacylglycerols, cholesteryl esters and phospholipids.

Medical Concepts/ Clinical Correlates

- **Pharmacological interventions (e.g. orlistat, olestra) to block fat digestion and/or absorption so as to prevent obesity are in recent use.**
- **Steatorrhea, characterized by the loss of lipids in feces, is commonly associated with impaired pancreatic function and biliary obstruction.**
- **Gastric ulcers are mainly caused by the bacterium *H. pylori*. The antibiotics that eliminate this bacterium are effective in the treatment.**
- **Acute pancreatitis is caused by auto-digestion of pancreas while chronic pancreatitis is associated with excessive consumption of alcohol.**

Degradation of triacylglycerols (fat)

Pancreatic lipase is the major enzyme that digests dietary fats. This enzyme preferentially cleaves fatty acids (particularly long chain, above 10 carbons) at positions 1 and 3 of triacylglycerols. The products are ***2-monoacylglycerol*** (formerly 2-monoglyceride) and ***free fatty acids*** (***Fig.8.6***).

Lipid esterase is a less specific enzyme present in pancreatic juice. It acts on monoacylglycerols, cholesteryl esters, vitamin esters etc. to liberate free fatty acids. The presence of bile acids is essential for the activity of lipid esterase.

Degradation of phospholipids

Phospholipases are enzymes responsible for the hydrolysis of phospholipids. Pancreatic juice is rich in phospholipase A_2 which cleaves the fatty acid at the 2nd position of phospholipids. The products are a free fatty acid and a lysophospholipid.

In the degradation of lecithin (phosphatidylcholine), the product lysolecithin is further hydrolysed by lysophospholipase to fatty acid and glycerylphosphoryl choline. The latter may be further degraded and absorbed or excreted in the feces.

An overview of the digestion of lipids is given in ***Fig.8.7***.

Absorption of lipids

The former and present theories to explain the absorption of lipids are briefly described hereunder

1. **Lipolytic theory put forth by Verzar :** According to this, ***fats*** are ***completely hydrolysed to glycerol and free fatty acids***. The latter are absorbed either as soaps or in association with bile salts.

2. **Partition theory proposed by Frazer :** This theory states that the ***digestion of triacylglycerols*** is ***partial and not complete***. The partially digested triacylglycerols, in association with bile salts, form emulsions. The lipids are taken up by the intestinal mucosal cells. As per this theory, resynthesis of lipids is not necessary for their entry into the circulation.

3. **Bergstrom theory :** This is a more recent and comprehensive theory to explain lipid absorption. It has almost replaced the earlier theories and is briefly described below.

The primary products obtained from the lipid digestion are ***2-monoacylglycerol, free fatty acids and free cholesterol***.

Role of bile salts in lipid absorption

Besides their participation in digestion, bile salts are ***essential for absorption of lipids***. Bile salts form ***mixed micelles*** with lipids. These micelles are smaller in size than the lipid emulsion droplets (utilized for digestion, described above). The micelles have a

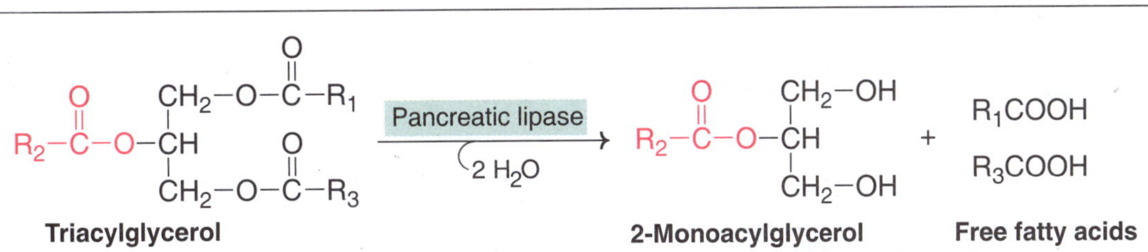

Fig.8.6 : *Enzymatic cleavage of dietary fat.*

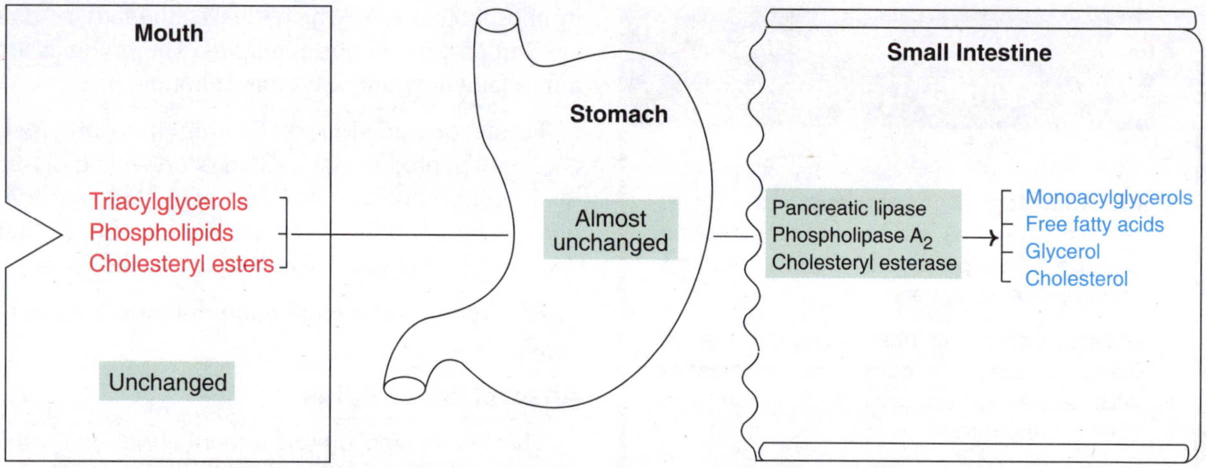

Fig.8.7 : *Overview of digestion of lipids.*

disk like shape with lipids (monoacylglycerol, fatty acids, cholesterol and phospholipids) at the interior and bile salts at the periphery. The hydrophilic groups of the lipids are oriented to the outside (close to the aqueous environment) and the hydrophobic groups to the inside. In this fashion, the bile salt micelles exert a solubilizing effect on the lipids.

Mechanism of lipid absorption

The ***mixed micelles*** serve as the major vehicles for the ***transport of lipids*** from the intestinal lumen to the membrane of the intestinal mucosal cells, the site of lipid absorption. The lipid components pass through the unstirred fluid layer and are absorbed through the plasma membrane by diffusion (***Fig.8.8***). Absorption is almost complete for monoacylglycerols and free fatty acids which are slightly water soluble. However, for water insoluble lipids, the absorption is incomplete. For instance, less than 40% of the dietary cholesterol is absorbed.

Synthesis of lipids in the intestinal mucosal cells

The ***fatty acids of short and medium chain length*** (< 10 carbons), after their absorption into the intestinal cells, ***do not undergo any modification***. They enter the portal circulation and are transported to the liver in a bound form to albumin.

The long chain fatty acids are activated by ***thiokinase*** (fatty acyl CoA synthetase) in the intestinal cells. The acyl CoA derivatives so formed combine with 2-monoacylglycerols to produce ***triacylglycerols***. These reactions are catalysed by a acyltransferases. Further, within the intestinal cells, cholesterol is converted to cholesteryl ester, and phospholipids are regenerated from the absorbed lysophospholipids. The

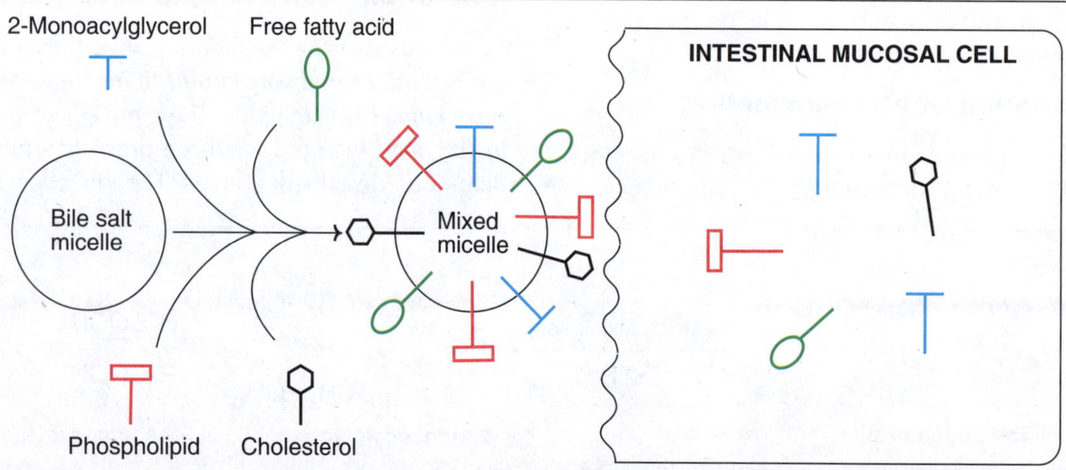

Fig.8.8 : *Absorption of lipids by intestinal mucosal cell.*

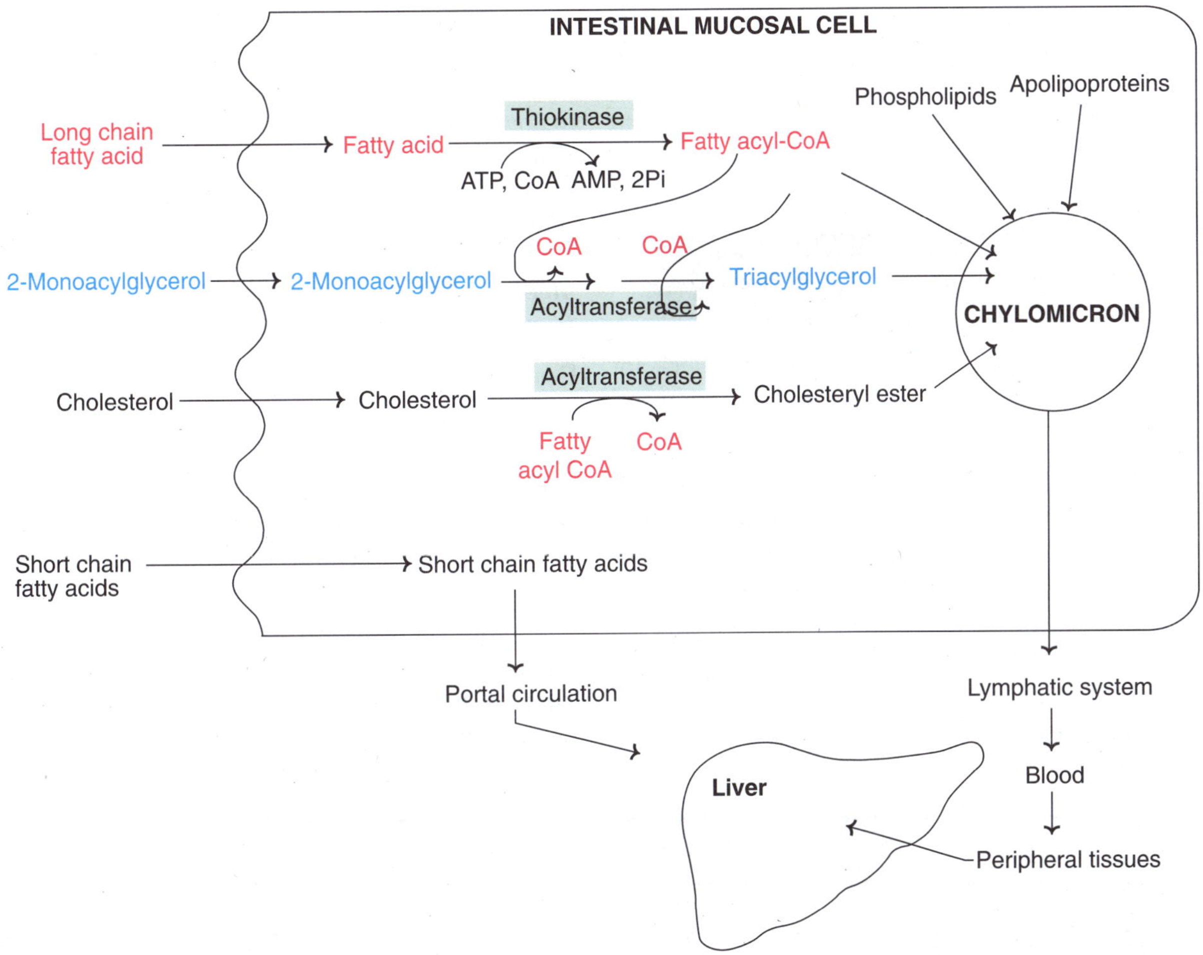

Fig.8.9 : *Formation and secretion of chylomicrons by intestinal mucosal cells.*

newly synthesized lipids are usually different from those consumed in the diet.

Secretion of lipids from the intestinal mucosal cells

The lipids that are resynthesized (described above) in the intestinal cells are hydrophobic in nature. They are put together as lipid droplets and surrounded by a thin layer consisting of mostly apolipoproteins and phospholipids (***Fig.8.9***). These particles are known as ***chylomicrons***.

Chylomicrons migrate to the plasma membrane of intestinal mucosal cells. They are released into the lymphatic vessels by exocytosis. Chylomicrons enter the large body veins via the thoracic duct. Blood from here flows to the lungs and then to the peripheral tissues (muscle, adipose tissue) and, finally, to the liver. Adipose tissue and muscle take up a large proportion of dietary lipids from chylomicrons for storage and transport.

Abnormalities of lipid digestion and absorption

The gastrointestinal tract possesses an efficient system for digestion and absorption of lipids. It can comfortably handle as much as 4 times the normal daily intake of lipids.

Steatorrhea : It is a condition characterized by the ***loss of lipids in the feces***. Steatorrhea may be due to

1. A defect in the secretion of bile or pancreatic juice into the intestine;

2. Impairment in the lipid absorption by the intestinal cells.

Steatorrhea is commonly seen in disorders associated with pancreas, biliary obstruction, severe liver dysfunction etc.

Cholesterol stones (gall stones)

Cholesterol stone formation in gall-bladder (gall stones) is a frequent health complication. It is found

more frequently in females than in males often in association with obesity. Cholesterol gall stones are formed when liver secretes bile (containing phospholipids, bile acids etc.), supersaturated with respect to cholesterol.

ABNORMALITIES RELATED TO DIGESTION AND ABSORPTION

The following are the major abnormalities (of interest to biochemists) concerned with digestion and absorption of food in the gastrointestinal tract.

Lactose intolerance, deficiency of sucrase, Hartnup's disease and steatorrhea have already been described. Peptic ulcer and pancreatitis are other important disorders associated with digestive system.

Peptic ulcers

Gastric and duodenal ulcers are collectively known as peptic ulcers. Ulceration occurs due to the autodigestion of mucosa by the gastric secretions (pepsin and HCl). In the patients of peptic ulcer, gastric HCl is always present in the pyloric regions of stomach and the duodenum. Gastic ulcers are mainly caused by the bacterium ***Helicobacter pylori*** which lives in the nutrient-rich gastric mucosa. *H. pylori* induces chronic inflammation in the stomach tissues, which gets exposed to acid damage. For this reason, the best mode of ***treatment for gastric ulcers is the use of antibiotics that eliminate H. pylori.***

Pancreatitis

Inflammation of the pancreas is known as pancreatitis. Acute pancreatitis is caused by the autodigestion of pancreas due to the unusual conversion of zymogens into the active enzymes by trypsin. In normal circumstances, this is prevented by trypsin inhibitor.

Acute pancreatitis is a life-threatening disorder. Measurement of serum ***amylase*** (highly elevated) is used ***in the diagnosis of pancreatitis***. Excessive ***consumption of alcohol*** over a long period is blamed as the prime cause of chronic pancreatitis.

SUMMARY

1. *Digestion is a process that converts complex foodstuffs into simpler ones which can be readily absorbed by the gastrointestinal tract.*
2. *Monosaccharides are the final absorbable products of carbohydrate digestion. Glucose is transported into the intestinal mucosal cells by a carrier mediated, Na^+-dependent energy requiring process.*
3. *Protein digestion begins in the stomach by pepsin, which is aided by gastric HCl. Pancreatic proteases (trypsin, chymotrypsin and elastase) and intestinal amino peptidases and dipeptidases complete the degradation of proteins to amino acids and some dipeptides.*
4. *The intestinal absorption of amino acids occurs by different transport systems. The uptake of amino acids is primarily a Na^+-dependent energy requiring process.*
5. *Digestion of lipids occurs in the small intestine. Emulsification of lipids, brought about by bile salts, is a prerequisite for their digestion. Pancreatic lipase aided by a colipase degrades triacylglycerol to 2-monoacylglycerol and free fatty acids. Cholesterol esterase and phospholipases, respectively, hydrolyse cholesteryl esters and phospholipids.*
6. *Lipid absorption occurs through mixed micelles, formed by bile salts in association with products of lipid digestion (2-monoacylglycerol, cholesterol and free fatty acids). In the intestinal mucosal cells, lipids are resynthesized from the absorbed components and packed as chylomicrons which enter the lymphatic vessels and then the blood.*

THE READER MUST REFER APPENDIX D (Page Nos : 422-431) FOR RELATED CLINICAL CASE STUDIES (Nos : 7, 8)

PLASMA PROTEINS

9

The ***plasma*** is the ***liquid medium of blood (55-60%)***, in which the cell components—namely erythrocytes, leukocytes, platelets—are suspended. If blood containing anticoagulants (e.g. heparin, potassium oxalate) is centrifuged, the plasma separates out as a supernatant while the cells remain at the bottom. The ***packed cell volume*** or ***hematocrit*** is about 45%.

The term serum is applied to the liquid medium which separates out after the blood clots (coagulates). Serum does not contain fibrinogen and other clotting factors. Thus, the main ***difference between plasma and serum is the presence or absence of fibrinogen***.

Importance of blood

The total volume of blood in an adult is around 4.5 to 5 liters. Blood performs several diversified functions. These include respiration, excretion, acid-base maintenance, water balance, transport of metabolites, hormones and drugs, body defense and coagulation.

Separation of plasma proteins

The total concentration of ***plasma proteins is about 6-8 g/dl***. The plasma is a complex mixture of proteins, and several techniques are employed to separate them. An age-old technique is based on the use of varying concentrations of ammonium sulfate or sodium sulfate. By this method, which is known as ***salting out process,*** the plasma proteins can be separated into three groups—namely albumin, globulins and fibrinogen.

Electrophoresis : This is the ***most commonly employed analytical technique*** for the separation of plasma (serum) proteins. The basic principles of electrophoresis are described in ***Chapter 41***. Paper or agar gel electrophoresis with vernol buffer (pH-8.6) separates plasma proteins into 5 distinct bands namely ***albumin, α_1, α_2, β and γ globulins*** (***Fig.9.1***).

Abnormal electrophoretic pattern

Electrophoresis of serum proteins is conveniently used for the diagnosis of certain diseases

1. **Multiple myeloma :** A sharp and distinct ***M band*** appears in the γ-globulin fraction.

2. **Acute infections :** α_1- and ***α_2-globulins*** are increased.

3. **Nephrotic syndrome :** Decreased albumin with sharp and prominent α_2-globulin.

Albumin/globulin (A/G) ratio : The albumin concentration of plasma is ***3.5 to 5.0 g/dl*** while that of total globulins is 2.5 to 3.5 g/dl. The ***normal A/G ratio is 1.2 to 1.5 : 1***. The A/G ratio is lowered either due to decrease in albumin or increase in globulins, as found in the following conditions

1. Decreased synthesis of albumin by liver—usually found in liver diseases and severe protein malnutrition.

2. Excretion of albumin into urine in kidney damage.

3. Increased production of globulins associated with chronic infections, multiple myelomas etc.

Components of plasma proteins

The important plasma proteins along with their characteristics (based on electrophoretic pattern) and major functions are given in ***Table 9.1***.

Some selected plasma proteins are discussed next.

TABLE 9.1 Important plasma proteins, characteristics and major functions

Protein	Plasma concentration	Molecular weight	Major function(s)
Albumin	3.5–5.0 g/dl	69,000	Osmotic, transport, nutritive and buffering
Prealbumin	25–30 mg/dl	61,000	Transports thyroxine to some extent
α_1-Globulins	0.3–0.5 g/dl	—	—
α_1-Antitrypsin	< 0.2 g/dl	54,000	Inhibitor of trypsin
α_1-Lipoproteins (HDL)	0.2–0.3 g/dl	—	Transports cholesterol and phospholipids
Orosomucoid	< 0.1 g/dl	44,000	Binds with progesterone
Retinol binding protein (RBP)	3–6 mg/dl	21,000	Transports vitamin A
Thyroxine binding globulin (TBG)	1–2 mg/dl	58,000	Transports thyroid hormones
Transcortin or cortisol binding protein (CBG)	3–4 mg/dl	52,000	Major transporter of steroid hormones (e.g. cortisol, corticosterone)
α_2-Globulins	0.4–0.8 g/dl	—	—
α_2-Macroglobulin	0.2–0.3 g/dl	800,000	Antitrypsin and antiplasmin activity
Haptoglobins (Hp 1-1; Hp 2-1 and Hp 2-2)	< 0.3 g/dl	90,000	Binds with plasma free hemoglobin and prevents its excretion
Prothrombin	< 0.02 g/dl	63,000	Participates in blood coagulation
Ceruloplasmin	< 0.03 g/dl	150,000	Transport of copper; oxidation of Fe^{2+} to Fe^{3+}.
β-Globulins	0.6–1.1 g/dl	—	—
β-Lipoproteins (LDL)	0.2–0.5 g/dl	—	Transports triacylglycerols and cholesterol
Transferrin	0.2–0.3 g/dl	76,000	Transports iron
Hemopexin	< 0.1 g/dl	57,000	Transports heme
Plasminogen	< 0.05 g/dl	140,000	Forms plasmin, involved in fibrinolysis
γ-Globulins	0.8–1.8 mg/dl	—	Antibody functions
(Immunoglobulins—IgG, IgA, IgM, IgD and IgE; refer ***Table 9.2*** for details)			
Fibrinogen	0.2–0.4 g/dl	340,000	Participates in blood coagulation

ALBUMIN

Albumin is the ***major constituent (60%) of plasma proteins*** with a ***concentration of 3.5-5.0 g/dl***. Human albumin has a molecular weight of 69,000 and consists of a single polypeptide chain of 585 amino acids with 17 disulfide bonds.

Synthesis of albumin

Albumin is exclusively synthesized by the ***liver***. For this reason, measurement of serum albumin concentration is conveniently ***used to assess liver function*** (synthesis decreased in liver diseases). Albumin has an ***half-life of 20 days***.

Functions of albumin

Plasma albumin performs ***osmotic***, ***transport*** and ***nutritive functions***

1. **Osmotic function :** Due to its high concentration and low molecular weight, albumin contributes to ***75-80% of the total plasma osmotic pressure*** (25 mm Hg). Thus, albumin plays a predominant role in maintaining blood volume and body fluid distribution. The edema observed in ***kwashiorkor***, a disorder of protein-energy malnutrition, is attributed to a drastic ***reduction in plasma albumin*** level.

2. **Transport functions :** Plasma albumin binds to several biochemically important compounds and

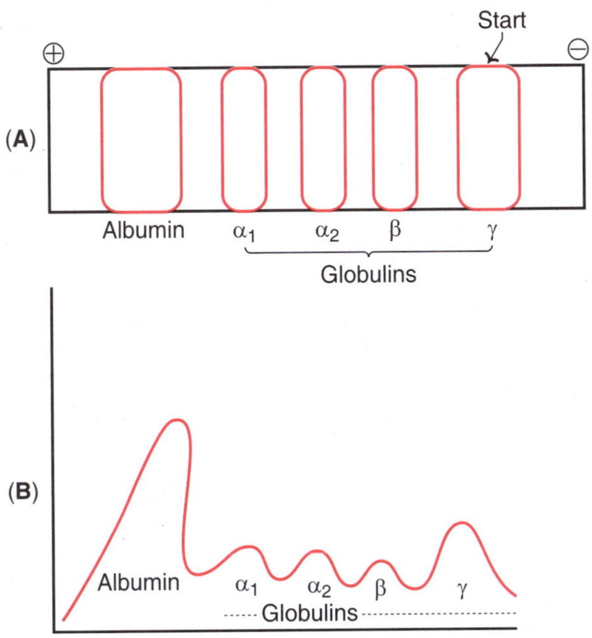

***Fig.9.1 :** Electrophoresis of plasma proteins—(**A**) Separated bands, (**B**) Densitometer scannning.*

transports them in the circulation. These include ***free fatty acids, bilirubin, steroid hormones, calcium and copper***.

[**Note :** Besides albumin, there are several other ***plasma transport proteins***. These include prealbumin, retinol binding protein, thyroxine binding protein, transcortin and others as stated in the functions of plasma proteins in ***Table 9.1***.]

3. **Nutritive functions :** Albumin serves as a ***source of amino acids*** for tissue protein synthesis to a limited extent, particularly in nutritional deprivation of amino acids.

4. **Buffering function :** Among the plasma proteins, ***albumin*** has the ***maximum buffering capacity***. However, the buffering action of albumin in plasma is not significant compared to bicarbonate buffer system.

Clinical significance of albumin

1. Albumin, binding to certain compounds in the plasma, ***prevents them from crossing the blood-brain barrier*** e.g. albumin-bilirubin complex, albumin-free fatty acid complex.

2. ***Hypoalbuminemia*** (lowered plasma albumin) is observed in ***malnutrition, nephrotic syndrome***.

The osmotic pressure of albumin significantly contributes to maintaining plasma volume, and fluid volume of interstitial fluids. In hypoalbuminemia (albumin <2g/dl), plasma osmotic pressure is decreased, leading to flow of water from plasma to interstitial compartment that results in ***edema of legs and other body parts***.

3. Albumin is excrected into urine (***albuminuria***) in nephrotic syndrome.

GLOBULINS

Globulins constitute several proteins that are separated into four distinct bands (***α_1, α_2, β and γ-globulins***) on electrophoresis (***Fig.9.1***). Globulins, in general, are bigger in size than albumin. They perform a variety of functions which include transport and immunity. In ***Table 9.1***, the important globulins are given, some of them are discussed here.

α_1-ANTITRYPSIN

α_1-Antitrypsin, more recently called as ***α-antiproteinase***, is a glycoprotein. It is a major constituent of α_1-globulin fraction of plasma proteins with a normal concentration of about 200 mg/dl. α_1-Antitrypsin is a serine protease inhibitor. It combines with trypsin, elastase and other protease enzymes and inhibits their activity.

Clinical significance of α_1-antitrypsin

α_1-Antitrypsin deficiency has been implicated in two diseases, namely, ***emphysema*** and ***α_1-AT deficiency liver disease.***

Emphysema (*Greek* : emphusan—to inflate) is a term used to represent the ***abnormal distension of lungs by air***. At least 5% of emphysema cases are due to the deficiency of α_1-AT. This is associated with ***lung infections*** (e.g. pneumonia) and increase in the activity of macrophages to release elastase that damages lung tissues. In the normal circumstances, elastase activity is inhibited by α_1-AT.

α_1-Antitrypsin deficiency and liver disease : This is due to the accumulation of a mutant α_1-AT which aggregates to form polymers. These polymers, in turn—by an unknown mechanism—cause liver damage (hepatitis) followed by accumulation of collagen resulting in fibrosis (cirrhosis).

HAPTOGLOBIN

Haptoglobin (***Hp***) is a plasma glycoprotein. Hp is an ***acute phase protein*** since its plasma concentration

Medical Concepts/ Clinical Correlates

- **Albumin is involved in osmotic function, transport of several compounds (fatty acids, steroid hormones), besides the buffering action.**
- **Hypoalbuminemia and albuminuria are observed in nephrotic syndrome.**
- **α_1-Antitrypsin deficiency has been implicated in emphysema (abnormal distension of lungs by air) which is more commonly associated with heavy smoking.**

is increased in several inflammatory conditions. In humans, Hp exists in 3 polymorphic forms, namely Hp 1-1, Hp 2-1 and Hp 2-2, synthesized by at least 2 separate genes. However, they do not exhibit any functional difference.

Functions of haptoglobin

Haptoglobin binds with the free hemoglobin (known as extra-corpuscular hemoglobin) that spills into the plasma due to hemolysis. The ***haptoglobin-hemoglobin (Hp-Hb) complex*** (mol. *wt.* 155,000) cannot pass through glomeruli of kidney while free Hb (mol. wt. 65,000) can. Haptoglobin, therefore, ***prevents the loss of free Hb into kidney***. The Hp-Hb complex is promptly removed by the hepatocytes and degraded.

CERULOPLASMIN

Ceruloplasmin is a blue coloured, ***copper—containing α_2-globulin***. Its plasma concentration is about 30 mg/dl. Ceruloplasmin binds with almost 90% of plasma copper (6 atoms of Cu bind to a molecule). This binding is rather tight and, as a result, copper from ceruloplasmin is not readily released to the tissues. Albumin carrying only 10% of plasma copper is the major supplier of copper to the tissues. Ceruloplasmin possesses oxidase activity. Ceruloplasmin is associated with ***Wilson's disease*** which is discussed under copper metabolism (***Chapter 19***).

ACUTE PHASE PROTEINS

Acute phase response refers to a non-specific response to the stimulus of infection, injury, various ***inflammatory conditions*** (affecting tissues/organs), cancers etc. This phase is associated with a characteristic pattern of changes in certain plasma proteins, collectively referred to as acute phase proteins e.g. α, ***antitrypsin, ceruloplasmin***, ***C-reactive protein (CRP)***. Estimation of CRP, a type of β-globulin in serum is important for the evaluation of acute phase response. Further, CRP is also useful for predicting the risk of heart diseases.

IMMUNOGLOBULINS

The higher vertebrates, including man, have evolved a defense system to protect themselves against the invasion of foreign substances—a virus, a bacterium or a protein. The defense strategies of the body are collectively known as ***immunity*** and are briefly described under immunology (***Chapter 42***). Immunoglobulins are discussed hereunder.

Immunoglobulins, a specialised group of proteins are mostly associated with ***γ-globulin fraction*** (on electrophoresis) of plasma proteins. Some immunoglobulins however, separate along with β and α-globulins. Therefore, it should be noted that γ-globulin and immunoglobulin are not synonymous. ***Immunoglobulin is a functional term while γ-globulin is a physical term.***

Structure of immunoglobulins

All the immunoglobulin (***Ig***) molecules basically consist of ***two identical heavy (H) chains*** (mol. wt. 53,000 to 75,000 each) and ***two identical light (L) chains*** (mol. wt. 23,000 each) held together by disulfide linkages and non-covalent interactions (***Fig.9.2***). Thus, immunoglobulin is a Y-shaped tetramer (H_2L_2). Each heavy chain contains approximately 450 amino acids while each light chain has 212 amino acids. The heavy chains of Ig are linked to carbohydrates, hence immunoglobulins are ***glycoproteins***.

Constant and variable regions : Each chain (L or H) of Ig has two regions (domains), namely the constant and the variable. The amino terminal half of the light chain is the variable region (V_L) while the carboxy terminal half is the constant region (C_L). As regards heavy chain, approximately

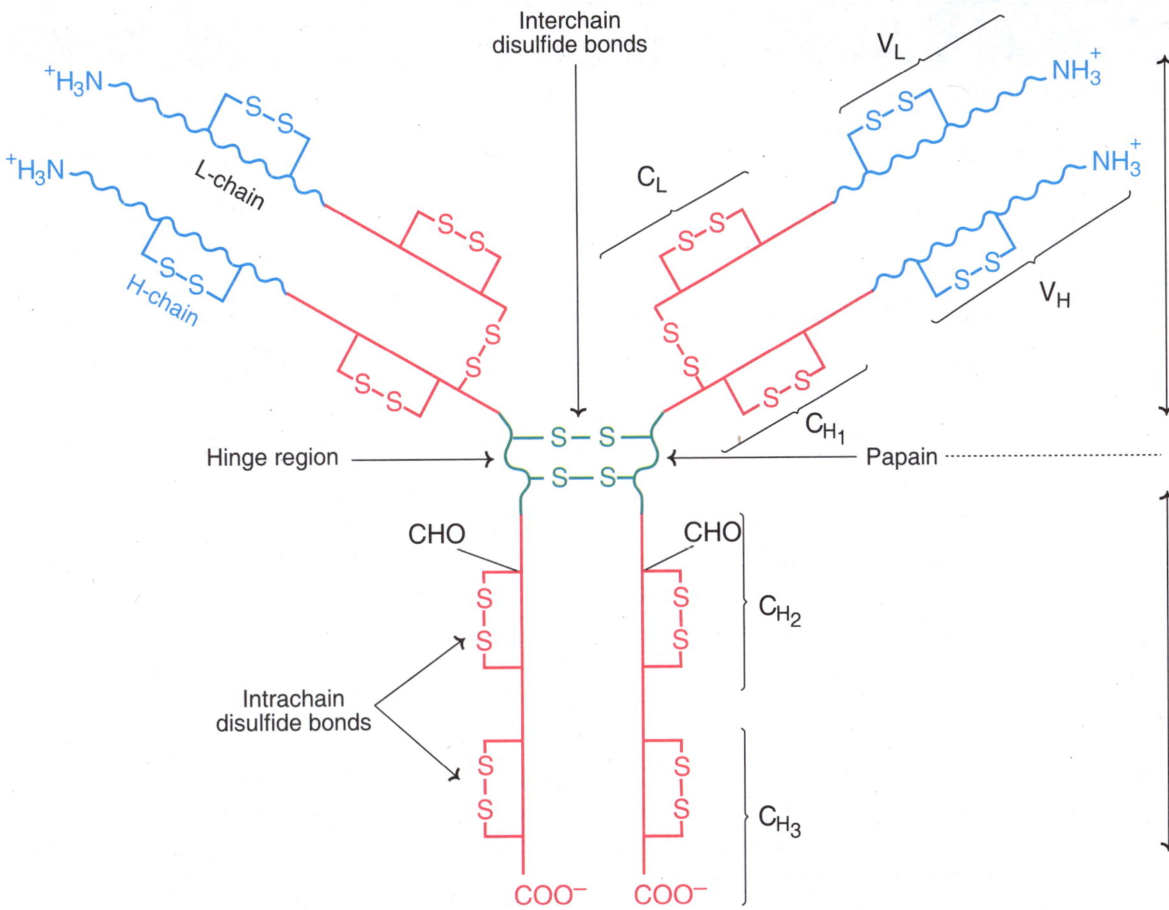

***Fig.9.2 :** Diagrammatic representation of human IgG molecule (V–variable region, C–constant region; CHO–carbohydrate; Each heavy chain is composed of four units—V_H, C_{H_1}, C_{H_2}, C_{H_3} while light chain consists of two units—V_L, C_L).*

one-quarter of the amino terminal region is variable (V_H) while the remaining three-quarters is constant (C_{H_1}, C_{H_2}, C_{H_3}). The amino acid sequence (with its tertiary structure) of variable regions of light and heavy chains is responsible for the specific binding of immunoglobulin (antibody) with antigen.

CLASSES OF IMMUNOGLOBULINS

Humans have five classes of immunoglobulins—namely ***IgG, IgA, IgM, IgD and IgE***—containing the heavy chains γ, α, μ, δ and ε, respectively. The type of heavy chain ultimately determines the class and the function of a given Ig.

Two types of light chains—namely ***kappa (κ) and lambda (λ)***—are found in immunoglobulins. They differ in their structure in C_L regions. An immunoglobulin (of any class) contains two κ or two λ light chains and never a mixture. The occurrence of κ chains is more common in human immunoglobulins than λ chains.

The characteristics of the 5 classes of human immunoglobulins are given in ***Table 9.2***.

Immunoglobulin G (IgG)

IgG is the ***most abundant*** (75–80%) class of immunoglobulins. IgG is composed of a single Y-shaped unit (monomer). It can traverse blood vessels readily. IgG is the only immunoglobulin that can cross the placenta and transfer the mother's immunity to the developing fetus. IgG triggers ***foreign cell destruction*** mediated by complement system.

Immunoglobulin A (IgA)

IgA occurs as a single (monomer) or double unit (dimer) held together by J chain. It is mostly ***found in the body secretions*** such as saliva, tears, sweat, milk and the walls of intestine. IgA is the most

TABLE 9.2 Characteristics of human immunoglobulins

Type	H-Chain	L-Chains	Molecular formula	Molecular weight	Percentage carbohydrate	Serum conc. mg/dl	Major function(s)
IgG	γ	κ or λ	$\gamma_2\kappa_2$ or $\gamma_2\lambda_2$	~150,000	3	800–1,500	Mostly responsible for humoral immunity
IgA	α	κ or λ	$(\alpha_2\kappa_2)_{1-3}$ or $(\alpha_2\lambda_2)_{1-3}$	$\sim(160,000)_{1-3}$	8	150–400	Protects the body surfaces
IgM	μ	κ or λ	$(\mu_2\kappa_2)_5$ or $(\mu_2\lambda_2)_5$	~900,000	12	50–200	Humoral immunity, serves as first line of defense
IgD	δ	κ or λ	$(\delta_2\kappa_2$ or $\delta_2\lambda_2)$	~180,000	13	1–10	B-cell receptor ?
IgE	ε	κ or λ	$\varepsilon_2\kappa_2$ or $\varepsilon_2\lambda_2$	~ 190,000	12	0.02–0.05	Humoral sensitivity and histamine release.

predominant antibody in the colostrum, the initial secretion from the mother's breast after a baby is born. The IgA molecules bind with bacterial antigens present on the body (outer epithelial) surfaces and remove them. In this way, IgA ***prevents the foreign substances*** from entering the body cells.

Immunoglobulin M (IgM)

IgM is the ***largest immunoglobulin*** composed of 5 Y-shaped units (IgG type) held together by a J polypeptide chain. Thus IgM is a pentamer. Due to its large size, IgM cannot traverse blood vessels, hence it is restricted to the blood stream. IgM is the first antibody to be produced in response to an antigen and is the most ***effective against invading microorganisms***.

Immunoglobulin D (IgD)

IgD is composed of a single Y-shaped unit and is present in a low concentration in the circulation. IgD molecules are present on the surface of B cells. Their function, however, is not known for certain. Some workers believe that IgD may function as ***B-cell receptor***.

Immunoglobulin E (IgE)

IgE is a single Y-shaped monomer. It is normally present in minute concentration in blood. IgE levels are elevated in individuals with ***allergies*** as it is associated with the body's allergic responses. The IgE molecules tightly bind with mast cells which release histamine and cause allergy.

MULTIPLE MYELOMA

Multiple myeloma, a ***plasma cell cancer***, constitutes about 1% of all cancers affecting the population. ***Females are more susceptible*** than males for this disorder, and it usually occurs in the age group 45-60 years.

Abnormal Ig production : Multiple myeloma is due to the malignancy of a single clone of plasma cells in the bone marrow. This results in the overproduction of abnormal immunoglobulins, ***mostly*** (75%) ***IgG*** and in some cases (25%) IgA or IgM. IgD type multiple myeloma found in younger adults is less common (<2%) but more severe. In patients of multiple myeloma, the synthesis of normal immunoglobulins is diminished causing depressed immunity. Hence recurrent infections are common in these patients.

Electrophoretic pattern : The plasma of multiple myeloma patients shows a characteristic pattern of electrophoresis. There is a sharp and distinct band (M band, for myeloma globulin) between β- and γ-globulins. Further, this ***M band*** almost replaces the γ-globulin band due to the diminished synthesis of normal γ-globulins.

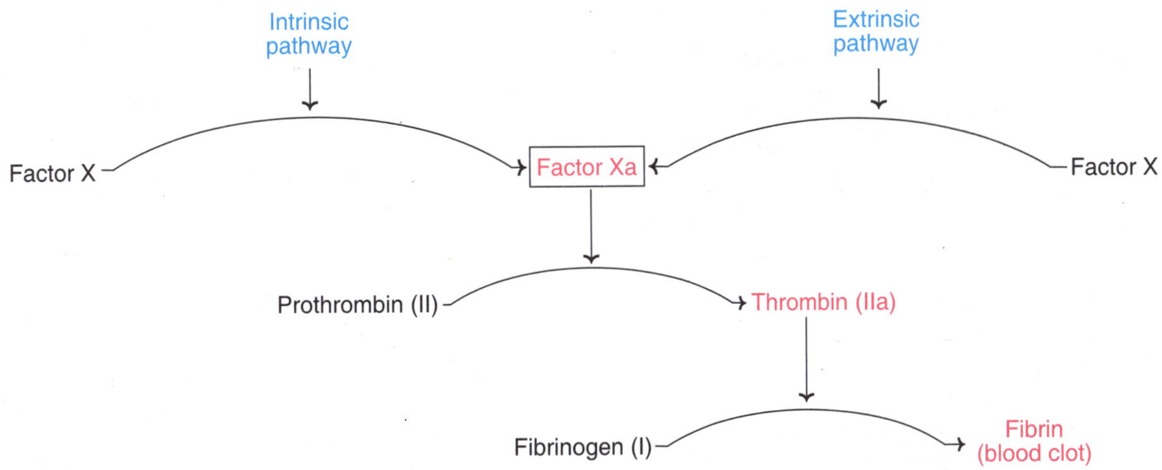

***Fig.9.3 :** Overview of blood clotting with the final common pathway.*

Bence Jones proteins : Henry Bence Jones first described them in 1847. These are the ***light chains (κ or λ) of immunoglobulins*** that are ***synthesized in excess***. Bence Jones proteins have a molecular weight of 20,000 or 40,000 (for dimer). In about 20% of the patients of multiple myeloma, Bence Jones proteins are excreted in urine which often damages the renal tubules.

The presence of Bence Jones proteins in urine can be detected by specific tests.

1. **Electrophoresis** of a concentrated urine is the best test to detect Bence Jones proteins in urine.

2. **The classical heat test** involves the ***precipitation*** of Bence Jones proteins when slightly acidified urine is ***heated to 40-50°C***, and redissolve on further heating urine to boiling point. They reappear again on cooling urine to ***about 70°C***.

3. **Bradshaw's test** involves layering of urine on concentrated HCl that forms a ***white ring*** of ***precipitate***, if Bence Jones proteins are present.

BLOOD CLOTTING

The term ***hemostasis*** is applied to the sequence of ***physiological responses to stop bleeding*** (loss of blood after an injury). This is carried out by blood clotting that involves two independent pathways

1. **The extrinsic pathway** is the initial process in clotting and involves the factors that are not present in the blood (hence the name).

2. **The intrinsic pathway** involves a series of reactions participated by the ***factors present in the blood***.

Strictly speaking, the extrinsic and intrinsic pathways are not independent, since they are coupled together. Further, the final reactions are identical for both pathways that ultimately lead to the activation of prothrombin to thrombin and the conversion of fibrinogen to fibrin clot (***Fig.9.3***).

The blood ***coagulation factors*** in human plasma along with their common names and molecular weights are listed in ***Table 9.3***. All but two of these factors are designated by a Roman numeral. It should, however, be noted that the numbers represent the ***order of their discovery and not the order of their action***. The cascade of blood clotting process is depicted in ***Fig.9.4***, and the salient features are discussed below. The active form of a factor is designated by a subscript **a.** The active clotting factors (with exception of fibrin) are serine proteases.

Conversion of fibrinogen to fibrin

Fibrinogen (factor I) is a soluble glycoprotein that constitutes 2-3% of plasma proteins (plasma concentration 0.3 g/dl). Fibrinogen consists of 6 polypeptide chains-two A α, two B β and two γ making the structure $(A\ \alpha)_2\ (B\ \beta)_2\ \gamma_2$.

Fibrinogen undergoes proteolytic cleavage catalysed by thrombin to release small ***fibrinopeptides (A and B)***. This results in the formation of fibrin monomers which can stick together to form hard clots (***Fig.9.5***).

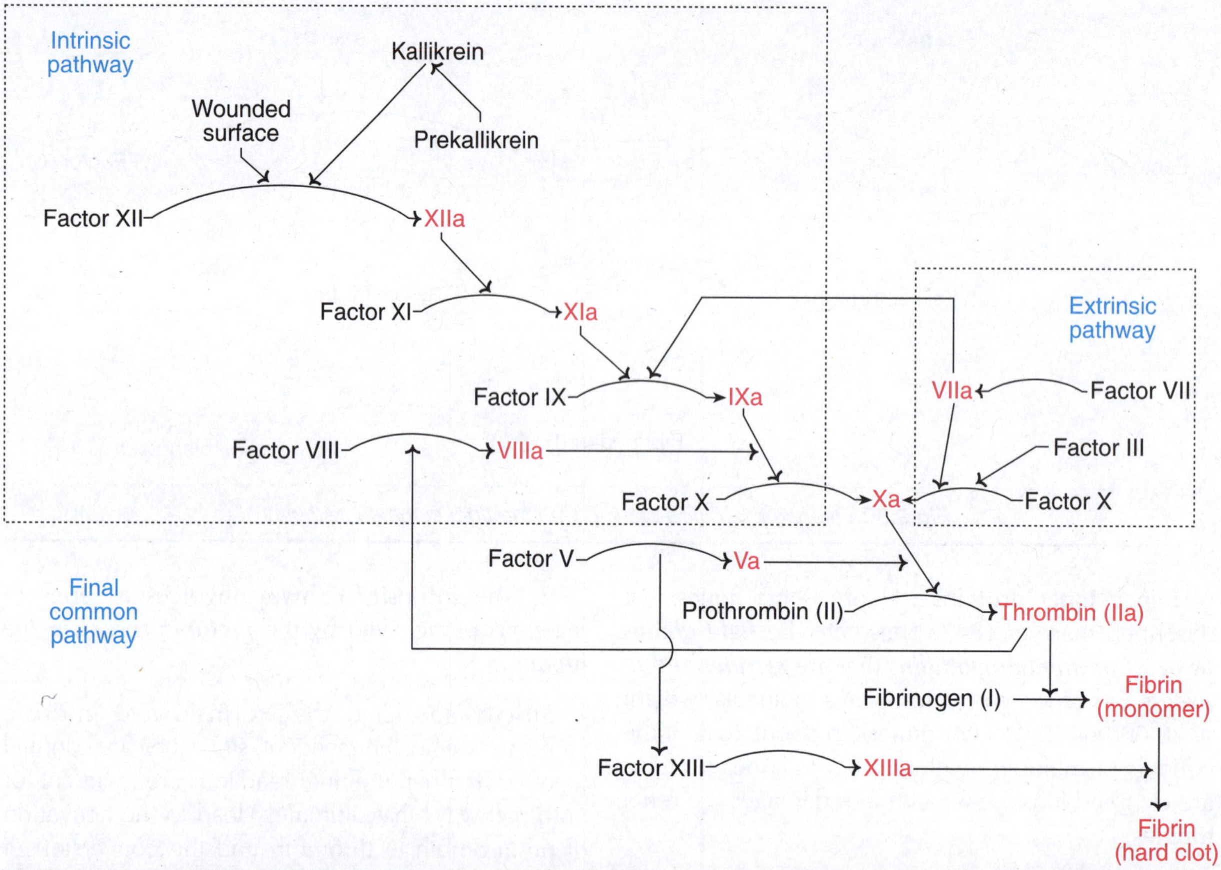

***Fig.9.4 :** The blood clotting cascade in humans (the active forms of the factors are represented in red with subscript 'a').*

TABLE 9.3 Blood coagulation factors in humans

Factor number	*Common name(s)*	*Subunit molecular weight*
I	Fibrinogen	340,000
II	Prothrombin	720,000
III	Tissue factor, thromboplastin	370,000
IV	Calcium (Ca^{2+})	—
V	Proaccelerin, labile factor	330,000
VII	Proconvertin, serum prothrombin conversion accelerator (SPCA)	50,000
VIII	Antihemophilic factor A, antihemophilic globulin (AHG)	330,000
IX	Christmas factor, antihemophilic factor B, Plasma thromboplastin component (PTC)	56,000
X	Staurt-Prower factor	56,000
XI	Plasma thromboplastin antecedent (PTA)	160,000
XII	Hageman factor	80,000
XIII	Fibrin-stabilizing factor (FSF), fibrinoligase, Liki Lorand factor	320,000
—	Prekallikrein	88,000
—	High molecular weight kininogen (HMK)	150,000

***Note :** The numbers represent the order of their discovery and not the order of their action. Factor Va was once referred to as factor VI, hence there is no factor VI.*

Medical Concepts/ Clinical Correlates

- Multiple myeloma, a plasma cell cancer disease of bone marrow, is characterized by overproduction of abnormal immunoglobulins (mostly IgG). Laboratory diagnosis of multiple myeloma can be made by the presence of a distinct M band on plasma/serum electrophoresis.
- Defects in clotting factors cause coagulation abnormalities such as hemophilia A (deficiency of factor VIII) and Christmas disease (deficiency of factor IX).
- Anticoagulants inhibit blood clotting. These include heparin, oxalate, fluoride, EDTA and citrate.
- Certain blood tests in the laboratory require unclotted blood. Anticoagulants are employed for collecting such specimens.

Conversion of prothrombin to thrombin

Prothrombin (II) is the inactive zymogen form of thrombin (IIa). The activation of prothrombin occurs on the platelets and requires the presence of factors Va and Xa, besides phospholipids and Ca^{2+}.

The extrinsic pathway

The extrinsic pathway is very rapid and occurs in response to ***tissue injury***. This pathway essentially involves the conversion of proconvertin (VII) to its active form (VIIa) and the generation factor Xa. The tissue factor (III), found to be necessary to accelerate the action VIIa on a factor X, is present in lung and brain.

The intrinsic pathway

The intrinsic pathway is rather slow. It involves the participation of a contact system (wounded surface) and a series of factors to generate factor Xa.

The Hageman factor (XII) is activated (XIIa) on exposure to activating wound surface containing collagen or platelet membranes. The formation of XIIa is accelerated by kallikrein and HMK. The activated Hageman factor (XIIa) activates factor XI. The XIa activate the Christmas factor (IX). The Christmas factor is also activated by active proconvertin (VIIa).

In the next step, the Staurt factor (X) is activated by Christmas factor (IXa) and this reaction requires the presence of antihemophilic factor (VIIIa), Ca^{2+} and phospholipids.

The extrinsic and intrinsic pathways lead to the formation of factor Xa which then participates in the final common pathway to ultimately result in the formation of fibrin clot.

Anticoagulants

Several substances, known as anticoagulants, are in use to ***inhibit the blood clotting***. Calcium is essentially required for certain reactions of blood coagulation. The substances which bind with Ca^{2+} are very effective as anticoagulants. These include ***oxalate***, ***fluoride***, ***EDTA*** and ***citrate***.

Heparin is an anticoagulant used ***to maintain normal hemostasis***. It is a heteropolysaccharide found in many tissues including mast cells in the endothelium of blood vessels. Heparin combines with antithrombin III which in turn, inhibits the clotting factors II, IX, X, XI, XII and kallikrein. Heparin can be administered to patients during and after surgery to retard blood clotting.

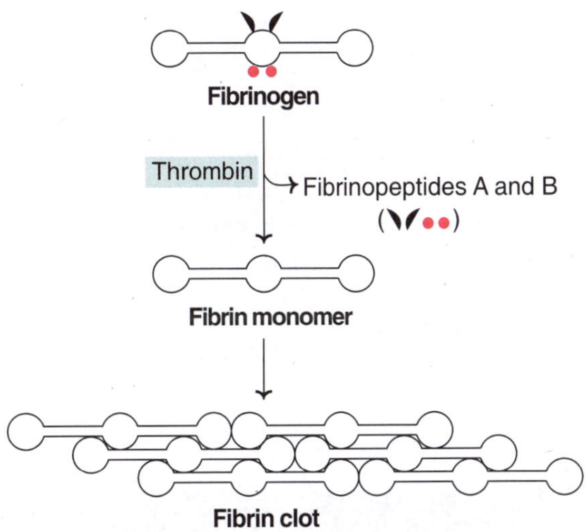

Fig.9.5 : Diagrammatic representation of fibrin clot formation from fibrinogen.

The blood contains another anticoagulant—namely ***protein C***—which is activated by thrombin. Active protein C hydrolyses and inactivates clotting factors V and VIII.

Warfarin, a vitamin K antagonist may be considered as an ***oral anticoagulant***. This acts by reducing the synthesis of certain clotting factors (II, VII, IX and X).

Fibrinolysis

The term fibrinolysis refers to the dissolution or lysis of blood clots. ***Plasmin*** is mostly responsible for the dissolution of fibrin clots. Plasminogen, synthesized in the kidney, is the inactive precursor of plasmin. Tissue plasminogen activator (TPA) and urokinase convert plasminogen to plasmin.

Streptokinase is a therapeutic fibrinolytic agent which activates plasminogen.

Abnormalities in blood clotting

Several abnormalities associated with blood clotting are known. These are due to defects in ***clotting factors*** which may be inherited or acquired.

Hemophilia A (classical hemophilia) : This is a sex-linked disorder transmitted by females affecting males. Hemophilia A is the most common clotting abnormality and is due to the ***deficiency of antihemophilic factor (VIII)***. The affected individuals have prolonged clotting time and suffer from internal bleeding (particularly in joints and gastrointestinal tract). Hemophilia A has gained importance due to the fact that the Royal families of Britain are among the affected individuals.

Hemophilia B (Christmas disease) : This is due to the deficiency of ***Christmas factor (IX)***. The clinical symptoms are almost similar to that found in hemophilia A.

SUMMARY

1. *The total concentration of plasma proteins is about 6-8 g/dl. Electrophoresis separates plasma proteins into 5 distinct bands, namely albumin, α_1, α_2, β and γ globulins.*
2. *Albumin is the major constituent (60%) of plasma proteins with a concentration 3.5 to 5.0 g/dl. It is exclusively synthesized in the liver. Albumin performs osmotic, transport and nutritive functions.*
3. *Haptoglobin (Hp) binds with free hemoglobin (Hb) that spills into the plasma due to hemolysis. The Hp-Hb complex cannot pass through the glomeruli, hence haptoglobin prevents the loss of free hemoglobin into urine.*
4. *Immunoglobulins are specialized proteins to defend the body against the foreign substances. They are mostly associated with γ-globulin fraction of plasma proteins.*
5. *Five classes of immunoglobulins—namely IgG, IgA, IgM, IgD and IgE—are found in humans. IgG is most abundant and is mainly responsible for humoral immunity. IgA protects body surfaces. IgM serves as a first line of defense for humoral immunity while IgE is associated with allergic reactions.*
6. *Blood clotting is the body's major defense mechanism against blood loss. The extrinsic and intrinsic pathways lead to the formation of factor Xa which then participates in the final common pathway to activate prothrombin to thrombin. Fibrinogen is then converted to fibrin clot.*

THE READER MUST REFER APPENDIX D (Page Nos : 422 - 431) FOR RELATED CLINICAL CASE STUDIES (No : 38)

HEMOGLOBIN AND PORPHYRINS

10

The structure, functions and abnormalities of hemoglobin, the synthesis and degradation of heme, the porphyrin containing compounds are discussed in this chapter.

HEMOGLOBIN

Hemoglobin (***Hb***) is the red blood pigment, exclusively found in ***erythrocytes*** (*Greek*: erythrose—red; kytos—a hollow vessel). The normal concentration of Hb in blood in ***males is 14–16 g/dl*** and in ***females 13–15 g/dl***. Hemoglobin performs two important biological functions concerned with respiration

1. Delivery of O_2 from the lungs to the tissues.
2. Transport of CO_2 and protons from tissues to lungs for excretion.

Structure of hemoglobin

Hemoglobin (mol. wt. 64,450) is a conjugated protein, containing ***globin***—the apoprotein part—and the ***heme***—the non-protein part (prosthetic group). Hemoglobin is a ***tetrameric allosteric protein*** (***Fig.10.1***).

Structure of globin : Globin consists of ***four polypeptide*** chains of two different primary structures (monomeric units). The common form of adult ***hemoglobin (HbA_1)*** is made up of ***two α-chains and two β-chains ($\alpha_2\beta_2$)***. Some authors consider hemoglobin consisting of two identical dimers—$(\alpha\beta)_1$ and $(\alpha\beta)_2$. Each α-chain contains 141 amino acids while β-chain contains 146 amino acids. Thus HbA_1 has a total of 574 amino acid residues. The four subunits of hemoglobin are held together by non-covalent interactions primarily hydrophobic, ionic and hydrogen bonds. Each subunit contains a heme group.

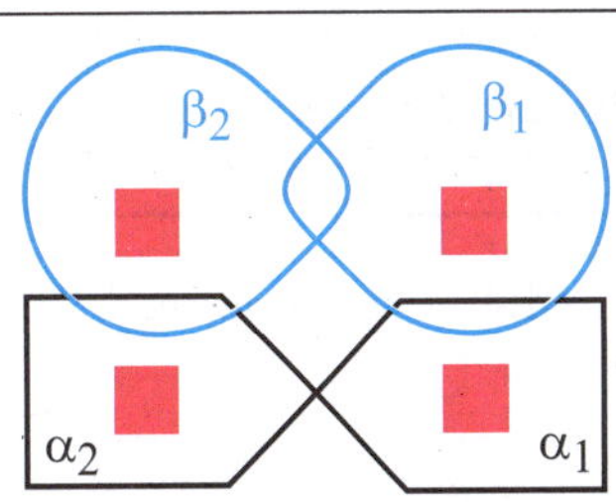

Fig.10.1 : *Diagrammatic representation of hemo-globin with 2α and 2β chains (Red–Heme).*

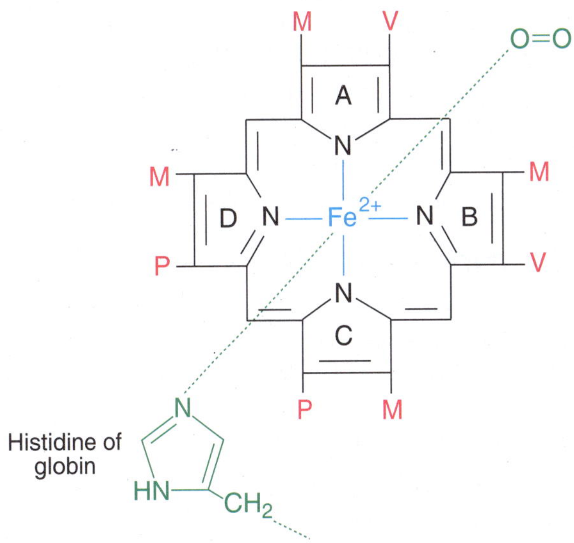

Fig.10.2 : *Structure of heme [M-Methyl ($-CH_3$); V-Vinyl ($-CH_2-CH_2-$); P-Propionyl ($-CH_2-CH_2-CH_2$).*

Structure of heme : The characteristic ***red colour*** of hemoglobin (ultimately blood) is ***due to heme***. Heme contains a porphyrin molecule namely ***protoporphyrin IX***, with iron at its center (***Fig.10.2***). Protoporphyrin IX consists of four ***pyrrole*** rings to

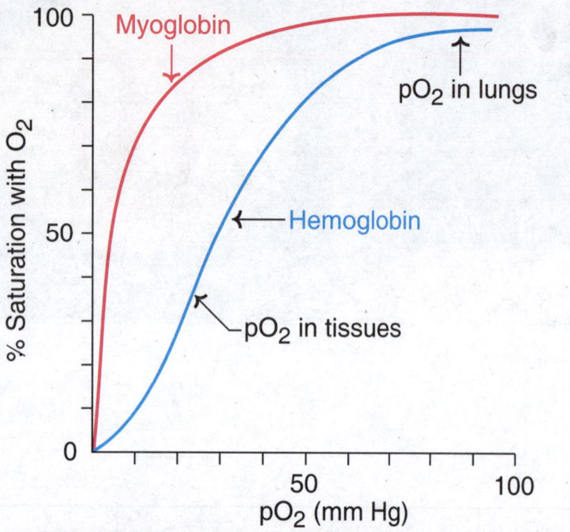

Fig.10.3 : *Oxygen dissociation curves of hemoglobin and myoglobin (pO_2–Partial pressure of oxygen).*

which four methyl, two propionyl and two vinyl groups are attached.

Myoglobin

Myoglobin (Mb) is monomeric oxygen binding ***hemoprotein*** found in ***heart*** and ***skeletal muscle.*** It has a single polypeptide (153 amino acids) chain with heme moiety. Myoglobin (mol. wt. 17,000) structurally resembles the individual subunits of hemoglobin molecule. For this reason, the more complex properties of hemoglobin have been conveniently elucidated through the study of myoglobin.

Myoglobin functions as a ***reservoir for oxygen.***

FUNCTIONS OF HEMOGLOBIN

Hemoglobin is largely responsible for the ***transport of O_2*** from lungs to tissues. It also helps to ***transport CO_2*** from the tissues to the lungs.

Binding of O_2 to hemoglobin

One molecule of hemoglobin (with four hemes) can bind with four molecules of O_2. This is in contrast to myoglobin (with one heme) which can bind with only one molecule of oxygen. In other words, each heme moiety can bind with one O_2.

Oxygen dissociation curve : The ***binding*** ability of ***hemoglobin with O_2*** at different partial pressures of oxygen (pO_2) can be measured by a graphic representation known as O_2 dissociation curve. The curves obtained for hemoglobin and myoglobin are depicted in ***Fig.10.3***.

It is evident from the graph that myoglobin has much higher affinity for O_2 than hemoglobin. Hence O_2 is bound more tightly with myoglobin than with hemoglobin. Further, pO_2 needed for half saturation (50% binding) of myoglobin is about 1 mm Hg compared to about 26 mm Hg for hemoglobin.

Transport of O_2 to the tissues

In the lungs, where the concentration of O_2 is high (hence high pO_2), the hemoglobin gets fully saturated (loaded) with O_2. Conversely, at the tissue level, where the O_2 concentration is low (hence low pO_2), the oxyhemoglobin releases (unloads) its O_2 for cellular respiration. This is often mediated by binding O_2 to myoglobin which serves as the immediate reservoir and supplier of O_2 to the tissues (***Fig.10.4***).

Transport of CO_2 by hemoglobin

In aerobic metabolism, for every molecule of O_2 utilized, one molecule of CO_2 is liberated. Hemoglobin actively participates in the transport of

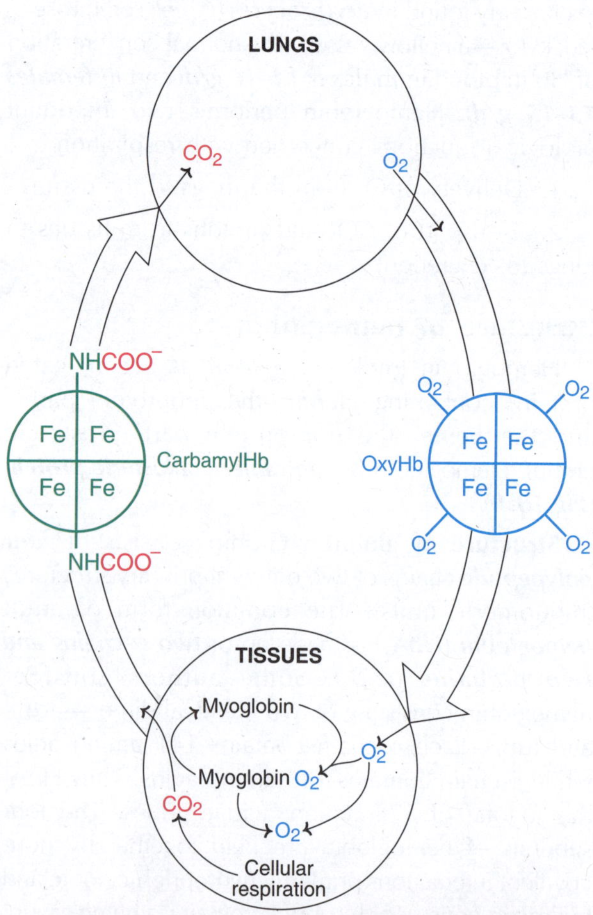

Fig.10.4 : *Diagrammatic representation of transport of O_2 and CO_2 by hemoglobin.*

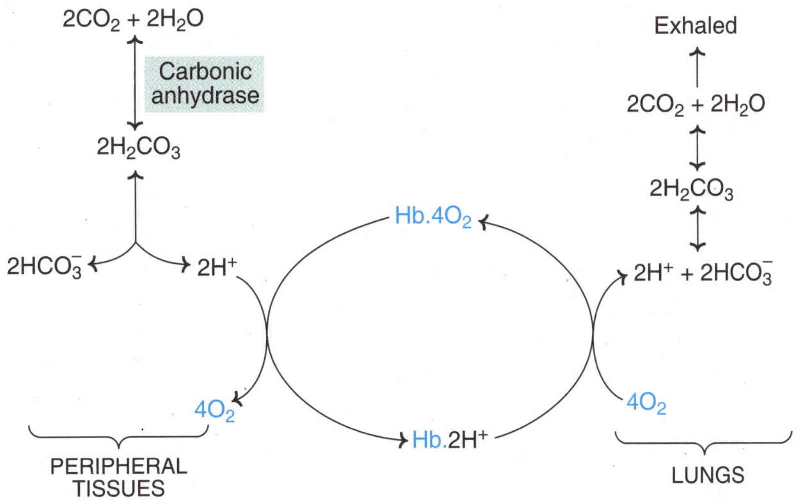

Fig.10.5 : *Transport of CO_2 through the mediation of hemoglobin.*

CO_2 from the tissues to the lungs. About 15% of CO_2 carried in blood directly binds with Hb. The rest of the tissue CO_2 is transported as bicarbonate (HCO_3^-).

Carbon dioxide molecules are bound to the uncharged α-amino acids of hemoglobin to form ***carbamyl hemoglobin*** as shown below

$$Hb - NH_2 + CO_2 \rightleftharpoons Hb - NH - COO^- + H^+$$

Hemoglobin also helps in the transport of CO_2 as bicarbonate, as explained below (***Fig.10.5***).

As the CO_2 enters the blood from tissues, the enzyme ***carbonic anhydrase*** present in erythrocytes catalyses the formation of carbonic acid ***(H_2CO_3)***. Bicarbonate (HCO_3^-) and proton (H^+) are released on dissociation of carbonic acid. Hemoglobin acts as a buffer and immediately binds with protons. It is estimated that for every 2 protons bound to Hb, 4 oxygen molecules are released to the tissues. In the lungs, binding O_2 to Hb results in the release of protons. The bicarbonate and protons combine to form carbonic acid. The latter is acted upon by carbonic anhydrase to release CO_2, which is exhaled.

BOHR EFFECT

The binding of oxygen to hemoglobin decreases with increasing H^+ concentration (lower pH) or when the hemoglobin is exposed to increased partial pressure of CO_2 (pCO_2). This phenomenon is known as Bohr effect. It is due to a change in the binding affinity of oxygen to hemoglobin. Bohr effect causes a ***shift in the oxygen dissociation curve to the right*** (***Fig.10.6***).

Bohr effect is primarily responsible for the release of O_2 from the oxyhemoglobin to the tissue. This is because of increased pCO_2 and decreased pH in the actively metabolising cells.

EFFECT OF 2,3-BISPHOSPHOGLYCERATE ON O_2 AFFINITY OF Hb

2,3-Bisphosphoglycerate (***2,3-BPG***; formerly, 2,3-diphosphoglycerate) is the most abundant organic phosphate in the erythrocytes. Its molar concentration

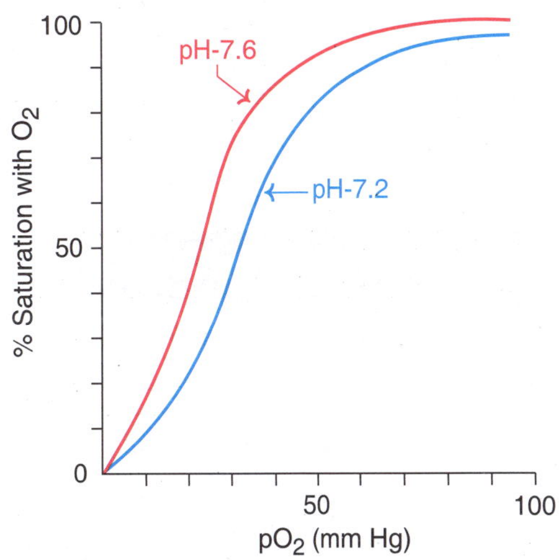

Fig.10.6 : *Effect of pH (Bohr effect) on oxygen dissociation curve (pO_2-Partial pressure of O_2).*

is approximately equivalent to that of hemoglobin. 2,3-BPG is produced in the erythrocytes from an intermediate (1,3-bisphosphoglycerate) of glycolysis. This short pathway referred to as ***Rapoport-Leubering cycle*** is described in carbohydrate metabolism (***Chapter 13***).

Binding of 2,3-BPG to deoxyhemoglobin

2,3-BPG regulates the binding of O_2 to hemoglobin. It specifically binds to deoxyhemoglobin (and not to oxyhemoglobin) and decreases the O_2 affinity to Hb. The effect of 2,3-BPG on Hb may be summarized as follows

$$\underset{\text{OxyHb}}{\text{Hb}\boxed{O_2}} + \text{2,3-BPG} \longrightarrow \underset{\text{DeoxyHb bound to 2,3-BPG}}{\text{Hb-2,3-BPG}} + \boxed{O_2}$$

The reduced affinity of O_2 to Hb facilitates the ***release of O_2*** at the partial pressure found in the tissues. This 2,3-BPG shifts the oxygen dissociation curve to the right (***Fig.10.7***).

Clinical significance of 2,3-BPG

Since the binding of 2,3-BPG with hemoglobin is primarily ***associated with the release of O_2 to the tissues***, this small molecule assumes a lot of biomedical significance. The erythrocyte levels of 2,3-BPG are related to tissue demands of oxygen supply.

1. **In hypoxia :** The concentration of ***2,3-BPG*** in erythrocytes ***is elevated in chronic hypoxic conditions*** associated with difficulty in O_2 supply. These include adaptation to high altitude, obstructive pulmonary emphysema (airflow in the bronchioles blocked) etc.

2. **In anemia :** 2,3-BPG levels in RBC are ***increased in severe anemia*** in order to cope up with the oxygen demands of the body.

HEMOGLOBIN DERIVATIVES

Hemoglobin (specifically heme) combines with different ligands and forms hemoglobin derivatives. The normal blood contains oxyHb and deoxyHb. Besides these, ***methemoglobin*** (metHb) and ***carboxyhemoglobin*** are the other important Hb derivatives. The Hb derivatives have characteristic colour and they can be detected by absorption spectra.

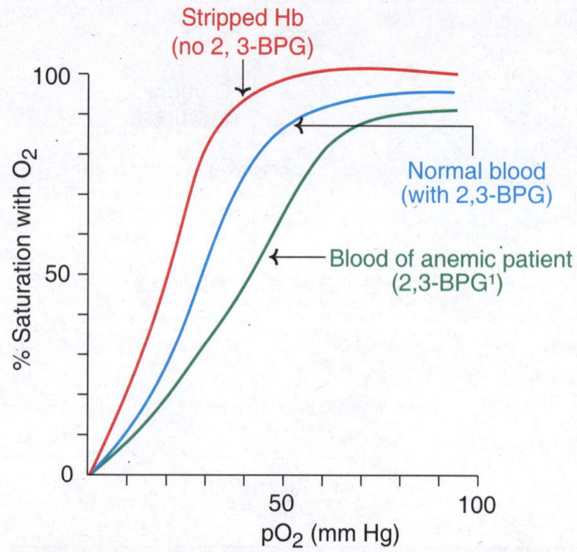

Fig.10.7 : *Effect of 2,3-bisphosphoglycerate (2,3-BPG) on oxygen dissociation curve (pO_2-Partial pressure of O_2).*

Methemoglobin

For the biological function of hemoglobin—to carry oxygen—the iron should remain in the ferrous (Fe^{2+}) state. Hemoglobin (Fe^{2+}) can be oxidized to methemoglobin (Fe^{3+}). In normal circumstances, however, molecular oxygen does not oxidize Hb, it only loosely binds to form oxyhemoglobin.

The oxidation of hemoglobin to methemoglobin (metHb) may be caused in the living system by ***H_2O_2, free radicals and drugs***. The methemoglobin (with Fe^{3+}) is unable to bind to O_2. Instead, a water molecule occupies the oxygen site in the heme of metHb.

In normal circumstances, the occasional oxidation of hemoglobin is corrected by the enzyme ***methemoglobin reductase*** present in erythrocytes.

Carboxyhemoglobin (COHb)

Carbon monoxide (CO) is a toxic industrial pollutant that can bind with Hb in the same manner as O_2 binds. However, CO has about 200 times more affinity than O_2 for binding with Hb.

Clinical manifestations of CO toxicity are observed when the COHb concentration exceeds 20%. The symptoms include headache, nausea, breathlessness, vomiting and irritability. Administration of O_2 through oxygen masks will help to reverse the manifestations of CO toxicity.

Medical Concepts/ Clinical Correlates

- Hemoglobin is primarily responsible for the delivery of O_2 from lungs to tissue and the transport of CO_2 from tissues to lungs.
- Increased erythrocyte 2,3-BPG levels in anemia and chronic hypoxia facilitate the release of more O_2 from the oxyhemoglobin to the tissues.
- Hemoglobin (Fe^{2+}) on oxidation by free radicals or drugs, forms methemoglobin (Fe^{3+}) which cannot transport O_2.
- Carboxyhemoglobin is produced when carbon monoxide binds to hemoglobin. The clinical manifestations of CO toxicity (> 20% COHb) include headache, nausea, breathlessness and vomiting.

ABNORMAL HEMOGLOBINS

Abnormal hemoglobins are the ***resultant of mutations in the genes that code for α or β chains of globin***. As many as 400 mutant hemoglobins are known. About 95% of them are due to alteration in a single amino acid of globin.

Hemoglobinopathies

It is a term used to describe the disorders caused by the synthesis of abnormal hemoglobin molecule or the production of insufficient quantities of normal hemoglobin or rarely both.

Sickle-cell anemia (HbS) and ***hemoglobin C disease (HbC)*** are the classical examples of abnormal hemoglobins. ***Thalassemias***, on the other hand, are caused by decreased synthesis of normal hemoglobin.

SICKLE-CELL ANEMIA OR SICKLE-CELL HEMOGLOBIN

Sickle-cell anemia (HbS) is the ***most common*** form of ***abnormal hemoglobins***. It is so named because the erythrocytes of these patients adopt a sickle shape (crescent like) at low oxygen concentration (***Fig.10.8***).

Occurrence of the disease

Sickle-cell anemia is largely confined to tropical areas of the world. It ***primarily occurs in the black population***. It is estimated that 1 in 500 newborn black infants in the USA are affected by sickle-cell anemia.

Molecular basis of HbS

The structure of hemoglobin (as described already) contains two α-and two β-globin chains. In case of sickle-cell anemia, the hemoglobin (HbS) has two normal α-globin chains and two abnormal (mutant) β-globin chains. This is due to a difference in a single amino acid. In HbS, ***glutamate at sixth position of β-chain is replaced by valine*** (Glu $\beta_6 \rightarrow$Val).

Abnormalities associated with HbS

Sickle-cell anemia is characterized by the following abnormalities

1. **Life-long hemolytic anemia :** The sickled ***erythrocytes*** are fragile and their ***continuous breakdown*** leads to life-long anemia.
2. **Tissue damage and pain :** The sickled cells block the capillaries resulting in poor blood supply to tissues. This leads to extensive damage and ***inflammation*** of certain tissues causing pain.
3. **Increased susceptibility to infection :** Hemolysis and tissue damage are accompanied by increased susceptibility to infection and diseases.
4. **Premature death :** Homozygous individuals of sickle-cell anemia die before they reach adulthood (<20 years).

Mechanism of sickling in sickle-cell anemia

Glutamate is a polar amino acid and it is replaced by a non-polar valine in sickle-cell hemoglobin. This

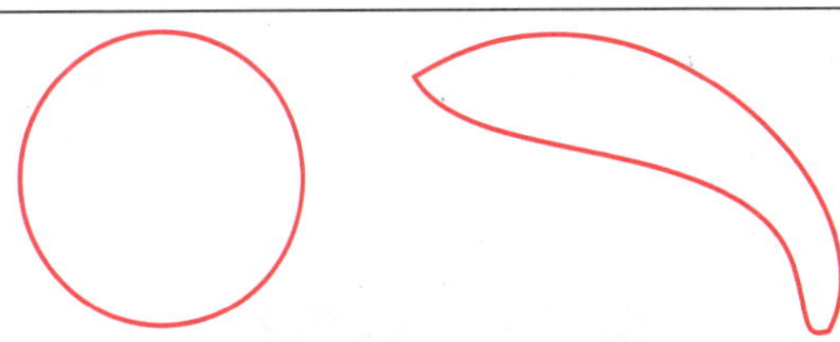

Fig.10.8 : *Erythrocytes : (A) From a normal person; (B) From a patient of sickel-cell anemia.*

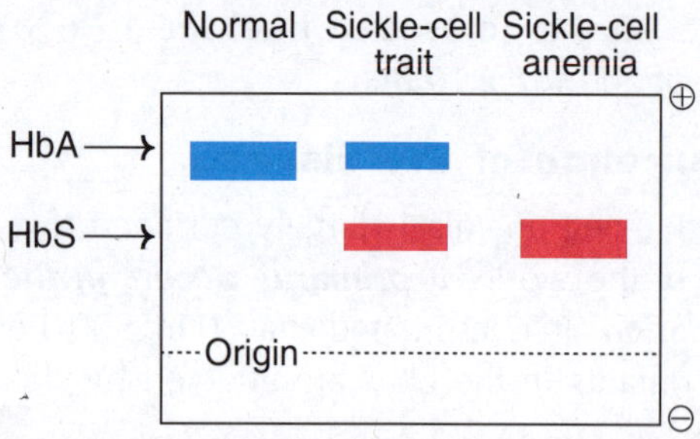

Fig. 10.9 : *Electrophoresis of hemoglobins at pH 8.6 (HbA–Normal adult hemoglobin; HbS–Sickle-cell hemoglobin).*

causes a marked ***decrease in the solubility of HbS*** in deoxygenated form (T-form). However, solubility of oxygenated HbS is unaffected. Sickling is due to polymerization of deoxyHbS.

Sickle-cell trait provides resistance to malaria

The incidence of sickle-cell disease coincides with the high incidence of malaria in tropical areas of the world (particularly among the black Africans).

Sickle-cell trait (heterozygous state with about 40% HbS) provides resistance to malaria which is a major cause of death in tropical areas. This is explained as follows

Malaria is a parasitic disease caused by *Plasmodium falciparum* in Africa. The malarial parasite spends a part of its life cycle in erythrocytes. ***Increased lysis of sickled cells*** (shorter life span of erythrocytes) ***interrupts the parasite cycle***.

Diagnosis of sickle-cell anemia

1. **Sickling test :** This is a simple microscopic examination of blood smear prepared by adding reducing agents such as sodium metabisulfite. Sickled erythrocytes can be ***detected under the microscope***.

2. **Electrophoresis :** When subjected to electrophoresis in alkaline medium (pH 8.6), sickle-cell hemoglobin (HbS) moves slowly towards anode (positive electrode) than does adult hemoglobin (HbA). The electrophoresis of hemoglobin obtained from lysed erythrocytes can be routinely used for the diagnosis of sickle-cell anemia and sickle-cell trait (***Fig.10.9***).

THALASSEMIAS

Thalassemias are a group of hereditary hemolytic disorders characterized by ***impairment/imbalance in the synthesis of globin chains of Hb***.

Thalassemias (*Greek*: thalassa-sea) mostly occur in the regions surrounding the Mediterranean sea, hence the name. These diseases, however, are also prevalent in Central Africa, India and the Far East.

Molecular basis of thalassemias

Hemoglobin contains 2α and 2β globin chains. The synthesis of individual chains is so coordinated that each α-chain has a β-chain partner and they combine to finally give hemoglobin ($\alpha_2\beta_2$). Thalassemias are characterized by a ***defect in the productiion of α-or β-globin chain***. There is however, no abnormality in the amino acids of the individual chains.

Thalassemias occur due to a variety of molecular defects

1. Gene deletion or substitution,
2. Underproduction or instability of mRNA,
3. Defect in the initiation of chain synthesis,
4. Premature chain termination.

α-Thalassemias

α-Thalassemias are caused by a ***decreased*** synthesis or total ***absence of α-globin chain of Hb***. There are four copies of α-globin gene, two on each one of the chromosome 16. Four types of α-thalassemias occur which depend on the number of missing α-globin genes.

1. **Silent carrier** state is due to loss of one of the four α-globin genes with no physical manifestations.

2. **α-Thalassemia trait** is caused by loss of two genes (both from the same gene pair or one from each gene pair). Minor anemia is observed.

3. **Hemoglobin H disease**, due to missing of three genes, is associated with moderate anemia.

4. **Hydrops fetalis** is the most severe form of α-thalassemias due to lack of all the four genes. The fetus usually survives until birth and then dies.

β-Thalassemias

Decreased synthesis or total ***lack of*** the formation of ***β-globin chain*** causes β-thalassemias. The production of α-globin chain continues to be normal, leading to the formation of α globin tetramer (α_4) that precipitate. This causes premature death of erythrocytes. There are mainly two types of β-thalassemias.

1. **β-Thalassemia minor :** This is an heterozygous state with a ***defect in only one of the two β-globin gene pairs on chromosome 11***. This disorder, also known as ***β-thalassemia trait***, is usually asymptomatic, since the individuals can make some amount of β-globin from the affected gene.

2. **β-Thalassemia major :** This is a homozygous state with a ***defect in both the genes responsible for β-globin synthesis***. The infants born with β-thalassemia major are healthy at birth since β-globin is not synthesized during the fetal development. They become severely anemic and die within 1-2 years.

PORPHYRINS

Porphyrins are cyclic compounds composed of 4 ***pyrrole rings*** held together by ***methenyl*** (=CH–) ***bridges*** (***Fig.10.10***). Metal ions can bind with nitrogen atoms of pyrrole rings to form complexes. Heme is an iron-containing porphyrin (See ***Fig.10.2***).

Presentation and nomenclature of porphyrins

The structure of porphyrins ($C_{20}H_{14}N_4$) has four ***pyrrole rings*** namely I, II, III and IV. Naturally occurring porphyrins contain substituent groups replacing the 8 hydrogen atoms of the porphyrin nucleus.

Hans Fischer, the father of porphyrin chemistry, proposed a shorthand model for presentation of porphyrin structures. Accordingly, each pyrrole ring is represented as a bracket. Thus porphyrin has 4 closed brackets with the 8 substituent positions.

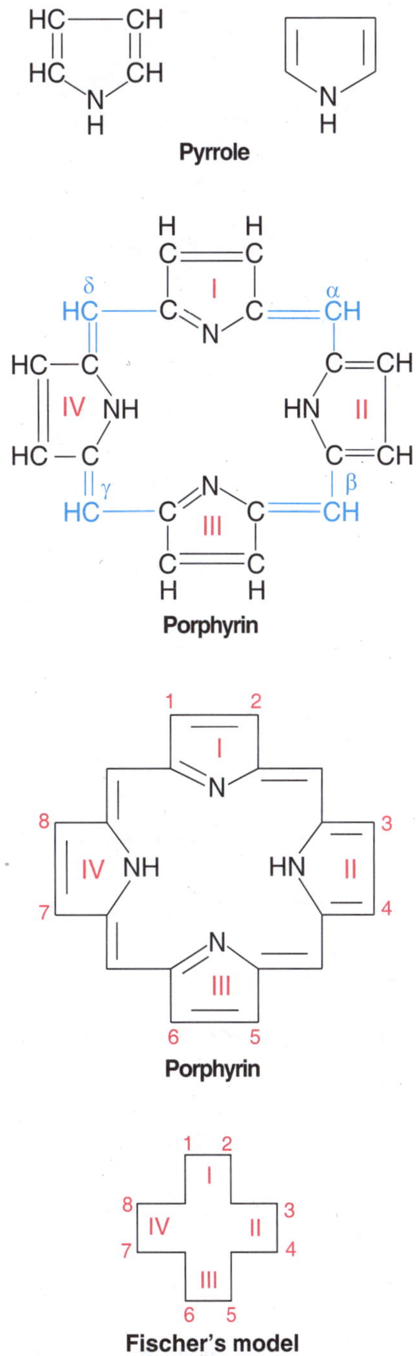

***Fig. 10.10 :** Structures of pyrrole and porphyrin [I-IV are pyrrole rings; 1-8 are substituent positions; α, β, γ, δ are methylene (–CH=) bridges.]*

BIOSYNTHESIS OF HEME

Heme is the most important porphyrin containing compound. It is primarily synthesized in the ***liver*** and the erythrocyte-producing cells of ***bone marrow*** (erythroid cells). Heme synthesis also occurs to some extent in other tissues. However, mature erythrocytes lacking mitochondria are a notable exception. Biosynthesis of heme occurs in the following stages (***Fig.10.11***).

1. **Formation of δ-aminolevulinate :** ***Glycine***, a non-essential amino acid and ***succinyl CoA***, an intermediate in the citric acid cycle, are the starting materials for porphyrin synthesis. Glycine combines with succinyl CoA to form δ-aminolevulinate (ALA). This reaction catalysed by a pyridoxal phosphate dependent ***δ-aminolevulinate synthase*** occurs in the mitochondria. It is a ***rate-controlling step in*** porphyrin synthesis.

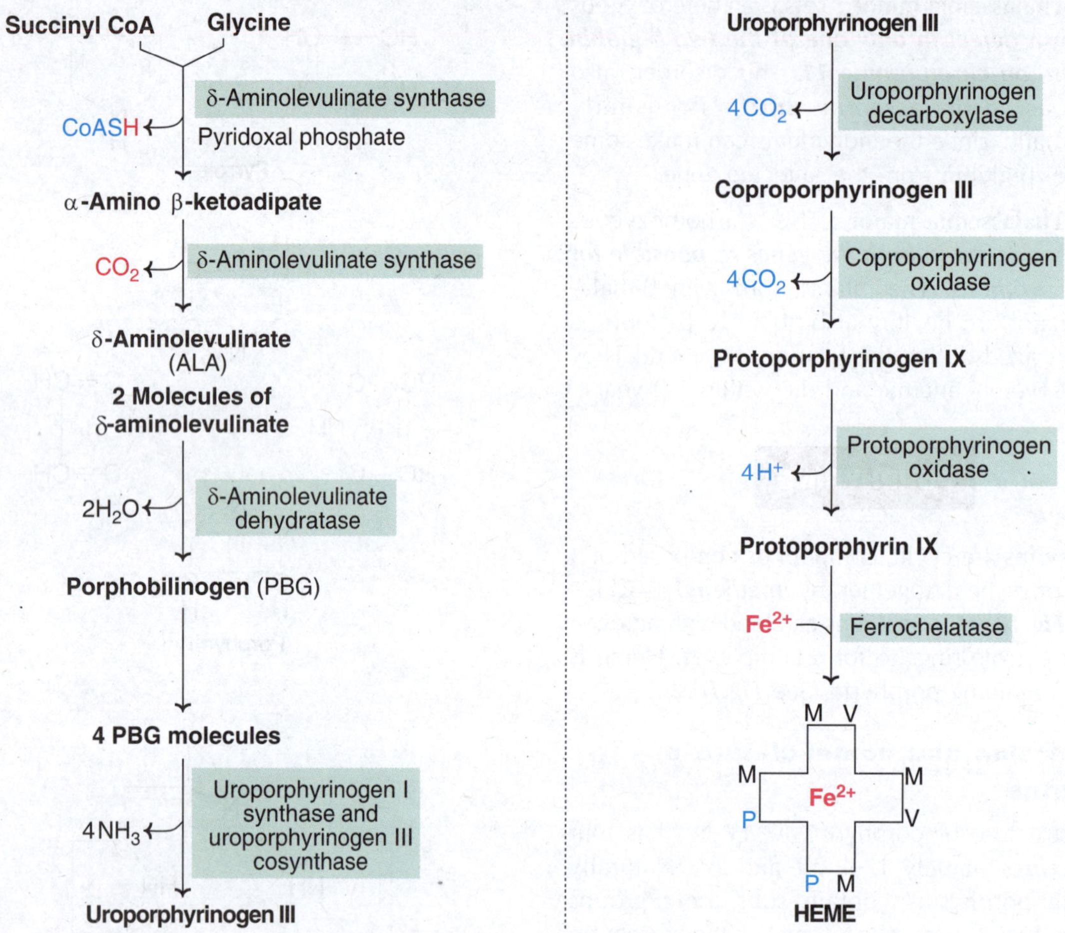

Fig. contd. next column

***Fig. 10.11 :** Biosynthesis of heme*
[A–Acetyl ($-CH_2-COO^-$); P–Propionyl ($-CH_2-CH_2-COO^-$); M–Methyl ($-CH_3$); V–Vinyl ($-CH=CH_2$)].

2. **Synthesis of porphobilinogen :** Two molecules of δ-aminolevulinate condense to form porphobilinogen (PBG) in the cytosol. This reaction is catalysed by a Zn-containing enzyme ***ALA dehydratase***. It is sensitive to inhibition by heavy metals such as lead.

3. **Formation of porphyrin ring :** Porphyrin synthesis occurs by condensation of four molecules of porphobilinogen. The four pyrrole rings in porphyrin are interconnected by methylene ($-CH_2$) bridges derived from α-carbon of glycine.

The interaction of two enzymes—namely uroporphyrinogen I synthase and uroporphyrinogen III cosynthase—results in condensation of porphobilinogen followed by ring closure and isomerization to produce uroporphyrinogen III.

In the series of reactions that follow (***Fig.10.11***), uroporphyrinogen III is finally converted to heme.

Regulation of heme synthesis

Regulation in the liver : The first committed step in heme biosynthesis, catalysed by δ-aminolevulinate (ALA) synthase, is regulatory. ***Heme*** or its oxidized product ***hemin*** (Fe^{3+}) controls this enzyme activity by three mechanisms

1. Feedback inhibition
2. Repression of ALA synthatase
3. Inhibition of transport of ALA synthase from cytosol to mitochondria (the site of action).

Regulation in the erythroid cells : The enzyme ALA synthase does not appear to control heme synthesis in erythroid cells. ***Uroporphyrinogen***

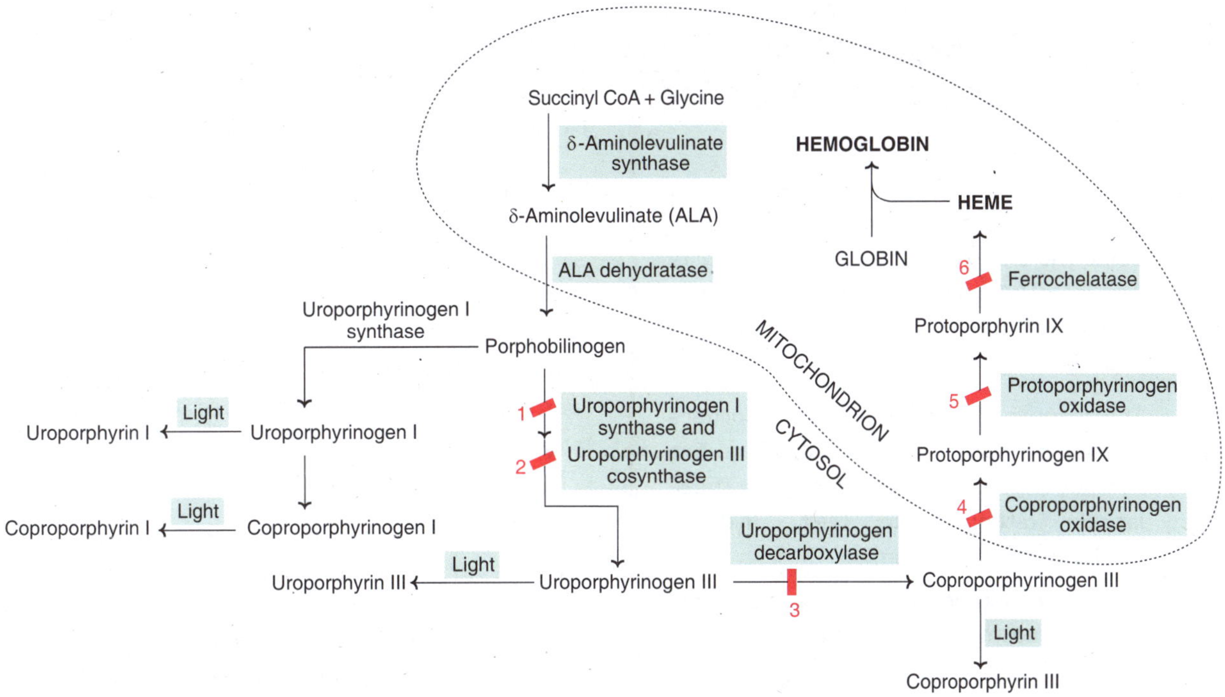

***Fig.10.12 :** Summary of heme synthesis along with porphyrias (1–Acute intermittent porphyria; 2–Congenital erythropoietic porphyria; 3–Porphyria cutanea tarda; 4–Hereditary coproporphyria; 5–Variegate porphyria; 6–Protoporphyria).*

synthase and ***ferrochelatase*** mostly regulate heme formation in these cells. Further, the cellular uptake of iron also influences heme synthesis.

PORPHYRIAS

Porphyrias are the metabolic disorders of heme synthesis, characterized by the ***increased excretion of porphyrins or porphyrin precursors***. Porphyrias are either inherited or acquired. They are broadly classified into two categories

1. **Erythropoietic :** Enzyme deficiency occurs in the erythrocytes.
2. **Hepatic :** Enzyme defect lies in the liver.

All the known porphyrias are inherited as autosomal dominant disorders. However, congenital erythropoietic porphyria is an exception, since it is autosomal recessive. The different types of porphyrias are described hereunder (***Fig.10.12, Table 10.1***).

I. Acute intermittent porphyria

This disorder occurs due to the deficiency of the enzyme ***uroporphyrinogen I synthase.*** Acute intermittent porphyria is characterized by increased excretion of porphobilinogen and δ-aminolevulinate. The urine gets darkened on exposure to air due to the conversion of porphobilinogen to porphobilin and porphyrin.

Acute intermittent porphyria is treated by administration of hematin which inhibits the enzyme ALA synthase and the accumulation of porphobilinogen.

II. Congenital erythropoietic porphyria

This disorder is due to a defect in the enzyme ***uroporphyrinogen III cosynthase***. Some workers, however, believe that congenital erythropoeitic porphyria is caused by an imbalance between the activities of uroporphyrinogen I synthase and uroporphyrinogen III cosynthase.

III. Porphyria cutanea tarda

This is also known as ***cutaneous hepatic porphyria*** and is the most common porphyria, usually associated with liver damage caused by alcohol overconsumption or iron overload. The partial deficiency of the enzyme ***uroporphyrinogen decarboxylase*** appears to be responsible for the occurrence of porphyria cutanea tarda.

IV. Hereditary coproporphyria

This disorder is due to a defect in the enzyme ***coproporphyrinogen oxidase***. As a result of this, coproporphyrinogen III and other intermediates (ALA and PBG) of heme synthesis prior to the blockade are excreted in urine and feces. The ***victims of*** hereditary coproporphyria are ***photosensitive***. They exhibit the clinical manifestations observed in the patients of acute intermittent porphyria.

V. Variegate porphyria

The enzyme ***protoporphyrinogen oxidase*** is defective in this disorder. Due to this blockade, protoporphyrin IX required for the ultimate synthesis of heme is not produced. Almost all the intermediates (porphobilinogen, coproporphyrin, uroporphyrin, protoporphyrin etc.) of heme synthesis accumulate in the body and are excreted in urine and feces. The urine of these patients is coloured, and they exhibit photosensitivity.

VI. Protoporphyria

This disorder, also known as ***erythropoietic protoporphyria***, is caused by a deficiency of the enzyme ***ferrochelatase***. Protoporphyrin IX accumulates in the tissues and is excreted into urine and feces.

Medical Concepts/ Clinical Correlates

- Sickle cell hemoglobin (HbS) causes hemolytic anemia, increased susceptibility to infection and premature death. However, HbS offers protection against malaria.
- Thalassemias are hemolytic disorders caused by impairment / imbalance in the synthesis of globin chains of Hb. These include α-thalassemia trait, hydrops fetalis and β-thalassemias.
- Administration of porphyrins can be used in the treatment of certain cancers by phototherapy.
- Abnormalities in heme synthesis cause porphyrias which may be erythropoietic (enzyme defect in RBC) or hepatic (enzyme defect in liver). Porphyrias are associated with elevated excretion of porphyrins, neuropsychiatric disturbances and cardiovascular abnormalities.

Table 10.1 A general summary of porphyrias

Type of porphyria	Enzyme defect	Characteristics
Hepatic		
Acute intermittent porphyria	Uroporphyrinogen I synthase	Abdominal pain, neuropsychiatric symptoms
Porphyria cutanea tarda	Uroporphyrinogen decarboxylase	Photosensitivity
Hereditary coproporphyria	Corpoporphyrinogen oxidase	Abdominal pain, photosensitivity, neuropsychiatric symptoms
Variegate porphyria	Protoporphyrinogen oxidase	Abdominal pain, photosensitivity, neuropsychiatric symptoms
Erythropoietic		
Congenital erythropoietic porphyria	Uroporphyrinogen III cosynthase	Photosensitivity, increased hemolysis
Protoporphyria	Ferrochelatase	Photosensitivity

Acquired (toxic) porphyria

The porphyrias, though not inherited, may be acquired due to the toxicity of several compounds. Exposure of the body to heavy metals (e.g. lead), toxic compounds (e.g. hexachlorobenzene) and drugs (e.g. griseofulvin) inhibits many enzymes in heme synthesis. These include ALA dehydratase, uroporphyrin I synthase and ferrochelatase.

DEGRADATION OF HEME TO BILE PIGMENTS

Erythrocytes have a life span of 120 days. At the end of this period, they are removed from the circulation. Erythrocytes are taken up and ***degraded by the macrophages*** of the reticuloendothelial (RE) system in the spleen and liver. The hemoglobin is cleaved to the protein part globin and non-protein heme. About 6 g of hemoglobin per day is broken down and resynthesized in an adult man (70 kg).

Fate of globin : The globin may be reutilized as such for the formation of hemoglobin or degraded to the individual amino acids.

Sources of heme : It is estimated that about 80% of the heme that is subjected for degradation comes from the erythrocytes and the rest (20%) comes from immature RBC, myoglobin and cytochromes.

Heme oxygenase : A complex microsomal enzyme namely heme oxygenase cleaves the methenyl bridges between the two pyrrole rings (A and B) to form ***biliverdin***. Simultaneously, ferrous iron (Fe^{2+}) is oxidized to ferric form (Fe^{3+}) and released. Heme promotes the activity of this enzyme.

Biliverdin is excreted in birds and amphibia while in mammals it is further degraded.

Biliverdin reductase : Biliverdin's methenyl bridges (between the pyrrole rings C and D) are reduced to methylene group to form ***bilirubin*** (yellow pigment). This reaction is catalysed by an NADPH dependent soluble enzyme, biliverdin reductase (***Fig 10.13***). ***One gram of hemoglobin on degradation finally yields about 35 mg bilirubin***. Approximately 250-350 mg of bilirubin is daily produced in human adults. The term bile pigments is used to collectively represent bilirubin and its derivatives.

Transport of bilirubin to liver : Bilirubin is lipophilic and therefore insoluble in aqueous solution. Bilirubin is transported in the plasma in a bound (non-covalently) form to albumin.

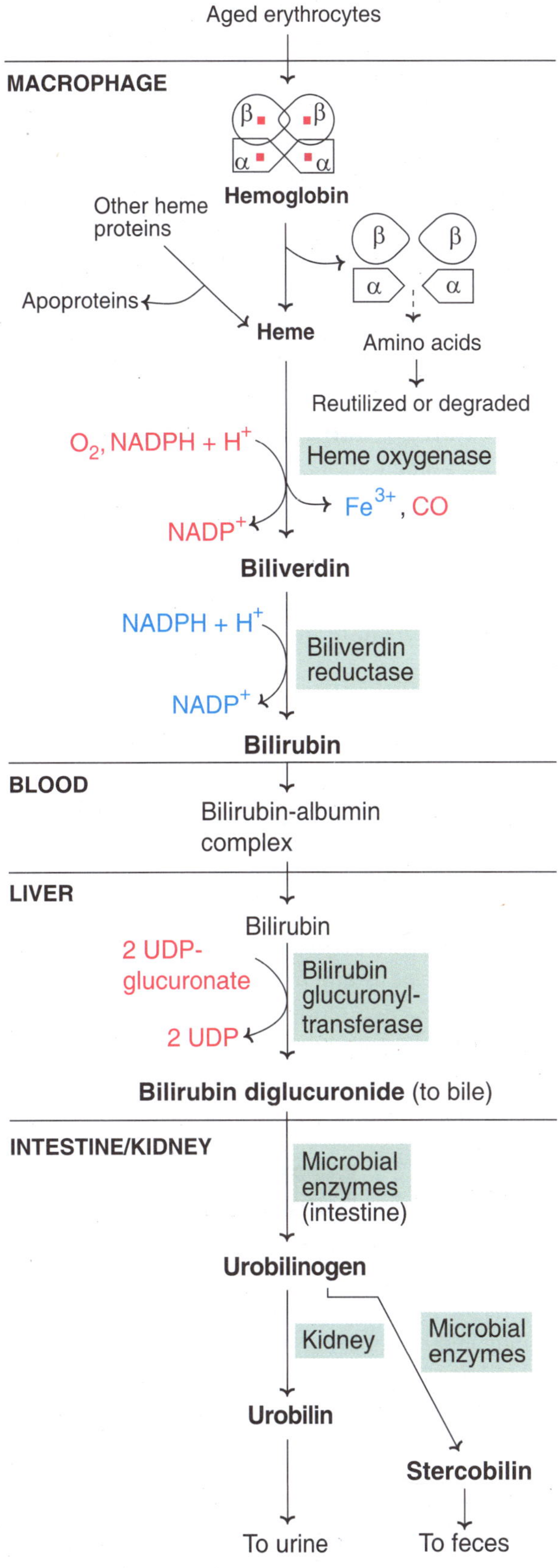

Fig.10.13 : Degradation of heme to bile pigments.

As the ***albumin-bilirubin complex*** enters the liver, bilirubin dissociates and is taken up by sinusoidal surface of the hepatocytes by a carrier mediated active transport.

Conjugation of bilirubin

In the liver, bilirubin is conjugated with two molecules of glucuronate supplied by UDP-glucuronate. This reaction, catalysed by ***bilirubin glucuronyltransferase*** results in the formation of a water soluble bilirubin diglucuronide (***Figs.10.13*** and ***10.14***). When bilirubin is in excess, bilirubin monoglucuronides also accumulate in the body.

Excretion of bilirubin into bile

Conjugated bilirubin is excreted into the bile canaliculi against a concentration gradient which then enters the bile. The transport of ***bilirubin diglucuronide*** is an active, energy-dependent and rate limiting process. This step is easily susceptible to any impairment in liver function.

Fate of bilirubin

Bilirubin glucuronides are hydrolysed (deconjugated) in the intestine by specific bacterial enzymes namely ***β-glucuronidases*** to liberate bilirubin. The latter is then converted to urobilinogen (colourless compound), a small part of which may be reabsorbed into the circulation. Urobilinogen can be converted to urobilin (an yellow colour compound) in the kidney and excreted. The characteristic ***yellow colour of urine is due to urobilin***.

A major part of urobilinogen is converted by bacteria to stercobilin which is excreted along with feces. The characteristic ***brown colour of feces is due to stercobilin***.

JAUNDICE

The normal ***serum total bilirubin*** concentration is in the range of ***0.2 to 1.0 mg/dl***. Of this, about 0.2-0.6 mg/dl is unconjugated while 0 to 0.2 mg/dl is conjugated bilirubin.

Jaundice (*French* : *Jaune*-yellow) is a clinical condition characterized by yellow colour of the white of the eyes (sclerae), skin and nail bed. It is caused by the deposition of bilirubin due to its elevated levels in the serum. The term ***hyperbilirubinemia*** is often used to represent the increased concentration of serum bilirubin. (***Note :*** For some more details on jaundice, ***refer Chapter 20***)

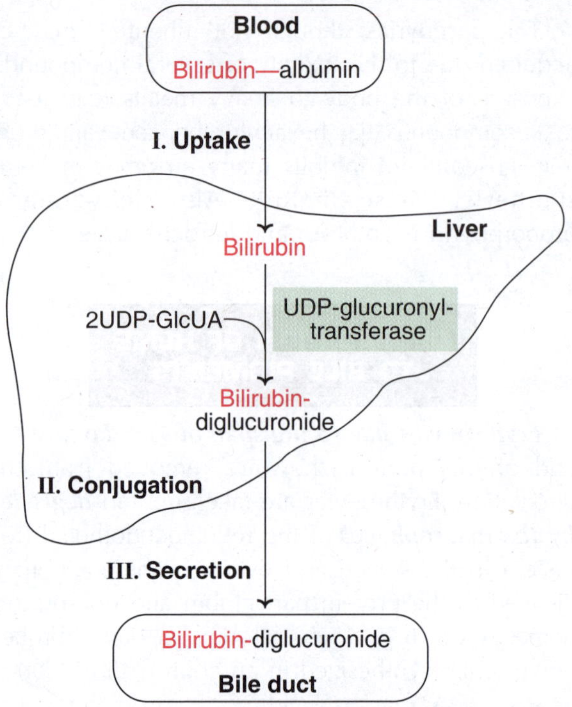

Fig.10.14 : *Summary of bilirubin metabolism (UDP-GlcUA—UDP-glucuronic acid).*

Classification of jaundice

Jaundice (also known as ***icterus***) may be more appropriately considered as a symptom rather than a disease. It is rather difficult to classify jaundice, since it is frequently caused due to multiple factors. For ***the sake of convenience to understand***, jaundice is classified into three major types—***hemolytic, hepatic and obstructive***.

1. **Hemolytic jaundice :** This condition is associated with ***increased hemolysis of erythrocytes*** (e.g. incompatible blood transfusion, malaria, sickle-cell anemia). This results in the overproduction of bilirubin beyond the ability of the liver to conjugate and excrete the same. It should, however be noted that liver possesses a large capacity to conjugate about 3.0 g of bilirubin per day against the normal bilirubin production of 0.3 g/day.

Hemolytic jaundice is characterized by

- ***Elevation*** in the serum ***unconjugated bilirubin***.
- Increased excretion of urobilinogen in urine.
- Dark brown colour of feces due to high content of stercobilinogen.

Medical Concepts/ Clinical Correlates

- **Jaundice is caused by elevated serum bilirubin (normal < 1.0 mg/dl) levels and is characterized by yellow coloration of white of the eyes, and skin.**
- **Phototherapy (by exposure to blue light) is used in to control severe cases of neonatal physiologic jaundice.**

2. **Hepatic (hepatocellular) jaundice :** This type of jaundice is caused by ***dysfunction of the liver*** due to damage to the parenchymal cells. This may be attributed to viral infection (viral hepatitis), ***poisons and toxins*** (chloroform, carbon tetrachloride, phosphorus etc.) ***cirrohosis*** of liver, cardiac failure etc. Among these, viral hepatitis is the most common.

Damage to the liver adversely affects the bilirubin uptake and its conjugation by liver cells. Hepatic jaundice is characterized by

- Increased levels of ***conjugated and unconjugated bilirubin in the serum***.
- Dark coloured urine due to the excessive excretion of bilirubin and urobilinogen.
- Increased activities of ***alanine transaminase*** (SGPT) and ***aspartate transaminase*** (SGOT) released into circulation due to damage to hapatocytes.
- The patients pass pale, clay coloured stools due to the absence of stercobilinogen.
- The affected individuals experience nausea and anorexia (loss of appetite).

3. **Obstructive (regurgitation) jaundice :** This is due to an obstruction in the bile duct that prevents the passage of bile into the intestine. The obstruction may be caused by gall stones, tumors etc.

Due to the blockage in bile duct, the conjugated bilirubin from the liver enters the circulation. Obstructive jaundice is characterized by

- Increased concentration of conjugated bilirubin in serum.
- Serum ***alkaline phosphatase*** is elevated as it is released from the cells of the damaged bile duct.
- Dark coloured urine due to elevated excretion of bilirubin and clay coloured feces due to absence of stercobilinogen.
- The patients experience nausea and gastro-intestinal pain.

JAUNDICE DUE TO GENETIC DEFECTS

There are certain hereditary abnormalities that cause jaundice.

Neonatal-physiologic jaundice

Physiological jaundice is ***not truly a genetic defect***. It is caused by ***increased hemolysis*** coupled with immature hepatic system for the uptake, conjugation and secretion of bilirubin. The activity of the enzyme ***UDP-glucuronyltransferase is low*** in the newborn. Further, there is a limitation in the availability of the substrate UDP-glucuronic acid for conjugation. The net effect is that in some infants the serum uncojugated bilirubin is highly elevated (may go beyond 25mg/dl), which can cross the blood-brain barrier. This results in ***hyperbilirubinemic toxic encephalopathy*** or ***kernicterus*** that causes mental retardation. The drug phenobarbital is used in the treatment of neonatal jaundice, as it can induce bilirubin metabolising enzymes in liver. In some neonates, blood transfusion may be necessary to prevent brain damage.

Phototherapy : Bilirubin can absorb blue light (420–470 nm) maximally. Phototherapy deals with the exposure of the jaundiced neonates to blue light. By a process called ***photoisomerization***, the toxic native unconjugated bilirubin gets converted into a non-toxic isomer namely ***lumirubin***. Lumirubin can be easily excreted by the kidneys in the unconjugated form (in contrast to bilirubin which cannot be excreted). Serum bilirubin is monitored every 12–24 hours, and phototherapy is continuously carried out till the serum bilirubin becomes normal (< 1 mg/dl).

Crigler-Najjar syndrome type I

This is also known as congenital non-hemolytic jaundice. It is a rare disorder and is due to a defect in the hepatic enzyme ***UDP-glucuronyltransferase***. Generally, the children die within first two years of life.

Crigler-Najjar syndrome type II

This is again a rare hereditary disorder and is due to a less severe defect in the bilirubin conjugation. It is believed that hepatic UDP-glucuronyltransferase that catalyses the addition of second glucuronyl group is defective. The serum bilirubin concentration is usually less than 20 mg/dl and this is less dangerous than type I.

Gilbert's disease : This is not a single disease but a combination of disorders. These include

1. A defect in the uptake of bilirubin by liver cells.
2. An impairment in conjugation due to reduced activity of UDP-glucuronyltransferase.
3. Decreased hepatic clearance of bilirubin.

SUMMARY

1. *Hemoglobin is responsible for the transport of O_2 from lungs to the tissues. Each heme (of Hb) can bind with one molecule of O_2 and this is facilitated by cooperative heme-heme interaction.*
2. *Hemoglobin actively participates in the transport of CO_2 from tissues to lungs. Increased partial pressure of CO_2 (pCO_2) accompanied by elevated H^+ decreases the binding of O_2 to Hb, a phenomenon known as Bohr effect.*
3. *Sickle-cell anemia (HbS) is a classical example of abnormal hemoglobins. It is caused when glutamate at 6th position of β-chain is replaced by valine.*
4. *Heme is the most important porphyrin compound, primarily synthesized in the liver from the precursors-glycine and succinyl CoA. Heme productioin is regulated by δ-aminolevulinate synthase.*
5. *Heme is degraded mainly to bilirubin, an yellow colour bile pigment. In the liver, it is conjugated to bilirubin diglucuronide, a more easily excretable form into bile.*
6. *Jaundice is a clinical condition caused by elevated serum bilirubin concentration (normal < 1.0 mg/dl). Jaundice is of three types-hemolytic (due to increased hemolysis), hepatic (due to impaired conjugation) and obstructive (due to obstruction in the bile duct).*

THE READER MUST REFER APPENDIX D (Page Nos : 422 - 431) FOR RELATED CLINICAL CASE STUDIES (Nos : 6, 16, 18, 19, 20)

BIOLOGICAL OXIDATION 11

For a better understanding of biological oxidation, it is worthwhile to have a basic knowledge of bioenergetics and the role of high-energy compounds in biological processes.

BIOENERGETICS

Bioenergetics or ***biochemical thermodynamics*** deals with the study of energy changes (transfer and utilization) in biochemical reactions. The reactions are broadly classified as ***exergonic*** (energy releasing) and ***endergonic*** (energy consuming). Bioenergetics is concerned with the initial and final states of energy component of the reactants and not the mechanism of chemical reactions.

Free energy

The energy actually ***available to do work*** (utilizable) is known as free energy. Changes in the free energy (ΔG) are valuable in predicting the feasibility of chemical reactions. The reactions can occur spontaneously if they are accompanied by decrease in free energy.

During a chemical reaction, heat may be released or absorbed. ***Enthalpy (ΔH)*** is a measure of the change in heat content of the reactants, compared to products.

Entropy (ΔS) represents a change in the randomness or disorder of reactants and products. Entropy attains a maximum as the reaction approaches equilibrium. The reactions of biological systems involve a temporary decrease in entropy.

The relation between the changes of free energy (ΔG), enthalpy (ΔH) and entropy (ΔS) is expressed as

$$\Delta G = \Delta H - T\Delta S$$

T represents the absolute temperature in Kelvin (K = 273 + °C).

The term ***standard free energy***, represented by $\Delta G°$ (note the superscript°), is often used. It indicates the ***free energy change*** when the reactants or products are at a concentration of 1 mol/l at pH 7.0.

Negative and positive ΔG

If free energy change (ΔG) is represented by a negative sign, there is a loss of free energy. The reaction is said to be ***exergonic***, and ***proceeds spontaneously***. On the other hand, a positive ΔG indicates that energy must be supplied to the reactants. The ***reaction cannot proceed spontaneously*** and is ***endergonic*** in character.

The ***hydrolysis of ATP*** is a classical example of ***exergonic reaction***

$$ATP + H_2O \longrightarrow ADP + Pi \ (\Delta G° = -7.3 \text{ Cal/mol})$$

The reversal of the reaction (ADP + Pi $\longrightarrow$ ATP) is endergonic and occurs only when there is a supply of energy of at least 7.3 Cal/mol ($\Delta G°$ is positive).

HIGH-ENERGY COMPOUNDS

The term high-energy compounds or ***energy rich compounds*** is usually applied to substances which possess sufficient free energy to liberate at least 7 Cal/mol at pH 7.0. Certain other compounds which liberate less than 7.0 Cal/mol (lower than ATP hydrolysis to ADP + Pi) are referred to as low-energy compounds. The list of important high-energy and low-energy metabolites is given in ***Table 11.1***.

All the high-energy compounds—when hydrolysed—liberate more energy than that of ATP. These include ***phosphoenol pyruvate, 1, 3-bisphosphoglycerate, phosphocreatine***, etc. Most of the high-energy compounds contain phosphate group (exception-acetyl CoA) hence they are called high-energy phosphate compounds.

Table 11.1 Standard free energy of hydrolysis of some important compounds

Compounds	DG° (Cal/mol)
High-energy phosphates	
Phosphoenol pyruvate	–14.8
Carbamoyl phosphate	–12.3
Cyclic AMP	–12.0
1,3-Bisphosphoglycerate	–11.8
Phosphocreatine	–10.3
Acetyl phosphate	–10.3
S-Adenosylmethionine	–10.0
Pyrophosphate	–8.0
Acetyl CoA	–7.7
ATP ⟶ ADP + Pi	–7.3
Low-energy phosphates	
ADP ⟶ AMP + Pi	–6.6
Glucose 1-phosphate	–5.0
Fructose 6-phosphate	–3.8
Glucose 6-phosphate	–3.3
Glycerol 3-phosphate	–2.2

ATP—the most important high-energy compound

Adenosine triphosphate (ATP) is a unique and the most important high-energy molecule in the living cells. It consists of an adenine, a ribose and a triphosphate moiety. ATP is a high-energy compound due to the presence of two phosphoanhydride bonds in the triphosphate unit (***Fig.11.1***). ATP serves as the ***energy currency of the cell*** as is evident from the ATP-ADP cycle.

ATP-ADP Cycle

The hydrolysis of ATP is associated with the release of large amount of energy.

$$ATP + H_2O \longrightarrow ADP + Pi + 7.3\ Cal.$$

The energy liberated is utilized for various processes like muscle contraction, active transport etc. ***ATP can also act as a donor of high-energy phosphate to low-energy compounds***, to make them energy rich. On the other hand, ***ADP can accept high-energy phosphate*** from the compounds possessing higher free energy content to form ATP.

ATP serves as an immediately available energy currency of the cell which is constantly being utilized and regenerated. This is represented by ATP-ADP cycle, the fundamental basis of energy exchange reactions in living system (***Fig.11.2***).

ATP acts as an ***energy link between the catabolism*** (degradation of molecules) and ***anabolism*** (synthesis) in the biological system.

Synthesis of ATP

ATP can be synthesized in two ways

1. **Oxidative phosphorylation :** This is the major source of ATP in aerobic organisms. It is ***linked with the mitochondrial electron transport chain*** (details described later).

2. **Substrate level phosphorylation :** ATP may be directly synthesized during ***substrate oxidation in the metabolism***. The high-energy compounds such as phosphoenolpyruvate and 1,3-bisphosphoglycerate (intermediates of glycolysis) and succinyl CoA (of citric acid cycle) can transfer high-energy phosphate to ultimately produce ATP.

Storage forms of high-energy phosphates

Phosphocreatine (creatine phosphate) stored in ***vertebrate muscle*** and brain is an energy-rich compound. In invertebrates, phosphoarginine (arginine phosphate) replaces phosphocreatine.

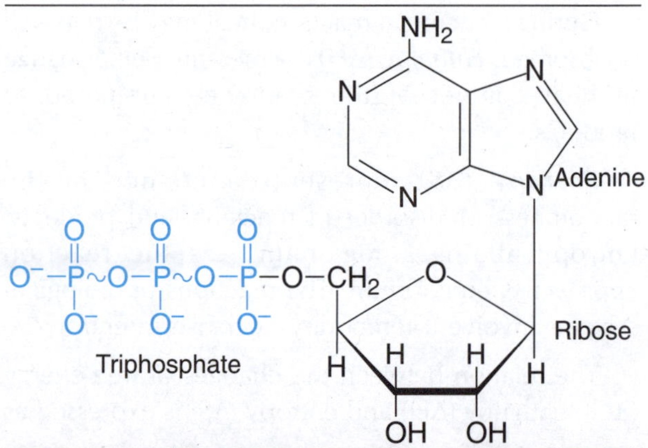

Fig.11.1 : Structure of ATP.

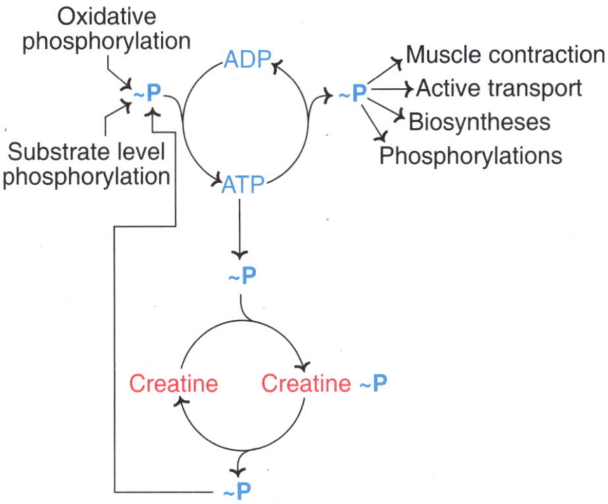

Fig.11.2 : *ATP-ADP cycle along with sources and utilization of ATP (Note that ~P does not exist in free form, but is only transferred).*

Redox pair	E_0 Volts
Succinate/α-ketoglutarate	–0.67
NAD^+/NADH	–0.32
$FMN/FMNH_2$ (enzyme bound)	–0.30
$FAD/FADH_2$	–0.22
Pyruvate/lactate	–0.19
Cytochrome b (Fe^{3+}/Fe^{2+})	+0.07
Coenzyme Q (ox/red)	+0.10
Cytochrome c_1 (Fe^{3+}/Fe^{2+})	+0.23
Cytochrome c (Fe^{3+}/Fe^{2+})	+0.25
Cytochrome a (Fe^{3+}/Fe^{2+})	+0.29
— O_2/H_2O	+0.82

Table 11.2 Standard redox potential (E_0) of some oxidation-reduction systems

BIOLOGICAL OXIDATION

Oxidation is defined as the ***loss of electrons and reduction as the gain of electrons***. This may be illustrated by the interconversion of ferrous ion (Fe^{2+}) to ferric ion (Fe^{3+}).

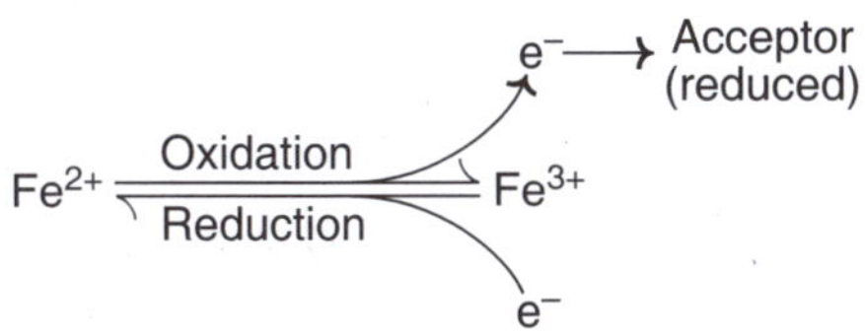

The electron lost in the oxidation is accepted by an acceptor which is said to be reduced. Thus the oxidation-reduction is a tightly coupled process.

Redox potential (E_0)

The ***oxidation-reduction*** potential or, simply, redox potential, is a ***quantitative measure of the tendency of a redox pair to lose or gain electrons***. The redox pairs are assigned specific ***standard redox potential*** (E_0 volts) at pH 7.0 and 25°C.

The redox potentials of some biologically important redox systems are given in ***Table 11.2***. The more negative redox potential represents a greater tendency (of reductant) to lose electrons. On the other hand, a more positive redox potential indicates a greater tendency (of oxidant) to accept electrons. ***The electrons flow from a redox pair with more negative E_0 to another redox pair with more positive E_0.***

ELECTRON TRANSPORT CHAIN

The energy-rich ***carbohydrates*** (particularly glucose), ***fatty acids*** and ***amino acids*** undergo a series of metabolic reactions and, finally, get ***oxidized to CO_2 and H_2O***. The reducing equivalents from various metabolic intermediates are transferred to coenzymes NAD^+ and FAD to produce, respectively, ***NADH and $FADH_2$***. The latter two reduced coenzymes pass through the electron transport chain (ETC) or ***respiratory chain*** and, finally, reduce oxygen to water. The passage of electrons through the ETC is associated with the loss of free energy. A part of this free energy is utilized to generate ATP from ADP and Pi (***Fig.11.3***).

Mitochondria —the power houses of cell

The mitochondria are the ***centres for metabolic oxidative reactions*** to generate reduced coenzymes (NADH and $FADH_2$) which, in turn, are utilized in ETC to liberate energy in the form of ATP. For this reason, mitochondrion is appropriately regarded as the ***power house of the cell***.

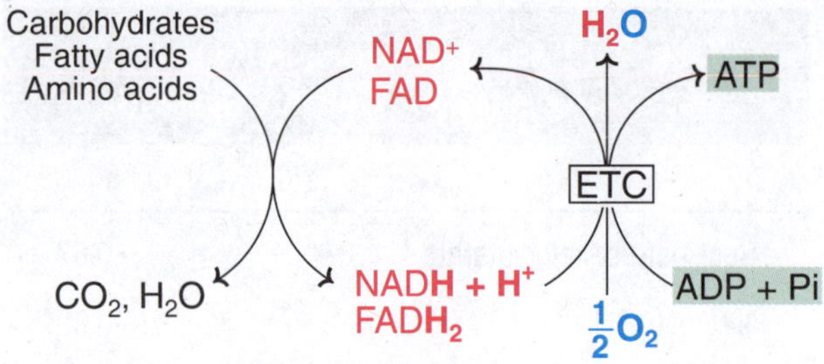

Fig.11.3 : *Overview of biological oxidation (ETC-Electron transport chain).*

Mitochondrial organization

The mitochondrion consists of ***five distinct parts***. These are the outer membrane, the inner membrane, the intermembrane space, the cristae and the matrix (***Fig.11.4***).

Inner mitochondrial membrane : The ***electron transport chain and ATP synthesizing system are located*** on the inner mitochondrial membrane which is a specialized structure, rich in proteins. It is impermeable to ions (H^+, K^+, Na^+) and small molecules (ADP, ATP). This membrane is highly folded to form ***cristae***. The surface area of inner mitochondrial membrane is greatly increased due to cristae. The inner surface of the inner mitochondrial membrane possesses specialized particles (that look like lollipops), the ***phosphorylating subunits*** which are the centres for ATP production.

Mitochondrial matrix : The interior ground substance forms the matrix of mitochondria. It is rich in the ***enzymes*** responsible for the ***citric acid cycle, β-oxidation of fatty acids*** and ***oxidation of amino acids***.

Components and reactions of the electron transport chain

There are five distinct carriers that participate in the electron transport chain (ETC). These carriers are sequentially arranged (***Fig.11.5***) and are responsible for the transfer of electrons from a given substrate to ultimately combine with proton and oxygen to form water.

I. Nicotinamide nucleotides

Of the two coenzymes NAD^+ and $NADP^+$ derived from the vitamin niacin, NAD^+ is more actively involved in the ETC. NAD^+ is reduced to ***NADH+H⁺*** by dehydrogenases with the removal of two hydrogen atoms from the substrate (AH_2). The substrates include glyceraldehyde 3-phosphate, pyruvate, isocitrate, α-ketoglutarate and malate.

$$AH_2 + NAD^+ \rightleftharpoons A + NADH + H^+$$

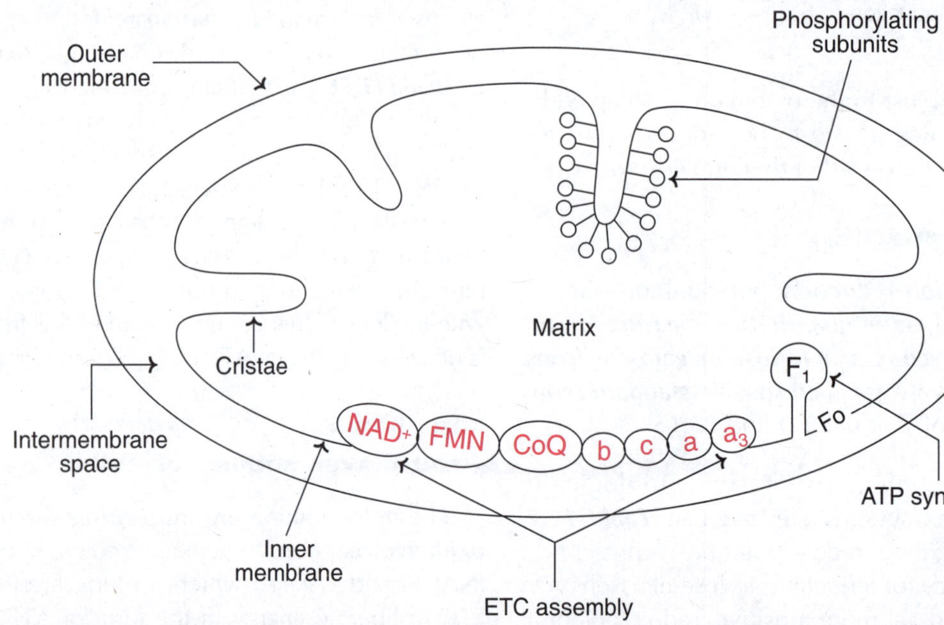

Fig.11.4 : *Structure of mitochondrion depicting electron transport chain (ETC) (F_0, F_1–Protein subunits).*

Medical Concepts/ Clinical Correlates

- **The supply of O_2 is very essential for the survival of life (exception—anaerobic bacteria).**
- **ATP, the energy currency of the cell, acts as a link between the catabolism and anabolism in the living system. The major production of body's ATP occurs in the mitochondria through oxidative phosphorylation coupled with respiration.**
- **Respiratory chain or electron transport chain (ETC) is blocked by site specific inhibitors such as rotenone, amytal, antimycin A, BAL, carbon monoxide and cyanide.**

II. Flavoproteins

The enzyme NADH dehydrogenase (NADH-coenzyme Q reductase) is a flavoprotein with FMN as the prosthetic group. The coenzyme FMN accepts two electrons and a proton to form $FMNH_2$. ***NADH dehydrogenase*** is a complex enzyme closely associated with non-heme iron proteins (NHI) or iron-sulfur proteins (FeS).

$$NADH + H^+ + FMN \longrightarrow NAD^+ + FMNH_2$$

III. Iron-sulfur proteins

The iron-sulfur (FeS) proteins exist in the oxidized (Fe^{3+}) or reduced (Fe^{2+}) state. About half a dozen FeS proteins connected with respiratory chain have been identified. However, the mechanism of action of iron-sulfur proteins in the ETC is not clearly understood.

IV. Coenzyme Q

Coenzyme Q is also known as ***ubiquinone*** since it is ubiquitous in living system. It is a quinone derivative with a variable ***isoprenoid*** side chain. The mammalian tissues possess a quinone with 10 isoprenoid units which is known as coenzyme Q_{10} (CoQ_{10}).

V. Cytochromes

The cytochromes are ***conjugated proteins*** containing ***heme group***. The latter consists of a porphyrin ring with iron atom. The heme group of cytochromes differ from that found in the structure of hemoglobin and myoglobin. The iron of heme in cytochromes is alternately oxidized (Fe^{3+}) and reduced (Fe^{2+}), which is essential for the transport of electrons in the ETC. This is in contrast to the heme iron of hemoglobin and myoglobin which remains in the ferrous (Fe^{2+}) state.

The ***electrons*** are ***transported*** from ***coenzyme Q to cytochromes*** (in the order) ***b, c_1, c, a and a_3***. The property of reversible oxidation-reduction of heme iron $Fe^{2+} \rightleftharpoons Fe^{3+}$ present in cytochromes allows them to function as effective carriers of electrons in ETC.

Cytochrome c (mol. wt. 13,000) is a small protein containing 104 amino acids and a heme group. It is a central member of ETC with an intermediate redox potential.

Cytochrome a and a_3 : The term ***cytochrome oxidase*** is frequently used to collectively represent cytochrome a and a_3 which is the terminal component of ETC. Cytochrome oxidase is the only electron carrier, the heme iron of which can directly react with molecular oxygen. Besides heme (with iron), this oxidase also contains copper that undergoes

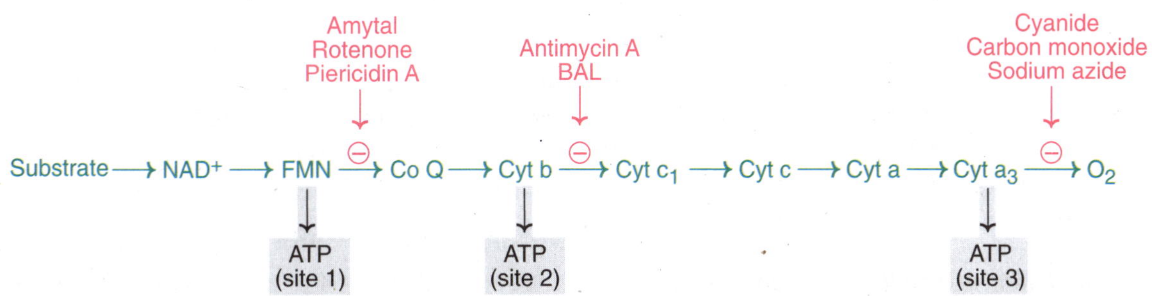

Fig.11.5 : Electron transport chain with sites of ATP synthesis and inhibitors (BAL-British antilewisite).

oxidation-reduction ($Cu^{2+} \rightleftharpoons Cu^{+}$) during the transport of electrons.

In the final stage of ETC, the transported electrons, the free protons and the molecular oxygen combine to produce water.

OXIDATIVE PHOSPHORYLATION

The transport of electrons through the ETC is linked with the release of free energy. The process of ***synthesizing ATP from ADP and Pi coupled with the electron transport chain*** is known as oxidative phosphorylation.

P : O Ratio

The P : O ratio refers to the ***number of inorganic phosphate*** molecules utilized for ATP generation for every ***atom of oxygen*** consumed.

(**Note :** ***Earlier, a P : O ratio of 3*** was assigned to ***NADH*** while it was ***2*** for ***$FADH_2$***. This was based on the belief that two protons were pumped at each complex (I, III and IV) resulting in the generation of one ATP, and thus a total of 3 for NADH and 2 for $FADH_2$. Studies in recent years suggest that this is not correct).

As per the recent evidence, the flow of 2 electrons (from NADH + H^+) through complexes I, III and IV results in the translocation of 10 protons from mitochondrial matrix. The synthesis of one ATP requires the involvement of 4 protons. Thus a total of 2.5 ATP (***P : O ratio 2.5***) are generated from ***NADH*** and ***1.5 ATP*** (***P : O ratio 1.5***) from ***$FADH_2$***.

It may be summarized that the oxidation of one ***NADH*** through ETC coupled with oxidative phosphorylation generates ***2.5 ATP*** while oxidation of one ***$FADH_2$*** yields ***1.5 ATP***. This difference is due to the fact that the coupling sites for NADH oxidation are 3 with 10 protons whereas for $FADH_2$ there are only two sites with 6 protons.

Sites of oxidative phosphorylation in ETC

The P : O ratio of 3 for NADH oxidation indicates that there are three reactions in the ETC that are exergonic to result in the synthesis of 3 ATP molecules (***See Fig.11.5***). The ***three sites of ATP formation*** in ETC are

1. ***Oxidation of $FMNH_2$*** by coenzyme Q.
2. ***Oxidation of cytochrome b*** by cytochrome c_1.
3. ***Cytochrome oxidase*** reaction.

Each one of the above reactions represents a ***coupling site*** for ATP production. There are only two coupling sites for the oxidation of $FADH_2$ (P : O ratio 2), since the first site is bypassed.

MECHANISM OF OXIDATIVE PHOSPHORYLATION

Several hypotheses have been put forth to explain the process of oxidative phosphorylation. The most important among them—namely, ***chemical coupling, conformational coupling*** and ***chemiosmotic***—are discussed below.

Chemical coupling hypothesis

According to chemical coupling hypothesis, during the course of electron transfer in respiratory chain, a series of ***phosphorylated high-energy intermediates*** are first produced which are utilized for the synthesis of ATP. These reactions are believed to be analogous to the substrate level phosphorylation that occurs in glycolysis or citric acid cycle. However, this hypothesis lacks experimental evidence, since all attempts, so far, to isolate any one of the high-energy intermediates have not been successful.

Chemiosmotic hypothesis

This mechanism, originally proposed by Peter Mitchell (1961), is now ***widely accepted***. It explains how the transport of electrons through the respiratory chain is effectively utilized to produce ATP from ADP + Pi.

Proton gradient : The inner mitochondrial membrane, as such, is impermeable to protons (H^+) and hydroxyl ions (OH^-). The transport of electrons through ETC is coupled with the ***translocation of protons (H^+)*** across the inner mitochondrial membrane (coupling membrane) from the matrix to the intermembrane space (***Fig.11.6***). The pumping of protons results in an ***electrochemical or proton gradient.*** This is due to the accumulation of more H^+ ions (low pH) on the outer side of the inner mitochondrial membrane than the inner side. The ***proton gradient*** developed due to the electron flow in the respiratory chain is ***sufficient to result in the synthesis of ATP from ADP and Pi*** (***Fig.11.7***).

Rotary motor model for ATP generation

Paul Boyer in 1964 proposed (Nobel Prize, 1997) that a conformational change in the mitochondrial membrane proteins leads to the synthesis of ATP.

Medical Concepts/ Clinical Correlates

- Cyanide is a popular suicidal agent that causes death by blocking ETC. Recently, hydroxycobalamin (binds with cyanide to form harmless cynocobalamin) is used as an antidote for cyanide poisoning.
- Uncoupling of respiration from oxidative phosphorylation under natural conditions assumes biological significance. The brown adipose tissue, rich in electron carriers, brings about oxidation uncoupled from phosphorylation. The presence of active brown adipose tissue in some individuals is believed to protect them from becoming obese. This is because the excess calories consumed by these people are burnt and liberated as heat instead of being stored as fat.

This hypothesis, considered as rotary motor/***engine driving model*** or ***binding change model***, is widely accepted for the generation of ATP.

The enzyme ATP synthase is F_0F_1, complex (of complex V) which is composed of many subunits. In response to the proton flux, conformational changes occur in the ATP synthase complex that finally lead to the generation of ATP.

Inhibitors of electron transport chain

Many site-specific inhibitors of ETC have contributed to the present knowledge of mitochondrial respiration. Selected examples of these inhibitors are given in ***Fig.11.5***. The inhibitors bind to one of the components of ETC and block the transport of electrons.

The synthesis of ATP (phosphorylation) is dependent on electron transport. Hence, all the site-specific inhibitors of ETC also inhibit ATP formation. Three possible sites of action for the inhibitors of ETC are identified

1. **NADH and coenzyme Q :** Fish poison ***rotenone***, barbituate drug ***amytal*** and antibiotic ***piercidin A*** inhibit at this site.
2. **Between cytochrome b and c_1 :** ***Antimycin A*** —an antibiotic, ***British antilewisite (BAL)***—an antidote used against war gas—are the two important inhibitors of the site between cytochrome b and c_1.
3. **Inhibitors of cytochrome oxidase :** Carbon monoxide, cyanide, hydrogen sulphide and azide effectively inhibit cytochrome oxidase. ***Carbon monoxide*** reacts with reduced form of the cytochrome while ***cyanide*** and ***azide*** react with oxidized form.

Cyanide poisoning

Cyanide is probably the most potent inhibitor of ETC. It binds to Fe^{3+} of cytochrome oxidase blocking mitochondrial respiration leading to cell death. Cyanide poisoning causes death due to tissue asphyxia (mostly of central nervous system). Cyanide is a popular suicidal agent commonly used by militants, terrorists and warriors.

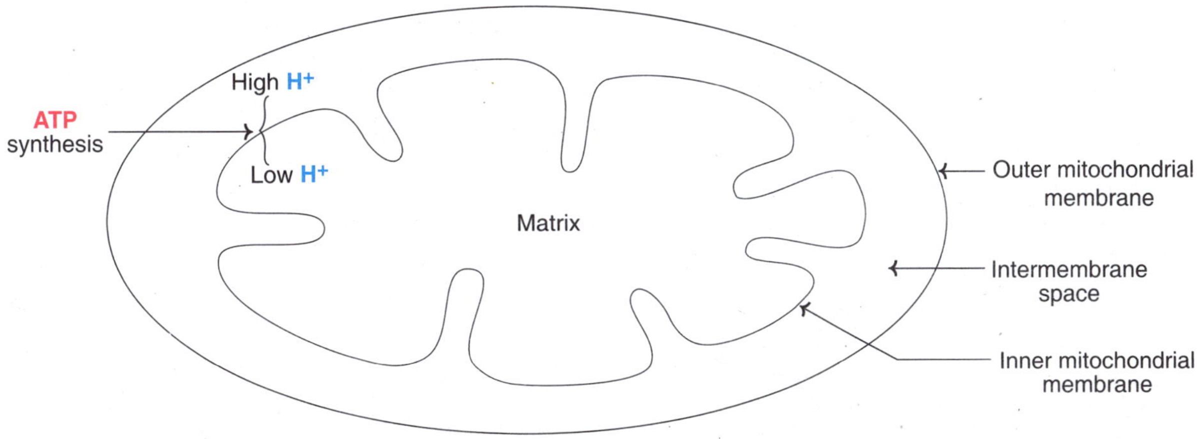

Fig.11.6 : *Outline of chemiosmotic hypothesis for oxidative phosphorylation.*

Cyanide containing compounds include ***hydrogen cyanide*** gas and certain cyanide salts e.g. ***potassium cyanide***. If cyanide is inhaled, it causes coma, seizures and cardiac arrest, and finally death in a few seconds. At lower doses, cyanide causes headache, weakness, giddiness, confusion and difficulty in breathing.

For the treatment of cyanide poisoning, specially designed ***antidote kits*** are available. The kits usually contain ***amyl nitrate*** and ***sodium nitrite***.

The nitrites convert hemoglobin into methemoglobin. Cyanide binds to methemoglobin forming cyano-methemoglobin. This results in the release of cyanide from cytochrome oxidase.

In recent years, ***hydroxycobalamin*** (a form of ***vitamin B_{12}***) is ***used as an antidote*** for cyanide poisoning. The reaction involves the ***conversion of cyanide to a harmless cyanocobalamin***.

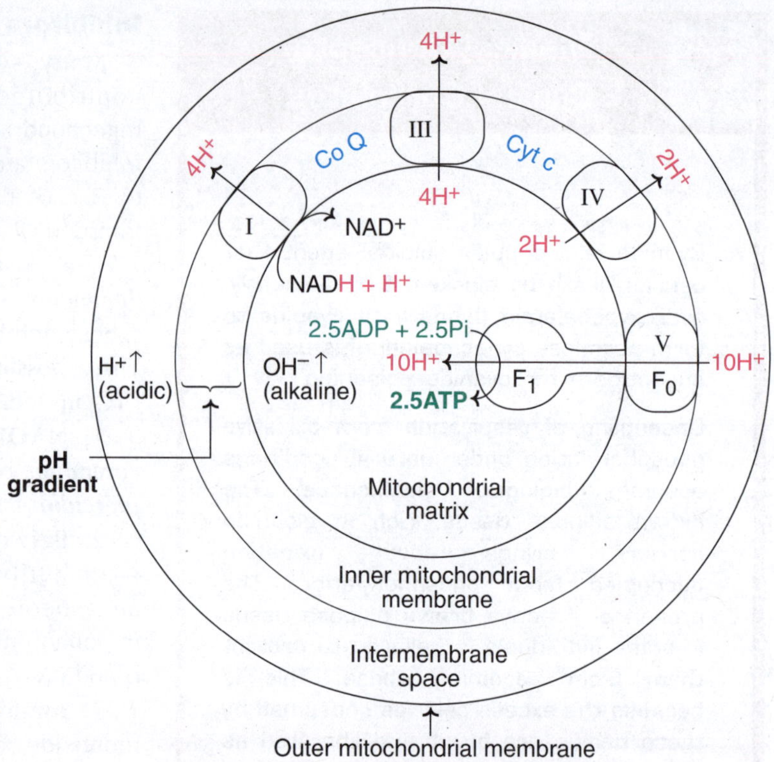

Fig.11.7 : *Diagrammatic representation of chemiosmotic hypothesis for oxidative phosphorylation (I, III, IV and V–Respiratory chain complexes; F_0 F_1–Protein subunits for phosphorylation).*

Inhibitors of oxidative phosphorylation

Uncouplers : The mitochondiral transport of electrons is tightly coupled with oxidative phosporylation (ATP synthesis). There are certain ***compounds that can uncouple*** (or delink) the ***electron transport from oxidative phosporylation***. Such compounds, known as uncouplers, increase the permeability of inner mitochondrial membrane to protons (H^+). Thus, an uncoupler allows the electron transport but blocks the establishment of proton gradient across the inner mitochondrial membrane. The result is that ATP synthesis does not occur. The energy linked with the transport of electrons is dissipated as heat. ***The uncouplers allow*** (often at accelerated rate) ***oxidation of substrates*** (via NADH or $FADH_2$) ***without ATP formation.***

The uncoupler, ***2,4-dinitrophenol (DNP)***, has been extensively studied. It is a small lipophilic molecule. DNP is a proton-carrier and can easily diffuse through the inner mitochondrial membrane.

Physiological uncouplers : Certain physiological substances which act as uncouplers at higher concentration have been identified. These include the hormone ***thyroxine*** and ***long chain free fatty acids***.

Significance of uncoupling

Uncoupling of respiration from oxidative phosphorylation under natural conditions assumes biological significance. The ***maintenance of body temperature*** is particularly important in hairless animals, hibernating animals and the animals adapted to cold. These animals possess a specialized tissue called ***brown adipose tissue*** in the upper back and neck portions. The mitochondria of brown adipose tissue are rich in electron carriers and are specialized to carry out an ***oxidation uncoupled from phosphorylation***. This causes liberation of heat when fat is oxidized in the brown adipose tissue. Brown adipose tissue may be considered as a site of ***non-shivering thermogenesis***. The presence of active brown adipose tissue in certain individuals is believed to protect them from becoming obese. The excess calories consumed by these people are burnt and ***liberated as heat***, instead of being stored as fat.

Thermogenin (or ***uncoupling protein***, UCP I) is a physiological uncoupler, located in the inner mitochondrial membrane of brown adipose tissue. It ***blocks the formation of ATP, and generates heat***. This assumes significant in the newborn, and during hibernation in animals.

Other inhibitors of oxidative phosphorylation

Oligomycin : This antibiotic prevents the mitochondrial oxidation as well as phosphorylation. It ***binds with the enzyme ATP synthetase*** and blocks the proton (H^+) channels.

Atractyloside : This is a plant toxin and inhibits oxidative phosphorylation by an indirect mechanism. Atractyloside inhibits adenine nucleotide carrier and, thus, ***blocks the adequate supply of ADP,*** thereby preventing phosphorylation.

Transport of reducing equivalents—shuttle pathways

The inner mitochondrial membrane is impermeable to NADH. Therefore, the NADH produced in the cytosol cannot directly enter the mitochondria. Two pathways—namely ***glycerol-phosphate shuttle*** and ***malate-aspartate shuttle***—are operative to do this job.

If glycerol-phosphate shuttle operates, the cytosolic NADH, through the mediation of mitochondrial $FADH_2$, enters ETC and generates 1.5ATP. In contrast, when malate-aspartate shuttle operates, 2.5ATP are produced from cytosolic NADH.

SUMMARY

1. *ATP is the energy currency of the cell. ATP-ADP cycle acts as a connecting energy link between catabolic and anabolic reactions.*
2. *Respiratory chain or electron transport chain (ETC) located in the inner mitochondrial membrane represents the final stage of oxidizing the reducing equivalents (NADH and $FADH_2$) derived from the metabolic intermediates to water.*
3. *ETC is organized into five distinct complexes. The complexes I to IV are electron carriers while complex V is responsible for ATP production. The components of ETC are arranged in the sequence*

$$NAD^+ \longrightarrow FMN \longrightarrow CoQ \longrightarrow Cyt\ b \longrightarrow Cyt\ c_1 \longrightarrow Cyt\ c \longrightarrow Cyt\ a + a_3 \longrightarrow O_2$$

4. *The process of synthesizing ATP from ADP and Pi coupled with ETC is known as oxidative phosphorylation. NADH oxidation with a P : O ratio 2.5 indicates that 2.5 ATP are synthesized while $FADH_2$ oxidation (P : O ratio 1.5) results in the production of 1.5 ATP.*
5. *Among the hypotheses put forth to explain the mechanism of oxidative phosphorylation, the chemiosmotic hypothesis (of Mitchell) is widely accepted. The flow of electrons through the ETC causes proton (H^+) gradient across the inner mitochondrial membrane which leads to the synthesis of ATP from ADP and Pi, catalysed by ATP synthetase.*
6. *NADH produced in the cytosol cannot directly enter mitochondria. Glycerol-phosphate shuttle (generates 1.5 ATP) and malate-asparate shuttle (generates 2.5 ATP) operate to overcome the difficulty.*

THE READER MUST REFER CHAPTER 34 FOR DETAILS ON FREE RADICALS AND ANTIOXIDANTS.

INTRODUCTION TO METABOLISM

12

Hundreds of reactions simultaneously take place in a living cell, in a well-organized and integrated manner. The entire spectrum of ***chemical reactions, occurring in the living system,*** are collectively referred to as ***metabolism.***

A ***metabolic pathway*** (or metabolic map) constitutes a series of enzymatic reactions to produce specific products. The term ***metabolite*** is applied to a substrate or an intermediate or a product in the metabolic reactions.

Metabolism is broadly divided into two categories (***Fig.12.1***).

1. **Catabolism :** The degradative processes concerned with the breakdown of complex molecules to simpler ones, with a concomitant release of energy.

2. **Anabolism :** The biosynthetic reactions involving the formation of complex molecules from simple precursors.

A clear demarcation between catabolism and anabolism is rather difficult, since there are several intermediates common to both the processes. The term ***amphibolism*** is used for reactions which are ***both catabolic and anabolic in nature***.

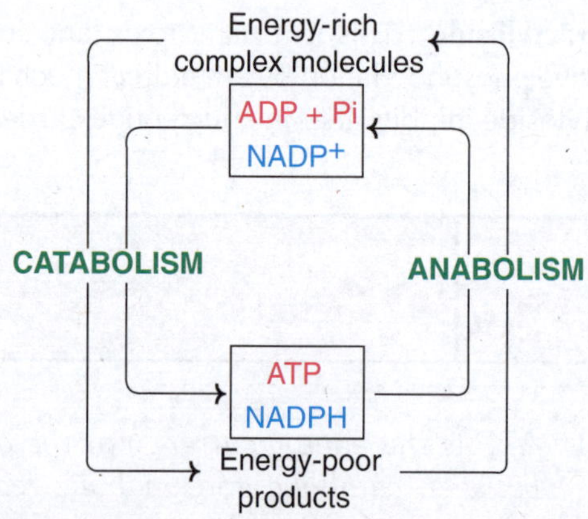

Fig.12.1 : *An outline of catabolism and anabolism.*

Catabolism

The very purpose of catabolism is to trap the energy of the biomolecules in the form of ATP and to generate the substances (precursors) required for the synthesis of complex molecules. Catabolism occurs in three stages (***Fig.12.2***).

1. **Conversion of complex molecules into their building blocks :** Polysaccharides are broken down to monosaccharides, lipids to free fatty acids and glycerol, proteins to amino acids.

2. **Formation of simple intermediates :** The building blocks produced in stage (1) are degraded to simple intermediates such as ***pyruvate and acetyl CoA***. These intermediates are not readily identifiable as carbohydrates, lipids or proteins. A small quantity of energy (as ATP) is captured in stage 2.

3. **Final oxidation of acetyl CoA :** Acetyl CoA is completely oxidized to CO_2, liberating NADH and $FADH_2$ that finally get oxidized ***to release*** large quantity of energy (as ***ATP***). ***Krebs cycle*** (or citric acid cycle) is the common metabolic pathway involved in the final oxidation of all energy-rich molecules. This pathway accepts the carbon compounds (pyruvate, succinate etc.) derived from carbohydrates, lipids or proteins.

Anabolism

For the synthesis of a large variety of complex molecules, the starting materials are relatively few. These include pyruvate, acetyl CoA and the

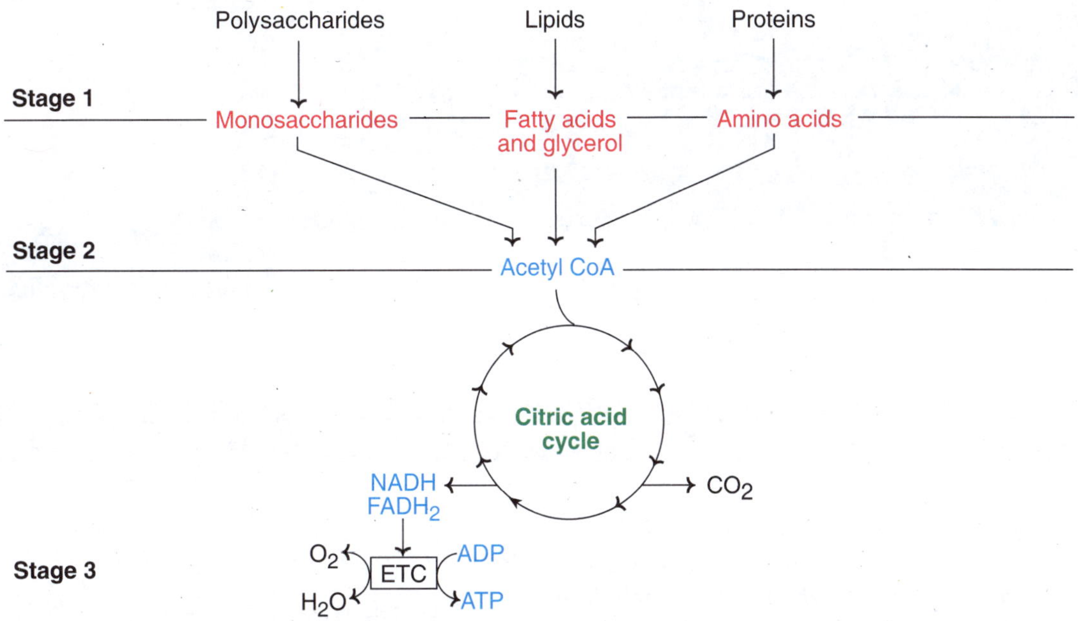

Fig.12.2 : *The three stages of catabolism (ETC–Electron transport chain).*

intermediates of citric acid cycle. Besides the availability of precursors, the anabolic reactions are dependent on the ***supply of energy*** (as ATP or GTP) and reducing equivalents (as NADPH + H^+).

The anabolic and catabolic pathways are not reversible and operate independently. As such, the metabolic pathways occur in specific cellular locations (mitochondria, microsomes etc.) and are controlled by different regulatory signals.

The terms—intermediary metabolism and energy metabolism—are also in use. ***Intermediary metabolism*** refers to the entire range of catabolic and anabolic reactions, not involving nucleic acids. ***Energy metabolism*** deals with the metabolic pathways concerned with the storage and liberation of energy.

Types of metabolic reactions

The biochemical reactions are mainly of four types

1. Oxidation-reduction.
2. Group transfer.
3. Rearrangement and isomerization.
4. Make and break of carbon-carbon bonds.

These reactions are catalysed by specific enzymes—more than 2,000 known so far.

Methods employed to study metabolism

The metabolic reactions do not occur in isolation. They are interdependent and integrated into specific series that constitute ***metabolic pathways.*** It is, therefore, not an easy task to study metabolisms. Fortunately, the ***basic metabolic pathways in most organisms are essentially identical.***

Several methods are employed to elucidate biochemical reactions and the metabolic pathways. These experimental approaches may be broadly divided into 3 categories

1. Use of whole organisms or its components.
2. Utility of metabolic probes.
3. Application of isotopes.

The actual methods employed may be either ***in vivo*** (in the living system) or ***in vitro*** (in the test tube) or, more frequently, both.

1. **Use of whole organism or its components :**

(a) **Whole organisms :** Glucose tolerance test (GTT).

(b) **Isolated organs, tissue slices, whole cells,** subcellular organelles etc., to elucidate biochemical reactions and metabolic pathways.

2. **Utility of metabolic probes :** Two types of metabolic probes are commonly used to trace out biochemical pathways. These are metabolic ***inhibitors*** and ***mutations.***

3. **Application of isotopes** (***Refer Chapter 40***).

METABOLISM OF CARBOHYDRATES

13

THE READER MUST REFER CHAPTER 36 FOR BLOOD GLUCOSE REGULATION, DIABETES MELLITUS ETC.

Carbohydrates are the ***major source of energy*** for the living cells. The monosaccharide ***glucose is the central molecule in carbohydrate metabolism*** since all the major pathways of carbohydrate metabolism are connected with it (*Fig.13.1*). Glucose is utilized as a source of energy, it is synthesized from non-carbohydrate precursors and stored as glycogen to release glucose as and when the need arises. The other monosaccharides important in carbohydrate metabolism are fructose, galactose and mannose.

The ***fasting blood glucose*** level in normal individuals is ***70-100 mg/dl*** (4.5-5.5 mmol/l) and it is very efficiently maintained at this level (***for details on the homeostasis of glucose, refer Chapter 36***). Liver plays a key role in monitoring and stabilizing blood glucose levels. Thus, ***liver*** may be appropriately considered as ***glucostat monitor***.

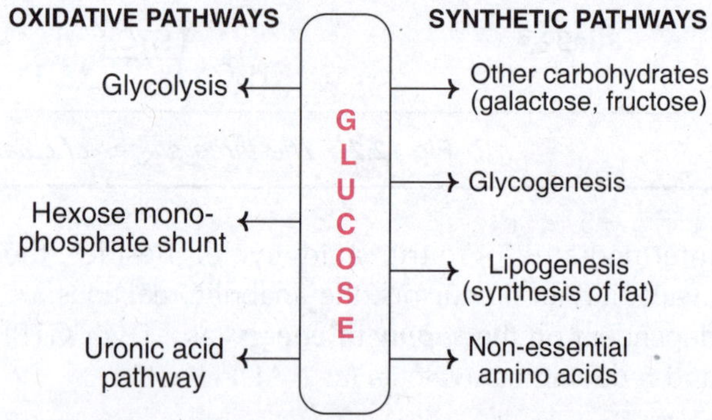

Fig.13.1 : *Overview of glucose metabolism. (**Note :** For majority of the pathways, glucose participates as glucose 6-phosphate).*

Entry of glucose into cells

Glucose concentration is very low in the cells compared to blood (for humans < 100 mg/dl). However, glucose does not enter the cells by simple diffusion. Two specific transport systems are recognized for the entry of glucose into the cells

1. **Insulin-independent transport system of glucose :** This is a carrier mediated uptake of glucose which is not dependent on the hormone insulin. This is ***operative in hepatocytes, erythrocytes and brain***.

2. **Insulin-dependent transport system :** This occurs in ***muscle and adipose tissue***.

GLYCOLYSIS

Glycolysis is derived from the *Greek* words (glycose—sweet or sugar; lysis—dissolution). It is a universal pathway in the living cells.

Glycolysis is defined as the sequence of reactions converting glucose (or glycogen) to pyruvate or lactate, with the production of ATP.

Salient features

1. Glycolysis (also known as ***Embden-Meyerhof-Parnas pathway*** (EMP pathway) takes place in all cells of the body. The ***enzymes*** of this pathway are present in the ***cytosomal fraction*** of the cell.

2. Glycolysis occurs in the absence of oxygen (anaerobic) or in the presence of oxygen (***aerobic***). ***Lactate*** is the end product under ***anaerobic condition***.

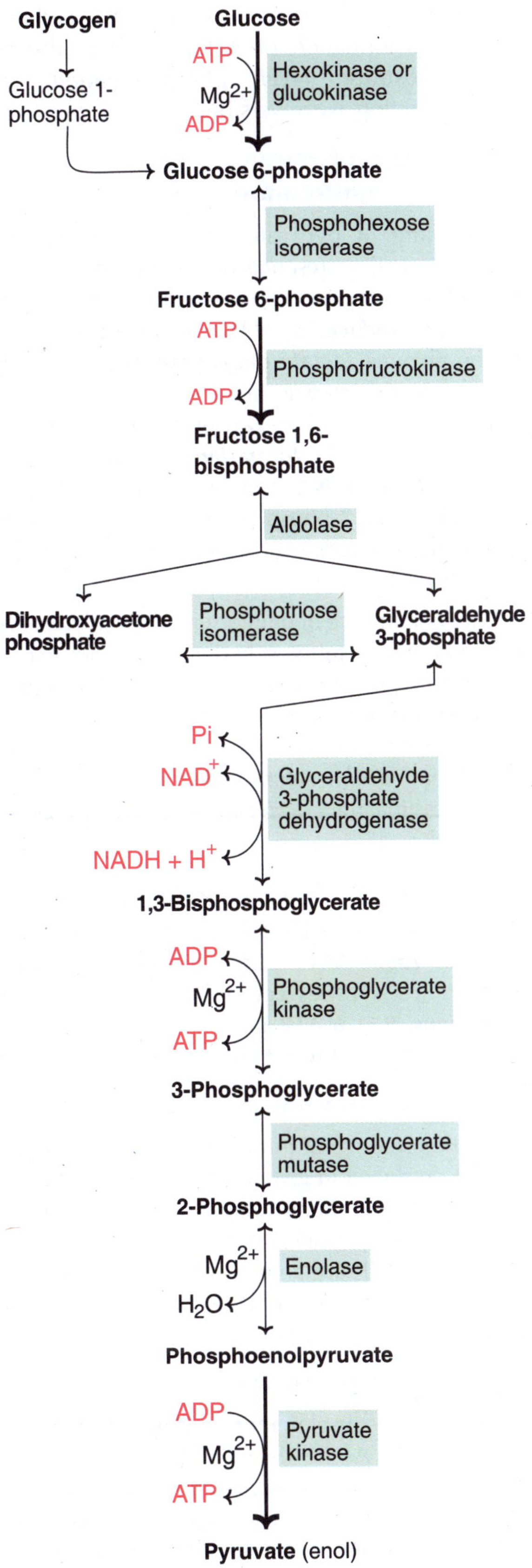

Fig. contd. next column

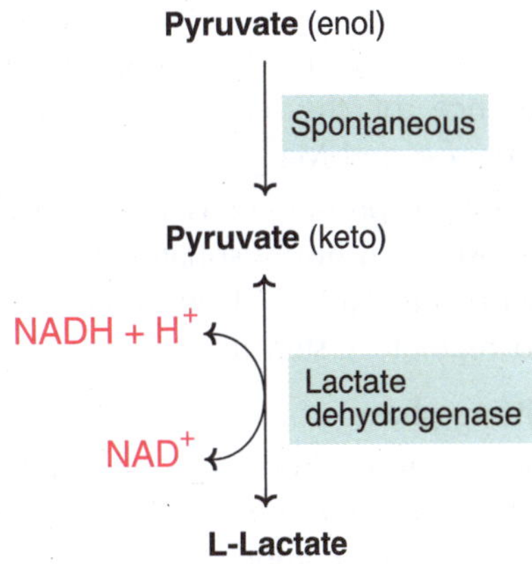

***Fig.13.2 :** The reactions in the pathway of glycolysis (The three steps, shown by thick lines, catalysed by hexokinase, phosphofructokinase and pyruvate kinase are irreversible).*

In the ***aerobic*** condition, ***pyruvate*** is formed, which is then oxidized to CO_2 and H_2O.

3. Glycolysis is a major pathway for ***ATP synthesis*** in tissues lacking mitochondria, e.g. ***erythrocytes, cornea, lens*** etc.

4. Glycolysis is very ***essential for brain*** which is dependent on glucose for energy. The glucose in brain has to undergo glycolysis before it is oxidized to CO_2 and H_2O.

5. Glycolysis (anaerobic) may be summarized by the net reaction

$$\text{Glucose} + 2\text{ADP} + 2\text{Pi} \longrightarrow 2\text{Lactate} + 2\text{ATP}$$

6. Reversal of glycolysis along with the alternate arrangements made at the irreversible steps will result in the synthesis of glucose (gluconeogenesis).

Reactions of glycolysis

The sequence of reactions of glycolysis are given in ***Fig.13.2***. The pathway can be divided into three distinct phases

A. Energy investment phase or priming stage

B. Splitting phase

C. Energy generation phase.

The sequence of reactions are discussed below.

A. **Energy investment phase**

1. Glucose is phosphorylated to glucose 6-phosphate by hexokinase or glucokinase (both are ***isoenzymes***). This is an irreversible reaction, dependent on ATP and Mg^{2+}.

2. Glucose 6-phosphate undergoes isomerization to give fructose 6-phosphate in the presence of the enzyme phosphohexose isomerase and Mg^{2+}.
3. Fructose 6-phosphate is phosphorylated to fructose 1,6-bisphosphate by phosphofructokinase (PFK). This is an irreversible and regulatory step in glycolysis.

B. **Splitting phase**

4. The six carbon fructose 1,6-bisphosphate is split (hence the name glycolysis) to two three-carbon compounds, glyceraldehyde 3-phosphate and dihydroxyacetone phosphate by the enzyme aldolase.
5. The enzyme phosphotriose isomerase catalyses the reversible interconversion of glyceraldehyde 3-phosphate and dihydroxyacetone phosphate. Thus, two molecules of glyceraldehyde 3-phosphate are obtained from one molecule of glucose.

C. **Energy generation phase**

6. Glyceraldehyde 3-phosphate dehydrogenase converts glyceraldehyde 3-phosphate to 1,3-bisphosphoglycerate. This step is involved in the formation of NADH + H^+ and a high energy compound 1,3-bisphosphoglycerate.
7. The enzyme phosphoglycerate kinase acts on 1,3-bisphosphoglycerate resulting in the synthesis of ATP and formation of 3-phosphoglycerate. This step is a good example of ***substrate level phosphorylation***, since ATP is synthesized from the substrate without the involvement of electron transport chain. ***Phosphoglycerate kinase reaction is reversible, a rare example among the kinase reactions.***
8. 3-Phosphoglycerate is converted to 2-phosphoglycerate by phosphoglycerate mutase.
9. The high energy compound phosphoenol pyruvate is generated from 2-phosphoglycerate by the enzyme ***enolase***. This enzyme requires Mg^{2+} or Mn^{2+} and is inhibited by ***fluoride.*** For ***blood glucose estimation*** in the laboratory, fluoride is added to the blood to ***prevent glycolysis*** by the cells, so that blood glucose is correctly estimated.
10. The enzyme pyruvate kinase catalyses the transfer of high energy phosphate from phosphoenol pyruvate to ADP, leading to the formation of ATP. This step also is a ***substrate level phosphorylation.*** This reaction is irreversible.

Conversion of pyruvate to lactate—significance

The fate of pyruvate produced in glycolysis depends on the presence or absence of oxygen in the cells. Under anaerobic conditions (lack of O_2), ***pyruvate is reduced by NADH to lactate*** in presence of the enzyme ***lactate dehydrogenase***. The NADH utilized in this step is obtained from the reaction catalysed by glyceraldehyde 3-phosphate dehydrogenase. The ***formation of lactate allows the regeneration of NAD***$^+$ which can be reused by glyceraldehyde 3-phosphate dehydrogenase so that glycolysis proceeds even in the absence of oxygen to supply ATP.

The occurrence of uninterrupted glycolysis is very essential in skeletal muscle during strenous exercise where oxygen supply is very limited. ***Glycolysis in the erythrocytes leads to lactate production***, since mitochondria—the centres for aerobic oxidation—are absent. ***Brain***, ***retina***, ***skin***, renal medulla and gastrointestinal tract ***derive most of their energy from glycolysis***.

Production of ATP in glycolysis

The details of ATP generation in glycolysis are given in ***Table 13.1***. Under ***anaerobic conditions, 2 ATP*** are synthesized while, under ***aerobic conditions, 8*** or ***6 ATP*** are synthesized—depending on the shuttle pathway that operates (***Chapter 11***).

Glycolysis and shuttle pathways

In the presence of mitochondria and oxygen, the NADH produced in glycolysis can participate in the shuttle pathways (***Refer Chapter 11***) for the synthesis of ATP. If the cytosolic NADH uses malate-aspartate shuttle, 2.5 ATP are generated from each molecule of NADH. This is in contrast to glycerolphosphate shuttle that produces only 1.5 ATP.

Cancer and glycolysis

Cancer cells display ***increased uptake of glucose, and glycolysis***. As the tumors grow rapidly, the blood vessels are unable to supply adequate oxygen, and thus a condition of hypoxia exists. Due to this, anaerobic glycolysis predominantly occurs to supply energy. The cancer cells get adapted to hypoxic glycolysis through the involvement of a transcription factor namely ***hypoxia-inducible transcription factor (HIF)***. HIF increases the synthesis of glycolytic enzymes and the glucose transporters. However, the cancer cells cannot grow and survive without proper vascularization. One of the modalities of ***cancer treatment is to use drugs that can inhibit vascularization of tumors***.

Irreversible steps in glycolysis

Most of the reactions of glycolysis are reversible. However, the three steps catalysed by the enzymes ***hexokinase*** (or glucokinase), ***phosphofructokinase*** and ***pyruvate kinase,*** are irreversible.

Regulation of glycolysis

The three enzymes namely hexokinase (and glucokinase), phosphofructokinase and pyruvate kinase, catalysing the ***irreversible reactions regulate glycolysis***.

Hexokinase is inhibited by glucose 6-phosphate. This enzyme prevents the accumulation of glucose 6-phosphate due to product inhibition. Glucokinase, which specifically phosphorylates glucose, is an ***inducible enzyme***. The substrate glucose, probably through the involvement of insulin, induces glucokinase.

Phosphofructokinase (PFK) is the most important regulatory enzyme in glycolysis. This enzyme catalyses the ***rate limiting committed step.*** PFK is an allosteric enzyme regulated by allosteric effectors. ATP, citrate and H^+ ions (low pH) are the most important allosteric inhibitors, whereas, fructose 2,6-bisphosphate, AMP and Pi are the allosteric activators.

Pyruvate kinase also regulates glycolysis. This enzyme is inhibited by ATP and activated by fructose 1,6-bisphosphate.

Pasteur effect

The ***inhibition of glycolysis by oxygen*** (aerobic condition) is known as Pasteur effect. This effect was discovered by Louis Pasteur, more than a century ago, while studying fermentation by yeast.

In the aerobic condition, the levels of glycolytic intermediates from fructose 1,6-bisphosphate onwards decrease while the earlier intermediates accumulate. This clearly indicates that Pasteur effect is due to the inhibition of the enzyme phosphofructokinase. The inhibitory effect of citrate and ATP (produced in the presence of oxygen) on phosphofructokinase explains the Pasteur effect.

TABLE 13.1 Generation of ATP in glucose metabolism

Pathway	*Enzyme (method of ATP synthesis)*	*Number of ATP synthesized*
Glycolysis	Glyceraldehyde 3-phosphate dehydrogenase (2 NADH, ETC, oxidative phosphorylation)	5
	Phosphoglycerate kinase (substrate level phosphorylation)	2
	Pyruvate kinase (substrate level phosphorylation)	2
	Two ATP are consumed in the reactions catalysed by hexokinase and phosphofructokinase	–2
	Net ATP synthesis in glycolysis in aerobic condition	***7***
	Pyruvate dehydrogenase (2 NADH, ETC, oxidative phosphorylation)	5
Citric acid cycle	Isocitrate dehydrogenase (2 NADH, ETC, oxidative phosphorylation)	5
	α-Ketoglutarate dehydrogenase (2 NADH, ETC, oxidative phosphorylation)	5
	Succinate thiokinase (substrate level phosphorylation)	2
	Succinate dehydrogenase (2 $FADH_2$, ETC, oxidative phosphorylation)	3
	Malate dehydrogenase (2 NADH, ETC, oxidative phosphorylation)	5
	Total ATP per mole of glucose under aerobic condition	***32***
	Total ATP per mole of glucose under anaerobic condition	2

Note : *Values represent ATP synthesized as per the new P:O ratios of 2.5 for NADH and 1.5 for $FADH_2$.*

Medical Concepts/ Clinical Correlates

- Glycolysis is an important source of energy supply for brain, retina, skin and renal medulla. The significance of glycolysis is its ability to generate ATP in the absence of oxygen.
- Skeletal muscle, during strenous exercise, requires the occurrence of uninterrupted glycolysis. This is due to the limited supply of oxygen.
- The occurrence of glycolysis is very much elevated in rapidly growing cancer cells.
- Glycolysis in erythrocytes is associated with 2,3-bisphosphoglycerate (2,3-BPG) production. In the presence of 2,3-BPG, oxyhemoglobin unloads more oxygen to the tissues.

Glycolysis and dental caries

Dental caries refers to the ***destruction or decalcification of hard teeth due to organic acids*** released by bacterial infections. The anaerobic bacteria (e.g. ***Streptococcus mutans, Lactobacillus*** sps) that colonize the oral cavity contribute to the development of dental caries. These bacteria grow optimally on refined and fermentable sugars (e.g. sucrose of chocolates, candies) by utilizing anaerobic glycolysis. ***Lactic acid and other acids*** (produced by bacteria) ***erode tooth enamel and dentin***, and dissolve hydroxyapatite matrix of teeth that ***results in cavity formation***. Low levels of ***fluoride***, from tooth pastes or when applied topically can inhibit the enzyme enolase and ***reduce glycolysis and thus tooth decay***. Further, fluoride integrates into tooth surface to form fluoroapatite which offers resistance to demineralization.

RAPOPORT-LEUBERING CYCLE

This is a supplementary pathway to glycolysis which is operative in the erythrocytes of man and other mammals. Rapoport-Leubering cycle is mainly concerned with the synthesis of ***2,3-bisphospho-***

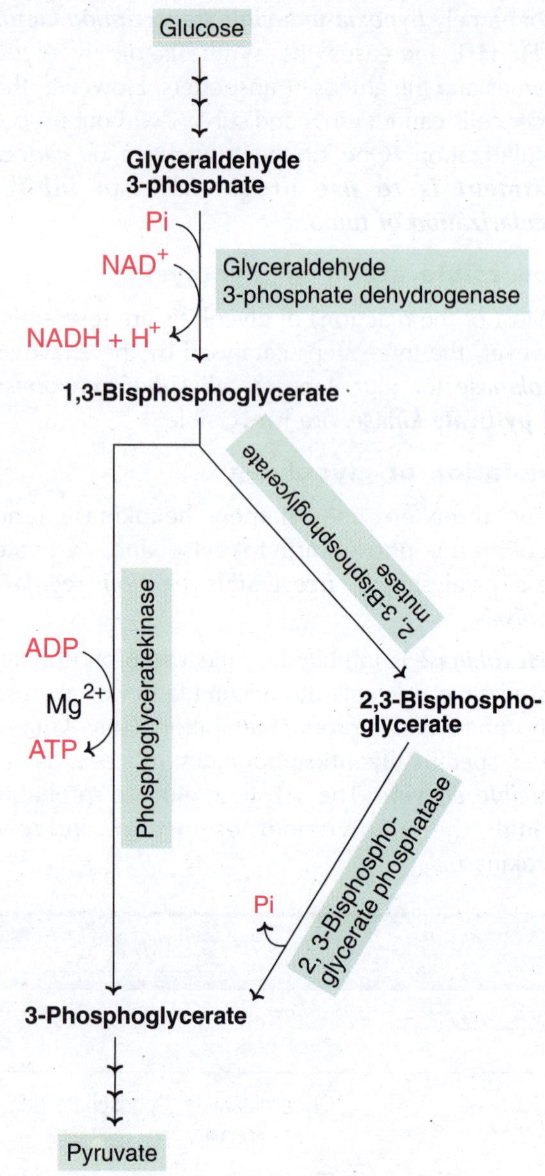

Fig.13.3 : *Rapoport-Leubering cycle for the synthesis of 2,3-bisphosphoglycerate (2,3-BPG).*

glycerate (2,3-BPG) in the RBC. 1,3-Bisphosphoglycerate (1,3-BPG) produced in glycolysis is converted to 2,3-BPG by the enzyme 2,3-bisphosphoglycerate mutase (***Fig.13.3***). 2,3-BPG is hydrolysed to 3-phosphoglycerate by bisphosphoglycerate phosphatase.

About 15-25% of the glucose converted to lactate in erythrocytes goes via 2,3-BPG synthesis.

Significance of 2,3-BPG

1. Production of 2,3-BPG allows the glycolysis to proceed without the synthesis of ATP. This is advantageous to erythrocytes since glycolysis occurs

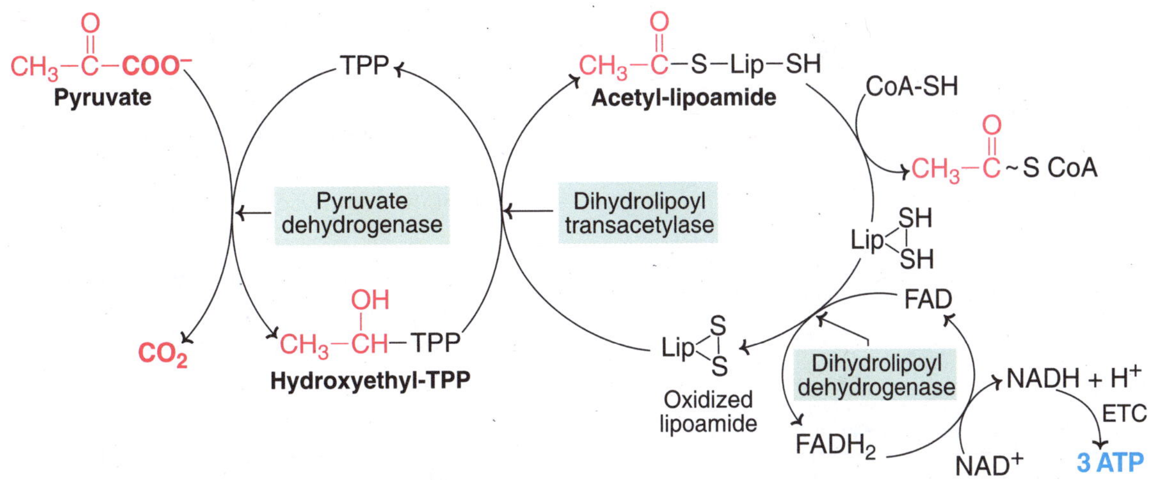

Fig.13.4 : *The mechanism of action of pyruvate dehydrogenase complex. (**Note :** The reaction involving the conversion of pyruvate to acetyl CoA requires five coenzymes—TPP, lipoamide, CoASH, FAD and NAD+).*

when the need for ATP is minimal. Rapoport-Leubering cycle is, therefore, regarded as ***a shunt pathway of glycolysis*** to dissipate or waste the energy not needed by erythrocytes.

2. 2,3-BPG, however, is not a waste molecule in RBC. It combines with hemoglobin (Hb) and reduces Hb affinity with oxygen. Therefore, in the presence of 2,3-BPG, ***oxyhemoglobin unloads more oxygen to the tissues.***

Increase in erythrocytes 2,3-BPG is observed in hypoxic condition, high altitude, fetal tissues, anemic conditions etc. In all these cases, 2,3-BPG will enhance the supply of oxygen to the tissues.

For a more detailed discussion on 2,3-BPG, refer ***Chapter 10***.

CONVERSION OF PYRUVATE TO ACETYL COA

Pyruvate is converted to acetyl CoA by ***oxidative decarboxylation***. This is an irreversible reaction, catalysed by a multienzyme complex, known as ***pyruvate dehydrogenase complex*** (***PDH***), which is found only in the mitochondria. High concentrations of PDH are found in cardiac muscle and kidney. The enzyme PDH requires five cofactors (coenzymes), namely—***TPP***, ***lipoamide***, ***FAD***, ***coenzyme A*** and ***NAD+*** (lipoamide contains lipoic acid linked to ε-amino group of lysine). The overall reaction of PDH is

$$\text{Pyruvate} + \text{NAD}^+ + \text{CoA} \xrightarrow{\text{PDH}} \text{Acetyl CoA} + \text{CO}_2 + \text{NADH} + \text{H}^+$$

Reactions of PDH complex

The sequence of reactions brought about by different enzymes of PDH complex in association with the coenzymes is depicted in ***Fig.13.4***. Pyruvate is decarboxylated to give hydroxyethyl TPP, catalysed by PDH (decarboxylase activity). Dihydrolipoyl transacetylase brings about the formation of acetyl lipoamide (from hydroxethyl-TPP) and then catalyses the transfer of acetyl group to coenzyme A to produce acetyl CoA. The cycle is complete when reduced lipoamide is converted to oxidized lipoamide by dihydrolipoyl dehydrogenase, transferring the reducing equivalents to FAD. $FADH_2$, in turn, transfers the reducing equivalents to NAD^+ to give NADH + H^+, which can pass through the respiratory chain to give 2.5 ATP (5 ATP from 2 moles of pyruvate formed from glucose) by oxidative phosphorylation.

A comparable enzyme with PDH is ***α-ketoglutarate dehydrogenase complex*** of citric acid cycle which catalyses the oxidative decarboxylation of α-ketoglutarate to succinyl CoA.

CITRIC ACID CYCLE

The citric acid cycle (***Krebs cycle*** or ***tricarboxylic acid—TCA cycle***) is the most important metabolic pathway for the energy supply to the body. About 65-70% of the ATP is synthesized in Krebs cycle. ***Citric acid cycle essentially involves the oxidation***

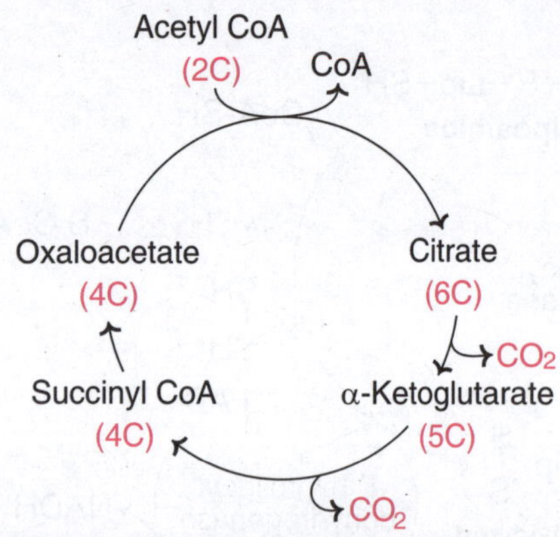

***Fig.13.5 :** An overview of Krebs cycle.*

of acetyl CoA to CO_2 and H_2O. This cycle utilizes about two-thirds of total oxygen consumed by the body. The name TCA cycle is used, since, at the outset of the cycle, tricarboxylic acids (citrate, cis-aconitate and isocitrate) participate.

TCA cycle—the central metabolic pathway

The citric acid cycle is the ***final common oxidative pathway for carbohydrates, fats and amino acids***. This cycle not only supplies energy but also provides many intermediates required for the synthesis of amino acids, glucose, heme etc. Krebs cycle is the most important central pathway ***connecting almost all the individual metabolic pathways*** (either directly or indirectly).

Location of TCA cycle

The enzymes of TCA cycle are located in ***mitochondrial matrix***, in close proximity to the electron transport chain.

TCA cycle—an overview

Krebs cycle basically involves the combination of a two carbon acetyl CoA with a four carbon oxaloacetate to produce a six carbon tricarboxylic acid, citrate. In the reactions that follow, the two carbons are oxidized to CO_2 and oxaloacetate is regenerated and recycled. ***Oxaloacetate is considered to play a catalytic role in citric acid cycle.*** An overview of Krebs cycle is depicted in ***Fig.13.5***.

Reactions of citric acid cycle

Oxidative decarboxylation of pyruvate to acetyl CoA by pyruvate dehydrogenase complex is discussed above. This step is a connecting link between glycolysis and TCA cycle. A few authors, however, describe the conversion of pyruvate to acetyl CoA along with citric acid cycle. The events of TCA cycle are described hereunder (***Fig.13.6***).

1. **Formation of citrate :** Krebs cycle proper starts with the condensation of acetyl CoA and oxaloacetate, catalysed by the enzyme citrate synthase.

2. **and** 3. **Citrate is isomerized to isocitrate** by the enzyme aconitase. This is achieved in a two stage reaction of dehydration followed by hydration through the formation of an intermediate—*cis*-aconitate.

4. **and** 5. **Formation of α-ketoglutarate :** The enzyme isocitrate dehydrogenase (ICD) catalyses the conversion (oxidative decarboxylation) of isocitrate to oxalosuccinate and then to α-ketoglutarate. The formation of NADH and the liberation of CO_2 occurs at this stage.

6. **Conversion of α-ketoglutarate to succinyl CoA** occurs through ***oxidative decarboxylation***, catalysed by α-ketoglutarate dehydrogenase complex. This enzyme is dependent on five cofactors—***TPP, lipoamide, NAD$^+$, FAD and CoA***. The mechanism of the reaction is analogous to the conversion of pyruvate to acetyl CoA (See ***Fig.13.4***).

7. **Formation of succinate :** Succinyl CoA is converted to succinate by succinate thiokinase. This reaction is coupled with the phosphorylation of GDP to GTP or ADP to ATP. This is a ***substrate level phosphorylation***.

Two ***isoenzymes*** of ***succinate thiokinase*** has been identified. One ***dependent on GDP*** is present in tissues like liver and kidney where gluconeogenesis occurs. The second succinate thiokinase isoenzyme is ***dependent on ADP*** and present in non-gluconeogenic tissues (***brain, heart***).

8. **Conversion of succinate to fumarate :** Succinate is oxidized by succinate dehydrogenase to fumarate. This reaction results in the production of $FADH_2$.

9. **Formation of malate :** The enzyme fumarase catalyses the conversion of fumarate to malate with the addition of H_2O.

10. **Conversion of malate to oxaloacetate :** Malate is then oxidized to oxaloacetate by malate dehydrogenase. The third and final synthesis of NADH occurs at this stage. The oxaloacetate is regenerated which can combine with another molecule of acetyl CoA and continue the cycle.

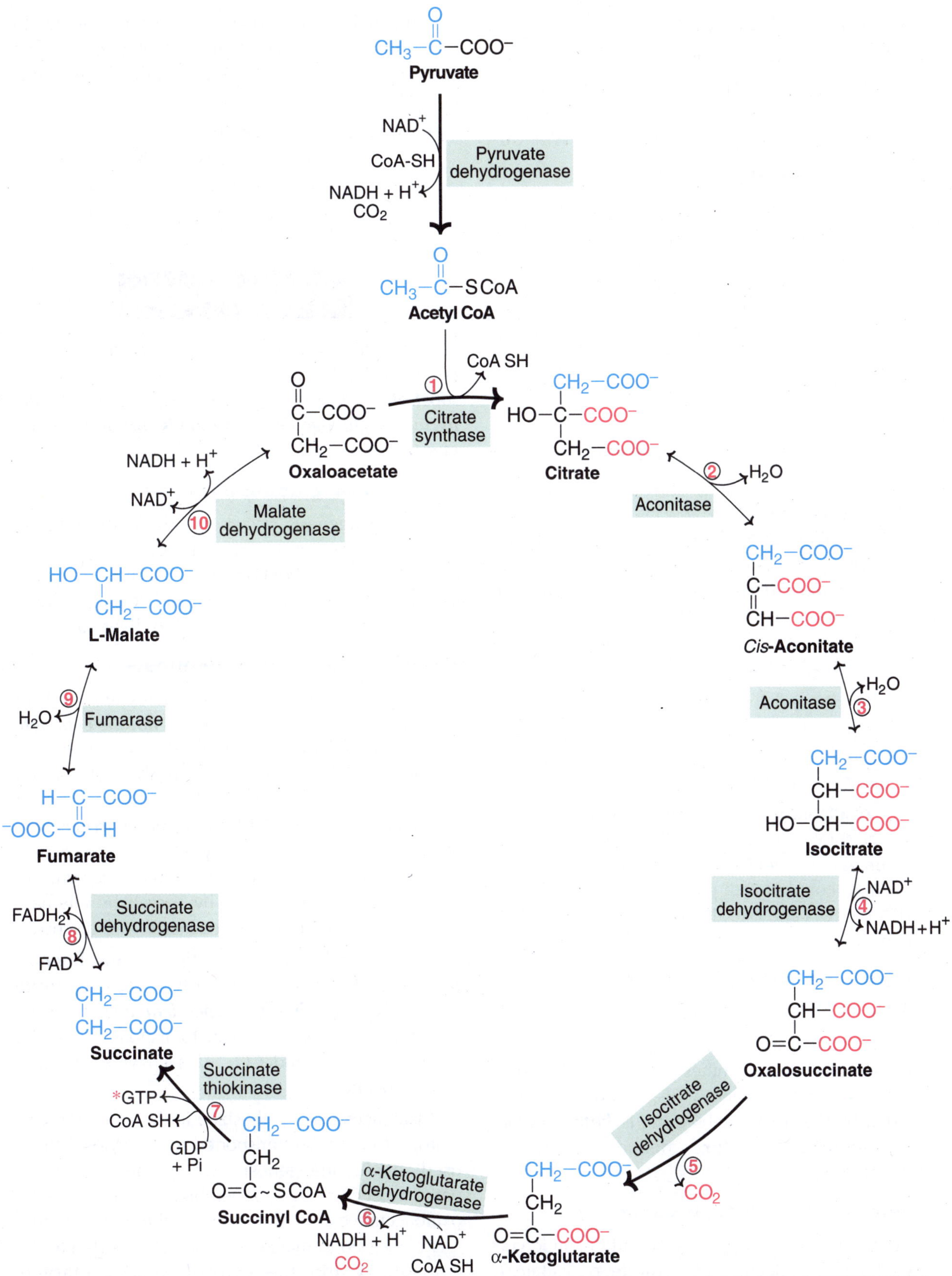

Fig.13.6 : *The citric acid (Krebs) cycle. (**Irreversible** reactions shown by **thick** arrows).*

Summary of TCA cycle

The events of Krebs cycle may be summarized as

$$\text{Acetyl CoA} + 3\ NAD^+ + FAD + GDP + Pi + 2H_2O \longrightarrow 2CO_2 + 3NADH + 3H^+ + FADH_2 + GTP + CoA$$

Energetics of citric acid cycle

During the process of oxidation of acetyl CoA via citric acid cycle, 4 reducing equivalents (3 as NADH and one as $FADH_2$) are produced. Oxidation of 1 NADH by electron transport chain coupled with oxidative phosphorylation results in the synthesis of 2.5 ATP, whereas $FADH_2$ leads to the formation of 1.5 ATP. Besides, there is one substrate level phosphorylation. A total of ***ten ATP*** are produced from one acetyl CoA.

Regulation of citric acid cycle

The cellular demands of ATP are crucial in controlling the rate of citric acid cycle. The regulation is brought about either by enzymes or the levels of ADP. Three enzymes—namely ***citrate synthase, isocitrate dehydrogenase*** and ***α-ketoglutarate dehydrogenase***—regulate citric acid cycle.

1. **Citrate synthase** is inhibited by ATP, NADH, acyl CoA and succinyl CoA.

2. **Isocitrate dehydrogenase** is activated by ADP and inhibited by ATP and NADH.

3. **α-Ketoglutarate dehydrogenase** is inhibited by succinyl CoA and NADH.

4. **Availability of ADP** is very important for the citric acid cycle to proceed. This is due to the fact that unless sufficient levels of ADP are available, oxidation (coupled with phosphorylation of ADP to ATP) of NADH and $FADH_2$ through electron transport chain stops.

Amphibolic nature of the citric acid cycle

The citric acid cycle provides various intermediates for the synthesis of many compounds (e.g. non-essential amino acids, porphyrins, heme) needed by the body. Krebs cycle is ***both catabolic and anabolic in nature***, hence regarded as ***amphibolic***.

Energetics of glucose oxidation

When a molecule of glucose (6 carbon) undergoes glycolysis, 2 molecules of pyruvate or lactate (3 carbon) are produced. Pyruvate is oxidatively decarboxylated to acetyl CoA (2 carbon) which enters the citric acid cycle and gets completely oxidized to CO_2 and H_2O. The overall process of glucose being completely oxidized to CO_2 and H_2O via glycolysis and citric acid cycle is given

$$C_6H_{12}O_6 + 6O_2 + 32ADP + 32Pi \longrightarrow 6CO_2 + 6H_2O + 32ATP$$

The enzymes of glucose metabolism responsible for generating ATP are given in ***Table 13.1***.

GLUCONEOGENESIS

The ***synthesis of glucose from non-carbohydrate compounds*** is known as gluconeogenesis. The major ***substrates***/precursors for gluconeogenesis are ***lactate***, ***pyruvate***, ***glucogenic amino acids***, ***propionate*** and ***glycerol.***

Location of gluconeogenesis

Gluconeogenesis occurs mainly in the ***cytosol***, although some precursors are produced in the mitochondria. Gluconeogenesis mostly takes place in ***liver*** (about 1kg glucose/day) and, to some extent, in ***kidney matrix*** (about one-tenth of liver capacity).

Reactions of gluconeogenesis

Gluconeogenesis closely resembles the reversed pathway of glycolysis, although it is ***not the complete reversal of glycolysis***. Essentially, 3 (out of 10) reactions of glycolysis are irreversible. The seven reactions are common for both glycolysis and gluconeogenesis (***Fig.13.7***). The ***three irreversible steps*** of glycolysis are catalysed by the enzymes, namely hexokinase, phosphofructokinase and pyruvate kinase. These three stages—by passed by alternate enzymes specific to gluconeogenesis—are discussed

1. **Conversion of pyruvate to phosphoenolpyruvate :** This takes place in two steps. ***Pyruvate carboxylase*** is a ***biotin—dependent*** mitochondrial enzyme that converts pyruvate to oxaloacetate in the presence of ATP and CO_2. This enzyme regulates gluconeogenesis.

Oxaloacetate is synthesized in the mitochondrial matrix. It has to be transported to the cytosol to be used in gluconeogenesis, where the rest of the pathway occurs. Due to membrane impermeability, oxaloacetate cannot diffuse out of the mitochondria. It is converted to malate and then transported to the cytosol. Within the cytosol, oxaloacetate is regenerated. The reversible conversion of oxaloacetate and malate is catalysed by malate

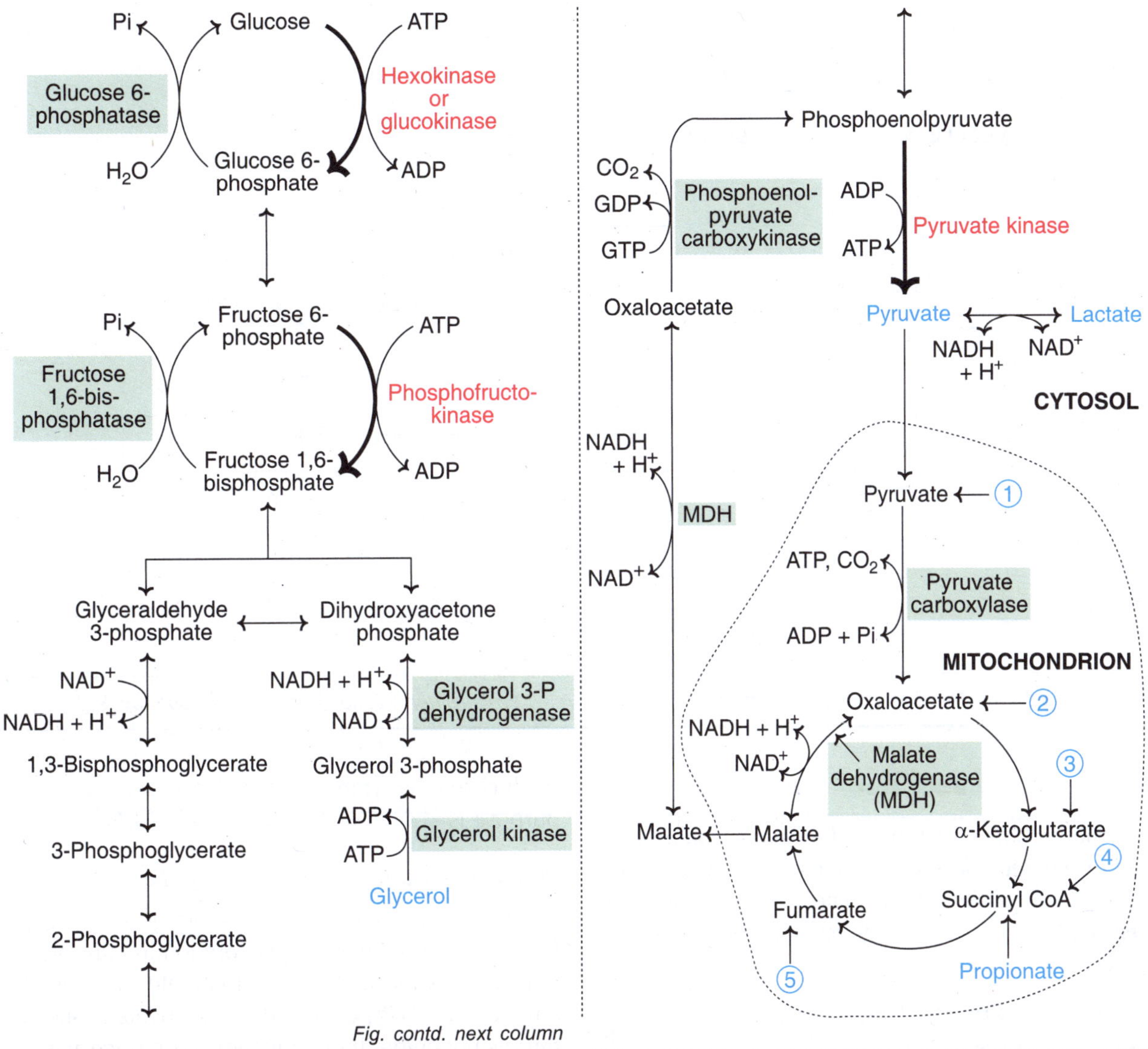

***Fig.13.7 :** The pathway of gluconeogenesis. [The enzymes catalysing irreversible steps in glycolysis are shown in red colour. The important enzymes participating in gluconeogenesis are shown in shaded green. The substrates for gluconeogenesis are in blue. The numbers represent the entry of glucogenic amino acids : (1) Alanine, glycine, serine, cysteine, threonine and tryptophan; (2) Aspartate and asparagine; (3) Arginine, glutamate, glutamine, histidine, proline; (4) Isoleucine methionine, valine; (5) Phenylalanine, tyrosine].*

dehydrogenase, an enzyme present in both mitochondria and cytosol.

In the cytosol, ***phosphoenolpyruvate carboxykinase*** converts oxaloacetate to phosphoenolpyruvate.

2. **Conversion of fructose 1,6-bisphosphate to fructose 6-phosphate :** Phosphoenolpyruvate undergoes the reversal of glycolysis until fructose 1,6-bisphosphate is produced. The enzyme ***fructose 1,6-bisphosphatase*** converts fructose 1,6-bisphosphate to fructose 6-phosphate.

3. **Conversion of glucose 6-phosphate to glucose :** ***Glucose 6-phosphatase*** catalyses the conversion of glucose 6-phosphate to glucose. The presence or absence of this enzyme in a tissue determines whether the tissue is capable of contributing glucose to the blood or not. It is mostly present in liver and kidney but ***absent in muscle***, brain and adipose tissue.

The ***overall summary of gluconeogenesis*** for the conversion of pyruvate to glucose is shown :

$$2\ \text{Pyruvate} + 4ATP + 2GTP + 2NADH + 2H^+ + 6H_2O \longrightarrow \text{Glucose} + 2NAD^+ + 4ADP + 2GDP + 6Pi + 6H^+$$

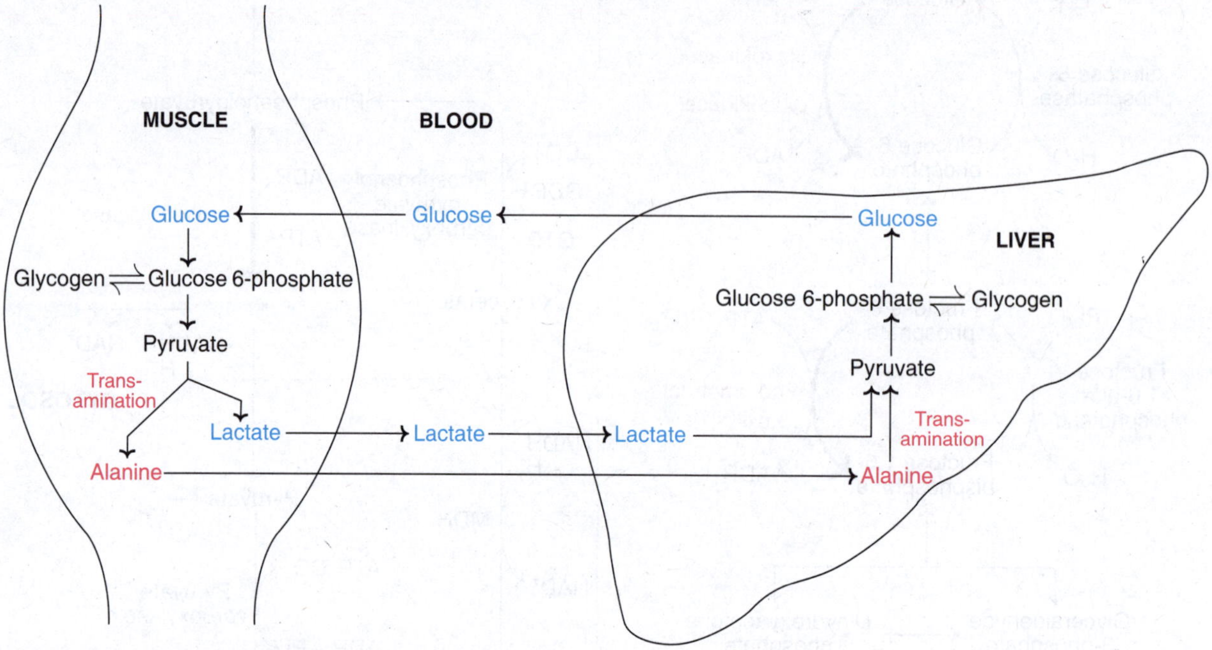

Fig.13.8 : *The Cori cycle (blue) and glucose-alanine (red) cycle (other reactions common for both cycles).*

Gluconeogenesis from amino acids

The carbon skeleton of ***glucogenic amino acids*** (all ***except leucine and lysine***) results in the formation of pyruvate or the intermediates of citric acid cycle (***Fig.13.7***) which, ultimately, result in the synthesis of glucose.

Gluconeogenesis from lactate (Cori cycle)

Lactate produced by active skeletal muscle is a ***major precursor for gluconeogenesis***. Under anaerobic conditions, pyruvate is reduced to lactate by lactate dehydrogenase (LDH)

$$\text{Pyruvate} + \text{NADH} + \text{H}^+ \xrightleftharpoons{\text{LDH}} \text{Lactate} + \text{NAD}^+$$

Lactate is a dead end in glycolysis, since it must be reconverted to pyruvate for its further metabolism.

The plasma membrane is freely permeable to lactate. Lactate is carried from the skeletal muscle through blood and handed over to liver, where it is oxidized to pyruvate. Pyruvate, so produced, is converted to glucose by gluconeogenesis, which is then transported to the skeletal muscle.

The cycle involving the ***synthesis of glucose in liver from the skeletal muscle lactate*** and the reuse of glucose thus synthesized by the muscle for energy purpose is known as Cori cycle (***Fig.13.8***).

Glucose-alanine (Cahill) cycle

There is a continuous transport of amino acids from muscle to liver, which predominantly occurs during starvation. Alanine dominates among the amino acids. It is postulated that pyruvate in skeletal muscle undergoes transamination to produce alanine. Alanine is transported to liver and used for gluconeogenesis. This cycle is referred to as glucose-alanine cycle (***Fig.13.8***).

Regulation of gluconeogenesis

The hormone ***glucagon*** and the availability of substrates mainly regulate gluconeogenesis, as discussed hereunder.

Influence of glucagon : This is a hormone, secreted by α-cells of the pancreatic islets. Glucagon ***stimulates gluconeogenesis.***

Availability of substrates : Among the various substrates, glucogenic amino acids have stimulating influence on gluconeogenesis. This is particularly important in a condition like diabetes mellitus (decreased insulin level) where amino acids are mobilized from muscle protein for the purpose of gluconeogenesis.

Alcohol inhibits gluconeogenesis

Ethanol oxidation in the liver to acetaldehyde by the enzyme alcohol dehydrogenase utilizes NAD^+. The excess NADH produced in the liver interferes with gluconeogenesis as illustrated below.

Ethanol + $NAD^+ \longrightarrow$ Acetaldehyde + NADH + H^+

Pyruvate + NADH + $H^+ \leftrightarrow$ Lactate + NAD^+

Oxaloacetate + NADH + $H^+ \leftrightarrow$ Malate + NAD^+

It is evident from the above reactions that pyruvate and oxaloacetate, the predominant ***substrates for gluconeogenesis, are made unavailable by alcohol intoxication***. This happens due to overconsumption of NAD^+ and excessive production of NADH by alcohol.

Alcohol consumption ***increases the risk of hypoglycemia*** (reduced plasma glucose) due to lowered gluconeogenesis. Hypoglycemia is frequently observed in diabetic patients (particularly on insulin treatment), and undernourished persons consuming alcohol.

Gluconeogenesis from fat?

It is often stated that glucose cannot be synthesized from fat. In a sense, it is certainly true, since the fatty acids (most of them being even chain), on oxidation, produce acetyl CoA which cannot be converted to pyruvate. Further, the two carbons of acetyl CoA disappear as 2 moles of CO_2 in TCA cycle. Therefore, even chain fatty acids cannot serve as precursors for glucose formation.

However, the ***glycerol*** released from lipolysis and the ***propionate*** obtained from the oxidation of odd chain fatty acids ***are good substrates*** for gluconeogenesis, as discussed above.

GLYCOGEN METABOLISM

Glycogen is the ***storage form of glucose in animals***, as is starch in plants. It is stored mostly in liver (6-8%) and muscle (1-2%). Due to more muscle mass, the quantity of glycogen in muscle (250 g) is about three times higher than that in the liver (75 g).

Functions of glycogen

The prime function of liver glycogen is to ***maintain the blood glucose levels***, particularly between meals. Liver glycogen stores increase in a well-fed state which are depleted during fasting. Muscle glycogen serves as a fuel reserve for the supply of ATP during muscle contraction.

GLYCOGENESIS

The ***synthesis of glycogen*** from glucose is glycogenesis (***Fig.13.9***). Glycogenesis takes place in the cytosol and requires ATP and UTP, besides glucose.

1. **Synthesis of UDP-glucose :** The enzymes hexokinase (in muscle) and glucokinase (in liver) convert glucose to ***glucose 6-phosphate***. Phosphoglucomutase catalyses the conversion of glucose 6-phosphate to glucose 1-phosphate. ***Uridine diphosphate glucose (UDPG)*** is synthesized from glucose 1-phosphate and UTP by UDP-glucose pyrophosphorylase.

2. **Requirement of primer to initiate glycogenesis :** A small fragment of pre-existing glycogen must act as a '***primer***' to initiate glycogen synthesis. It is recently found that in the absence of glycogen primer, a specific protein—namely '***glycogenin***'—can accept glucose from UDPG.

3. **Glycogen synthesis by glycogen synthase :** Glycogen synthase is responsible for the ***formation of 1,4-glycosidic linkages***. This enzyme transfers the glucose from UDP-glucose to the non-reducing end of glycogen to form α-1,4 linkages.

4. **Formation of branches in glycogen :** The formation of branches is brought about by the action of a branching enzyme, namely ***glucosyl α-4-6 transferase***. This enzyme transfers a small fragment of five to eight glucose residues from the non-reducing end of glycogen chain (by breaking α-1,4 linkages) to another glucose residue where it is linked by α-1, 6 bond. This leads to the formation of a new non-reducing end, besides the existing one. Glycogen is further elongated and branched, respectively, by the enzymes glycogen synthase and glucosyl 4-6 transferase.

The overall reaction of the glycogen synthesis for the addition of each glucose residue is

$$(Glucose)_n + Glucose + 2ATP \longrightarrow (Glucose)_{n+1} + 2\ ADP + Pi$$

Of the two ATP utilized, one is required for the phosphorylation of glucose while the other is needed for conversion of UDP to UTP.

GLYCOGENOLYSIS

The ***degradation of stored glycogen*** in liver and muscle constitutes glycogenolysis. The pathway for the synthesis and degradation of glycogen are not reversible. An independent set of enzymes present in the cytosol carry out glycogenolysis. Glycogen is degraded by breaking α-1,4- and α-1,6-glycosidic bonds (***Fig.13.10***).

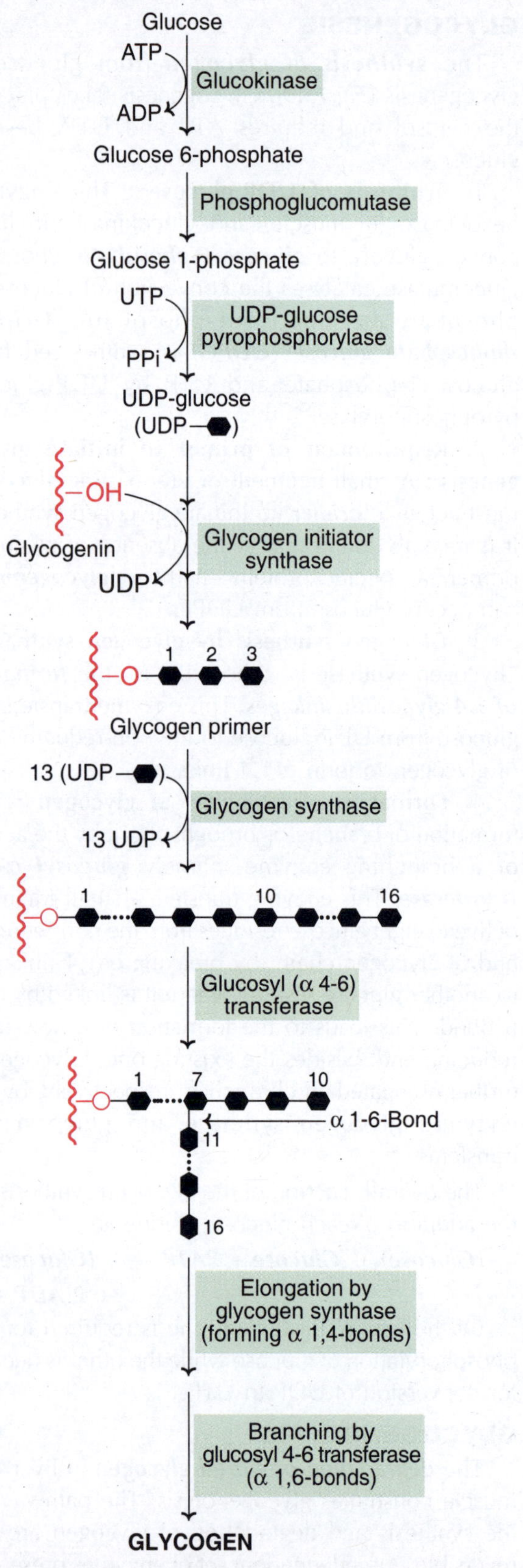

Fig.13.9 : *Glycogen synthesis from glucose (glycogenesis).*

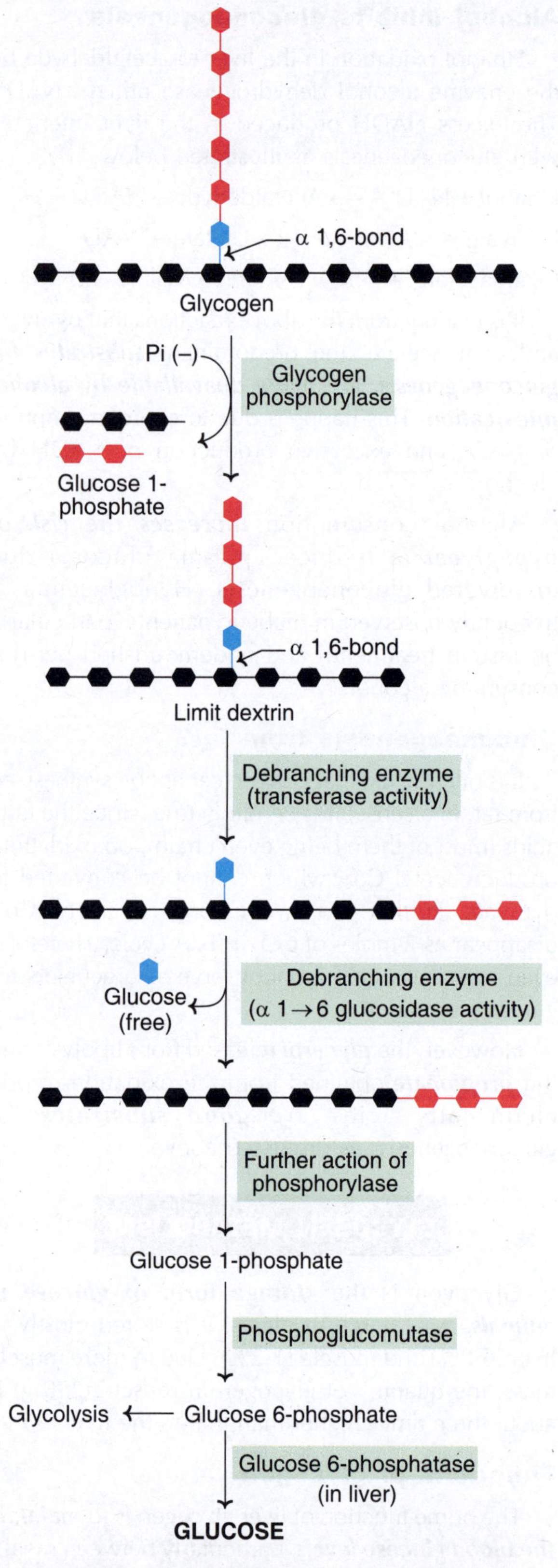

Fig.13.10 : *Glycogen degradation to glucose-glycogenolysis. (The ratio of glucose 1-phosphate to glucose is 8 : 1).*

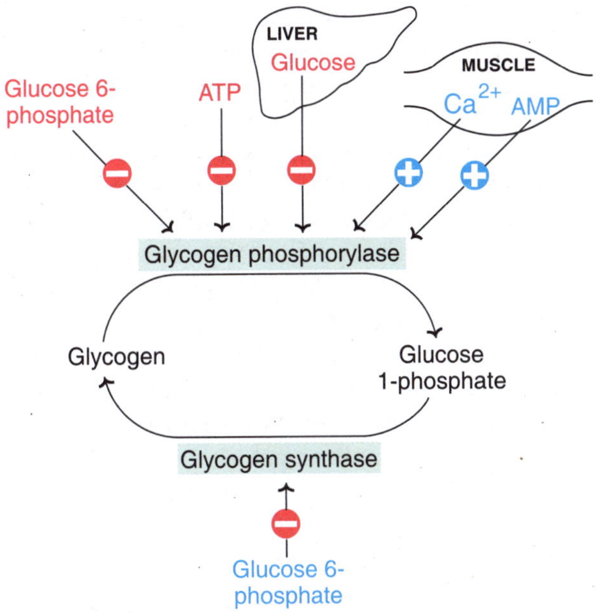

Fig.13.11 : *Allosteric regulation of glycogenolysis and glycogenesis (⊖ : Inhibition; ⊕: Activation).*

1. **Action of glycogen phosphorylase :** The α-1, 4 glycosidic bonds are cleaved sequentially by the enzyme glycogen phosphorylase to yield glucose 1-phosphate. This process—called ***phosphorolysis***—continues until four glucose residues remain on either side of branching point (α-1,6-glycosidic link). The glycogen so formed is known as ***limit dextrin*** which cannot be further degraded by phosphorylase.

2. **Action of debranching enzyme :** The branches of glycogen are cleaved by two enzyme activities present on a single polypeptide called ***debranching enzyme,*** hence it is a ***bifunctional enzyme.***

Glycosyl 4 : 4 transferase activity removes a fragment of three or four glucose residues attached at a branch and transfers them to another chain. Here, one α-1,4-bond is cleaved and the same α-1,4 bond is made, but the places are different.

Amylo α-1,6-glucosidase breaks the α-1,6 bond at the branch with a single glucose residue and ***releases a free glucose molecule.***

The remaining molecule of glycogen is again available for the action of phosphorylase and debranching enzyme to repeat the reactions stated in 1 and 2.

3. **Formation of glucose 6-phosphate and glucose :** Through the combined action of glycogen phosphorylase and debranching enzyme, glucose 1-phosphate and free glucose in ratio of 8 : 1 are produced. Glucose 1-phosphate is converted to glucose 6-phosphate by the enzyme phosphoglucomutase.

The fate of glucose 6-phosphate depends on the tissue. The liver, kidney and intestine contain the enzyme ***glucose 6-phosphatase*** that cleaves glucose 6-phosphate to glucose. This enzyme is ***absent in muscle and brain, hence free glucose cannot be produced*** from glucose 6-phosphate in these tissues. Therefore, liver is the major glycogen storage organ to provide glucose into the circulation to be utilised by various tissues.

Regulation of glycogenesis and glycogenolysis

A good coordination and regulation of glycogen synthesis and its degradation is essential to maintain the blood glucose levels. Glycogenesis and glycogenolysis are, respectively, controlled by the enzymes ***glycogen synthase*** and ***glycogen phosphorylase***. Regulation of these enzymes is accomplished by three mechanisms

1. Allosteric regulation
2. Hormonal regulation

1. **Allosteric regulation of glycogen metabolism :** In a well-fed state, the availability of ***glucose 6-phosphate*** is high which allosterically ***activates glycogen synthase*** for more glycogen synthesis. On the other hand, glucose 6-phosphate and ATP allosterically inhibit glycogen phosphorylase. Free ***glucose*** in liver also acts as an allosteric ***inhibitor of glycogen phosphorylase*** (***Fig.13.11***).

2. **Hormonal regulation of glycogen metabolism :** The hormones, through a complex series of reactions, bring about covalent modification, namely phosphorylation and dephosphorylation of enzyme proteins which, ultimately control glycogen synthesis or its degradation.

Regulation of glycogen synthesis by cAMP : The inhibition of glycogen synthesis brought by epinephrine (also norepinephrine) and glucagon occurs through cAMP by converting active glycogen synthase *'a'* to inactive synthase *'b'*.

Regulation of glycogen degradation by cAMP : Glycogen phosphorylase exists in two forms, an active *'a'* form and inactive *'b'* form. The hormones like epinephrine and glucagon bring about glycogenolysis by their action on glycogen phosphorylase through

Medical Concepts/ Clinical Correlates

- Glycogen storage diseases—characterized by deposition of normal or abnormal type of glycogen in one or more tissues—result in muscular weakness, or even death.
- The occurrence of HMP shunt (NADPH production) in the RBC is necessary to maintain the integrity of erythrocyte membrane and to prevent the accumulation of methemoglobin.
- Deficiency of glucose 6-phosphate dehydrogenase results in hemolysis of RBC, causing hemolytic anemia. The subjects of G6PD deficiency are, however, resistant to malaria.
- The conversion of glucose to fructose is impaired in diabetes mellitus, causing accumulation of sorbitol. This compound has been implicated in the development of cataract, nephropathy, peripheral neuropathy etc.
- Severe cases of galactosemia are associated with the development of cataract, believed to be due to the accumulation of galactitol.

cAMP. These hormones promote glycogenolysis by converting inactive phosphorylase *'b'* to active phosphorylase *'a'*.

The overall effect of hormones in glycogen metabolism is that ***an elevated glucagon or epinephrine level increases glycogen degradation whereas an elevated insulin results in increased glycogen synthesis.*** (Insulin lowers cAMP levels in liver).

FUTILE (SUBSTRATE) CYCLES

The ***synthesis and degradative pathways*** of metabolism (particularly reactions involving phosphorylation and dephosphorylation utilizing ATP) are ***well regulated*** and subjected to fine tuning and rapid response to meet the body demands, ***with minimal wastage of energy and metabolites***. Thus, glycolysis and gluconeogenesis (breakdown of glucose to pyruvate, and conversion of pyruvate to glucose), glycogenolysis and glycogenesis operate in a selective fashion to suit the cellular demands. If on the other hand, the ***synthesis and degradative metabolic pathways*** of a particular substance (say gluconeogenesis and glycolysis related to glucose) ***operate to the same extent simultaneously***, this ***would result in futile cycles***. However, futile cycles, consuming energy (ATP) are ***wasteful metabolic exercises***. They are minimally operative due to a well coordinated metabolic machinery. The enzymes, namely ***protein kinases*** and ***phosphatases*** serve as ***molecular switches***. They can be ***turned on or off, as per the metabolic needs of the cells***.

Recent studies suggest that futile cycles may have some physiological (operative in arousal of hibernating animals) and pathological (malignant hyperthermia due to loss of control)significance.

GLYCOGEN STORAGE DISEASES

The ***metabolic defects*** concerned with the ***glycogen synthesis*** and ***degradation*** are collectively referred to as glycogen storage diseases (***GSD***) ***or glycogenosis***. These disorders are due to defects in the enzymes which may be either generalized (affecting all tissues) or tissue-specific. The inherited disorders are characterized by deposition of normal or abnormal type of glycogen in one or more tissues.

von Gierke's disease (type I)

The incidence of type I glycogen storage disease is 1 per 200,000 persons. It is transmitted by ***autosomal recessive trait***. This disorder results in various biochemical manifestations.

1. **Fasting hypoglycemia :** Due to the defect in the enzyme ***glucose 6-phosphatase***, enough free glucose is not released from the liver into blood.

2. **Lactic acidemia :** Glucose is not synthesized from lactate produced in muscle and liver. Lactate level in blood increases and the pH is lowered (acidosis).

3. **Hyperlipidemia :** There is a blockade in gluconeogenesis. Hence more fat is mobilized to meet energy requirements of the body. This results in increased plasma free fatty acids and ketone bodies.

4. **Hyperuricemia :** Glucose 6-phosphate that accumulates is diverted to pentose phosphate pathway, leading to increased synthesis of ribose phosphates which increase the cellular levels of phosphoribosyl pyrophosphate and enhance the metabolism of purine nucleotides to uric acid. Elevated plasma levels of uric acid (hyperuricemia) are often associated with ***gouty arthritis*** (painful joints).

The other important glycogen storage disorders (with enzyme defect) are ***Pompe's disease*** (acid maltase), ***Cori's disease*** (debranching enzyme), ***Anderson's disease*** (branching enzyme), ***McArdle's disease*** (muscle glycogen phosphorylase) and ***Her's disease*** (liver glycogne phosphorylase).

HEXOSE MONOPHOSPHATE SHUNT

Hexose monophosphate pathway or ***HMP shunt*** is also called ***pentose phosphate pathway*** or phosphogluconate pathway or ***Dickens-Horecker pathway***. ***This is an alternative pathway to glycolysis and TCA cycle for the oxidation of glucose.*** However, HMP shunt is more anabolic in nature, since it is concerned with the biosynthesis of NADPH and pentoses.

Location of the pathway

The enzymes of HMP shunt are located in the ***cytosol***. The tissues such as ***liver, adipose tissue, adrenal gland, erythrocytes, testes*** and ***lactating mammary gland***, are highly active in HMP shunt. Most of these tissues are involved in the biosynthesis of fatty acids and steroids which are dependent on the supply of NADPH.

Reactions of HMP shunt

The sequence of reactions of HMP shunt (***Fig.13.12***) is divided into two phases—***oxidative*** and ***non-oxidative.***

1. **Oxidative phase :** ***Glucose 6-phosphate dehydrogenase (G6PD)*** is an NADP-dependent enzyme that converts glucose 6-phosphate to 6-phosphogluconolactone. The latter is then hydrolysed by the gluconolactone hydrolase to 6-phosphogluconate. The next reaction involving the synthesis of NADPH is catalysed by 6-phosphogluconate dehydrogenase to produce 3 keto 6-phosphogluconate which then undergoes decarboxylation to give ribulose 5-phosphate.

2. **Non-oxidative phase :** The non-oxidative reactions are concerned with the interconversion of three, four, five and seven carbon monosaccharides. Ribulose 5-phosphate is acted upon by an epimerase to produce xylulose 5-phosphate while ribose 5-phosphate ketoisomerase converts ribulose 5-phosphate to ribose 5-phosphate.

The enzyme ***transketolase*** catalyses the transfer of two carbon moiety from xylulose 5-phosphate to ribose 5-phosphate to give a 3-carbon glyceraldehyde 3-phosphate and a 7-carbon ***sedoheptulose 7-phosphate.*** Transketolase is dependent on the coenzyme ***thiamine pyrophosphate (TPP)*** and Mg^{2+} ions. Transaldolase brings about the transfer of a 3-carbon fragment (active dihydroxyacetone) from sedoheptulose 7-phosphate to glyceraldehyde 3-phosphate to give fructose 6-phosphate and four carbon erythrose 4-phosphate. Transketolase acts on xylulose 5-phosphate and transfers a 2-carbon fragment (glyceraldehyde) from it to erythrose 4-phosphate to generate fructose 6-phosphate and glyceraldehyde 3-phosphate.

Fructose 6-phosphate and glyceraldehyde 3-phosphate can be further catabolised through glycolysis and citric acid cycle. Glucose may also be synthesized from these two compounds.

For the complete oxidation of glucose 6-phosphate to $6CO_2$, we have to start with 6 molecules of glucose 6-phosphate. Of these 6,5 moles are regenerated with the production of 12 NADPH.

The ***overall reaction*** may be represented as

$$6 \text{ Glucose 6-phosphate} + 12\ NADP^+ + 6H_2O \longrightarrow 6CO_2 + 12\ NADPH + 12H^+ + 5 \text{ Glucose 6-phosphate.}$$

Significance of HMP shunt

HMP shunt is unique in generating two important products—***pentoses*** and ***NADPH***—needed for the biosynthetic reactions and other functions.

A. Importance of pentoses

In the HMP shunt, hexoses are converted into pentoses, the most important being ribose 5-phosphate. This pentose or its derivatives are useful for the ***synthesis of nucleic acids*** (RNA and DNA) and many ***nucleotides*** such as ATP, NAD^+, FAD and CoA.

B. Importance of NADPH

1. NADPH is required for the reductive ***biosynthesis of fatty acids and steroids***, hence HMP shunt is more active in the tissues concerned with lipogenesis, e.g. adipose tissue, liver etc.

2. NADPH is used in the synthesis of certain amino acids involving the enzyme ***glutamate dehydrogenase.***

3. There is a continuous production of H_2O_2 in the living cells which can chemically damage unsaturated lipids, proteins and DNA. This is, however, prevented to a large extent through ***antioxidant reactions*** involving NADPH. Glutathione mediated reduction of H_2O_2 is given in next page.

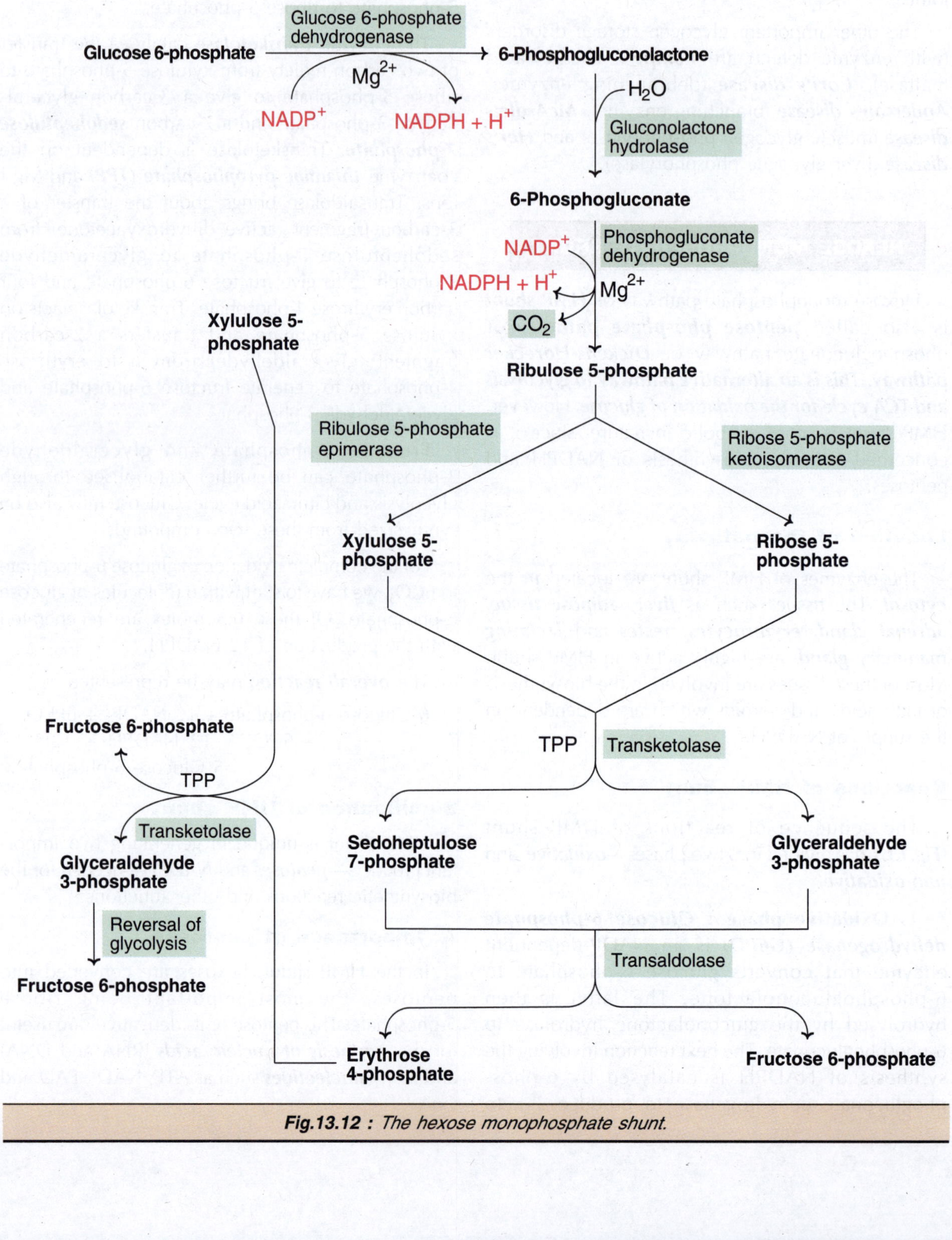

***Fig.13.12 :** The hexose monophosphate shunt.*

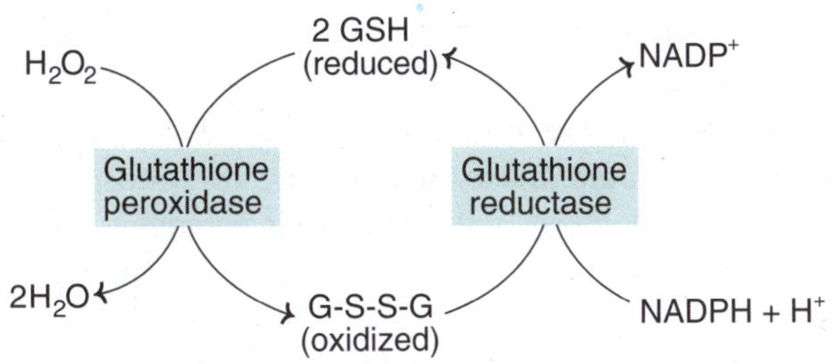

Glutathione (reduced, GSH) detoxifies H_2O_2, peroxidase catalyses this reaction. NADPH is responsible for the regeneration of reduced glutathione from the oxidized one.

4. Microsomal cytochrome P_{450} system (in liver) brings about the ***detoxification of drugs*** and foreign compounds by hydroxylation reactions involving NADPH.

5. **Special functions of NADPH in RBC :** NADPH produced in erythrocytes has special functions to perform. It maintains the concentration of reduced glutathione (reaction explained in 3) which is essentially required to preserve the ***integrity of RBC membrane.*** NADPH is also necessary to keep the ferrous iron (Fe^{2+}) of hemoglobin in the reduced state so that accumulation of methemoglobin (Fe^{3+}) is prevented.

Glucose 6-phosphate dehydrogenase (G6PD) deficiency

G6PD deficiency is an ***inherited sex-linked trait***. Although the deficiency occurs in all the cells of the affected individuals, it is more severe in RBC.

HMP shunt is the only means of providing NADPH in the erythrocytes. Decreased activity of G6PD impairs the synthesis of NADPH in RBC. This results in the accumulation of methemoglobin and peroxides in erythrocytes leading to ***hemolysis.***

Clinical manifestations in G6PD deficiency : Most of the patients with G6PD deficiency do not usually exhibit clinical symptoms. Some of them, however, develop ***hemolytic anemia*** if they are administered oxidant drugs or exposed to a severe infection. The drugs such as primaquine (antimalarial), acetanilide (antipyretic), sulfamethoxazole (antibiotic) or ingestion of fava beans (***favism***) produce ***hemolytic jaundice*** in these patients. Severe infection results in the generation of free radicals (in macrophages) which can enter RBC and cause hemolysis (due to decreased NADPH and reduced GSH).

G6PD deficiency and resistance to malaria : It is interesting to note that G6PD deficiency is associated with resistance to malaria (caused by *Plasmodium falciparum*). This is explained from the fact that the ***parasites that cause malaria are dependent on HMP shunt and reduced glutathione for their optimum growth in RBC***. Therefore, G6PD deficiency—which is seen frequently in Africans—protects them from malaria, a common disease in this region. It is regarded as an adaptability of the people living in malaria-infected regions of the world.

URONIC ACID PATHWAY

This is an ***alternative oxidative pathway for glucose*** and is also known as ***glucuronic acid pathway.*** It is concerned with the synthesis of glucuronic acid, pentoses and vitamin, ascorbic acid (except in primates and guinea pigs).

Importance of UDP-glucuronate

UDP-glucuronate is the metabolically active form of glucuronate which is utilized for conjugation with many substances like bilirubin, steroid hormones and certain drugs.

Synthesis of ascorbic acid in some animals

L-Gulonate is the precursor for the synthesis of ascorbic acid (vitamin C) in many animals. The enzyme ***L-gulonolactone oxidase***—which converts gulonate to ascorbic acid—is absent in man, other primates and guinea pigs. Therefore, vitamin C has to be supplemented in the diet for these animals.

METABOLISM OF GALACTOSE

The specific enzyme, namely ***galactokinase***, phosphorylates galactose to galactose 1-phosphate. This reacts with UDP-glucose in an exchange reaction to form UDP-galactose in presence of the enzyme galactose 1-phosphate uridyltransferase. UDP-galactose is an active donor of galactose for many synthetic reactions involving the formation of compounds like lactose, glycosaminoglycans, glycoproteins, cerebrosides and glycolipids. UDP-galactose can be converted to UDP-glucose by UDP hexose 4-epimerase (***Fig.13.13***). In this way, ***galactose can enter the metabolic pathways of glucose***. It may be noted that ***galactose is not an essential nutrient*** since UDP-glucose can be converted to UDP-galactose by the enzyme UDP hexose 4-epimerase.

Disorders of galactose metabolism

Classical galactosemia : This is due to the deficiency of the enzyme ***galactose 1-phosphate uridyltransferase.*** It is a rare congenital disease in infants, inherited as an autosomal recessive disorder. The salient features of galactosemia are listed.

1. Galactose metabolism is impaired leading to ***increased galactose levels*** in circulation (galactosemia) and urine (galactosuria).

2. The accumulated galactose is diverted for the production of ***galactitol*** (dulcitol) by the enzyme aldose reductase (the same enzyme that converts glucose to sorbitol). Galactitol (like sorbitol, discussed later) has been implicated in the ***development of cataract***.

3. The accumulation of galactose 1-phosphate and galactitol in various tissues like ***liver***, nervous tissue, ***lens*** and ***kidney*** leads to ***impairment in their function***.

4. High levels of galactose 1-phosphate in liver results in the depletion of inorganic phosphate (***sequestering of phosphate***) for other metabolic functions. Galactose 1-phosphate inhibits glycogen phosphorylase resulting in ***hypoglycemia***.

5. The clinical symptoms of galactosemia are—loss of weight (in infants) hepatosplenomegaly, jaundice, mental retardation etc. In severe cases, ***cataract***, ***amino aciduria*** and albuminuria are also observed.

Diagnosis : Early detection is possible (biochemical diagnosis) by measuring the activity of galactose 1-phosphate uridyltransferase in erythrocytes.

Treatment : The therapy includes the supply of diet deprived of galactose and lactose.

Galactokinase deficiency : The defect in the enzyme ***galactokinase,*** responsible for phosphorylation of galactose, will also result in galactosemia and galactosuria. Here again galactose is shunted to the formation of galactitol. Generally, galactokinase-deficient individuals do not develop hepatic and renal complications.

METABOLISM OF FRUCTOSE

Fructose is mostly phosphorylated by fructokinase to fructose 1-phosphate. Fructokinase has been identified in liver, kidney and intestine. Hexokinase, which phosphorylates various monosaccharides, can also act on fructose to produce fructose 6-phosphate. However, hexokinase has low affinity (high K_m) for fructose, hence this is a minor pathway.

Fructose 1-phosphate is cleaved to glyceraldehyde and dihydroxyacetone phosphate (DHAP) by ***aldolase B.*** This is in contrast to fructose 6-phosphate which is converted to fructose 1, 6-bisphosphate and split by ***aldolase A*** (details in glycolysis—***Fig.13.2***). Glyceraldehyde is phosphorylated by the enzyme triokinase to glyceraldehyde 3-phosphate which, along with DHAP, enters glycolysis or gluconeogenesis.

Sorbitol pathway/polyol pathway

Polyol pathway (so termed since sorbitol is a polyhydroxy sugar) basically involves the conversion

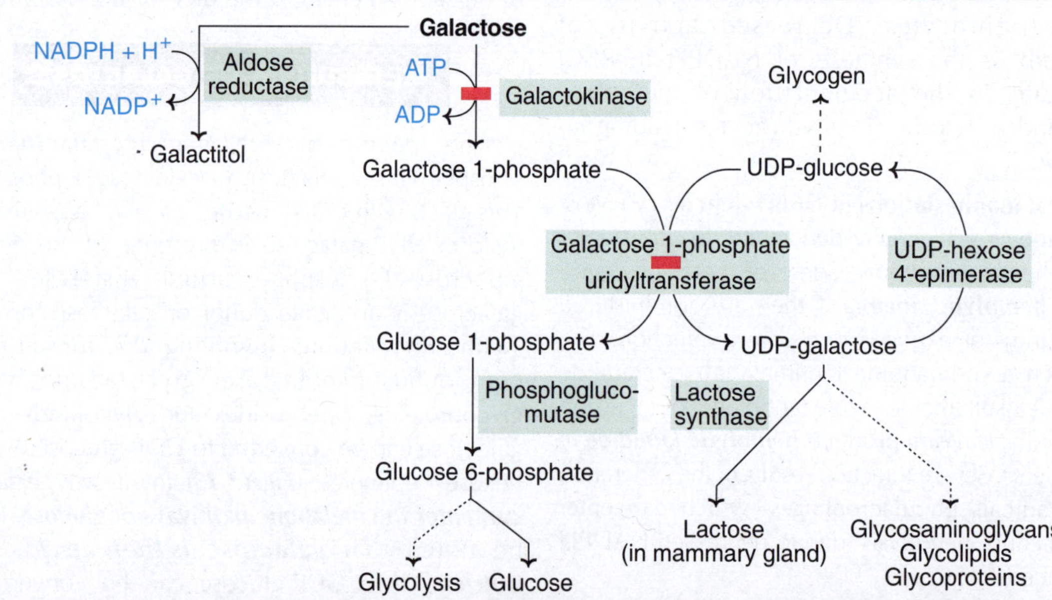

Fig. 13.13 : Metabolism of galactose (defects in galactosemia are indicated by dark blocks).

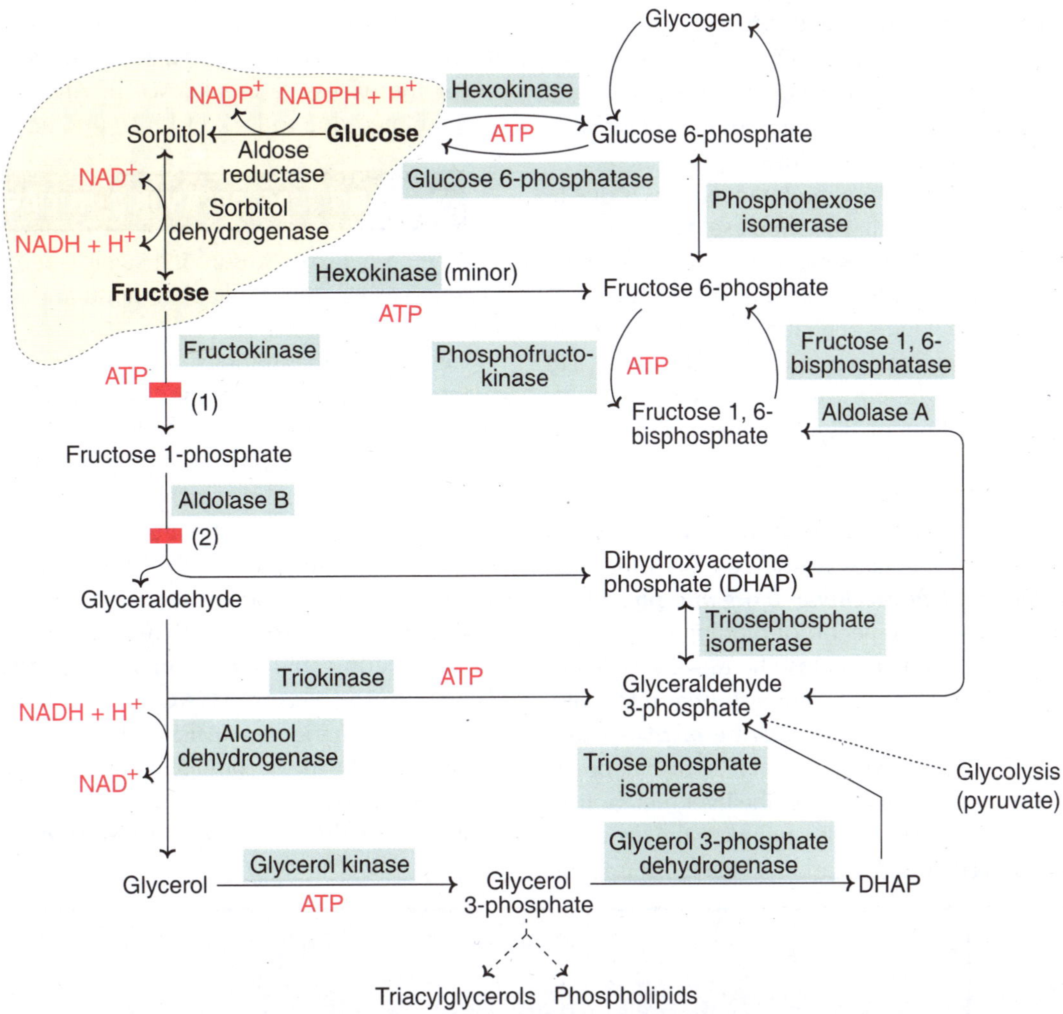

***Fig.13.14 :** Metabolism of fructose (Metabolic defects 1-Essential fructosuria; 2-Fructose intolerance).*

of glucose to fructose via sorbitol (***Fig.13.14***). This pathway is absent in liver. Sorbitol pathway is directly related to glucose concentration, and is ***higher in uncontrolled diabetes.***

The enzyme aldose reductase reduces glucose to sorbitol (glucitol) in the presence of NADPH. Sorbitol is then oxidized to fructose by sorbitol dehydrogenase and NAD^+. Aldose reductase is absent in liver but found in many tissues like lens and retina of the eye, kidney, placenta, Schwann cells of peripheral nerves, erythrocytes and seminal vesicles. The enzyme sorbitol dehydrogenase is present in liver, seminal vesicle, spleen and ovaries. Fructose is a preferred carbohydrate for energy needs of sperm cells due to the presence of sorbitol pathway.

Sorbitol pathway in diabetes mellitus

In uncontrolled diabetes (hyperglycemia), large amounts of glucose enter the cells which are not dependent on insulin. Significantly, the cells with increased intracellular glucose levels in diabetes (lens, retina, nerve cells, kidney etc.) possess high activity of aldose reductase and sufficient supply of NADPH. This results in a rapid and ***efficient conversion of glucose to sorbitol***. The enzyme sorbitol dehydrogenase, however, is either low in activity or absent in these cells, hence ***sorbitol is not converted to fructose***. Sorbitol cannot freely pass through the cell membrane, and ***accumulates in the cells*** where it is produced. Sorbitol—due to its hydrophilic nature—causes strong osmotic effects leading to swelling of the cells. Some of the ***pathological changes associated with diabetes*** (like cataract formation, peripheral neuropathy, nephropathy etc.) are ***believed to be due to the accumulation of sorbitol***, as explained above.

Defects in fructose metabolism

1. **Essential fructosuria :** Due to the deficiency of the enzyme hepatic ***fructokinase***, fructose is not converted to fructose 1-phosphate. Treatment involves the restriction of dietary fructose.

2. **Hereditary fructose intolerance :** This is due to the absence of the enzyme ***aldolase B.*** Hereditary fructose intolerance causes intracellular accumulation of fructose 1-phosphate, severe hypoglycemia, vomiting, hepatic failure and jaundice. Early detection and intake of diet free from fructose and sucrose, are advised to overcome fructose intolerance.

3. **Consumption of high fructose :** Fructose is rapidly converted to fructose 1-phosphate by fructokinase. The activity of the enzyme aldolase B is relatively less, and, due to this, fructose 1-phosphate accumulates in the cell. This leads to the ***depletion of intracellular inorganic phosphate*** (Pi) levels. The phenomenon of binding of Pi to the organic molecules (like fructose here)—that leads to the less availability of Pi for the essential metabolic functions—is known as ***sequestering of phosphate.*** Due to the decreased availability of Pi, which happens in overconsumption of fructose, the liver metabolism is adversely affected. This includes the lowered synthesis of ATP from ADP and Pi. High consumption of fructose over a long period is associated with ***increased uric acid in blood leading to gout***. This is due to the excessive breakdown of ADP and AMP (accumulated due to lack of Pi) to uric acid.

METABOLISM OF AMINO SUGARS

When a hydroxyl group of a sugar is replaced by an amino group, the resultant compound is an amino sugar.

The important amino sugars are ***glucosamine***, ***galactosamine***, ***mannosamine***, ***sialic acid*** etc. They are essential components of glycosaminoglycans, glycolipids (gangliosides) and glycoproteins. They are also found in some oligosaccharides and certain antibiotics. It is estimated that about 20% of the glucose is utilized for the synthesis of amino sugars, which mostly occurs in the connective tissue.

Mucopolysaccharidoses : The lysosomal storage diseases, caused by ***defects in the degradation of glycosaminoglycans (GAGs)*** are known as mucopolysaccharidoses (***MPS***). These diseases are characterized by the accumulation of GAGs in various tissues that may result in skeletal deformities and mental retardation e.g., ***Hurler's syndrome*** or MPS I (enzyme defect L-iduronidase).

SUMMARY

1. *Carbohydrates are the major source of energy for the living cells. Glucose (normal fasting blood level 70-100 mg/dl) is the central molecule in carbohydrate metabolism.*
2. *Glucose is oxidized in glycolysis, either in anaerobic (2 ATP formed) or aerobic (7 ATP formed) conditions, resulting in the formation of 2 moles of lactate or pyruvate, respectively.*
3. *Acetyl CoA is produced from pyruvate which is completely oxidized in citric acid cycle, the final common oxidative pathway for all foodstuffs. The complete oxidation of one mole of glucose generates 32 ATP.*
4. *Gluconeogenesis is the synthesis of glucose from noncarbohydrate precursors like amino acids (except leucine and lysine), lactate, glycerol, propionate etc.*
5. *Hexose monophosphate shunt (HMP shunt) is the direct oxidative pathway of glucose. HMP shunt assumes significance since it generates NADPH and pentoses, respectively required for the synthesis of lipids and nucleic acids.*
6. *Galactosemia is mostly due to the defect in the enzyme galactose 1-phosphate uridyltransferase. This results in the diversion of galactose to produce galactitol which has been implicated in the development of cataract.*

THE READER MUST REFER CHAPTER 36 FOR BLOOD GLUCOSE REGULATION, DIABETES MELLITUS ETC.

THE READER MUST REFER APPENDIX D (Page Nos : 422 - 431) FOR RELATED CLINICAL CASE STUDIES (Nos : 8, 9, 30, 31, 32, 36, 40)

METABOLISM OF LIPIDS

14

Lipids are indispensable for cell structure and function. Due to their hydrophobic and non-polar nature, lipids differ from rest of the body compounds and are unique in their action.

Triacylglycerols—the body fuel reserve

Lipids constitute about 15-20% of the body weight in humans. Triacylglycerols ***(formerly triglycerides)*** are the most abundant lipids comprising 85-90% of body lipids. Most of the triacylglycerols (TG; also called ***neutral fat*** or depot fat) are stored in the ***adipose tissue*** and serve as energy reserve of the body. This is in contrast to carbohydrates and proteins which cannot be stored to a significant extent for energy purposes. Fat also acts as an insulating material for maintaining the body temperature of animals.

Mobilization of fat from adipose tissue

Triacylglycerol (TG) is the stored fat in the adipose tissue. The enzyme, namely ***hormone-sensitive triacylglycerol lipase***, removes the fatty acid either from carbon 1 or 3 of the triacylglycerol to form diacylglycerol. The other two fatty acids of TG are cleaved by additional lipases specific for diacylglycerol and monoacylglycerol. The complete degradation of triacylglycerol to glycerol and free fatty acids is known as ***lipolysis*** (***Fig.14.1***).

Fate of glycerol : The adipose tissue lacks the enzyme glycerol kinase, hence glycerol produced in lipolysis cannot be phosphorylated here. It is transported to liver where it is activated to ***glycerol 3-phosphate***. The latter may be used for the synthesis of triacylglycerols and phospholipids. Glycerol 3-phosphate may also enter glycolysis by getting converted to dihydroxyacetone phosphate.

Fate of free fatty acids : The fatty acids released by lipolysis in the adipocytes enter the circulation and are transported in a bound form to albumin. The free fatty acids enter various tissues and are utilized for the energy. About ***95% of the energy obtained from fat comes from the oxidation of fatty acids***. Certain tissues, however, cannot oxidize fatty acids, e.g. brain, erythrocytes.

FATTY ACID OXIDATION

The fatty acids in the body are mostly oxidized by ***β-oxidation***. β-Oxidation may be defined as the ***oxidation of fatty acids on the β-carbon atom***. This results in the sequential removal of a two carbon fragment, acetyl CoA.

Fatty acid oxidation—stages and tissues

The β-oxidation of fatty acids involves three stages

I. ***Activation*** of fatty acids occurring in the cytosol;

II. ***Transport*** of fatty acids into mitochondria;

III. ***β-Oxidation proper*** in the mitochondrial matrix.

Fatty acids are oxidized by most of the tissues in the body. However, ***brain, erythrocytes and adrenal medulla cannot utilize fatty acids for energy requirement***.

I. Fatty acid activation

Fatty acids are activated to acyl CoA by ***thiokinases*** or ***acyl CoA synthetases.*** In this reaction, ***two high energy phosphates are utilized***, since ATP is converted to pyrophosphate (PPi). The enzyme ***inorganic pyrophosphatase*** hydrolyses PPi to phosphate (Pi).

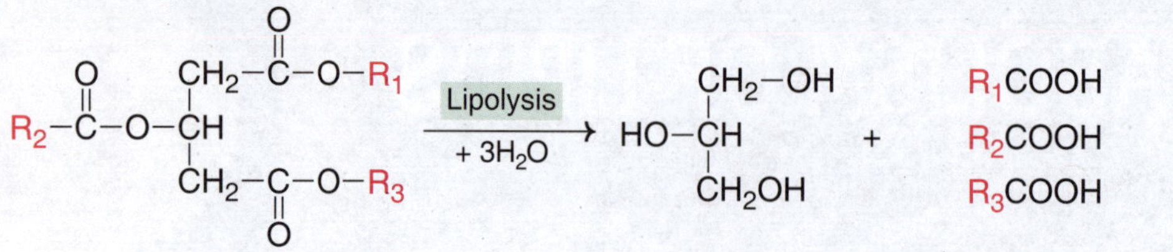

Triacylglycerol Glycerol Free fatty acids (3)

Fig.14.1 : Complete hydrolysis (lipolysis) of triacylglycerol.

The immediate elimination of PPi makes this reaction totally irreversible.

II. Transport of acyl CoA into mitochondria

The inner mitochondrial membrane is impermeable to fatty acids. A specialized ***carnitine carrier system*** (***carnitine shuttle***) operates to transport activated fatty acids from cytosol to the mitochondria.

III. β-Oxidation proper

Each cycle of β-oxidation, liberating a two carbon unit-acetyl CoA, occurs in a sequence of four reactions (***Fig.14.2***).

1. **Oxidation :** Acyl CoA undergoes dehydrogenation by an FAD-dependent flavoenzyme, acyl CoA dehydrogenase. A double bond is formed between α and β carbons (i.e., 2 and 3 carbons).

2. **Hydration :** Enoyl CoA hydratase brings about the hydration of the double bond to form β-hydroxyacyl CoA.

3. **Oxidation :** β-Hydroxyacyl CoA dehydrogenase catalyses the second oxidation and generates NADH. The product formed is β-ketoacyl CoA.

4. **Cleavage :** The final reaction in β-oxidation is the liberation of a 2 carbon fragment, ***acetyl CoA*** from acyl CoA. This occurs by a thiolytic cleavage catalysed by ***β-ketoacyl CoA thiolase*** (or simply thiolase).

The new acyl CoA, containing two carbons less than the original, reenters the β-oxidation cycle. The process continues till the fatty acid is completely oxidized.

The scheme of fatty acid oxidation discussed above corresponds to saturated (no double bond) and even carbon fatty acids. This occurs most predominantly in biological system.

Oxidation of palmitoyl CoA

The ***summary of β-oxidation*** of palmitoyl CoA is shown in the next column

$$\text{Palmitoyl CoA} + 7\ \text{CoASH} + 7\ \text{FAD} + 7\ \text{NAD}^+ + 7H_2O \longrightarrow 8\ \text{Acetyl CoA} + 7\ FADH_2 + 7\text{NADH} + 7H^+$$

Palmitoyl CoA undergoes 7 cycles of β-oxidation to yield 8 acetyl CoA. Acetyl CoA can enter citric acid cycle and get completely oxidized to CO_2 and H_2O.

Energetics of β-oxidation

The ultimate aim of fatty acid oxidation is to generate energy. The energy (net ***106 ATP***) obtained from the complete oxidation of ***palmitic acid*** (16 carbon) is given in ***Table 14.1***, and ***Fig.14.3***.

Oxidation of odd carbon chain fatty acids

The β-oxidation of saturated fatty acids containing odd number of carbon atoms proceeds in the same manner, as described above for even carbon fatty acids. The only difference is that in the last and final β-oxidation cycle, a three-carbon fragment is left behind (in place of 2 carbon unit for saturated fatty acids). This compound is ***propionyl CoA*** which is converted to succinyl CoA and metabolised.

Oxidation of unsaturated fatty acids

Due to the presence of double bonds, the unsaturated fatty acids are not reduced to the same extent as saturated fatty acids. Therefore, ***oxidation of unsaturated fatty acids***, in general, provides ***less energy than*** that of saturated fatty acids.

Most of the reactions involved in the oxidation of unsaturated fatty acids are the same as found in the β-oxidation of saturated fatty acids. However, the presence of double bonds poses problem for β-oxidation to proceed. This is overcome by two additional enzymes—an ***isomerase*** and an ***reductase***.

α-Oxidation of fatty acids

β-Oxidation is the most predominant pathway for fatty acid degradation. However, the ***removal of one carbon unit at a time*** by the oxidation of α-carbon

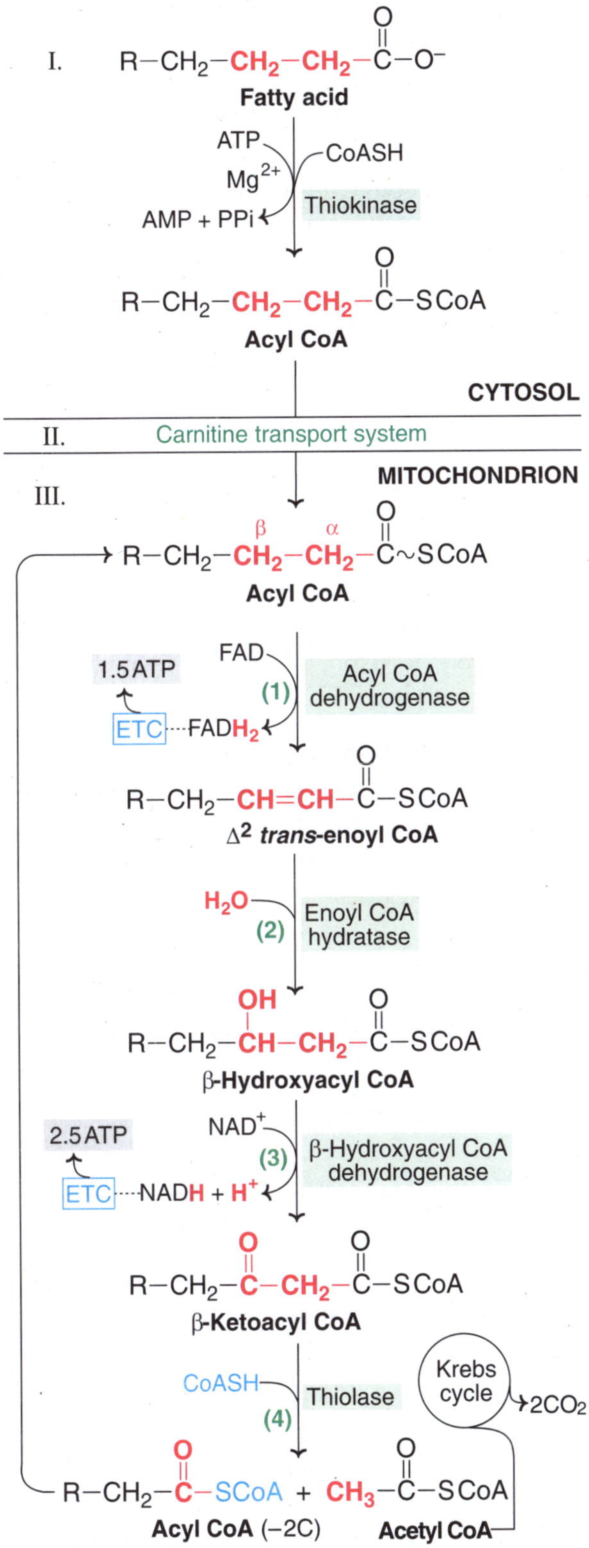

Fig.14.2 : *β-Oxidation of fatty acids : Palmitoyl CoA (16 carbon) undergoes seven cycles to yield 8 acetyl CoA [I–Activation; II–Transport; III–β Oxidation proper—* ***(1)*** *Oxidation,* ***(2)*** *Hydration,* ***(3)*** *Oxidation and* ***(4)*** *Cleavage].*

atom of fatty acid is known. α-Oxidation does not involve the binding of fatty acid to coenzyme A and no energy is produced.

Refsum's disease is a rare but severe ***neurological disorder***. The patients of this disease accumulate large quantities of an unusual fatty acid, ***phytanic acid.*** It is derived from phytol, a constituent of ***chlorophyll***. Hence it is found mostly in plant foods. However, it is also present in milk lipids and animal fats. Phytanic acid ***cannot undergo β-oxidation*** due to the presence of a methyl group on carbon-3. This

TABLE 14.1 Energetics of palmitic acid oxidation

Mechanism	*ATP yield*
I. **β-Oxidation 7 cycles**	
7 $FADH_2$ [oxidized by electron transport chain (ETC), each $FADH_2$ gives 1.5 ATP]	***10.5***
7 NADH (oxidized by ETC, each NADH liberates 2.5 ATP)	***17.5***
II. **From 8 acetyl CoA**	
Oxidized by citric acid cycle, each acetyl CoA provides 10 ATP	80
Total energy from one mole of palmitoyl CoA	***108***
Energy utilized for activation (formation of palmitoyl CoA)	–2
Net yield for one molecule of palmitate	***106***

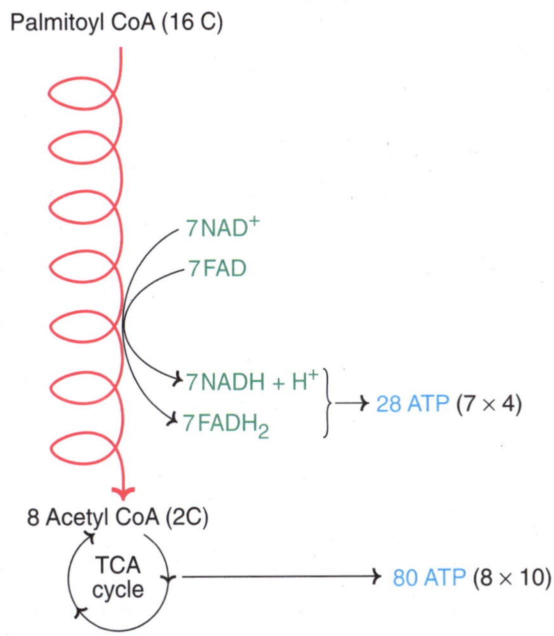

Fig.14.3 : *An overview of oxidation of palmitic acid.*

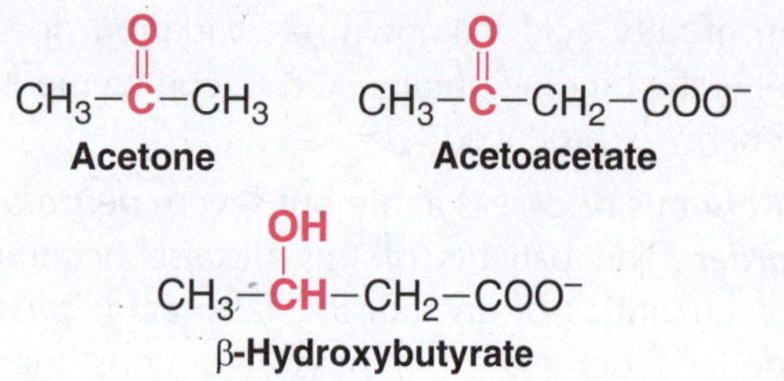

Fig.14.4 : *Structures of ketone bodies.*

fatty acid undergoes ***initial α-oxidation*** and this is followed by β-oxidation.

Refsum's disease is caused by a defect in the α-oxidation due to the ***deficiency of the enzyme phytanic acid α-oxidase***. The result is that phytanic acid cannot be converted to a compound that can be degraded by β-oxidation. The patients should not consume diet containing chlorophyll (i.e., all green leafy vegetables).

Oxidation of fatty acids and metabolic water

Fatty acid oxidation (even other forms of aerobic respiration) is accompanied by the production of water, referred to as metabolic water. For instance, when one molecule of palmitic acid is oxidized, it releases ***16 molecules of water***. This metabolic water has great significance in some animals. ***Camel*** can store ***lipids*** in its ***hump*** which is a ***good source of water***, besides ***energy*** supply. For this reason, camel can travel in deserts for long periods even without food and water supply.

KETONE BODIES

The compounds namely ***acetone, acetoacetate*** and ***β-hydroxybutyrate*** (or 3-hydroxybutyrate) are known as ketone bodies (***Fig.14.4***). Only the first two are true ketones while β-hydroxybutyrate does not possess a keto (C=O) group. ***Ketone bodies are water-soluble*** and ***energy yielding***. Acetone, however, is an exception since it cannot be metabolised.

Ketogenesis

The synthesis of ketone bodies occurs in the ***liver***. The enzymes for ketone body synthesis are located in the ***mitochondria matrix***. Acetyl CoA, formed by oxidation of fatty acids, pyruvate or some amino acids, is the ***precursor for ketone bodies***. Ketogenesis occurs through the following reactions (***Fig.14.5***).

1. Two moles of acetyl CoA condense to form acetoacetyl CoA. This reaction is catalysed by thiolase.

2. Acetoacetyl CoA combines with another molecule of acetyl CoA to produce β-hydroxy β-methyl glutaryl CoA (HMG CoA). ***HMG CoA synthase***, catalysing this reaction, ***regulates the synthesis of ketone bodies.***

3. HMG CoA lyase cleaves HMG CoA to produce acetoacetate and acetyl CoA.

4. Acetoacetate can undergo spontaneous decarboxylation to form acetone.

5. Acetoacetate can be reduced by a dehydrogenase to β-hydroxybutyrate.

The carbon skeleton of some amino acids (ketogenic) is degraded to acetoacetate or acyl CoA and, therefore, to ketone bodies, e.g. leucine, lysine, phenylalanine etc.

Utilization of ketone bodies

The ketone bodies, ***being water soluble***, are ***easily transported from the liver to various tissues.***

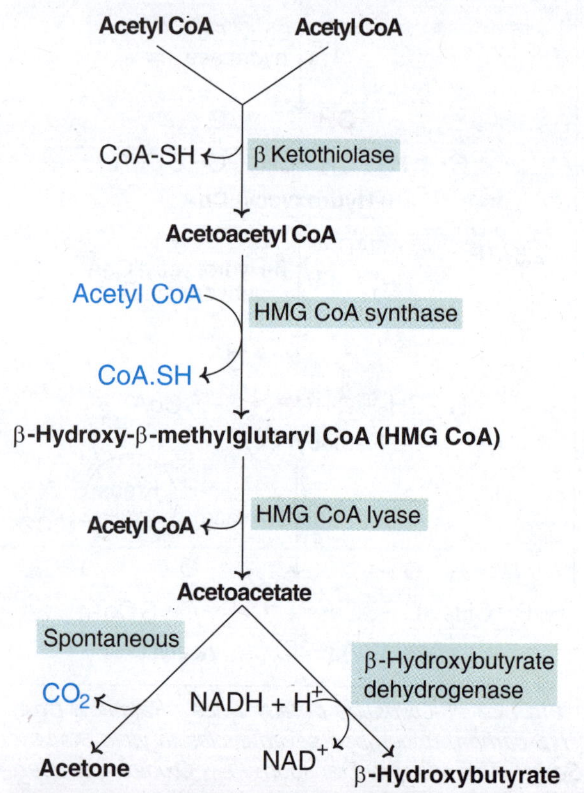

Fig.14.5 : *Synthesis of ketone bodies (ketogenesis).*

Medical Concepts/ Clinical Correlates

- An adult human body contains about 10–11 kg of fat reserve corresponding to about 100,000 Cal. This can meet the energy requirements for several weeks of food deprivation in man.
- Refsum's disease is due to a defect in α-oxidation of fatty acids. The patients are advised not to consume diets containing chlorophyll.
- Ketosis is commonly associated with uncontrolled diabetes mellitus and starvation. Diabetic ketoacidosis is dangerous—may result in coma or even death. Starvation, however, is not accompanied by ketoacidosis.
- The lack of the ability of the organisms to introduce double bonds in fatty acids beyond C_9 and C_{10} makes linoleic and linolenic acids essential to mammals.

The two ketone bodies—acetoacetate and β-hydroxybutyrate serve as an important source of energy for the ***peripheral tissues*** such as skeletal muscle, cardiac muscle, renal cortex etc. The tissues which lack mitochondria (e.g. erythrocytes) however, cannot utilize ketone bodies. The production of ketone bodies and their utilization become more significant when glucose is in short supply to the tissues, as observed in ***starvation*** and ***diabetes mellitus***.

During prolonged ***starvation***, ketone bodies are the major ***fuel source for the brain*** and other parts of central nervous system.

Reactions of ketone bodies : β-Hydroxybutyrate is first converted to acetoacetate (reversal of synthesis) and metabolised. Acetoacetate is activated to acetoacetyl CoA by a mitochondrial enzyme ***thiophorase*** (***Fig.14.6***). The coenzyme A is donated by succinyl CoA, an intermediate in citric acid cycle. ***Thiophorase is absent in liver***, hence ketone bodies are not utilized by the liver. Thiolase cleaves acetoacetyl CoA to two moles of acetyl CoA.

The summary of ketone body synthesis, utilization and excretion is depicted in ***Fig.14.7***.

Overproduction of ketone bodies

In normal individuals, there is a constant production of ketone bodies by liver and their utilization by extrahepatic tissues. The concentration of ketone bodies in blood is maintained around 1 mg/dl. Their excretion in ***urine*** is very low and undetectable by routine tests (***Rothera's test***).

When the rate of synthesis of ketone bodies exceeds the rate of utilization, their concentration in blood increases, this is known as ***ketonemia.*** This is followed by ***ketonuria***—excretion of ketone bodies in urine. The overall picture of ketonemia and ketonuria is commonly referred to as ***ketosis.*** Smell of acetone in breath is a common feature in ketosis. ***Ketosis is most commonly associated with starvation and severe uncontrolled diabetes mellitus.***

Starvation : Starvation is accompanied by ***increased degradation of fatty acids*** (from the fuel reserve triacylglycerol) to meet the energy needs of the body. This causes an ***overproduction*** of acetyl CoA and consequently ***ketone bodies.***

Diabetes mellitus : Diabetes mellitus is associated with insulin deficiency. This results in

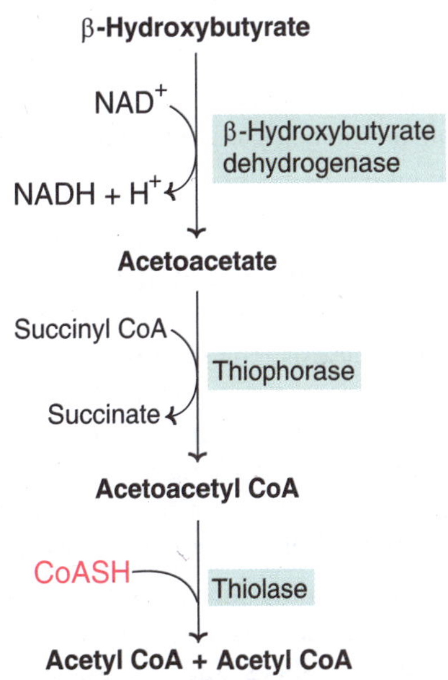

Fig.14.6 : Metabolism (utilization) of ketone bodies to acetyl CoA.

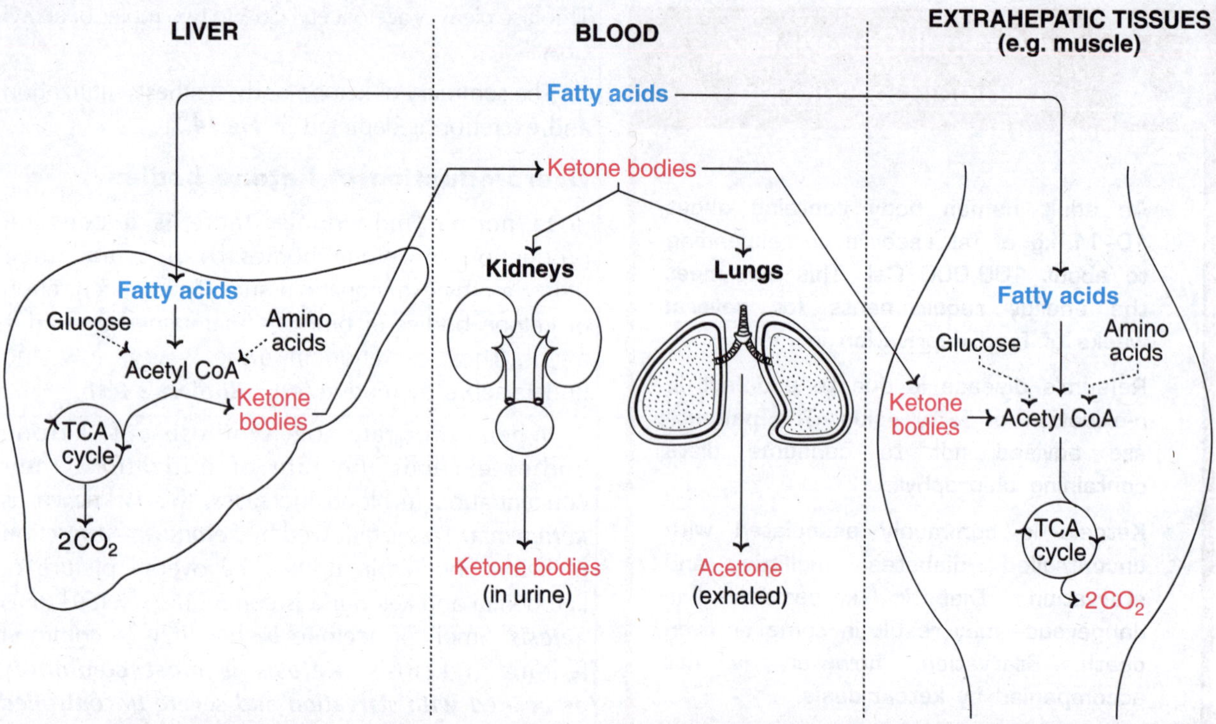

Fig.14.7 : *Summary of ketone body synthesis, utilization and excretion.*

impaired carbohydrate metabolism and ***increased lipolysis***, both of them ultimately leading to the accumulation of acetyl CoA and its conversion to ***ketone bodies***.

Ketogenic and antiketogenic substances

The ketogenic substances (promote ketogenesis) include fatty acids and certain amino acids (***leucine***, ***lysine***, tyrosine etc.). The antiketogenic substances (inhibit ketogenesis) are glucose, glycerol and glucogenic amino acids (e.g. ***glycine***, ***alanine***, serine, glutamate etc.)

Ketoacidosis

Both acetoacetate and β-hydroxybutyrate are strong acids. Increase in their concentration in blood would cause acidosis. The carboxyl group has a p^{Ka} around 4. Therefore, the ketone bodies in the blood dissociate and release H^+ ions which lower the pH. Diabetic ketoacidosis is dangerous—may result in coma, and even death, if not treated. Ketosis due to starvation is not usually accompanied by ketoacidosis.

Treatment of ketoacidosis : Rapid treatment of ***diabetic ketoacidosis*** is required to correct the metabolic abnormalities and the associated water and electrolyte imbalance. Administration of ***insulin*** is necessary to stimulate uptake of glucose by tissues and inhibition of ketogenesis.

BIOSYNTHESIS OF FATTY ACIDS

The ***dietary carbohydrates and amino acids***, when consumed ***in excess***, can be converted to fatty acids and ***stored as triacylglycerols***. ***De novo*** (new) synthesis of fatty acids occurs predominantly in liver, kidney, adipose tissue and lactating mammary glands. The ***enzyme machinery*** for fatty acid production is located in the ***cytosomal fraction*** of the cell. Acetyl CoA is the source of carbon atoms while NADPH provides the reducing equivalents and ATP supplies energy for fatty acid formation. The fatty acid synthesis may be learnt in 2 stages

I. Conversion of acetyl CoA to malonyl CoA.

II. Reactions of fatty acid synthase complex.

I. Formation of malonyl CoA

Acetyl CoA is carboxylated to malonyl CoA by the enzyme ***acetyl CoA carboxylase.*** This is an ATP-dependent reaction and requires ***biotin*** for CO_2 fixation (***Fig.14.8***). Acetyl CoA carboxylase is a regulatory enzyme in fatty acid synthesis (details given later).

II. Reactions of fatty acid synthase complex

The remaining reactions of fatty acid synthesis are catalysed by a multifunctional enzyme known as ***fatty acid synthase (FAS) complex.*** In eukaryotic cells,

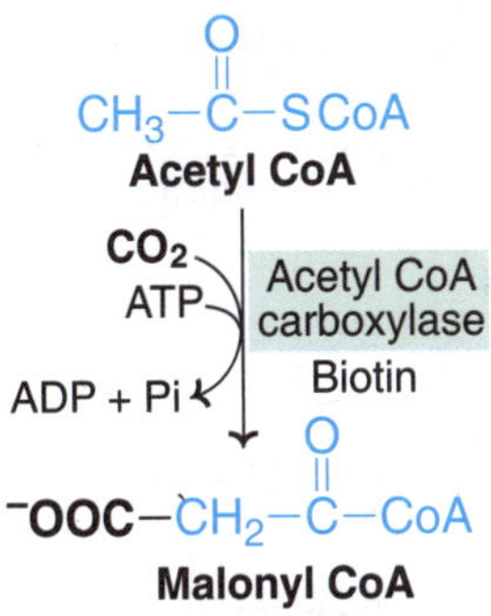

Fig.14.8 : Conversion of acetyl CoA to malonyl CoA.

including man, the fatty acid synthase exists as a dimer with two identical units. Each monomer possesses the activities of seven different enzymes and an ***acyl carrier protein (ACP)*** bound to 4′-phosphopantetheine. Fatty acid synthase functions as a single unit catalysing all the seven reactions. Dissociation of the synthase complex results in loss of the enzyme activities.

The sequence of reactions for the extra—mitochondrial synthesis of fatty acids (palmitate) is depicted in ***Fig.14.9***, and described below

1. The two carbon fragment of acetyl CoA is transferred to ACP of fatty acid synthase, catalysed by the enzyme, ***acetyl CoA-ACP transacylase***. The acetyl unit is then transferred from ACP to cysteine residue of the enzyme. Thus ACP site falls vacant.

2. The enzyme ***malonyl CoA-ACP transacylase*** transfers malonate from malonyl CoA to bind to ACP.

3. The acetyl unit attached to cysteine is transferred to malonyl group (bound to ACP). The malonyl moiety loses CO_2 which was added by ***acetyl CoA carboxylase.*** Thus, CO_2 is never incorporated into fatty acid carbon chain.

4. β-Ketoacyl ACP reductase reduces ketoacyl group to hydroxyacyl group. The reducing equivalents are supplied by NADPH (from HMP shunt).

5. β-Hydroxyacyl ACP undergoes dehydration. A molecule of water is eliminated and a double bond is introduced between α and β carbons.

6. A second NADPH-dependent reduction, catalysed by ***enoyl-ACP reductase*** occurs to produce acyl-ACP. The four-carbon unit attached to ACP is butyryl group.

The carbon chain attached to ACP is transferred to cysteine residue and the reactions 2-6 are repeated 6 more times. Each time, the fatty acid chain is lengthened by a two-carbon unit (obtained from malonyl CoA). At the end of 7 cycles, the fatty acid synthesis is complete and a 16-carbon fully saturated fatty acid—namely palmitate—bound to ACP is produced.

7. The enzyme palmitoyl thioesterase separates palmitate from fatty acid synthase. This completes the synthesis of palmitate.

Summary of palmitate synthesis

Of the 16 carbons present in palmitate, only two come from acetyl CoA directly. The remaining 14 are from malonyl CoA which, in turn, is produced by acetyl CoA. The overall reaction of palmitate synthesis is summarized

8 Acetyl CoA + 7 ATP + 14 NADPH

+ 14 H^+ ⟶ Palmitate + 8 CoA + 7 ADP

+ 7 Pi + $6H_2O$

Fatty acid synthase complex

The diagrammatic representation of the model for fatty acid synthase (***FAS***) ***multienzyme complex*** is depicted in ***Fig.14.10***.

Fatty acid synthase is a ***dimer*** composed of two identical subunits (monomers). Each subunit contains the activities of 7 enzymes of FAS and an ACP with 4′-phosphopantetheine —SH group. The two subunits lie in antiparallel (head-to-tail) orientation. The —SH group of phosphopantetheine of one subunit is in close proximity to the —SH of cysteine residue (of the enzyme ketoacyl synthase) of the other subunit.

Each monomer of FAS contains all the enzyme activities of fatty acid synthesis. But ***only the dimer is functionally active***. This is because the functional unit consists of half of each subunit interacting with the complementary half of the other. Thus, the FAS structure has both functional division and subunit division (***Fig.14.10***). ***The two functional subunits of FAS independently operate and synthesize two fatty acids simultaneously***.

SYNTHESIS OF TRIACYLGLYCEROLS

Triacylglycerol (TG) synthesis mostly occurs in ***liver*** and ***adipose tissue***, and to a lesser extent in other tissues (***Fig.14.11***). Fatty acids and glycerol must be activated prior to the synthesis of triacylglycerols. Conversion of fatty acids to acyl CoA by thiokinase is already described (refer β-oxidation).

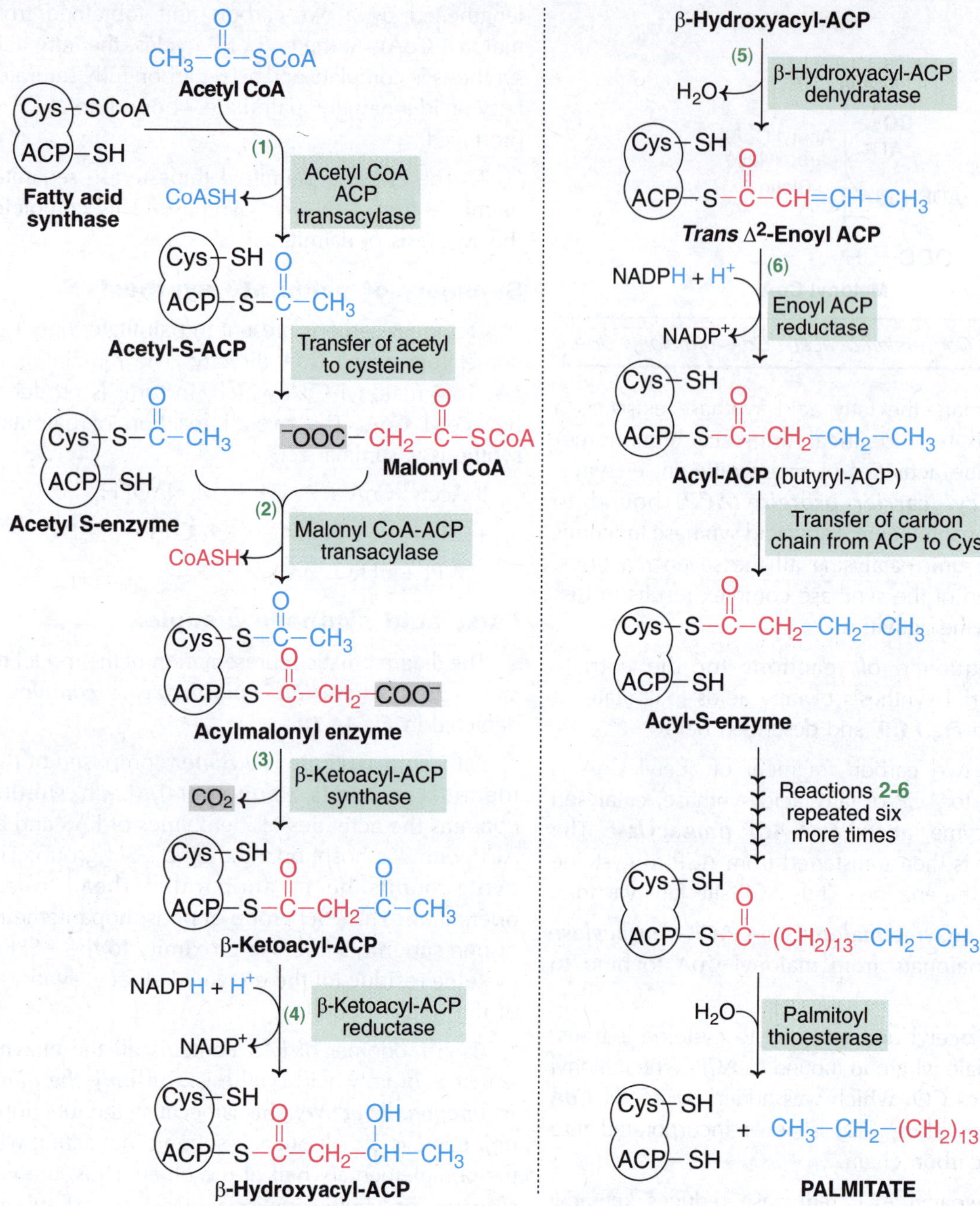

Fig. contd. next column

***Fig.14.9 :** Biosynthesis of long chain fatty acid-palmitate. (Cys–Cysteine; ACP–Acyl carrier protein; The pathway repeats 7 times to produce palmitate; the first two carbons at the methyl end are directly from acetyl CoA, the rest of the carbons come from malonyl CoA).*

Synthesis of glycerol 3-phosphate

Two mechanisms are involved for the synthesis of glycerol 3-phosphate

1. In the liver, glycerol is activated by glycerol kinase. This enzyme is absent in adipose tissue.

2. In both liver and adipose tissue, glucose serves as a precursor for glycerol 3-phosphate. Dihydroxyacetone phosphate (DHAP) produced in glycolysis is reduced by glycerol 3-phosphate dehydrogenase to glycerol 3-phosphate.

Addition of acyl groups to form TG

Glycerol 3-phosphate acyltransferase catalyses the transfer of an acyl group to produce lysophosphatidic acid. DHAP can also accept acyl group, ultimately

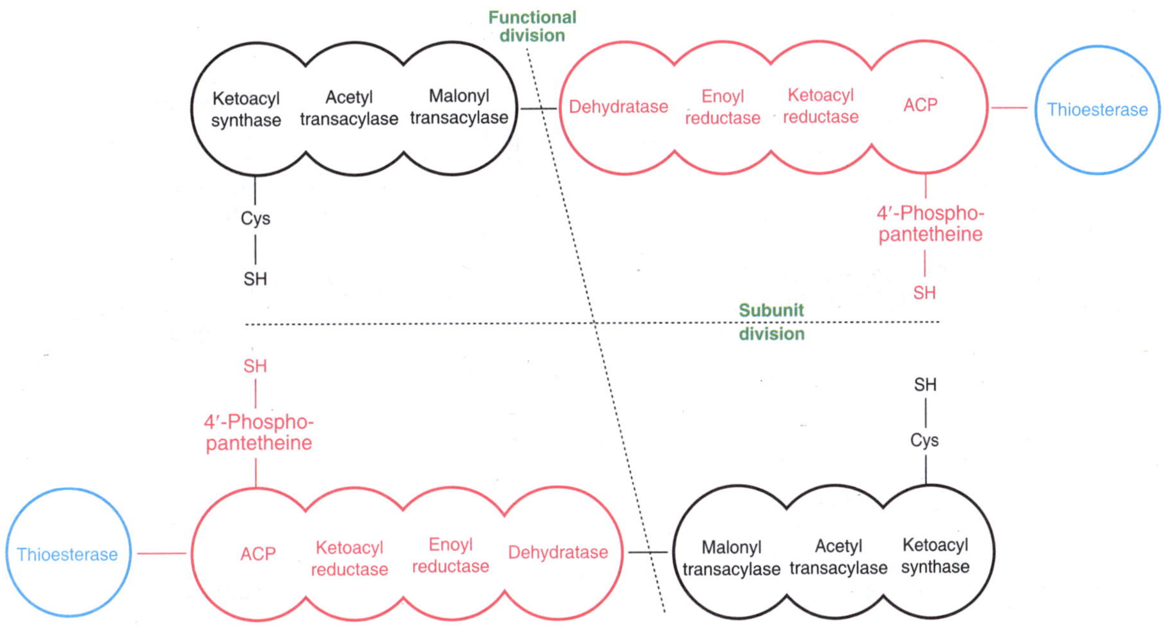

Fig.14.10 : *Fatty acid synthase multienzyme complex (ACP–Acyl carrier protein; FAS has two identical subunits which organize into two functional subunits to simultaneously synthesize two fatty acids).*

resulting in the formation of lysophosphatidic acid. Another acyl group is added to lysophosphatidic acid to form phosphatidic acid (1,2-diacylglycerol phosphate). The enzyme phosphatase cleaves off phosphate of phosphatidic acid to produce diacylglycerol. Incorporation of another acyl group finally results in synthesis of triacylglycerol.

The ***three fatty acids found in triacylglycerol are not of the same type***. A saturated fatty acid is usually present on carbon 1, an unsaturated fatty acid is found on carbon 2, and carbon 3 may have either.

The intermediates of TG synthesis phosphatidic acid and diacylglycerol are also utilized for phospholipid synthesis (described later).

METABOLISM OF PHOSPHOLIPIDS

Phospholipids are a specialized group of lipids performing a variety of functions. These include the membrane structure and functions, involvement in blood clotting and supply of arachidonic acid for the synthesis of prostaglandins (for details ***refer Chapter 32***).

Synthesis of phospholipids

Phospholipids are synthesized from phosphatidic acid and 1,2-diacylglycerol, intermediates in the production of triacylglycerols (***Fig.14.12***). Phospholipid synthesis occurs in the smooth endoplasmic reticulum.

1. **Formation of lecithin and cephalin :** Choline and ethanolamine first get phosphorylated and then combine with CTP to form, respectively, CDP-choline and CDP-ethanolamine.

Phosphatidylcholine (lecithin) is synthesized when CDP-choline combines with 1,2-diacylglycerol. Phosphatidyl ethanolamine (cephalin) is produced when CDP-ethanolamine reacts with 1,2-diacylglycerol. Phosphatidyl ethanolamine can be converted to phosphatidylcholine on methylation.

2. **Synthesis of phosphatidylserine :** Phosphatidyl ethanolamine can exchange its ethanolamine group with free serine to produce phosphatidylserine. The latter, on decarboxylation, gives the former.

3. **Formation of phosphatidylinositol :** CDP-diacylglycerol produced from phosphatidic acid combines with inositol to form phosphatidylinositol (***PI***). This phospholipid contains ***arachidonic acid*** on carbon 2 of glycerol which serves as a substrate for ***prostaglandin synthesis***. Further, Pi is important for ***signal transmission across membranes***.

4. **Synthesis of phosphatidyl glycerol and cardiolipin :** CDP-diacylglycerol combines with glycerol 3-phosphate to form phosphatidyl glycerol 3-phosphate, which then forms phosphatidylglycerol. The later combines with another molecule of

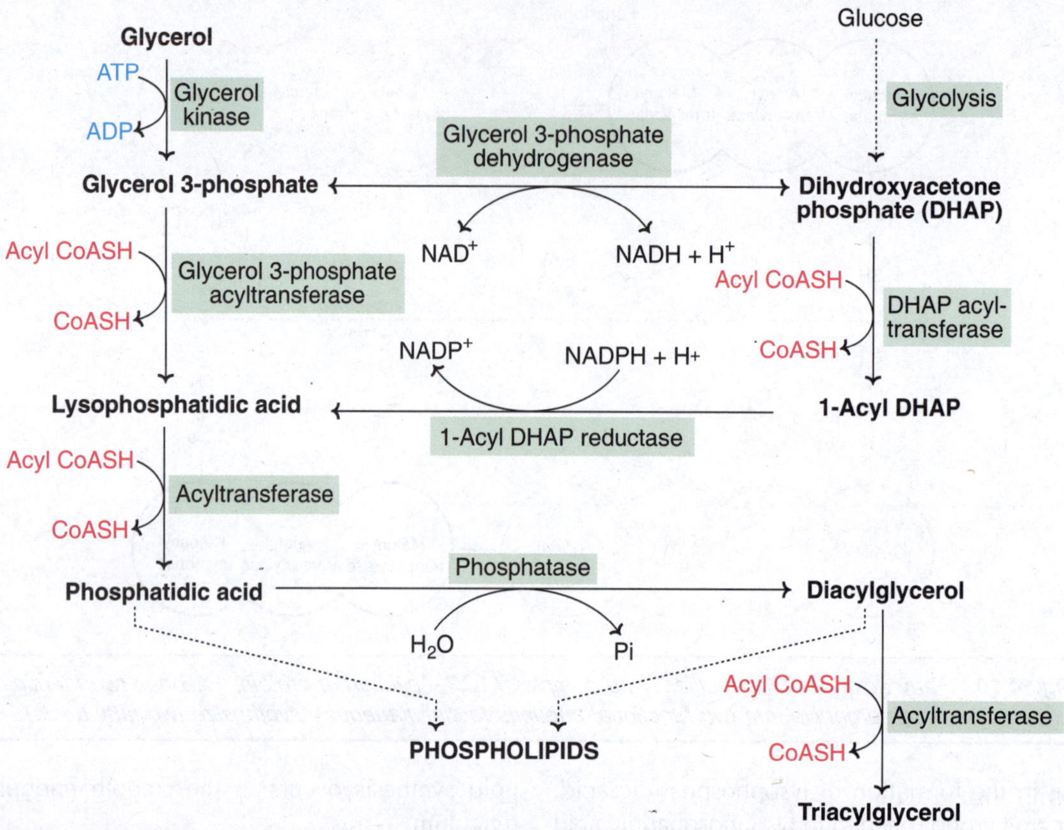

Fig.14.11 : Synthesis of triacylglycerol.

phosphatidylglycerol to finally produce ***cardiolipin*** (***Fig.14.12***).

5. **Synthesis of sphingomyelins :** These are phospholipids containing a complex amino alcohol, sphingosine, instead of glycerol. Palmitoyl CoA and serine combine and undergo a sequence of reactions to produce sphingosine which is then acylated to produce ***ceramide.*** Sphingomyelin is synthesized when ceramide combines with CDP-choline (***Fig.14.13***).

Role of LCAT in lecithin metabolism

Lecithin-cholesterol acyltransferase (LCAT) is a plasma enzyme, synthesized in the liver. LCAT activity is associated with apo A_1 of HDL. This enzyme esterifies cholesterol by transferring acyl group from the second position of lecithin

$$\text{Lecithin + Cholesterol} \xrightarrow{\text{LCAT}} \text{Lysolecithin + Cholesterol ester}$$

The above reaction is responsible for the ***reverse cholesterol transport mediated by HDL*** (more details given under lipoprotein metabolism).

Degradation of sphingomyelins

The enzyme ***sphingomyelinase*** of lysosomes hydrolyses sphingomyelins to ceramide and phosphorylcholine (***Fig.14.25***). Ceramide formed can be further degraded to sphingosine and free fatty acid.

Niemann-Pick disease : It is an inherited disorder due to a defect in the enzyme ***sphingomyelinase***. This causes accumulation of sphingomyelins in liver and spleen, resulting in the enlargement of these organs. Victims of Niemann-Pick disease suffer from ***severe mental retardation***, and death may occur in early childhood.

Farber's disease : A ***defect*** in the enzyme ***ceramidase*** results in Farber's disease. This disorder is characterized by ***skeletal deformation***, subcutaneous nodules, dermatitis and mental retardation. It is fatal in early life.

METABOLISM OF CHOLESTEROL

Cholesterol is found exclusively in animals, hence it is often called as ***animal sterol***. The total body

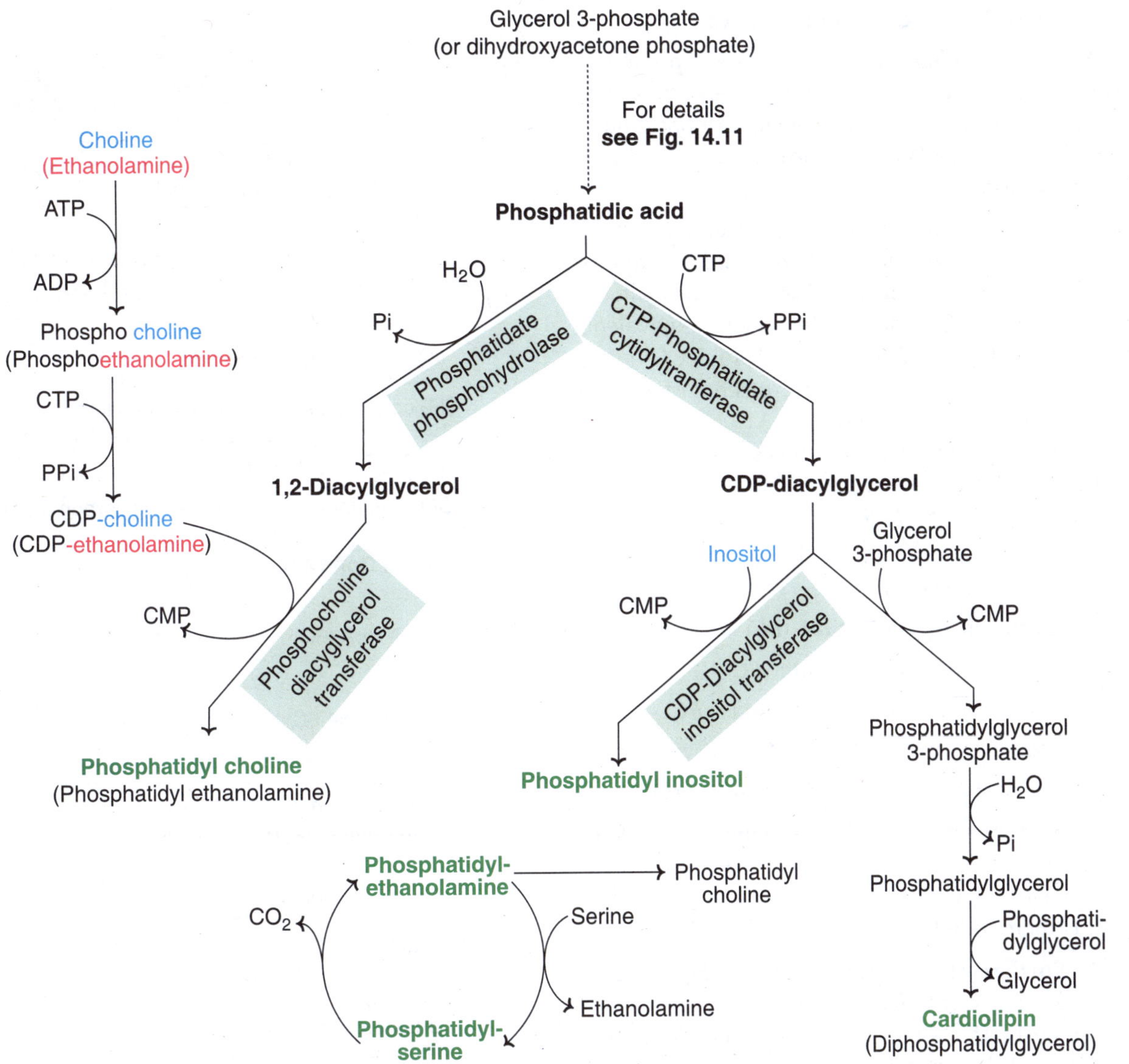

Fig.14.12 : *Biosynthesis of phospholipids.*

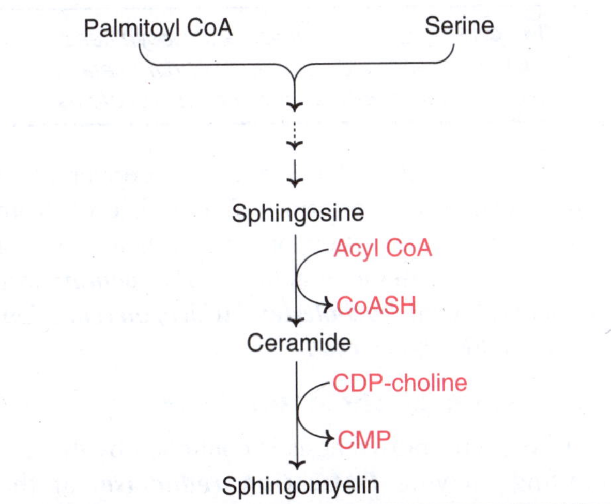

Fig.14.13 : *An outline of the synthesis of sphingomyelin.*

content of cholesterol in an adult man weighing 70 kg is about 140 g i.e., around 2 g/kg body weight. Cholesterol is ***amphipathic*** in nature, since it possesses both hydrophilic and hydrophobic regions in the structure.

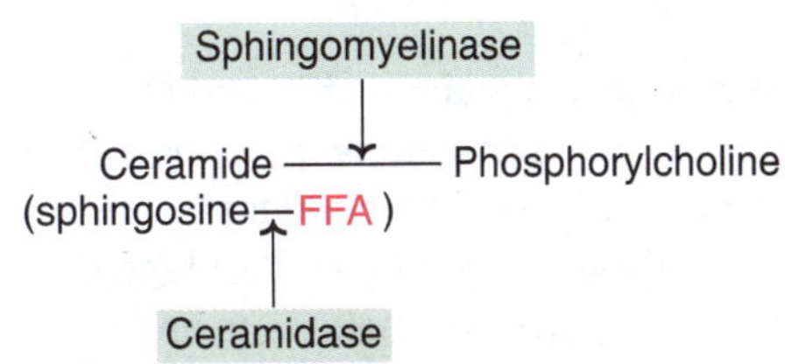

Fig.14.14 : *Site of action of sphingomyelinase and ceramidase on sphingomyelin.*

CHOLESTEROL BIOSYNTHESIS

About ***1 g of cholesterol is synthesized per day*** in adults. Almost all the tissues of the body participate in cholesterol biosynthesis. The largest contribution is made by ***liver*** (50%), ***intestine*** (15%), skin, adrenal cortex, reproductive tissue etc.

The enzymes involved in cholesterol synthesis are found in the ***cytosol*** and ***microsomal fractions*** of the cell. Acetate of ***acetyl CoA*** provides all the carbon atoms in cholesterol. The reducing equivalents are supplied by ***NADPH*** while ***ATP*** provides energy. For the production of one mole of cholesterol, 18 moles of acetyl CoA, 36 moles of ATP and 16 moles of NADPH are required. The sources of carbon atoms and the key intermediates of cholesterol formation are depicted in ***Fig.14.15***.

The synthesis of cholesterol may be learnt in 5 stages (details in ***Fig.14.16***)

1. Synthesis of ***HMG CoA (6C)***
2. Formation of ***mevalonate (6C)***
3. Production of ***isoprenoid units (5C)***
4. Synthesis of ***squalene (30C)***
5. Conversion of squalene to ***cholesterol (27C)***.

1. **Synthesis of β-hydroxy β-methylglutaryl CoA (HMG CoA) :** Two moles of acetyl CoA condense to form acetoacetyl CoA. Another molecule of acetyl CoA is then added to produce HMG CoA.

2. **Formation of mevalonate :** ***HMG CoA reductase*** is the ***rate limiting enzyme*** in cholesterol biosynthesis. This enzyme is present in endoplasmic reticulum and catalyses the reduction of HMG CoA to mevalonate. The reducing equivalents are supplied by NADPH.

3. **Production of isoprenoid units :** In a three-step reaction catalysed by kinases, mevalonate is converted to 3-phospho 5-pyrophosphomevalonate which on decarboxylation forms isopentenyl pyrophosphate (IPP). The latter isomerizes to dimethylallyl pyrophosphate (DPP). Both IPP and DPP are 5-carbon isoprenoid units.

4. **Synthesis of squalene :** IPP and DPP condense to produce a 10-carbon geranyl pyrophosphate (GPP). Another molecule of IPP condenses with GPP to form a 15-carbon farnesyl pyrophosphate (FPP). Two units of farnesyl pyrophosphate unite and get reduced to produce a 30-carbon squalene.

5. **Conversion of squalene to cholesterol :** Squalene undergoes hydroxylation and cyclization

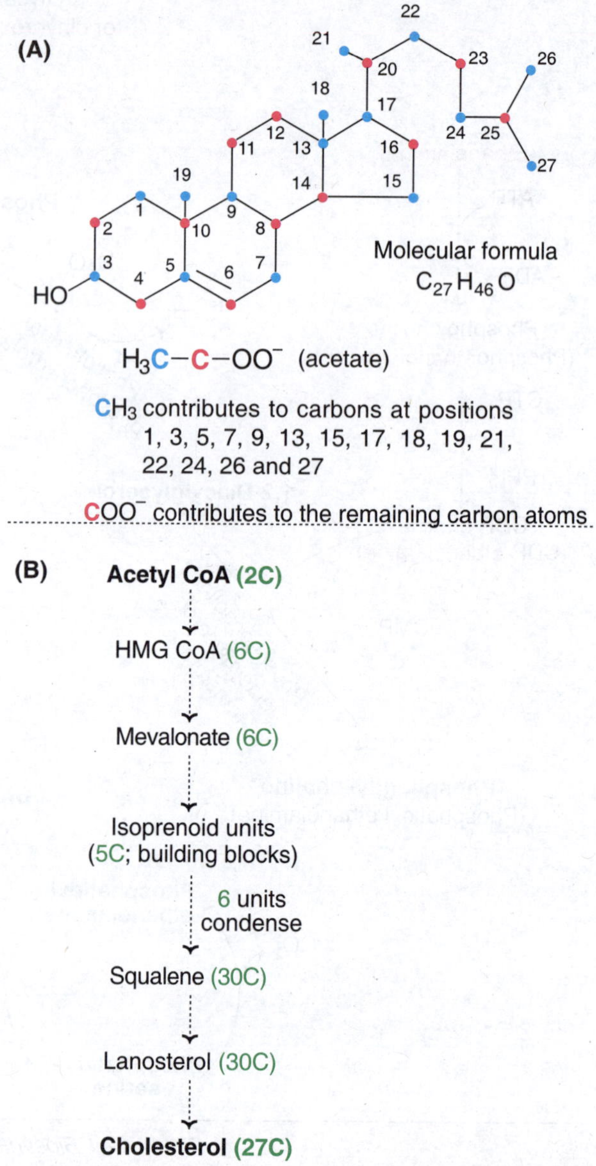

Fig.14.15 : *Outline of cholesterol biosynthesis— (A) Derivation of carbon atoms from acetate, (B) Key intermediates with the carbon atoms.*

utilizing O_2 and NADPH and gets converted to lanosterol. The formation of cholesterol from lanosterol is a multistep process with a series of about 19 enzymatic reactions. ***The penultimate product is 7-dehydrocholesterol which, on reduction, finally yields cholesterol.***

Regulation of cholesterol synthesis

Cholesterol biosynthesis is controlled by the rate limiting enzyme ***HMG CoA reductase,*** at the beginning of the pathway. HMG CoA reductase is subjected to different metabolic controls.

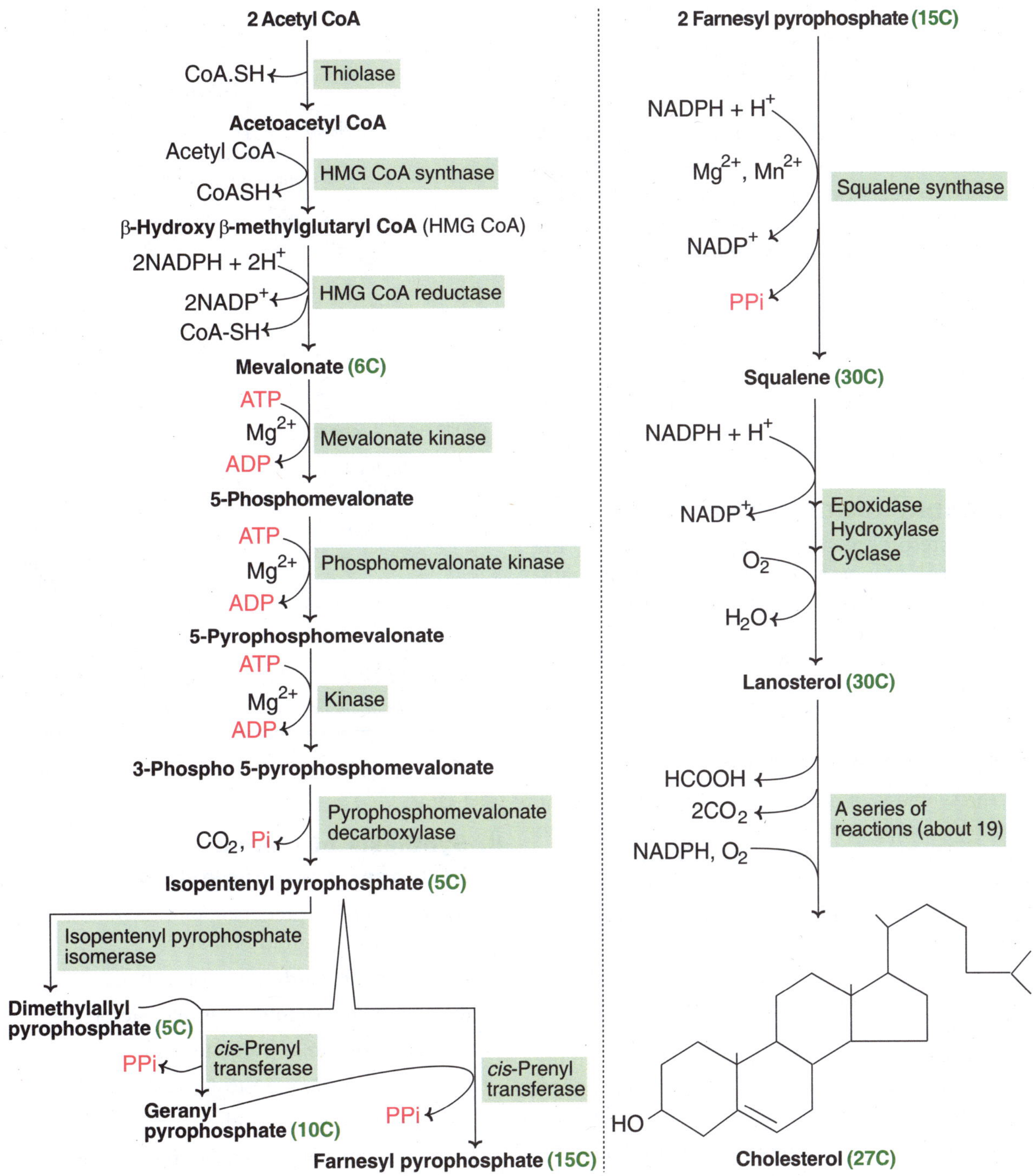

Fig. contd. next column

Fig.14.16 : *Biosynthesis of cholesterol.*

1. **Feedback control :** The end product cholesterol controls its own synthesis by a feedback mechanism. Increase in the cellular concentration of cholesterol reduces the synthesis of the enzyme HMG CoA reductase. This is achieved by ***decreasing the transcription*** of the gene responsible for the production of HMG CoA reductase.

2. **Hormonal regulation :** The enzyme HMG CoA reductase exists in two interconvertible forms. The dephosphorylated form of HMG CoA is more active while the phosphorylated form is less active. ***Glucagon*** and ***glucocorticoids*** favour the formation of inactive HMG CoA reductase (phosphorylated form) and, thus, decrease cholesterol synthesis. On

the other hand, ***insulin*** and ***thyroxine*** increase cholesterol production by enhancing the formation of active HMG CoA reductase (dephosphorylated form).

3. **Inhibition by drugs :** The drugs ***compactin*** and ***lovastatin*** (mevinolin) are fungal products. They are used to decrease the serum cholesterol level in patients with hypercholesterolemia. Compactin and lovastatin are ***competitive inhibitors*** of the enzyme ***HMG CoA reductase*** and, therefore, reduce cholesterol synthesis. About 50 to 60% decrease in serum cholesterol level has been reported by a combined use of these two drugs.

4. HMG CoA reductase activity is inhibited by ***bile acids***. ***Fasting*** also ***reduces the activity of this enzyme***.

DEGRADATION OF CHOLESTEROL

The steroid nucleus (ring structure) of the ***cholesterol cannot be broken down*** in humans. Cholesterol (50%) is converted to ***bile acids***, excreted in feces, serves as a precursor for the synthesis of ***steroid hormones***, ***vitamin D***, ***coprostanol*** and ***cholestanol***. The latter two are the fecal sterols, besides cholesterol.

I. Synthesis of bile acids

The synthesis of primary bile acids takes place in the liver and involves a series of reactions (***Fig.14.17***). The step catalysed by ***7 α-hydroxylase*** is inhibited by bile acids and this is the ***rate limiting*** reaction. Cholic acid and chenodeoxycholic acid are the primary bile acids and the former is found in the largest amount in bile. On conjugation with glycine or taurine, conjugated bile acids (glycocholic acid, taurocholic acid etc.) are formed which are more efficient in their function as surfactants. In the bile, the conjugated bile acids exist as sodium and potassium salts which are known as ***bile salts***.

In the intestine, a portion of primary bile acids undergo deconjugation and dehydroxylation to form ***secondary bile acids*** (deoxycholic acid and lithocholic acid). These reactions are catalysed by bacterial enzymes in the intestine.

Cholelithiasis : Bile salts and phospholipids are responsible for keeping the cholesterol in bile in a soluble state. Due to their deficiency (particularly bile salts), cholesterol crystals precipitate in the gall bladder often resulting in cholelithiasis—***cholesterol gall stone disease***. Cholelithiasis may be due to defective absorption of bile salts from the intestine,

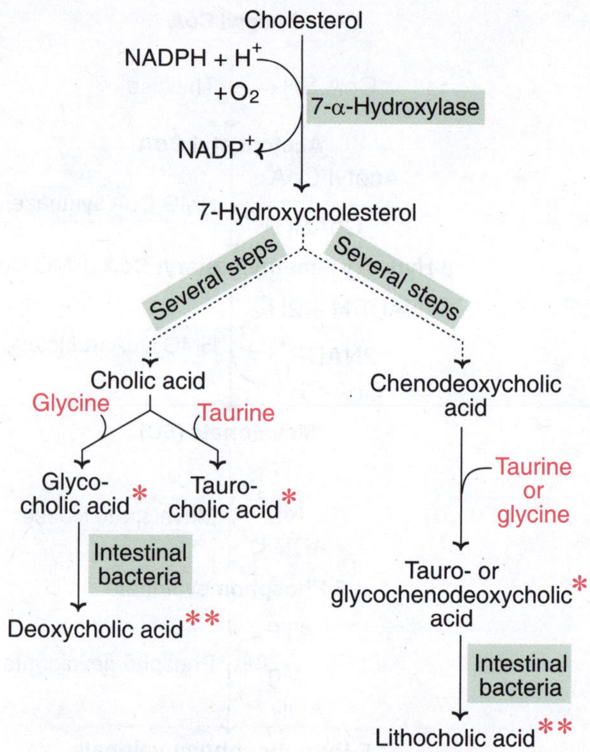

Fig.14.17 : *Outline of bile acid synthesis (*–Primary bile acids, **–Secondary bile acids).*

impairment in liver function, obstruction of biliary tract etc.

II. Synthesis of steroid hormones from cholesterol

Cholesterol is the precursor for the synthesis of all the five classes of steroid hormones

(a) ***Glucocorticoids*** (e.g. cortisol)

(b) ***Mineralocorticoids*** (e.g. aldosterone)

(c) ***Progestins*** (e.g. progesterone)

(d) ***Androgens*** (e.g. testosterone)

(e) ***Estrogens*** (e.g. estradiol).

A brief outline of steroid hormonal synthesis is given in ***Fig.14.18*** and more details are discussed under '***Hormones***' (***Chapter 19***).

III. Synthesis of vitamin D

7-Dehydrocholesterol, an intermediate in the synthesis of cholesterol, is converted to cholecalciferol (vitamin D_3) by ultraviolet rays in the skin.

A brief summary of prominent sources and the major pathways for utilization of cholesterol with the liver as the central metabolic organ is depicted in ***Fig.14.19***.

Medical Concepts/ Clinical Correlates

- **Niemann-Pick disease, caused by a defect in the enzyme sphingomyelinase, results in the accumulation of sphingomyelins in liver and spleen.**
- **About a dozen glycolipid storage diseases are known. These include Gaucher's disease and Krabbe's disease.**
- **Hypercholesterolemia is associated with atherosclerosis and coronary heart diseases. Consumption of polyunsaturated fatty acids and fiber decreases cholesterol in circulation. Drugs—such as lovastatin, cholestyramine, compactin and clofibrate—reduce plasma cholesterol.**
- **It has been proved that there is no significant correlation between dietary cholesterol and serum cholesterol. Thus, consumption of cholesterol-rich foods (eggs, butter, sea foods etc.) are now considered safe.**

Transport of cholesterol

Cholesterol is present in the plasma lipoproteins in two forms

1. About 70-75% of it is in an esterified form with long chain fatty acids.

2. About 25-30% of it as free cholesterol. This form of cholesterol readily exchanges between different lipoproteins and also with the cell membranes.

Role of LCAT : High density lipoproteins (HDL) and the enzyme ***lecithin-cholesterol acyltransferase (LCAT)*** are responsible for the transport and elimination of cholesterol from the body.

The cholesterol (cholesteryl) ester forms an integral part of HDL. In this manner, the cholesterol from the peripheral tissues is trapped in HDL, by a reaction catalysed by LCAT and then transported to liver for degradation and excretion. This mechanism is commonly known as ***reverse cholesterol transport.***

Plasma cholesterol— biomedical importance

In healthy individuals, the total plasma cholesterol is in the range of 150-200 mg/dl. Cholesterol level increases with increasing age (in women particularly after menopause), and also in pregnancy.

Plasma cholesterol is associated with different lipoprotein fractions (LDL, VLDL and HDL).

In adults, the normal LDL-cholesterol is about 70-200 mg/dl while HDL-cholesterol is around 30-60 mg/dl. Elevation in plasma HDL-cholesterol is beneficial to the body, since it protects the body from atherosclerosis and coronary heart diseases (CHD). On the other hand, increase in LDL-cholesterol is harmful to the body as it may lead to various complications, including CHD.

HYPERCHOLESTEROLEMIA

Increase in plasma cholesterol (***>200 mg/dl***) concentration is known as hypercholesterolemia and is observed in many disorders

1. **Diabetes mellitus :** Due to ***increased cholesterol synthesis*** since the availability of acetyl CoA is increased.

2. **Hypothyroidism (myxoedema) :** This is believed to be due to ***decrease in the HDL receptors*** on hepatocytes.

3. **Obstructive jaundice :** Due to an ***obstruction in the excretion*** of cholesterol through bile.

4. **Nephrotic syndrome :** Increase in plasma globulin concentration is the characteristic feature of nephrotic syndrome. Cholesterol elevation is due to an increase in plasma lipoprotein fractions in this disorder.

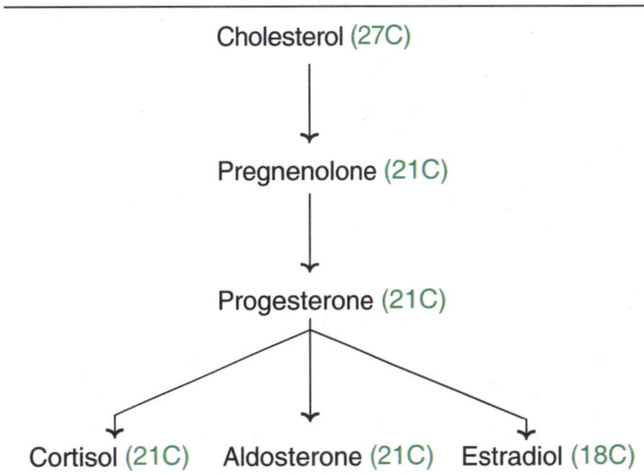

***Fig.14.18 :** Outline of steroid hormone synthesis from cholesterol (Numbers in the brackets represent the number of carbon atoms).*

Hypercholesterolemia is also associated with ***atherosclerosis*** and ***coronary heart disease.***

Bad cholesterol and good cholesterol : Cholesterol is a natural metabolite performing a wide range of functions (membrane structure, precursor for steroid hormones, bile acids). The usages good and bad to cholesterol, although inappropriate, are still in use. The cholesterol in high concentration, present in ***LDL***, is considered ***bad*** due to its involvement in altherosclerosis and related complications. Thus, LDL may be regarded as ***l***ethally ***d***angerous ***l***ipoprotein. Small dense LDL (***sdLDL***) is considered to be the ***most dangerous fraction of LDL*** associated with CHD. On the other hand, ***HDL cholesterol*** is ***good*** since its high concentration counteracts atherogenesis. HDL may be considered as ***h***ighly ***d***esirable ***l***ipoprotein.

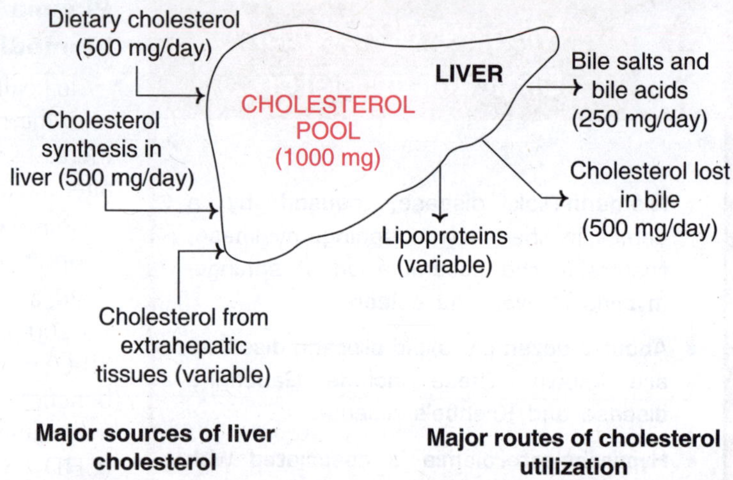

Fig.14.19 : *Summary of major sources of liver cholesterol and its utilization (values given in brackets are variable).*

Affects of lifestyles on serum cholesterol level : Individual lifestyles and habits certainly influence serum cholesterol, and thus play a significant role in the development coronary heart disease. The parametres such as ***high blood pressure***, ***emotional stress***, ***smoking***, drinking of soft water (against hard water), coffee drinking, lack of exercise, obesity (particlarly of abdomen) ***elevate serum cholesterol level***.

Control of hypercholesterolemia

Several measures are advocated to lower the plasma cholesterol (particularly LDl) level

1. **Consumption of PUFA :** Dietary intake of polyunsaturated fatty acids (PUFA) ***reduces the plasma cholesterol level***. PUFA will help in transport of cholesterol by LCAT mechanism (described earlier) and its excretion from the body. The oils with rich PUFA content include cottonseed oil, soya bean oil, sunflower oil, corn oil, fish oils etc. Ghee and coconut oil are poor sources of PUFA.

2. **Dietary cholesterol :** Dietary cholesterol ***influence*** on plasma cholesterol is ***minimal***. However, avoidance of cholesterol-rich foods is advocated, and a dietary intake of <300 mg/day is advised. Certain drugs (e.g. ezetimibe) inhibit intestinal cholesterol absorption.

3. **Plant sterols :** Certain plant sterols and their esters (e.g. sitostanol esters) ***reduce*** plasma cholesterol levels. They inhibit the intestinal absorption of dietary cholesterol.

4. **Dietary fiber :** Fiber present in vegetables ***decreases the cholesterol absorption*** from the intestine.

5. **Avoiding high carbohydrate diet :** Diets rich in carbohydrates (e.g. sucrose) should be avoided to control hypercholesterolemia.

6. **Impact of lifestyles :** Elevation in plasma cholesterol is obseved in people with smoking, abdominal obesity, lack of exercise, stress, high blood pressure, consumption of soft water etc. Therefore, adequate ***changes in the lifestyles*** will bring down plasma cholesterol.

7. **Moderate alcohol cosumption :** The beneficial effects of moderate alcohol intake are masked by the ill effects of chronic alcoholism. ***Red wine*** is particularly ***beneficial*** due to its antioxidants, besides low alcohol content.

8. **Use of drugs :** Drugs such as ***lovastatin*** which inhibit HMG CoA reductase and decrease cholesterol synthesis are used. Statins currently in use include atorvastatin, simvastatin, fluvastatin and pravastatin. ***Statins*** are usually ***taken at night to ensure maximum effect*** (HMG CoA reductase activity at peak about 6 hours after dark). Certain drugs—***cholestyramine*** and ***colestipol***—bind with bile acids and decrease their intestinal reabsorption. This helps in the conversion of more cholesterol to bile acids and its excretion through feces. ***Clofibrate*** increases the activity of lipoprotein lipase and reduces the plasma cholesterol and triacylglycerols.

Medical Concepts/ Clinical Correlates

- High density lipoproteins—in association with lecithin-cholesterol acyltransferase (LCAT)—are responsible for the transport and elimination of cholesterol from the body.
- Hyperlipoproteinemias are a group of disorders caused by the elevation of one or more of plasma lipoprotein fractions.
- Excessive accumulation of triacylglycerols causes fatty liver which can often be prevented by the consumption of lipotropic factors (choline, betaine, methionine).

Dietary cholesterol—safe to consume!

In the dietary guidelines, cholesterol was treated as a nutrient of concern and its daily intake was restricted to not more than 300 mg. The US Government in 2015 has removed ***cholesterol from the list of nutrient of concern taking a U-turn on its earlier recommendations***.

The earlier warning (since 1970) world over was to stay away from high cholesterol foods to avoid heart diseases and clogging of arteries. Research findings repeatedly indicated that there is ***no significant relationship between consumption of dietary cholesterol and serum cholesterol***. Further, majority of people that have heart attacks were found to have normal cholesterol. Thus, ***cholesterol has become the biggest medical scam of all time.***

With the recent developments, ***consumption of cholesterol—rich foods such as eggs, sea foods***, animal fats and oils, fish and fish oils, cheese etc. are now ***considered safe*** (with regard to cholesterol).

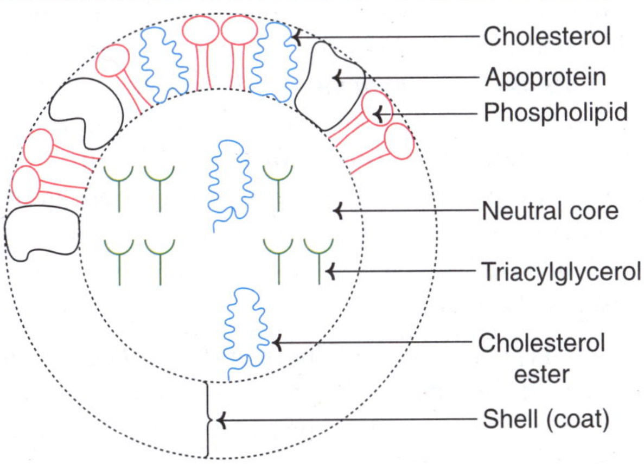

Fig.14.20 : A general structure of lipoprotein complex. (Note : For the sake of clarity, only a part of the shell and core are filled with the constituents).

Cholesterol is a friend, not an enemy

Cholesterol is a vital and universal ***component of every cell*** membrane on earth. Further, cholesterol is the precursor for the ***synthesis of steroid hormones, bile acids, vitamin D*** etc. Thus, there is no life without cholesterol. Therefore, cholesterol should be treated as one of our best friends, not an evil.

The focus now is on dietary transfat and sugars as culprits in the causation of heart diseases.

LIPOPROTEINS

Lipoproteins are ***molecular complexes*** that consist of ***lipids and proteins*** (conjugated proteins). They function as transport vehicles for lipids in blood plasma. Lipoproteins deliver the lipid components (cholesterol, triacylglycerol etc.) to various tissues for utilization.

Structure of lipoproteins

A lipoprotein basically consists of a neutral lipid core (with triacylglycerol and/or cholesteryl ester) surrounded by a coat shell of phospholipids, apoproteins and cholesterol (***Fig.14.20***). The polar portions (amphiphilic) of phospholipids and cholesterol are exposed on the surface of lipoproteins so that lipoprotein is soluble in aqueous solution.

Classification of lipoproteins

Five major classes of lipoproteins are identified in human plasma, based on their ***separation by electrophoresis*** (***Fig.14.21***).

1. **Chylomicrons :** They are synthesized in the intestine and transport exogenous (dietary) triacylglycerol to various tissues. They consist of highest (99%) quantity of lipid and lowest (1%) concentration of protein.

2. **Very low density lipoproteins (VLDL) :** They are produced in liver and intestine and are responsible for the transport of endogenously synthesized triacylglycerols.

Table 14.2 Classification and characteristics of hyperlipoproteinemias (hyperlipidemias)

Hyperlipoproteinemia Type	Increased plasma lipoprotein(s)	Increased plasma lipid (most)	Probable metabolic defect	Risk of atherosclerosis	Suggested treatment
I	Chylomicrons	Triacylglycerols	Deficiency of lipoprotein lipase	May increase	Low fat diet
IIa	LDL	Cholesterol	Deficiency of LDL receptors	Very high (mostly in coronary artery)	Low cholesterol fat diet; cholestyramine
IIb	LDL and VLDL	Triacylglycerols and cholesterol	Overproduction of apo-B	— do —	— do —
III	IDL	Triacylglycerols and cholesterol	Abnormality in apo-E	Very high (mostly in peripheral vessels)	Low fat and low caloric diet; clofibrate
IV	VLDL	Triacylglycerols	Overproduction of TG	May or may not increase	Low fat and low caloric diet; niacin
V	Chylomicrons and VLDL	Triacylglycerols	—	— do —	— do —

3. **Low density lipoproteins (LDL) :** They are formed from VLDL in the blood circulation. They transport cholesterol from liver to other tissues.

4. **High density lipoproteins (HDL) :** They are mostly synthesized in liver. Three different fractions of HDL (1,2 and 3) can be identified by ultracentrifugation. HDL particles transport cholesterol from peripheral tissues to liver (reverse cholesterol transport).

5. **Free fatty acids—albumin :** Free fatty acids in the circulation are in a bound form to albumin. Each molecule of albumin can hold about 20-30 molecules of free fatty acids. This lipoprotein cannot be separated by electrophoresis.

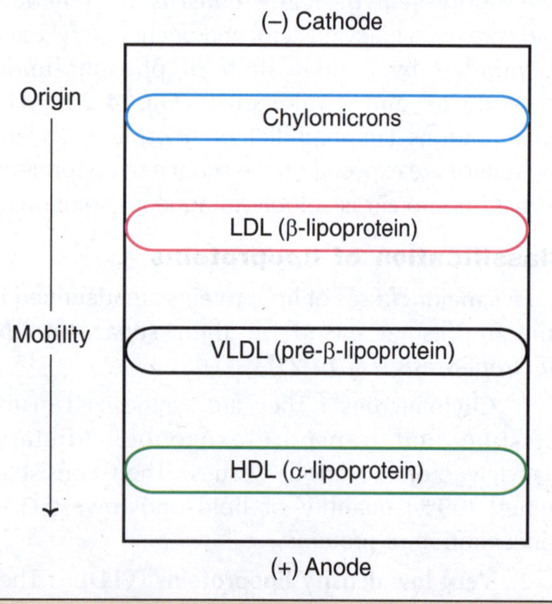

***Fig.14.21 :** Electrophoresis of plasma (serum) lipoproteins.*

DISORDERS OF PLASMA LIPOPROTEINS

Inherited disorders of lipoproteins are encountered in some individuals resulting in ***primary*** hyper- or hypolipoproteinemias. These are due to ***genetic defects*** in lipoprotein metabolism and transport. The ***secondary*** acquired lipoprotein disorders are due to some other diseases (e.g. diabetes mellitus, nephrotic syndrome, atherosclerosis, hypothyrodism etc.), resulting in abnormal lipoprotein pattern which often resembles the primary inherited condition.

Hyperlipoproteinemias

Elevation in one or more of the lipoprotein fractions constitutes hyperlipoproteinemias. These disorders may be either primary or secondary. Some authors use ***hyperlipidemias*** or ***dyslipidemias*** instead of hyperlipoproteinemias. ***Fredrickson's classification of hyperlipoproteinemias***—based on the electrophoretic patterns of plasma lipoproteins —is widely accepted to understand these disorders. It is given in ***Table 14.2***, and briefly discussed hereunder.

1. **Type I :** This is due to familial ***lipoprotein lipase deficiency.*** The enzyme defect causes increase in plasma chylomicron and triacylglycerol levels.

2. **Type IIa :** This is also known as hyperbetalipoproteinemia and is caused by ***a defect in LDL receptors.*** Secondary type IIa hyperlipoproteinemia

Medical Concepts/ Clinical Correlates

- **Atherosclerosis is characterized by hardening of arteries due to the accumulation of lipids and other compounds. The probable causes of atherosclerosis include hyperlipoproteinemias, diabetes mellitus, obesity, high consumption of saturated fat, lack of exercise and stress.**
- **Atherosclerosis and coronary heart disease are directly correlated with plasma cholesterol and LDL, inversely with HDL. Elevation of plasma lipoprotein a suggests increased risk of CHD.**
- **Alcoholism is associated with fatty liver, hyperlipidemia and atherosclerosis.**

is observed in association with diabetes mellitus, hypothyroidism, nephrotic syndrome etc. This disorder is characterized by hypercholesterolemia.

3. **Type IIb :** Both LDL and VLDL increase along with elevation in plasma cholesterol and triacylglycerol. This is believed to be due to ***overproduction of apo B***.

4. **Type III :** This is commonly known as broad beta disease and characterized by the appearance of a broad β-band corresponding to ***intermediate density lipoprotein*** (IDL) on electrophoresis.

5. **Type IV :** This is due to overproduction of endogenous triacylglycerols with a concomitant ***rise in VLDL***. Type IV disorder is usually associated with obesity, alcoholism, diabetes mellitus etc.

6. **Type V :** Both ***chylomicrons and VLDL*** are elevated. This is mostly a secondary condition, due to disorders such as obesity, diabetes and excessive alcohol consumption etc.

FATTY LIVER

The normal concentration of ***lipid*** (mostly phospholipid) ***in liver is around 5%***. Liver is not a storage organ for fat, unlike adipose tissue. However, in certain conditions, lipids—especially the ***triacylglycerols—accumulate*** excessively in liver, resulting in fatty liver. Fatty liver may occur due to two main causes.

1. Increased synthesis of triacylglycerols
2. Impairment in lipoprotein synthesis.

1. **Increased triacylglycerol synthesis :** Mobilization of free fatty acids from adipose tissue and their influx into liver is much higher than their utilization. This leads to the overproduction of triacylglycerols and their accumulation in liver. ***Diabetes mellitus***, ***starvation***, ***alcoholism*** and ***high fat diet*** are associated with increased mobilization of fatty acids that often cause fatty liver. Alcohol also inhibits fatty acid oxidation and, thus, promotes fat synthesis and its deposition.

2. **Impaired synthesis of lipoproteins :** The synthesis of very low density lipoproteins (VLDL) actively takes place in liver. VLDL formation requires phospholipids and apoprotein B. Fatty liver caused by impaired lipoprotein synthesis may be due to :

- A defect in phospholipid synthesis;
- A block in apoprotein formation;
- A failure in the formation/secretion of lipoprotein.

Among the three causes, fatty liver due to impairment in phospholipid synthesis has been studied in some detail. This is usually associated with the dietary ***deficiency of lipotropic factors*** such as choline, betaine, inositol etc. (more details given later). Deficiency of essential fatty acids leads to a decreased formation of phospholipids. Further, excessive consumption of cholesterol competes with essential fatty acids and impairs phospholipid synthesis.

Certain ***chemicals*** (e.g. puromycin, ethionine, carbon tetrachloride, chloroform, lead, phosphorus etc.) that ***inhibit protein synthesis cause fatty liver***. This is due to a blockade in the synthesis of apoprotein B required for VLDL production.

Lipoprotein synthesis and their secretion require ATP. Decrease in the availability of ATP, sometimes found in pyridoxine and pantothenic acid deficiency, impairs lipoprotein formation. The action of ethionine in the development of fatty liver is believed to be due to a reduction in the availability of ATP. Ethionine competes with methionine and traps the available adenosine (as adenosylethionine)—thereby reducing ATP levels.

Deficiency of vitamin E is associated with ***fatty liver***. ***Selenium*** acts as a ***protective*** agent in such a condition.

Endocrine factors : Certain hormones like ACTH, ***insulin***, ***thyroid hormones***, adrenocorticoids ***promote deposition of fat in liver***.

Alcoholism and fatty liver : The most common cause of fatty liver and cirrhosis is chronic alcoholism. Due to ***impairment in citric acid cycle in alcoholism*** (described later), acetyl CoA accumulates ***leading to fatty acid synthesis and TG accumulation in liver***.

Non-alcoholic fatty liver disease (NAFLD) : Uncontrolled diabetes and consumption of high fat diet are also associated with fatty liver.

LIPOTROPIC FACTORS

These are the substances the ***deficiency of which causes fat*** (triacylglycerol) ***to accumulate in liver.*** This may happen despite the fatty acid synthesis and uptake by liver being normal.

Important lipotropic factors

These include ***choline***, ***betaine***, ***methionine*** and ***inositol***. ***Folic acid***, ***vitamin*** B_{12}, glycine and serine also serve as lipotropic factors to some extent.

Action of lipotropic factors

Choline and ***inositol*** are ***components of phospholipids*** and, hence, required for their synthesis. The other lipotropic factors are directly or indirectly concerned with transmethylation reactions and, ultimately, the synthesis of choline. Severe ***protein deficiency*** (e.g. kwashiorkor) causes fatty liver. This is due to a defect in the synthesis of choline as a result of insufficient amino acid (particularly methionine) supply. In other words the ***non-availability of methyl groups may lead to fatty liver***.

Choline deficiency and fatty liver

Several explanations are offered to understand choline deficiency causing fatty liver :

1. Decreased phospholipid synthesis
2. Impaired formation of lipoprotein membrane
3. Reduced synthesis of carnitine due to insufficient supply of methyl groups
4. Impairment in fatty acid oxidation.

ATHEROSCLEROSIS

Atherosclerosis (*Greek*: athere—mush) is a complex disease characterized by thickening or ***hardening of arteries due to the accumulation of lipids*** (particularly cholesterol, free, and esterified) collagen, fibrous tissue, proteoglycans, calcium deposits etc. in the inner arterial wall. Atherosclerosis is a progressive disorder that narrows and ultimately blocks the arteries. ***Coronary arteries***—the arteries supplying blood to heart—are the most commonly affected leading to myocardial infarction or heart attacks.

Causes of atherosclerosis and CHD : The development of atherosclerosis and the risk for the coronary heart disease (CHD) is directly correlated with plasma cholesterol and LDL (bad cholesterol). On the other hand, plasma HDL is inversely correlated with CHD (good cholesterol).

Disorders that may cause atherosclerosis

Certain diseases are associated with atherosclerosis. These include ***diabetes mellitus***, ***hyperlipoproteinemias***, ***nephrotic syndrome***, ***hypothyroidism*** etc. Many other factors like obesity, high consumption of saturated fat, excessive smoking, ***lack of physical exercise***, ***hypertension***, stress etc., are the probable causes of atherosclerosis.

Relation between HDL and CHD

The increased levels of plasma HDL (good cholesterol) are correlated with a low incidence of cardiovascular disorders. ***Women*** have higher HDL and are ***less prone to heart diseases*** compared to men. This is attributed to ***estrogens*** in women. Strenuous physical exercise, moderate alcohol intake, consumption of unsaturated fatty acids (vegetable and fish oils), reduction in body weight—all tend to increase HDL levels and reduce the risk cardiovascular diseases.

Lipoprotein a and CHD

Lipoprotein a (Lp-a) is almost identical in structure to LDL. Lp-a contains an additional apoprotein, apo-a. Lp-a inhibits fibrinolysis. Recent studies have shown that elevation of lipoprotein-a in the plasma (>30 mg/dl) suggests ***increased risk of CHD***. It is hypothesized that elevated Lp-a reduces the breakdown of blood clots by interfering with plasminogen activation. This results in intravascular thrombosis, and increased risk of heart attacks. ***Indians*** have ***higher levels of Lp-a*** compared to Western population.

Antioxidants and atherosclerosis

Antioxidants, in general, decrease the oxidation of LDL. There is some evidence, based on the epidemiological studies that taking of antioxidants (***vitamins E and C, or β-carotene***) reduces the risk of atherosclerosis, and CHD.

ALCOHOL METABOLISM

Walker has rightly said 'alcohol can be a food, a drug or a poison depending on the dose.' In small quantities, alcohol relieves tension and anxiety. Unfortunately, consumption of alcohol seldom ends with small doses, hence the beneficial effects are over-shadowed by the harmful effects.

Alcohol (***ethanol or ethyl alcohol***) is readily absorbed by the stomach and intestine. Consequently, less than 2% of the alcohol consumed is excreted through lungs, urine and sweat.

Alcohol gets oxidized in the liver by alcohol dehydrogenase to acetaldehyde.

$$\underset{\textbf{Alcohol}}{CH_3{-}CH_2{-}OH} \xrightarrow[NAD^+ \;\; NADH + H^+]{\text{Alcohol dehydrogenase}} \underset{\textbf{Acetaldehyde}}{CH_3CHO}$$

Besides ADH, microsomal ethanol oxidizing system (MEOS) is also involved in the metabolism of alcohol. ***Aldehyde,*** produced by the action of either ADH or MEOS, is ***responsible for the manifestations of alcohol***.

Biochemical changes in alcoholism

The metabolism of alcohol (by both dehydrogenases) involves the consumption of NAD^+, and consequently a high $NADH/NAD^+$ ratio. This is mostly responsible for the metabolic alterations observed in alcoholism. Some of them are listed.

1. High concentration of NADH favours the conversion of pyruvate to lactate which may lead to ***lactic acidosis***.
2. ***Hypoglycemia*** due to reduced gluconeogenesis is observed. This happens as a result of decreased availability of pyruvate and oxaloacetate (the latter gets converted to malate by high NADH).
3. Acetaldehyde interferes with the functioning of neurotransmitters, with an overall effect of neurological depression.
4. ***Acetaldehyde causes headache, nausea***, tachycardia, reduced blood pressure etc.

Effects of chronic alcoholism

Chronic alcoholism is associated with ***cirrhosis of liver***, ***neurodegenerative changes***, ***cardiomyopathy***, ***diuresis, impotence etc***.

SUMMARY

1. *Triacylglycerols (TG) are the highly concentrated form of energy, stored in adipose tissue. Hormone-sensitive lipase hydrolyses TG to free fatty acids.*
2. *Fatty acids are activated (acyl CoA) and transported by carnitine to mitochondria where they get oxidized (mostly β type) to liberate energy. Complete oxidation of one mole palmitate liberates 106 ATP.*
3. *Excessive utilization of fatty acids occurs in uncontrolled diabetes mellitus and starvation. This results in the overproduction of ketone bodies (in liver), namely acetone, acetoacetic acid and β-hydroxy butyric acid. The last two ketone bodies serve as energy source for peripheral tissues.*
4. *Fatty acid biosynthesis occurs from acetyl CoA in the cytosol through the involvement of a multienzyme complex associated with acyl carrier protein (ACP). The reducing equivalents ($NADPH + H^+$) are supplied mostly by HMP shunt.*
5. *Cholesterol is synthesized from acetyl CoA in a series of reactions involving HMG CoA, mevalonate, isoprenoid units and squalene as the intermediates. Cholesterol serves as a precursor for bile acids, steroid hormones and vitamin D.*
6. *Lipoproteins are the transport vehicles for lipids in the plasma. Lipoprotein disorders are associated with abnormalities in their plasma levels. Elevation in LDL and VLDL—in association with cholesterol and TG—poses a serious health problem with increased risk of atherosclerosis and CHD.*

THE READER MUST REFER APPENDIX D (Page Nos : 422 - 431) FOR RELATED CLINICAL CASE STUDIES (Nos : 10, 36)

METABOLISM OF AMINO ACIDS

15

Proteins are the ***most abundant organic compounds*** and constitute a major part of the body dry weight (10-12 kg in adults). They perform a wide variety of static (structural) and dynamic (enzymes, hormones, clotting factors, receptors etc.) functions. About half of the body protein (predominantly collagen) is present in the supportive tissue (skeleton and connective) while the other half is intracellular.

The proteins on degradation (proteolysis) release individual amino acids. Amino acids are not just the structural components of proteins. Each of the 20 naturally occurring amino acids undergoes its own metabolism and performs specific functions. Some of the amino acids also serve as precursors for the synthesis of many biologically important compounds (e.g. melanin, serotonin, creatine etc.), ***Protein metabolism is more appropriately learnt as metabolism of amino acids.***

AMINO ACID POOL

An adult has about ***100 g of free amino acids*** which represent the amino acid pool of the body. The amino acid pool may be an oversimplification of the facts, since there is no single compartment—rather, several compartments exist.

Glutamate and ***glutamine*** together constitute ***about 50%***, and essential amino acids about 10% of the body pool (100 g). The concentration of intracellular amino acids is always higher than the extracellular amino acids. Amino acids enter the cells against a concentration gradient by active transport.

The amino acid pool (100 g) of the body is maintained by the sources that contribute (input) and the metabolic pathways that utilize (output) the amino acids (***Fig.15.1***).

I. Sources of amino acid pool

Turnover of body protein, intake of dietary protein and the synthesis of non-essential amino acids contribute to the body amino acid pool.

(a) **Protein turnover :** The protein present in the body is in a dynamic state. It is estimated that about ***300-400 g of protein per day*** is constantly degraded and synthesized which represents body protein turnover.

(b) **Dietary protein :** There is a regular loss of nitrogen from the body due to degradation of amino acids. In healthy adults, it is estimated that about 30-50 g of protein is lost everyday from the body. This amount of ***protein (30-50 g/day)*** must, therefore, be supplied daily in the diet to ***maintain nitrogen balance***. The RDA for protein is 1 g/kg body weight/day.

(c) **Synthesis of non-essential amino acids :** ***Ten*** out of the 20 naturally occurring amino acids ***can be synthesized*** by the body which contributes to the amino acid pool.

II. Utilization of amino acids from body pool

(a) Most of the ***body proteins (300-400 g/day)*** degraded are synthesized from the amino acid pool. These include enzymes, hormones, immunoproteins, contractile proteins etc.

(b) Many important ***nitrogenous compounds*** (porphyrins, purines, pyrimidines, etc.) are produced from the amino acids.

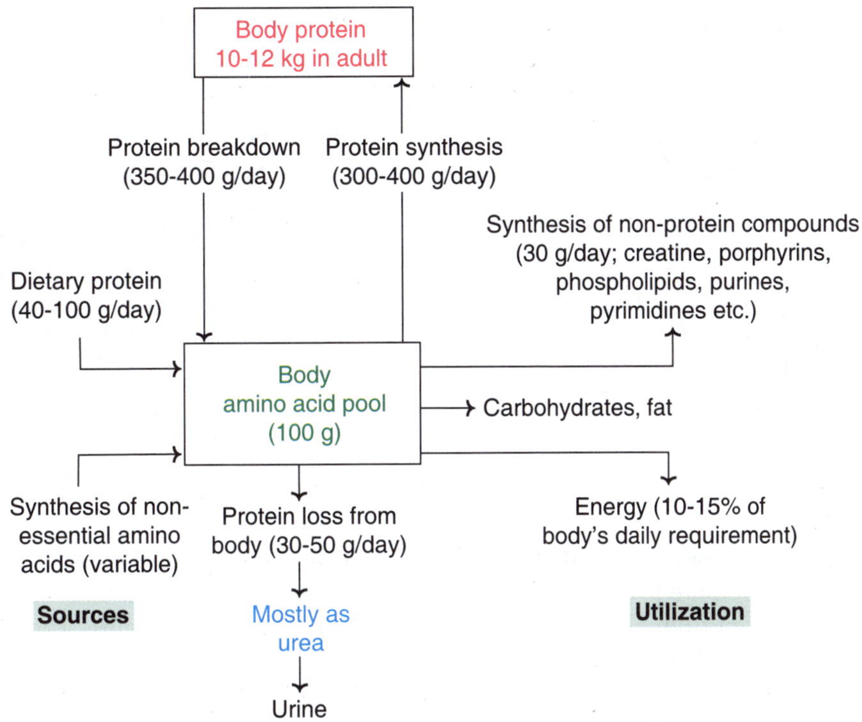

***Fig.15.1 :** Overview of body's amino acid pool—sources and utilization.*

(c) Generally, ***about 10-15% of body energy*** requirements are met from the amino acids.

(d) The amino acids are ***converted to carbohydrates and fats***.

METABOLISM OF AMINO ACIDS— GENERAL ASPECTS

The amino acids undergo certain common reactions like ***transamination*** followed by ***deamination*** for the liberation of ***ammonia***. The amino group of the amino acids is utilized for the formation of ***urea*** which is an excretory ***end product of protein metabolism***. The carbon skeleton of the amino acids is first converted to keto acids (by transamination) which meet one or more of the following fates.

1. Utilized to ***generate energy***.
2. Used for the ***synthesis of glucose***.
3. Diverted for the ***formation of fat or ketone bodies***.
4. Involved in the ***production of non-essential amino acids***.

A general picture of amino acid metabolism is depicted in ***Fig.15.2***.

TRANSAMINATION

The ***transfer of an amino ($-NH_2$) group from an amino acid to a keto acid*** is known as transamination (***Fig.15.3***). This process involves the interconversion of a pair of amino acids and a pair

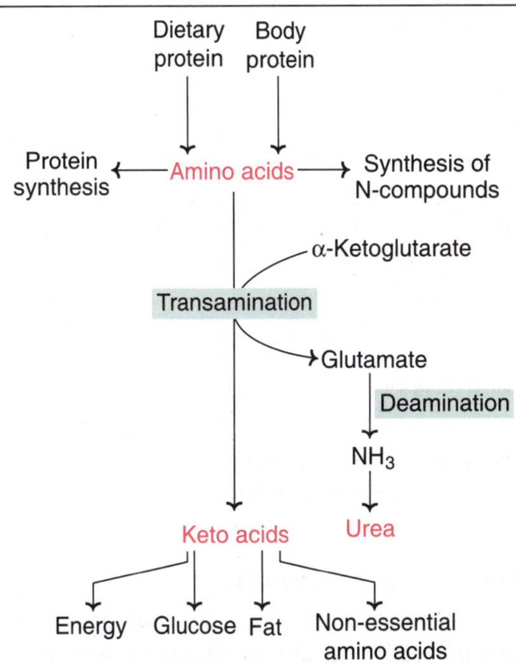

***Fig.15.2 :** An overview of amino acid metabolism.*

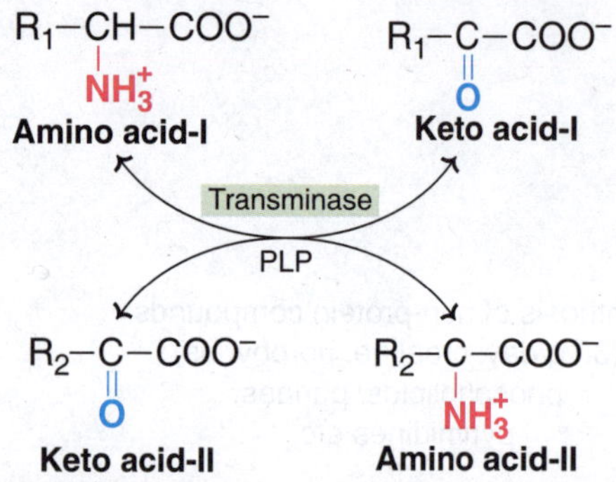

Fig.15.3 : Transamination reaction.

of keto acids, catalysed by a group of enzymes called ***transaminases*** (recently ***aminotransferases***).

Salient features of transamination

1. All transaminases require ***pyridoxal phosphate*** (***PLP***), a coenzyme derived from vitamin B_6.

2. There is ***no free NH_3 liberated***, only the transfer of amino group occurs.

3. Transamination is ***reversible***.

4. It involves both catabolism (degradation) and anabolism (synthesis) of amino acids. Transamination is ultimately responsible for the ***synthesis of non-essential amino acids***.

5. Transamination diverts the excess amino acids towards ***energy generation***.

6. The amino acids undergo transamination to finally concentrate nitrogen in glutamate. ***Glutamate*** is the only amino acid that undergoes oxidative deamination to a significant extent to ***liberate free NH_3 for urea synthesis***.

7. All amino acids except lysine, threonine, proline and hydroxyproline participate in transamination.

DEAMINATION

The ***removal of amino group*** from the amino acids as NH_3 is deamination. It results in the liberation of ammonia ***for urea synthesis***. Deamination may be either oxidative or non-oxidative.

I. Oxidative deamination

Oxidative deamination is the ***liberation of free ammonia*** from the amino group of amino acids ***coupled with oxidation***. This takes place mostly in liver and kidney. The purpose of oxidative deamination is to provide NH_3 for urea synthesis and α-keto acids for a variety of reactions, including energy generation.

Role of glutamate dehydrogenase : In the process of transamination, the amino groups of most amino acids are transferred to α-ketoglutarate to produce glutamate. Thus, ***glutamate*** serves as a ***'collection centre' for amino groups*** in the biological system. Glutamate rapidly undergoes oxidative deamination, catalysed by glutamate dehydrogenase (GDH) to liberate ammonia. This enzyme is unique in that it can utilize ***either NAD^+ or $NADP^+$ as a coenzyme***. Conversion of glutamate to α-ketoglutarate occurs through the formation of an intermediate, α-iminoglutarate (***Fig.15.4***).

Oxidative deamination by amino acid oxidases : L-Amino acid oxidase and D-amino acid oxidase are flavoproteins, possessing FMN and FAD, respectively. They act on the corresponding amino acids (L or D) to produce α-keto acids and NH_3. In this reaction, oxygen is reduced to H_2O_2, which is later decomposed by catalase (***Fig.15.5***).

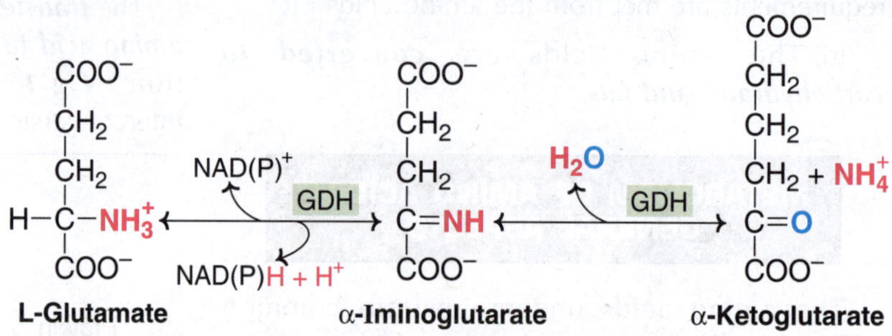

Fig.15.4 : Oxidation of glutamate by glutamate dehydrogenase (GDH).

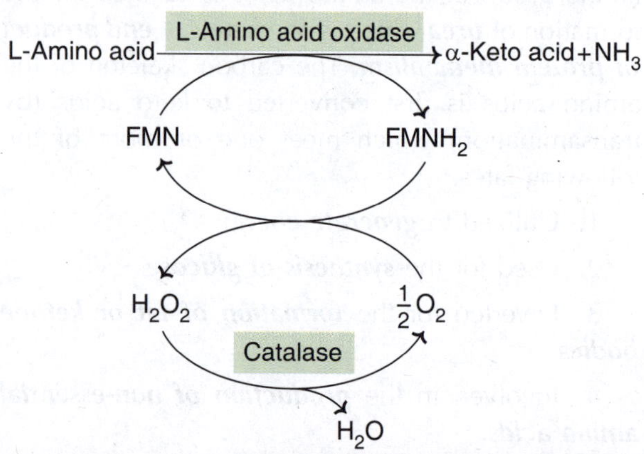

Fig.15.5 : Oxidative deamination of amino acids.

II. Non-oxidative deamination

Some of the amino acids can be deaminated to liberate NH_3 without undergoing oxidation

(a) **Amino acid dehydrases :** Serine, threonine and homoserine are the hydroxy amino acids. They undergo non-oxidative deamination catalysed by PLP-dependent dehydrases (dehydratases).

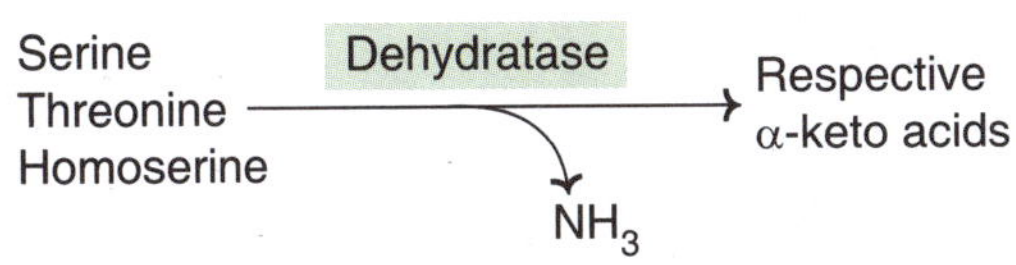

(b) **Deamination of histidine :** The enzyme histidase acts on histidine to liberate NH_3 by a non-oxidative deamination process.

METABOLISM OF AMMONIA

Ammonia is constantly being liberated in the metabolism of amino acids (mostly) and other nitrogenous compounds. At the physiological pH, ***ammonia exists as ammonium (NH_4^+) ion.***

I. Formation of ammonia

The production of NH_3 occurs from the ***amino acids*** (transamination and deamination), ***biogenic amines, amino group of purines and pyrimidines*** and by the action of intestinal bacteria (urease) on urea.

II. Disposal of ammonia

The organisms, during the course of evolution, have developed different mechanisms for the disposal of ammonia from the body. The animals in this regard are of three different types

(a) **Ammoniotelic :** The aquatic animals dispose off ***NH_3 into the surrounding water.***

(b) **Uricotelic :** Ammonia is converted mostly to ***uric acid*** e.g. reptiles and birds.

(c) **Ureotelic :** The mammals including man convert NH_3 to urea. ***Urea*** is a non-toxic and soluble compound, hence easily excreted.

III. Toxicity of ammonia

Even a marginal elevation in the blood ammonia concentration is ***harmful to the brain.*** Ammonia, when it accumulates in the body, results in ***slurring of speech and blurring of the vision*** and causes tremors. It may lead to coma and, finally death, if not corrected.

Hyperammonemia : Elevation in plasma NH_3 (***reference range 15-45 µg/dl***) level may be genetic or acquired. Impairment in urea synthesis due to a defect in any one of the five enzymes is described in urea synthesis. All these disorders lead to hyperammonemia and cause ***hepatic coma*** and ***mental retardation***. The acquired hyperammonemia may be due to hepatitis, alcoholism etc. where the urea synthesis becomes defective, hence NH_3 accumulates.

UREA CYCLE

Urea is the end product of protein metabolism (***amino acid metabolism***). The nitrogen of amino acids, converted to ammonia (as described above), is toxic to the body. It is converted to urea and detoxified. As such, urea accounts for ***80-90% of the nitrogen containing substances excreted in urine***.

Urea is ***synthesized in liver*** and transported to kidneys for excretion in urine. Urea cycle is the ***first metabolic cycle*** that was elucidated by Hans Krebs and Kurt Henseleit (1932) or Ornithine cycle, hence it is known as ***Krebs-Henseleit cycle***. The individual reactions, however, were described in more detail later on by Ratner and Cohen.

Urea has ***two amino ($-NH_2$) groups***, one derived ***from NH_3*** and the other ***from aspartate.*** Carbon atom is supplied by CO_2. Urea synthesis is a five-step cyclic process, with five distinct enzymes. The first ***two enzymes*** are present in ***mitochondria*** while the ***rest*** are localized in ***cytosol***. The details of urea cycle are described (***Figs. 15.6*** and ***15.7***).

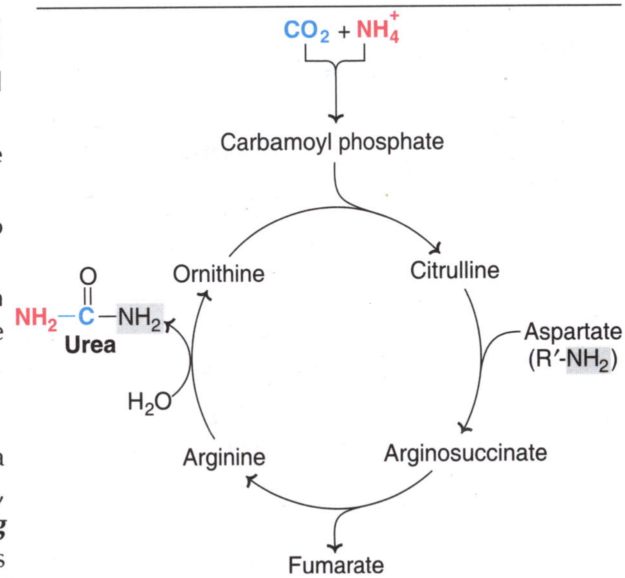

***Fig.15.6 :** Outline of urea cycle. (**Note :** In the synthesis of urea one amino group comes from ammonium ion while the other is from aspartate; carbon is derived from CO_2. This is represented in colours/shades.)*

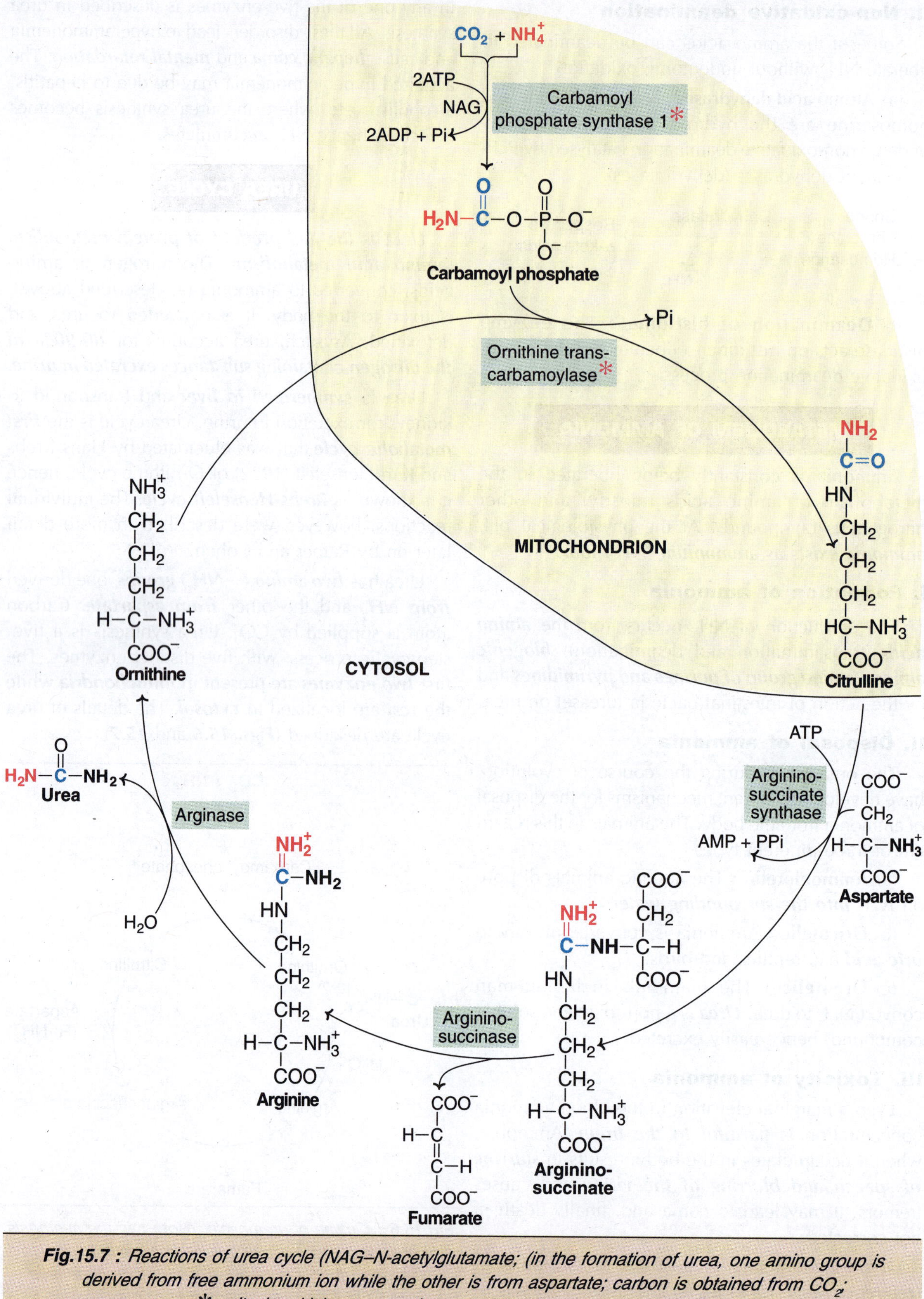

Fig.15.7 : *Reactions of urea cycle (NAG–N-acetylglutamate; (in the formation of urea, one amino group is derived from free ammonium ion while the other is from aspartate; carbon is obtained from CO_2; * mitochondrial enzymes, the rest of the enzymes are cytosomal).*

1. **Synthesis of carbamoyl phosphate :** Carbamoyl phosphate synthase I (***CPS I***) of mitochondria catalyses the condensation of NH_4^+ ions with CO_2 to form carbamoyl phosphate. This step consumes two ATP and is ***irreversible***, and ***rate-limiting***. CPS I requires ***N-acetylglutamate*** for its activity.

2. **Formation of citrulline :** Citrulline is synthesized from carbamoyl phosphate and ornithine by ***ornithine transcarbamoylase***. Ornithine is regenerated and used in urea cycle.

3. **Synthesis of argininosuccinate :** ***Argininosuccinate synthase*** condenses citrulline with aspartate to produce argininosuccinate. The second amino group of urea is incorporated in this reaction. This step requires ATP which is cleaved to AMP and pyrophosphate (PPi).

4. **Cleavage of argininosuccinate :** ***Argininosuccinase*** cleaves argininosuccinate to give arginine and fumarate. Arginine is the immediate precursor for urea.

5. **Formation of urea :** ***Arginase*** is the fifth and final enzyme that cleaves arginine to yield urea and ornithine. Ornithine, so regenerated, enters mitochondria for its reuse in the urea cycle.

Overall reaction and energetics

The urea cycle is irreversible and consumes 4 ATP. Two ATP are utilized for the synthesis of carbamoyl phosphate. One ATP is converted to AMP and PPi to produce arginosuccinate which equals to 2 ATP. Hence ***4 ATP*** are actually ***consumed***.

$$NH_4^+ + CO_2 + \text{Aspartate} + 3\text{ATP} \longrightarrow \text{Urea} + \text{Fumarate} + 2\ \text{ADP} + 2\ \text{Pi} + \text{AMP} + \text{PPi}$$

Regulation of urea cycle

The first reaction catalysed by ***carbamoyl phosphate synthase I*** (CPS I) is ***rate limiting*** reaction or committed step in urea synthesis. CPS I is allosterically activated by N-acetylglutamate (NAG).

Metabolic disorders of urea cycle

Metabolic defects associated with each of the five enzymes of urea cycle have been reported (***Table 15.1***). All the disorders invariably lead to a build-up in blood ammonia ***(hyperammonemia)***, leading to toxicity. Other metabolites of urea cycle also accumulate which, however, depends on the specific enzyme defect. The clinical symptoms associated with defect in urea cycle enzymes include vomiting, lethargy, irritability, ataxia and mental retardation.

TABLE 15.1 Metabolic defects in urea cycle

Defect	*Enzyme involved*
Hyperammonemia type I	Carbamoyl phosphate synthase I
Hyperammonemia type II	Ornithine transcarbamoylase
Citrullinemia	Argininosuccinate synthase
Argininosuccinic aciduria	Argininosuccinase
Hyperargininemia	Arginase

Blood urea—clinical importance

In healthy people, the ***normal blood urea*** concentration is ***10-40 mg/dl***.

Blood urea estimation is widely used as a screening test for the evaluation of ***kidney*** (renal) ***function***. It is estimated in the laboratory either by urease method or diacetyl monoxime (DAM) procedure. Elevation in blood urea may be broadly classified into three categories.

1. **Pre-renal :** This is associated with ***increased protein breakdown***, leading to a negative nitrogen balance, as observed after major surgery, prolonged fevers, diabetic coma, thyrotoxicosis etc. In leukemia and bleeding disorders also, blood urea is elevated.

2. **Renal :** In renal disorders like ***acute glomerulonephritis***, chronic nephritis, nephrosclerosis, polycystic kidney, blood urea is increased.

3. **Post-renal :** Whenever there is an ***obstruction*** in the ***urinary tract*** (e.g. tumors, stones, enlargement of prostate gland etc.), blood urea is elevated. This is due to increased reabsorption of urea from the renal tubules.

METABOLISM OF INDIVIDUAL AMINO ACIDS

In the preceding pages, the general aspects of amino acid metabolism have been discussed. A summary of the biologically important or ***specialized products*** obtained from or ***contributed by the amino acids*** is given in the ***Table 15.2***. The metabolism of individual amino acids with special emphasis on the specialized products obtained from them is described.

TABLE 15.2 A summary of the specialized products formed/contributed by amino acids

Amino acid	*Specialized product(s)*
Glycine	Creatine, glutathione, heme, purines, conjugated bile acids.
Tyrosine	Thyroxine, triiodothyronine, epinephrine, norepinephrine, dopamine, melanin.
Tryptophan	NAD^+, $NADP^+$ (coenzymes of niacin), serotonin, melatonin.
Methionine	Active methionine, creatine, epinephrine, polyamines.
Cysteine	Glutathione, taurine, coenzyme A, active sulfate.
Histidine	Histamine
Arginine	Creatine, nitric oxide
Lysine	Carnitine
Glutamate	γ-Amino butyric acid, glutathione, γ-carboxyglutamate.
Glutamine	Purines, pyrimidines, amino sugars.
Aspartate	Purines, pyrimidines
Serine	Phosphatidylserine, sphingomyelins, choline.
β-Alanine	Coenzyme A

GLYCINE

Glycine (Gly, G) is a ***non-essential***, optically inactive and ***glycogenic*** (precursor for glucose) amino acid. It is indispensable for chicks. The outline of glycine metabolism is depicted in ***Fig.15.8***. Glycine is actively involved in the ***synthesis of*** many specialized products (***heme***, ***purines***, ***creatine*** etc.) in the body, besides its incorporation into proteins, synthesis of serine and glucose and participation in one-carbon metabolism.

Synthesis of glycine

Glycine is synthesized ***from serine*** by the enzyme serine hydroxymethyl transferase which is dependent on tetrahydrofolate (THF). Glycine can also be obtained from threonine, catalysed by threonine aldolase.

Degradation of glycine

Glycine undergoes oxidative deamination by glycine synthase to liberate NH_4^+, CO_2 and one-carbon fragment as N^5, N^{10}-methylene THF. This provides a major route for glycine breakdown in mammals. This reaction is reversible and, therefore, glycine can be generated from one-carbon unit (methylene fragment of THF).

Glycine is reversibly converted to serine by THF dependent serine hydroxymethyl transferase. Pyruvate produced from serine by serine dehydratase, serves as a precursor for glucose. Serine is degraded to glyoxylate which undergoes transamination to give back glycine. Glyoxylate is also converted to oxalate, an excretory product and formate which enters one-carbon pool.

Synthesis of specialized products

1. **Formation of purine ring :** The ***entire molecule of glycine*** is utilized for the formation of positions 4 and 5 of carbon and position 7 of nitrogen of purines.

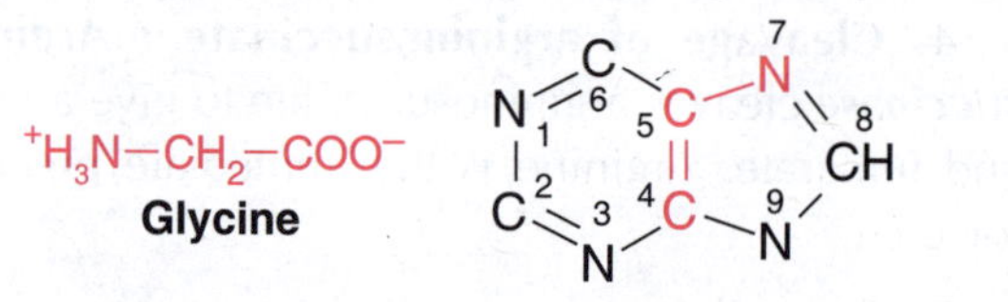

2. **Synthesis of glutathione :** Glutathione is a ***tripeptide*** (γ-glutamyl-cysteinyl-glycine) and requires glycine for its formation.

3. **Conjugation reactions :** As a conjugating agent, glycine performs two important functions

(a) The bile acids – cholic acid and chenodeoxy cholic acid – are conjugated with glycine.

Cholic acid + Glycine ⟶ Glycocholic acid

Chenodeoxycholic acid + Glycine ⟶ Glycochenodeoxy cholic acid

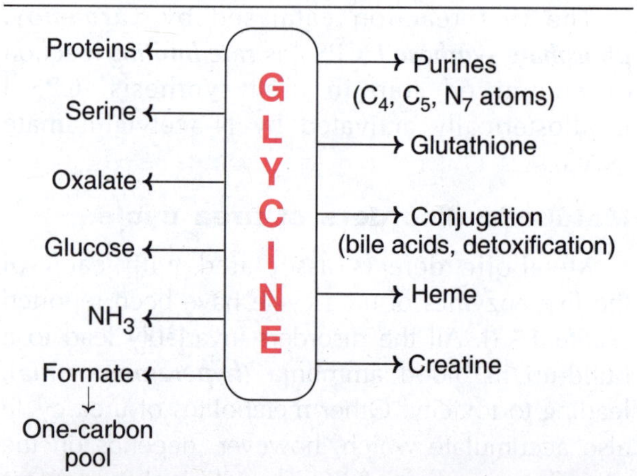

Fig.15.8 : Overview of glycine metabolism.

Medical Concepts/ Clinical Correlates

- The amino acids are mainly utilized for protein biosynthesis, production of specialized products (creatine, porphyrin, amines, purines, pyrimidines) and generation of energy.
- Ammonia accumulation in blood is toxic to brain causing slurring of speech, blurring of vision, tremors and even death. Mammals convert NH_3 to urea, a non-toxic excretory product. Metabolic defects in urea cycle enzymes result in hyperammonemia.
- Hyperammonemia is sometimes associated with hepatic coma and liver failure.
- Blood urea estimation is commonly used to assess renal function. Elevation of blood urea level (normal 10–40 mg/dl) is associated with several disorders which may be pre-renal (burns), renal (acute glomerulonephritis) and post-renal (tumors or stones in the urinary tract).
- Estimation of serum creatinine (normal < 1 mg/dl) is considered to be a more reliable indicator for the evaluation of kidney function.

(b) Glycine is important for **detoxification** of benzoic acid (commonly used as a food preservative) to hippuric acid.

4. **Synthesis of heme :** Glycine condenses with succinyl CoA to form δ-amino levulinate which serves as a precursor for heme synthesis (details given in porphyrin metabolism—***Chapter 10***).

$$\text{Succinyl CoA} + \text{Glycine} \xrightarrow{\text{ALA synthase}} \delta\text{-Amino levulinate (ALA)}$$

5. **Biosynthesis of creatine :** Creatine is present in the tissues (muscle, brain, blood etc.) as the high energy compound, phosphocreatine and as free creatine. Three amino acids—***glycine***, ***arginine*** and ***methionine***—are required for creatine formation (***Fig.15.9***). The first reaction occurs in the kidney. It involves the transfer of guanidino group of arginine to glycine, catalysed by arginine-glycine transamidinase to produce guanidoacetate (glycocyamine). ***S Adenosylmethionine (active methionine)*** donates methyl group to glycocyamine to produce creatine. This reaction occurs in liver. Creatine is reversibly phosphorylated to phosphocreatine (creatine phosphate) by creatine kinase. It is stored in muscle as high energy phosphate.

Creatinine is an anhydride of creatine. It is formed by spontaneous cyclization of creatine or creatine phosphate. Creatinine is excreted in urine.

Creatine and creatinine—clinical importance : The normal concentration of serum creatinine in man is 0.6–1.0 mg/dl.

Estimation of ***serum creatinine*** (along with blood urea) is used as a ***diagnostic test to assess kidney function.*** Serum creatinine concentration is not influenced by endogenous and exogenous factors, as is the case with urea. Hence, some workers consider ***serum creatinine*** as a ***more reliable indicator of renal function.***

PHENYLALANINE AND TYROSINE

Phenylalanine (Phe, F) and tyrosine (Tyr, Y) are structurally related ***aromatic amino acids***.

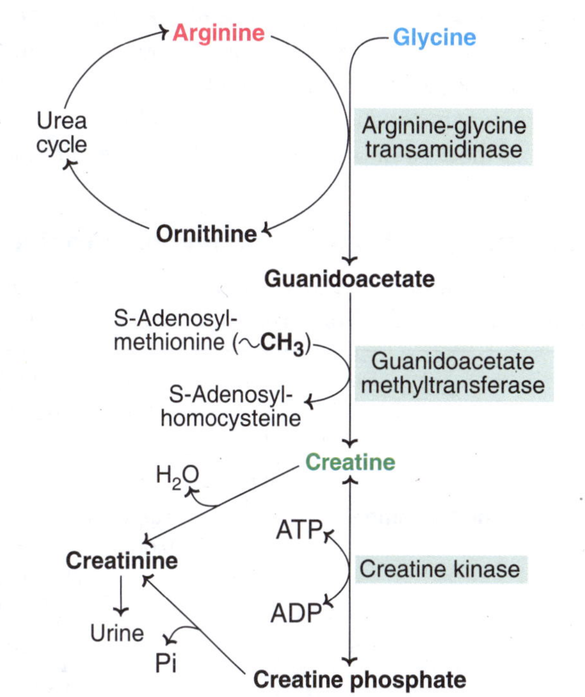

Fig.15.9 : *Metabolism of creatine.*

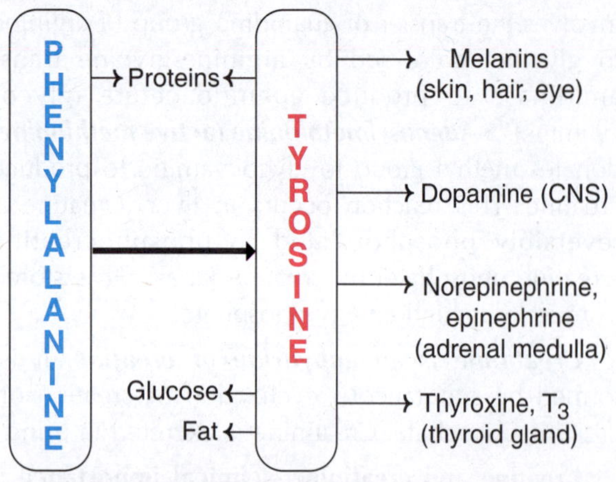

Fig.15.10 : Overview of phenylalanine and tyrosine metabolism (CNS–Central nervous system; T_3–Triiodothyronine).

Phenylalanine is an ***essential amino*** acid while tyrosine is non-essential. Besides its incorporation into proteins, the only function of phenylalanine is its conversion to tyrosine. For this reason, ingestion of tyrosine can reduce the dietary requirement of phenylalanine. This phenomenon is referred to as ***'sparing action' of tyrosine on phenylalanine.***

The predominant metabolism of phenylalanine occurs through tyrosine. Tyrosine is incorporated into proteins and is involved in the synthesis of a variety of biologically important compounds—***epinephrine, norepinephrine, dopamine*** (catecholamines), ***thyroid hormones***—and the pigment ***melanin*** (***Fig.15.10***). During the course of degradation, phenylalanine and tyrosine are converted to metabolites which can serve as precursors for the synthesis of glucose and fat. Hence, these amino acids are both ***glucogenic*** and ***ketogenic***.

Conversion of phenylalanine to tyrosine

Phenylalanine is hydroxylated at para-position by ***phenylalanine hydroxylase*** to produce tyrosine (p-hydroxy phenylalanine). This is an irreversible reaction and requires the participation of a specific coenzyme ***biopterin*** (containing pteridine ring) which is structurally related to folate. The active form of biopterin is ***tetrahydrobiopterin*** (H_4-biopterin). In the phenylalanine hydroxylase reaction, tetrahydrobiopterin is oxidized to dihydrobiopterin (H_2-biopterin). Tetrahydrobiopterin is then regenerated by an NADPH-dependent dihydrobiopterin reductase (***Fig.15.11***).

The enzyme phenylalanine hydroxylase is present in the liver. Due to a defect in ***phenylalanine hydroxylase***, the conversion of phenylalanine to tyrosine is blocked resulting in the disorder ***phenylketonuria*** (***PKU***).

Degradation of tyrosine (Phenylalanine)

The metabolism of phenylalanine and tyrosine is considered together. The sequence of the reactions in the degradation of these amino acids depicted in ***Fig.15.12***, is described hereunder

1. As phenylalanine is ***converted to tyrosine*** (details in ***Fig.15.11***), a single pathway is responsible for the degradation of both these amino acids, which occurs mostly in liver.

2. Tyrosine first undergoes transamination to give p-hydroxyphenylpyruvate. This reaction is catalysed by ***tyrosine transaminase*** (PLP dependent).

3. p-Hydroxyphenylpyruvate hydroxylase catalyses oxidative decarboxylation as well as hydroxylation of the phenyl ring of p-hydroxyphenylpyruvate to produce ***homogentisate***.

4. ***Homogentisate oxidase*** cleaves the benzene ring of homogentisate to form 4-maleylacetoacetate. Molecular oxygen is required for this reaction to break the aromatic ring.

5. Maleylacetoacetate undergoes isomerization to form 4-fumaryl acetoacetate and this reaction is catalysed by ***maleylacetoacetate isomerase***.

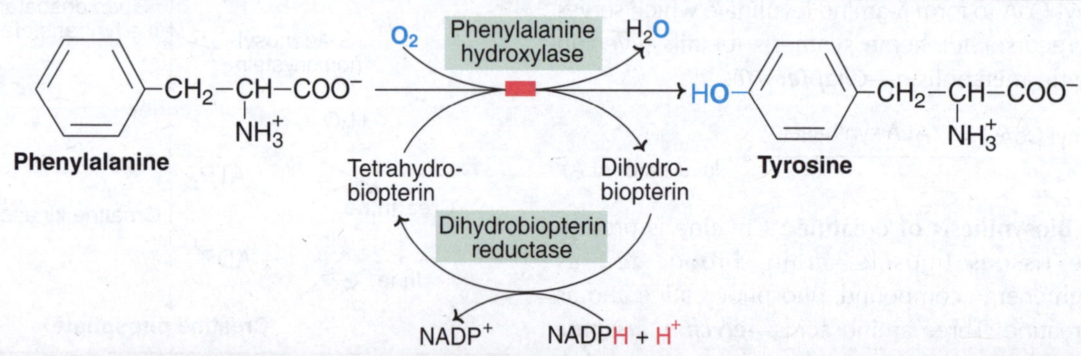

Fig.15.11 : Synthesis of tyrosine from phenylalanine (▬—Block in phenylketonuria).

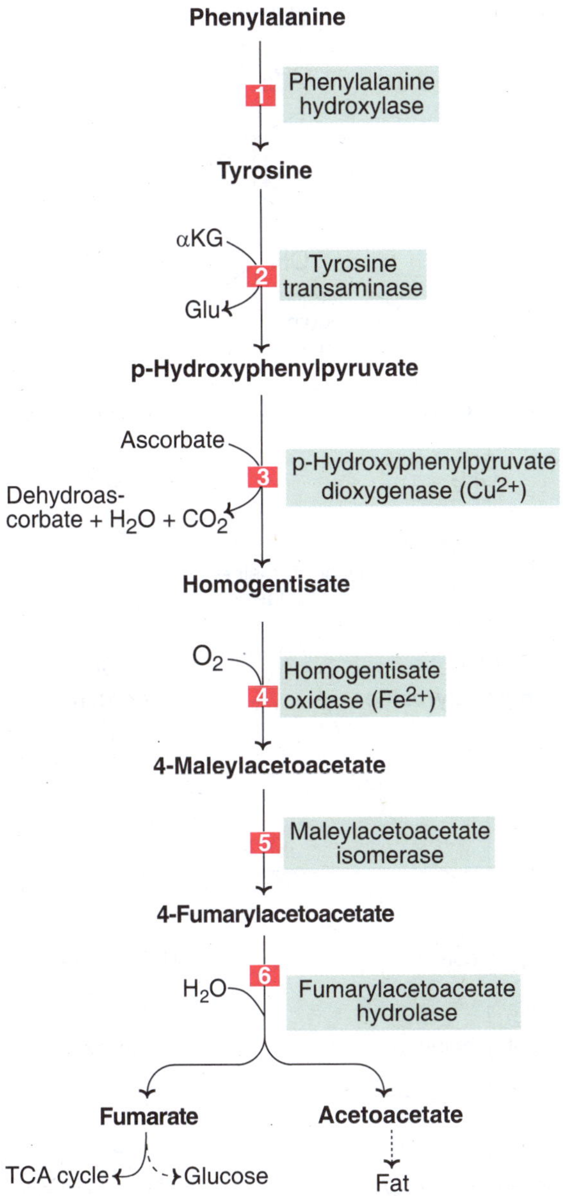

Fig.15.12 : *Tyrosine metabolism—degradative pathway [αKG—α-Ketoglutarate; Glu-Glutamate; The circled numbers indicate metabolic defects (1) Phenylketonuria; (2) Tyrosinemia type II; (3) Neonatal tyrosinemia; (4) Alkaptonuria; (5) and (6) Tyrosinosis (tyrosinemia, type I)].*

6. ***Fumaryl acetoacetase*** (fumaryl acetoacetate hydrolase) brings about the hydrolysis of fumaryl acetoacetate to liberate fumarate and acetoacetate.

Fumarate is an intermediate of citric acid cycle and can also serve as precursor for gluconeogenesis. Acetoacetate is a ketone body from which fat can be synthesized. Phenylalanine and tyrosine are, therefore, both glucogenic and ketogenic.

The inborn errors of phenylalanine and tyrosine metabolism are indicated in ***Fig.15.12***. Some details on these disorders are given later.

Synthesis of melanin

Melanin (*Greek* : melan—black) is the ***pigment of skin, hair and eye***. The synthesis of melanin occurs in melanosomes present in melanocytes, the pigment-producing cells. Tyrosine is the precursor for melanin and only one enzyme, namely ***tyrosinase*** (a copper-containing oxygenase), is involved in its formation. Tyrosinase hydroxylates tyrosine to form ***3,4-dihydroxyphenylalanine*** (***DOPA***). DOPA can act as a cofactor for tyrosinase. The next reaction is also catalysed by tyrosinase in which DOPA is converted to dopaquinone. It is believed that the subsequent couple of reactions occur spontaneously, forming leucodopachrome followed by 5,6-dihydroxyindole (***Fig.15.13***). The oxidation of 5,6-dihydroxyindole to indole 5,6-quinone is catalysed by tyrosinase, and DOPA serves as a cofactor. This reaction, inhibited by tyrosine regulates the synthesis of melanin. ***Melanochromes*** are formed from indole quinone,

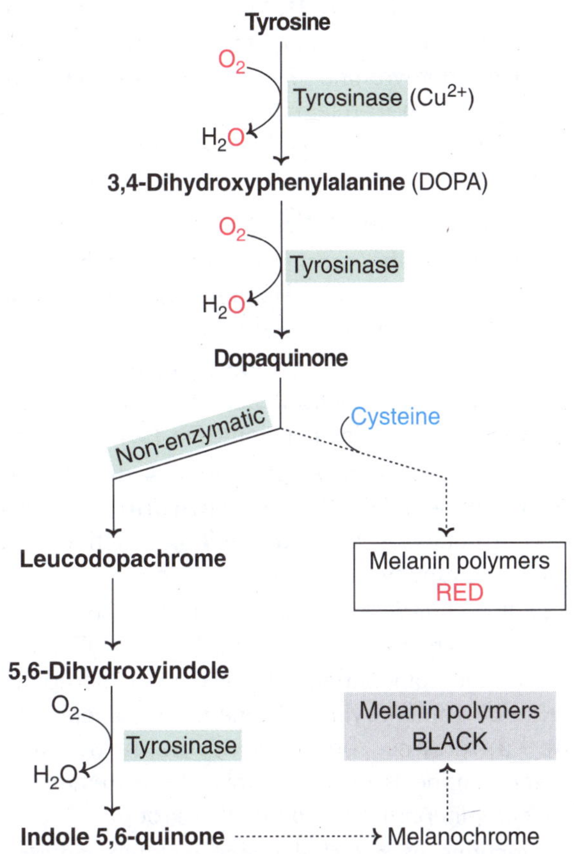

Fig.15.13 : *Metabolism of tyrosine—biosynthesis of melanin (Defect in tyrosinase causes albinism).*

which on ***polymerization*** are converted to black melanin.

Melanin—the colour pigment : The ***skin colour*** of the individual is ***determined by the relative concentrations of black an red melanins***. This, in turn, is dependent on many factors, both genetic and enviornmental. These include the activity of tyrosinase, the density of melanocytes, availability of tyrosine etc.

The presence of ***moles*** on the body represents a localized severe ***hyperpigmentation*** due to hyperactivity of melanocytes. On the other hand, localized absence or degeneration of melanocytes results in ***white patches*** on the skin commonly known as ***leucoderma***. ***Albinism*** is an ***inborn error*** with generalized lack of melanin synthesis (details described later).

Biosynthesis of thyroid hormones

Thyroid hormones—***thyroxine*** (tetraiodothyronine) and ***triiodothyronine***—are synthesized from the tyrosine residues of the protein ***thyroglobulin*** and activated iodine (***Fig.15.14***). Iodination of tyrosine ring occurs to produce mono-and diiodotyrosine from which triiodothyronine (T_3) and thyroxine (T_4) are synthesized. The protein thyroglobulin undergoes proteolytic breakdown to release the free hormones, T_3 and T_4.

Biosynthesis of catecholamines

Tyrosine is the precursor for the synthesis of catecholamines, namely ***dopamine***, ***norepinephrine*** (noradrenaline) and ***epinephrine*** (adrenaline).

The conversion of tyrosine to catecholamines occurs in adrenal medulla and central nervous system involving the following reactions (***Fig.15.15***).

Tyrosine is hydroxylated to 3,4-dihydroxyphenylalanine (DOPA) by ***tyrosine hydroxylase***. This enzyme catalyses the ***rate limiting*** reaction and requires tetrahydrobiopterin as a coenzyme.

DOPA undergoes PLP-dependent decarboxylation to give dopamine which, in turn, is hydroxylated to produce norepinephrine. Methylation of norepinephrine by S-adenosylmethionine gives epinephrine. The difference between epinephrine and norepinephrine is only a methyl group (remember that ***nor***epinephrine has ***no*** methyl group).

Functions of catecholamines : Norepinephrine and epinephrine ***regulate carbohydrate and lipid metabolisms***. They stimulate the degradation of triacylglycerol and glycogen. They cause an increase

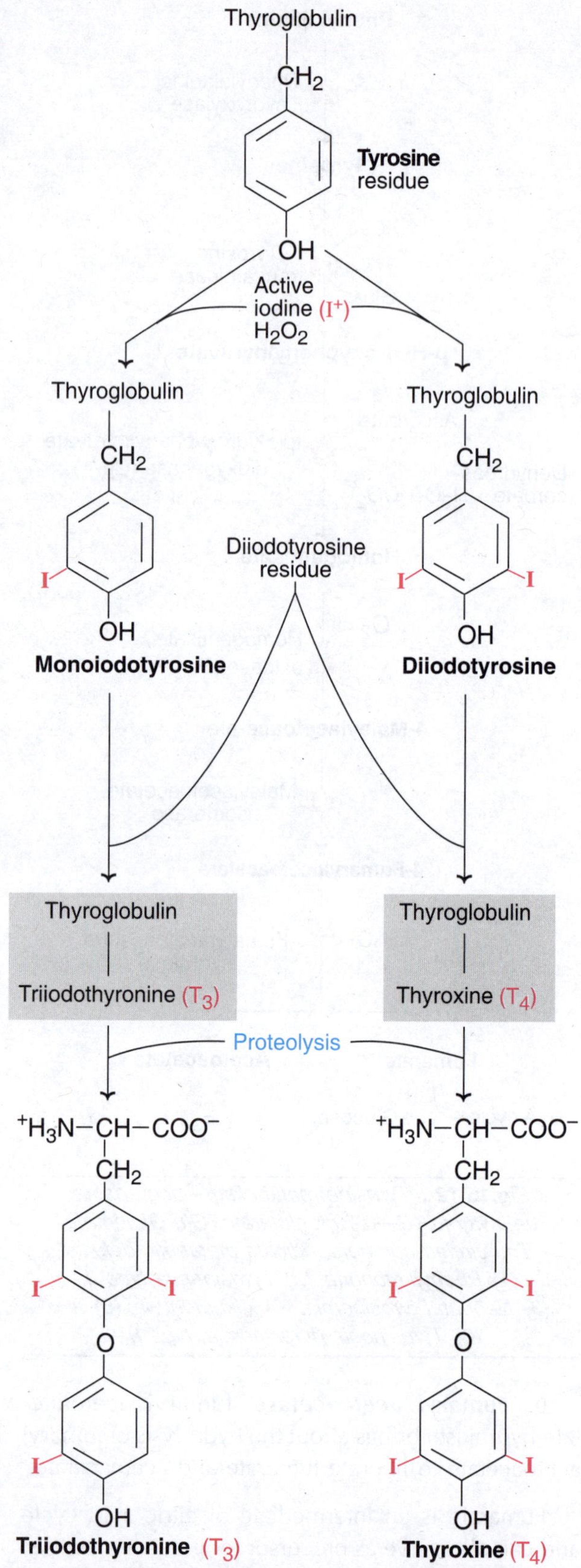

Fig.15.14 : Metabolism of tyrosine-synthesis of thyroid hormones.

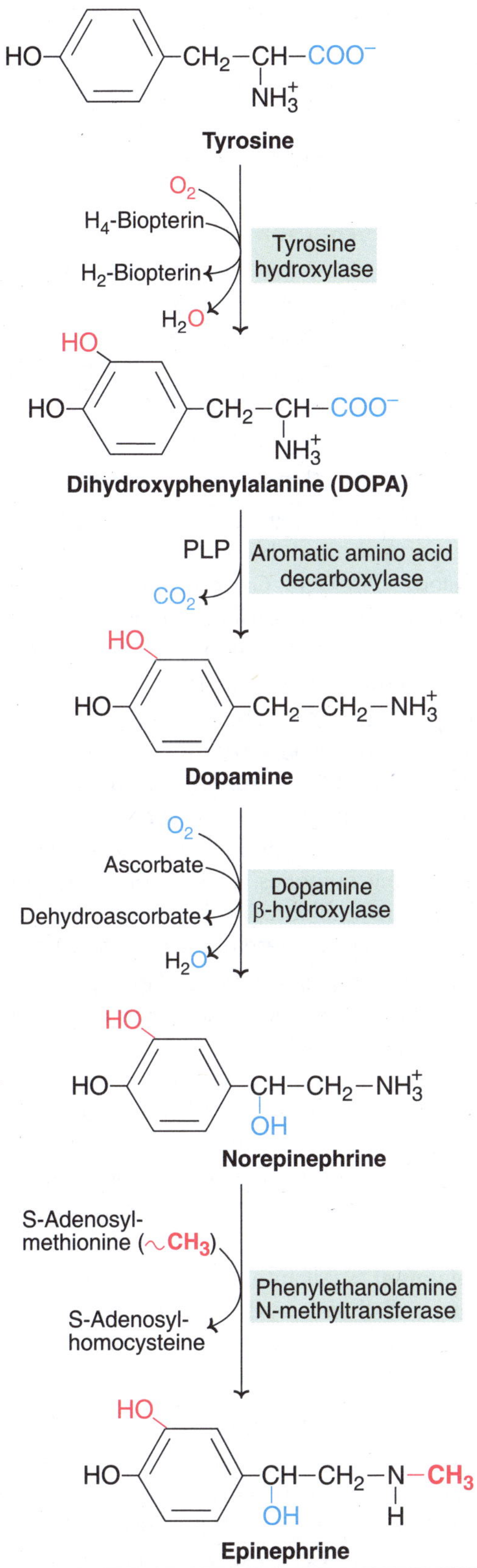

Fig.15.15 : Metabolism of tyrosine-synthesis of catecholamines (dopamine, norepinephrine, epinephrine; PLP–pyridoxal phosphate).

in the ***blood pressure***. Dopamine and norepinephrine serve as ***neurotransmitters in the brain*** and autonomic nervous system.

Dopamine and Parkinson's disease

Parkinson's disease is a common ***disorder in many elderly people***, with about 1% of the population above 60 years being affected. It is characterized by muscular rigidity, tremors, expressionless face, lethargy, involuntary movements etc.

Biochemical basis : The exact biochemical cause of this disorder has not been identified. Parkinson's disease is, however, ***linked with a decreased production of dopamine***. The disease is due to degeneration of certain parts of the brain (substantia nigra and coeruleus locus), leading to the impairment in the synthesis of dopamine.

Treatment : Dopamine cannot enter the brain, hence its administration is of no use. ***DOPA*** (levodopa or L-dopa) is used in the treatment of Parkinson's disease. In the brain, DOPA is decarboxylated to dopamine which alleviates the symptoms of this disorder. Unfortunately, dopamine synthesis occurs in various other tissues and results in side-effects such as nausea, vomiting, hypertension etc.

Administration of ***dopa analogs***—that inhibit dopa decarboxylase (in various tissues) but not enter brain (due to blood-brain barrier)—are found to be effective. ***Carbidopa*** and ***α-methyldopa*** (dopa analogs) are administered along with dopa for the treatment of Parkinson's disease.

DISORDERS OF TYROSINE (PHENYLALANINE) METABOLISM

Several enzyme defects in phenylalanine/tyrosine degradation leading to metabolic disorders are known. In ***Fig.15.12***, the ***deficient enzymes*** and the respective ***inborn errors*** are ***depicted*** and they are briefly discussed here.

Phenylketonuria

Phenylketonuria (PKU) is the most common metabolic disorder in amino acid metabolism. The incidence of PKU is 1 in 10,000 births. It is due to the ***deficiency of the hepatic enzyme***, ***phenylalanine hydroxylase*** (See ***Fig.15.11***).

Phenylalanine metabolism in PKU : Phenylketonuria primarily causes the accumulation of phenylalanine in tissues and blood, and results in its increased excretion in urine. Due to disturbances in the routine metabolism, phenylalanine is diverted to alternate pathways, resulting in ***the excessive production of phenylpyruvate, phenylacetate,***

Medical Concepts/ Clinical Correlates

- **Melanin—the pigment of skin, hair and eyes—is produced from tyrosine. Lack of melanin synthesis (mostly due to a deficiency of tyrosinase) causes albinism.**
- **Defect in melanin synthesis, due to disappearance of melanocytes from hair roots, results in greying of hair.**
- **Parkinson's disease—a common disorder of the elderly—is linked with decreased synthesis of dopamine. It is characterized by muscular rigidity, tremors, lethargy etc.**

phenyllactate and ***phenylacetylglutamine***. All these metabolites are excreted in urine in high concentration in PKU.

The name phenylketonuria is coined due to the fact that the metabolite phenylpyruvate is a ***keto*** acid ($C_6H_5CH_2$–***CO***–COO^-) excreted in urine in high amounts.

Clinical manifestations of PKU : Mental retardation, failure to walk or talk, failure of growth, seizures and tremor are the characteristic findings in PKU. If untreated, the patients show very low IQ (below 50). The biochemical basis of mental retardation in PKU is not well understood. Hypopigmentation that causes light skin colour, fair hair, blue eyes etc. is also observed.

Treatment of PKU : The maintenance of ***plasma phenylalanine concentration within the normal range*** (1-2 mg/dl) is a challenging task in the treatment of PKU. This is done by selecting foods with low ***phenylalanine content*** and/or feeding synthetic amino acid preparations, low in phenylalanine.

Tyrosinemia type II

This disorder—also known as ***Richner-Hanhart syndrome***, is due to a defect in the enzyme ***tyrosine transaminase.*** The result is a blockade in the routine degradative pathway of tyrosine. Accumulation and excretion of tyrosine and its metabolites—namely p-hydroxyphenylpyruvate, p-hydroxyphenyllactate, p-hydroxyphenylacetate, N-acetyltyrosine—and tyramine are observed.

Tyrosinemia type II is characterized by ***skin*** (dermatitis) and ***eye lesions*** and, rarely, mental retardation. A disturbed self-coordination is seen in these patients.

Neonatal tyrosinemia

The absence of the enzyme ***p-hydroxyphenylpyruvate dioxygenase*** causes neonatal tyrosinemia. This is mostly a temporary condition and usually responds to ascorbic acid. It is explained that the substrate inhibition of the enzyme is overcome by the presence of ascorbic acid.

Alkaptonuria (Black urine disease)

Alkaptonuria has great historical importance. It was first described by Lusitanus in 1649 and characterized in 1859. Garrod conceived the idea of inborn errors of metabolism from his observation on alkaptonuria. The prevalance of this autosomal recessive disorder is 1 in 25,000.

Enzyme defect : The defective enzyme in alkaptonuria is ***homogentisate oxidase*** in tyrosine metabolism (***Fig.15.12***). ***Homogentisate accumulates*** in tissues and blood, and is ***excreted*** into urine. Homogentisate, on standing, gets oxidized to the corresponding quinones, which polymerize to give black or brown colour. For this reason, the ***urine*** of alkaptonuric patients resembles ***coke in colour.***

Biochemical manifestations : Homogentisate gets oxidized by polyphenol oxidase to benzoquinone acetate which undergoes polymerization to produce a pigment called ***alkapton.*** Alkapton deposition occurs in connective tissue, bones and various organs (nose, ear etc.) resulting in a condition known as ***ochronosis.*** Many alkaptonuric patients suffer from arthritis and this is believed to be due to the deposition of pigment alkapton (in the joints), produced from homogentisate.

Treatment : Alkaptonuria is ***not a dangerous disorder*** and, therefore, does not require any specific treatment. However, consumption of protein diet with relatively low phenylalanine content is recommended.

Tyrosinosis or tyrosinemia type I

This is due to the deficiency of the enzymes ***fumarylacetoacetate hydroxylase*** and/or ***maleylacetoacetate isomerase***. Tyrosinosis is a rare but serious disorder. It causes liver failure, rickets, renal tubular dysfunction and polyneuropathy. Tyrosine, its metabolites and many other amino acids are excreted in urine.

In acute tyrosinosis, the infant exhibits diarrhea, vomiting and 'cabbage-like' odor. Death may even occur due to liver failure within one year. For the treatment, diets low in tyrosine, phenylalanine and methionine are recommended.

Albinism

Albinism (*Greek*: albino—white) is an inborn error, due to the ***lack of*** synthesis of the pigment ***melanin***. It is an autosomal recessive disorder with a frequency of 1 in 20,000.

Biochemical basis : The colour of skin and hair is controlled by a large number of genes. About 150 genes have been identified in mice. The melanin synthesis can be influenced by a variety of factors. Many possible causes (rather explanations) for albinism have been identified

1. Deficiency or ***lack of the enzyme tyrosinase***.
2. Decrease in melanosomes of melanocytes.
3. Impairment in melanin polymerization.
4. Lack of protein matrix in melanosomes.

The most common cause of albinism is a ***defect in tyrosinase***, the enzyme most responsible for the synthesis of melanin (***See Fig.15.13***).

Clinical manifestations : The most important function of melanin is the protection of the body from sun radiation. Lack of melanin in ***albinos*** makes them ***sensitive to sunlight***. Increased susceptibility to ***skin cancer*** (carcinoma) is observed. ***Photophobia*** (intolerance to light) is associated with lack of pigment in the eyes. However, there is no impairment in the eyesight of albinos.

TRYPTOPHAN

Tryptophan (Trp, W) was the first to be identified as an essential amino acid. It contains an indole ring and chemically it is α-amino β-indole propionic acid. Tryptophan is both ***glucogenic*** and ***ketogenic*** in nature. It is a precusor for the synthesis of important compounds, namely ***NAD⁺*** and ***NADP⁺*** (coenzymes of niacin), ***serotonin*** and ***melatonin (Fig.15.16)***.

The metabolism of tryptophan is divided into

I. Kynurenine (kynurenine-anthranilate) pathway;

II. Serotonin pathway.

I. Kynurenine pathway

This pathway mostly occurs in liver leading to oxidation of tryptophan and the synthesis of NAD^+ and $NADP^+$ (***Fig.15.17***).

Tryptophan pyrrolase or oxygenase cleaves the five-membered ring of the indole nucleus to produce formylkynurenine. Formamidase hydrolyses formylkynurenine and liberates formate which enters the one carbon pool. Kynurenine formed in this reaction is a branch point with different fates. In the prominent pathway, kynurenine undergoes NADPH-dependent hydroxylation to give 3-hydroxykynurenine.

Kynureninase, a pyridoxal phosphate (PLP)-dependent enzyme, acts on the 3-hydroxykynurenine and splits off alanine. Tryptophan is glucogenic, since alanine is a good precursor for glucose. The enzyme kynureninase is very sensitive to vitamin B_6 deficiency. Due to the lack of PLP, kynureninase reaction is blocked and 3-hydroxykynurenine is diverted to form xanthurenate. Elevated excretion of ***xanthurenate*** serves as an ***indication of B_6 deficiency***. The enzyme kynurenine hydroxylase is inhibited by estrogen, hence ***women***, in general, are ***more susceptible to pellagra***.

3-Hydroxyanthranilate is cleaved by an oxidase to form an intermediate, 2-amino 3-carboxy muconate semialdehyde. This compound has three fates.

1. It undergoes spontaneous cyclization to form quinolinate for NAD^+ synthesis.

2. Picolinate carboxylase decarboxylates the intermediate which cyclizes to produce picolinate.

3. The intermediate undergoes decarboxylation, catalysed by amino carboxysemialdehyde decarboxylase to produce 2-aminomuconate semialdehyde that enters glutarate pathway. The semialdehyde is converted to 2-aminomuconate by a dehydrogenase. The amino muconate, in a series of reactions involving reduction, deamination, decarboxylation etc., is converted to glutaryl CoA and finally to acetyl CoA. The latter is either completely oxidized via TCA cycle or converted to fat (hence tryptophan is ketogenic).

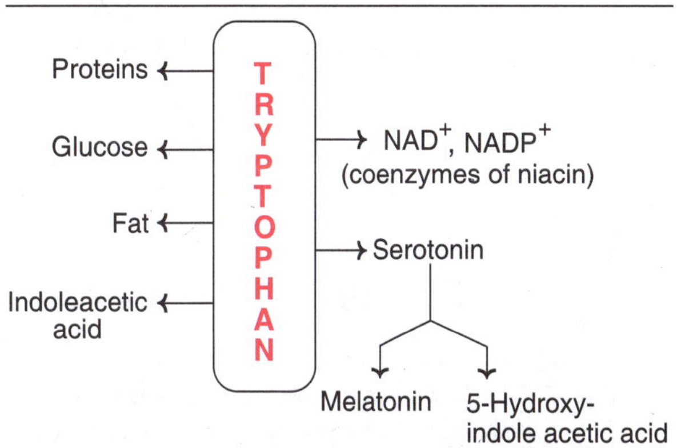

Fig.15.16 : Overview of tryptophan metabolism.

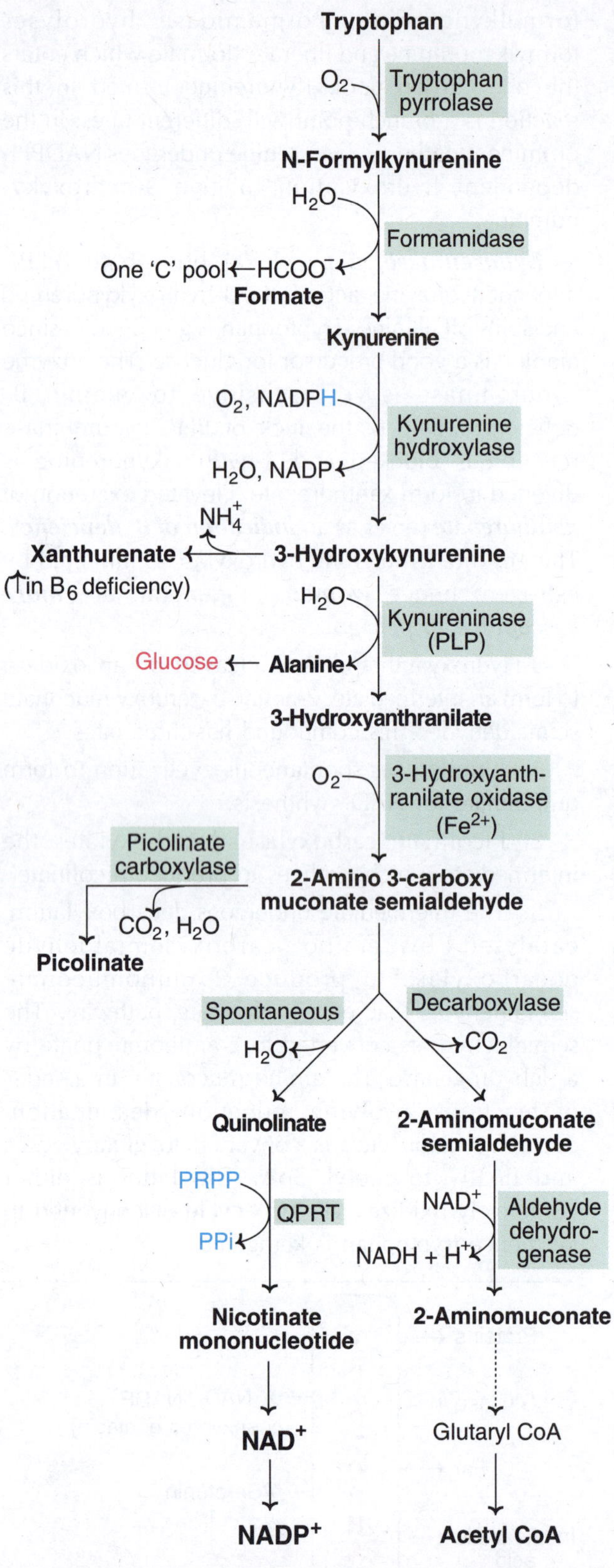

Fig.15.17 : Metabolism of tryptophan—kynurenine pathway (PLP–Pyridoxal phosphate; QPRT–Quinolinate phosphoribosyl transferase; PRPP–Phosphoribosyl pyrophosphate).

NAD⁺ Pathway : Tryptophan is not a precursor for the synthesis of free niacin. Quinolinate undergoes decarboxylation and is converted to nicotinate mononucleotide by the enzyme quinolinate phosphoribosyl transferase (QPRT). The synthesis of NAD⁺ and NADP⁺ from nicotinate mononucleotide is similar to that from niacin as described in the chapter on vitamins (***Refer Fig.7.12***).

Conversion of tryptophan to indole acetate : Tryptophan undergoes deamination and decarboxylation to produce indolepyruvate and tryptamine, respectively. Both these compounds are converted to indoleacetate (***Fig.15.18***) and excreted in urine.

II. Serotonin pathway

Serotonin or ***5-hydroxytryptamine*** (5HT) is a ***neurotransmitter***, synthesized from tryptophan.

Synthesis of serotonin : In mammals, the ***largest amount*** of serotonin is synthesized ***in the intestinal cells***. Tryptophan is first hydroxylated at 5th carbon by tryptophan hydroxylase. This enzyme requires tetrahydrobiopterin as a cofactor. 5-Hydroxytryptophan is decarboxylated by aromatic amino acid decarboxylase (PLP-dependent) to give serotonin (***Fig.15.18***).

Degradation of serotonin : ***Monoamine oxidase*** (MAO) degrades serotonin to 5-hydroxyindoleacetate (5HIA) which is excreted in urine.

Functions of serotonin : Serotonin is a ***neurotransmitter*** and performs a variety of functions.

1. Serotonin is a powerful ***vasoconstrictor*** and results in smooth muscle contraction in bronchioles and arterioles.

2. It is closely involved in the ***regulation of cerebral activity*** (excitation).

3. Serotonin, regarded as a ***happiness hormone***, ***controls the behavioural patterns***, sleep, blood pressure and body temperature.

Malignant carcinoid syndrome : Serotonin is produced by argentaffin cells of gastrointestinal tract. When these cells undergo uncontrolled growth, they develop into a tumor called malignant carcinoid or ***argentaffinomas.*** The patients exhibit symptoms like respiratory distress, sweating, hypertension etc.

Normally, about 1% of the tryptophan is utilized for serotonin synthesis. In case of carcinoid syndrome, very high amount (up to

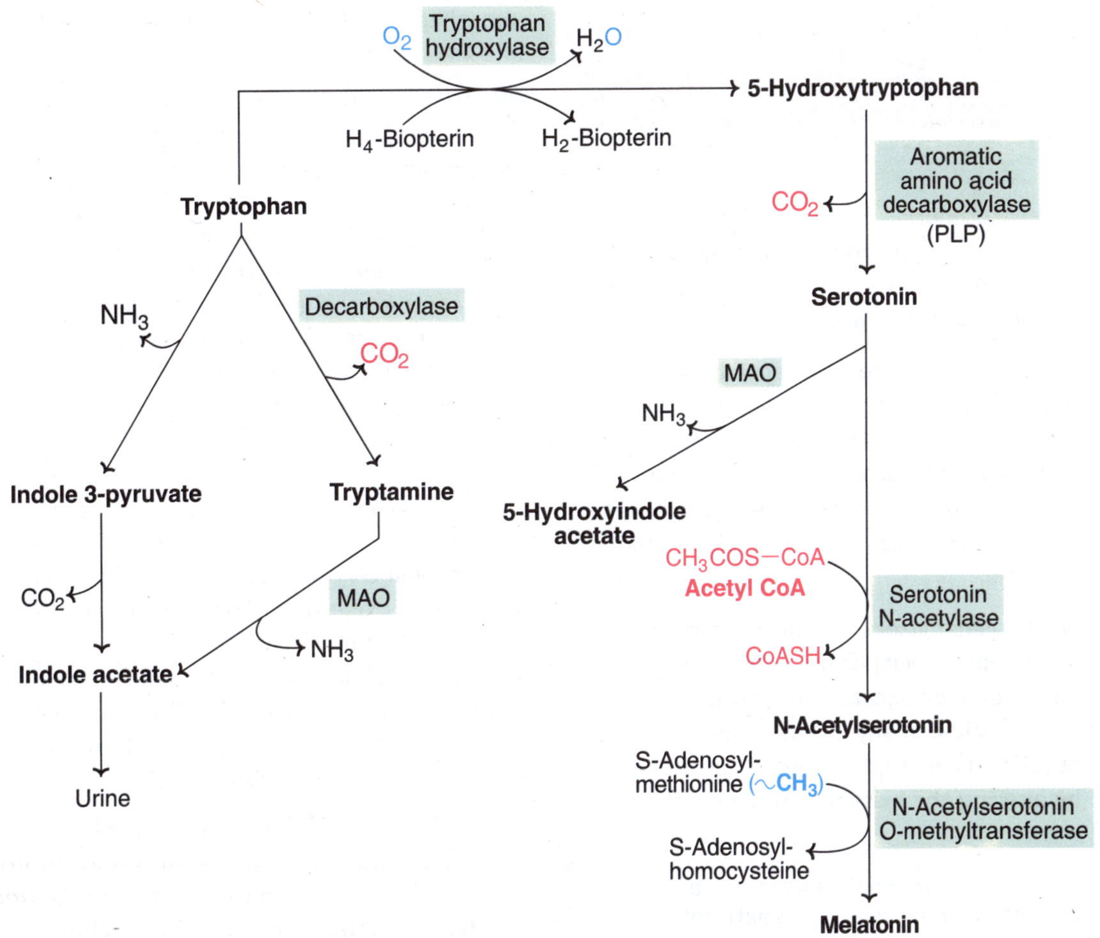

Fig.15.18 : Metabolism of tryptophan—serotonin and melatonin synthesis (PLP–Pyridoxal phosphate; MAO–Monoamine oxidase).

60%) of tryptophan is diverted for serotonin production. This disturbs the normal tryptophan metabolism and ***impairs the synthesis of NAD⁺ and NADP⁺***. Hence, the patients of carcinoid syndrome develop symptoms of ***pellagra*** (niacin deficiency). Further, ***negative nitrogen balance*** is also observed.

Diagnosis : The excretion of ***5-hydroxy indole acetate in urine*** is tremendously ***elevated*** (upto 500mg/day against normal <5 mg/day) in carcinoid syndrome. The estimation of 5 HIA in urine is used for the diagnosis of this disorder.

Melatonin

Melatonin is a hormone, mostly synthesized by the ***pineal gland.*** Serotonin—***produced from tryptophan***—is acted upon by serotonin N-acetylase (the rate limiting enzyme), to give N-acetylserotonin. The latter undergoes methylation, S-adenosylmethionine being the methyl group donor to produce melatonin or N-acetyl 5-methoxyserotonin (***Fig.15.18***).

Functions of melatonin

1. Melatonin is involved in ***circadian rhythms*** or diurnal variations (24 hr cyclic process) of the body. It plays a significant role ***in sleep and wake process.***

2. Melatonin also performs a ***neurotransmitter*** function.

SULFUR AMINO ACIDS

The sulfur-containing amino acids are ***methionine, cysteine and cystine***. Among these, only ***methionine*** is ***essential***. It serves as a precursor for the synthesis of cysteine and cystine which are, therefore, non-essential.

METABOLISM OF METHIONINE

Methionine (or sulfur amino acids) metabolism may be divided into three parts (***Fig.15.19***).

1. Utilization of methionine for transmethylation reactions.

Medical Concepts/ Clinical Correlates

- **Phenylketonuria, due to a defect in the enzyme phenylalanine hydroxylase, is characterized by failure of growth, seizures and mental retardation (low IQ).**
- **Alkaptonuria causes the accumulation of homogentisate which undergoes oxidation followed by polymerization to produce the pigment alkapton. Deposition of alkapton in connective tissue, causes ochronosis which is associated with arthritis.**
- **Neurotransmitters namely serotonin, dopamine, epinephrine, nonepinephrine and γ-aminobutyrate are derived from amino acids. Many drugs used to treat psychiatric and neurologic disorders act by altering the metabolism of neurotransmitters.**
- **Malignant carcinoid syndrome, a tumor of argentaffin cells of gastrointestinal tract, is characterized by tremendously increased production of serotonin. This disorder can be diagnosed by the elevated levels of 5-hydroxyindoleacetate in urine.**

2. Conversion of methionine to cysteine and cystine.

3. Degradation of cysteine and its conversion to specialized products.

Transmethylation

The ***transfer of methyl group ($-CH_3$) from active methionine to an acceptor*** is known as transmethylation. Methionine has to be activated to ***S-adenosylmethionine*** (***SAM***) or active methionine to donate the methyl group.

Synthesis of S-adenosylmethionine

The synthesis of S-adenosylmethionine occurs by the transfer of an adenosyl group from ATP to sulfur atom of methionine. This reaction is catalysed by methionine S-adenosyltransferase. Three high energy phosphates ***(3 ATP) are consumed in the formation of SAM***.

Functions of S-adenosylmethionine

S-Adenosylmethionine is highly reactive due to the presence of a positive charge. The enzymes involved in the transfer of methyl group are collectively known as ***methyltransferases***. ***S-Adenosylmethionine transfers the methyl group to an acceptor*** and gets itself converted to S-adenosylhomocysteine. The loss of free energy in this reaction makes the methyl transfer essentially irreversible. S-Adenosylhomocysteine is hydrolysed to homocysteine and adenine. Homocysteine can be remethylated to methionine by N^5-methyl tetrahydrofolate. In this manner, methionine can be regenerated for reuse. It should be noted that there is no net synthesis of methionine in the S-adenosylmethionine cycle (***Fig.15.20***). Hence, methionine is an essential amino acid.

S-Adenosylmethionine (carbon fragment) is also involved in the synthesis of polyamines (spermidine, spermine). The most important transmethylation reactions are listed in ***Table 15.3***.

Significance of transmethylation

1. Transmethylation is of great biological significance since ***many compounds become functionally active*** only after methylation.

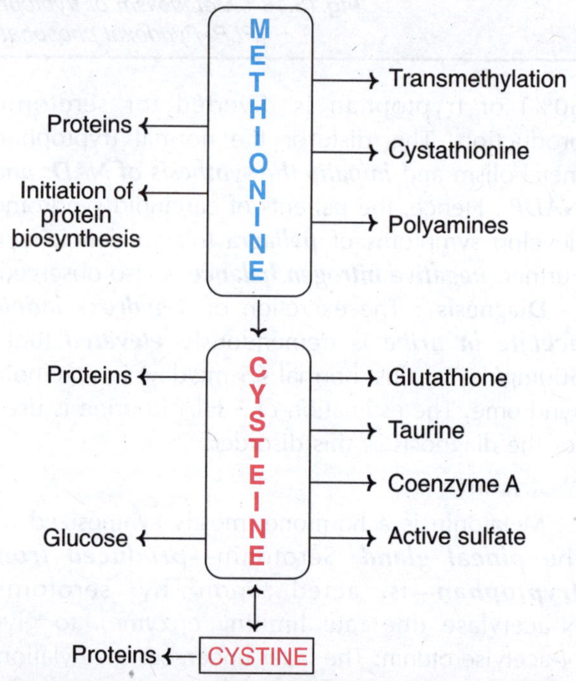

Fig.15.19 : Overview of the metabolism of sulfur amino acids.

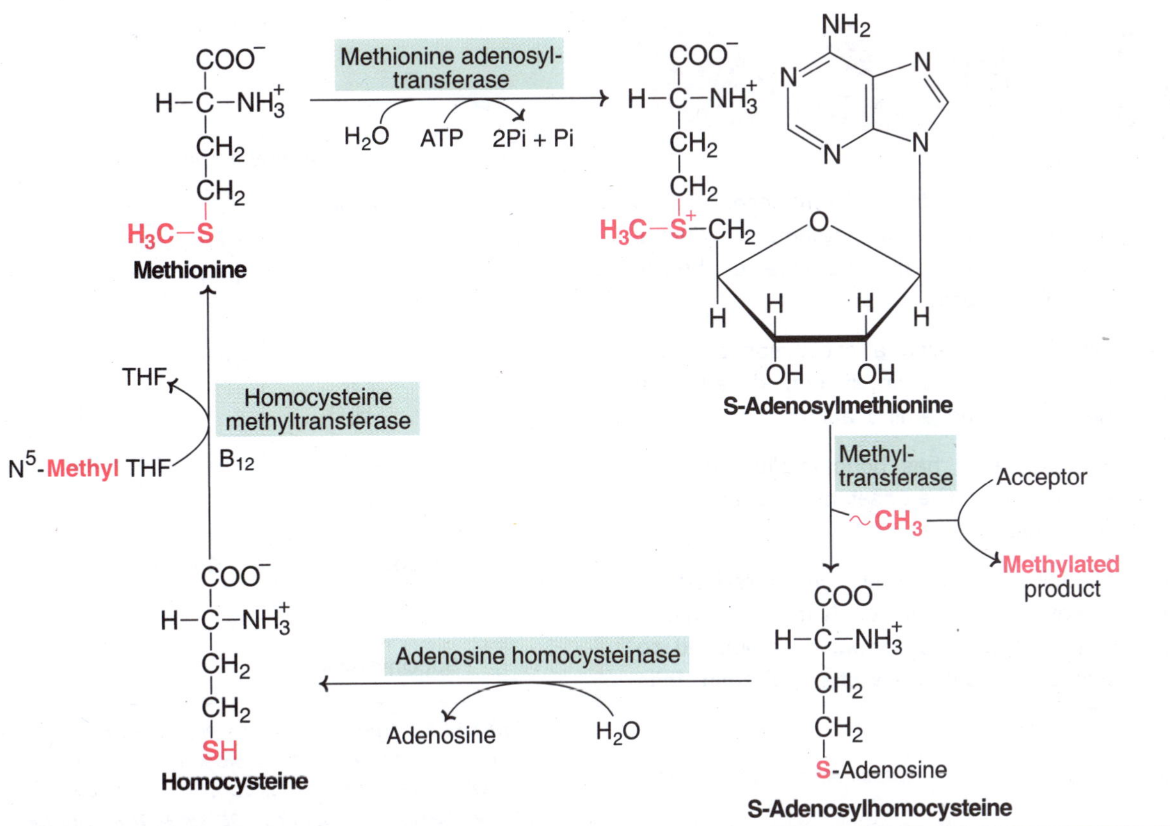

Fig.15.20 : S-Adenosylmethionine cycle—synthesis, utilization and regeneration.

2. Protein (amino acid residues) methylation helps to control protein turnover. In general, ***methylation protects*** the ***proteins from immediate degradation***.

Synthesis of cysteine

Homocysteine formed from methionine is a precursor for the synthesis of cysteine (***Fig.15.21***). Homocysteine condenses with serine to form cystathionine. This reaction is catalysed by a PLP-dependent cystathionine synthase. The enzyme cystathioninase (PLP-dependent) cleaves and deaminates cystathionine to cysteine and α-ketobutyrate. The sum of the reactions catalysed by cystathionine synthase and cystathioninase is a good example of ***transsulfuration*** (transfer of sulfur from one compound to another). It should be noted that only the ***sulfur atom of cysteine*** comes from homocysteine (originally methionine) while the ***rest of the molecule is from serine***.

Homocysteine and heart attacks

Homocysteine is an intermediate in the synthesis of cysteine from methionine (***Fig.15.21***), ***Elevation*** in ***plasma homocysteine*** (normal < 15 μ mol/l) has been ***implicated in coronary artery diseases***, although the mechanism is not known. It is believed that

TABLE 15.3 Transmethylation reactions

S-Adenosylmethionine ($\sim CH_3$)	*S-Adenosylhomocysteine*

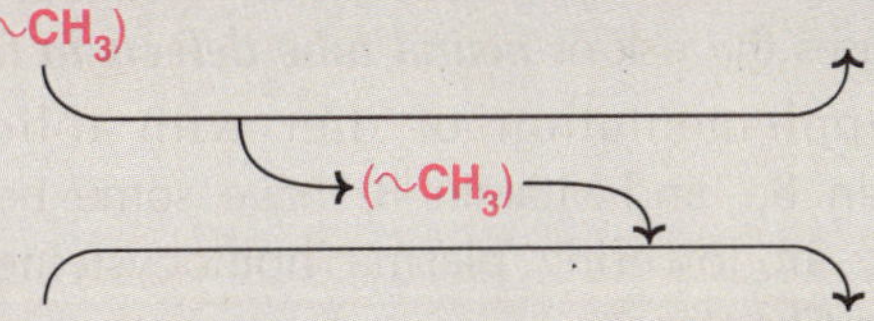

Methyl group acceptor	*Methylated product*
Guanidoacetate	Creatine
Norepinephrine	Epinephrine
Epinephrine	Metanephrine
Ethanolamine	Choline
Nicotinamide	N-Methylnicotinamide
Acetyl serotonin	Melatonin
Phosphatidylethanolamine	Phosphatidylcholine
Serine	Choline

Medical Concepts/ Clinical Correlates

- **Melatonin, produced from serotonin, is involved in circadian rhythms or diurnal variations, i.e., maintenance of body's biological clock.**
- **Tryptophan, being a precursor of serotonin and melatonin, is considered as a sleep inducing amino acid.**
- **Homocysteine has been implicated as a risk factor in the onset of coronary heart diseases.**
- **Histidine loading test, characterized by elevated excretion of N-formiminoglutamate (FIGLU), is commonly employed to assess the deficiency of the vitamin, folic acid.**

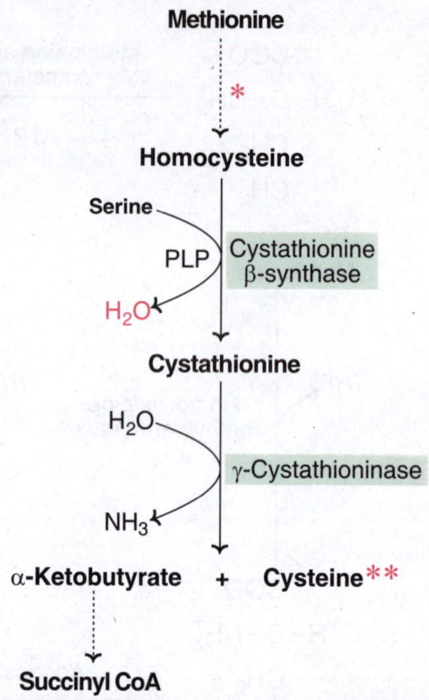

Fig.15.21 : Synthesis of cysteine from methionine (–For reactions of methionine conversion to homocysteine refer Fig. 15.20) (**–In cysteine synthesis, only sulfur is obtained from homocysteine, the rest of the molecule is from serine).*

homocysteine reacts with collagen to produce reactive free radicals, besides interfering with collagen cross links. Homocysteine is also involved in the aggregation of LDL particles. All this leads to an increased tendency for ***atherogenesis***, and consequently heart complications. ***Hyperhomocysteinemia*** in pregnancy increases the risk of ***neural tube defects in fetus***.

Supplementation of diet with folic acid, vitamin B_{12} and vitamin B_6 have some beneficial affects in lowering plasma homocysteine levels (***Refer Chapter 7***).

Degradation of cysteine

Cystine and cysteine are interconvertible by an NAD^+-dependent cystine reductase (***Fig.15.22***). Cysteine on decarboxylation produces ***mercaptoethanolamine*** which is involved in the biosynthesis of ***coenzyme A*** from the vitamin pantothenic acid (***Fig.15.22***).

The enzyme cysteine dioxygenase oxidizes cysteine to cysteine sulfinate which, on further oxidation, is converted to cysteic acid. The latter undergoes decarboxylation to produce ***taurine*** which conjugates with ***bile acids***. Cysteic acid can also be degraded to pyruvate.

Cysteine sulfinate cleaves off alanine to produce sulfite which is converted to sulfate and excreted in urine. Some amount of sulfate condenses with ATP to form ***active sulfate*** or ***3′-phosphoadenosine 5′-phosphosulfate*** (***PAPS***). Active sulfate (PAPS) is utilized for the synthesis of ***mucopolysaccharides*** (sulfation), besides being used in ***detoxification***. Sulfate is also a structural component of some proteins, lipids etc.

Inborn errors of sulfur amino acid metabolism

Cystinuria (cystine-lysinuria) : Cystinuria is one of the most common inherited diseases with a frequency of 1 in 7,000. It is primarily characterized by increased excretion of cystine (25-40 times normal). Elevation in the urinary output of lysine, arginine and ornithine is also observed. A specific carrier system exists in kidney tubules for the reabsorption of amino acids, namely **c**ysteine, **o**rnithine, **a**rginine and **l**ysine (remember ***COAL*** to recall). In cystinuria, this ***carrier system*** becomes ***defective*** leading to the excretion of all these four amino acids in urine.

Cystine is relatively insoluble and its increased concentration leads to its precipitation and formation of cystine stones in kidney and urinary tract.

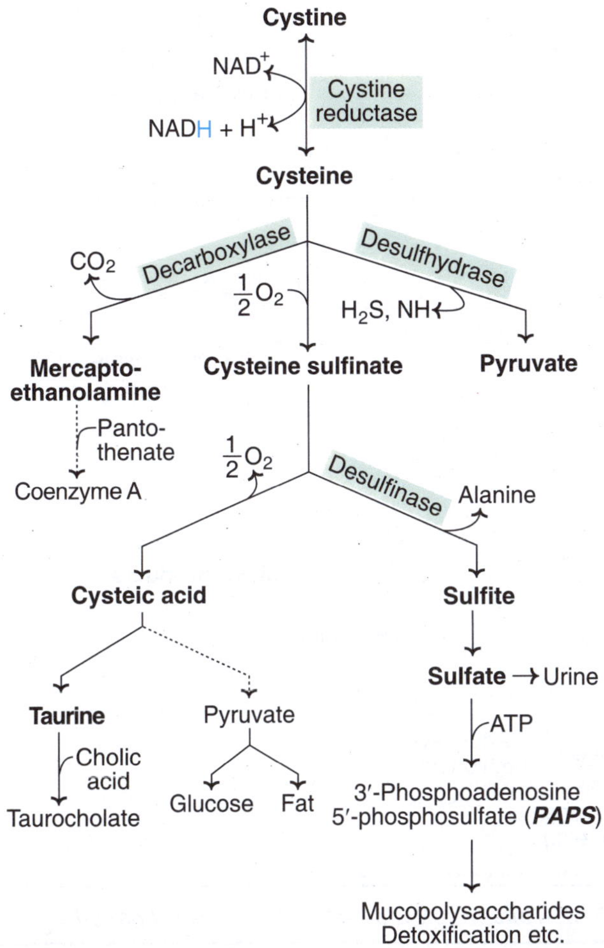

Fig. 15.22 : Metabolism of cysteine and cystine.

Cystinosis (cystine storage disease) : Cystine crystals are deposited in many tissues and organs of reticuloendothelial system throughout the body. These include spleen, lymph nodes, liver, kidney, bone marrow etc. A ***defect in the lysosomal function*** is said to be the primary cause of this disorder. In fact, cystine is accumulated in the lysosomes of various tissues. Impairment in renal function is commonly seen in cystinosis. It is characterized by generalized amino aciduria.

Homocystinurias : Homocystinurias are a group of ***metabolic disorders*** characterized by the accumulation and ***increased urinary excretion*** of ***homocysteine and S-adenosylmethionine***. Plasma concentration of methionine is increased.

Homocystinuria type I has been more thoroughly investigated. It is due to a defect in the enzyme ***cystathionine synthase.*** Accumulation of homocysteine results in various complications—thrombosis, osteoporosis and, very often, mental retardation. Further, the deficiency of cystathionine is associated with damage to endothelial cells which might lead to atherosclerosis.

ONE-CARBON METABOLISM

Amino acid metabolism is particularly important for the transfer or exchange of one-carbon units. The following ***one-carbon fragments*** are encountered in the biological reactions, which constitute one-carbon pool

Methyl	($-CH_3$)
Hydroxymethyl	($-CH_2OH$)
Methylene	($=CH_2$)
Methenyl	($-CH=$)
Formyl	($-CH=O$)
Formimino	($-CH=NH$)

Tetrahydrofolate (THF) is a versatile coenzyme that ***actively participates in one-carbon metabolism.*** With regard to the transfer of methyl groups from S-adenosylmethionine, vitamin B_{12} is also involved besides THF ***(Refer Fig.7.23***; Folic acid in ***Chapter** 7*).

I. Generation of one-carbon units

Many compounds (particularly amino acids) act as donors of one-carbon fragments

1. The formate released from glycine and tryptophan metabolism combines with THF to form N^{10}-formyl THF.
2. Histidine contributes formimino fragment to produce N^5-formimino THF.
3. When serine is converted to glycine, N^5, N^{10}-methylene THF is formed. This is the most predominant entry of one carbon units into one carbon pool.
4. Choline and betaine contribute to the formation of N^5-methyl THF.

The ***different derivatives of THF carrying one-carbon units are interconvertible***, and this is metabolically significant for the continuity of one-carbon pool (***Fig.15.23***).

II. Utilization of one-carbon moieties

One-carbon fragments from THF are used for the synthesis of a wide variety of compounds. These include ***purines***, ***formylmethionine tRNA*** (required for initiation of protein synthesis), ***glycine, pyrimidine nucleotide*** (thymidylate) etc.

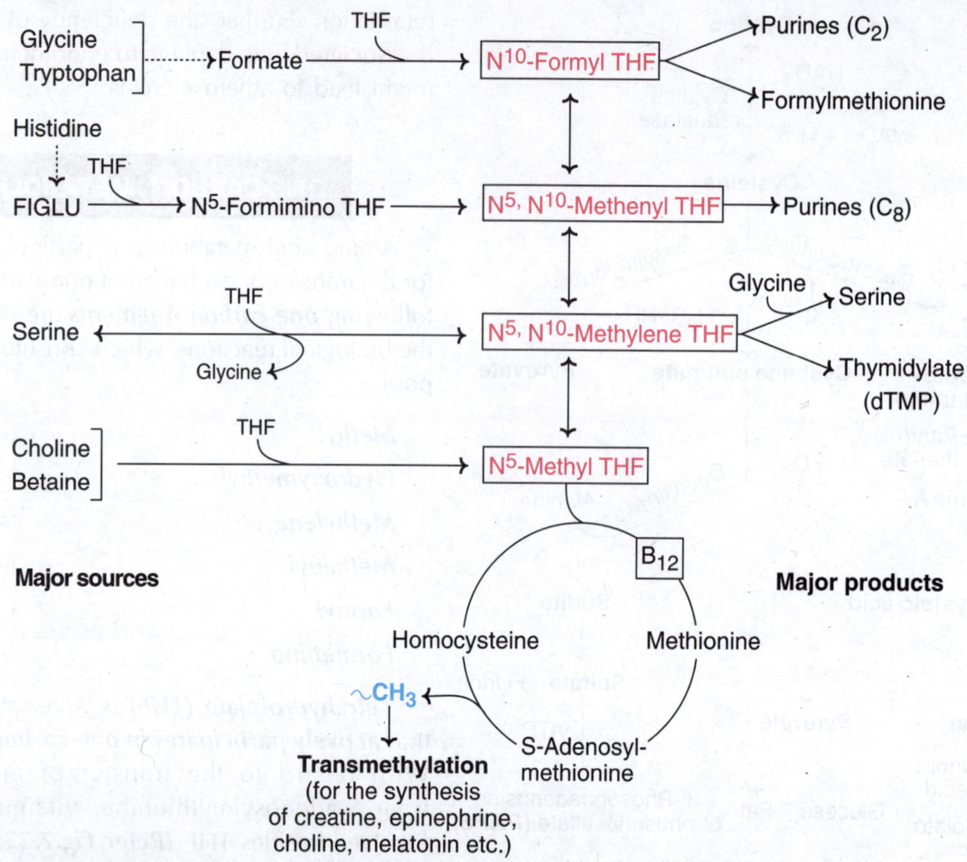

Fig. 15.23 : *Summary of one-carbon metabolism (THF–Tetrahydrofolate; ~CH_3–Active methyl group).*

BRANCHED CHAIN AMINO ACIDS

Valine, leucine and ***isoleucine*** are the branched chain and ***essential*** amino acids. These three amino acids initially undergo a common pathway and then diverge to result in different end products. Based on the products obtained from the carbon skeleton, the branched chain amino acids are either glycogenic or ketogenic

Valine – ***glycogenic***
Leucine – ***ketogenic***
Isoleucine – ***glycogenic and ketogenic.***

The first three metabolic reactions are common to the branched chain amino acids (***Fig.15.24***).

1. **Transamination :** The three amino acids undergo a reversible transamination to form their respective keto acids.

2. **Oxidative decarboxylation :** ***α-Keto acid dehydrogenase*** is a complex mitochondrial enzyme. It is ***comparable*** in function to ***pyruvate dehydrogenase complex*** and employs 5 coenzymes—TPP, lipoamide, FAD, coenzyme A and NAD^+—for its action. α-Keto acid dehydrogenase catalyses oxidative decarboxylation of the keto acids to the corresponding acyl CoA thioesters. This is a regulatory enzyme in the catabolism of branched amino acids.

3. **Dehydrogenation :** The dehydrogenation is similar to that in fatty acid oxidation. FAD is the coenzyme and there is an incorporation of a double bond. It is now believed that there are two enzymes responsible for dehydrogenation.

After the initial three common reactions, the metabolism of branched chain amino acids diverges and takes independent routes. In a series of reactions that follow, valine is converted to propionyl CoA, a precursor for glucose. Leucine produces acetyl CoA and acetoacetate, the substrates for fatty acid synthesis. Isoleucine is degraded to propionyl CoA and acetyl CoA. Thus, valine is glycogenic and leucine is ketogenic while isoleucine is both glycogenic and ketogenic.

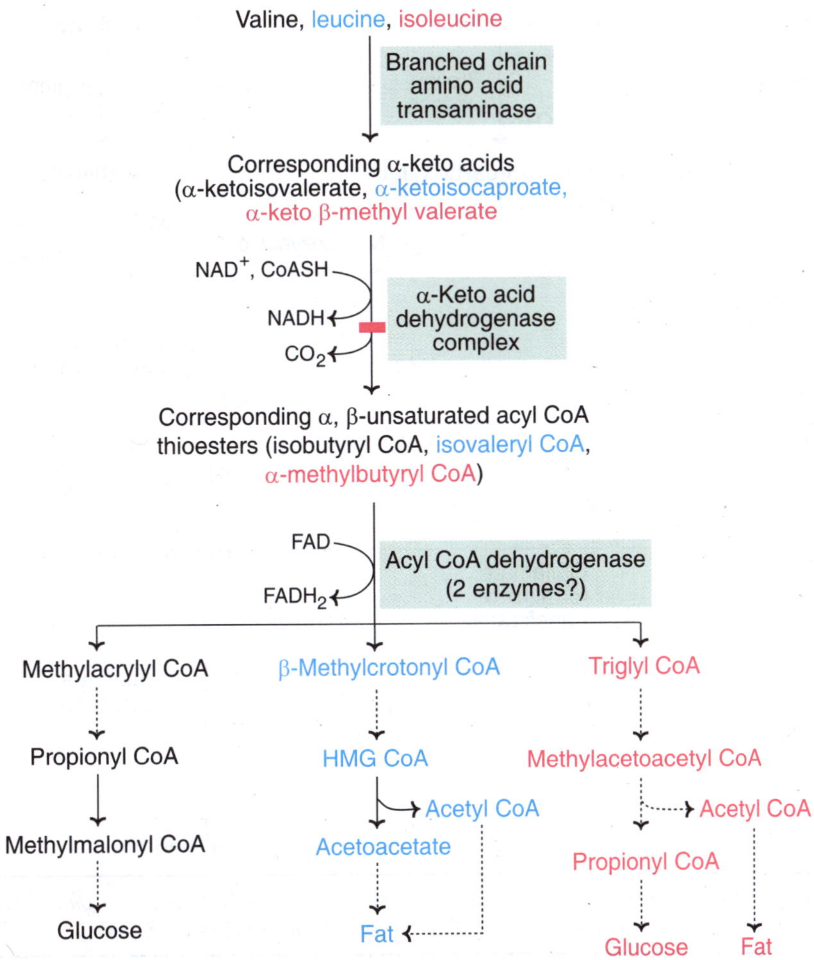

***Fig. 15.24 :** Summary of branched chain amino acid metabolism (▬–Defect in α-keto acid dehydrogenase results in maple syrup urine disease; HMG CoA–β-Hydroxy β-methylglutaryl CoA; Valine is glycogenic, leucine is ketogenic, while isoleucine is both).*

Metabolic defects of branched chain amino acids

1. **Maple syrup urine disease :** This is a metabolic disorder of branched chain amino acids. The urine of the affected individuals ***smells like*** maple syrup or ***burnt sugar***—hence the name.

Enzyme defect : Maple syrup urine disease is due to a defect in the enzyme ***branched chain α-keto acid dehydrogenase.*** This causes a blockade in the conversion of α-keto acids to the respective acyl CoA thioesters. The plasma and urine concentrations of branched amino acids and their keto acids are highly elevated. This disease is also known as ***branched chain ketonuria.***

Biochemical complications and symptoms :

- Accumulation of branched chain amino acids causes an ***impairment in transport and function of other amino acids.***
- Protein biosynthesis is reduced.
- Branched chain amino acids competitively inhibit glutamate dehydrogenase.
- The disease results in ***acidosis, lethargy, convulsions***, mental retardation, coma and, finally, death within one year after birth.

Diagnosis and treatment : An early diagnosis by enzyme analysis—preferably within the first week of life—is ideal. Estimation of urinary branched amino acids and keto acids will also help in diagnosis.

The treatment is to feed a diet with low (or no) content of branched amino acids. The plasma levels of branched amino acids should be constantly monitored for adjusting their dietary intake.

HISTIDINE, PROLINE AND ARGININE

The metabolism of histidine, proline and arginine is considered together, as all the

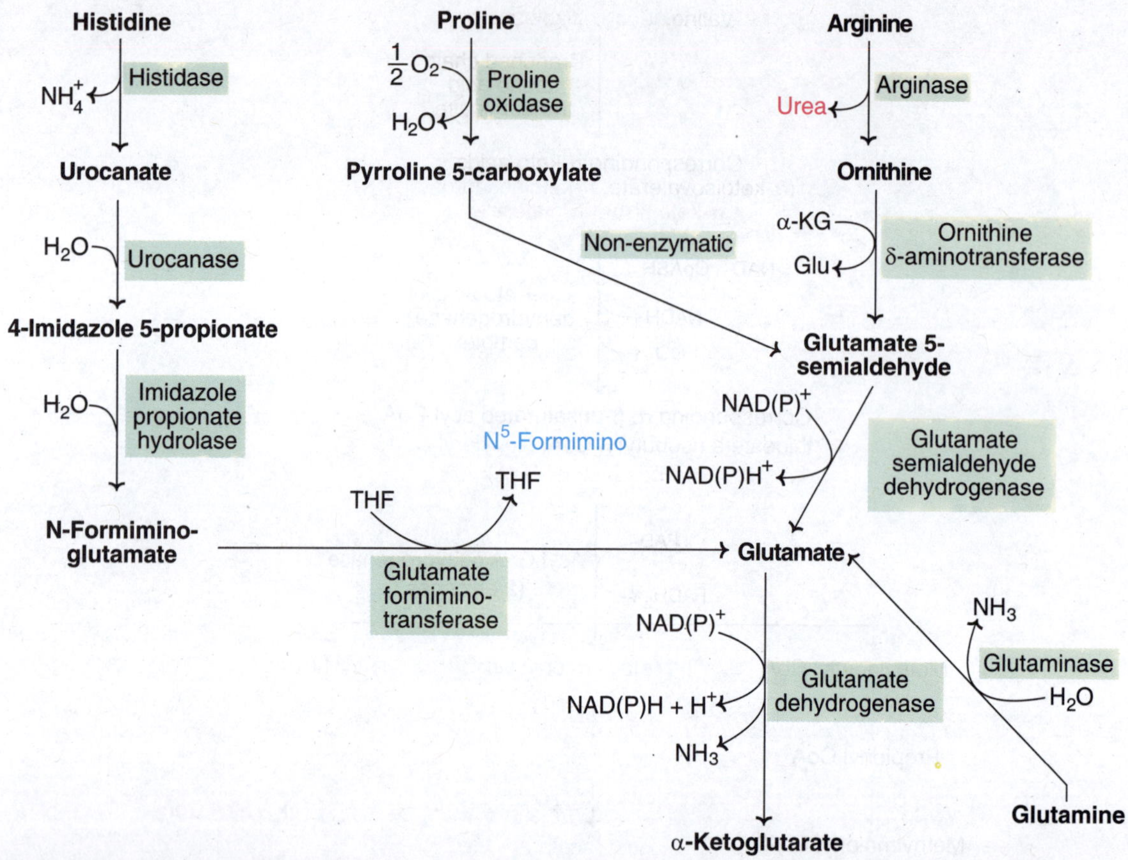

***Fig.15.25 :** Metabolism of histidine, proline, arginine, glutamate and glutamine (THF–Tetrahydrofolate; α-KG–α-Ketoglutarate; Glu–Glutamate).*

three are converted to glutamate and metabolized (***Fig.15.25***).

Histidine

The metabolism of histidine is important for the generation of ***one-carbon*** unit, namely formimino group. The enzyme histidase acts on histidine to split off ammonia. Urocanate formed in this reaction is acted upon by urocanase to produce 4-imidazole 5-propionate. Imidazole ring of the product is cleaved by a hydrolase to give N-formiminoglutamate (FIGLU). Tetrahydrofolate (THF) takes up the formimino group to form N^5-formimino THF, and glutamate is liberated. ***Deficiency of folate*** blocks this reaction and causes ***elevated excretion of FIGLU*** in urine. Histidine loading test is commonly employed to assess folate deficiency.

Histidine, on decarboxylation, gives the corresponding amine—***histamine.*** Histamine regulates HCl secretion by gastric mucosa. Excessive production of histamine causes asthma and allergic reactions.

Proline

Proline is oxidized to pyrroline 5-carboxylate which undergoes a non-enzymatic conversion to glutamate 5-semialdehyde. The latter is converted to glutamate and then transaminated to α-ketoglutarate. The five carbons of proline are converted to α-ketoglutarate.

Hyperprolinemia type I : It is due to a defect in the enzyme ***proline oxidase*** (proline dehydrogenase).

Arginine

Arginine is cleaved by arginase to liberate urea and produce ornithine. Ornithine undergoes transamination of δ-amino group to form glutamate γ-semialdehyde which is converted to glutamate. ***Hyperargininemia*** is an inborn error in arginine metabolism due to a defect in the enzyme ***arginase***.

Nitric oxide (NO) : Arginine is the substrate for the production of nitric oxide (NO), a wonder molecule with a wide range of functions. The enzyme ***nitric oxide synthase*** cleaves the nitrogen from the

Medical Concepts/ Clinical Correlates

- **Nitric oxide (NO), synthesized from arginine, is involved in several biological functions—vasodilation, platelet aggregation, neurotransmission and bactericidal action. NO is used in the treatment of chronic obstructive pulmonary disease (COPD).**
- **γ-Aminobutyric acid (GABA), produced from glutamate, is an inhibitory neurotransmitter. Low levels of GABA result in convulsions.**
- **The carbon skeleton of amino acids may be converted to glucose (glycogenic) or fat (ketogenic), besides being responsible for the synthesis of non-essential amino acids.**
- **Polyamines (spermine, putrescine) are involved in the synthesis of DNA, RNA and proteins and, thus, they are essential for cell growth and differentiation.**

guanidino group of arginine to form NO. NO has a very short half-life (about 5 seconds).

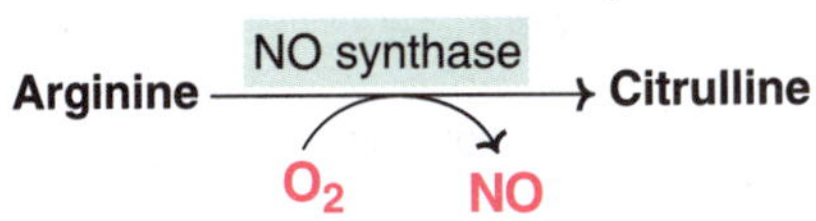

Functions of NO : Nitric oxide acts as a mediator for several biological functions.

1. NO functions as a vasodilator and causes relaxation of smooth muscles.
2. It is a key molecule in the ***regulation of*** blood flow and the ***blood pressure*** (inhibitors of NO synthesis markedly raise blood pressure).
3. NO acts as an inhibitor of platelet aggregation and adhesion.
4. It functions as a messenger molecule of the nervous system (***neurotransmitter***).
5. NO mediates the bactericidal actions of macrophages.
6. It is involved in the ***erection of penis.***

Mechanism of action : Nitric oxide promotes the synthesis of cGMP. It is believed that some of the actions of NO are mediated through cGMP and protein kinase G.

GLUTAMATE AND GLUTAMINE

Glutamate and glutamine are ***non-essential*** glycogenic amino acids. Both of them play a predominant role in the amino acid metabolism and are directly involved in the final transfer of amino group for urea synthesis.

The amino acids—histidine, proline and arginine —are converted in their metabolism to glutamate (***Fig.15.25***). α-Ketoglutarate—an intermediate in TCA cycle—serves as an immediate precursor for glutamate formation. Glutamate—besides being converted to glutamine—is involved in the synthesis of certain specialized products (***Fig.15.26***).

1. **Glutathione** is a tripeptide that contains glutamate. Its formation is described under glycine metabolism.
2. **N-Acetylglutamate** is an allosteric regulator of carbamoyl phosphate synthase I, the first enzyme in urea synthesis.
3. Glutamate is present in the ***clotting factors*** (II, VII, IX, X) as γ-carboxyglutamate and is involved in coagulation.

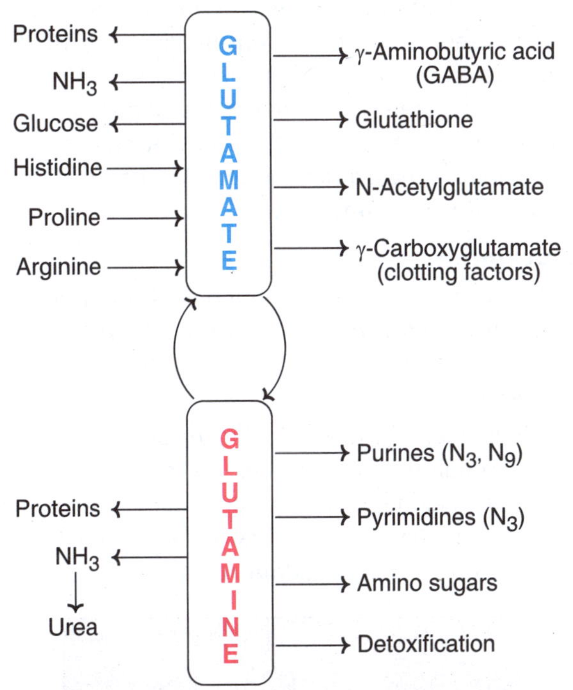

***Fig.15.26 :** Overview of glutamate and glutamine metabolism.*

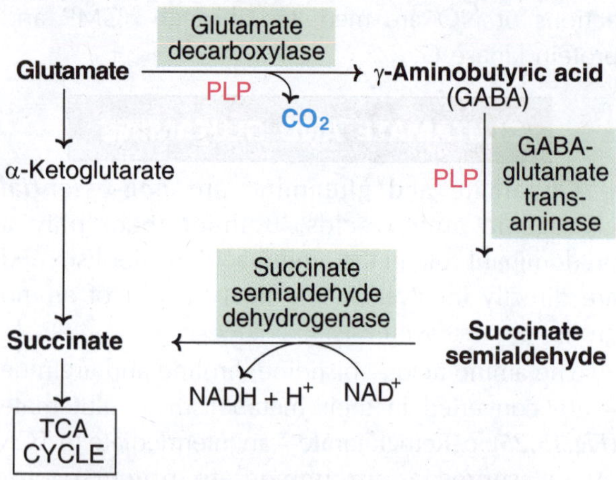

***Fig. 15.27 :** Metabolism of γ-aminobutyric acid (PLP–Pyridoxal phosphate).*

4. **γ-Aminobutyric acid (GABA) :** Glutamate is decarboxylated to GABA—mostly in the brain. This is a sensitive and irreversible reaction, ***catalysed by a pyridoxal phosphate-dependent glutamate decarboxylase.***

GABA undergoes transamination followed by oxidation to form succinate which enters TCA cycle (***Fig.15.27***). The reactions involving the fate of GABA constitute a bypass route for glutamate to enter TCA cycle, which is known as ***GABA shunt.***

Functions of GABA : It is one of the major ***inhibitory neurotransmitters in the brain***. GABA regulates the activity of neurons by discouraging the transmission signals. It is believed that GABA opens chloride channels and increases the permeability of post-synaptic membranes. Thus GABA functions as an ***inhibitory neurotransmitter. Decreased GABA levels will cause convulsions.***

Glutamine

Glutamine is a versatile amino acid. ***Ammonia*** is temporarily ***stored in the form of glutamine.*** Glutamine is freely diffusible and, hence, easily transported.

Glutamine is the donor of nitrogen atom for purine and pyrimidine synthesis. It is the chief source of ammonia in kidneys. The NH_3 production is elevated in acidosis to maintain ***acid-base balance.*** Glutamine also takes part in ***conjugation reactions.***

FATE OF CARBON SKELETON OF AMINO ACIDS

The metabolic reactions of individual amino acids are described above. After the removal of amino groups, the carbon skeleton of amino acids is converted to intermediates of TCA cycle or their precursors. The ***carbon skeleton*** finally has one or more of the following fates

1. Oxidation via TCA cycle to produce ***energy*** (about 10-15% of body needs).
2. Synthesis of ***glucose.***
3. Formation of lipids—fatty ***acids and ketone bodies.***
4. Synthesis of ***non-essential amino acids.***

The carbon skeletons of the 20 standard amino acids (or the amino acids of proteins) are degraded to one of the following seven products—***pyruvate, α-ketoglutarate, succinyl CoA, fumarate, oxaloacetate, acetyl CoA*** and ***acetoacetate***. Some authors use the term amphibolic (*Greek*: amphiboles—uncertain) intermediates to these compounds due to their multiple metabolic functions.

The amino acids are classified into two groups, based on the nature of the metabolic end products of carbon skeleton.

TABLE 15.4 Classification of amino acids based on the fate of carbon skeleton

Glycogenic (glucogenic)	*Glycogenic and ketogenic*	*Ketogenic*
Alanine	Phenylalanine*	Leucine*
Arginine*	Isoleucine*	Lysine*
Aspartate	Tyrosine	
Cysteine	Tryptophan*	
Glutamine		
Glutamate		
Glycine		
Histidine*		
Hydroxyproline		
Methionine*		
Proline		
Serine		
Threonine*		
Valine*		

** Essential amino acids; (Helpful tips to recall—ketogenic amino acids start with letter 'L'; PITT for glyco-and ketogenic amino acids; rest of the 20 amino acids are only glycogenic).*

TABLE 15.5 Inborn errors of amino acid metabolism

Disorder	Metabolic defect (enzyme/other)
I. Defects in urea synthesis —See *Table 15.1*	
II. Phenylalanine and tyrosine	
1. Phenylketonuria	Phenylalanine hydroxylase
2. Tyrosinemia type II	Tyrosine transaminase
3. Neonatal tyrosinemia	p-Hydroxy phenylpyruvate dioxygenase
4. Alkaptonuria	Homogentisate oxidase
5. Tyrosinosis (tyrosinemia type I)	Maleyl acetoacetate isomerase or fumaryl acetoacetate hydrolase
6. Albinism	Tyrosinase
III. Sulfur amino acids (methionine, cysteine and cystine)	
7. Cystinuria	Defect in renal reabsorption
8. Cystinosis	Impairment in cystine utilization (defect in lysosomal function)
9. Homocystinuria type I	Cystathionine synthetase
10. Homocystinuria type II	N^5, N^{10}-Methylene THF reductase
11. Homocystinuria type III	N^5-Methyl THF-homocysteine methyltransferase
12. Cystathioninuria	Cystathioninase
IV. Glycine	
13. Glycinuria	Defect in renal reabsorption
14. Primary hyperoxaluria	Glycine transaminase
V. Tryptophan	
15. Hartnup's disease	Defective intestinal absorption
VI. Branched chain amino acids (valine, leucine and isoleucine)	
16. Maple syrup urine disease	Branched chain α-keto acid dehydrogenase
17. Intermittent branched chain ketonuria	Variant of the above enzyme (less severe)
18. Hypervalinemia	Valine transaminase
19. Isovaleric acidemia	Isovaleryl CoA dehydrogenase
VII. Histidine	
20. Histidinemia	Histidase
VIII. Proline	
21. Hyperprolinemia type I	Proline oxidase

1. **Glycogenic (glucogenic) amino acids :** These are the amino acids whose carbon skeleton is finally degraded to pyruvate or one of the intermediates of TCA cycle (α-ketoglutarate, succinyl CoA, fumarate and oxaloacetate). These intermediates serve as good substrates for gluconeogenesis leading to the ***formation of glucose*** or glycogen.

2. **Ketogenic amino acids :** The amino acids whose carbon skeleton is ***metabolized to acetyl CoA or acetoacetate*** can be ***converted to fat*** (i.e., fatty acids or ketone bodies). Acetoacetate is a ketone body (besides acetone and β-hydroxybutyrate).

Some of the amino acids are both glycogenic and ketogenic since they serve as precursors for glucose as well as fat.

The classification of amino acids (glycogenic, ketogenic, or both) is given in ***Table 15.4***. For the various products obtained from the carbon skeleton of glycogenic amino acids and their connection with the citric acid cycle refer ***Fig.13.7***.

TABLE 15.6 A summary of the biogenic amines and their functions

Amino acid	Amine	Function(s)
Serine	Ethanolamine	Forms choline
Glutamate	γ-Aminobutyric acid	Inhibitory neurotransmitter
Histidine	Histamine	Vasodilator, promotes gastric HCl and pepsin synthesis
Phenylalanine	Dopamine	For the synthesis of norepinephrine and epinephrine
Tyrosine	Tyramine	Vasoconstrictor (increases blood pressure)
Tryptophan	Tryptamine	Elevates blood pressure
	Serotonin	Stimulates cerebral activity
	Melatonin	Circadian rhythms
Cysteine	Taurine (taurocholic acid)	Constituent of bile acid

Inborn errors of amino acid metabolism—a summary

Several inherited disorders are associated with amino acid metabolism. The details of these metabolic disorders are described in the respective amino acids. ***Table 15.5*** gives a summary of the inborn errors of amino acid metabolism.

BIOGENIC AMINES

In general, the ***decarboxylation of amino acids*** or their derivatives results in the formation of amines.

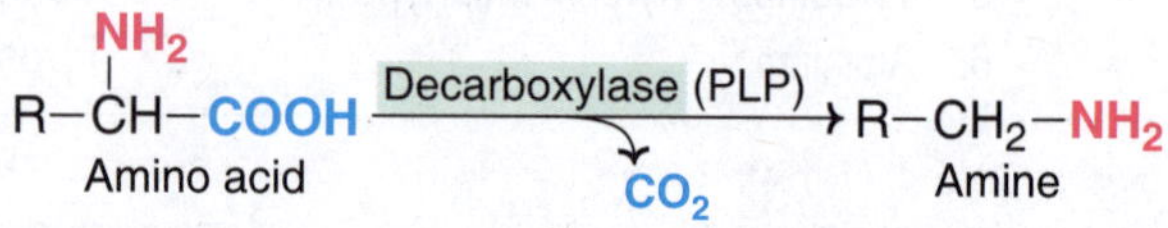

A summary of the biogenic amines derived from different amino acids and their major functions are given in ***Table 15.6***.

SUMMARY

1. *The amino acids undergo transamination and deamination to liberate ammonia for the synthesis of urea, the end product of protein metabolism. Besides being present as structural components of proteins, amino acids participate in the formation of several biologically important compounds.*
2. *Glycine is involved in the synthesis of creatine, heme, purines, glutathione etc.*
3. *Phenylalanine is hydroxylated to tyrosine, which is a precursor for the production of skin pigment (melanin), catecholamines (dopamine, epinephrine and norepinephrine) and thyroid hormones (T_3 and T_4).*
4. *Tryptophan is converted to NAD^+ and $NADP^+$, the coenzymes of niacin, serotonin (a neurotransmitter) and melatonin.*
5. *The active methionine (SAM) is a donor of methyl group (transmethylation) for the synthesis of many biological compounds (epinephrine, choline, methylcytosine etc.).*
6. *The carbon skeleton of amino acids is involved either in the synthesis of glucose (glycogenic) or fat (ketogenic), or both – glucose and fat.*

THE READER MUST REFER APPENDIX D (Page Nos : 422 - 431) FOR RELATED CLINICAL CASE STUDIES (Nos : 11, 12)

INTEGRATION OF METABOLISM

16

Metabolism is a continuous process, with thousands of reactions, simultaneously occurring in the living cell. However, biochemists prefer to present metabolism in the form of reactions and metabolic pathways. This is done for the sake of convenience in presentation and understanding.

In the preceeding three ***chapters (13-15)***, we have learnt the ***metabolism of carbohydrates, lipids and amino acids***. We shall now consider the organism as a whole and integrate the metabolism with particular reference to energy demands of the body.

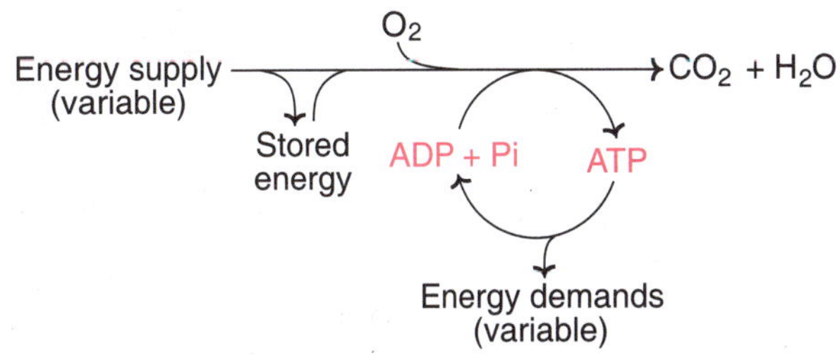

Fig.16.1 : *A summary of body's energy supply and demands. (**Note :** ATP serves as the energy currency).*

Energy demand and supply

The organisms possess variable energy demands, hence the supply (input) is also equally variable. The consumed metabolic fuel may be burnt (oxidized to CO_2 and H_2O) or stored to meet the energy requirements as per the body needs. ***ATP*** serves as the ***energy currency of the cell*** in this process (***Fig.16.1***).

The humans possess enormous capacity for food consumption. It is estimated that one can consume as much as 100 times his/her basal requirements! Obesity, a disorder of overnutrition mostly prevalent in affluent societies, is primarily a consequence of overconsumption.

Integration of major metabolic pathways of energy metabolism

An overview of the interrelationship between the important metabolic pathways, concerned with fuel metabolism depicted in ***Fig.16.2***, is briefly described here. For detailed information on these metabolic pathways, the reader must refer the respective chapters.

1. **Glycolysis :** The degradation of glucose to pyruvate (lactate under anaerobic condition) generates 7 ATP. Pyruvate is converted to acetyl CoA.

2. **Fatty acid oxidation :** Fatty acids undergo sequential degradation with a release of 2-carbon fragment, namely acetyl CoA. The energy is trapped in the form of NADH and $FADH_2$.

3. **Degradation of amino acids :** Amino acids, particularly when consumed in excess than required for protein synthesis, are degraded and utilized to meet the fuel demands of the body. The glucogenic amino acids can serve as precursors for the synthesis of glucose via the formation of pyruvate or intermediates of citric acid cycle. The ketogenic amino acids are the precursors for acetyl CoA.

4. **Citric acid cycle :** Acetyl CoA is the key and common metabolite, produced from different fuel sources (carbohydrates, lipids, amino acids). ***Acetyl***

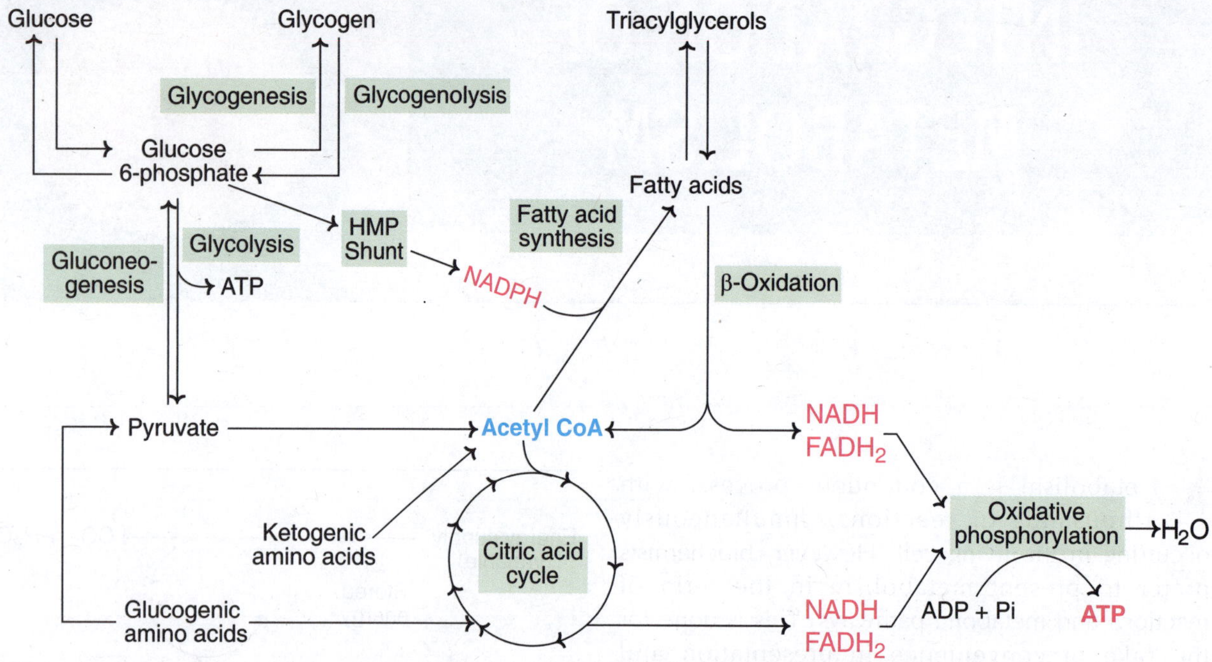

***Fig.16.2 :** An overview of integration of metabolic pathways of energy metabolism (HMP shunt—Hexose monophosphate shunt).*

CoA enters citric acid cycle and gets oxidized to CO_2. Thus, citric acid cycle is the ***final common metabolic pathway for the oxidation of all foodstuffs.*** Most of the energy is trapped in the form of NADH and $FADH_2$.

5. **Oxidative phosphorylation :** The NADH and $FADH_2$, produced in different metabolic pathways, are finally oxidized in the electron transport chain (ETC). The ETC is coupled with oxidative phosphorylation ***to generate ATP.***

6. **Hexose monophosphate shunt :** This pathway is primarily concerned with the liberation of ***NADPH*** and ***ribose sugar***. NADPH is utilized for the biosynthesis of several compounds, including fatty acids. Ribose is an essential component of nucleotides and nucleic acids (note—DNA contains deoxyribose).

7. **Gluconeogenesis :** The synthesis of glucose from non-carbohydrate sources constitutes gluco-neogenesis. Several compounds (e.g. pyruvate, glycerol, amino acids) can serve as precursors for gluconeogenesis.

8. **Glycogen metabolism :** Glycogen is the storage form of glucose, mostly found in liver and muscle. It is degraded (glycogenolysis) and synthesized (glycogenesis) by independent pathways. Glycogen effectively serves as a ***fuel reserve*** to meet body needs, ***for a brief period*** (between meals).

Regulation of metabolic pathways

The metabolic pathways, in general, are controlled by four different mechanisms

1. The ***availability of substrates***
2. Covalent ***modification of enzymes***
3. ***Allosteric regulation***
4. Regulation of ***enzyme synthesis.***

The details of these regulatory processes are discussed under the individual metabolic pathways, in the respective chapters.

ORGAN SPECIALIZATION AND METABOLIC INTEGRATION

The ***various tissues and organs of the body work*** in a ***well coordinated manner to meet its metabolic demands***. The major organs along with their most important metabolic functions, in a well-fed absorptive state (usually 2-4 hours after food consumption), are described (***Fig.16.3***).

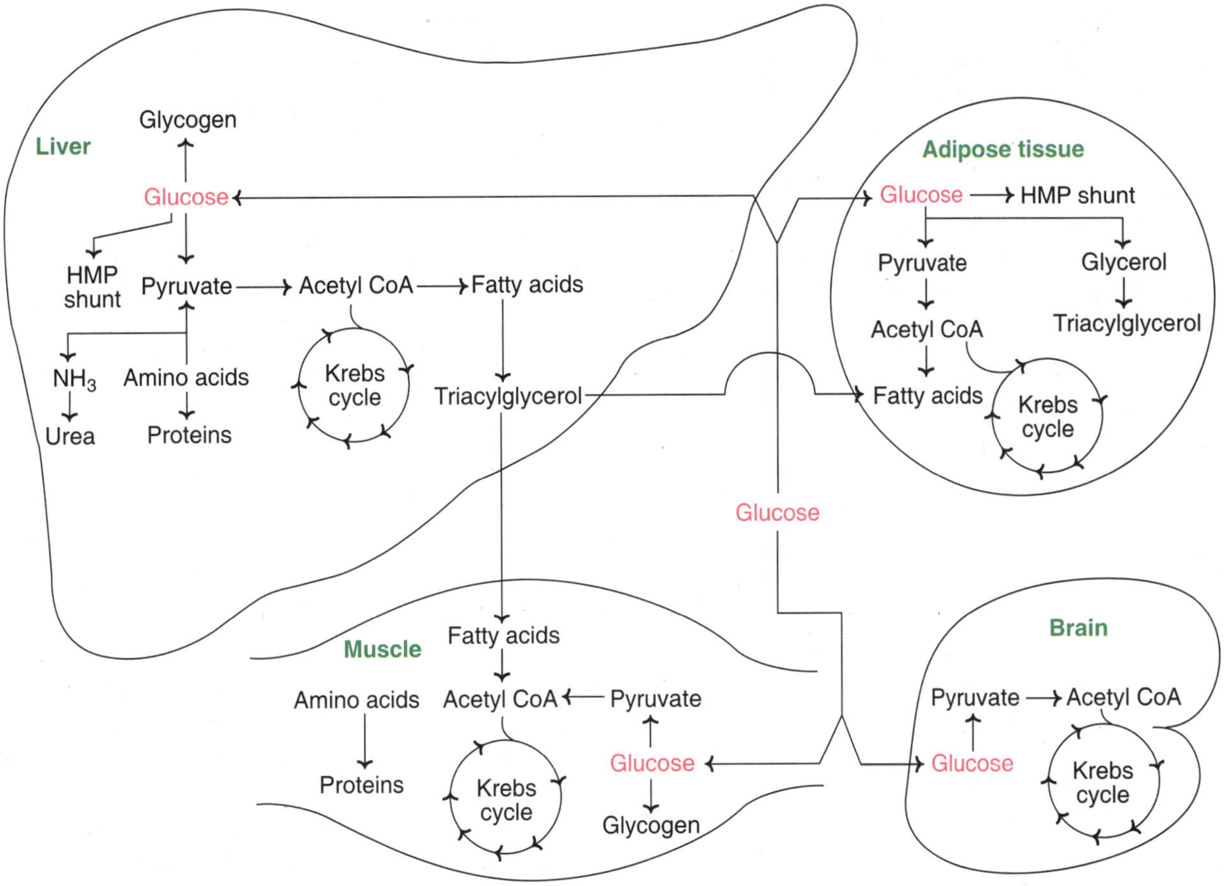

***Fig.16.3 :** Metabolic interrelationship among the major tissues in a well fed state (HMP shunt–Hexose monophosphate shunt).*

Liver

The liver is specialized to serve as the body's ***central metabolic clearing house.*** It processes and distributes the nutrients to different tissues for utilization. After a meal, the liver takes up the carbohydrates, lipids and most of the amino acids, processes them and routes to other tissues. The major metabolic functions of liver, in an absorptive state, are listed

1. **Carbohydrate metabolism** : Increased glycolysis, glycogenesis and hexose monophosphate shunt and decreased gluconeogenesis.

2. **Lipid metabolism** : Increased synthesis of fatty acids and triacylglycerols.

3. **Protein metabolism** : Increased degradation of amino acids and protein synthesis.

Adipose tissue

Adipose tissue is regarded as the ***energy storage tissue.*** As much as 15 kg. of triacylglycerol (equivalent to 135,000 Cal) is stored in a normal adult man. The major metabolic functions of adipose tissue in an absorptive state are listed here.

1. **Carbohydrate metabolism** : The uptake of glucose is increased. This follows an increase in glycolysis and hexose monophosphate shunt.

2. **Lipid metabolism** : The synthesis of fatty acids and triacylglycerols is increased. The degradation of triacylglycerols is inhibited.

Skeletal muscle

The ***metabolism*** of skeletal muscle is rather ***variable*** depending on its needs. For instance, the resting muscle of the body utilizes about 30% of body's oxygen consumption. However, during strenuous exercise, this may be as high as 90%. The important metabolic functions of skeletal muscle in an absorptive state are listed

1. **Carbohydrate metabolism** : The uptake of glucose is higher and glycogen synthesis is increased.

Medical Concepts/ Clinical Correlates

- **Biochemists, for their convenience, learn body chemical processes in terms of individual metabolic reactions and pathways, although thousands of reactions simultaneously occur in a living cell.**
- **The metabolic pathways in various tissues and organs are well coordinated to meet the demands of the body.**
- **Liver is appropriately regarded as the body's 'central metabolic clearing house' while adipose tissues constitute the energy (fat) storehouse.**

2. **Lipid metabolism :** Fatty acids taken up from the circulation are also important fuel sources for the skeletal muscle.

3. **Protein metabolism :** Incorporation of amino acids into proteins is higher.

Brain

The human brain constitutes about 2% of the body's weight. But it ***utilizes*** as much as **20% *of the oxygen*** consumed by the body. Being a vital organ, special priority is given to the metabolic needs of the brain.

1. **Carbohydrate metabolism :** In an absorptive state, glucose is the only fuel source to the brain. About 120 g of glucose is utilized per day by an adult brain. This constitutes about 60% of the glucose consumed by the body at rest. It is estimated that about 50% of the energy consumed by brain is utilized by plasma membrane Na^+ - K^+-ATPase to maintain membrane potential required for nerve impulse transmission.

2. **Lipid metabolism :** The free fatty acids cannot cross the blood-brain barrier, hence their contribution for the supply of energy to the brain is insignificant. Further, in a fed state, ketone bodies are almost negligible as fuel source to the brain. However, brain predominantly ***depends on ketone bodies during prolonged starvation***.

METABOLISM IN STARVATION

Starvation may be due to food scarcity or the desire to rapidly lose weight or certain clinical conditions (e.g. surgery, burns etc.). Starvation is a metabolic stress which imposes certain metabolic compulsions on the organism. The metabolism is reorganized to meet the new demands of starvation.

Glucose is the fuel of choice for brain and muscle. Unfortunately, the carbohydrate reserve of the body is so low that it cannot meet the energy requirements even for a day. The fuel stores (or energy reserves) of a 70 kg normal man are given in ***Table 16.2***. Triacylglycerol (fat) of adipose tissue is the predominant energy reserve of the body. The survival time of an individual on starvation is mostly dependent on his/her fat stores. And for this reason, obese individuals can survive longer than lean individuals without consuming food.

Protein is basically a structural constituent, mostly present in the muscle. However, during starvation, protein can also meet the fuel demands of the body. It is estimated that about $^1/_3$rd of the body's protein can be utilized towards energy needs without compromising the vital functions.

Table 16.2 Energy reserves of a normal 70 kg man

Energy source (main storage tissue)	*Weight (kg)*	*Energy equivalent (in Cal)*
Triacylglycerol (adipose tissue)	15	135,000
Protein (muscle)	6	24,000
Glycogen (muscle, liver)	0.2	800

METABOLISM OF NUCLEOTIDES

17

Nucleotides consist of a nitrogenous base, a pentose and a phosphate. The pentose sugar is D-ribose in ribonucleotides of RNA while in deoxyribonucleotides (deoxynucleotides) of DNA, the sugar is 2-deoxy D-ribose. Nucleotides participate in almost all the biochemical processes, either directly or indirectly. They are the structural components of nucleic acids (DNA, RNA), coenzymes, and are involved in the regulation of several metabolic reactions.

BIOSYNTHESIS OF PURINE RIBONUCLEOTIDES

Many compounds contribute to the purine ring of the nucleotides (***Fig.17.1***).

1. N_1 of purine is derived from amino group of aspartate.
2. C_2 and C_8 respectively arise from formate of N^{10}-formyl THF and from N^5,N^{10}-methenyl THF.
3. N_3 and N_9 are obtained from amide group of glutamine.
4. C_4, C_5 and N_7 are contributed by glycine.
5. C_6 directly comes from CO_2.

It should be remembered that purine bases are not synthesized as such, but they are formed as ribonucleotides. ***The purines are built upon a pre-existing ribose 5-phosphate***. Liver is the major site for purine nucleotide synthesis. Erythrocytes, polymorphonuclear leukocytes and brain cannot produce purines.

The pathway for the synthesis of ***inosine monophosphate*** (IMP or inosinic acid), the '***parent*** ***purine nucleotide***' is given in ***Fig.17.2***. The reactions are briefly described below

1. Ribose 5-phosphate, reacts with ATP to form phosphoribosyl pyrophosphate (PRPP).
2. Glutamine transfers its amide nitrogen to PRPP to replace pyrophosphate and produce 5-phosphoribosylamine. The enzyme ***PRPP glutamyl amidotransferase*** is controlled by ***feedback inhibition*** of nucleotides (IMP, AMP and GMP). This reaction is the '***committed step***' in purine nucleotide biosynthesis.
3. Phosphoribosylamine reacts with glycine in the presence of ATP to form glycinamide ribosyl 5-phosphate or glycinamide ribotide (GAR).
4. N^{10}-Formyl tetrahydrofolate donates the formyl group and the product formed is formylglycinamide ribosyl 5-phosphate.

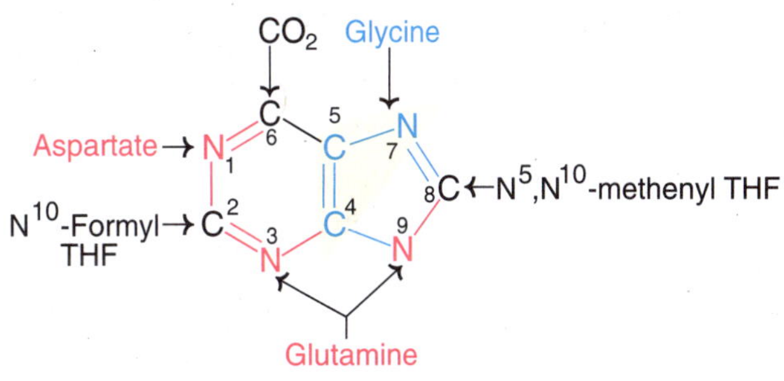

Fig.17.1 : *The sources of individual atoms in purine ring.*

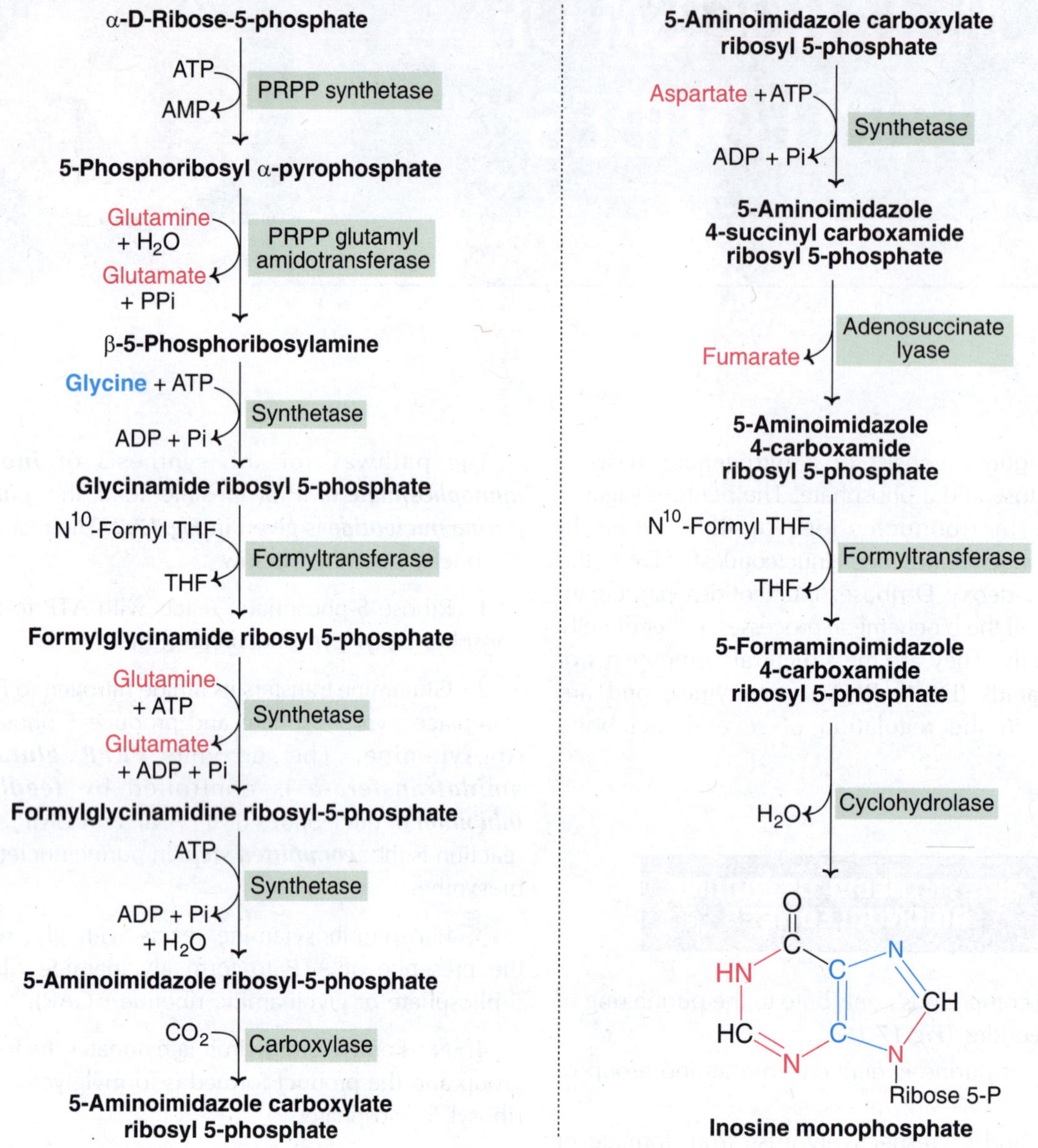

Fig. contd. next column

Fig.17.2 : *The metabolic pathway for the synthesis of inosine monophosphate, the parent purine nucleotide (PRPP–Phosphoribosyl pyrophosphate; PPi–Pyrophosphate).*

5. Glutamine transfers the second amido amino group to produce formylglycinamidine ribosyl 5-phosphate.

6. The imidazole ring of the purine is closed in an ATP dependent reaction to yield 5-aminoimidazole ribosyl 5-phosphate.

7. Incorporation of CO_2 (carboxylation) occurs to yield aminoimidazole carboxylate ribosyl 5-phosphate. This reaction ***does not require*** the vitamin ***biotin*** and/or ATP which is the case with most of the ***carboxylation reactions***.

8. Aspartate condenses with the product in reaction 7 to form aminoimidazole 4-succinyl carboxamide ribosyl 5-phosphate.

9. Adenosuccinate lyase cleaves off fumarate and only the amino group of aspartate is retained to yield aminoimidazole 4-carboxamide ribosyl 5-phosphate.

10. N^{10}-Formyl tetrahydrofolate donates a one-carbon moiety to produce formaminoimidazole 4-carboxamide ribosyl 5-phosphate.

11. The final reaction catalysed by cyclohydrolase leads to ring closure with an elimination of water

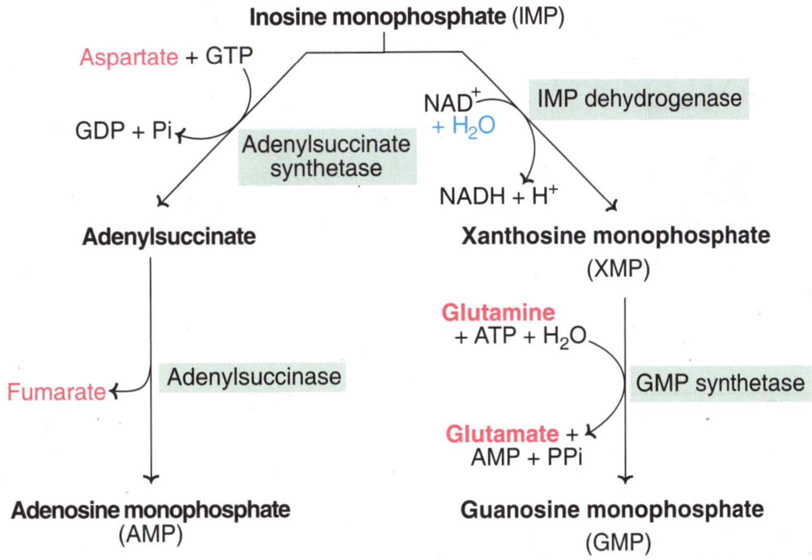

Fig.17.3 : *Synthesis of AMP and GMP from inosine monophosphate.*

molecule. The product obtained is inosine monophosphate (IMP) from which other purine nucleotides can be synthesized.

Inhibitors of purine synthesis

Inhibition of microbial growth : Folic acid (as THF) is essential for the synthesis of purine nucleotides (reactions 4 and 10). ***Sulfonamides*** are the structural analogs of para-aminobenzoic acid (PABA) (see ***page no. 81*** for structures). These sulfa drugs can be used to ***inhibit the synthesis of folic acid by microorganisms***. This indirectly reduces the synthesis of purines and, therefore, the nucleic acids (DNA and RNA). Sulfonamides have no influence on humans, since folic acid is not synthesized, and is supplied through diet.

Treatment of cancer : The structural ***analogs of folic acid*** (e.g. ***methotrexate***) are widely ***used to control cancer***. They inhibit the synthesis of purine nucleotides (reactions 4 and 10) and, thus, nucleic acids. Both these reactions are concerned with the transfer of one-carbon moiety (formyl group). These inhibitors also affect the proliferation of normally growing cells. This causes many ***side-effects including anemia, baldness, scaly skin etc***.

Synthesis of AMP and GMP from IMP

Inosine monophosphate is the immediate precursor for the formation of AMP and GMP (***Fig.17.3***). Aspartate condenses with IMP in the presence of GTP to produce adenylsuccinate which, on cleavage, forms AMP.

For the synthesis of GMP, IMP undergoes NAD^+ dependent dehydrogenation to form xanthosine monophosphate (XMP).

Glutamine then transfers amide nitrogen to XMP to produce GMP.

Formation of purine nucleoside diphosphates and triphosphates

The nucleoside monophosphates (AMP and GMP) have to be converted to the corresponding di-and triphosphates to participate in most of the metabolic

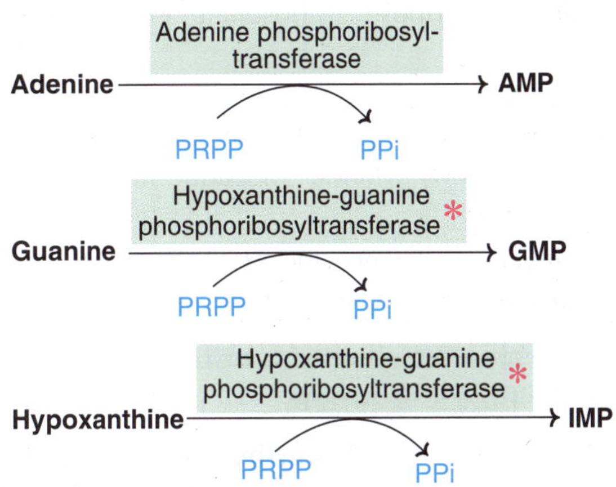

Fig.17.4 : *Salvage pathways of purine nucleotide synthesis (PRPP–Phosphoribosyl pyrophosphate; PPi–Inorganic pyrophosphate; AMP–Adenosine monophosphate; GMP–Guanosine monophosphate; IMP–Inosine monophosphate; * Deficiency of HGPRT causes Lesch-Nyhan syndrome).*

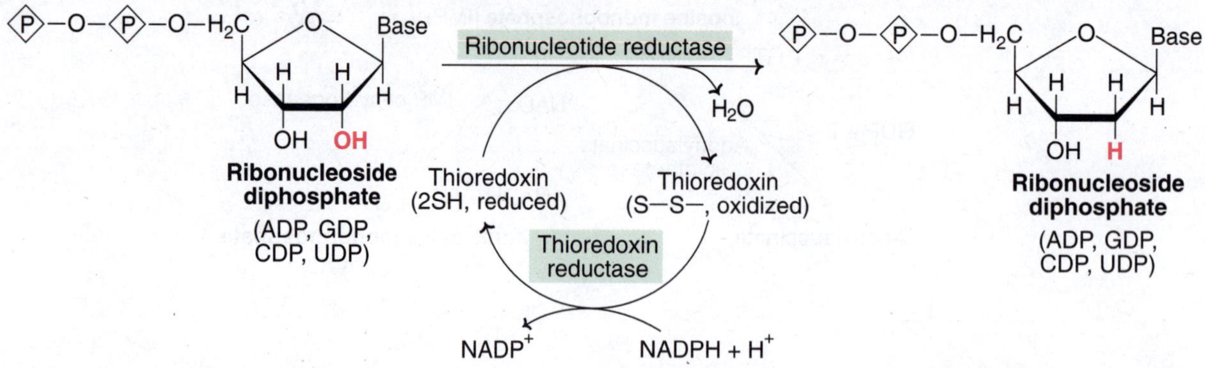

***Fig.17.5** : Formation of deoxyribonucleotides from ribonucleotides.*

reactions. This is achieved by the transfer of phosphate group from ATP, catalysed by nucleoside monophosphate (NMP) kinases and nucleoside diphosphate (NDP) kinases.

Salvage pathway for purines

The free purines (adenine, guanine and hypoxanthine) are formed in the normal turnover of nucleic acids (particularly RNA) and also obtained from the dietary sources. The ***purines*** can be ***directly converted to the corresponding nucleotides*** and this process is known as 'salvage pathway' (***Fig.17.4***).

Adenine phosphoribosyl transferase catalyses the formation of AMP from adenine. Hypoxanthine-guanine phosphoribosyl transferase (HGPRT) converts guanine and hypoxanthine, respectively, to GMP and IMP. Phosphoribosyl pyrophosphate (PRPP) is the donor of ribose 5-phosphate in the salvage pathway.

The salvage pathway is particularly important in certain tissues such as erythrocytes and brain where *de novo* (a new) synthesis of purine nucleotides is not operative.

A ***defect*** in the enzyme ***HGPRT*** causes ***Lesch-Nyhan syndrome*** (details given later).

Conversion of ribonucleotides to deoxyribonucleotides

The synthesis of purine and pyrimidine deoxyribonucleotides occurs from ribonucleotides ***by a reduction at the C_2 of ribose moiety*** (***Fig. 17.5***). This reaction is catalysed by ***ribonucleotide reductase***.

Supply of reducing equivalents : The enzyme ribonucleotide reductase itself provides the hydrogen atoms needed for reduction from its sulfhydryl groups. The reducing equivalents, in turn, are supplied by ***thioredoxin***, a monomeric protein with two cysteine residues.

NADPH-dependent thioredoxin reductase converts the oxidized thioredoxin to reduced form which can be recycled again and again. ***Thioredoxin*** thus serves as a ***protein cofactor in an enzymatic reaction***.

DEGRADATION OF PURINE NUCLEOTIDES

The ***end product*** of purine metabolism in humans is ***uric acid***. The sequence of reactions in purine nucleotide degradation is given in ***Fig.17.6***.

1. The nucleotide monophosphates (AMP, IMP and GMP) are converted to their respective nucleoside forms (adenosine, inosine and guanosine) by the action of ***nucleotidase***.

2. The amino group, either from AMP or adenosine, can be removed to produce IMP or inosine, respectively.

3. Inosine and guanosine are, respectively, converted to hypoxanthine and guanine (purine bases) by purine nucleoside phosphorylase.

4. Guanine undergoes deamination by guanase to form ***xanthine***.

5. ***Xanthine oxidase*** is an important enzyme that converts hypoxanthine to xanthine, and xanthine to uric acid. This enzyme contains FAD, molybdenum and iron.

Uric acid (2,6,8-trioxypurine) is the final ***excretory product of purine metabolism*** in humans.

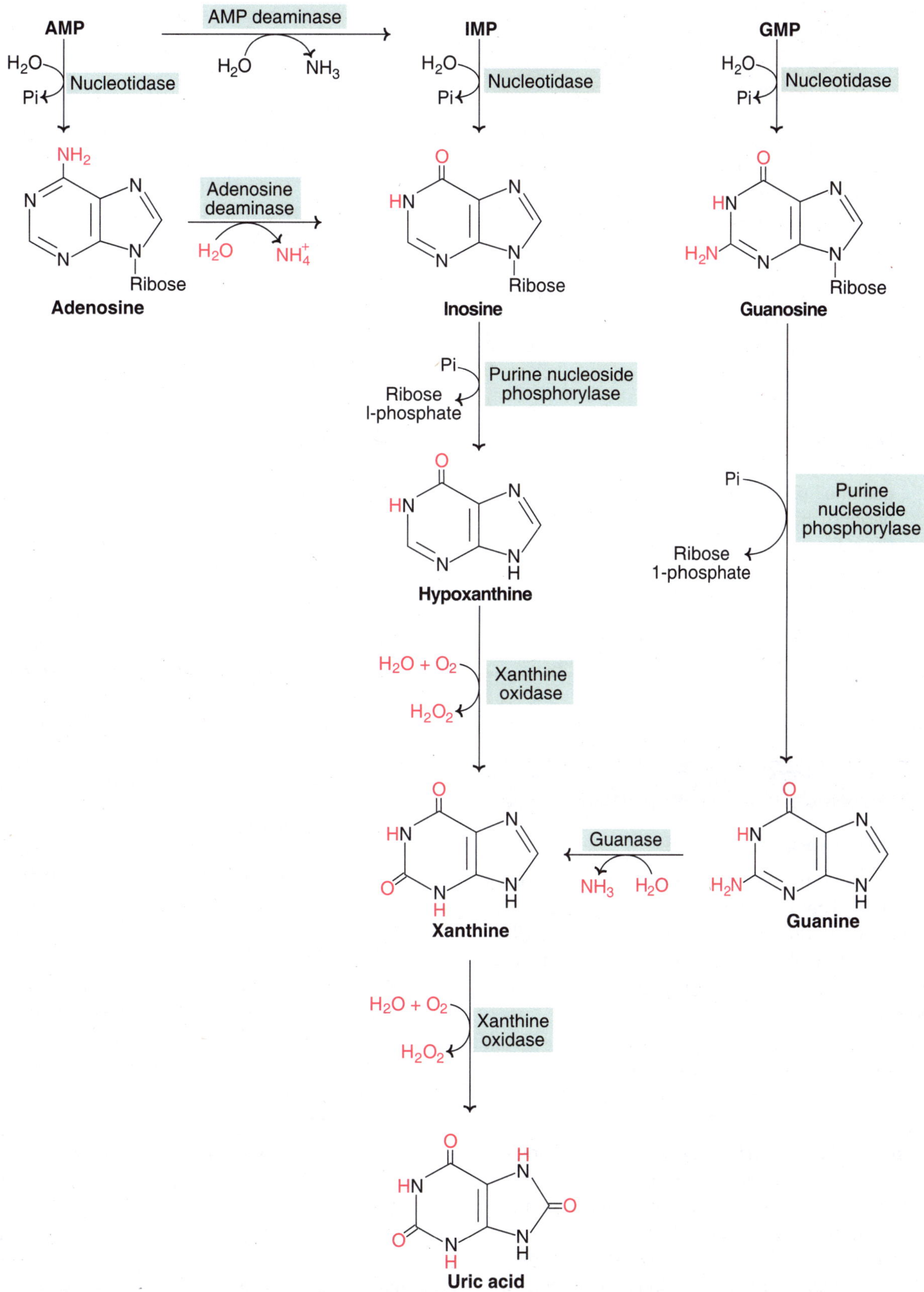

***Fig.17.6 :** Degradation of purine nucleotides to uric acid (AMP–Adenosine monophosphate; IMP–Inosine monophosphate; GMP–Guanosine monophosphate).*

Medical Concepts/ Clinical Correlates

- Folic acid is essential for the synthesis of purine nucleotides. Folic acid analogs (methotrexate) are employed to control cancer.
- Gout is the disorder associated with the overproduction of uric acid, the end product of purine metabolism. Allopurinol is the drug of choice for the treatment of gout.
- Lesch-Nyhan syndrome is caused by a defect in the enzyme hypoxanthine-guanine phosphoribosyltransferase. The patients have an irresistible urge to bite their fingers and lips.
- Orotic aciduria, a metabolic defect in pyrimidine biosynthesis, is characterized by anaemia and retarded growth, besides the excretion of orotic acid in urine.

DISORDERS OF PURINE METABOLISM

Hyperuricemia and gout

Uric acid is the end product of purine metabolism in humans. The normal concentration of ***uric acid in the serum*** of adults is in the ***range of 3-7 mg/dl***. In women, it is slightly lower (by about 1 mg) than in men. The daily excretion of uric acid is about 500-700 mg.

Hyperuricemia refers to an ***elevation in the serum uric acid*** concentration. This is sometimes associated with increased uric acid excretion (uricosuria).

Gout is a metabolic disease associated with ***overproduction of uric acid***. At the physiological pH, uric acid is found in a more soluble form as sodium urate. In severe hyperuricemia, crystals of sodium urate get deposited in the soft tissues, particularly in the joints. Such deposits are commonly known as ***tophi***. This causes inflammation in the joints resulting in a painful ***gouty arthritis***.

The ***prevalence*** of gout is about ***3 per 1,000*** persons, ***mostly*** affecting ***males***. Post-menopausal women, however, are as susceptible as men for this disease. Gout is of two types—primary and secondary.

1. **Primary gout :** It is an inborn error of metabolism due to ***overproduction of uric acid***. This is mostly related to increased synthesis of purine nucleotides. Lack of feedback regulation of purine nucleotide synthesis, and ***Lesch-Nyhan syndrome*** result in primary gout.

2. **Secondary gout :** Secondary hyperuricemia is due to various diseases causing increased synthesis or decreased excretion of uric acid. Increased degradation of nucleic acids (hence more uric acid formation) is observed in various cancers (leukemias, polycythemia, lymphomas, etc.) psoriasis and increased tissue breakdown (trauma, starvation etc.).

Treatment of gout

The drug of choice for the treatment of primary gout is ***allopurinol***. This is a structural analog of hypoxanthine that competitively inhibits the enzyme ***xanthine oxidase***. Further, allopurinol is oxidized to alloxanthine by xanthine oxidase. Alloxanthine, in turn, is a more effective inhibitor of xanthine oxidase. This type of inhibition is referred to as ***suicide inhibition***.

Besides the drug therapy, ***restriction in dietary intake of purines and alcohol is advised***. Consumption of plenty of water will also be useful.

The ***anti-inflammatory drug colchicine*** is used for the treatment of gouty arthritis. Other anti-inflammatory drugs—such as ***phenylbutazone***, indomethacin, oxyphenbutazone, corticosteroids— are also useful.

Lesch-Nyhan syndrome

This disorder is due to the ***deficiency of hypoxanthine-guanine phosphoribosyltransferase*** (***HGPRT***), an enzyme of purine salvage pathway (***See Fig.17.4***). It was first described in 1964 by Michael Lesch (a medical student) and William L. Nyhan (his teacher).

Lesch-Nyhan syndrome is a sex-linked metabolic disorder since the structural gene for HGPRT is located on the X-chromosome. It ***affects*** only the ***males*** and is characterized by excessive uric acid production (often gouty arthritis), and neurological abnormalities such as mental retardation, aggressive behavior, learning disability etc. The patients of this disorder have an ***irresistible urge to bite their fingers and lips, often causing self-mutilation***.

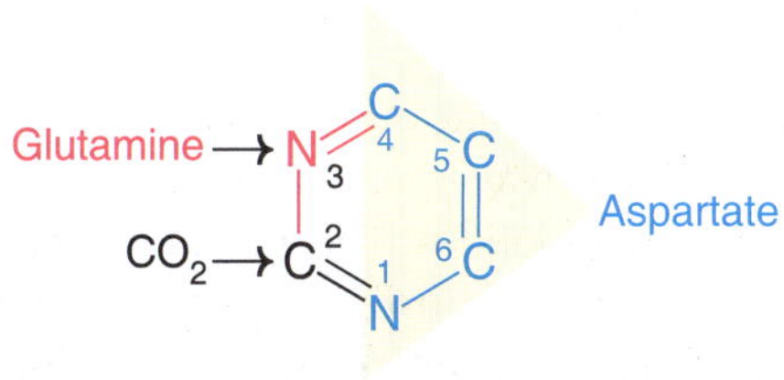

Fig.17.7 : *Sources of individual atoms in pyrimidine ring.*

BIOSYNTHESIS OF PYRIMIDINE RIBONUCLEOTIDES

The synthesis of pyrimidines is a much simpler process compared to that of purines. ***Aspartate, glutamine*** (amide group) and CO_2 ***contribute to atoms*** in the formation of ***pyrimidine ring*** (***Fig.17.7***). Pyrimidine ring is first synthesized and then attached to ribose 5-phosphate. This is in contrast to purine nucleotide synthesis wherein purine ring is built upon a pre-existing ribose 5-phosphate. The pathway of pyrimidine synthesis is depicted in ***Fig.17.8***, and the salient features are described below.

Glutamine transfers its amido nitrogen to CO_2 to produce carbamoyl phosphate. This reaction is ATP-dependent and is catalysed by cytosomal enzyme ***carbamoyl phosphate synthetase II (CPS II)***.

CPS II is activated by ATP and PRPP and inhibited by UTP. Carbamoyl phosphate synthetase I (CPS I) is a mitochondrial enzyme which synthesizes carbamoyl phosphate from ammonia and CO_2 and, in turn urea (***Refer*** protein metabolism, ***Chapter 15***, for more details). Prokaryotes have only one carbamoyl phosphate synthetase which is responsible for the biosynthesis of arginine and pyrimidines.

Carbamoyl phosphate condenses with aspartate to form carbamoyl aspartate. This reaction is catalysed by aspartate transcarbamoylase. Dihydroorotase catalyses the pyrimidine ring closure with a loss of H_2O.

The next step in pyrimidine synthesis is an NAD^+ dependent dehydrogenation, leading to the formation of orotate.

Ribose 5-phosphate is now added to orotate to produce orotidine monophosphate (OMP). This reaction is catalysed by orotate phosphoribosyl-transferase, an enzyme comparable with HGPRT in its function. OMP undergoes decarboxylation to uridine monophosphate (UMP).

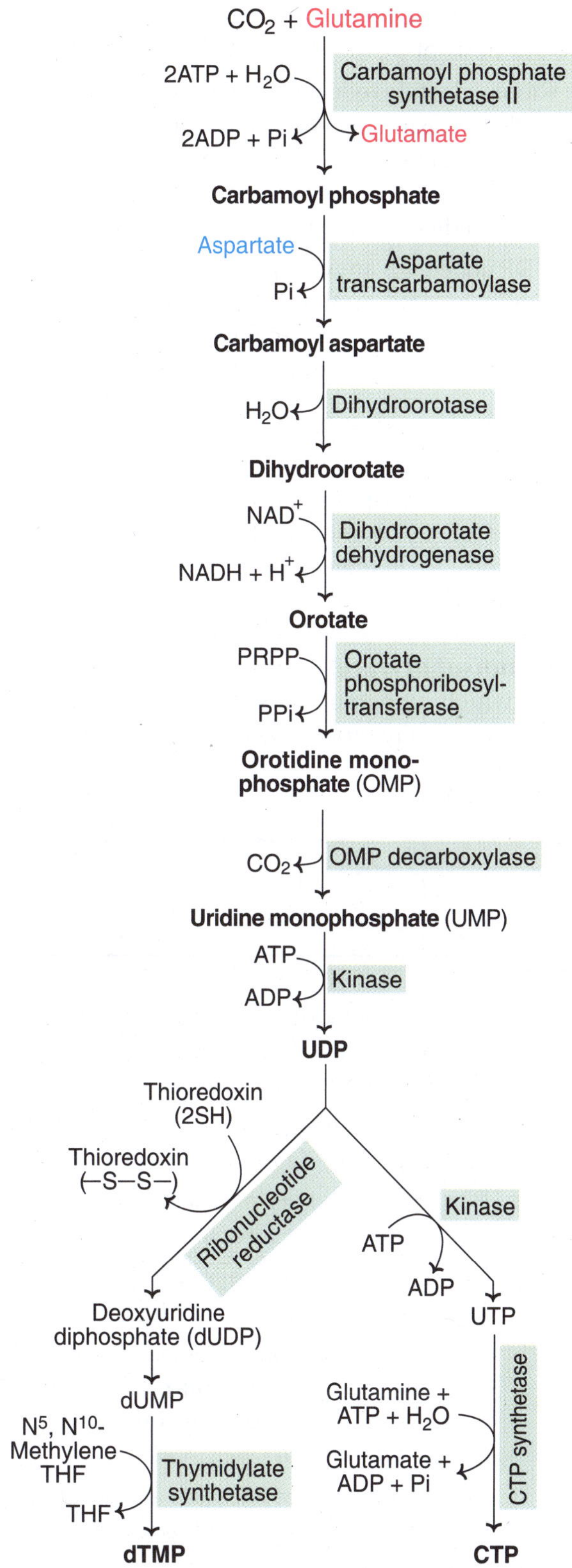

Fig.17.8 : *Metabolic pathway for the synthesis of pyrimidine nucleotides.*

By an ATP-dependent kinase reaction, UMP is converted to UDP which serves as a precursor for the synthesis of dUMP, dTMP, UTP and CTP.

Ribonucleotide reductase converts UDP to dUDP by a thioredoxin-dependent reaction. Thymidylate synthetase catalyses the transfer of a methyl group from N^5, N^{10}-methylene tetrahydrofolate to produce deoxythymidine monophosphate (dTMP).

UDP undergoes an ATP-dependent kinase reaction to produce UTP. Cytidine triphosphate (CTP) is synthesized from UTP by amination. CTP synthetase is the enzyme and glutamine provides the nitrogen.

Degradation of pyrimidine nucleotides

The pyrimidine nucleotides undergo similar reactions (dephosphorylation, deamination and cleavage of glycosidic bond) like that of purine nucleotides to liberate the nitrogenous bases—cytosine, uracil and thymine. The bases are then degraded to highly soluble products—***β-alanine*** and ***β-aminoisobutyrate.*** These are the amino acids which undergo transamination and other reactions to finally produce acetyl CoA and succinyl CoA.

Salvage pathway

The pyrimidines (like purines) can also serve as precursors in the salvage pathway to be converted to the respective nucleotides. This reaction is catalysed by pyrimidine phosphoribosyltransferase which utilizes PRPP as the source of ribose 5-phosphate.

Disorders of pyrimidine metabolism

Orotic aciduria : This is a rare metabolic disorder characterized by the excretion of orotic acid in urine, severe anemia and retarded growth. It is due to the deficiency of the enzymes ***orotate phosphoribosyl transferase*** and ***OMP decarboxylase*** of pyrimidine synthesis (***Fig.17.8***). Both these enzyme activities are present on a single protein as domains (bifunctional enzyme).

Biosynthesis of nucleotide coenzymes

The nucleotide coenzymes FMN, FAD, NAD^+ $NADP^+$ and coenzyme A are synthesized from the B-complex vitamins. Their formation is described under the section on vitamins (***Chapter 7***).

SUMMARY

1. *Purine nucleotides are synthesized in a series of reactions starting from ribose 5-phosphate. Glycine, glutamine, aspartate, formate and CO_2 contribute to the synthesis of purine ring.*
2. *The synthesis of purine nucleotides also occurs from free purines by a salvage pathway. The defect in the enzyme HGPRT causes Lesch-Nyhan syndrome.*
3. *Purine nucleotides are degraded to uric acid, the excretory product in humans. Uric acid serves as a natural antioxidant in the living system.*
4. *Uric acid in many animal species (other than primates) is converted to more soluble forms such as allantoin, allantoic acid etc., and excreted.*
5. *Gout is a metabolic disease associated with overproduction of uric acid. This often leads to the accumulation of sodium urate crystals in the joints, causing painful gouty arthritis. Allopurinol, an inhibitor of xanthine oxidase, is the drug used for the treatment of gout.*
6. *Pyrimidine nucleotides are synthesized from the precursors aspartate, glutamine and CO_2, besides ribose 5-phosphate.*

THE READER MUST REFER APPENDIX D (Page Nos : 422 - 431) FOR RELATED CLINICAL CASE STUDIES (Nos : 13, 14, 15)

MINERAL METABOLISM

18

The ***mineral (inorganic) elements*** constitute ***only a small proportion of the body weight***. There is a wide variation in their body content. For instance, calcium constitutes about 2% of body weight while cobalt about 0.00004%.

General functions

Minerals perform several vital functions which are ***absolutely essential for the very existence of the organism***. These include ***calcification of bone, blood coagulation***, neuromuscular irritability, ***acid-base equilibrium***, fluid balance and osmotic regulation.

Certain minerals are ***integral components*** of biologically important compounds such as ***hemoglobin (Fe), thyroxine (I), insulin (Zn)*** and ***vitamin B_{12} (Co)***. Sulfur is present in thiamine, biotin, lipoic acid and coenzyme A. Several minerals participate as cofactors for enzymes in metabolism (e.g. Mg, Mn, Cu, Zn, K). Some elements are essential constituents of certain enzymes (e.g. Co, Mo, Se).

Classification

The minerals are classified as principal elements and trace elements.

The seven ***principal elements*** (***macroelements***) constitute 60-80% of the body's inorganic material. These are ***calcium, phosphorus, magnesium, sodium, potassium, chloride*** and ***sulfur***. The principal elements are required in amounts greater than 100 mg/day.

The ***trace elements*** (***microelements***) are required in amounts less than 100 mg/day. They are subdivided into three categories

1. **Essential trace elements :** ***Iron, copper, iodine, manganese, zinc, molybdenum, cobalt, fluorine, selenium and chromium.***

2. **Possibly essential trace elements :** Nickel, vanadium, cadmium and barium.

3. **Non-essential trace elements :** Aluminium, lead, mercury, boron, silver, bismuth etc.

A summary of the major characteristics of principal elements and trace elements is respectively given in ***Tables 18.1*** and ***18.2***. The individual elements are described next.

CALCIUM

Calcium is the ***most abundant among the minerals in the body***. The total content of calcium in an adult man is about 1 to 1.5 kg. As much as ***99%*** of it is ***present in the bones and teeth***. A small fraction (1%) of the calcium, found outside the skeletal tissue, performs a wide variety of functions.

Biochemical functions

1. **Development of bones and teeth :** Calcium, along with phosphate, is required for the formation (of ***hydroxyapatite***) and physical strength of skeletal tissue. Bones which are in a dynamic state serve as ***reservoir of Ca***. Osteoblasts are responsible for bone formation while osteoclasts result in demineralization.

2. **Muscle contraction :** Ca^{2+} interacts with troponin C to ***trigger muscle contraction***. Calcium also activates ATPase, increases the interaction between actin and myosin.

TABLE 18.1 A summary of the major characteristics of principal elements (macroelements)

Element	*Major functions*	*Deficiency disease/symptoms*	*Recommended dietary allowance*	*Major sources*
Calcium	Constituent of bones and teeth; muscle contraction, nerve transmission	Rickets; osteomalacia, osteoporosis	0.8–1.0 g/d	Milk and milk products, leafy vegetables, beans
Phosphorus	Constituent of bones and teeth; in the formation of high energy phosphates, nucleic acids, nucleotide coenzymes.	Rickets, osteomalacia	0.8–1.0 g/d	Milk, cereals, leafy vegetables
Magnesium	Constituent of bones and teeth; cofactor for enzymes e.g. kinases.	Neuromuscular weakness, irritation	300–350 mg/d	Cereals, vegetables, fruits, milk
Sodium	Chief cation of extracellular fluids; acid-base balance, osmotic pressure; nerve and muscle function	Almost unknown on normal diet	5–10 g/d	Table salt, salt added foods
Potassium	Chief cation of intracellular fluids; acid-base balance; osmotic pressure; muscle function	Muscular weakness, mental confusion	3–4 g/d	Fruits, nuts, vegetables
Chloride	Regulation of acid-base balance; formation of HCl	Almost unknown on normal diet	5–10 g/d	Table salt
Sulfur	Constituent of sulfur containing amino acids, certain vitamins (thiamine, biotin) and other compounds (heparin, chondroitin sulfate).	Almost unknown	—	Sulfur containing amino acids

3. **Blood coagulation :** Several reactions in the cascade of blood clotting process are dependent on ***Ca^{2+}(factor IV)***.

4. **Nerve transmission :** Ca^{2+} is ***necessary*** for the transmission of nerve impulse.

5. **Membrane integrity and permeability :** Ca^{2+} ***influences*** the membrane structure and transport of water and several ions across it.

6. **Activation of enzymes :** Ca^{2+} is needed for the direct activation of enzymes such as ***lipase*** (pancreatic), ***ATPase*** and succinate dehydrogenase.

7. **Calmodulin mediated action of Ca^{2+} :** Calmodulin (mol. wt. 17,000) is a calcium binding regulatory protein. ***Ca-calmodulin complex activates certain enzymes*** e.g. adenylate cyclase, Ca^{2+} dependent protein kinases.

8. **Calcium as intracellular messenger :** Certain hormones exert their action through the mediation of Ca^{2+} (instead of cAMP). Calcium is regarded as a ***second messenger*** for such hormonal action e.g. epinephrine in liver glycogenolysis. Calcium serves as a third messenger for some hormones e.g. antidiuretic hormone (ADH) acts through cAMP, and then Ca^{2+}.

9. **Release of hormones :** The release of certain hormones (insulin, PTH, calcitonin) from the endocrine glands is facilitated by Ca^{2+}.

Dietary requirements

Adult men and women	—	***800 mg/day***
Women during pregnancy, lactation and post-menopause	—	1.5 g/day
Children (1-18 yrs.)	—	0.8–1.2 g/day
Infants (<1 year)	—	300–500 mg/day

TABLE 18.2 A summary of the major characteristics of trace elements (microelements)

Element	Major functions	Deficiency disease/symptoms	Recommended dietary allowance	Major sources
Iron	Constituent of heme e.g. hemoglobin, myoglobin, cytochromes; involved in O_2 transport and biological oxidation.	Hypochromic, microcytic anemia	10–15 mg/d	Organ meats (liver, heart), leafy vegetables, iron cookware
Copper	Constituent of enzymes e.g. cytochrome C oxidase, catalase, tyrosinase; in iron transport.	Anemia, Menke's disease	2–3 mg/d	Organ meats, cereals, leafy vegetables
Iodine	Constituent of thyroxine and triiodothyronine	Cretinism, goiter, myxedema	150–200 μg/d	Iodized salt, sea foods
Manganese	Cofactor for enzymes e.g. arginase, pyruvate carboxylase; glycoprotein synthesis.	Almost unknown	2–9 mg/d	Cereals, leafy vegetables
Zinc	Cofactor for enzymes e.g. alcohol dehydrogenase, carbonic anhydrase, lactate dehydrogenase.	Growth retardation, poor wound healing, hypogonadism	10–15 mg/d	Meat, fish, milk
Molybdenum	Constituent of enzymes e.g. xanthine oxidase	Almost unknown	75–250 μg/d	Vegetables
Cobalt	Constituent of vitamin B_{12}, required for the formation of erythrocytes	Pernicious anemia (as in vitamin B_{12} deficiency)	5–8 μg/d	Foods of animal origin
Fluorine	Helps in the proper formation of bones and teeth	Dental caries, osteoporosis	2–4 mg/d	Drinking water
Selenium	Involved in antioxidant function along with vitamin E; constituent of glutathione peroxidase and selenocysteine	Muscular degeneration, cardiomyopathy	50–200 μg/d	Organ meats, sea foods
Chromium	Promotes insulin function (as glucose tolerance factor)	Impaired glucose tolerance	10–100 μg/d	Brewer's yeast, meat, whole grains

Sources

Best sources	—	***Milk and milk products***
Good sources	—	Beans, leafy vegetables, fish, cabbage, egg yolk.

Absorption

The absorption of calcium mostly occurs in the duodenum by an energy dependent active process. It is influenced by several factors.

Factors promoting Ca absorption

1. Vitamin D (through its active form ***calcitriol***) induces the synthesis of calcium binding protein in the intestinal epithelial cells and ***promotes Ca absorption***.

2. ***Parathyroid hormone enhances Ca absorption*** through the increased synthesis of calcitriol.

3. ***Acidity*** (low pH) is more ***favourable*** for Ca absorption.

4. ***Lactose promotes*** calcium uptake by intestinal cells.

5. The amino acids ***lysine*** and ***arginine facilitate*** Ca absorption.

Factors inhibiting Ca absorption

1. ***Phytates*** and ***oxalates*** form insoluble salts and interfere with Ca absorption.

2. High content of dietary phosphate results in the formation of insoluble calcium phosphate and prevents Ca uptake. The dietary ratio of ***Ca and P—between 1 : 2 and 2 : 1—is ideal*** for optimum Ca absorption by intestinal cells.

3. The free fatty acids react with Ca to form insoluble calcium soaps. This is particularly observed when the fat absorption is impaired.

4. ***Alkaline condition*** (high pH) is ***unfavourable*** for Ca absorption.

5. ***High content of dietary fiber interferes*** with Ca absorption.

Plasma calcium

Most of the blood Ca is present in the plasma since the blood cells contain very little of it. The normal concentration of plasma or serum Ca is ***9-11 mg/dl*** (***4.5-5.5 mEq/l***). About half of this (5 mg/dl) is in the ionized form which is functionally the most active. At least 1 mg/dl serum Ca is found in association with citrate and/or phosphate. The other half of serum Ca (4-5mg/dl) is bound to proteins, mostly albumin and, to a lesser extent, globulin. Ionized and citrate (or phosphate) bound Ca is diffusible from blood to the tissues while protein bound Ca is non-diffusible. In the usual laboratory determination of serum Ca, all the three fractions are measured together.

FACTORS REGULATING PLASMA Ca LEVEL

As already stated, calcium is almost exclusively present in blood plasma (or serum). The hormones—***calcitriol, parathyroid hormone*** (PTH) and ***calcitonin*** are the major factors that regulate the plasma calcium (***homeostasis of Ca; Fig.18.1***) within a narrow range (9-11 mg/dl).

Calcitriol

The physiologically active form of vitamin D is a hormone, namely calcitriol or ***1,25-dihydroxycholecalciferol (1,25 DHCC)***. The synthesis of calcitriol and its wide range of biochemical actions are described under ***Vitamins*** (***Chapter 7***).

Calcitriol induces the synthesis of a specific calcium binding protein in the intestinal cells. This protein increases the intestinal absorption of calcium as well as phosphate. Thus blood Ca level is increased by calcitriol (the active vitamin D). Furthermore, calcitriol ***stimulates calcium uptake by osteoblasts of bone and promotes calcification*** or mineralization (deposition of calcium phosphate) and remodelling.

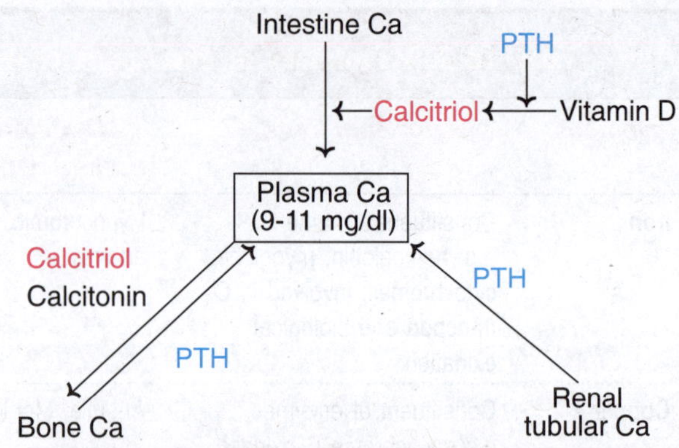

Fig.18.1 : *Overview of calcium homeostasis (PTH–Parathyroid hormone).*

Parathyroid hormone

Parathyroid hormone (PTH) is secreted by two pairs of parathyroid glands that are closely associated with thyroid glands. Parathyroid hormone (mol. wt. 95,000) is a single chain polypeptide, containing 84 amino acids. It is originally synthesized as preproPTH which is degraded to proPTH and, finally, to active PTH. The rate of formation (by degradation of proPTH) and the secretion of PTH are promoted by low Ca^{2+} concentration. Thus, the ***release of PTH*** from parathyroid glands is ***under the negative feedback regulation of serum*** Ca^{2+}.

Mechanism of action of PTH : PTH binds to a membrane receptor protein on the target cell and activates adenylate cyclase to liberate cAMP. This, in turn, increases intracellular calcium that promotes the phosphorylation of proteins (by kinases) which, finally brings about the biological actions. PTH has 3 independent tissues—bone, kidneys and intestine—to exert its action. The ***prime function of PTH is to elevate serum calcium level.***

Action on the bone : PTH causes decalcification or ***demineralization of bone***, a process carried out by ***osteoclasts***. This is brought out by PTH stimulated increased activity of the enzymes pyrophosphatase and collagenase. These enzymes result in bone resorption. Demineralization ultimately leads to an increase in the blood Ca level. The action of PTH on bone is quantitatively very significant to maintain Ca homeostasis. It must, however, be noted that this is

Medical Concepts/ Clinical Correlates

- **Serum calcium level is increased (normal 9–11 mg/dl) in hyperparathyroidism. This condition is also associated with elevated urinary excretion of Ca and P, often leading to stone formation.**
- **Tetany, caused by a drastic reduction in serum Ca, is characterized by neuromuscular irritability and convulsions.**
- **Rickets is due to defective calcification of bones. This may be caused by deficiency of Ca and P or vitamin D or both.**

being done at the expense of loss of Ca from bone, particularly in dietary Ca deficiency.

Action on the kidney : ***PTH increases the Ca reabsorption*** by kidney tubules. This is the most rapid action of PTH to elevate blood Ca levels. However, quantitatively, this is less important compared to the action of PTH on bone. PTH promotes the production of calcitriol (1,25 DHCC) in the kidney by stimulating 1-hydroxylation of 25-hydroxycholecalciferol.

Action on the intestine : The action of PTH on the intestine is indirect. It ***increases the intestinal absorption of Ca*** by promoting the synthesis of calcitriol.

Calcitonin (CT)

Calcitonin is a peptide containing 32 amino acids. It is secreted by parafollicular cells of thyroid gland. The ***action of CT on calcium metabolism is antagonistic to that of PTH***. Thus, calcitonin promotes calcification by increasing the activity of osteoblasts. Further, calcitonin decreases bone resorption and increases the excretion of Ca into urine. CT, therefore, has a ***decreasing influence on blood calcium***.

Importance of Ca : P ratio

The ratio of plasma Ca : P is important for calcification of bones. The product of Ca ↔ P (in mg/dl) in children is around 50 and in adults around 40. This product is less than 30 in rickets.

Excretion of calcium

Calcium is excreted partly through the kidneys and mostly through the intestine. The renal threshold for serum Ca is 10 mg/dl. Calcium gets excreted into urine beyond this concentration. Ingestion of excess protein causes increased calcium excretion in urine. This is mainly due to an increase in the acidity of urine as a result of high protein diet.

Calcium in the teeth

The teeth calcium is not subjected to regulation as observed for bone calcium. Thus the adult teeth, once formed, ***do not undergo decalcification*** to meet the body needs of calcium. However, proper calcification of teeth is important in the growing children.

DISEASE STATES

The blood Ca level is maintained within a narrow range by the ***homeostatic control***, most predominantly by PTH. Hence ***abnormalities in Ca metabolism*** are mainly ***associated with alterations in PTH***.

Hypercalcemia

The serum Ca level (***normal 9–11 mg/dl***) is elevated in hypercalcemia. Hypercalcemia is associated with ***hyperparathyroidism*** caused by increased activity of parathyroid glands. Decrease in serum phosphate (due to increased renal losses) and increase in alkaline phosphatase activity are also found in hyperparathyroidism. Elevation in the urinary excretion of Ca and P, often resulting in the formation of urinary calculi, is also observed in these patients.

The symptoms of hypercalcemia include lethary, muscle weakness, loss of appetite, constipation, nausea, increased myocardial contractility and susceptibility to fractures.

Hypocalcemia

Hypocalcemia is a more serious and life threatening condition. It is characterized by a fall in the serum Ca to ***below 7 mg/dl***, causing ***tetany***. The symptoms of tetany include neuromuscular irritability, spasms and convulsions.

Hypoparathyroidism is associated with a decrease in serum Ca and an increase in serum phosphate, besides the reduced urinary excretion of both Ca and P.

Rickets

Rickets is a disorder of defective calcification of bones. This may be due to a ***low levels of vitamin D*** in the body or due to a dietary deficiency of Ca and P—or both. The concentration of serum Ca and P may be low or normal. An ***increase*** in the activity of ***alkaline phosphatase*** is a characteristic feature of rickets.

Osteoporosis

Osteoporosis is characterized by ***demineralization of bone*** resulting in the progressive loss of bone mass.

Occurrence : The ***elderly people (over 60 yr.)*** of both sexes are ***at risk*** for osteoporosis. However, it more predominantly occurs in the postmenopausal women. Osteoporosis results in ***frequent bone fractures*** which are a major cause of disability among the elderly.

Etiology : The etiology of osteoporosis is largely unknown, but it is believed that several causative factors may contribute to it. The ability to produce ***calcitriol*** from vitamin D is ***decreased with age***, particularly in the postmenopausal women. Immobilized or sedentary individuals tend to decrease bone mass while those on regular exercise tend to increase bone mass. ***Deficiency of sex hormones*** (in women) has been implicated in the development of osteoporosis.

Treatment : ***Estrogen*** administration along with ***calcium supplementation*** (in combination with vitamin D) to postmenopausal women reduces the risk of fractures. Higher dietary intake of Ca (about 1.5 g/day) is recommended for elderly people.

PHOSPHORUS

An adult body contains about ***1 kg phosphate*** and it is found in every cell of the body. Most of it (about 80%) occurs in combination ***with Ca*** in the ***bones*** and ***teeth***. About 10% of body P is found in muscles and blood in association with proteins, carbohydrates and lipids. The remaining 10% is widely distributed in various chemical compounds.

Biochemical functions

1. Phosphorus is ***essential*** for the development of ***bones and teeth***.

2. It plays a central role for the formation and utilization of ***high-energy phosphate compounds*** e.g. ***ATP***, GTP, creatine phosphate etc.

3. Phosphorus is required for the formation of ***phospholipids***, phosphoproteins and nucleic acids (DNA and RNA).

4. It is an essential component of several ***nucleotide coenzymes*** e.g. NAD^+, $NADP^+$, pyridoxal phosphate, ADP, AMP.

5. Several ***proteins and enzymes*** are activated by phosphorylation.

6. ***Phosphate buffer*** system is important for the maintenance of pH in the blood (around 7.4) as well as in the cells.

Dietary requirements

The recommended dietary allowance (RDA) of phosphate is based on the intake of calcium. The ratio of ***Ca : P of 1 : 1 is recommended*** (i.e. 800 mg/day) for an adult. For infants, however, the ratio is around 2 : 1, which is based on the ratio found in human milk.

Sources

Milk, cereals, leafy vegetables, meat, eggs.

Absorption

Phosphate absorption occurs from jejunum

1. Calcitriol promotes phosphate uptake along with calcium.

2. Absorption of phosphorus and calcium is optimum when the dietary Ca : P is between 1 : 2 and 2 : 1.

Serum phosphate

The phosphate level of the whole blood is around 40 mg/dl while serum contains about ***3-4 mg/dl***. This is because the RBC and WBC have very high content of phosphate.

The serum phosphate may exist as free ions (40%) or in a complex form (50%) with cations such as Ca^{2+}, Mg^{2+}, Na^+, K^+. About 10% of serum phosphate is bound to proteins.

Excretion

About 500 mg phosphate is excreted in urine per day. The renal threshold is 2 mg/dl. The reabsorption of phosphate by renal tubules is inhibited by PTH.

Disease states

1. Serum ***phosphate*** level is ***increased in hypoparathyroidism*** and decreased in hyperparathyroidism.

2. In severe renal diseases, serum phosphate content is elevated causing acidosis.

3. Vitamin D deficient rickets is characterized by decreased serum phosphate (1–2 mg/dl).

4. Renal rickets is associated with low serum phosphate levels and increased alkaline phosphatase activity.

TEETH

The chemical composition of tooth closely resembles that of bone. Thus, tooth contains organic matrix composed of ***protein*** (around 5%) and inorganic material (up to 95%) — calcium phosphate in the form of ***hydroxyapatite*** [$Ca_{10}(PO_4)_6(OH)_2$], and ***fluroapatite***. The tooth is organized into 3 distinct layers—***enamel***, ***dentine*** and ***cementum***. The enamel is the hard upper surface of tooth covering the dentine which in turn is connected by cementum to gum.

Development of teeth

Vitamin D (active form–***calcitriol***) is ***required for*** the ***calcification teeth*** i.e., for the absorption of calcium and phosphate from the intestine, and their deposition in tooth (calcification), Vitamins C and A are also necessary for proper development of teeth and calcification process (also see calcium in teeth, discussed already).

Dental caries

Dental caries is the ***destruction (decay) or decalcification of hard teeth due to organic acids***, released by bacterial infections (often *Streptococcus mutans*). These organic acids are mostly produced by the action of microorganisms on carbohydrates, prodominantly sucrose (source-chocolates), particularly during the formation of teeth in children.

The incidence of dental caries is higher in the deficiency of fluoride. It is known that fluoride prevents dental caries by forming acid-resistant protective layer of ***fluoroapatite*** with hydroxyapatite. (See page 223 for ***dental fluorosis***).

MAGNESIUM

The adult body contains about 20 g magnesium, 70% of which is found in bones in combination with calcium and phosphorus. The remaining 30% occurs in the soft tissues and body fluids.

Biochemical functions

1. Magnesium is required for the ***formation of bones and teeth***.

2. Mg^{2+} serves as a ***cofactor for several enzymes*** requiring ATP e.g. ***hexokinase***, glucokinase, phosphofructokinase, adenylate cyclase.

3. Mg^{2+} is necessary for proper ***neuromuscular function***. Low Mg^{2+} levels lead to neuromuscular irritability.

Dietary requirements

Adult man	—	***350 mg/day***
Adult woman	—	300 mg/day

Sources

Cereals, nuts, beans, vegetables (cabbage, cauliflower), meat, milk, fruits.

Serum Mg

Normal serum concentration of Mg is 2–3 mg/dl. It is present in the ionized form (60%), in combination with other ions (10%) and bound to proteins (30%).

Disease states

1. Magnesium deficiency causes ***neuromuscular irritation, weakness and convulsions***. These symptoms are similar to that observed in tetany (Ca deficiency) which are relieved only by Mg. Malnutrition, alcoholism and cirrhosis of liver may lead to Mg deficiency.

2. Low levels of Mg may be observed in uremia, rickets and abnormal pregnancy.

SODIUM

Sodium is the ***chief cation in the extracellular fluid***. About 50% of body sodium is present in the bones, 40% in the extracellular fluid and the remaining (10%) in the soft tissues.

Biochemical functions

1. In association with chloride and bicarbonate, sodium ***regulates the body's acid-base balance***.

2. Sodium is required for the ***maintenance of osmotic pressure*** and fluid balance.

3. It is necessary for the nomal ***muscle irritability*** and cell permeability.

4. Sodium is involved in the intestinal ***absorption of glucose***, galactose and amino acids.

5. It is necessary for initiating and maintaining heart beat.

Dietary requirements

For normal individuals, the requirement of sodium is about ***5-10 g/day*** which is mainly consumed as NaCl. For persons with a family history of hypertension, the daily NaCl intake should be less than 5 g. For patients of hypertension, around 1 g/day is recommended. It may be noted that 10 g of NaCl contains 4 g of sodium. The daily consumption of Na is generally higher than actually required due to its flavour.

Sources

The ***common salt (NaCl)*** used in the cooking medium is the major source of sodium. The ingested foods also contribute to sodium. The good sources of sodium include bread, whole grains, leafy vegetables, nuts, eggs and milk.

Plasma sodium

In the plasma (serum), the normal concentration of sodium is ***135-145 mEq/l***. Sodium is an ***extracellular cation,*** therefore, the blood cells contain much less (35 mEq/l). The mineralocorticoids, secreted by adrenal cortex, influence sodium metabolism. A decrease in plasma sodium and an increase in its urinary excretion are observed in adrenocortical insufficiency.

Excretion

Kidney is the major route of sodium excretion from the body. As much as 800 g Na/day is filtered by the glomeruli, 99% of this is reabsorbed by the renal tubules by an active process. This is ***controlled by aldosterone***.

Disease states

1. **Hyponatremia :** This is a condition in which the ***serum sodium level falls*** below the normal. Hyponatremia may occur due to diarrhea, vomiting, chronic renal diseases, adrenocortical insufficiency ***(Addison's disease)***. Administration of salt free fluids to patients may also cause hyponatremia. This is due to overhydration. Decreased serum sodium concentration is also observed in edema which occurs in cirrhosis or congestive heart failure.

The manifestations of hyponatremia include reduced blood pressure and circulatory failure.

2. **Hypernatremia :** This condition is characterized by an ***elevation in the serum sodium level***. The symptoms include increase in blood volume and blood pressure. It may occur due to hyperactivity of adrenal cortex ***(Cushing's syndrome)***, prolonged administration of cortisone, ACTH and/or sex hormones. Loss of water from the body causing dehydration, as it occurs in diabetes insipidus, results in hypernatremia.

In ***edema***, along with ***water***, ***sodium*** concentration in the body is also ***elevated***. Administration of diuretic drugs increases the urinary output of water along with sodium. In the patients of ***hypertension*** and congestive cardiac failure salt ***(Na^+) restriction*** is advocated.

POTASSIUM

Potassium is the principal ***intracellular cation***. It is equally important in the extracellular fluid for specific functions.

Biochemical functions

1. Potassium maintains ***intracellular osmotic pressure***.

2. It is required for the regulation of ***acid-base balance***, and water balance in the cells.

3. The enzyme pyruvate kinase (of glycolysis) is dependent on K^+ for optimal activity.

4. Potassium is required for the transmission of nerve impulse.

5. Adequate intracellular concentration K^+ is necessary for proper biosynthesis of proteins by ribosomes.

6. Extracellular K^+ influences cardiac muscle activity.

Dietary requirements

About ***3-4 g/day***.

Sources

Banana, orange, pineapple, potato, beans, chicken, liver. Tender coconut water is a rich source of potassium.

Plasma potassium

The plasma (serum) concentration of potassium is ***3.4-5.0 mEq/l***. The whole blood contains much higher level of K^+ (50 mEq/l), since it is predominantly an intracellular cation. Care should, therefore, be taken to avoid hemolysis of RBC for the estimation of serum K^+.

Excretion

Potassium is mainly excreted through urine. The maintenance of body acid-base balance influences K^+ excretion. Aldosterone increases excretion of potassium.

Medical Concepts/ Clinical Correlates

- **Osteoporosis is a bone disorder of the elderly, characterized by demineralization resulting in a progressive loss of bone mass. It is the major cause of bone fractures and disability in the old people.**
- **Decreased levels of serum Na (hyponatremia) are observed in diarrhea and vomiting, besides Addison's disease, while increased serum Na (hypernatremia) is found in Cushing's syndrome.**
- **Iron deficiency anemia is the most prevalent nutritional disorder worldover. It is most commonly observed in pregnant and lactating women.**

Disease states

Serum potassium concentration is maintained within a narrow range. Either high or low concentrations are dangerous since potassium affects the contractility of heart muscle.

Hypokalemia : ***Decrease*** in the concentration of serum potassium is observed due to overactivity of adrenal cortex ***(Cushing's syndrome)***, prolonged cortisone therapy, intravenous administration of K^+-free fluids, treatment of diabetic coma with insulin, prolonged diarrhea and vomiting.

The symptoms of hypokalemia include irritability, muscular weakness, tachycardia, cardiomegaly and cardiac arrest. Changes in the ECG are observed (flattened waves with inverted T wave).

Hyperkalemia : ***Increase*** in the concentration of ***serum potassium*** is observed in renal failure, adrenocortical insufficiency ***(Addison's disease)***, diabetic coma, severe dehydration, intravenous administration of fluids with excessive potassium salts.

The manifestations of hyperkalemia include depression of central nervous system, mental confusion, numbness, bradycardia with reduced heart sounds and, finally, cardiac arrest.

CHLORINE

Chlorine is a constituent of sodium chloride. Hence, the metabolism of ***chloride*** and sodium are intimately related.

Biochemical functions

1. Chloride is involved in the regulation of ***acid-base equilibrium, fluid balance*** and osmotic pressure. These functions are carried out by the interaction of chloride with Na^+ and K^+.
2. Chloride is necessary for the formation of ***HCl in the gastric juice***.
3. Chloride shift involves the active participation of Cl^-.
4. The enzyme ***salivary amylase*** is activated by chloride.

Dietary requirements

The daily requirement of chloride as ***NaCl is 5-10 g***. Adequate intake of sodium will satisfy the chloride requirement of the body.

Sources

Common salt as cooking medium, whole grains, leafy vegetables, eggs and milk.

Plasma chloride

The normal plasma concentration of chloride is ***95-105 mEq/l***. Cerebrospinal fluid (CSF) contains higher level of Cl^- (125 mEq/l). This is due to the fact that protein content is low in CSF and, therefore, Cl^- is higher in order to maintain Donnan membrane equilibrium.

Excretion

There exists a parallel relationship between excretion of chloride and sodium. The renal threshold for Cl^- is about 110 mEq/l.

Disease states

1. **Hypochloremia :** A reduction in the serum Cl^- level may occur due to vomiting, diarrhea, respiratory alkalosis, Addison's disease and excessive sweating.
2. **Hyperchloremia :** An increase in serum Cl^- concentration may be due to dehydration, respiratory acidosis and Cushing's syndrome.

IRON

The ***total content*** of iron in an adult body is ***3-5 g***. About 70% of this occurs in the erythrocytes

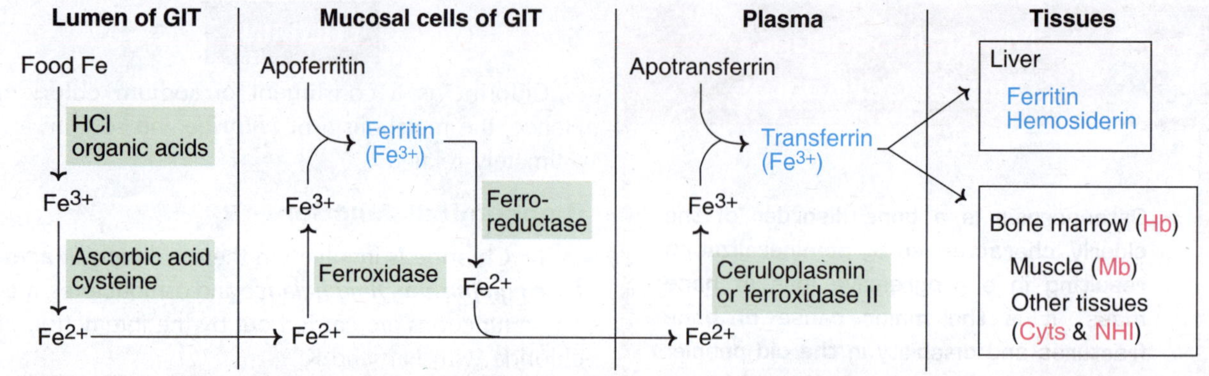

Fig.18.2 : *Iron absorption and transport (GIT–Gastrointestinal tract; Hb–Hemoglobin; Mb–Myoglobin; Cyts–Cytochromes; NHI–Non-heme iron).*

of blood as a constituent of hemoglobin. At least 5% of body iron is present in myoglobin of muscle. Heme is the most predominant iron-containing substance. It is a constituent of several proteins/ enzymes ***(hemoproteins)***—hemoglobin, myoglobin, cytochromes, xanthine oxidase, catalase, tryptophan pyrrolase, peroxidase. Certain other proteins contain ***non-heme iron*** e.g. transferrin, ferritin, hemosiderin.

Biochemical functions

1. Iron mainly exerts its functions through the compounds in which it is present. ***Hemoglobin*** and ***myoglobin*** are required for the ***transport of O_2 and CO_2***.

2. Cytochromes and certain non-heme proteins are necessary for electron transport chain and oxidative phosphorylation.

3. Peroxidase, the lysosomal enzyme, is required for phagocytosis and killing of bacteria by neutrophils.

4. Iron is associated with effective immuno-competence of the body.

Dietary requirements

Adult man — ***10 mg/day***

Menstruating woman — ***18 mg/day***

Pregnant and lactating woman — 40 mg/day

Sources

Rich sources — ***Organ meats*** (liver, heart, kidney).

Good sources — Leafy vegetables, pulses, cereals, fish, apples, dried fruits, molasses.

Poor sources — ***Milk***, wheat, polished rice.

Absorption, transport and storage

Iron is mainly absorbed in the stomach and duodenum. In normal people, ***about 10% of dietary iron is usually absorbed***.

Iron is mostly found in the foods in ferric form (Fe^{3+}), bound to proteins or organic acids. In the acid medium provided by gastric HCl, the Fe^{3+} is released from foods. Reducing substances such as ascorbic acid (vitamin C) and cysteine convert ferric iron (Fe^{3+}) to ferrous form (Fe^{2+}). Iron in the ***ferrous form is soluble and readily absorbed (Fig.18.2).***

Factors affecting Fe absorption

1. ***Acidity***, ***ascorbic acid*** and cysteine ***promote*** iron absorption.

2. In ***iron deficiency anemia***, Fe absorption is ***increased*** to 2-10 times that of normal.

3. Small peptides and amino acids favour iron uptake.

4. ***Phytate*** (found in cereals) and ***oxalate*** (found in leafy vegetables) ***interfere*** with Fe absorption.

5. A diet with high phosphate content decreases Fe absorption while low phosphate promotes.

6. Impaired absorption of iron is observed in malabsorption syndromes such as steatorrhea.

7. In patients with partial or total surgical removal of stomach and/or intestine, iron absorption is severely impaired.

Mucosal block theory

Homeostasis of iron in the body is maintained by ***regulating iron absorption***, and not by its excretion. When iron stores in the body are less, its absorption is increased. On the other hand, if iron stores are adequate, then absorption is decreased. This unique

phenomenon of regulating iron absorption is referred to as mucosal block theory. It is interesting to note that no other nutrient in the body is regulated in this fashion.

Iron in the mucosal cells : The iron (Fe^{2+}) entering the mucosal cells by absorption is oxidized to ferric form (Fe^{3+}) by the enzyme ferroxidase. Fe^{3+} then combines with ***apoferritin to form ferritin*** which is the temporary storage form of iron. From the mucosal cells, iron may enter the blood stream (which mainly depends on the body needs) or lost when the cells are desquamated.

Transport of Fe in the plasma

The iron liberated from the ferritin of mucosal cells enters the plasma in ferrous state. Here, it is oxidized to ferric form by a copper-containing protein, ***ceruloplasmin*** which possesses ***ferroxidase activity***. Another cuproprotein ferroxidase II also helps for the conversion of Fe^{2+} to Fe^{3+}.

Ferric iron then binds with a specific iron-binding protein, namely ***transferrin*** or siderophilin (a glycoprotein with mol. wt. 90,000). Each transferrin molecule can bind with two atoms of ferric iron (Fe^{3+}). The plasma transferrin (concentration 250 mg/dl) can bind with 400 mg of iron/dl plasma. This is known as ***total iron binding capacity (TIBC)*** of plasma.

Storage of iron

Iron is ***stored in liver, spleen*** and bone marrow in the form of ***ferritin***. In the mucosal cells, ferritin is the temporary storage form of iron. A molecule of apoferritin (mol. wt. 500,000) can combine with 4,000 atoms of iron. The maximum iron content of ferritin on weight basis is around 25%.

Hemosiderin is another ***iron storage protein*** which can hold about 35% of iron by weight. Hemosiderin accumulates in the body (spleen, liver) when the supply of iron is in excess of body demands.

Iron is a one-way substance

Iron metabolism is unique as it operates in a closed system. It is very efficiently ***utilized and reutilized by the body***. Further, iron losses from the body are minimal (<1 mg/day) which may occur through bile, sweat, hair loss etc. Iron is ***not excreted into urine***. Thus, iron differs from the vitamins or other organic and inorganic substances which are either inactivated or excreted during the course of metabolic function. Hence, iron is appropriately regarded as a one-way substance.

Iron entry into the body is controlled at the absorption level, depending on the body needs. Thus the periodical blood loss in menstruating women increases its requirements. Increased iron demands are also observed in pregnancy, lactation, and in growing children.

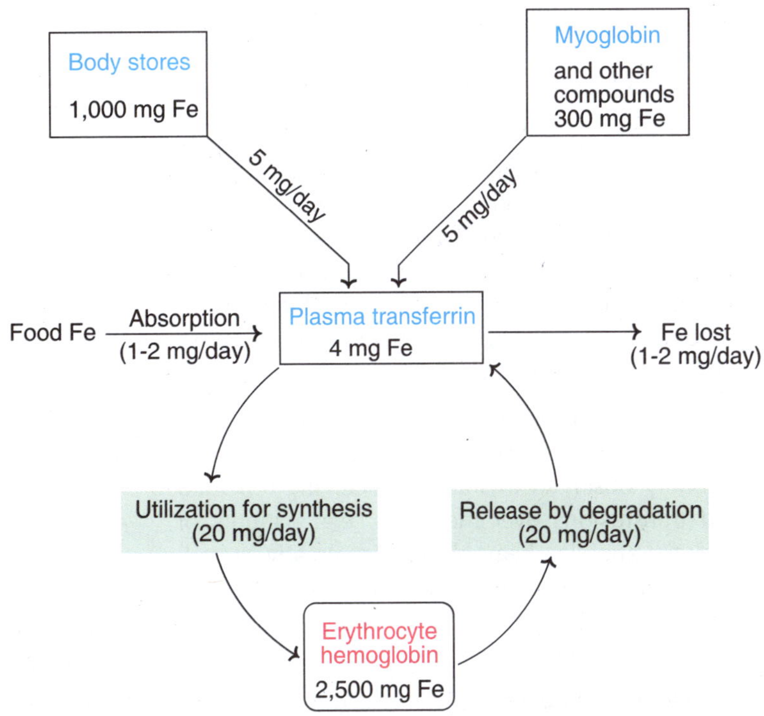

Fig.18.3 : *A general overview of iron metabolism.*

Overview of iron metabolism

A general overview of iron metabolism is depicted in ***Fig.18.3***. It shows the distribution of iron in the body and its efficient reutilization. It may be noted that about 1-2 mg of iron is absorbed per day to replace the loss.

Disease states

1. **Iron deficiency anemia :** This is the ***most prevalent nutritional disorder worldover***, including the well developed countries (e.g. USA). Several factors may contribute to iron deficiency anemia. These include inadequate intake or defective absorption of iron, chronic blood loss, repeated pregnancies and hookworm infections.

Strict ***vegetarians are more prone*** for iron deficiency anemia. This is due to the presence of inhibitors of iron absorption in the vegetarian foods, besides a relatively low content of iron.

Iron deficiency anemia mostly occurs in growing children, adolescent girls, pregnant and lactating women. It is characterized by ***microcytic hypochromic anemia*** with reduced blood hemoglobin levels (***<12 g/dl***). The other manifestations include apathy (dull and inactive), sluggish metabolic activities, retarded growth and loss of appetite.

2. **Hemosiderosis :** This is a less common disorder and is due to ***excessive iron in the body***. It is commonly observed in subjects receiving repeated blood transfusions over the years, e.g. patients of hemolytic anemia, hemophilia. As already stated, iron is a one-way compound, once it enters the body, it cannot escape. Excessive iron is deposited as ferritin and hemosiderin.

3. **Hemochromatosis :** This is a rare disease in which ***iron is directly deposited in the tissues*** (liver, spleen, pancreas and skin). Hemosiderosis is sometimes accompanied by hemochromatosis. Bronzed-pigmentation of the skin, cirrhosis of liver, pancreatic fibrosis are the manifestations of this disorder. Hemochromatosis causes a condition known as ***bronze diabetes***.

COPPER

The body contains about 100 mg copper distributed in different organs. It is involved in several important functions.

Biochemical functions

1. Copper is an essential constituent of several ***enzymes***. These include ***cytochrome oxidase***, ***catalase***, ***tyrosinase***, superoxide dismutase, monoamine oxidase, ascorbic acid oxidase, ALA synthase, phenol oxidase and uricase. Due to its presence in a wide variety of enzymes, Copper is involved in many metabolic reactions.

2. Copper is necessary for the ***synthesis of hemoglobin*** (Cu is a constituent of ALA synthase).

3. Lysyl oxidase (a copper-containing enzyme) is required for the conversion of certain lysine residues of ***collagen and elastin*** to allysine which are necessary for cross-linking these structural proteins.

4. Ceruloplasmin serves as ferroxidase and is involved in the conversion of iron from Fe^{2+} to Fe^{3+} in which form iron (transferrin) is ***transported in plasma***.

5. Hemocyanin, a copper protein complex in invertebrates, functions like hemoglobin for O_2 transport.

Dietary requirements

Adults — ***2-3 mg/day***

Infants and children — 0.5-2 mg/day

Sources

Liver, kidney, meat, egg yolk, cereals, nuts and green leafy vegetables. Milk is a poor source.

Plasma copper

The copper concentration of plasma is about 100-200 µg/dl. Most of this (95%) is tightly bound to ceruloplasmin while a small fraction (5%) is loosely held to albumin. Normal concentration of serum ***ceruloplasmin*** is 25-50 mg/dl.

Disease states

1. **Copper deficiency :** Severe deficiency of copper causes demineralization of bones, demyelination of neural tissue, anemia, fragility of arteries, myocardial fibrosis, hypopigmentation of skin, greying of hair.

2. **Wilson's disease (hepatolenticular degeneration) :** It is a rare disorder of abnormal copper metabolism and is characterized by the following manifestations.

- Copper is ***deposited*** in abnormal amounts ***in liver and lenticular nucleus of brain***. This may lead to hepatic cirrhosis and brain necrosis.
- ***Low levels of copper and ceruloplasmin in plasma*** with increased excretion of copper in urine.
- Copper deposition in kidney causes renal damage. This leads to increased excretion of amino acids, glucose, peptides and hemoglobin in urine.
- Intestinal ***absorption of copper*** is ***very high***, about 4-6 times higher than normal.

Treatment : Administration of ***penicillamine***, a naturally occurring copper chelating agent, is used for the treatment of Wilson's disease.

IODINE

The total body contains about 20 mg iodine, most of it (80%) being present in the ***thyroid gland***. Muscle, salivary glands and ovaries also contain some amount of iodine.

Biochemical functions

The only known function of iodine is its requirement for the synthesis of thyroid hormones namely, ***thyroxine*** (***T_4***) and ***triiodothyronine*** (***T_3***). These hormones are involved in several biochemical functions (***Chapter 19***). Functionally, T_3 is more active than T_4.

Dietary requirements

Adults	—	100-150 μg/day
Pregnant women	—	200 μg/day

Sources

Seafoods, drinking water, vegetables, fruits (grown on seaboard). High altitudes are deficient in iodine content in water as well as soil.

Absorption, storage and excretion

Iodine as iodide is mainly absorbed from the small intestine. Normally, about 30% of dietary iodine is taken up by the intestinal cells. Iodine absorption also occurs through skin and lungs.

About 80% of body's iodine is stored in the organic form as ***iodothyroglobulin*** (a glycoprotein) in the thyroid gland. This protein contains thyroxine, diiodotyrosine and triiodothyronine in different proportions.

Plasma iodine

The normal concentration of plasma iodine is ***4-10 μg/dl***. Most of this is present as ***protein bound iodine (PBI)*** and represents the iodine contained in the circulating thyroid hormones. PBI level decreases in hypothyroidism and increases in hyperthyroidism. RBC do not contain iodine.

Disease states

The disorders of iodine metabolism—***simple goiter*** and ***toxic goiter***—are discussed in detail under thyroid hormones (***Chapter 19***).

ZINC

The total content of zinc in an adult body is about 2 g. ***Prostate gland is very rich*** in Zn (100 μg/g). Zinc is mainly an intracellular element.

Biochemical functions

1. Zn is an essential component of several enzymes e.g. ***carbonic anhydrase***, alcohol dehydrogenase, alkaline phosphatase, carboxypeptidase, superoxide dismutase (cytosolic).

2. The storage and ***secretion of insulin*** from the β-cells of pancreas requires Zn.

3. Zn is necessary to ***maintain the normal levels of vitamin A*** in serum.

Dietary requirements

Zinc requirement for an adult is 10-15 mg/day. It is increased (by about 50%) in pregnancy and lactation.

Sources

Meat, fish, eggs, milk, beans, nuts.

COBALT

Cobalt is only important as a constituent of ***vitamin B_{12}***. Cobalt content of vitamin B_{12} is about 4% by weight. The functions of cobalt are the same as that of vitamin B_{12} (***Chapter 7***). Administration of cobalt stimulates the production of the hormone erythropoietin, which promotes erythropoiesis.

FLUORINE

Fluoride is mostly found in bones and teeth. The beneficial effects of fluoride in trace amounts are overshadowed by its harmful effects caused by excess consumption.

Biochemical functions

1. Fluoride ***prevents the development of dental caries***. It forms a protective layer of acid resistant ***fluoroapatite*** with hydroxyapatite of the enamel and prevents the tooth decay by bacterial acids. Further, fluoride inhibits the bacterial enzymes and reduces the production of acids.

> **Medical Concepts/ Clinical Correlates**
>
> - **Wilson's disease is due to disturbed copper metabolism. It is characterized by abnormal deposition of copper in liver and brain, besides the low levels of plasma copper and ceruloplasmin.**
> - **Endemic goitre, due to dietary iodine deficiency, is very common. Consumption of iodized salt is advocated to overcome this problem.**
> - **Fluorosis is caused by an excessive intake of fluoride. The manifestations include mottling of enamel and discoloration of teeth. In the advanced stages, hypercalcification of limb bones and ligaments of spine get calcified, ultimately crippling the individual.**

2. Fluoride is necessary for the ***proper development of bones***.

3. It inhibits the activities of certain enzymes. Sodium fluoride inhibits enolase (of glycolysis) while fluoroacetate inhibits aconitase (of citric acid cycle).

Dietary requirements and sources

An intake of less than 2 ppm of fluoride will meet the daily requirements. Drinking water is the main source.

Disease states

1. **Dental caries :** It is clearly established that drinking water containing ***less than 0.5 ppm*** of fluoride is associated with the development of dental caries in children.

2. **Fluorosis :** ***Excessive*** intake of fluoride is harmful to the body. An intake above 2 ppm (***particularly > 5 ppm***) in children causes ***mottling of enamel and discoloration of teeth***. The teeth are weak and become rough with characteristic brown or yellow patches on their surface. These manifestations are collectively referred to as ***dental fluorosis***.

An intake of fluoride above 20 ppm is toxic and causes pathological changes in the bones. Hypercalcification, increasing the density of the bones of limbs, pelvis and spine, is a characteristic feature. Even the ligaments of spine and collagen of bones get calcified. Neurological disturbances are also commonly observed. The manifestations described here constitute ***skeletal fluorosis***. In the advanced stages, the individuals are crippled and cannot perform their daily routine work due to stiff joints. This condition of advanced fluorosis is referred to as ***genu valgum***.

The fluoride content of water in some parts of Andhra Pradesh, Punjab and Karnataka is quite high. Fluorosis is prevalent in these regions, causing concern to government and health officials.

SELENIUM

Selenium was originally identified as an element that causes toxicity to animals (alkali disease) in some parts of USA, containing large amounts of Se in the soil. Later work, however, has shown that Se in smaller amounts is biologically important.

Biochemical functions

1. Selenium, along ***with vitamin E***, prevents the development of hepatic necrosis and muscular dystrophy.

2. Se is involved in maintaining ***structural integrity of biological membranes***.

3. Se is an integral component of the enzyme ***glutathione peroxidase*** which provides defense against peroxide damage to the cells.

4. Se ***prevents lipid peroxidation*** and protects the cells against the free radicals, including superoxide (O_2^-).

5. Se protects animals from carcinogenic chemicals. However, the precise role of Se in humans with regard to cancer prevention is not clearly identified.

6. Se binds with certain heavy metals (Hg, Cd) and protects the body from their toxic effects.

Requirements and sources

A daily intake of 50-200 μg of Se has been recommended for adults. The good sources of Se are organ meats (liver, kidney) and sea foods.

Toxicity

Selenosis is the toxicity due to very excessive intake of Se. The manifestations of selenosis include weight loss, emotional disturbances, diarrhea, hair loss and garlic odor in breath. The compound ***dimethyl selenide*** is responsible for the ***garlic odor***.

CHROMIUM

The total human body contains about 6 mg chromium. The Cr content of blood is about 20 μg/dl. Cr performs several biochemical functions.

1. In association with insulin, ***Cr promotes the utilization of glucose***. However, Cr alone is ineffective.

2. Cr lowers the total serum cholesterol level.

3. It is involved in lipoprotein metabolism. Cr decreases serum low density lipoproteins (LDL) and increases high density lipoproteins (HDL) and, thus, promotes health.

4. It is believed that Cr participates in the transport of amino acids into the cells (heart and liver).

The dietary requirement of Cr is not known. It is estimated that an adult man consumes about 10 to 100 mg/day. The good sources of Cr include brewer's yeast, grains, cereals, cheese and meat.

Chromium deficiency causes disturbances in carbohydrate, lipid and protein metabolisms. Excessive intake of Cr results in toxicity, leading to liver and kidney damage.

SUMMARY

1. *The minerals or inorganic elements are required for normal growth and maintenance of the body. They are classified as principal elements and trace elements. There are seven principal elements—Ca, P, Mg, Na, K, Cl and S. The trace elements include Fe, Cu, I, Zn, Mn, Mo, Co, F, Se and Cr.*
2. *Calcium is required for the development of bones and teeth, muscle contraction, blood coagulation, nerve transmission etc. Absorption of Ca from the duodenum is promoted by vitamin D, PTH and acidity while it is inhibited by phytate, oxalate, free fatty acids and fiber.*
3. *Sodium, potassium and chlorine are involved in the regulation of acid-base equilibrium, fluid balance and osmotic pressure in the body. Sodium is the principal extracellular cation (serum level 135-145 mEq/l), while potassium is the chief intracellular cation (serum level 3.5-5.0 mEq/l).*
4. *Iron is mainly required for O_2 transport and cellular respiration. Absorption of iron is promoted by ascorbic acid, cysteine, acidity and small peptides while it is inhibited by phytate, oxalate and high phosphate.*
5. *Copper is an essential constituent of several enzymes (e.g catalase, cytochrome oxidase, tyrosinase). Iodine is important as a component of thyroid hormones (T_4 and T_3) while cobalt is a constituent of vitamin B_{12}. Zinc is necessary for the storage and secretion of insulin and maintenance of normal vitamin A levels in serum.*
6. *Fluorine in trace amounts (<2 ppm) prevents dental caries while its higher intake leads to fluorosis. Selenium is assigned an antioxidant role as it protects the cells from free radicals. Chromium promotes the utilization of glucose and reduces serum cholesterol.*

THE READER MUST REFER APPENDIX D (Page Nos : 422-431) FOR RELATED CLINICAL CASE STUDIES (Nos. 6, 17, 29, 33)

HORMONES

19

The living body possesses a remarkable communication system to coordinate its biological functions. This is achieved by two distinctly organized functional systems.

1. The nervous system coordinates the body functions through the transmission of electrochemical impulses.

2. The ***endocrine system acts*** through a wide range of ***chemical messengers*** known as hormones.

Hormones are conventionally defined as organic substances, produced in small amounts by specific tissues (endocrine glands), secreted into the blood stream to control the metabolic and biological activities in the target cells. Hormones may be regarded as the ***chemical messengers*** involved in the transmission of information from one tissue to another and from cell to cell. The major endocrine organs in human body are depicted in ***Fig.19.1***.

CLASSIFICATION OF HORMONES

Hormones may be classified in many ways based on their characteristics and functions. Two types of classification are discussed here

I. Based on the chemical nature

The hormones can be categorized into three groups considering their chemical nature.

1. **Protein** or **peptide hormones** e.g. ***insulin***, glucagon, antidiuretic hormone, oxytocin.

2. **Steroid hormones** e.g. glucocorticoids, mineralocorticoids, ***sex hormones***.

3. **Amino acid derivatives** e.g. ***epinephrine***, norepinephrine, thyroxine (T_4), triiodothyronine (T_3).

II. Based on the mechanism of action

Hormones are classified into two broad groups (I and II) based on the location of the receptors to which they bind and the signals used to mediate their action.

1. **Group I hormones :** These hormones ***bind to intracellular receptors*** to form receptor-hormone complexes (the intracellular messengers) through which their biochemical functions are mediated. Group I hormones are lipophilic in nature and are derivatives of cholesterol (exception—T_3 and T_4). e.g. estrogens, androgens, glucocorticoids, calcitriol, T_3, T_4.

2. **Group II hormones :** These hormones bind to cell surface (plasma membrane) receptors and stimulate the release of certain molecules, namely the ***second messengers*** which, in turn, perform the biochemical functions. Thus, hormones themselves are the first messengers.

Group II hormones are subdivided into three categories based on the chemical nature of the second messengers.

(a) The second messenger is ***cAMP*** e.g. ACTH, FSH, LH, PTH, glucagon, calcitonin.

(b) The second messenger is ***phosphatidylinositol/calcium*** e.g. TRH, GnRH, gastrin, CCK.

(c) The second messenger is ***unknown*** e.g. growth hormone, insulin, oxytocin, prolactin.

The principal human hormones, their classification based on the mechanism of action, and major functions are given in ***Table 19.1***.

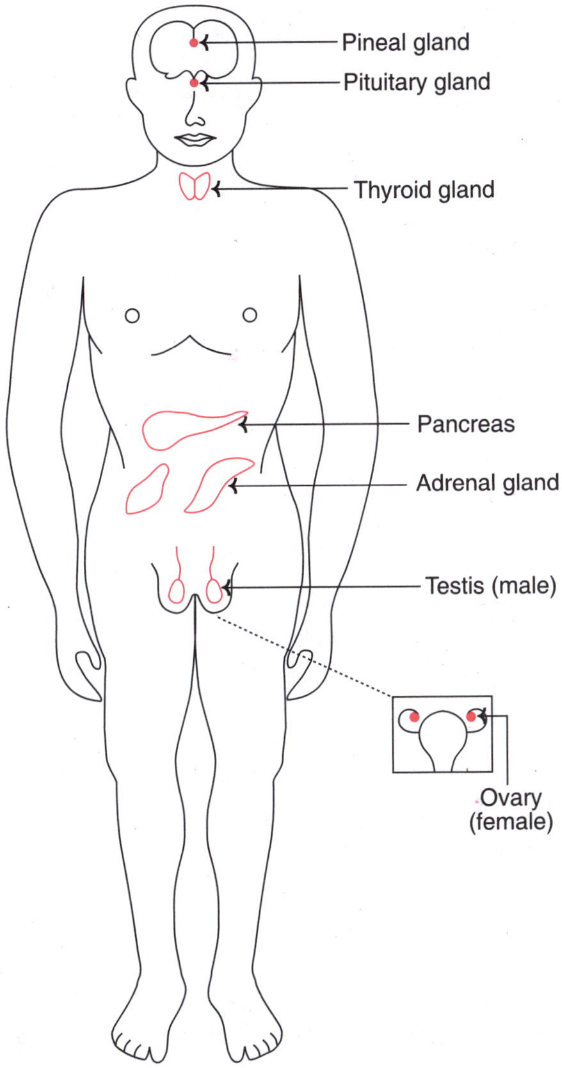

Fig. 19.1 : Diagrammatic representation of major endocrine glands.

Mechanism of action of group I hormones

These hormones are lipophilic in nature and can easily pass across the plasma membrane. They act through the ***intracellular receptors*** located either in the cytosol or the nucleus. The hormone-receptor complex binds to specific regions on the DNA called hormone responsive element (HRE) and causes ***increased expression of specific genes*** (***Fig.19.2***). It is believed that the interaction of hormone receptor complex with HRE promotes initiation and, to a lesser extent, elongation and termination of RNA synthesis (transcription). The ultimate outcome is the production of specific proteins (translation) in response to hormonal action.

Mechanism of action of group II hormones

These ***hormones*** are considered as the ***first messengers***. They exert their action through mediatory molecules, collectively called ***second messengers***.

cAMP—THE SECOND MESSENGER

Cyclic AMP (cAMP, cyclic adenosine 3′,5′-monophosphate) is a ubiquitous nucleotide. It consists of adenine, ribose and a phosphate (linked by 3′,5′ linkage). cAMP acts as a second messenger for a majority of ***polypeptide hormones***.

The membrane-bound enzyme adenylate cyclase converts ATP to cyclic AMP. cAMP is hydrolysed by phosphodiesterase to 5′-AMP (***Fig.19.3***).

Adenylate cyclase system

A series of events occur at the membrane level that influence the activity of adenylate cyclase leading to the ***synthesis of cAMP***. This process is mediated by ***G-proteins***, so designated due to their ability to bind to guanine nucleotides.

Action of cAMP—a general view

Once produced, cAMP performs its role as a second messenger in eliciting biochemical responses (***Fig.19.4***). cAMP activates protein kinase ***A*** (A stands for c**A**MP). This enzyme is a heterotetramer consisting of 2 regulatory subunits (R) and 2 catalytic subunits (C).

cAMP binds to inactive protein kinase and causes the dissociation of R and C subunits.

$$\underset{}{4cAMP} + \underset{(inactive)}{R_2C_2} \longrightarrow \underset{(inactive)}{R_2(4\ cAMP)} + \underset{(active)}{2C}$$

The active subunit (C) catalyses phosphorylation of proteins (transfer of phosphate group to serine and threonine residues). It is the phosphoprotein that ultimately causes the biochemical response.

It should, however, be remembered that cAMP does not act on all protein kinases. e.g. protein kinase C.

Dephosphorylation of proteins : A group of enzymes called protein phosphatases hydrolyse and remove the phosphate group added to proteins.

Degradation of cAMP : cAMP undergoes rapid hydrolysis, catalysed by the ***enzyme phospho-***

Table 19.1 Principal human hormones—classification (by mechanism of action), origin and major functions

Hormone(s)	Origin	Major Function(s)
Group I. HORMONES THAT BIND TO INTRACELLULAR RECEPTORS		
Estrogens	Ovaries and adrenal cortex	Female sexual characteristics, menstrual cycle.
Progestins	Ovaries and placenta	Involved in menstrual cycle and maintenance of pregnancy.
Androgens	Testes and adrenal cortex	Male sexual characteristics, spermatogenesis.
Glucocorticoids	Adrenal cortex	Affect metabolisms, suppress immune system.
Mineralocorticoids	Adrenal cortex	Maintenance of salt and water balance.
Calcitriol (1,25-DHCC)	Kidney (final form)	Promotes absorption of Ca^{2+} from intestine, kidney and bone.
Thyroid hormones (T_3, T_4)	Thyroid	Promote general metabolic rate.
Group II. HORMONES THAT BIND TO CELL SURFACE RECEPTORS		
A. The second messenger is cAMP		
Adrenocorticotropic hormone (ACTH)	Anterior pituitary	Stimulates the release of adrenocorticosteroids.
Follicle stimulating hormone (FSH)	Anterior pituitary	In females, stimulates ovulation and estrogen synthesis. In males, promotes spermatogenesis.
Luteinizing hormone (LH)	Anterior pituitary	Stimulates synthesis of estrogens and progesterone and causes ovulation. Promotes androgen synthesis by testes.
Chorionic gonadotropin (hCG)	Anterior pituitary	Stimulates progesterone release from placenta.
Thyroid stimulating hormone (TSH)	Anterior pituitary	Promotes the release of thyroid hormones (T_3, T_4).
β-Endorphins and enkephalins	Anterior pituitary	Natural endogenous analgesics (pain relievers).
Antidiuretic hormone (ADH)	Posterior pituitary (stored)	Promotes water reabsorption by kidneys.
Glucagon	Pancreas	Increases blood glucose level, stimulates glycogenolysis and lipolysis.
Parathyroid hormone (PTH)	Parathyroid	Increases serum calcium, promotes Ca^{2+} release from bone.
Calcitonin	Thyroid	Lowers serum calcium. Decreases Ca^{2+} uptake by bone and kidney.
Epinephrine	Adrenal medulla	Increases heart rate and blood pressure. Promotes glycogenolysis in liver and muscle and lipolysis in adipose tissue.
Norepinephrine	Adrenal medulla	Stimulates lipolysis in adipose tissue.
B. The second messenger is phosphatidyl inositol/calcium		
Thyrotropin-releasing hormone (TRH)	Hypothalamus	Promotes TSH release.
Gonadotropin-releasing hormone (GnRH)	Hypothalamus	Stimulates release of FSH and LH.
Gastrin	Stomach	Stimulates gastric HCl and pepsinogen secretion.
Cholecystokinin (CCK)	Intestine	Stimulates contraction of gall bladder and secretion of pancreatic enzymes.
C. The second messenger is unknown/unsettled		
Growth hormone (GH)	Anterior pituitary	Promotes growth of the body (bones and organs).
Prolactin (PRL)	Anterior pituitary	Growth of mammary glands and lactation.
Oxytocin	Posterior pituitary (stored)	Stimulates uterine contraction and milk ejection.
Insulin	Pancreas	Lowers blood glucose (hypoglycemic effect), promotes protein synthesis and lipogenesis.
Somatomedins (insulin-like growth factors, IGF-I, IGF-II)	Liver	Growth related functions of GH are mediated. Stimulates growth of cartilage.

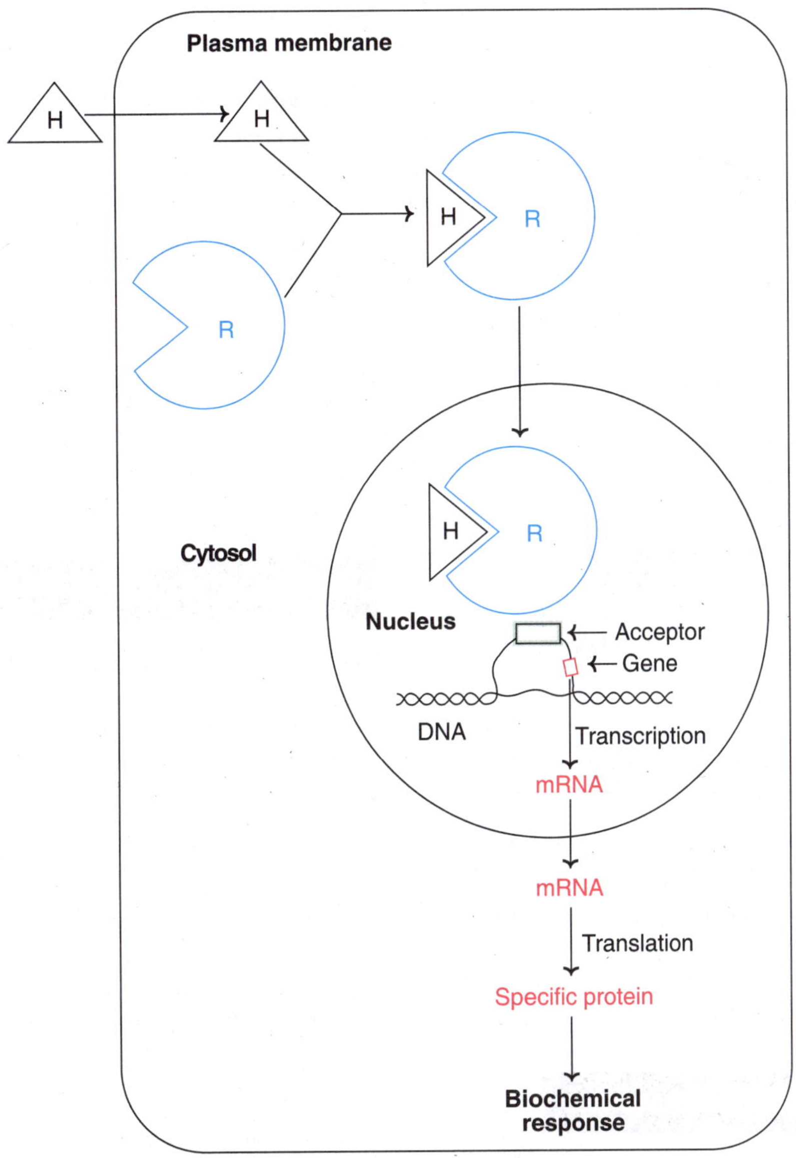

Fig.19.2 : *Mechanism of action of steroid hormones (H–Hormone; R–Receptor; HR–Hormone-receptor complex).*

diesterase, to 5′ AMP which is inactive. Hence, the effect of cAMP will be shortlived if the hormone stimulating adenylate cyclase is removed. Caffeine and theophylline (methylxanthine derivatives) can inhibit phosphodiesterases and increase the intracellular levels of cAMP.

HYPOTHALAMIC AND PITUITARY HORMONES

The pituitary gland or hypophysis (weighing about 1 g) is located below the hypothalamus of the brain. It consists of two distinct parts— the anterior pituitary (adenohypophysis) and the posterior pitutitary (neurohypophysis) connected by pars intermedia (***Fig.19.5***). The latter is almost absent in humans, although found in lower organisms.

Hypothalamus is a specialized center in the brain that functions as a ***master coordinator of hormonal action***. In response to the stimuli of central nervous system, hypothalamus liberates certain releasing factors or hormones. These factors stimulate or inhibit the release of corresponding tropic hormones from the anterior pituitary. Tropic hormones stimulate the target endocrine tissues to secrete the hormones they

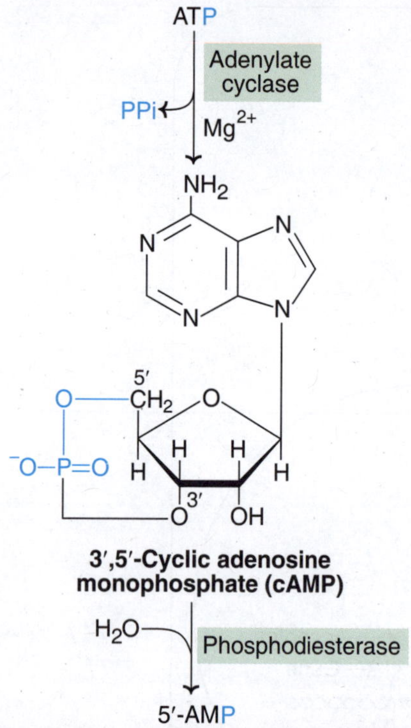

Fig.19.3 : Synthesis and degradation of cAMP.

synthesize. The relationship between hypothalamus and pituitary with endocrine glands is illustrated in ***Fig.19.6***. ***In general, the hormonal system is under feedback control***. For instance, adrenocorticotropic hormone (ACTH) inhibits the release of corticotropin releasing hormone (CRH).

HYPOTHALAMIC HORMONES

Hypothalamus produces at least six releasing factors or hormones.

1. **Thyrotropin-releasing hormone (TRH) :** It is a tripeptide consisting of glutamate derivative (pyroglutamate), histidine and proline. TRH stimulates anterior pituitary to release thyroid-stimulating hormone (TSH or thyrotropin) which, in turn, stimulates the release of thyroid hormones (T_3 and T_4).

2. **Corticotropin-releasing hormone (CRH) :** It stimulates anterior pituitary to release adrenocorticotropic hormone (ACTH) which in turn, acts on adrenal cortex to liberate adrenocorticosteroids. CRH contains 41 amino acids.

3. **Gonadotropin-releasing hormone (GnRH) :** It is a decapeptide. GnRH stimulates anterior pituitary to release *gonadotropins,* namely luteinizing hormone (LH) and follicle stimulating hormone (FSH).

4. **Growth hormone-releasing hormone (GRH)** with 44 amino acids stimulates the release of growth hormone (GH or somatotropin) which promotes growth.

5. **Growth hormone release-inhibiting hormone (GRIH) :** It contains 14 amino acids and is also known as somatostatin. GRIH inhibits the release of growth hormone from the anterior pituitary.

6. **Prolactin release-inhibiting hormone (PRIH) :** It is believed to be a dopamine and/or a small peptide that inhibits the release of prolactin (PRL) from anterior pituitary.

ANTERIOR PITUITARY HORMONES

Anterior pituitary or adenohypophysis is truly the ***master endocrine organ***. It produces several hormones that influence—either directly or indirectly—a variety of biochemical processes in the body. The hormones of adenohypophysis are broadly classified into three categories.

I. The growth hormone-prolactin group.

II. The glycoprotein hormones.

III. The pro-opiomelanocortin peptide family.

I. The growth hormone-prolactin group

Growth hormone (GH), prolactin (PRL) and chorionic somatomammotropin (CS; placental lactogen) are protein hormones with many striking similarities in their structure.

Growth hormone (GH) : The growth hormone (or ***somatotropin***) is produced by somatotropes, a special group of acidophilic cells of anterior pituitary.

Regulation of GH release : Two hypothalamic factors play a prominent role in the release of growth hormones. These are the growth hormone-releasing hormone (GRH) that stimulates and the growth hormone release-inhibiting hormone (GRIH, ***somatostatin***) that inhibits. This, in turn, is regulated by a feedback mechanism.

Growth hormone production is influenced by many factors such as sleep, stress (pain, cold, surgery), exercise, food intake etc. It is observed that the largest increase in the production of GH occurs after the onset of sleep. This supports the ***adage "If you don't sleep, you won't grow."***

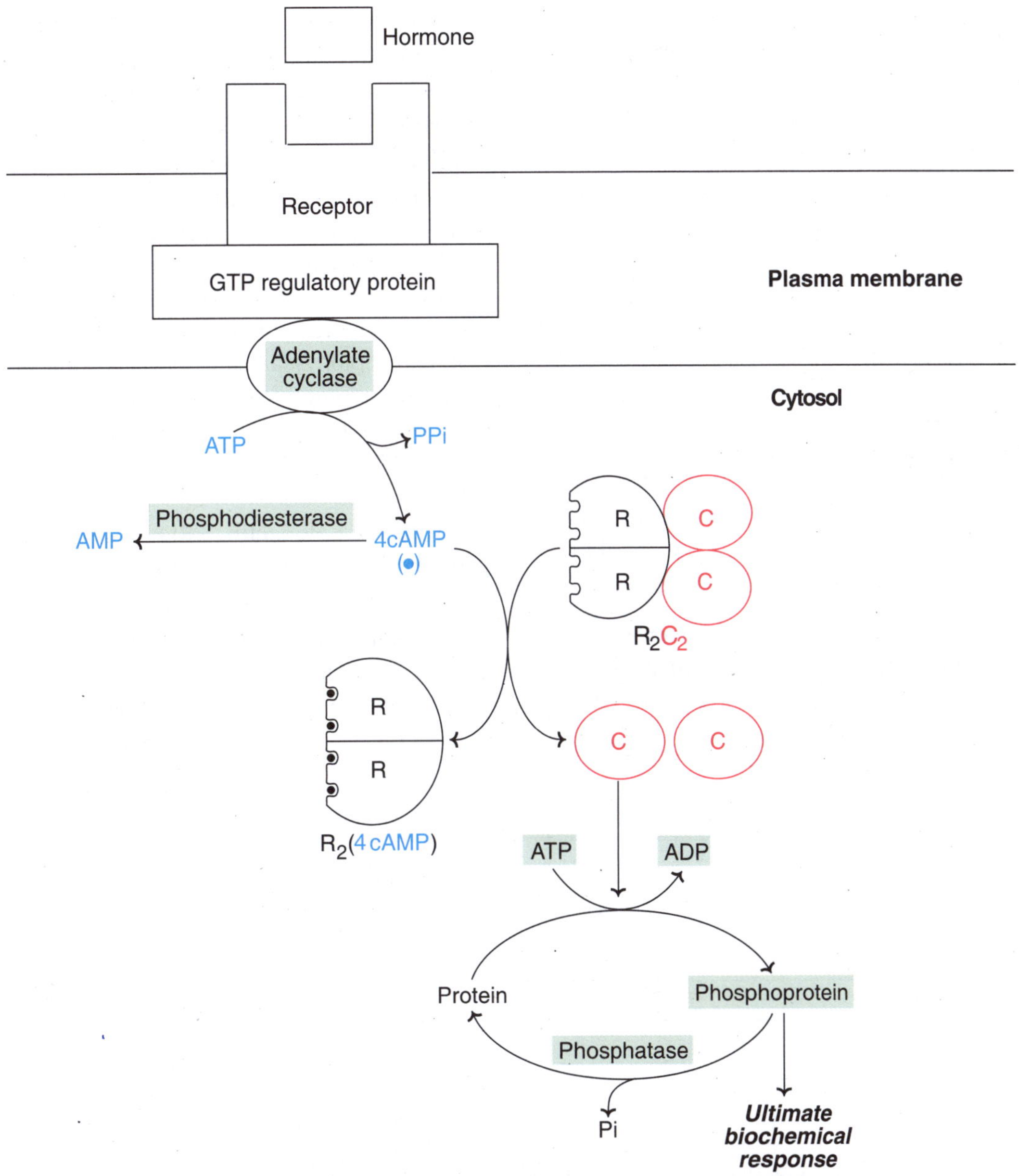

Fig.19.4 : *Overview of synthesis and action of cAMP (R_2C_2–cAMP dependent protein kinase A; R_2–Regulatory subunits; C_2–Catalytic subunits; C–Active catalytic unit of R_2C_2).*

Biochemical functions of GH : Growth hormone promotes growth and also influences the normal metabolisms (protein, carbohydrate, lipid and mineral) in the body.

1. **Effects on growth :** As is obvious from the name, GH is ***essential for the growth***. The growth-related effects of GH are mediated through ***insulin like growth factor I (IGF-I)*** which is also known as ***somatomedin C*** (formerly sulfation factor), produced by liver.

2. **Effects on protein metabolism :** Growth hormone has an ***anabolic effect*** on protein metabolism. It promotes the uptake of amino acids into the tissues and increases the protein synthesis. The overall effect of GH is a ***positive nitrogen balance*** that leads to increase in body weight.

3. **Effects on carbohydrate metabolism :** Growth hormone is antagonistic to insulin and ***causes hyperglycemia***. GH increases gluconeogenesis, decreases glucose utilization, impairs glycolysis and reduces the tissue uptake of glucose.

4. **Effects on lipid metabolism :** Growth hormone ***promotes lipolysis*** in the adipose tissue and increases

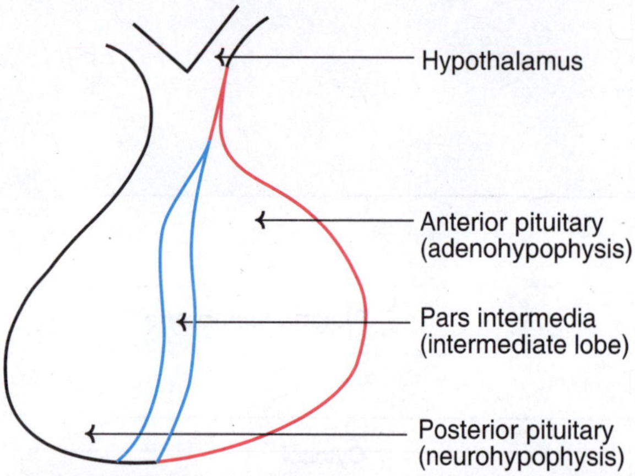

Fig.19.5 : *A diagrammatic view of pituitary gland.*

the circulatory levels of free fatty acids and their oxidation. It increases ketogenesis, particularly in diabetes.

5. **Effects on mineral metabolism :** Growth hormone ***promotes bone mineralization*** and its growth, as clearly observed in the growing children.

Abnormalities of GH production

Deficiency of GH : Impairment in the secretion of growth hormone in the growing age causes ***dwarfism***. The other deficiency metabolic effects are not that serious in nature.

Overproduction of GH : Excessive production of GH causes ***gigantism in children*** and ***acromegaly in adults***. This usually occurs in the acidophil tumor of pituitary gland. Gigantism is characterized by increased growth of long bones and this is observed before the epiphyseal plates close. Acromegaly occurs after epiphyseal closure and is characterized by increase in the size of hands, facial changes (enlarged nose, protruding jaw), excessive hair, thickening of skin etc.

Prolactin : Prolactin (PRL) is also called lactogenic hormone, luteotropic hormone, mammotropin or luteotropin.

Biochemical functions of PRL : Prolactin is primarily concerned with the initiation and maintenance of lactation in mammals. PRL increases

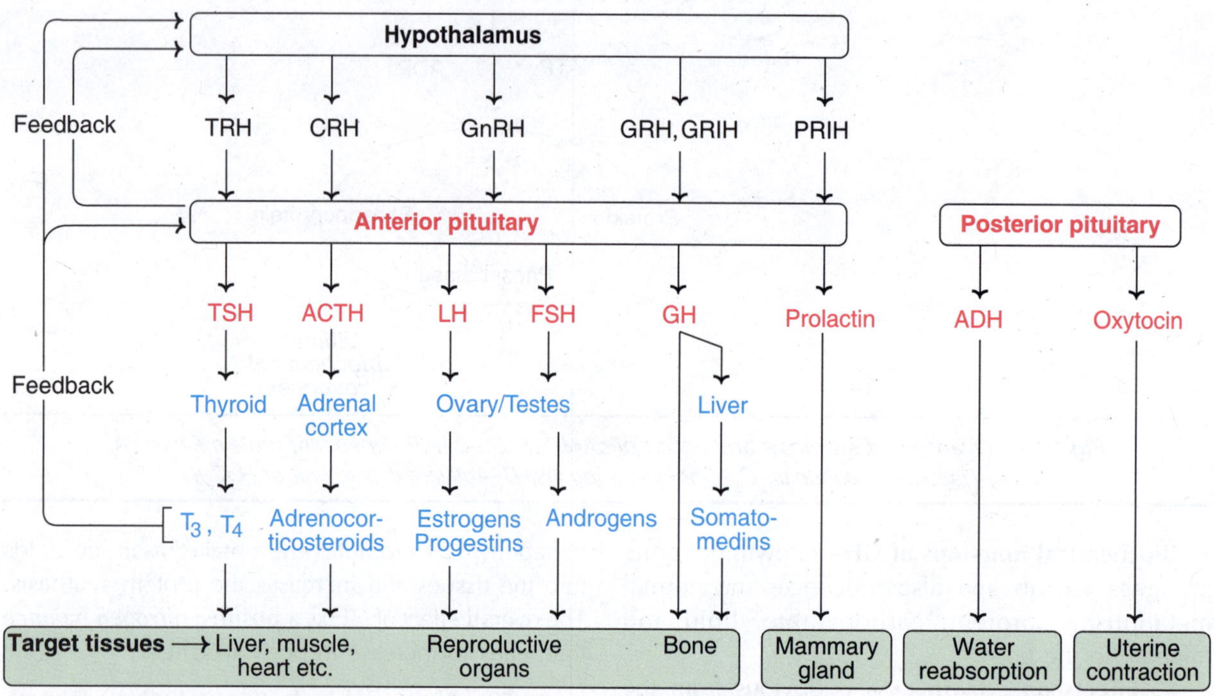

Fig. 19.6 : *Hormonal hierarchy relationships between hypothalamus and pituitary with other endocrine glands [TRH–Thyrotropin releasing hormone; CRH–Corticotropin releasing hormone; GnRH–Gonadotropin releasing hormone; GRH–Growth hormone releasing hormone; GRIH–Growth hormone release inhibiting hormone; TSH–Thyroid stimulating hormone; ACTH–Adrenocorticotropic hormone; LH–Luteinizing hormone; FSH–Follicle stimulating hormone; GH–Growth hormone; ADH–Antidiuretic hormone; T_3–Triiodothyronine; T_4–Tetraiodothyronine (thyroxine)].*

the levels of several enzymes involved in carbohydrate and lipid metabolism. PRL promotes HMP shunt, increases lipid biosynthesis and stimulates lactose production in mammary glands.

Prolactin promotes the growth of corpus luteum (hence also known as luteotropic hormone) and stimulates the production of progesterone.

II. The glycoprotein hormones

The following four hormones are glycoprotein in nature and possess certain structural similarities, despite their functional diversity.

1. Thyroid stimulating hormone (TSH)
2. Follicle stimulating hormone (FSH)
3. Luteinizing hormone (LH)
4. Human chorionic gonadotropin (hCG).

The last three hormones (2-4) are collectively referred to as ***gonadotropins*** due to their involvement in the function of gonads. The hormone hCG is produced by human placenta and not by pituitary. However, due to its structural resemblance with other hormones, it is also considered here.

1. **Thyroid stimulating hormone (TSH) :** TSH is a dimer (αβ) glycoprotein with a molecular weight of about 30,000.

Regulation of TSH production : The release of TSH from anterior pituitary is controlled by a feedback mechanism. This involves the hormones of thyroid gland (T_3 and T_4) and thyrotropin-releasing hormone (TRH) of hypothalamus.

Functions of TSH : The biochemical effects of TSH on thyroid gland are briefly discussed here. TSH binds with plasma membrane receptors and stimulates adenylate cyclase with a consequent increase in cAMP level. TSH, through the mediation of cAMP, exerts the following effects.

(a) ***Promotes the uptake of iodide*** (iodide pump) from the circulation by thyroid gland.

(b) Enhances the ***conversion of iodide*** (I^-) to ***active iodide*** (I^+), a process known as organification.

(c) Increases the proteolysis of thyroglobulin to ***release*** T_3 ***and*** T_4 into the circulation.

TSH increases the synthesis of proteins, nucleic acids and phospholipids in thyroid gland.

Gonadotropins : The follicle-stimulating hormone (***FSH***), luteinizing hormone (***LH***) and human chorionic gonadotropin (***hCG***) are commonly known as gonadotropins. All three are glycoproteins.

The release of FSH and LH from the anterior pituitary is controlled by gonadotropin-releasing hormone (GnRH) of hypothalamus.

2. **Biochemical functions of FSH :** In females, FSH stimulates follicular growth, increases the weight of the ovaries and enhances the production of estrogens.

In males, FSH stimulates testosterone production, required for spermatogenesis. FSH also promotes growth of seminiferous tubules.

3. **Biochemical functions of LH :** Luteinizing hormone stimulates the production of progesterone from corpus luteum cells in females and testosterone from Leydig cells in males. LH and FSH are collectively responsible for the development and maintenance of secondary sexual characters in males.

4. **Human chorionic gonadotropin (hCG) :** hCG is a glycoprotein (mol. wt. 100,000), produced by syncytiotrophoblast cells of placenta. The structure of hCG closely resembles that of LH.

The levels of hCG in plasma and urine increase almost immediately after the implantation of fertilized ovum. The detection of ***hCG in urine*** is conveniently used for the ***early detection*** (within a week after missing the menstrual cycle) ***of pregnancy***.

III. The pro-opiomelanocortin (POMC) peptide family

This family consists of the hormones— adreno-corticotropic hormone (***ACTH***), lipotropin (***LPH***) and melanocyte stimulating hormone (***MSH***) and several (about 24) neuromodulators such as endorphins and enkephalins.

The synthesis of POMC family is very interesting. All the members of POMC are produced from a single gene of the anterior and intermediate lobes of pituitary. It is fascinating that a single polypeptide — pro-opiomelanocortin—is the precursor (approximately 285 amino acids) that contains multiple hormones. The name ***pro-opiomelano-cortin*** is derived since it is a ***pro***hormone to ***opio***ids, ***melano***cyte-stimulating hormone and ***corti***cotropi***n***.

Products of POMC : The pituitary multihormone precursor is synthesized as pre-proopiomelano-cortin from which POMC is formed. The POMC consists of 3 peptide groups.

1. ACTH that can give rise to α-MSH and corticotropin like intermediate lobe peptide (CLIP).

Medical Concepts/ Clinical Correlates

- Growth hormone deficiency causes dwarfism while its excessive production results in gigantism (in children) or acromegaly (in adults).
- Identification of hCG in urine is employed for the early detection of pregnancy.
- Cushing's syndrome is due to over-production of ACTH that results in the increased synthesis of adrenocortico-steroids. The symptoms of this syndrome include hypertension, edema and negative nitrogen balance.

2. ***β-Lipotropin*** (β-LPH) that can produce γ-LPH, β-MSH and β-endorphin. The latter yields γ- and α-endorphins.

3. An N-terminal peptide that forms γ-MSH.

The products obtained from POMC are depicted in ***Fig.19.7***. These products undergo many modifications such as glycosylation, acetylation etc.

1. **Adrenocorticotropic hormone (ACTH) :** ACTH is a polypeptide with 39 amino acids and a molecular weight of 4,500. This hormone is primarily concerned with the growth and functions of adrenal cortex.

Regulation of ACTH production : The release of ACTH from the anterior pituitary is under the regulation of hypothalamic hormone, namely corticotropin releasing hormone (CRH).

Biochemical functions of ACTH

(a) ACTH promotes the conversion of cholesterol to pregnenolone in the adrenal cortex.

(b) It enhances RNA and protein synthesis and thus promotes adrenocortical growth.

(c) ACTH increases lipolysis by activating lipase of adipose tissue.

Overproduction of ACTH : *Cushing's syndrome* is caused by an excessive production of ACTH which may be due to a tumor. This syndrome is characterized by hyperpigmentation and increased production of adrenocorticosteroids. The associated symptoms include negative nitrogen balance, impaired glucose tolerance, hypertension, edema, muscle atrophy etc.

2. **β-Lipotropin (β-LPH) :** β-LPH is derived from POMC and contains 93 carboxy terminal amino acids. This polypeptide consists of γ-LPH and β-endorphin from which β-MSH and γ-endorphin are, respectively, formed. γ-Endorphin can be converted to α-endorphin and then to enkephalins (***Fig.19.7***). β-LPH is found only in the pituitary and not in other tissues since it is rapidly degraded.

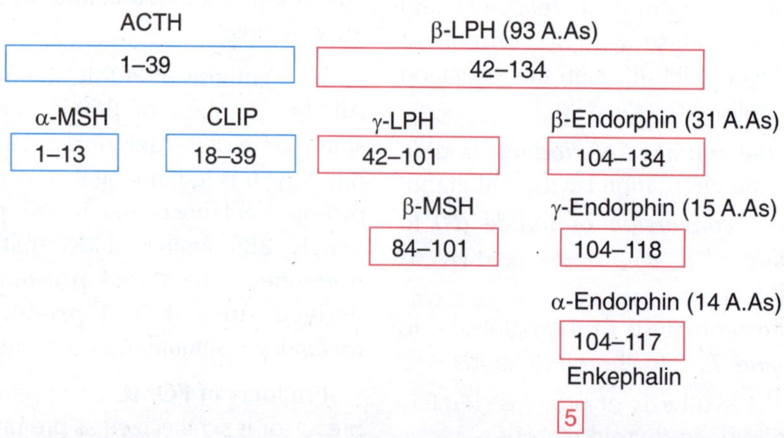

***Fig.19.7 :** The members of the pro-opiomelanocortin (POMC) family derived from POMC cleavage. (Numbers in blocks represent amino acids in sequence; In the brackets are the number of amino acids–AAs; (ACTH–Adrenocorticotropic hormone; LPH–Lipotropin; MSH–Melanocyte-stimulating hormone; CLIP–Corticotropin like intermediate lobe peptide).*

The biochemical functions of β-LPH, as such, are limited. It promotes lipolysis and increases the mobilization of fatty acids. The most important function of β-LPH is its precursor role for the formation of β-endorphin and enkephalins.

Endorphins and enkephalins : These are the ***natural analgesics*** that control pain and emotions. They were discovered after an unexpected finding of opiate receptors in the human brain.

Synthesis : Endorphins and enkephalins are produced from β-endorphin which, in turn, is derived from POMC (***Fig.19.7***). β-Lipotropin has 31 amino acids while its modified products α and γ-endorphins have 15 and 14 amino acids, respectively. Methionine enkephalin (Tyr–Gly–Gly–Phe–***Met***) and leucine enkephalin (Tyr-Gly-Gly-Phe-***Leu***) are the two important pentapeptide derivatives of β-endorphin.

Biochemical actions : Endorphins and enkephalins are ***peptide neurotransmitters*** that produce opiate-like effects on the central nervous system, hence they are also known as ***opioid-peptides***. They bind to the same receptors as the morphine opiates and are believed to ***control*** the ***endogenous pain perception***. Endorphins and enkephalins are more potent (20-30 times) than ***morphine in their function*** as analgesics.

It is ***believed that the pain relief*** through ***acupuncture and placebos is mediated through opioid peptides***.

3. **Melanocyte-stimulating hormone (MSH) :** Three types of MSH (α, β and γ) are present in the precursor POMC molecule. In humans, γ MSH is important while in some animals α and β are functional. The activity of γ-MSH is contained in the molecule γ-LPH or its precursor β-LPH (***Fig.19.7***).

The functions of MSH has been clearly established in some animals. MSH ***promotes the synthesis of skin pigment melanin*** (melanogenesis) and disperses melanin granules that ultimately leads to darkening of the skin. In humans, MSH does not appear to play any role in melanin synthesis.

POSTERIOR PITUITARY HORMONES

Two hormones namely ***oxytocin*** and ***antidiuretic hormone*** (ADH, vasopressin) are produced from the posterior pituitary (neurohypophysis). Both of them are nonapeptides (9 amino acids). Their structures are depicted in ***Fig.19.8***.

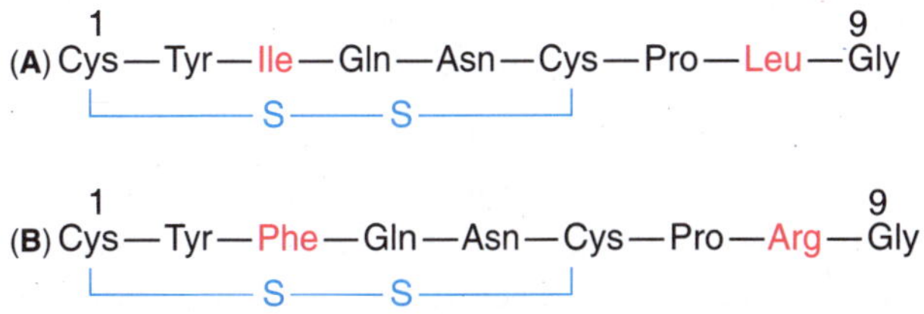

Fig.19.8 : Structures of (A) Human oxytocin and (B) Human antidiuretic hormone (ADH).

Oxytocin

The release of oxytocin from posterior pituitary gland is caused by the neural impulses of nipple stimulation. The other stimuli responsible for oxytocin release include vaginal and uterine distention.

Biochemical functions

1. **Effect on uterus :** Oxytocin causes the contraction of pregnant uterus (smooth muscles) and induces labor.

2. **Effect on milk ejection :** In mammals, oxytocin causes contraction of myoepithelial cells (look like smooth muscle cells) of breast. This stimulates the squeezing effect, causing milk ejection from the breast.

3. Oxytocin synthesized in the ovary appears to ***inhibit the synthesis of steroids***.

Antidiuretic hormone (ADH)

The release of ADH (also called ***vasopressin***) is mostly controlled by osmoreceptors (of hypothalamus) and baroreceptors (of heart). Any increase in the osmolarity of plasma stimulates ADH secretion.

Biochemical functions : ADH is primarily concerned with the ***regulation of water balance*** in the body. It stimulates kidneys ***to retain water*** and, thus, increases the blood pressure.

In the absence of ADH, the urine output would be around 20 l/day. ADH acts on the distal convoluted tubules of kidneys and causes water reabsorption with a result that the urine output is around 0.5-1.5 l/day.

Mechanism of action : ADH stimulates adenylate cyclase causing production of cAMP. Water reabsorption is promoted by cAMP. Inhibitors of adenylate cyclase (e.g. calcium) inhibit the activity of ADH. This supports the view that ADH action is mostly mediated through cAMP.

Diabetes insipidus : This disorder is characterized by the excretion of large volumes of dilute urine ***(polyuria)***. It may be due to insufficient levels of ADH or a defect in the receptors of target cells.

THYROID HORMONES

Thyroid gland (weighs about 30 g in adults) is located on either side of the trachea below the larynx. It produces two principal hormones (***Fig.19.9***)—***thyroxine*** (T_4; 3,5,3′,5′-tetraiodothyronine) and ***3,5,3′-triiodothyronine*** (T_3)—which regulate the metabolic rate of the body. Thyroid gland also secretes calcitonin, a hormone concerned with calcium homeostasis (discussed under calcium metabolism, ***Chapter 18***).

Biosynthesis of thyroid hormones

Iodine is essential for the synthesis of thyroid hormones. More than half of the body's total iodine content is found in the thyroid gland.

Tyrosine (of thyroglobulin) is first iodinated at position 3 to form monoiodotyrosine (MIT) and then at position 5 to form diiodotyrosine (DIT). Two molecules of DIT couple to form thyroxine (T_4). One molecule of MIT, when coupled with one molecule of DIT, triiodothyronine (T_3) is produced (***Refer Fig.15.14***).

Biochemical functions of thyroid hormones

Triiodothyronine (T_3) is ***about four times more active*** in its biological functions ***than*** thyroxine (T_4). The following are the biochemical functions attributed to thyroid hormones (T_3 and T_4).

1. **Influence on the metabolic rate :** Thyroid hormones ***stimulate the metabolic activities*** and increases the oxygen consumption in most of the tissues of the body (exception—brain, lungs, testes and retina).

Na^+-K^+ ATP pump : This is an energy dependent process which consumes a major share of cellular ATP. Na^+-K^+ ATPase activity is directly correlated to thyroid hormones and this, in turn, with ATP utilization. ***Obesity*** in some individuals is attributed to a ***decreased energy utilization*** and heat production ***due to diminished Na^+-K^+ ATPase activity***.

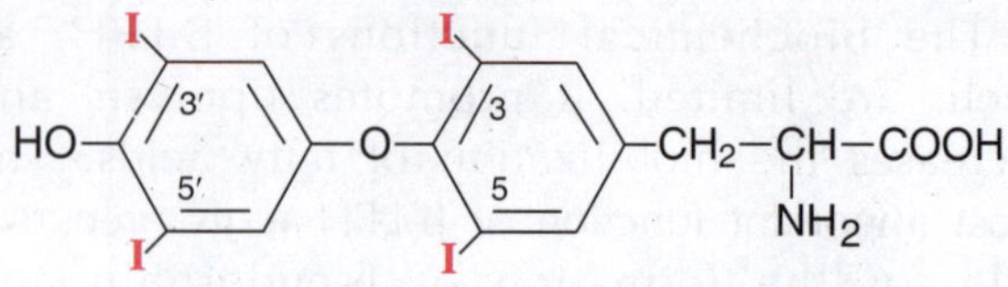

3, 5, 3′, 5′-Tetraiodothyronine (thyroxine, T_4)

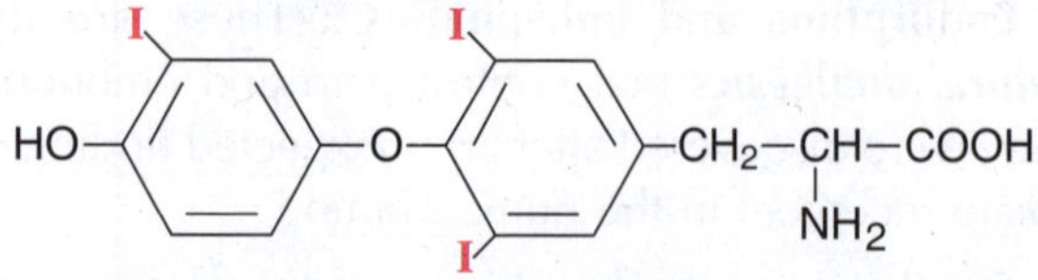

3,5,3′-Triiodothyronine (T_3)

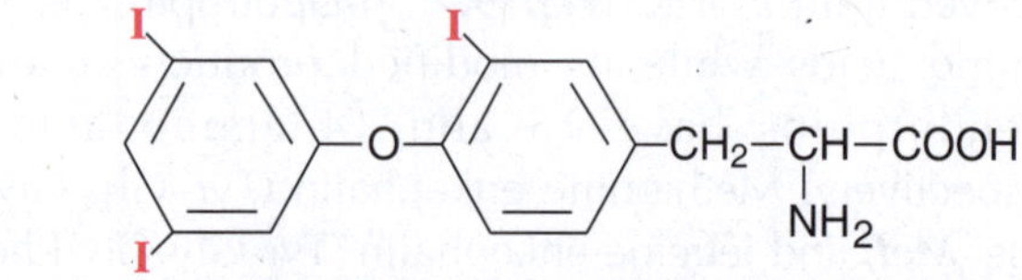

3, 3′, 5′-Triiodothyronine (reverse T_3, rT_3)

Fig.19.9 : *Structures of thyroid hormones (Refer* **Fig. 15.14** *for their biosynthesis).*

2. **Effect on protein synthesis :** Thyroid hormones act like steroid hormones in promoting protein synthesis by acting at the transcriptional level (activate DNA to produce RNA). Thyroid hormones, thus, function as anabolic hormones and cause ***positive nitrogen balance and promote growth and development***.

3. **Influence on carbohydrate metabolism :** Thyroid hormones promote intestinal absorption of glucose and its utilization. These hormones ***increase gluconeogenesis and glycogenolysis***, with an overall effect of enhancing blood glucose level (hyperglycemia).

4. **Effect on lipid metabolism :** Lipid turnover and utilization are stimulated by thyroid hormones. ***Hypothyroidism*** is associated with ***elevated plasma cholesterol levels*** which can be reversed by thyroid hormone administration.

Regulation of T_3 and T_4 synthesis

The synthesis of thyroid hormones is controlled by ***feedback regulation***. T_3 appears to be more actively involved than T_4 in the regulation process. The production of thyroid stimulating hormone (TSH) by pituitary and thyrotropin releasing hormone (TRH) by hypothalamus are inhibited by T_3 and, to a lesser degree, by T_4. The increased synthesis of TSH and TRH occurs in response to decreased circulatory levels of T_3 and T_4.

Medical Concepts/ Clinical Correlates

- Endorphins and enkephalins are the natural painkillers in the brain. It is believed that the pain relief through acupuncture and placebos is mediated through these compounds.
- Deficiency of ADH causes diabetes insipidus, a disorder characterized by excretion of large volumes of dilute urine.
- Thyroid hormones directly influence $Na^+ - K^+$ ATP pump which consumes a major share of cellular ATP. Obesity in some individuals is attributed to decreased energy utilization (heat production) due to diminished $Na^+ - K^+$ ATPase activity.

Abnormalities of thyroid function

Among the endocrine glands, ***thyroid is the most susceptible for hypo- or hyperfunction***.

Three abnormalities associated with thyroid functions are known.

Goiter : Any abnormal increase in the size of the thyroid gland is known as goiter. Enlargement of thyroid gland is mostly to compensate the decreased synthesis of thyroid hormones and is associated with ***elevated TSH***. Goiter is primarily due to a failure in the autoregulation of T_3 and T_4 synthesis. This may be ***caused by deficiency or excess of iodide***.

Simple endemic goiter : This is due to iodine deficiency in the diet. It is mostly found in the geographical regions away from sea coast where the water and soil are low in iodine content. Consumption of iodized salt is advocated to overcome the problem of endemic goiter. In certain cases, administration of thyroid hormone is also employed.

Hyperthyroidism : This is also known as ***thyrotoxicosis*** and is associated with overproduction of thyroid hormones. Hyperthyroidism is characterized by increased metabolic rate (***higher BMR***) nervousness, irritability, anxiety, rapid heart rate, loss of weight despite increased appetite, weakness, diarrhea, sweating, sensitivity to heat and often protrusion of eyeballs (exopthalmos).

Hyperthyroidism is caused by ***Grave's disease*** (particularly in the developed countries) or due to increased intake of thyroid hormones. Grave's disease is due to elevated ***thyroid stimulating IgG*** also known as long acting thyroid stimulator (LATS) which activates TSH and, thereby, increases thyroid hormonal production.

Hypothyroidism : This is due to an impairment in the function of thyroid gland that often causes decreased circulatory levels of T_3 and T_4. Disorders of pituitary or hypothalamus also contribute to hypothyroidism. Women are more susceptible than men. Hypothyroidism is characterized by ***reduced BMR***, slow heart rate, weight gain, sluggish behaviour, constipation, sensitivity to cold, dry skin etc.

Hypothyroidism in children is associated with physical and mental retardation, collectively known as ***cretinism***. Early diagnosis and proper treatment are essential. Hypothyroidism in adult causes ***myxoedema,*** characterized by bagginess under the eyes, puffiness of face, slowness in physical and mental activities.

Thyroid hormonal administration is employed to treat hypothyroidism.

Laboratory diagnosis of thyroid function

Measurement of ***basal metabolic rate*** (***BMR***) was once used to reflect thyroid activity. The estimation of serum ***protein bound iodine*** (***PBI***), representing the circulating thyroid hormones, was employed for a long time to assess thyroid function. The normal serum PBI concentration is 3-8 μg/100 ml.

Hypothyroidism is associated with decreased PBI and hyperthyroidism with increased PBI.

In recent years, more sensitive and reliable tests have been developed to assess thyroid activity. The concentration of free T_3 and T_4, and ***TSH*** are measured (by RIA or ELISA) and their serum normal concentrations are

Free triiodothyronine (T_3)	—	80–220 ng/dl
Free thyroxine (T_4)	—	0.8–2.4 ng/dl
Total thyroxine (T_4)	—	5–12 μg/dl
Thyroid stimulating hormone (TSH)	—	5–10 μU/ml

Radioactive iodine uptake (RAIU) and scanning of thyroid gland are also used for diagnosis.

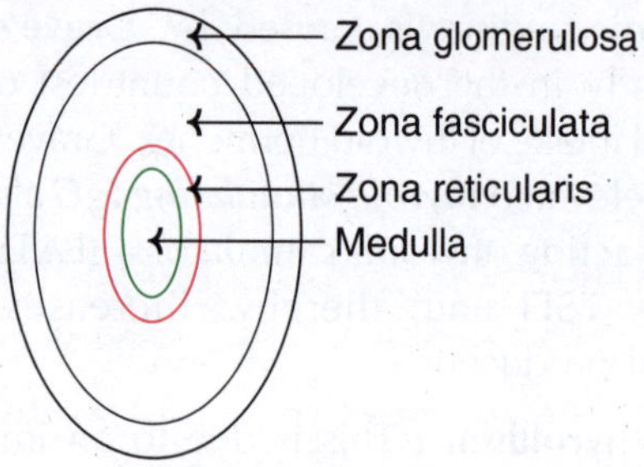

Fig.19.10 : Adrenal gland with zones (3) and medulla.

Thyroid activity and serum cholesterol

Serum cholesterol level is ***increased in hypothyroidism*** and decreased in hyperthyroidism. Unfortunately, cholesterol estimation will be of no value in the assessment of thyroid function. This is due to the fact that serum cholesterol level is elevated in many other disorders (diabetes, obstructive jaundice, nephrotic syndrome etc.). However, ***cholesterol estimation may be utilized for monitoring thyroid therapy***.

HORMONES OF ADRENAL CORTEX

The adrenal glands are two small organs (each weighing about 10 g), located above the kidneys. Each adrenal consists of two distinct tissues—an outer cortex (with 3 zones) and inner medulla (***Fig.19.10***).

As many as 50 steroid hormones (namely adrenocorticosteroids), produced by adrenal cortex, have been identified. However, only a few of them possess biological activity.

Adrenocorticosteroids are classified into three groups according to their dominant biological action. However, there is some overlap in their functions.

1. **Glucocorticoids :** These are 21-carbon steroids, produced mostly by zona fasciculata. They affect glucose (hence the name), amino acid and fat metabolism in a manner that is opposite to the action of insulin. ***Cortisol*** (also known as hydrocortisone) is the most important glucocorticoid in humans. Corticosterone is predominant in rats.

2. **Mineralocorticoids :** These are also 21-carbon containing steroids produced by zona glomerulosa. They regulate water and electrolyte balance. ***Aldosterone*** is the most prominent mineralocorticoid.

3. **Androgens and estrogens :** The innermost adrenal cortex zona reticularis produces small quantities of androgens (19-carbon) and ***estrogens*** (18-carbon). These hormones affecting ***sexual development*** and functions are mostly produced by gonads. Dehydroepiandrosterone—a precursor for androgens—is synthesized in adrenal cortex.

Synthesis of adrenocorticosteroids

Cholesterol undergoes cleavage with an elimination of a 6-carbon fragment to form pregnenolone. ***Pregnenolone is the common precursor for the synthesis of all steroid hormones.***

Conversion of cholesterol to pregnenolone is catalysed by cytochrome P_{450} side chain cleavage enzyme. This reaction is promoted by ACTH. The enzymes—hydroxylases, dehydrogenases/ isomerases and lyases associated with mitochondria or endoplasmic reticulum—are responsible for the synthesis of steroid hormones. The metabolic pathway for the formation of major adrenocorticosteroids is given in ***Fig.19.11***.

Biochemical functions of adrenocorticosteroids

1. **Glucocorticoid hormones :** The important glucocorticoids are—***cortisol***, ***cortisone*** and ***corticosterone***. They bring about several biochemical functions in the body.

(a) **Effects on carbohydrate metabolism :** Glucocorticoids ***promote*** the synthesis of glucose (***gluconeogenesis***). This is brought about by increasing the substrates (particularly amino acids) and enhancing the synthesis of phosphoenolpyruvate carboxy kinase, the rate limiting enzyme in gluconeogenesis.

The overall influence of glucocorticoids on carbohydrate metabolism is to increase blood glucose concentration. The biological actions of glucocorticoids generally oppose that of insulin.

(b) **Effects on lipid metabolism :** Glucocorticoids ***increase the circulating free fatty acids***. This is caused by two mechanisms.

(i) Increased breakdown of storage triacylglycerol (lipolysis) in adipose tissue.

(ii) Reduced utilization of plasma free fatty acids for the synthesis of triacylglycerols.

(c) **Effects on protein and nucleic acid metabolism :** Glucortiocoids exhibit both

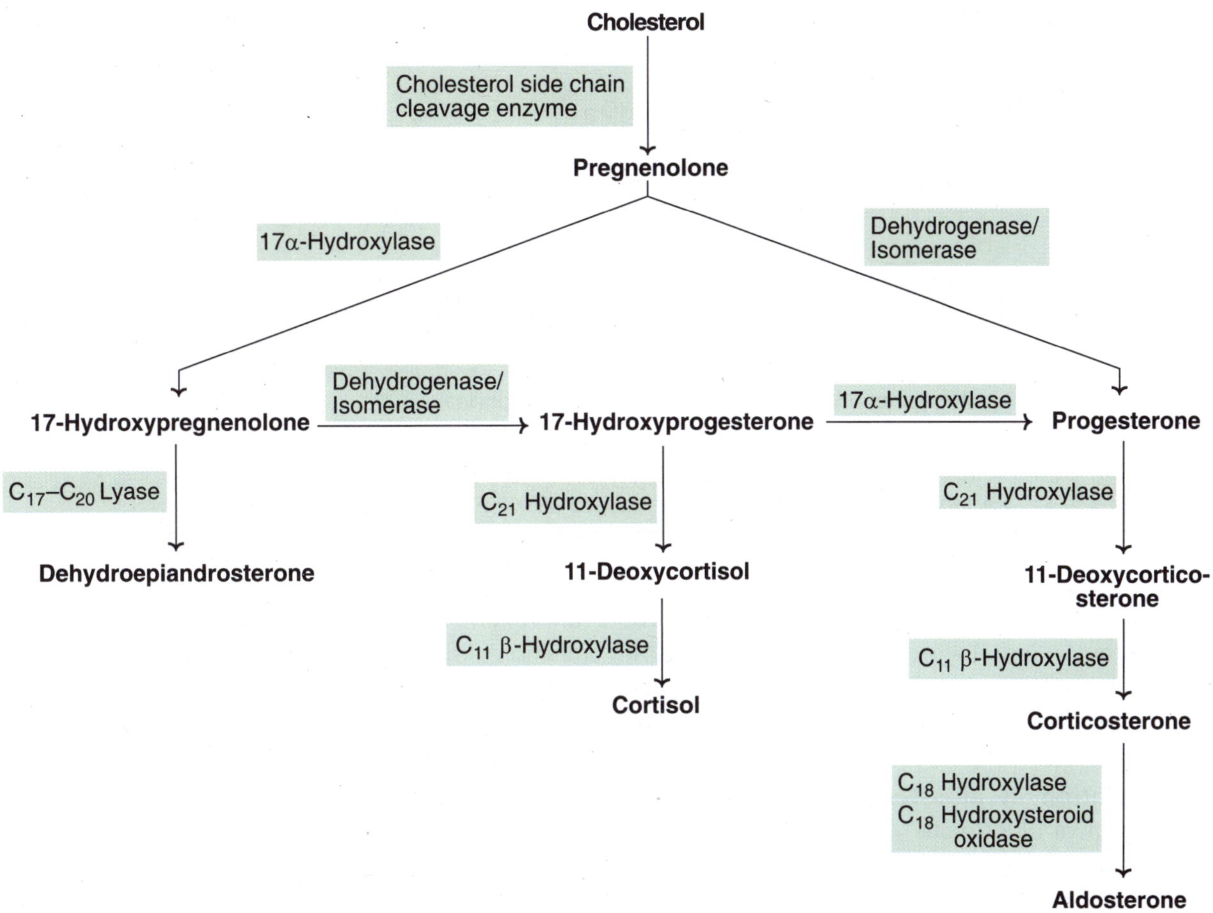

Fig.19.11 : *Biosynthesis of major adrenocorticosteroids.*

catabolic and anabolic effects on protein and nucleic acid metabolism. They promote transcription (RNA synthesis) and protein biosynthesis in liver. These anabolic effects of glucocorticoids are caused by the stimulation of specific genes.

Glucocorticoids (particularly at high concentration) cause catabolic effects in extrahepatic tissues (e.g. muscle, adipose tissue, bone etc.). This results in enhanced degradation of proteins.

(d) **Effects on water and electrolyte metabolism :** The influence of glucocorticoids on water metabolism is ***mediated through antidiuretic hormone (ADH)***. Deficiency of glucocorticoids causes increased production of ADH. ADH decreases glomerular filtration rate causing water retention in the body.

(e) **Effects on the immune system :** Glucocorticoids (particularly cortisol), in high doses, suppress the host immune response. The steroid hormones act at different levels—damaging lymphocytes, impairment of antibody synthesis, suppression of inflammatory response etc.

(f) **Other physiological effects of glucocorticoids :** Glucocorticoids are involved in several physiological functions.

(i) ***Stimulate the fight and flight response*** (to face sudden emergencies) of catecholamines.

(ii) Increase the production of gastric HCl and pepsinogen.

(iii) Inhibit the bone formation, hence the subjects are at a risk for osteoporosis.

Mechanism of action of glucocorticoids : Glucocorticoids bind to specific receptors on the target cells and bring about the action. These hormones mostly act at the transcription level and control the protein synthesis.

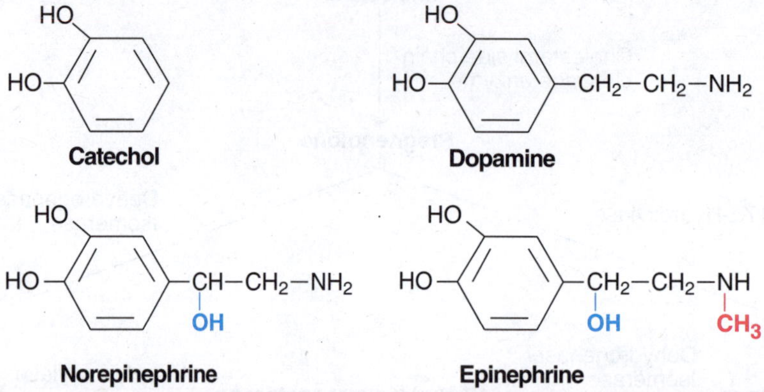

***Fig.19.12 :** Catecholamines (dopamine, norepinephrine and epinephrine) produced by adrenal medulla (Refer* **Fig. 15.15** *for biosynthesis).*

2. **Mineralocorticoid hormones :** The most active and potent mineralocorticoid is ***aldosterone***. It promotes Na^+ reabsorption at the distal convoluted tubules of kidney. Na^+ retention is accompanied by corresponding excretion of K^+, H^+ and NH_4^+ ions.

Regulation of aldosterone synthesis : The production of aldosterone is regulated by different mechanisms. These include renin-angiotensin, potassium, sodium and ACTH.

Mechanism of aldosterone action : Aldosterone acts like other steroid hormones. It binds with specific receptors on the target tissue and promotes transcription and translation.

Metabolism of adrenocorticosteroids : The steroid hormones are metabolized in the liver and excreted in urine as conjugates of glucuronides or sulfates.

The urine contains mainly two steroids—17-hydroxysteroids and 17-ketosteroids—derived from the metabolism of glucocorticoids and mineralocorticoids. Androgens synthesized by gonads also contribute to the formation of ***17-ketosteroids.***

Urinary 17-ketosteroids estimated in the laboratory are expressed in terms of dehydroepiandrosterone and their normal excretion is in the range of 0.2–2.0 mg/day.

Abnormalities of adrenocortical function

Addison's disease : ***Impairment in adrenocortical function results in Addison's disease.*** This disorder is characterized by decreased blood glucose level (hypoglycemia), loss of weight, loss of appetite (anorexia), muscle weakness, impaired cardiac function, low blood pressure, decreased Na^+ and increased K^+ level in serum, increased susceptibility to stress etc.

Cushing's syndrome : ***Hyperfunction of adrenal cortex may*** be due to long term pharmacological use of steroids or tumor of adrenal cortex or tumor of pituitary. Cushing's syndrome is characterized by hyperglycemia (due to increased gluconeogenesis), fatigue, muscle wasting, edema, osteoporosis, negative nitrogen balance, hypertension, moon-face etc.

Assessment of adrenocortical function

The adrenocortical function can be assessed by measuring plasma cortisol (5-15 μg/dl at 9.00 AM), plasma ACTH, urinary 17-ketosteroids etc.

HORMONES OF ADRENAL MEDULLA

Adrenal medulla is an extension of sympathetic nervous system. It produces two important hormones—***epinephrine*** (formerly adrenaline) and ***norepinephrine*** (formerly noradrenaline). Both these hormones are catecholamines since they are amine derivatives of ***catechol nucleus*** (dihydroxylated phenyl ring). Epinephrine is a methyl derivative of norepinephrine. Dopamine is another catecholamine, produced as an intermediate during the synthesis of epinephrine. Norepinephrine and dopamine are important neurotransmitters in the brain and autonomic nervous system. The structures of the three catecholamines are given in ***Fig.19.12.***

Synthesis of catecholamines

The amino acid tyrosine is the precursor for the synthesis of catecholamines. The pathway is described under tyrosine metabolism (***Chapter 15, Fig.15.15***). Catecholamines are produced in response to fight, fright and flight. These include the emergencies like shock, cold, fatigue, emotional conditions like anger etc.

Biochemical functions of catecholamines

Catecholamines cause diversified biochemical effects on the body. The ultimate goal of their action is to mobilize energy resources and prepare the individuals ***to meet emergencies*** (e.g. shock, cold, low blood glucose etc.).

1. **Effects on carbohydrate metabolism :** Epinephrine and norepinephrine in general ***increase the degradation of glycogen*** (glycogenolysis), synthesis of glucose (gluconeogenesis) and decrease glycogen formation (glycogenesis).

The overall effect of catecholamines is to ***elevate blood glucose levels*** and make it available for the brain and other tissues to meet the emergencies.

2. **Effects on lipid metabolism :** Both epinephrine and norepinephrine ***enhance the breakdown of triacylglycerols*** (lipolysis) in adipose tissue. This causes increase in the free fatty acids in the circulation which are effectively utilized by the heart and muscle as fuel source.

The metabolic effects of catecholamines are mostly related to the increase in adenylate cyclase activity causing elevation in cyclic AMP levels (refer carbohydrate and lipid metabolisms for more details).

3. **Effects on physiological functions :** In general, catecholamines (most predominantly epinephrine) ***increase cardiac output, blood pressure and oxygen consumption***. They cause smooth muscle relaxation in bronchi, gastrointestinal tract and the blood vessels supplying skeletal muscle. On the other hand, catecholamines stimulate smooth muscle contraction of the blood vessels supplying skin and kidney. Platelet aggregation is inhibited by catecholamines.

Metabolism of catecholamines

Catecholamines are rapidly inactivated and metabolized. The enzymes—catechol-O methyltransferase (COMT) and ***monoamine oxidase (MAO)***, found in many tissues act on catecholamines. The metabolic products metanephrine and ***vanillylmandelic acid (VMA)*** are excreted in urine.

Abnormalities of catecholamine production

Pheochromocytomas : These are the tumors of adrenal medulla. The diagnosis of pheochromocytoma is possible only when there is an excessive production of epinephrine and norepinephrine that causes severe hypertension. In the individuals affected by this disorder, the ratio of norepinephrine to epinephrine is increased. The measurement of ***urinary VMA*** (normal <8 mg/day) is helpful in the ***diagnosis*** of pheochromocytomas.

HORMONES OF GONADS

The gonads (testes in males, ovaries in females) perform closely related dual functions.

1. Synthesize ***sex hormones***;
2. Produce germ cells.

The steroid sex hormones are responsible for growth, development, maintenance and regulation of reproductive system. Sex hormones are essentially required for the development of germ cells.

The ***sex hormones*** are categorized into three groups

1. **Androgens** or ***male sex hormones*** which are C-19 steroids.

2. **Estrogens** or ***female sex hormones*** which are C-18 steroids. Ring A of steroid nucleus is phenolic in nature and is devoid of C-19 methyl group.

3. **Progesterone** is a C-21 steroid produced during the luteal phase of menstrual cycle and also during pregnancy.

ANDROGENS

The male sex hormones or androgens are produced by the Leydig cells of the testes and to a minor extent by the adrenal glands in both the sexes. Ovaries also produce small amounts of androgens.

Biosynthesis of androgens

Cholesterol is the precursor for the synthesis of androgens. It is first converted to pregnenolone which then forms androstenedione by two pathways—either through progesterone or through 17-hydroxypregnenolone. Testosterone is produced from androstenedione (***Fig.19.13***). The production of androgens is under the control of LH and FSH.

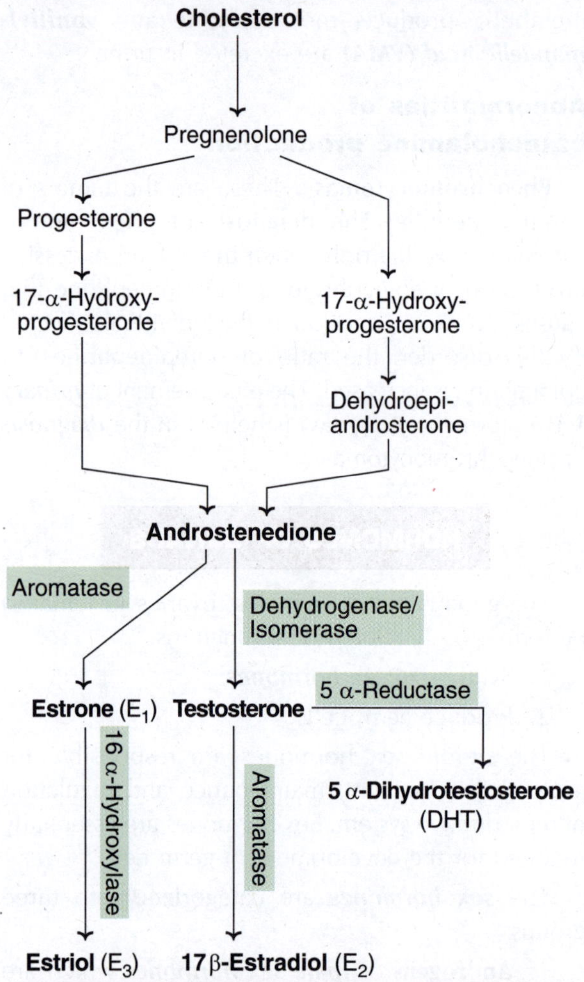

***Fig.19.13 :** Biosynthesis of steroid sex hormones from cholesterol*
*(**Note :** Male and female sex hormones are given together).*

Active form of androgen : The primary product of testes is ***testosterone***. However, the active hormone in many tissues is not testosterone but its metabolite ***dihydrotestosterone*** (DHT). Testosterone, on reduction by the enzyme 5 α-reductase, forms DHT. This conversion mostly occurs in the peripheral tissues. Some workers consider testosterone as a prohormone and dihydrotestosterone, the more potent form as the hormone.

Physiological and biochemical functions of androgens

1. **Sex related physiological functions :** The androgens, primarily DHT and testosterone, influence :

   (a) Growth, development and maintenance of male reproductive organs.

   (b) Sexual differentiation and secondary sexual characteristics.

   (c) Spermatogenesis.

   (d) Male pattern of aggressive behavior.

2. **Biochemical functions :** Many specific biochemical effects of androgens that ultimately influence the physiological functions stated above are identified. Androgens are anabolic in nature.

   (a) **Effects on protein metabolism :** Androgens promote RNA synthesis (transcription) and protein synthesis (translation). Androgens cause positive nitrogen balance and increase the muscle mass.

   (b) **Effects on carbohydrate and fat metabolisms :** Androgens increase glycolysis fatty acid synthesis and citric acid cycle.

   (c) **Effects on mineral metabolism :** Androgens promote mineral deposition and bone growth before the closure of epiphyseal cartilage.

ESTROGENS

Estrogens are predominantly ovarian hormones, synthesized by the follicles and corpus luteum of ovary. These hormones are responsible for maintenance of menstrual cycle and reproductive process in women.

Synthesis of estrogens

Estrogen synthesis occurs from the precursor cholesterol (***Fig.19.13***). Estrogens are produced by aromatization (formation of aromatic ring) of androgens. The ovary produces estradiol (E_2) and estrone (E_1) while the placenta synthesizes these two steroid hormones and estriol (E_3). The synthesis of estrogens is under the control of LH and FSH.

Physiological and biochemical functions of estrogens

1. **Sex-related physiological functions :** The estrogens are primarily concerned with

   (a) Growth, development and maintenance of female reproductive organs.

   (b) Maintenance of menstrual cycles.

   (c) Development of female sexual characteristics.

2. **Biochemical functions :** Estrogens are involved in many metabolic functions.

Medical Concepts/ Clinical Correlates

- **Catecholamines are produced in response to fight, fright and flight. The ultimate goal of catecholamine function is to mobilize energy resources and prepare the individual to meet emergencies such as shock, cold, fatigue, anger etc.**
- **Pheochromocytomas are the tumors of adrenal medulla, characterized by excessive production of epinephrine and norepinephrine, associated with severe hypertension.**
- **The low incidence of atherosclerosis and coronary heart disease in the women during reproductive age is due to estrogens.**

(a) **Lipogenic effect :** Estrogens increase lipogenesis in adipose tissue and, for this reason, women have relatively more fat (about 5%) than men.

(b) **Hypocholesterolemic effect :** Estrogens lower the plasma total cholesterol. The LDL fraction of lipoproteins is decreased while the HDL fraction is increased. This explains the low incidence of atherosclerosis and coronary heart diseases in the women during reproductive age.

(c) **Anabolic effect :** Estrogens in general promote transcription and translation. The synthesis of many proteins in liver is elevated e.g. transferrin, ceruloplasmin.

(d) **Effect on bone growth :** Estrogens like androgens promote calcification and bone growth. It is believed that decalcification of bone in the postmenopausal women leading to osteoporosis is due to lack of estrogens.

(e) **Effect on transhydrogenase :** Transhydrogenase is an enzyme activated by estrogen. It is capable of ***transferring reducing equivalents from NADPH to NAD$^+$***. The NADH so formed can be oxidized. It is explained that in the women ***after menopause, due to deficiency of estrogens***, the transhydrogenase activity is low. This results in the diversion of NADPH towards lipogenesis—causing ***obesity***.

PROGESTERONE

Progesterone is synthesized and secreted by corpus luteum and placenta. Progesterone, as such, is an intermediate in the formation of steroid hormones from cholesterol (***See Fig.19.13***). LH controls the production of progesterone.

Biochemical functions of progesterone

1. Progesterone is essentially required for the implantation of fertilized ovum and maintenance of pregnancy.

2. It promotes the growth of glandular tissue in uterus and mammary gland.

3. Progesterone increases the body temperature by 0.5–1.5 °F. The exact mechanism of this thermogenic effect is not clearly known. The measurement of temperature was used as an indicator for ovulation.

THE MENSTRUAL CYCLE

The occurrence of menstrual cycle is a good example of coordination among the hormonal functions. In humans, the menstrual cycle is under the control of FSH, LH, estrogens and progesterone. The cycle normally varies between 25 and 35 days in length, with a mean of 28 days. The menstrual cycle can be divided into two phases—follicular phase and luteal phase (***Fig.19.14***).

1. **Follicular phase :** Follicular stimulating hormone (FSH) causes the ***development and maturation of ovarian follicles***. As the follicle enlarges, estradiol progessively rises and reaches its peak value 24 hours before LH and FSH attain their respective maximum levels. LH surge or peak initiates ovulation-release of ovum from the ruptured follicles. The levels of progesterone are low during follicular phase.

2. **Luteal phase :** After the ovulation occurs, the ruptured follicles form corpus luteum and start producing progesterone and estradiol. The predominant hormone of luteal phase is ***progesterone*** which ***prepares the endometrium of uterus for implantation of the fertilized ovum***. LH maintains the corpus for a few days. In the absence of

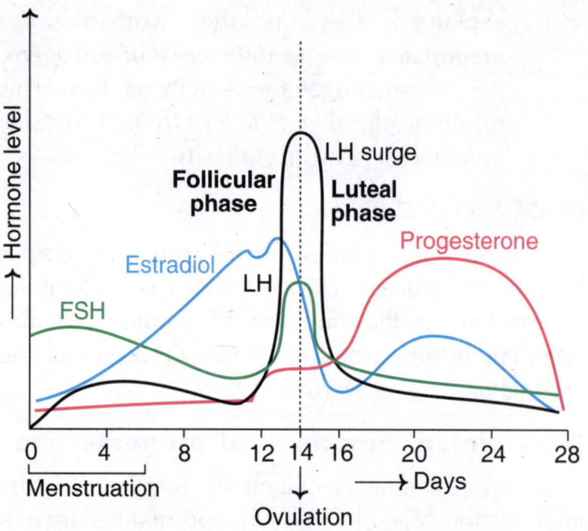

Fig.19.14 : Hormonal pattern in women during mestrual cycle (FSH–Follicle stimulating hormone; LH–Luteinizing hormone).

implantation, the corpus luteum regresses and sheds endometrium causing menstruation. And another new cycle begins.

MENOPAUSE

The ***menstrual cycles*** which begin in the women after puberty, continue till the age of 45-50 years. The cycles ***cease*** around this age which coincides with the loss of ovarian function. The progesterone and estrogen levels are very low in these women. However, the concentration of LH and FSH are elevated due to lack of feedback inhibition by estrogens.

Postmenopausal women are susceptible to two complications associated with insufficient levels of sex hormones.

1. **Atrophy of secondary sex tissues :** Mainly the epithelial tissue of vagina and lower urinary tract.

2. **Osteoporosis :** Decreased density of bones and increased susceptibility to fractures.

GASTROINTESTINAL HORMONES

The digestion and absorption of nutrients (***Chapter 8***) is a complicated process which is regulated by the autonomic nervous system. This occurs in association with peptide hormones of gastrointestinal tract (GIT).

The specialized cells lining the GIT are responsible for the production of GIT hormones. Hence GIT may be considered as the largest mass of cells that secrete hormones. A large number of GIT hormones have been identified. However, only four GIT hormones have been well characterised.

1. **Gastrin :** This hormone contains 17 amino acids and is produced by gastric mucosa. It ***stimulates the secretion of gastric HCl and pepsinogen*** (proenzyme of pepsin). The release of gastrin is stimulated by vagus nerve of stomach and partially digested proteins. HCl and certain other hormones inhibit gastrin release.

2. **Secretin :** It resembles glucagon in many ways. Secretin is synthesized by the mucosa of the upper small intestine. It is released in response to the presence of HCl in chyme in the duodenum which is passed on from the stomach. Secretin ***stimulates pancreatic cells to produce bicarbonate (HCO_3^-)*** in order to neutralize HCl.

3. **Cholecystokinin (CCK) :** It is produced by the upper part of small intestine. The secretion of CCK is stimulated by the products of protein and lipid digestion, namely peptides, amino acids, mono- or diacylglycerols, fatty acids and glycerol.

4. **Gastric inhibitory peptide (GIP) :** It contains 43 amino acids and is produced by duodenal mucosa. The release of GIP is stimulated by the presence of glucose in the gut. The most important function of GIP is to stimulate the release of insulin from pancreas.

GIT hormones show certain structural relations and may be considered under two families.

(a) **Gastrin family :** Some of the C-terminal amino acids are identical. This family includes gastrin and CCK.

(b) **Secretin family :** Secretin, GIP and glucagon are structurally related, hence may be considered under this family.

Besides the hormones described above, several other hormones (in hundreds!) from the GIT have been identified. These hormones are often known as ***candidate hormones,*** since their biological functions are yet to be precisely identified. The candidate hormones include ***vasoactive intestinal peptide*** (VIP), ***motilin, enteroglucagon, substance P, neurotensin, somatostatin*** and ***enkephalins***.

Mechanism of action of GIT hormones

Many of the GIT hormones have receptor sites specific for their action. At least two distinct mechanisms have been identified through which these hormones act.

1. ***Production of cAMP*** through the activation of adenylate cyclase e.g. secretin, VIP etc.

2. ***Stimulation of intracellular Ca^{2+}*** usually mediated through the metabolism of phosphatidylinositol e.g. gastrin, CCK.

Both these mechanisms ultimately influence the enzyme secretions/other biological effects.

Other hormones

Besides the hormones discussed above, there are a few other important hormones which are not referred to in this chapter. ***Insulin and glucagon*** are described under diabetes mellitus (***Chapter 36***) while ***parathyroid hormone*** and ***calcitonin*** are discussed under calcium metabolism (***Chapter 18***) These hormones are not given here to avoid repetition.

SUMMARY

1. *Hormones are the organic substances, produced in minute quantities by specific tissues (endocrine glands) and secreted into the blood stream to control the biological activities in the target cells. They may be regarded as the chemical massengers involved in the regulation and coordination of body functions.*
2. *Hypothalamus is the master coordinator of hormonal action as it liberates certain releasing factors or hormones (TRH, CRH, GRH, GRIH) that stimulate or inhibit the corresponding trophic hormones from the anterior pituitary.*
3. *Thyroid gland produces two principal hormones—thyroxine (T_4) and triiodothyronine (T_3)—which are primarily concerned with the regulation of the metabolic activity of the body. Goiter is a disorder caused by enlargement of thyroid gland and is mainly due to iodine deficiency in the diet.*
4. *Adrenal cortex synthesizes glucocorticoids (e.g. cortisol) that influence glucose, amino acid and fat metabolism, and mineralocorticoids (e.g. aldosterone) that regulate water and electrolyte balance.*
5. *Adrenal medulla produces two important hormones—epinephrine and norepinephrine (catecholamines). They influence diversified biochemical functions with an ultimate goal to mobilize energy resources and prepare the individual to meet emergencies (shock, anger, fatigue etc.)*
6. *The steroid sex hormones, primarily androgens in males and estrogens in females, are, respectively synthesized by the testes and ovaries. These hormones are responsible for growth, development, maintenance and regulation of reproductive system in either sex.*

THE READER MUST REFER CHAPTER 36 FOR INSULIN AND GLUCAGON; CHAPTER 18 FOR PARATHYROID HORMONE AND CALCITONIN

THE READER MUST REFER APPENDIX D (Page Nos : 422 - 431) FOR RELATED CLINICAL CASE STUDIES (Nos : 27, 28, 34, 35)

ORGAN FUNCTION TESTS

20

Each organ of the body has to perform its biochemical functions to keep the body, as a whole, in a healthy state. This is possible only when the cells of the organ are intact in structure and function. Any abnormality in the tissue, caused by exogenous or endogenous factors, will seriously impair the organ function which, in turn, influences the health of the organism.

Based on the functional capabilities of the organs, ***specific biochemical investigations*** have been developed ***in the laboratory***, to assess their function. In this chapter, the biochemical investigations to measure the functioning of liver, kidney and stomach are discussed. The tests to evaluate the function of endocrine organs are discussed elsewhere (***Chapter 19***).

LIVER FUNCTION TESTS

Liver performs several diversified functions. It is the central organ of body's metabolism.

Functions of liver

1. **Metabolic functions :** Liver actively participates in carbohydrate, lipid, protein, mineral and vitamin metabolisms.

2. **Excretory functions :** Bile pigments, bile salts and cholesterol are excreted in the bile into intestine.

3. **Protective functions and detoxification :** Kupffer cells of liver perform phagocytosis to eliminate foreign compounds. Ammonia is detoxified to urea. Liver is responsible for the metabolism of xenobiotics (detoxification).

4. **Hematological functions :** Liver participates in the formation of blood (particularly in the embryo), synthesis of plasma proteins (including blood clotting factors), and destruction of erythrocytes.

5. **Storage functions :** Glycogen, vitamins A, D and B_{12} and trace element iron are stored in liver.

The above list is an oversimplification and inconclusive with regard to the role of liver in the body.

Tests to assess liver function

The liver function tests (***LFT***) are the biochemical investigations to assess the capacity of the liver to carry out any of the functions it performs. LFT will ***help to detect the abnormalities and the extent of liver damage***.

The major liver function tests commonly employed in the laboratory are described.

BILIRUBIN

Bilirubin is a bile pigment, and is the excretory end product of heme degradation. It is conjugated in the liver to form bilirubin diglucuronide, and excreted in bile. The details of bilirubin metabolism are discussed elsewhere (***Chapter 10***).

Serum bilirubin

The normal concentration of ***serum bilirubin*** is in the range of ***0.2-1.0 mg/dl***. Of this, the conjugated bilirubin (diglucuronide 75%; monoglucuronide 25%) is about 0.2-0.4 mg/dl, while the unconjugated bilirubin is 0.2-0.6 mg/dl.

van den Bergh reaction

This is a specific reaction ***to identify*** the increase in ***serum bilirubin*** (above the reference level). ***Normal serum gives a negative van den Bergh reaction.***

Direct and indirect reactions : Bilirubin as such is insoluble in water while the conjugated bilirubin is soluble. van den Bergh reagent reacts with ***conjugated bilirubin*** and gives a purple colour immediately (normally within 30 seconds). This is referred to as a ***direct positive*** van den Bergh reaction. ***Addition of methanol*** (or alcohol) ***dissolves the unconjugated bilirubin*** which then gives the van den Bergh reaction (normally within 30 minutes) positive and this is referred to as ***indirect positive***. If the serum contains both unconjugated and conjugated bilirubin in high concentration, the purple colour is produced immediately (direct positive) which is further intensified by the addition of alcohol (indirect positive). This type of reaction is known as ***biphasic***.

van den Bergh reaction and jaundice : This reaction is highly useful in understanding the nature of jaundice. This is due to the fact that the type of jaundice is characterized by increased serum concentration of unconjugated bilirubin (hemolytic), conjugated bilirubin (obstructive) or both of them (hepatic). Therefore, the response of van den Bergh reaction can differentiate the jaundice as follows

Indirect positive — ***Hemolytic jaundice***
Direct positive — ***Obstructive jaundice***
Biphasic — ***Hepatic jaundice***

Bilirubin in urine

The ***conjugated bilirubin***, being water soluble, is ***excreted in urine***. This is in contrast to unconjugated bilirubin which is not excreted. Bilirubin in urine can be detected by Fouchet's test or Gmelin's test.

SERUM ENZYMES DERIVED FROM LIVER

Liver cells contain several enzymes which may be released into the circulation in liver damage. Measurement of ***selected enzymes in serum is often used to assess the liver function***.

Transaminases or aminotransferases

The activities of two enzymes—namely serum glutamate pyruvate transaminase (SGPT; recently called as ***alanine transaminase***—ALT) and serum glutamate oxaloacetate transaminase (SGOT; recently known as ***aspartate transaminase***—AST) — are widely used to assess the liver function. ALT is a cytoplasmic enzyme while AST is found in both cytoplasm and mitochondria. The activity of these enzymes is low in normal serum (ALT 10-45 U/l; AST 10-50U/l). Serum ALT and AST are increased in liver damage. However, ***alanine transaminase*** is more sensitive and ***reliable for the assessment of LFT***.

Alkaline phosphatase

Alkaline phosphatase (ALP) is mainly derived from bone and liver (the cells lining the bile canaliculi). A rise in serum ALP (normal 3-13 KA units/dl), usually associated with elevated serum bilirubin is an indicator of biliary obstruction (***obstructive/ posthepatic jaundice***). ALP is also elevated in cirrhosis of liver and hepatic tumors.

γ-Glutamyl transpeptidase

This is a microsomal enzyme widely distributed in body tissues, including liver. Measurement of γ-glutamyl transpeptidase (***GGT***) activity provides a ***sensitive index to assess liver abnormality***. The activity of this enzyme almost parallels that of transaminases in hepatic damage. Serum GGT is highly elevated (normal ***5-50 U/l***) in ***biliary obstruction*** and ***alcoholism***. Further, several drugs (e.g. phenytoin) induce (liver synthesis) and increase this enzyme in circulation.

JAUNDICE

Jaundice (*French* : jaune—yellow) is characterized by ***yellow coloration of sclera (of eyes)*** and skin. This is due to the ***elevated serum bilirubin*** level, usually ***beyond 2 mg/dl*** (***normal <1 mg/dl***).

The metabolism of heme to produce bilirubin and its conjugated derivatives and the types of jaundice have already been described. The reader must refer this (***Chapter 10***) now. The biochemical changes and the related parameters for the differential diagnosis of the three types of jaundice (hemolytic, obstructive and hepatic) are given in ***Table 20.1***.

The pattern of rise in the serum alanine transaminase, aspartate transaminase and bilirubin in acute viral hepatitis is depicted in ***Fig.20.1***. It may be noted that the transaminase activities (more predominantly ALT) are elevated much before the bilirubin starts increasing.

Choice of liver functions tests

The choice of biochemical tests to measure liver functions mostly depends on the purpose

Medical Concepts/ Clinical Correlates

- **The impairment in the functions of any organ in the body will adversely influence the health of the organism. Organ function tests are the laboratory tools to biochemically evaluate the working of a given organ.**
- **Acute viral hepatitis is associated with elevated alanine transaminase (predominantly), aspartate transaminase and bilirubin.**
- **Increase in serum γ-glutamyl transpeptidase is observed in biliary obstruction and alcoholism.**

of the investigation. Frequently, a combination of laboratory investigations are employed in LFT. These include serum bilirubin (conjugated and unconjugated), alanine transaminase, aspartate transaminase, alkaline phosphatase, γ-glutamyl transpeptidase and proteins (albumin, globulins).

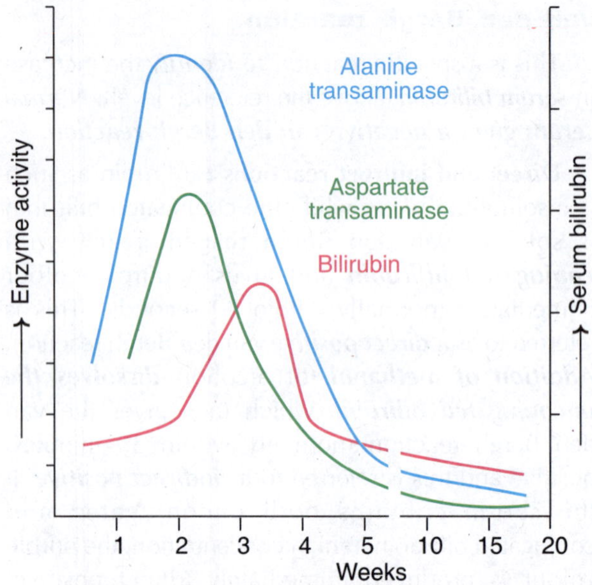

***Fig. 20.1 :** Pattern of rise in serum enzymes and bilirubin in viral hepatitis.*

KIDNEY (RENAL) FUNCTION TESTS

The kidneys are the vital organs of the body, performing the following major functions.

1. **Maintenance of homeostasis :** The kidneys are largely responsible for the regulation of ***water, electrolyte and acid-base balance*** in the body.

2. **Excretion of metabolic waste products :** The end products of protein and nucleic acid metabolism are eliminated from the body. These include ***urea, creatinine, creatine, uric acid***, sulfate and phosphate.

3. **Retention of substances vital to body :** The kidneys reabsorb and retain several substances of biochemical importance in the body e.g. ***glucose, amino acids*** etc.

4. **Hormonal functions :** The kidneys also function as endocrine organs by producing hormones. e.g., ***erythropoietin, calcitriol***.

TABLE 20.1 Biochemical changes for the differential diagnosis of three types of jaundice

Parameter	*Hemolytic jaundice (prehepatic jaundice)*	*Obstructive jaundice (post-hepatic jaundice)*	*Hepatic jaundice (intrahepatic jaundice)*
Serum bilirubin	Unconjugated bilirubin ↑	Conjugated bilirubin ↑	Both ↑
van den Bergh reaction	Indirect positive	Direct positive	Biphasic
Serum enzymes	ALT, AST and ALP →	ALP ↑↑, ALT and AST marginal ↑	ALT and AST ↑↑, ALP marginal ↑
Bilirubin in urine	Not excreted	Excreted	Excreted
Urobilinogen in urine	Excretion ↑	→ or ↓	→ or ↓

ALT : Alanine transaminase; AST : Aspartate transaminase; ALP : Alkaline phosphatase; ↑ : Increase; ↓ : Decrease; → : Normal.

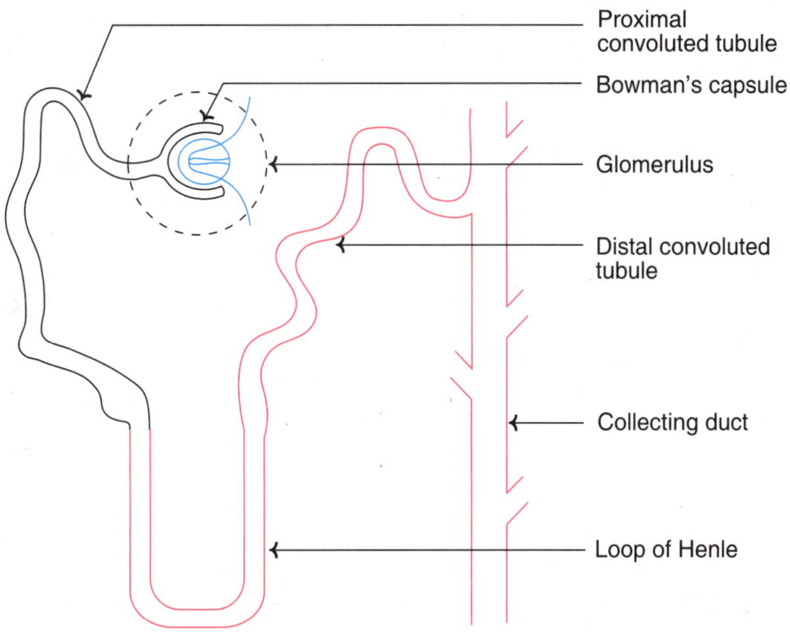

Fig.20.2 : Diagrammatic representation of a nephron.

The formation of urine

Nephron is the functional unit of kidney. Each kidney is composed of approximately one million nephrons. The structure of a nephron, as depicted in ***Fig.20.2***, consists of a ***Bowman's capsule*** (with blood capillaries), proximal convoluted tubule (PCT), loop of Henle, distal convoluted tubule (DCT) and collecting tubule.

The blood supply to kidneys is relatively large. About 1200 ml of blood (650 ml plasma) passes through the kidneys, every minute. From this, about ***120-125 ml is filtered per minute by the kidneys*** and this is referred to as ***glomerular filtration rate*** (GFR). With a normal GFR (120-125 ml/min), the glomerular filtrate formed in an adult is about 175-180 litres per day, out of which only 1.5 litres is excreted as urine. Thus, more than 99% of the glomerular filtrate is reabsorbed by the kidneys.

The process of urine formation basically involves two steps—glomerular filtration and tubular reabsorption.

Tests to assess renal function

The kidney function tests may be divided into four groups.

1. **Glomerular function tests :** All the clearance tests (inulin, creatinine, urea) are included in this group.

2. **Tubular function tests :** Urine concentration or dilution test, urine acidfication test.

3. **Analysis of blood/serum :** Estimation of blood urea, serum creatinine, protein and electrolytes are often useful to assess renal function.

4. **Urine examination :** Simple routine examination of urine for volume, pH, specific gravity, osmolality and presence of certain abnormal constituents (proteins, blood, ketone bodies, glucose etc.) also helps, of course to a limited degree, to assess kidney functioning.

Some of the important rėnal function tests are discussed in the following pages.

CLEARANCE TESTS

The clearance tests, measuring the ***glomerular filtration rate (GFR)*** are the most useful in assessing the renal function.

Clearance, in general, is defined as the ***volume of plasma that would be completely cleared of a substance per minute***. In other words, ***clearance of a substance refers to the milliliters of plasma which contains the amount of that substance excreted by kidney per minute.*** Clearance (C), expressed as ml/minute, can be calculated by using the formula

$$C = \frac{U \times V}{P}$$

where U = Concentration of the substance in urine.

V = Volume of urine in ml excreted per minute.

P = Concentration of the substance in plasma.

Care should be taken to express the concentrations of plasma and urine in the same units (mmol/l or mg/dl).

The clearance of a given substance is determined by its mode of excretion. The maximum rate at which the plasma can be cleared of any substance is equal to the GFR. This can be easily calculated by measuring the clearance of a plasma compound which is freely filtered by the glomerulus and is neither absorbed nor secreted in the tubule. ***Inulin*** (a plant carbohydrate, composed of fructose units) and ^{51}Cr-EDTA satisfy this criteria. Inulin is intravenously administered ***to measure GFR***.

In practice, however, measurement of clearance for the substances already present in the blood is preferred. The two compounds, namely ***creatinine and urea***, are commonly employed for this purpose. ***Creatinine clearance (~145 ml/min) is marginally higher than the GFR*** as it is secreted by the tubules. On the other hand, ***urea clearance (~75 ml/min) is less than the GFR***, since it is partially reabsorbed by the tubules.

Creatinine clearance test

Creatinine is an excretory product derived from creatine phosphate (largely present in muscle). The ***excretion of creatinine is rather constant*** and is not influenced by body metabolism or dietary factors. As already stated, creatinine is filtered by the glomeruli and only marginally secreted by the tubules. The value of creatinine clearance is close to GFR, hence its measurement is a sensitive and good approach to assess the renal glomerular function. Creatinine clearance may be defined as ***the volume (ml) of plasma that would be completely cleared of creatinine per minute***.

Procedure : In the traditional method, creatinine content of a 24 hr urine collection and the plasma concentration in this period are estimated. The creatinine clearance (C) can be calculated as follows :

$$C = \frac{U \times V}{P}$$

where U = Urine concentration of creatinine

V = Urine output in ml/min (24 hr urine volume divided by $24 \leftrightarrow 60$)

P = Plasma concentration of creatinine.

As already stated, creatinine concentration in urine and plasma should be expressed in the same units (mg/dl or mmol/l).

Reference values : The normal range of creatinine clearance is around ***120-145 ml/min***. These values are slightly lower in women.

Diagnostic importance : A ***decrease in creatinine clearance value (<75% normal)*** serves as ***sensitive indicator of a decreased GFR***, due to renal damage. This test is useful for an early ***detection of impairment of kidney function***, often before the clinical manifestations are seen.

Urea clearance test

Urea is the end product of protein metabolism. After being filtered by the glomeruli, it is partially reabsorbed by the renal tubules. Hence, urea clearance is less than the GFR and, further, it is influenced by the protein content of the diet. For these reasons, ***urea clearance is not as sensitive as creatinine clearance*** for assessing renal function. Despite this fact, several laboratories traditionally use this test.

Urea clearance is defined as the volume (ml) of plasma that would be completely cleared of urea per minute. It is calculated by the formula

$$C_m = \frac{U \times V}{P}$$

where C_m = Maximum urea clearance

U = Urea concentration in urine (mg/ml)

V = Urine excreted per minute in ml

P = Urea concentration in plasma (mg/ml).

The above calculation is applicable if the ***output of urine is more than 2 ml per minute.*** This is referred to as ***maximum urea clearance***, and the normal value is around 75 ml/min.

Standard urea clearance : It is observed that the urea clearance drastically changes when the volume of urine is less than 2 ml/min. This is known as ***standard urea clearance (C_s)*** and the normal value is around 54 ml/min. It is calculated by a modified formula

$$C_s = \frac{U \times \sqrt{V}}{P}$$

Diagnostic importance : A urea clearance value below 75% of the normal is viewed seriously, since it is an indicator of renal damage. Blood urea

Medical Concepts/ Clinical Correlates

- A combination of laboratory investigations—instead of a single one—are commonly employed in assessing organ function. Kidney function can be accurately assessed by clearance tests, measuring glomerular filtration rate. A reduction in clearance reflects renal damage.
- Cystatin C is a more reliable and better biomarker of kidney function than creatinine.
- Zollinger-Ellison syndrome, a tumor of gastrin secreting cells of the pancreas, is associated with increased gastric HCl production.

level as such is found to increase only when the clearance falls below 50% normal. As already stated, creatinine clearance is better indicator of renal function.

Analysis of blood (or serum)

Estimation of ***serum creatinine*** and blood urea are often used to ***assess the overall kidney function***, although these tests are less sensitive than the clearance tests. Serum ***creatinine*** is a ***better indicator*** than urea in this regard. The diagnostic importance of urea and creatinine estimations are discussed elsewhere (***Refer Chapter 15***).

Cystatin C

Cystatin C is mainly used as a ***biomarker of kidney function***. Normally, serum concentration of cystatin C is in the range of ***0.8-1.2 mg/dl***.

Choice of renal function tests

In general, the assessment of kidney function starts with the routine urine examination, followed by serum creatinine and/or blood urea estimations and, finally, the specific tests to measure the tubular and glomerular functions (clearance tests).

GASTRIC FUNCTION TESTS

There are several tests for gastric function evaluation, some of the important ones are discussed hereunder.

Fractional test meal

This is rather old and not used these days. Fractional test meal (***FTM***) involves the collection of stomach contents by Ryle's tube in fasting. This is followed by a gastric stimulation, giving a test meal (rice gruel, black coffee etc.) The stomach contents are aspirated by ***Ryle's tube*** at different time periods (usually every 15 min for 2 hrs.) The samples are analysed for free and total acidity in the laboratory. The results are normally represented by a graph.

Pentagastrin stimulation test

Pentagastrin is a synthetic peptide which stimulates the gastric secretion in a manner similar to the natural gastrin.

Basal acid output (BAO) refers to the acid output (millimol per hour) under the basal conditions i.e. basal secretion.

Maximal acid output (MAO) represents the acid output (millimol per hour) after the gastric stimulation by pentagastrin i.e. maximum secretion.

In normal individuals, the BAO is 4-10 mmol/hr while the MAO is 20-50 mmol/hr.

Augmented histamine test meal

Histamine is a powerful stimulant of gastric secretion. The basal gastric secretion is collected for one hour. Histamine (0.04 mg/kg body weight) is administered subcutaneously and the gastric contents are aspirated for the next one hour (at 15 minute intervals). The acid content is measured in all these samples.

Abnormalities of gastric function

Increased gastric HCl secretion is found in ***Zollinger-Ellison syndrome*** (a tumor of gastrin secreting cells of the pancreas), chronic duodenal ulcer, gastric cell hyperplasia, excessive histamine production etc.

A ***decrease in gastric HCl is observed in gastritis, gastric carcinoma, pernicious anemia*** etc.

OTHER ORGAN FUNCTION TESTS

PANCREATIC FUNCTION TESTS

The pancreas is a specialized organ with exocrine and endocrine functions. The endocrine functions are discussed under the topic diabetes mellitus (***Chapter 36***).

The exocrine functions involve the synthesis of pancreatic juice containing several enzymes (for the digestion of foodstuffs) and bicarbonate. The major enzymes of pancreatic juice are trypsin, chymotrypsin, elastase, carboxypeptidase, amylase and lipase.

Pancreatic enzymes in serum : ***Serum amylase*** and ***lipase measurements*** are commonly employed to assess the pancreatic function. Both these enzyme activities are elevated in acute pancreatitis, obstruction in the intestine and/or pancreatic duct.

THYROID FUNCTION TESTS

Thyroid gland produces two principal hormones—thyroxine (T_4) and triiodothyronine which regulate the metabolic rate of the body. The laboratory tests employed for the diagnosis of thyroid function are described in the ***Chapter 19*** on ***hormones***.

SUMMARY

1. *The liver function can be evaluated by the tests based on its excretory function (serum bilirubin), serum enzymes (transaminases), metabolic capability (galactose tolerance test) and synthetic functions (prothrombin time).*
2. *Serum bilirubin (normal < 1mg/dl) is derived from heme degradation. van den Bergh reaction is a specific test to identify the increased serum bilirubin. Conjugated bilirubin gives a direct positive test while the unconjugated bilirubin gives an indirect positive test.*
3. *The serum enzymes—namely alanine transaminase (ALT), aspartate transaminase (AST), alkaline phosphatase (ALP) and γ-glutamyltranspeptidase (GGT)—are frequently used for LFT. Increase in the activities of these enzymes indicates an impairment in liver function.*
4. *Jaundice is due to elevated serum bilirubin level (>2 mg/dl). The three types of jaundice (hemolytic, obstructive and hepatic) can be differentially diagnosed by biochemical tests. Thus, unconjugated bilirubin (indirect positive) is increased in hemolytic jaundice, conjugated bilirubin (direct positive) in obstructive jaundice and both of them (biphasic) are increased in hepatic jaundice.*
5. *The renal (kidney) function is usually assessed by evaluating either the glomerular (clearance tests) or tubular function (urine concentration test). This is often guided by blood analysis (for urea, creatinine) and/or urine examination.*
6. *The clearance is defined as the volume of the plasma that would be completely cleared of a substance per minute. Inulin clearance represents glomerular filtration rate (GFR). Creatinine clearance and urea clearance tests are often used to assess renal function. A decrease in their clearance is an indication of renal damage.*

THE READER MUST REFER CHAPTER 19 FOR THYROID FUNCTION TESTS AND ADRENOCORTICAL FUNCTION TESTS.

THE READER MUST REFER APPENDIX D (Page Nos : 422 - 431) FOR RELATED CLINICAL CASE STUDIES (Nos : 18, 19, 20, 21, 23, 25)

WATER, ELECTROLYTE AND ACID-BASE BALANCE

21

The ***organism*** possesses tremendous capacity to survive against odds and ***maintain homeostasis***. This is particularly true with regard to water, electrolyte and acid-base status of the body. These three are interrelated, hence they are considered together for the discussion. Kidney actively participates in the regulation of water, electrolyte and acid-base balance. The general functions of kidney have already been described (***Chapter 20***).

WATER AND LIFE

Water is ***the solvent of life***. Undoubtedly, water is more important than any other single compound to life. It is involved in several body functions.

Functions of water

1. Water provides the ***aqueous medium*** to the organism which is ***essential for*** the various biochemical reactions to occur.

2. Water directly ***participates as a reactant*** in several metabolic reactions.

3. It serves as a vehicle for ***transport of solutes***.

4. Water is closely associated with the regulation of ***body temperature***.

Distribution of water

Water is the major body constituent. An adult human contains about 60% water (men 55-70%, women 45-60%). The women and obese individuals have relatively less water which is due to the higher content of stored fat in an anhydrous form.

A 70 kg normal man contains about 42 litres of water. This is distributed in intracellular (inside the cells 28l) and extracellular (outside the cells 14l) compartments, respectively known as ***intracellular fluid*** (ICF) and ***extracellular fluid*** (ECF). The ECF is further divided into interstitial fluid (10.5l) and plasma (3.5l). The distribution of water in an adult man is given in ***Table 21.1***.

WATER TURNOVER AND BALANCE

The body possesses tremendous capacity to regulate its water content. In a healthy individual, this is achieved by balancing the daily water intake and water output.

Water intake

Water is supplied to the body by ***exogenous*** and ***endogenous*** sources.

Exogenous water : Ingested water and beverages, water content of solid foods-constitute the exogenous source of water. Water intake is highly variable which may range from 0.5-5 litres. It largely depends on the social habits and climate. In general, people

TABLE 21.1 Distribution of water in an adult man, weighing 70 kg.

Compartment	% Body weight	Volume (l)
Total	60	42
Intracellular fluid (ICF)	40	28
Extracellular fluid (ECF)	20	14
Interstitial fluid	15	10.5
Plasma	5	3.5

living in hot climate drink more water. Ingestion of water is mainly controlled by a ***thirst centre*** located in the ***hypothalamus***.

Endogenous water : The ***metabolic water*** produced within the body is the endogenous water. This water (300-350 ml/day) is ***derived from the oxidation of foodstuffs***. It is estimated that 1 g each of carbohydrate, protein and fat, respectively, yield 0.6 ml, 0.4 ml and 1.1 ml of water. On an average, about 125 ml of water is generated for 1,000 Cal consumed by the body.

Water output

Water losses from the body are variable. There are four distinct routes for the elimination of water from the body—urine, skin, lungs and feces.

Urine : This is the ***major route*** for water loss from the body. In a healthy individual, the urine output is about ***1-2 l/day***.

Hormonal regulation of urine production : It is indeed surprising to know that about 180 litres of water is filtered by the glomeruli into the renal tubules everyday. However, most of this is reabsorbed and only 1-2 litres is excreted as urine. Water excretion by the kidney is tightly controlled by ***vasopressin*** also known as ***antidiuretic hormone (ADH)*** of the posterior pituitary gland. The secretion of ADH is regulated by the osmotic pressure of plasma. An increase in osmolality promotes ADH secretion that leads to an increased water reabsorption from the renal tubules (less urine output).

Diabetes insipidus is a disorder characterized by the ***deficiency of ADH*** which results in an ***increased loss of water from the body***.

Skin : Loss of water (450 ml/day) occurs through the body surface by perspiration. This is an unregulated process by the body which mostly depends on the atmospheric temperature and humidity.

Lungs : During respiration, some amount of water (about 400 ml/day) is lost through the expired air.

The loss of water by perspiration (via skin) and respiration (via lungs) is collectively referred to as ***insensible water loss***.

Feces : Most of the water entering the gastrointestinal tract is reabsorbed by the intestine. About 150 ml/day is lost through feces in a healthy individual. Fecal loss of water is tremendously increased in diarrhea.

A summary of the water intake and output in the body is depicted in ***Fig.21.1***. It may be noted that water balance of the body is regulated predominantly by controlling the urine output.

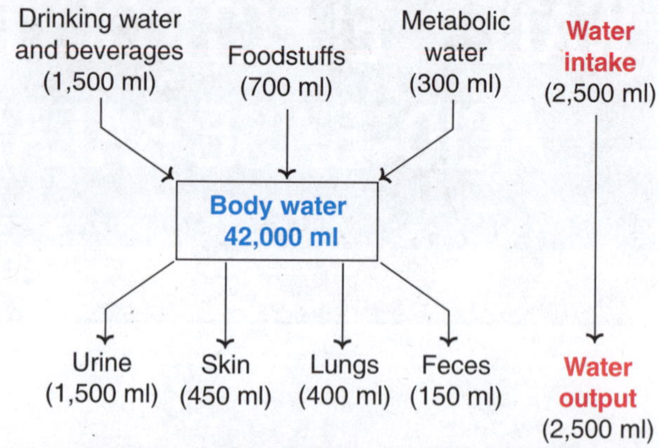

Fig.21.1 : *Water balance in the body, represented by daily intake and output (values are variable).*

The abnormalities associated with water balance —dehydration and overhydration—will be described, following a discussion on electrolyte balance.

ELECTROLYTE BALANCE

Electrolytes are the compounds which readily dissociate in solution and exist as ions i.e. positively and negatively charged particles. For instance, NaCl does not exist as such, but it exists as cation (Na^+) and anion (Cl^-). The concentration of ***electrolytes*** are ***expressed as milliequivalents (mEq/l)*** rather than milligrams.

Electrolyte composition of body fluids

Electrolytes are well distributed in the body fluids in order to maintain the osmotic equilibrium and water balance. The total concentration of cations and anions in each body compartment (ECF or ICF) is equal to maintain electrical neutrality.

There is a marked difference in the concentration of electrolytes (cations and anions) between the extracellular and intracellular fluids. ***Na^+*** is the principal ***extracellular cation*** while ***K^+*** is the ***intracellular cation***. This difference in the concentration is essential for the cell survival which is maintained by ***Na^+ – K^+ pump*** (for details, ***Refer Chapter 33***).

Osmolality of plasma

Osmolality is a measure of the solute particles present in the fluid medium. The osmolality of plasma is in the range of ***285-295 milliosmoles/kg***. Sodium and its associated anions make the largest contribution (~90%) to plasma osmolality. Osmolality is generally measured by osmometer.

For practical purposes, ***plasma osmolality*** can be computed from the concentrations (mmol/l) of Na^+, K^+, urea and glucose as follows

$$2(Na^+) + 2(K^+) + \text{Urea} + \text{Glucose}$$

The factor 2 is used for Na^+ and K^+ ions to account for the associated anion concentration (assuming complete ionization of the molecules). Since plasma Na^+ is the most predominant contributor to osmolality, the above calculation is further simplified as follows

$$\underset{(\text{mmol/kg})}{\textit{Plasma osmolality}} = 2 \times \underset{(\text{mmol/l})}{\textit{Plasma } Na^+}$$

Regulation of electrolyte balance

Electrolyte and water balance are regulated together, and the kidneys play a predominant role in this regard. The regulation is mostly achieved through the hormones aldosterone, ADH and renin-angiotensin.

Aldosterone : It is a ***mineralocorticoid*** produced by adrenal cortex. Aldosterone increases Na^+ reabsorption by the renal tubules at the expense of K^+ and H^+ ions. The net effect is the retention of Na^+ in the body.

Antidiuretic hormone (ADH) : An increase in the plasma osmolality (mostly due to Na^+) stimulates hypothalamus to release ADH. ADH effectively ***increases water reabsorption*** by renal tubules.

Renin-angiotensin : The ***secretion of aldosterone is controlled by renin-angiotensin system***. Decrease in the blood pressure (due to a fall in ECF volume) is sensed by juxtaglomerular apparatus of the nephron which secrete renin. Renin acts on angiotensinogen to produce angiotensin I. The latter is then converted to angiotensin II which stimulates the release of aldosterone.

The relation between renin, angiotensin and aldosterone in the regulation of Na^+ balance is depicted in ***Fig.21.2***. Aldosterone and ADH coordinate with each other to maintain the normal fluid and electrolyte balance.

Dehydration

Dehydration is a condition characterized by ***water depletion in the body***. It may be due to insufficient intake or excessive water loss or both. Dehydration is generally classified into two types.

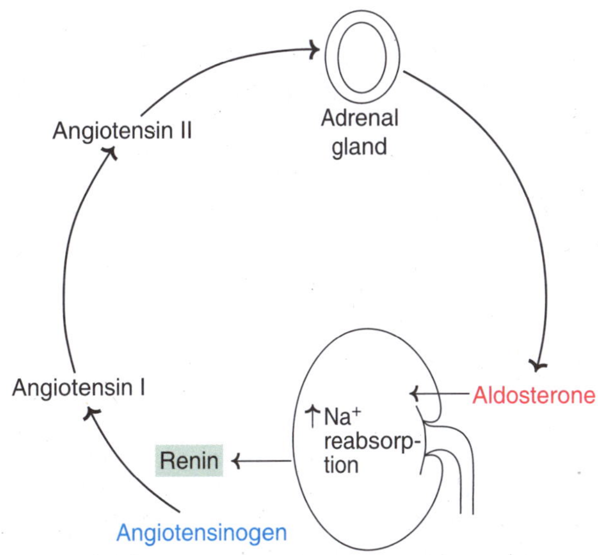

Fig.21.2 : *Hormonal regulation of Na^+ balance by the kidney.*

1. Due to loss of water alone.
2. Due to deprivation of water and electrolytes.

Osmotic imbalance and dehydration in cholera

Cholera is transmitted through water and foods, contaminated by the bacterium ***Vibrio cholerae***. This bacterium produces a toxin which stimulates the intestinal cells to secrete various ions (Cl^-, Na^+, K^+, HCO_3^- etc.) into the intestinal lumen. These ions collectively raise the osmotic pressure and suck the water into lumen. This results in diarrhea with a heavy loss of water (5–10 liters/day). If not treated in time, the victims of cholera will die due to dehydration and loss of dissolved salts. Thus, cholera and other forms of severe diarrhea are the major killers of young children in many developing countries.

Oral rehydration therapy (ORT) is commonly used to treat cholera and other diarrheal diseases.

Overhydration

Overhydration or water intoxication is caused by excessive ***retention of water in the body***. This may occur due to excessive intake of large volumes of salt free fluids, renal failure, overproduction of ADH etc.

ACID-BASE BALANCE

The ***normal pH*** of the ***blood*** is maintained in the narrow range of ***7.35-7.45***, i.e. slightly alkaline.

Maintenance of blood pH is an important homeostatic mechanism of the body. In normal circumstances, the ***regulation is so effective that the blood pH varies very little***.

Production of acids by the body

The metabolism of the body is accompanied by an overall production of acids. These include the ***volatile acids like carbonic acid*** (most predominent, about 20,000 mEq/day) or ***non-volatile acids*** (about 80 mEq/day) such as ***lactic acid***, ***sulfuric acid***, ***phosphoric*** acid etc. Carbonic acid is formed from the metabolic product CO_2 while lactic acid is produced in anaerobic metabolism. These acids add up H^+ ions to the blood. A ***diet rich in animal proteins results in more acid production*** by the body that ultimately leads to the excrètion of ***urine*** which is profoundly ***acidic***.

Production of bases by the body

The formation of basic compounds in the body, in the normal circumstances, is negligible. Some amount of bicarbonate is generated from the organic acids such as lactate and citrate. A vegetarian diet has a tendency for a net production of bases. For this reason, a ***vegetarian diet has an alkalizing effect on the body***.

MAINTENANCE OF BLOOD pH

The body has developed three lines of defense to regulate the body's acid-base balance and maintain the blood pH (around 7.4).

I. ***Blood buffers*** II. ***Respiratory mechanism***
III. ***Renal mechanism***.

I. Blood buffers

A buffer may be defined as a solution of a weak acid (HA) and its salt (BA) with a strong base. The buffer resists the change in pH by the addition of acid or alkali and the buffering capacity is dependent on the absolute concentration of salt and acid. It should be borne in mind that the buffer cannot remove H^+ ions from the body. It temporarily acts as a shock absorbant to reduce the free H^+ ions. The H^+ ions have to be ultimately eliminated by the renal mechanism (described later).

The blood contains 3 buffer systems.

1. Bicarbonate buffer 2. Phosphate buffer
3. Protein buffer.

1. **Bicarbonate buffer system :** Sodium bicarbonate and carbonic acid ($NaHCO_3 - H_2CO_3$) is the most ***predominant buffer*** system of the extracellular fluid, particularly the plasma. Carbonic acid dissociates into hydrogen and bicarbonate ions.

$$H_2CO_3 \rightleftharpoons H^+ + HCO_3^-$$

By the law of mass action, at equilibrium

$$K_a = \frac{[H^+][HCO_3^-]}{[H_2CO_3]}$$

(K_a = Dissociation constant of H_2CO_3).

The equation for bicarbonate buffer may be written as

$$pH = pK_a + \log\frac{[HCO_3^-]}{[H_2CO_3]}.$$

The general equation referred to as ***Henderson-Hasselbalch equation*** for any buffer is written as

$$pH = pK_a + \log\frac{[Base]}{[Acid]}.$$

Blood pH and the ratio of HCO_3^- to H_2CO_3 : The plasma bicarbonate (HCO_3^-) concentration is around 24 mmol/l (range 22-26 mmol/l). Carbonic acid is a solution of CO_2 in water. Its concentration is given by the product of pCO2 (arterial partial pressure of CO_2 = 40 mm Hg) and the solubility constant of CO_2 (0.03).

Thus $H_2CO_3 = 40 \times 0.03 = 1.2$ mmol/l.

At a ***blood pH 7.4, the ratio of bicarbonate to carbonic acid is 20 : 1***. Thus, the bicarbonate concentration is much higher (20 times) than carbonic acid in the blood. This is referred to as ***alkali reserve*** and is responsible for the effective buffering of H^+ ions, generated in the body. In normal circumstances, the concentration of bicarbonate and carbonic acid determines the pH of blood.

2. **Phosphate buffer system :** Sodium dihydrogen phosphate and disodium hydrogen phosphate ($NaH_2PO_4 - Na_2HPO_4$) constitute the phosphate buffer. It is ***mostly an intracellular buffer*** and is of less importance in plasma due to its low concentration.

3. **Protein buffer system :** The ***plasma proteins and hemoglobin*** together constitute the protein buffer

Medical Concepts/ Clinical Correlates

- Sodium is the principal extracellular cation while K^+ is intracellular. The maintenance of the differential concentration of these electrolytes is essential for the survival of life which is brought about by Na^+-K^+ pump.
- The body metabolism is accompanied by the production of acids such as carbonic acid, sulfuric acid, phosphoric acid etc.
- Vegetarian diet has an alkalizing effect on the body. This is attributed to the formation of organic acids such as sodium lactate which can deplete H^+ ions by combining with them.

system of the blood. The buffering capacity of proteins is dependent on the pK of ionizable groups of amino acids.

II. Respiratory mechanism for pH regulation

Respiratory system provides a ***rapid mechanism for the maintenance of acid-base balance***. This is achieved by regulating the concentration of carbonic acid (H_2CO_3) in the blood i.e. the denominator in the bicarbonate buffer system.

The large volumes of CO_2 produced by the cellular metabolic activity endanger the acid-base equilibrium of the body. But in normal circumstances, all of this CO_2 is eliminated from the body in the expired air via the lungs, as summarized below.

$$H_2CO_3 \xleftrightarrow{\text{Carbonic anhydrase}} CO_2 + H_2O$$

The rate of respiration (or the rate of removal of CO_2) is controlled by a respiratory centre, located in the medulla of the brain. This centre is highly sensitive to changes in the pH of blood. Any decrease in blood pH causes hyperventilation to blow off CO_2, thereby reducing the H_2CO_3 concentration. Simultaneously, the H^+ ions are eliminated as H_2O.

Respiratory control of blood pH is rapid but only a short term regulatory process, since hyperventilation cannot proceed for long.

III. Renal mechanism for pH regulation

The role of kidneys in the maintenance of acid-base balance of the body (blood pH) is highly significant. The ***renal mechanism*** tries to provide a ***permanent solution to the acid-base disturbances***.

The kidneys regulate the blood pH by maintaining the alkali reserve, besides excreting or reabsorbing the acidic or basic substances, as the situation demands.

Urine pH normally lower than blood pH : The pH of urine is normally acidic (~6.0). This clearly indicates that the ***kidneys have contributed to the acidification of urine***, when it is formed from the blood plasma (pH 7.4). In other words, the H^+ ions generated in the body in the normal circumstances, are eliminated by acidified urine.

Carbonic anhydrase and renal regulation of pH : The enzyme carbonic anhydrase is of central importance in the renal regulation of pH which occurs by the following mechanisms.

1. Excretion of H^+ ions
2. Reabsorption of bicarbonate
3. Excretion of titratable acid
4. Excretion of ammonium ions.

1. **Excretion of H^+ ions :** Kidney is the ***only route through which the H^+ can be eliminated from the body***. H^+ excretion occurs in the proximal convoluted tubules (renal tubular cells) and is coupled with the regeneration of HCO_3^-. The process depicted in ***Fig.21.3***, occurs as follows.

Carbonic anhydrase catalyses the production of carbonic acid (H_2CO_3) from CO_2 and H_2O in the renal tubular cell. H_2CO_3 then dissociates to H^+ and HCO_3^-. The H^+ ions are secreted into the tubular lumen in exchange for Na^+. The Na^+ in association with HCO_3^- is reabsorbed into the blood. This is an effective mechanism to eliminate acids (H^+) from the body with a simultaneous generation of HCO_3^-. The latter adds up to the alkali reserve of the body. The H^+ combines with a non-carbonate base and is excreted in urine.

2. **Reabsorption of bicarbonate :** This mechanism is primarily ***responsible to conserve the blood HCO_3^-***, with a simultaneous excretion of H^+ ions. The normal urine is almost free from HCO_3^-.

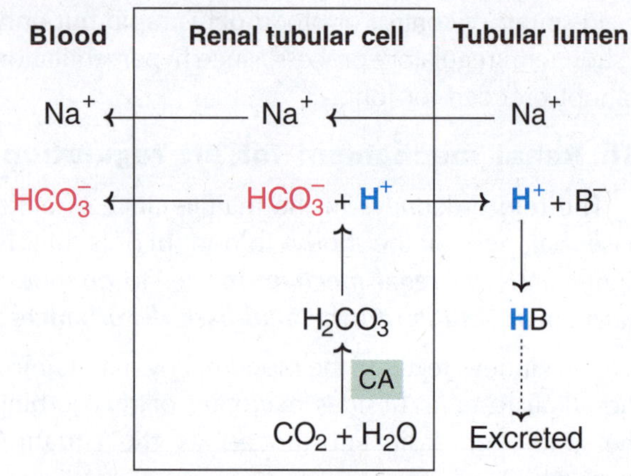

Fig.21.3 : Renal regulation of blood pH–Excretion of H^+ ions (CA–Carbonic anhydrase).

3. **Excretion of titratable acid :** Titratable acidity is a ***measure of acid excreted into urine by the kidney***. This can be estimated by titrating urine back to the normal pH of blood (7.4). In quantitative terms, titratable acidity refers to the number of milliliters of N/10 NaOH required to titrate 1 liter of urine to pH 7.4. Titratable acidity ***reflects the H^+ ions excreted into urine which resulted in a fall of pH from 7.4*** (that of blood). The excreted H^+ ions are actually buffered in the urine by phosphate buffer.

4. **Excretion of ammonium ions :** This is another mechanism to buffer H^+ ions secreted into the tubular fluid. The ***H^+ ion combines with NH_3 to form ammonium ion (NH_4^+)***. The renal tubular cells deamidate glutamine to glutamate and NH_3. This reaction is ***catalysed by*** the enzyme ***glutaminase***. The NH_3, liberated in this reaction, diffuses into the tubular lumen where it combines with H^+ to form NH_4^+. ***Ammonium*** ions cannot diffuse back into tubular cells and, therefore, are ***excreted into urine***.

NH_4^+ is a major urine acid. It is estimated that about half to two-thirds of body acid load is eliminated in the form of NH_4^+ ions.

Carbon dioxide—the central molecule of pH regulation

As is observed from the foregoing discussion, CO_2 is of central importance in the acid-base balance of the body. It has the ability to combine with H_2O to from H_2CO_3 which can dissociate to HCO_3^- and H^+. A summary of the interaction between the lungs, erythrocytes and kidneys in handling CO_2 to maintain pH of the blood is depicted in ***Fig.21.4***. The CO_2 generated by aerobic metabolism may be exhaled via lungs, or converted to HCO_3^- by erythrocytes and kidneys to add up to the alkali reserve of the body.

DISORDERS OF ACID-BASE BALANCE

The body has developed an efficient system for the maintenance of acid-base equilibrium with a result that the pH of blood is almost constant (7.4). The ***blood pH compatible to life is 6.8-7.8***, beyond which life cannot exist.

For a better understanding of the disorders of acid-base balance, the ***Henderson-Hasselbalch equation*** must be frequently consulted.

$$pH = pK_a + \log \frac{[HCO_3^-]}{[H_2CO_3]}$$

It is evident from the above equation that the blood pH (H^+ ion concentration) is dependent on the relative concentration (ratio) of bicarbonate (HCO_3^-) and carbonic acid (H_2CO_3).

The acid-base disorders are mainly classified as

1. **Acidosis—a decline in blood pH**
   (a) ***Metabolic acidosis***—due to a decrease in bicarbonate.
   (b) ***Respiratory acidosis***—due to an increase in carbonic acid.
2. **Alkalosis—a rise in blood pH**
   (a) ***Metabolic alkalosis***—due to an increase in bicarbonate.

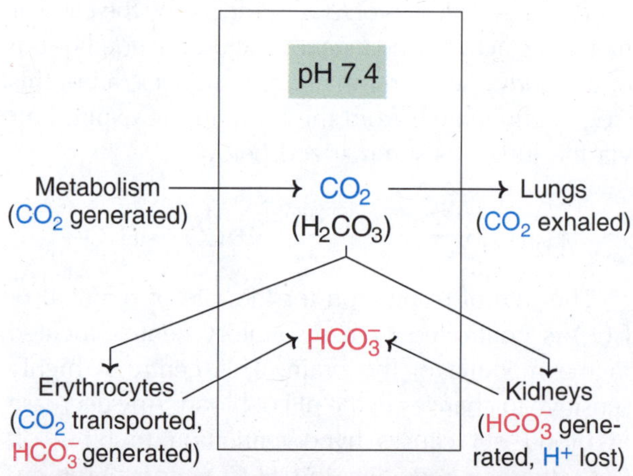

Fig.21.4 : Carbon dioxide—the central molecule of blood pH regulation.

(b) ***Respiratory alkalosis***—due to a decrease in carbonic acid.

The four acid-base disorders referred to above are primarily due to alterations in either bicarbonate or carbonic acid. It may be observed that the metabolic acid-balance disorders are caused by a direct alteration in bicarbonate concentration while the respiratory disturbances are due to a change in carbonic acid level (i.e. CO_2).

Clinical causes of acid-base disorders

The most important clinical causes/disease states that result in acid-base disorders are listed in ***Table 21.2***. Metabolic acidosis could occur due to ***diabetes mellitus (ketoacidosis)***, ***lactic acidosis***, ***renal failure*** etc. Respiratory acidosis is common in severe asthma and cardiac arrest. Vomiting and hypokalemia may result in metabolic alkalosis while hyperventilation and ***severe anemia*** may lead to respiratory alkalosis.

Compensation of acid-base disorders

To counter the acid-base disturbances, the ***body gears up*** its homeostatic mechanism and makes every attempt ***to restore the pH to normal level (7.4)***. This is referred to as compensation which may be partial or full. Sometimes the acid-base disorders may remain uncompensated.

The principal acid-base disturbances, along with the blood concentration of HCO_3^- and H_2CO_3, in acute and compensated states are given in the ***Table. 21.3***.

TABLE 21.2 Major clinical causes of acid-base disorders

Metabolic acidosis	*Respiratory acidosis*
Diabetes mellitus (ketoacidosis)	Severe asthma Pneumonia Cardiac arrest
Renal failure	Obstruction in airways
Lactic acidosis	Chest deformities
Severe diarrhea Renal tubular acidosis	Depression of respiratory center (by drugs e.g. opiates)
Metabolic alkalosis	***Respiratory alkalosis***
Severe vomiting	Hyperventilation
Hypokalemia	Anemia High altitude
Intravenous administration of bicarbonate	Salicylate poisoning

For the acute metabolic disorders (due to changes in HCO_3^-), respiratory compensation sets in and regulates the H_2CO_3 (i.e. CO_2) by hyper- or hypoventilation. As regards acute respiratory disorders (due to changes in H_2CO_3), the renal compensation

TABLE 21.3 Acid-base disorders along with the concentrations of bicarbonate (HCO_3^-) and carbonic acid (H_2CO_3) in plasma

Disorder	*Blood pH*	*[HCO_3^-]*	*[H_2CO_3]*
Metabolic acidosis			
Acute	↓	↓	→
Compensated (by ↑ ventilation)	↘ or →	↓	↓
Respiratory acidosis			
Acute	◊	→	↑
Compensated (HCO_3^- retained by kidney)	↘ or →	↑	↑
Metabolic alkalosis			
Acute	↑	↑	→
Compensated (by ↓ ventilation)	↗ or →	↑	↑
Respiratory alkalosis			
Acute	↑	→	↓
Compensated (↑HCO_3^- excretion by kidney)	↗ or →	↓	↓

↑: Increased; ↓: Decreased; →: Normal; ↘: Marginally decreased; ↗: Marginally increased.

Medical Concepts/ Clinical Correlates

- **The blood pH is maintained by blood buffers, respiratory and renal mechanisms.**
- **Carbon dioxide is the central molecule of acid-base regulation.**
- **Disturbances in acid-base regulation result in acidosis (decreased blood pH) or alkalosis (raised blood pH).**
- **Uncontrolled diabetes mellitus is associated with metabolic acidosis, commonly referred to as ketoacidosis (due to the overproduction of ketone bodies).**

occurs to maintain the HCO_3^- level, by increasing or decreasing its excretion.

Anion gap

For a better understanding of acid-base disorders, adequate knowledge of anion gap is essential. The total concentration of cations and anions (expressed as mEq/l) is equal in the body fluids. This is required to maintain electrical neutrality.

The commonly measured electrolytes in the plasma are Na^+, K^+, Cl^- and HCO_3^-. Na^+ and K^+ together constitute about 95% of the plasma cations. Cl^- and HCO_3^- are the major anions, contributing to about 80% of the plasma anions. The remaining 20% of plasma anions (not normally measured in the laboratory) include proteins, phosphate, sulfate, urate and organic acids.

Anion gap is defined as ***the difference between the total concentration of measured cations (Na^+ and K^+) and that of measured anion (Cl^- and HCO_3^-).*** The anion gap (A^-) in fact ***represents the unmeasured anions in the plasma*** which may be calculated as follows, by substituting the normal concentration of electrolytes (mEq/l).

$$Na^+ + K^+ = Cl^- + HCO_3^- + A^-$$
$$136 + 4 = 100 + 25 + A^-$$
$$A^- = 15 \text{ mEq/l}$$

The anion gap in a healthy individual is around 15 mEq/l (range 8-18 mEq/l). Acid-base disorders are often associated with alterations in the anion gap.

Metabolic acidosis

The primary defect in metabolic acidosis is a ***reduction in bicarbonate concentration*** which leads to a fall in blood pH. The bicarbonate concentration may be decreased due to its utilization in buffering H^+ ions, loss in urine or gastrointestinal tract or failure to be regenerated.

The most important cause of metabolic acidosis is due to an ***excessive production of organic acids*** which combine with $NaHCO_3^-$ and deplete the alkali reserve.

$$NaHCO_3^- + \text{Organic acids} \longrightarrow \text{Na salts of organic acids} + CO_2$$

Metabolic acidosis is commonly seen in ***severe uncontrolled diabetes mellitus*** which is associated with excessive production of acetoacetic acid and β-hydroxybutyric acid (both are organic acids).

Compensation of metabolic acidosis : The acute metabolic acidosis is usually compensated by ***hyperventilation of lungs***. This leads to an increased elimination of CO_2 from the body (hence $H_2CO_3^-$). but respiratory compensation is only short-lived. Renal compensation sets in within 3-4 days and the H^+ ions are excreted as NH_4^+ ions.

Respiratory acidosis

The primary defect in respiratory acidosis is due to a ***retention of CO_2*** ($H_2CO_3\uparrow$). There may be several causes for respiratory acidosis which include depression of the respiratory centre (overdose of drugs), pulmonary disorders (bronchopneumonia) and breathing air with high content of CO_2.

The renal mechanism comes for the rescue to compensate respiratory acidosis. ***More HCO_3^- is generated and retained by the kidneys*** which adds up to the alkali reserve of the body. The excretion of ***titratable acidity*** and NH_4^+ is ***elevated in urine***.

Metabolic alkalosis

The primary abnormality in metabolic alkalosis is an ***increase in HCO_3^- concentration***. This may occur due to excessive vomiting (resulting in loss of H^+) or an excessive intake of sodium bicarbonate for therapeutic purposes (e.g. control of gastric acidity).

The respiratory mechanism initiates the compensation by hypoventilation to retain CO_2 (hence $H_2CO_3\uparrow$). This is slowly taken over by renal mechanism which excretes more HCO_3^- and retains H^+.

Respiratory alkalosis

The primary abnormality in respiratory alkalosis is a ***decrease in H_2CO_3 concentration***. This may occur due to prolonged hyperventilation resulting in increased exhalation of CO_2 by the lungs. Hyperventilation is observed in conditions such as hysteria, hypoxia, raised intracranial pressure, excessive artificial ventilation and the action of certain drugs (salicylate) that stimulate respiratory centre.

The ***renal mechanism tries to compensate by increasing the urinary excretion of HCO_3^-***.

Mixed acid-base disorders

Sometimes, the ***patient may have two or more acid-base disturbances occurring simultaneously***. In such instances, both HCO_3^- and H_2CO_3 are altered. In general, if the biochemical data (of blood gas analysis) cannot be explained by a specific acid-base disorder, it is assumed that a mixed disturbance is occurring. Many a times, ***compensatory mechanisms may lead to mixed acid-base disorders***.

SUMMARY

1. *Water is the solvent of life and constitutes about 60% of the total body weight, distributed in intracellular and extracellular fluids.*
2. *Electrolytes are distributed in the intracellular and extracellular fluids to maintain the osmotic equilibrium and water balance, Na^+ is the principal extracellular cation while K^+ is the intracellular cation.*
3. *The normal pH of blood is maintained in the narrow range of 7.35–7.45. The metabolism of the body is accompanied by an overall production of acids. The body has developed three lines of defense (blood buffers, respiratory and renal mechanisms) to regulate the acid-base balance and maintain the blood pH.*
4. *Among the blood buffers, bicarbonate buffer (with a ratio of HCO_3^- to H_2CO_3 as 20 : 1) is the most important in regulating blood pH. The respiratory system regulates the concentration of carbonic acid by controlling the elimination of CO_2 via lungs.*
5. *The renal (kidney) mechanism regulates blood pH by excreting H^+ and NH_4^+ ions besides the reabsorption of HCO_3^-. The pH of urine is normally acidic which indicates that the kidneys have contributed to the acidification of urine.*
6. *The acid-base disorders are classified as acidosis (metabolic or respiratory) and alkalosis (metabolic or respiratory), respectively, due to a rise or fall in blood pH. The metabolic disturbances are associated with alterations in HCO_3^- concentration while the respiratory disorders are due to changes in H_2CO_3 (i.e. CO_2).*

THE READER MUST REFER APPENDIX D (Page Nos : 422-433) FOR RELATED CLINICAL CASE STUDIES (Nos : 36, 37)

TISSUE PROTEINS AND BODY FLUIDS

22

The body possesses a vast number of proteins designed with specific structures to perform specialized functions. A selected few of the most important proteins that are intimately connected with the tissue structure and functions are briefly described in this chapter. In addition, the body fluids are also discussed.

CONNECTIVE TISSUE PROTEINS

The connective tissue or ***extracellular matrix (ECM)*** refers to the complex material surrounding the mammalian cells in tissues. The major protein components of ECM include collagen, elastin, fibrillin, fibronectin, and proteoglycans. Besides these proteins, the structural proteins namely keratins are also described.

COLLAGEN

Collagen is the ***most abundant protein in mammals***, comprising approximately one-third of the total body protein. Collagen is the predominant component of the connective tissue, although its distribution varies in different tissues. For instance, collagen forms 90% of the organic matrix of bones, 85% of tendons, 70% of skin, and 4% of liver.

Functions of collagen

1. Being a major component of the connective tissue, collagen ***gives strength***, ***support*** and ***shape to the tissues***.

2. Collagen contributes to proper alignment of cells, which in turn helps in cell proliferation, and their differentiation to different tissues and organs.

3. Collagen (that is exposed in blood vessels) contributes to thrombus formation.

Types of collagen

Collagen is not a single homogeneous protein, but a group of structurally related and genetically distinct proteins. In humans, ***at least 19 different types of collagens***, composed of 30 distinct polypeptide chains (encoded by separate genes), have been identified. The types of collagen are numbered (by Roman numerals) as I, II...XIX.

Structure of collagen

In principle, all types of collagen are ***triple helical structures***. The triple helix may occur throughout the molecule, or only a part of the molecule.

Type I mature collagen, containing about 1000 amino acids (for each polypeptide chain) possesses triple helical structure throughout the molecule. It is composed of three similar polypeptide chains twisted around each other to form a rod like molecule of 1.4 nm diameter, and about 300 nm length (***Fig.22.1***). The amino acid composition of collagen is unique. Approximately $^{1}/_{3}$rd of the amino acids are contributed by glycine i.e. every third amino acid in collagen is glycine. Hence, the repetitive amino acid sequence of collagen is represented by $(\textbf{\textit{Gly-X-Y}})_n$, where X and Y represent other amino acids.

Abnormalities associated with collagen

The biosynthesis of collagen is a complex process, involving at least ***30 genes*** (in humans), and about ***8 post-translational modifications***. Expectedly, many inherited diseases due to gene mutations, linked with collagen formation have been identified. A couple of them are listed in the next page.

- **Ehlers-Danlos syndrome**–a group of inherited disorders characterized by ***hyperextensibility of skin***, and abnormal tissue fragility.
- **Osteogenesis imperfecta**–characterized by ***abnormal fragility of bones*** due to decreased formation of collagen.

ELASTIN

Elastin is another important (besides collagens) connective tissue protein. It is mainly responsible for the extensibility and elasticity of tissues. Elastin is found in large quantities in lungs, arterial blood vessels, elastic ligments etc.

Williams syndrome is a genetic disease due to impairment in elastin synthesis. The connective tissue and central nervous system are affected.

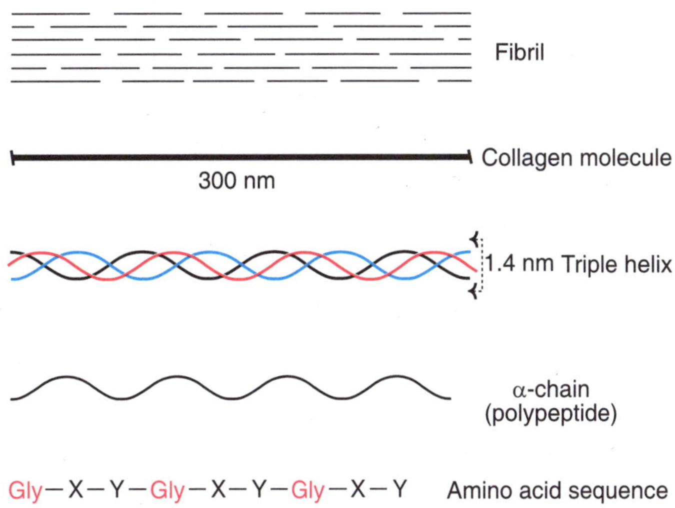

***Fig.22.1 :** A diagrammatic representation of the structure of collagen and fibril (X and Y represent amino acids other than glycine)*

FIBRILLIN

Fibrillin is a structural ***component of myofibrils*** found in various tissues.

Marfan syndrome is a genetic disorder due to a mutation in the gene for fibrillin. It is characterized by ***hyperextensibility of joints*** and skeletal system. Consequently, the patients of Marfan syndrome are tall, and have long digits. These patients may also have cardiovascular complications. Some researchers believe that Abraham Lincoln was a victim of Marfan syndrome.

PROTEOGLYCANS

Proteoglycans are ***conjugated proteins containing glycosaminoglycans*** (GAGs). Several proteoglycans with variations in core proteins and GAGs are known e.g. syndecan, betaglycan, aggrecan, fibromodulin. For more information on the structure and functions of proteoglycans ***Refer Chapter 2***. GAGs, the components of proteoglycans, are affected in a group of genetic disorders namely mucopolysaccharidoses (***Chapter 13***).

KERATINS

Keratins are structural ***proteins*** found ***in hair***, skin, nails, horns etc. The 3 polypeptides of keratin form α-helical structure and are held together by disulfide bonds. The toughness and strength of keratin are directly related to the number of disulfide bonds. Thus, the harder keratin possesses more disulfide bonds. The mechanical strength of the hair is attributed to disulfide bonds.

CONTRACTILE PROTEINS

The proteins that are involved in the ***movement of body organs*** (e.g. muscle, heart, lung) are regarded as contractile proteins. It is worthwhile to understand the basic structure of muscle before learning the contractile proteins.

STRUCTURE OF MUSCLE

Muscle is the ***single largest tissue*** of the human body. Muscle constitutes about 20% of body mass at birth, 40% in young adults and 30% in aged adults. Three types of muscles are found in vertebrates-skeletal, cardiac and smooth. The skeletal and cardiac muscles are ***striated*** while the smooth muscles are ***non-striated***.

The structure of striated muscle is represented in ***Fig.22.2***. It is composed of bundles of multinucleated muscle fibre cells. Each cell is surrounded by an electrically excitable plasma membrane, the sarcolemma. The muscle fibre cells are long which may extend the entire length of the muscle. The intracellular fluid of fibre cells is the sarcoplasm (i.e. cytoplasm) into which the ***myofibrils*** are embedded. The sarcoplasm is rich in glycogen, ATP, creatine phosphate, and the enzymes of glycolysis.

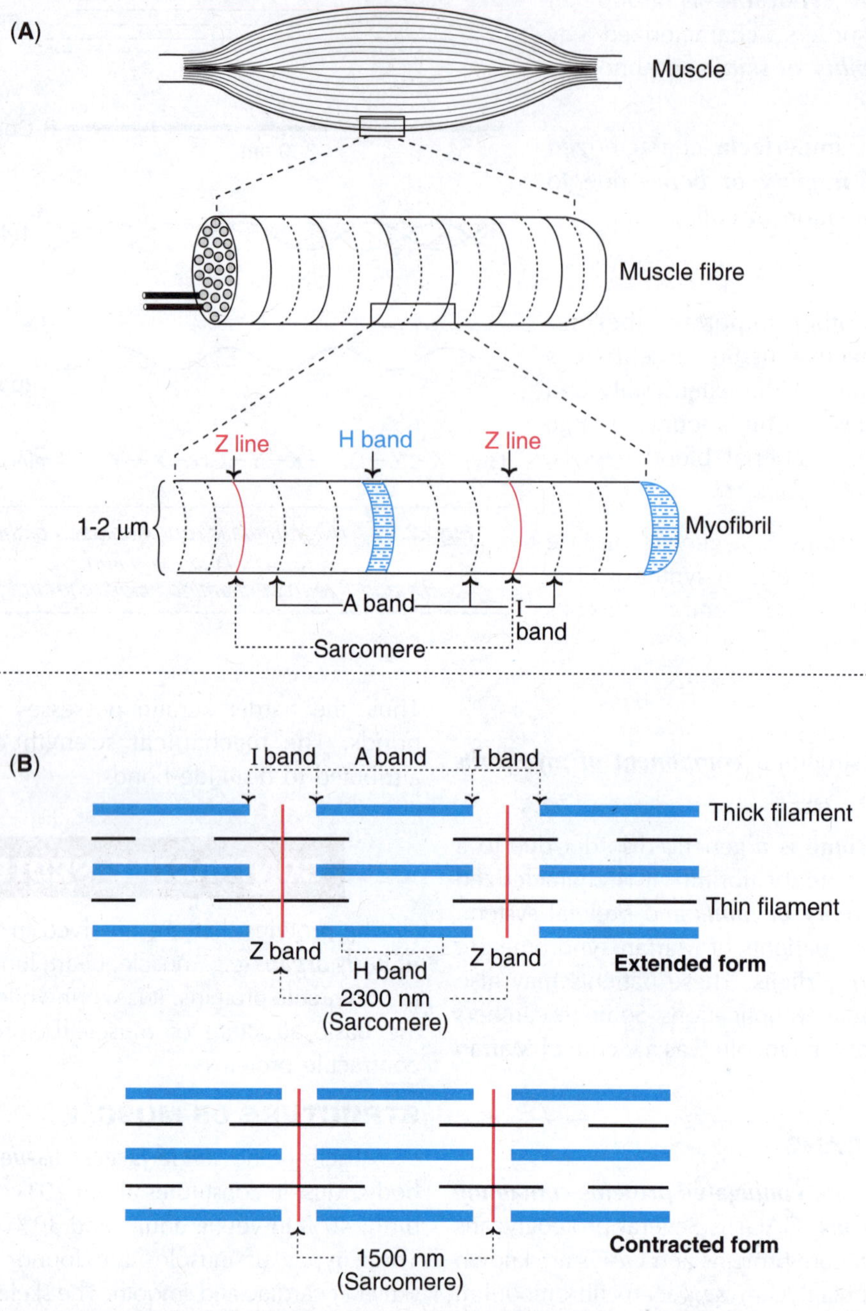

***Fig.22.2 :** (**A**) Structure of myofibril of a straited muscle (**B**) Arrangement of filaments of myofibril in extended and contracted form. (**Note** : The length of sarcomere is reduced from 2300 nm to 1500 nm during contraction)*

When the myofibril is examined under electron microscope, alternating dark bands (anisotropic or A bands) and light bands (isotropic or I bands) are observed. The less dense central region of A band is referred to as H band (or H line). A narrow and dense Z line bisects the I band. The region of the muscle fibre between two Z lines is termed as sarcomere. ***Sarcomere is the functional unit of muscle.***

In the electron microscopy, it is further observed that the myofibrils are composed of thick and thin longitudinal filaments. The thick filaments contain the protein myosin, and are confined to A band. The thin filaments lie in the I band, and can extend into A band (but not into H line). These thin filaments contain the proteins actin, tropomyosin and troponin.

During the course of ***muscle contraction***, the thick and thin filaments slide over each other (***sliding filament model*** of muscle contraction). Consequently, the H bands and I bands shorten. However, there is no change in the length of thick and thin filaments. The length of sarcomere which is around 2300 nm in an extended form of myofibril is reduced to 1500 nm in a contracted form (***Fig.22.2B***).

MUSCLE PROTEINS

More than 20% of the muscle mass is composed of proteins. This is largely contributed by structural proteins namely actin, myosin, and the actin cross-linking proteins, tropomyosin and troponin. Muscle also contains other proteins –myoglobin, collagen, enzymes etc.

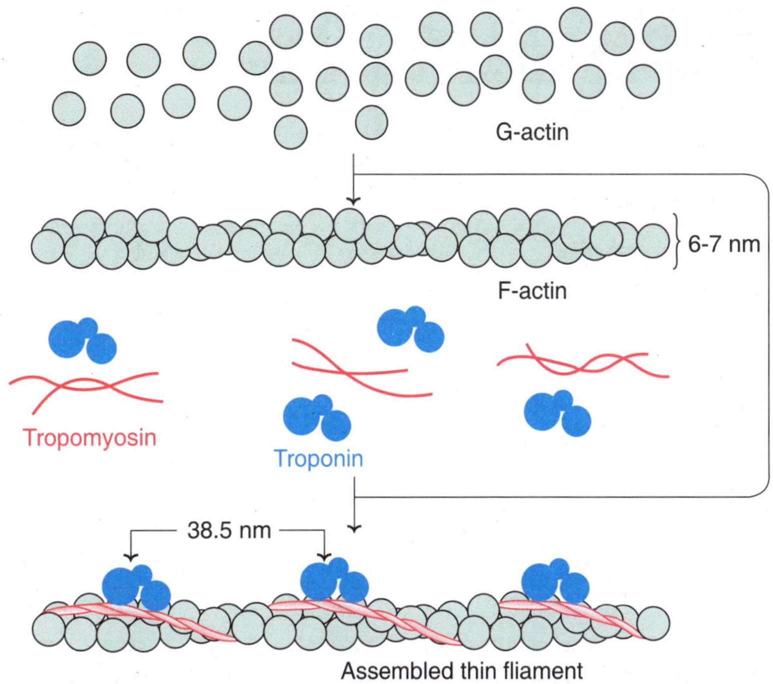

Fig.22.3 : *A diagrammatic representation of the thin filament of a sarcomere.*

ACTIN

Actin is a major constituent of thin filaments of sarcomere. It exists in two forms – monomeric ***G-actin*** (i.e. ***g***lobular actin) and polymeric ***F-actin*** (i.e. ***f***ilament actin). G-actin constitutes about 25% of the muscle proteins by weight. In the presence of Mg^{2+} ions, G-actin polymerizes (non-covalently) to form an insoluble double helical F-actin with a thickness of 6-7 nm (***Fig.22.3***).

Tropomyosin and troponin : These two are cross-linking proteins found in association with actin. Although, minor in terms of mass, they are important in terms of their function. Tropomyosin, composed of two chains, attaches to F-actin in the grooves (***Fig.22.3***). Troponin consists of three polypeptide chains – troponin T (TpT binding to ***t***ropomyosin), tropinin I (TpI that ***i***nhibits F-actin myosin interaction) and troponin C (TpC, ***c***alcium binding polypeptide). TpC is comparable to calmodulin.

MYOSINS

Myosins are actually a family of proteins with about 15 members. The myosin that is predominantly present in muscle is myosin II.

In terms of quantity, myosin constitutes approximately 55% of muscle protein, and is found in thick filaments. Myosin is composed of six polypeptide chains (hexamer). It contains one pair of heavy (H) chains, and two pairs of light (L) chains.

PROTEIN MISFOLDING AND DISEASES

The process of protein folding is complex and has been briefly described in ***Chapter 25***. Sometimes, improperly folded proteins may be formed (either spontaneous or by gene mutations). Such misfolded proteins usually get degraded within the cell. However, as the individuals age, the misfolded proteins accumulate and cause a number of diseases. Prion diseases and amyloidosis are two groups of diseases due to protein misfolding.

Prion diseases

The term prion represents ***pro***teinous ***in***fectious agents. ***Prion proteins*** (***PrP***) are the altered forms of three-dimensional structures of normal proteins. However, no differences in the primary structure (i.e. amino acid sequence) and post-translational modifications are observed.

Some examples of prion diseases are listed

- Transmissible sponsiform encephalopathies in humans.
- Scrapie in sheep.
- ***Mad cow disese*** in cattle.

Table 22.1 Composition of milk in different species

Constituent	Human	Cow	Buffalo	Goat
Water	87.6	87.2	83.5	87.0
Total solids (g/dl)	12.4	12.8	16.5	13.0
Carbohydrates (g/dl)	7.5	4.4	5.4	4.6
Lipids (g/dl)	3.8	3.8	6.5	3.5
Proteins (g/dl)	1.1	3.3	4.3	3.7
Calcium (mg/dl)	35	150	160	175
Magnesium (mg/dl)	2.2	13	10	8
Phosphorus (mg/dl)	16	100	100	70
Sodium (mg/dl)	15	60	60	50
Potassium (mg/dl)	55	140	130	85

Kuru is an interesting prion disease. It was first reported in Papau New Guinea in the tribal people who practice ***cannibalism*** (they eat the brains of the dead people).

As of now, there is no treatment for prion diseases. Transmissible spongiform encephalopathies are invariably fatal in humans.

Amyloidosis

The term amyloids is used to refer to the ***altered proteins*** (with ***β-sheets***) that ***accumulate in the body***, particularly in the nervous system. Amyloids are formed by protein misfolding or due to gene mutations. They are not infectious agents as prion proteins. However, as the age advances, amyloids accumulate, and they have been implicated in many degenerating diseases. A total of at least 15 different proteins are involved in amyloidosis.

Alzheimer's disease is a ***neurodegenerative disorder***, affecting about 5-10% of the people above 60 years of age. It is characterized by memory loss, confusion, hallucinations, personality changes with abnormal behaviour. As the disease progresses, the patient may enter a vegetative state, and may die after 10 years after the onset of the disease manifestations. The accumulation of amyloids (in the form of amyloid plaque) has been clearly demonstrated in the patients of Alzheimer's disease.

BODY FLUIDS

The specialized fluids of the body are ***milk***, ***cerebrospinal fluid***, ***amniotic fluid***, ***aqueous humor***, ***sweat and tears***. In a broader perspective, blood, plasma and serum are also biological fluids. Their biochemical importance is discussed elsewhere (Hemoglobin, ***Chapter 10***; Plasma proteins, ***Chapter 9***; Diagnostic enzymes, ***Chapter 6***; Acid-base balance, ***Chapter 21***). Urine is an excretory biological fluid.

MILK

Milk is secreted by mammary glands. It is almost a ***complete natural food***. Milk is the only food for the offsprings of mammals on their birth.

COMPOSITION OF MILK

The major constituents of milk in different species—human, cow, buffalo and goat are given in ***Table 22.1***. Water is the major constituent, with a concentration in the range of 83–87%, depending on the species. The remaining 13–17% is made up of solids–carbohydrates, lipids, proteins, minerals and vitamins.

Carbohydrates in milk

Milk contains the disaccharide ***lactose*** which imparts sweetness. Human milk has a higher concentration of lactose (7.5%) compared to milk of other species. Thus, human milk is sweet enough for the babies to relish.

Lipids in milk

The lipids in the milk are dispersed as small globules. Milk fat is mainly composed of ***triacylglycerols***. Mono- and diacylglycerols are also present in trace quantities. The fatty acids found in milk (i.e. in TG) are mostly medium or short chain, and saturated e.g. palmitic acid, myristic acid, stearic acid, lauric acid and butyric acid. Oleic acid, an unsaturated fatty acid, is also present.

Proteins in milk

The major milk proteins are ***casein*** (about 80%) and ***lactalbumin***. Small concentrations of enzymes (proteases, lipase, xanthine oxidase, lysozyme) and immunoglobulins are also found.

Milk casein (a phosphoprotein) is almost a complete protein (next to egg albumin), containing all the essential amino acids. It is present in milk in the form of aggregates called micelles. The ***white colour of milk*** is due to the dispersion of ***calcium caseinate micelles***.

Whey proteins : If milk is acidified, casein gets precipitated at isoelectric point (pH 4.7). The supernatant fluid contains whey proteins (20% of milk proteins). These include lactalbumin, lactoglobulin and various enzymes.

Minerals in milk

Milk is rich in calcium, magnesium, phosphorus, sodium, potassium and chlorine. However, milk is a ***poor source of iron*** and ***copper***.

Vitamins in milk

Both fat soluble and water soluble vitamins are found in good concentration in milk. However milk is ***deficient in vitamin C***.

Calorific value of milk

Due to variability in the nutrient composition (carbohydrates, fats and proteins), the calorific value of milk from different species varies. Thus, human milk can provide about 70 Cal/100 ml, while for ***buffalo milk***, it is ***around 95 Cal/100 ml***.

CEREBROSPINAL FLUID (CSF)

Cerebrospinal fluid is a clear, colourless liquid formed within the cavities (ventricles) of brain and around the spinal cord. CSF originates in the choroid plexus (as an ultrafiltrate of plasma) and returns to blood through arachnoid villi. About 500 ml of CSF is formed everyday. However, at any given time, there is about 120–150 ml CSF in the system. Further, CSF is completely replaced about three times a day.

Functions of CSF

As the brain has no lymphatic system, CSF drains into the ventricular system and moves into spaces surrounding the brain and spinal cord. The major functions of CSF are listed.

- CSF serves as a hydraulic ***shock absorber***. It can diffuse the force from a hard blow to the skull that might otherwise cause severe injury.
- It helps in the regulation of intracranial pressure.

Collection of CSF

Cerebrospinal fluid is usually collected by a spinal puncture for the purpose of biochemical analysis. The puncture is performed in the lumbar region, between the third and fourth, or between the fourth and fifth lumber vertebrae.

The sterile lumbar puncture (spinal tap) is carried out in a side lying (lateral) position with head fixed into the chest and knees. This position helps to increase the space between the lumbar vertebrae so that the needle can be inserted with ease. A sitting position of the patient with head flexed to chest can also be used for lumbar puncture.

Composition of CSF in health and disease

The normal composition of cerebrospinal fluid is given in the ***Table 22.2***. From the diagnostic point of view, the total cell count of lymphocytes (Reference : 0-5-10^6/1), protein concentration (15–45 mg/dl) and glucose concentration (45–85 mg/dl) are important.

In the ***Table 22.3***, the major alterations in the CSF in the disease states are given. The total ***cell count*** and ***protein content*** are ***increased*** while ***glucose*** concentration is ***reduced in tuberculosis meningitis***.

TABLE 22.2 Normal composition of cerebrospinal fluid

Parameter	Description/concentration
Volume	90–150 ml
Appearance	Clear and colourless
Specific gravity	1.006–1.008
Osmolality	280–290 mOsm/kg
Total cell count (lymphocytes)	0–5 × 10^6/l
pH	7.3–7.4
Protein	15–45 mg/dl
A/G ratio (albumin/globulin)	8 : 1
Glucose	45–85 mg/dl
Chloride	118–130 mEq/l
Calcium	2.1–2.7 mEq/l
Sodium	145–155 mEq/l
Potassium	2.0–3.5 mEq/l

- **Improper formation of collagen is associated with certain genetic diseases e.g. Ehlers-Danlos syndrome (abnormal tissue fragility), osteogenesis imperfecta (abnormal fragility of bones).**
- **Protein misfolding results in prion diseases (e.g. mad cow disease) and amyloidosis (Alzheimer's disease).**
- **Biochemical analysis of CSF is useful for the diagnosis of certain diseases – tuberculosis meningitis (increased protein and decreased glucose concentrations).**
- **Amniotic fluid is analysed to assess fetal maturity (creatinine concentration > 1.6 mg/dl), lung maturity (lecithin-sphingomyelin ratio > 2 : 1) and for the prenatal diagnosis of congenital disorders (e.g. hemophilia, Down's syndrome).**

In case of brain tumors, there is no change in total cell count while the protein concentration may be marginally increased.

The colour and appearance of CSF is sometimes a guiding factor in the disease diagnosis. For instance, CSF is opalescent and slightly yellow coloured in tuberculosis meningitis.

AMNIOTIC FLUID

Amniotic fluid is a ***liquid produced by the membranes and the fetus***. It ***surrounds the fetus*** throughout pregnancy. The volume of amniotic fluid increases with the gestational age. Thus, the volume increases from 30 ml (at 2 weeks of gestation) to 350 ml (at 20 weeks), and thereafter to 500–1000 ml. Amniotic fluid is almost clear with some desquamated fetal cells and a little lipid.

Functions of amniotic fluid

- It provides physical ***protection to the fetus***.
- Amniotic fluid is a medium for the ***exchange of*** various ***chemicals***.

Diagnostic importance of amniotic fluid

Assessment of fetal maturity : Fetal maturity can be assessed by cytological staining of fat cells, and estimation of creatinine concentration (> 1.6 mg/dl indicates fetal maturity).

Lung maturity : The fetal lung maturity is evaluated by measuring ***lecithin–sphingomyelin (L/S) ratio***. A L/S ratio of ***2 : 1*** or more ***indicates lung maturity***. If L/S ratio is less than 1.2 : 1, it is better to delay the induced delivery until the lung has become more mature.

Diagnosis of congenital disorders : Amniotic fluid analysis is useful for the ***prenatal diagnosis*** of congenital disorders. Some of the important ones are listed.

- Chromosomal disorders such as ***Down's syndorme***.
- Metabolic disorders e.g. cystic fibrosis.
- Sex-linked disorders e.g. hemophilia.

Table 22.3 Changes in cerebrospinal fluid in the disease states

Disease	*Colour and appearance*	*Total cell count*	*Protein*	*Glucose*
Normal	Clear and colourless	0–5 ↔ 10^6/l	15–45 mg/dl	45–85 mg/dl
Tuberculosis meningitis	Opalescent and slightly yellow	Increased	Increased	Relatively low
Bacterial meningitis	Opalescent and turbid	Markedly increased	Markedly increased	Markedly decreased
Brain tumour	Clear and colourless	No change	Increased	Low
Subarachnoid hemorrhage	Slightly blood colour	RBC and WBC present	Increased	Almost normal

AQUEOUS HUMOR

Aqueous humor is the ***fluid that fills the anterior chamber of the eye***. This fluid is responsible for maintaining the intraocular tension. Aqueous humor, secreted by the ciliary body, enters the anterior chamber. Blockade in the flow of aqueous humor causes ***glaucoma due to increased intraocular pressure***.

PLEURAL FLUID

Pleural fluid is the filtrate of plasma, and is present in a minimal quantity in pleural cavity. The amount of pleural fluid increases in disease states due to ***pleural effusion***.

Pleural fluid may be ***transudate*** or ***exudate***, based on the composition. If the ratio of protein content between pleural fluid and plasma is less than 0.5, it is transudate. If this ratio is more than 0.5, the pleural fluid is exudate.

Transudate is fluid-like, with ***specific gravity <1.015***. It will not clot, and is not associated with inflammation. Transudate is observed in nephrotic syndrome (due to low osmotic pressure of plasma), and congestive cardiac failure (increased hydrostatic pressure).

Exudate is viscous in nature ***with specific gravity >1.015***. It may clot, and is associated with inflammation or malignancy. Exudate accumulates in various infections (tuberculosis, pneumonia, rheumatoid arthritis) and some cancers (e.g. lung cancer).

Biochemical analysis of exudate is useful in certain disease states

- Amylase activity increased in pancreatitis.
- Rheumatoid factor elevated in rheumatoid arthritis.
- Carcinoembryonic antigen (CEA) increased in malignancy.

SUMMARY

1. *The major proteins of connective tissue are collagen, elastin, fibrillin, and proteoglycans. Among these, collagen is the most abundant, constituting one-third of the total body proteins.*

2. *Type I mature collagen is a triple helical structure i.e. contains three polypeptide chains. The repetitive amino acid sequence of collagen is* $(Gly\text{-}X\text{-}Y)_n$. *Glycine constitutes about 1/3 rd of the amino acids while X and Y represent other amino acids.*

3. *Muscle is the single largest tissue of the human body (30–40% of body weight). It is composed of fibre cells into which myofibrils are embedded. Each myofibril contains alternating A and I bands. Sarcomere is the functional unit of muscle.*

4. *Actin, myosin, tropomyosin and troponin are the major contractile proteins found in muscles. The muscle contraction and relaxation occur due to the active involvement of these proteins. ATP is the immediate source of energy for muscle contraction.*

5. *Milk is almost a complete food with various nutrients—carbohydrates, lipids, proteins, vitamins and minerals. However, milk is deficient in vitamin C, iron and copper.*

6. *Cerebrospinal fluid is an ultrafiltrate of plasma. In the disease, tuberculosis meningitis, the total cell count and protein concentration are increased, while glucose concentration is decreased in CSF.*

THE READER MUST REFER APPENDIX D (Page Nos : 422 - 431) FOR RELATED CLINICAL CASE STUDIES (No : 39)

NUTRITION

23

Whether a man eats for living or lives for eating, food is his prime concern. ***Nutrition*** may be defined as the ***utilization of food by living organisms.*** Biochemists have largely contributed to the science of nutrition.

Methodology in nutrition : Most of the existing knowledge on nutrition is originally derived from ***animal experimentation***. This is despite the fact that there may exist several differences in the biochemical composition between man and animals! For instance, some animals can synthesize ascorbic acid while man cannot do so.

Study of human nutrition : The study of nutrition may be logically divided into three areas—***ideal nutrition***, ***undernutrition*** and ***overnutrition***. Ideal nutrition is the concern of everyone. Undernutrition is the prime concern of developing countries while overnutrition is a serious concern of developed countries.

Macronutrients and micronutrients

The nutrients necessary to support growth and maintenance of living organisms are broadly divided into two groups—macronutrients and micronutrients. The macronutrients forming the ***main bulk of food are carbohydrates, proteins and fats***. The ***micronutrients*** are ***required in small quantities*** and they include the various ***vitamins*** and ***minerals***.

Essential nutrients

A nutrient is said to be essential when it is ***required for normal physiological functions*** of the body and ***cannot be synthesized***, and therefore must be supplied through the diet. In case of humans, there are approximately ***8 amino acids, 2 fatty acids, 13 vitamins and 15 minerals*** that are considered ***as essential nutrients***. Besides these, there are certain ***conditionally essential nutrients***, as they are indispensable in developmental stages, and in pathological states.

NUTRITION AND ENERGY SUPPLY

Food is the fuel source of the body. The ingested food undergoes metabolism to liberate energy required for the vital activities of the body.

Energy content of foods

The calorific value (energy content) of a food is calculated from the heat released by the total combustion of food in a calorimeter.

Unit of heat : ***Calorie*** is the unit of heat. One calorie represents the amount of ***heat required to rise the temperature of one gram of water by 1°C*** (i.e. from 15° to 16°C). A calorie is too small a unit. Therefore, it is more conveniently expressed as kilocalories (1,000 times calorie) which is represented by kcal or simply ***Cal*** (with capital ***'C'***).

Calorie value of foods : The energy values of the three principal foodstuffs—carbohydrate, fat and protein—measured in a bomb calorimeter and in the body are given in the ***Table 23.1***. The energy values

Table 23.1 Calorific values of foodstuffs

Foodstuff	Energy value (Cal/g)	
	In bomb calorimeter	In the body
Carbohydrate	4.1	4
Fat	9.4	9
Protein	5.4	4
Alcohol	7.1	7

of carbohydrates, fats and proteins (when utilized in the body) respectively, are 4, 9 and 4 Cal/g.

Alcohol is a recent addition to the calorie (***7 Cal/g***) contribution, as it is a significant dietary component for some people.

It must be noted that the nutrients, namely ***vitamins and minerals***, have ***no calorific value***, although they are involved in several important body functions, including the generation of energy from carbohydrates, fats and proteins.

UTILIZATION OF ENERGY IN MAN

Man consumes energy to meet the fuel demands of the three ongoing processess in the body.

1. Basal metabolic rate
2. Specific dynamic action
3. Physical activity.

Besides the above three, additional energy supply is needed during growth, pregnancy and lactation.

BASAL METABOLIC RATE

Basal metabolism or basal metabolic rate (BMR) is defined as the ***minimum amount of energy required by the body*** to maintain life ***at complete physical and mental rest*** in the post-absorptive state (i.e. 12 hours after the last meal). The term ***resting metabolic rate (RMR)*** is in recent use for BMR.

Under the basal conditions, although the body appears to be at total rest, several functions within the body continuously occur. These include working of heart and other organs, conduction of nerve impulses, reabsorption by renal tubules, gastrointestinal motility and ion transport across membranes (***Na^+-K^+ pump*** consumes about 50% of basal energy).

Normal values of BMR : ***For adult man 35–38 Cal/sq. m/hr; for adult women 32-35 Cal/sq.m/hr.*** A BMR value between –15% and +20% is considered as normal.

Some authors continue to represent BMR as Cal/day. ***For an adult man BMR is around 1,600 Cal/day, while for an adult woman around 1,400 Cal/day.*** This is particularly important for easily calculating energy requirements per day.

Factors affecting BMR

1. **Surface area :** The BMR is ***directly proportional to the surface area***.

2. **Sex :** ***Men*** have ***marginally higher*** (about 5%) BMR than women. This is due to the higher proportion of lean muscle mass in men.

3. **Age :** In infants and growing children, with lean muscle mass, the BMR is higher. In ***adults, BMR decreases*** at the rate of about 2% per decade of life.

4. **Physical activity :** BMR is ***increased*** in persons (notably athletes) ***with regular exercise***. This is mostly due to increase in body surface area.

5. **Hormones :** ***Thyroid hormones*** (T_3 and T_4) have a ***stimulatory effect*** on the metabolism of the body and, therefore, BMR. Thus, BMR is raised in hyperthyroidism and reduced in hypothyroidism. In fact, the measurement of BMR was earlier used to assess thyroid function.

6. **Environment :** In cold climates, the BMR is higher compared to warm climates.

7. **Disease states :** BMR is elevated in various infections, leukemias, polycythemia, cardiac failure, hypertension etc. In Addison's disease (adrenal insufficiency), BMR is marginally lowered.

SPECIFIC DYNAMIC ACTION

The phenomenon of the ***extra heat production by the body***, over and above the calculated caloric value, when a given ***food is metabolized*** by the body, is known as specific dynamic action (SDA). It is also known as ***calorigenic action*** or ***thermogenic action*** or thermic action (effect) of food.

SDA for different foods : SDA for protein, fat and carbohydrate are 30%, 13% and 5%, respectively. Thus, ***proteins*** possess the ***highest SDA*** while ***carbohydrates*** have the ***lowest***.

SDA for mixed diet : For a regularly consumed mixed diet, the SDA is ***around 10%***.

Significance of SDA : For the utilization of foods by the body, certain amount of energy is consumed from the body stores. This is actually an expenditure by the body for the utilization of foodstuffs. It is the highest for proteins (30%) and lowest for carbohydrates (5%), and for a mixed diet around 10%. It is, therefore, essential that an additional 10% calories should be added to the total energy needs (of the body) towards SDA. And the diet should be planned, accordingly. [SDA is quite comparable to the handling charges levied by a bank for an outstation cheque].

Mechanism of SDA : The exact cause of SDA is not known. It is generally believed that SDA of foods is due to the energy required for digestion, absorption, transport, metabolism and storage of foods in the body.

PHYSICAL ACTIVITY OF THE BODY

The physical activity of the individual is highly variable. The amount of energy needed for this depends mainly on the duration and intensity of

Medical Concepts/ Clinical Correlates

- The dietary essential nutrients for humans are approximately 8 amino acids, 2 fatty acids, 13 vitamins and 15 minerals.
- The body at total rest (physical and mental) requires energy to meet the basal requirements such as working of heart, conduction of nerve impulse, membrane transport etc.
- Non-exercise physical activity (NEPA) is important in keeping the body in order. Defect in NEPA significantly contributes to obesity.
- Adequate intake of dietary fiber prevents constipation, eliminates bacterial toxins, reduces GIT cancers, improves glucose tolerance and reduces plasma cholesterol.

muscular activity. The expenditure of energy for the various physical activities has been calculated (***Table 23.2***).

Energy requirements of man

As already stated, the three factors—basal metabolic rate, specific dynamic action and physical activity—determine the energy needed by the body. In an individual with light work, ***about 60% of the calories are spent towards BMR, about 30% for physical activity and about 10% to take care of the SDA***.

The daily requirement of energy is rather variable which depends on the BMR (in turn depends on age, sex, body size etc.) and physical activity. As per some rough calculation, caloric requirements of adults per day (***Cal/day***) are in the following ranges.

Light work	—	2,200 –2,500
Moderate work	—	***2,500 –2,900***
Heavy work	—	2,900 –3,500
Very heavy work	—	3,500–4,000

Non-exercise activity thermogenesis (NEAT)

NEAT, also called as ***non-exercise physical activity (NEPA)*** refers to the ***energy expended for all our routine physical activities other than sleeping*** or ***sports-like exercises***. These include the energy used for walking to work, climbing steps, performing yard work, kitchen work and fidgeting.

Normally, about 60–70% of calories are spent towards BMR, 10–15% towards SDA, and at least 15–30% towards NEAT. In individuals, with very sedentary life, NEAT may be as low as 15% while in active individuals, it may go up to 50%.

NEAT plays an important role in keeping the body in order. ***Defect in NEAT significantly contributes to obesity***.

NUTRITIONAL IMPORTANCE OF CARBOHYDRATES

Dietary carbohydrates are the ***chief source of energy***. They contribute to 60-70% of total caloric requirement of the body. Incidentally, carbohydrate rich foods cost less.

From the nutritional point of view, carbohydrates are grouped into 2 categories.

1. Carbohydrates ***utilized*** by the body—starch, glycogen, sucrose, lactose, glucose, fructose etc.
2. Carbohydrates ***not utilized*** (***not digested***) by the body—cellulose, hemicellulose, pectin, gums etc.

Table 23.2 Type of physical activity and energy expenditure (over and above BMR, about 65 Cal/hr).

Physical activity	*Energy requirement (Cal/hr)*
Sitting (quietly)	25
Standing (quietly)	30
Writing/eating/reading	30
Car driving	60
Typing	75
Household work (dish washing)	80
Walking (slow)	130
Sexual intercourse	140
Cycling (slow)	150
Running (moderate)	500
Swimming	600
Walking upstairs	800

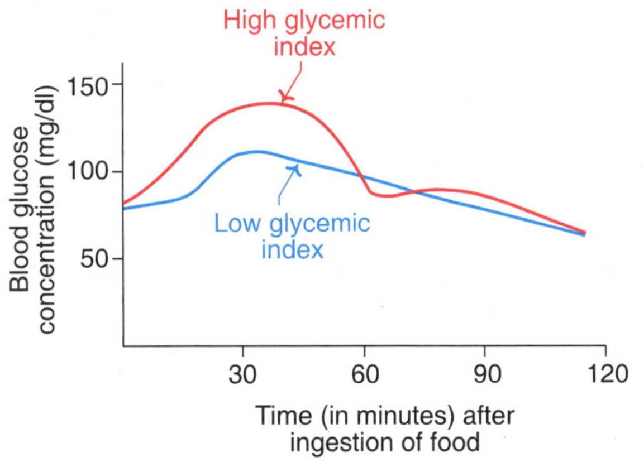

***Fig.23.1 :** The glycemic index curve after the ingestion of two different foods.*

Functions of carbohydrates

1. **Major sources of energy :** Carbohydrates are the principal source of energy, supplying 60–80% of the caloric requirements of the body.

2. **Protein sparing action :** Proteins perform a specialized function of body building and growth. The wasteful expenditure of proteins to meet the energy needs of the body should be curtailed. Carbohydrates come to the rescue and spare the proteins from being misused for caloric purpose.

3. **Absolute requirement by brain :** The brain and other parts of central nervous system are dependent on glucose for energy.

4. **Required for the oxidation of fat :** Acetyl CoA is the product formed in fatty acid oxidation. For its further oxidation via citric acid cycle, acetyl CoA combines with oxaloacetate, the latter is predominantly derived from carbohydrates. It may therefore be stated '***Fat burns in a fuel of carbohydrate***'.

5. **Synthesis of non-essential amino acids :** The intermediates of carbohydrate metabolism, mainly the keto acids (e.g. pyruvic acid), serve as precursors for the synthesis of non-essential amino acids.

6. **Synthesis of fat :** Excess consumption of carbohydrates leads to the formation of fat which is stored in the adipose tissue.

Glycemic index

There are variations in the increase and fall of blood glucose levels after the ingestion of different carbohydrate containing foods. These quantitative differences are assayed by glycemic index which ***measures*** the ***time course of post-prandial glucose concentrations*** from a graph. Glycemic index may be defined as the area under the blood glucose curve after the ingestion of a food compared with the area under the blood glucose curve after taking the same amount of carbohydrate as glucose. It is expressed as percentage.

$$\text{Glycemic index} = \frac{\text{Area under the blood glucose curve after ingestion of test meal}}{\text{Area under the curve after ingestion of glucose}} \times 100$$

A graphic representation of high and low glycemic indices is depicted in ***Fig.23.1***.

The glycemic index of a ***complex carbohydrate (i.e. starch) is lower than that of a refined carbohydrate (i.e. glucose)***. This is explained on the basis of slow digestion and absorption of complex carbohydrates. Further, the glycemic index of carbohydrate is usually lower when it is combined with protein, fat or fiber.

Sources of carbohydrates

Carbohydrates are abundant in several naturally occurring foods. These include table sugar (99%), ***cereals (60–80%), pulses*** (50–60%), roots and tubers (20–40%) and bread (50–60%).

Requirement of carbohydrates

In a well balanced diet, at ***least 40% of the caloric*** needs of the body should be met from carbohydrates.

FIBER IN NUTRITION

The ***complex carbohydrates*** that are ***not digested*** by the human enzymes are collectively referred to as dietary fiber. ***Soluble fibers***, mostly found in fruits and legumes, dissolve in water and form gels (e.g., ***pectins***, gums, mucilages). ***Insoluble fibers***, present in vegetables and grains, adsorb water and swell up (e.g. ***cellulose hemicellulose***, lignin). Certain fibers (e.g. pectins, gums) are digestible by intestinal bacterial ezymes. It may be stated that once regarded as a nutritional waste, a lot of importance is now given to dietary fiber in human health.

Beneficial effects of fiber

1. **Prevents constipation :** Fiber can absorb 10–15 times its own weight of water, by drawing fluid into the lumen of the intestine. This increases ***bowel motility***, and prevents constipation, besides decreasing the risk of hemorrhoids and diverticulosis.

2. **Eliminates bacterial toxins :** Fiber also ***adsorbs toxic compounds*** produced by intestinal bacteria and helps in their easier expulsion.

3. **Decreases GIT cancers :** The ***lower*** incidence of cancers of gastrointestinal tract (e.g. colon and rectum) ***in vegetarians*** compared to non-vegetarians is attributed to dietary fiber.

4. **Improves glucose tolerance :** Fiber improves glucose tolerance by the body. This is mainly done by a diminished rate of glucose absorption from the intestine.

5. **Reduces plasma cholesterol level :** Fiber decreases the absorption of dietary cholesterol from the intestine. Further, fiber binds with the bile salts and reduces their enterohepatic circulation. Thus, degradation of cholesterol to bile salts and its disposal from the body is increased.

6. **Satiety value :** Dietary fiber adds to the weight of the foodstuff ingested and gives a sensation of stomachfulness, giving satiety without consumption of excess calories.

Sources of dietary fiber

Fruits, leafy vegetables, vegetables, whole wheat legumes, rice bran etc. are rich sources of fiber.

NUTRITIONAL IMPORTANCE OF LIPIDS

Triacylglycerols (fats and oils) are the concentrated dietary source of fuel, contributing 15-50% of the body energy requirements.

Major nutritional functions of lipids

Dietary lipids have two major nutritive functions.

1. Supply triacylglycerols that normally constitute about 90% of dietary lipids which is a concentrated source of fuel to the body.

2. Provide essential fatty acids and fat soluble vitamins (A, D, E and K).

ESSENTIAL FATTY ACIDS

The unsaturated fatty acids which the ***body cannot synthesize*** and, therefore, must be ***consumed in the diet*** are referred to as essential fatty acids (EFA).

The fatty acids—***linoleic*** and ***linolenic acid***—cannot be synthesized by humans. In a strict sense, only these two are essential fatty acids. Arachidonic acid can be synthesized from linoleic acid in some animal species, including man. However, the conversion efficiency of linoleic acid to arachidonic acid is not clearly known in man. And for this reason, some nutritionists recommend that it is better to include ***some amount of arachidonic acid*** also in the diet.

Functions of EFA

1. Essential fatty acids are the ***structural components of biological membranes***.

2. Participate in the ***transport and utilization of cholesterol***.

3. ***Prevent fat*** accumulation in the ***liver***.

4. Required for the ***synthesis of prostaglandins***.

5. Maintain proper growth and reproduction of the organisms.

Deficiency of EFA

The EFA deficiency in humans is characterized by a scaly dermatitis on the posterior and lateral parts of limbs and buttocks. This condition is referred to as ***phrynoderma*** or ***toad skin***. Poor wound healing and hair loss is also observed in EFA deficiency.

EFA content of foods

The essential fatty acids, more frequently called ***polyunsaturated fatty acids (PUFA)***, are predominantly present in vegetable oils and fish oils. The rich vegetable sources include sunflower oil, cotton seed oil, corn oil, soyabean oil etc.

The fats of animal origin (exception—fish) contain less PUFA e.g. butter, fat of meat, pork and chicken.

Dietary intake of EFA

Nutritionists recommend that at least 30% of the dietary fat should contain PUFA. Very high intake of PUFA (i.e. totally replacing saturated fatty acids) may not be advisable. This is due to the fact that excess PUFA, unless accompanied by ***antioxidants*** (vitamin E, carotenes), is believed to be injurious to the cells due to the overproduction of free radicals.

ω-3 and ω-6 fatty acids

These are long chain PUFA with double bond beginning at 3rd (ω-3) and 6th (ω-6) position from the methyl end. ω-3 fatty acids (e.g. linolenic acid, ***docosahexaenoic acid (DHA), eicosapentaenoic acid (EPA)*** and ω-6 fatty acids (e.g. arachidonic acid) are known to ***reduce serum cholesterol*** and ***triacylglycerols***, thereby decrease the tendency for thrombosis, lower blood presure and reduce the risk of CHD. In recent years, ω-3 fatty acids are included in infant formulas to ***promote brain development***.

Requirement of dietary fat

The recommended fat intake is around 20–30% of the daily calorie requirement, containing about 50% of PUFA.

NUTRITIONAL IMPORTANCE OF PROTEINS

Proteins have been traditionally regarded as '***body-building foods***'. However, 10–15% of the total body energy is derived from proteins.

Functions of proteins

1. Proteins are the ***fundamental basis of cell structure and its function***.

2. All the ***enzymes***, several hormones, immunoglobulins, transport carriers etc., are proteins.

3. Proteins are involved in the maintenance of ***osmotic pressure***, clotting of blood, muscle contraction etc.

4. During starvation, proteins (amino acids) serve as the major suppliers of energy.

Essential amino acids

The nutritional importance of proteins is based on the content of essential amino acids. The details of the essential amino acids are described under chemistry of proteins (***Chapter 4***). There are ten essential amino acids—***arginine, valine, histidine, isoleucine, leucine, lysine, methionine, phenylalanine, tryptophan and threonine*** (code to recall—***AV HILL MP TT***). Of these two—namely arginine and histidine—are semiessential.

NITROGEN BALANCE

Dietary protein is almost an exclusive source of nitrogen to the body. Therefore, the term nitrogen balance truly represents the protein (16% of which is nitrogen) utilization and its loss from the body.

Nitrogen balance is determined by comparing the intake of nitrogen (chiefly by proteins) and the excretion of nitrogen (mostly undigested protein in feces; urea and ammonia in urine). A normal healthy adult is in a ***nitrogen equilibrium*** since the daily ***dietary intake*** (I) ***is equal to the loss*** through ***urine (U)***, ***feces (F)*** and ***sweat (S)***.

$$I = U + F + S$$

Thus, an individual is said to be in a nitrogen balance if the intake and output of nitrogen are the same (***Fig.23.2***). There are two other situations—a positive and a negative nitrogen balance.

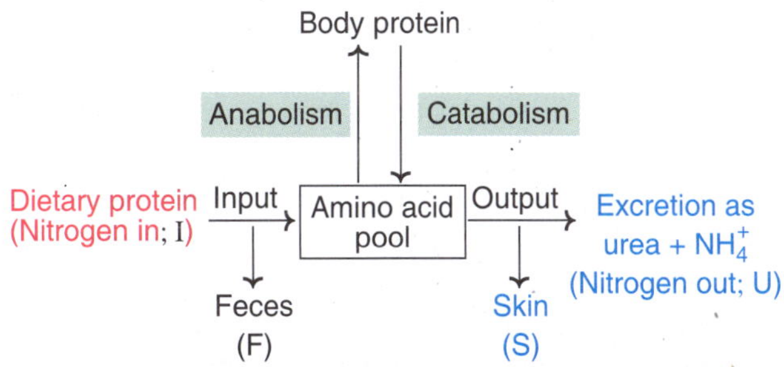

Fig.23.2 : *Overview of nitrogen balance (At equilibrium N input = N output; For positive N balance, N input > N output; for negative N balance N input < N output).*

Positive nitrogen balance : This is a state in which the ***nitrogen intake is higher than the output.*** Some amount of nitrogen is retained in the body causing a net increase in the body protein. Positive nitrogen balance is observed in growing children, pregnant women or during recovery after serious illness.

Negative nitrogen balance : This is a situation in which the ***nitrogen output is higher than the input***. The result is that some amount of nitrogen is lost from the body depleting the body protein. Prolonged negative nitrogen balance may even lead to death. This is sometimes observed in children suffering from kwashiorkor or marasmus.

Negative nitrogen balance may occur due to inadequate dietary intake of protein (deficiency of even a single essential amino acid) or destruction of tissues or serious illness. In all these cases, the body adapts itself and increases the breakdown of tissue proteins causing loss of nitrogen from the body.

Assessment of nutritive value of proteins

The quality of the proteins depends on the composition of essential amino acids. Several laboratory methods are in use to assess the nutritive value of proteins.

The data obtained on the nutritive value of proteins (from major food sources) employing four different methods — ***protein efficiency ratio (PER)***, ***biological value (BV)***, ***net protein utilization (NPU)*** and ***chemical score*** are given in ***Table 23.3***.

Mutual supplementation of proteins

As is observed from the ***Table 23.3***, the ***animal proteins are superior*** in their nutritive value compared to the proteins of vegetable origin. Further, some of the essential amino acids are limiting in certain foods. For instance, ***rice*** and ***wheat*** proteins are ***limiting in lysine*** and ***threonine*** while the protein of Bengal gram is limited in sulfur-containing amino acids (methionine and cystine).

It is fortunate that humans (worldover) have the habit of consuming a mixed diet, with different foods, simultaneously. This helps to overcome the deficiency of certain essential amino acids in one food by being supplemented from the others. This phenomenon is referred to as ***mutual supplementation***. For instance, an Indian diet with cereals (wheat, rice) is taken along with pulses (dal). The limitation of lysine and threonine in cereal proteins is overcome by their supplementation from dal proteins. Simultaneously, the limitation of sulfur-containing amino acids in dal is also compensated by the cereals which are rich in them.

The nutritive value of protein of a particular food can be enhanced by appropriate combination with other foods. Due to the consumption of mixed diets, ***dietary deficiency of essential amino acids is most uncommon***. Further, the principle of mixed diet takes care to supply adequate quantities of essential amino acids to the people subsisting on pure vegetarian diets. It has to be remembered that the effect of mutual supplementation in proteins is best observed with the same meal (or at least on the same day).

Requirement of proteins

The requirement of protein is dependent on its nutritive value, caloric intake and physiological states (growth, pregnancy, lactation) of the individual. For an ***adult, 0.8-1.0 g protein/kg body weight/day*** is adequate. The requirement should be nearly double for growing children, pregnant, and lactating women.

Dietary sources of proteins

The protein content of foods is variable, cereals have 6-12%; pulses 18-22%; meat 18-25%, egg 10-14%; milk 3-4% and leafy vegetables 1-2%. In general, the ***animal proteins*** are ***superior*** than vegetable proteins as the dietary source.

NUTRITIONAL IMPORTANCE OF VITAMINS AND MINERALS

The nutritional aspects including metabolism, biochemical functions, dietary sources, requirements and associated disorders for vitamins (***Chapter 7***) and for minerals (***Chapter 19***) have already been discussed in much detail.

RECOMMENDED DIETARY ALLOWANCES (RDA)

The recommended dietary/daily allowances (RDA) represents the ***quantities of the nutrients to be provided in the diet daily*** for maintaining good health and physical efficiency of the body. It must be remembered that RDA is not the minimum amount to just meet the body needs, but allowance is given for a safe margin.

Factors affecting RDA

1. **Sex :** The RDA for ***men*** is about ***20% higher than*** that for ***women***. Iron is an exception as the requirement is greater in menstruating women.

2. **Age :** In general, the nutrient requirement is ***higher in growing age***. For instance, the protein requirement for a growing child is about 2 g/kg body wt/day compared to 1 g/kg body wt/day for adults.

RDA for adult man

The details of RDA for each of the nutrients in relation to age, sex and physiological status is

Table 23.3 Nutritive value of food proteins, assessed by PER, BV, NPU and chemical score

Source of protein	*PER*	*BV*	*NPU*	*Chemical score*	*Limiting amino acid(s)*
Egg	4.5	94	90	100	Nil
Milk	3.0	84	75	65	S-Containing amino acids
Fish	3.0	85	70	60	Tryptophan
Meat	2.7	75	76	70	S-Containing amino acids
Rice	2.2	68	60	60	Lysine, threonine
Wheat	1.5	58	47	42	Lysine, threonine
Bengal gram	1.7	58	47	45	S-Containing amino acids
Red gram	1.5	57	46	45	S-Containing amino acids
Groundnut	1.7	55	45	44	Lysine, threonine, S-amino acids

PER–Protein efficiency ratio; BV–Biological value; NPU–Net protein utilization; S–Sulfur.

Medical Concepts/ Clinical Correlates

- In general, vegetable oils are good sources for essential fatty acids while animal proteins are superior for the supply of essential amino acids.
- The recommended dietary allowance (RDA) of nutrients depends on the sex and age, besides pregnancy and lactation in the women.
- Kwashiorkor and marasmus, the two extreme forms of protein-energy malnutrition in infants and pre-school children, are highly prevalent in developing countries.
- Use of oral contraceptives may lead to deficiencies of vitamins B_6, B_{12}, and folic acid.

described in the respective chapters. For a quick recapitulation, the RDA of macronutrients (carbohydrate, fat and protein) and selected micronutrients (vitamins and minerals) for an adult man weighing 70 kg are given in ***Table 23.4***.

BALANCED DIET

After discussing the nutritional aspects of dietary ingradients and their RDA, it is worthwhile to formulate a diet for man. A balanced diet or ***prudent diet*** is defined as the diet which ***contains different types of foods, possessing the nutrients—carbohydrates, fats, proteins, vitamins and minerals—in a proportion to meet the requirements of the body***. A balanced diet invariably supplies a little more of each nutrient than the minimum requirement to withstand the short duration of leanness and keep the body in a state of good health.

The basic composition of balanced diet is highly variable, as it differs from country to country, depending on the availability of foods. Social and cultural habits, besides the economic status, age, sex and physical activity of the individual largely influence the intake of diet.

Table 23.4 Recommended dietary allowance (RDA) of important nutrients for an adult man, weighing 70 kg.

Nutrient(s)	*RDA*
Carbohydrates	400 g
Fats	70 g
Proteins	56 g*
Essential fatty acids	4 g
Vitamin A	1,000 µg**
Vitamin D	5 µg***
Vitamin E	10 µg
Vitamin K	70 µg
Ascorbic acid	60 mg
Thiamine	1.5 mg
Riboflavin	2 mg
Niacin	20 mg
Pyridoxine	2 mg
Folic acid	150 µg
Cobalamin	2 µg
Calcium	800 mg
Phosphorus	800 mg
Iron	10 mg

* *0.8 g/kg body weight/day;* ** *Retinol equivalents;* *** *As cholecalciferol*

The Nutrition Expert Group, constituted by the Indian Council of Medical Research has recommended the composition of balanced diets for Indians. This is done taking into account the commonly available foods in India.

NUTRITIONAL DISORDERS

While the people of developing countries suffer from ***undernutrition***, ***overnutrition*** is the major concern of the developed countries. Some of the important nutritional diseases are discussed hereunder.

PROTEIN-ENERGY MALNUTRITION

Protein-energy malnutrition (***PEM***)—sometimes called ***protein-calorie malnutrition (PCM)***—is the most common nutritional disorder of the developing countries. PEM is widely prevalent in the infants and pre-school children. ***Kwashiorkor*** and ***marasmus*** are the two extreme forms of protein-energy malnutrition.

Kwashiorkor

Kwashiorkor literally means ***sickness of the deposed child*** i.e. a disease the child gets when the next baby is born.

Occurrence and causes : Kwashiorkor is predominantly found in children between 1-5 years of age. This is primarily due to ***insufficient intake of proteins***, as the diet of a weaning child mainly consists of carbohydrates.

Clinical symptoms : The major clinical manifestations of kwashiorkor include ***stunted growth***, ***edema*** (particularly on legs and hands), diarrhea, discoloration of hair and skin, anemia, apathy and moonface.

Biochemical manifestations : Kwashiorkor is associated with a ***decreased plasma albumin*** concentration (<***2 g/dl*** against normal 3–4.5 g/dl), fatty liver, deficiency of K^+ due to diarrhea. Edema occurs due to lack of adequate plasma proteins to maintain water distribution between blood and tissues. Disturbance in the metabolism of carbohydrate, protein and fat is also observed. Several vitamin deficiencies occur.

Treatment : Ingestion of ***protein-rich foods*** or the dietary combinations to provide about 3–4 g of protein/kg body weight/day will control kwashiorkor. The treatment can be monitored by measuring plasma albumin concentration, disappearance of edema and gain in body weight.

Marasmus

Marasmus literally means '***to waste***'. It mainly occurs in children under one year age. Marasmus is predominantly due to the ***deficiency of calories.*** This is usually observed in children given watery gruels (of cereals) to supplement the mother's breast milk.

The symptoms of marasmus include ***growth retardation***, ***muscle wasting*** (emaciation), anemia and weakness. A marasmic child does not show edema or decreased concentration of plasma albumin. This is a major difference to distinguish marasmus from kwashiorkor.

OTHER NUTRITIONAL DISORDERS

There are several other nutritional disorders which have been discussed elesewhere. These include ***obesity*** and ***atherosclerosis*** (***Chapter 14***); vitamin deficiency disorders—***xerophthalmia***, ***rickets***, ***beri-beri***, ***pellagra***, ***scurvy*** and ***pernicious anemia*** (***Chapter 7***); ***goiter*** and other disorders of minerals (***Chapter 18***). The biochemical ramifications of starvation are discussed along with the integration of metabolism (***Chapter 16***).

OBESITY

Obesity is an abnormal increase in the body weight due to excessive fat deposition.

Nutritional basis

Men and women are considered as obese if their weight due to fat (in adipose tissue), respectively, exceeds more than 20% and 25% of body weight. Obesity is basically a disorder of excess calorie intake, in simple language—***overeating***. It has to be remembered that every 7 calories of excess consumption leads to 1 g fat deposit and increase in body weight. Over-eating—coupled with ***lack of physical exercise***—contribute to obesity.

Body mass index (BMI)

Clinical obesity is represented by body mass index. BMI is calculated as the weight (in kilograms) divided by the height (in meters^2).

$$BMI\ (kg/m^2) = \frac{Weight\ (kg)}{[height\ (m)^2]}$$

Healthy ***reference range*** for BMI is between ***18.5–24.9 kg/m²***.

- Grade I obesity or overweight – BMI 25–30 kg/m^2
- Grade II or clinical obesity – BMI > 30 kg/m^2
- Grade III or morbid obesity – BMI > 40 kg/m^2

Obesity is associated with many health complications e.g. type II diabetes, CHD, hypertension, stroke, arthritis, gall bladder disease.

In recent years, the ***ratio between waist and hip sizes*** (for men < 0.9 and for women < 0.85) is considered as ***more effective than BMI***, particularly with regard to the risk of heart diseases. The lower is the waist to hip ratio, the lower the risk for health complications, and therefore better is the health.

Obesity and adipose tissue

Obesity is due to an increase in both the number and size of adipocytes (of adipose tissue). There are two types of adipose tissues

1. **White adipose tissue :** The fat is mostly stored and this tissue is metabolically less active.

2. **Brown adipose tissue :** The stored fat is less but the tissue is metabolically very active.

Brown adipose tissue possesses high proportion of mitochondria and cytochromes but low activity of ATP synthase. This is an active centre for the oxidation of fat and glucose and is responsible for the ***diet-induced thermogenesis***.

The peculiarity of mitochondria of brown adipose tissue is that the oxidation and phosphorylation are not coupled. Mitochondrial oxidation produces ***more heat*** and ***less ATP***. A specific protein—namely ***thermogenin***—has been isolated in the inner membrane of these mitochondria. Thermogenin functions like an uncoupler and dissipates the energy in the form of heat, and thus blocks the formation of ATP.

Brown adipose tissue is mostly found in hibernating animals, and the animals exposed to cold, besides the newborn. In adult humans, though not a prominent tissue, it is located in the thoracic region. It is significant to note that ***brown adipose tissue is almost absent in obese persons***. Some individuals are fortunate to have active brown adipose tissue. They eat and liberate it as heat, and therefore do not become obese.

Pharmacological treatment of obesity : In recent years, synthetic lipids such as ***Olestra*** and Orlistat are used to treat obesity. They taste like natural lipids but cannot be digested, and excreted unchanged.

METABOLIC SYNDROME

Metabolic syndrome (MS) is a cluster of different conditions that adversely affect the health. The ***components*** contributing to MS include ***abdominal obesity, insulin resistance, dyslipidemia, elevated blood pressure, overnutrition, sedentary lifestyles*** etc. As per WHO criteria, metabolic syndrome has the following characteristics

1. **Insulin resistance** – identified either as type 2 diabetes or elevated fasting blood glucose (>100 mg/dl) or impaired glucose tolerance.
2. And ***any two*** of the following
   (i) ***Hypertension*** (> = 140/90 mm Hg)
   (ii) ***Dyslipidemia*** (serum TG > = 150 mg/dl or HDL cholesterol <40 mg/dl in men or <50 mg/dl in women.
   (iii) ***BMI*** >30 kg/m^2 or waist : hip ratio of >0.9 in men or 0.85 in women.

Metabolic syndrome can be ***managed by healthy habits*** and change in lifestyles—restricted balanced diet, adequate intake of fiber, and antioxidants, exercise, avoiding smoking, stress-free life etc.

SUMMARY

1. *The calorific values of carbohydrates, fats and proteins respectively are 4, 9 and 4 Cal/g. These three nutrients (macronutrients) supply energy to the body to meet the requirements of basal metabolic rate, specific dynamic action and physical activity.*
2. *Basal metabolic rate (BMR) represents the minimum amount of energy required by the body to maintain life at complete physical and mental rest, in the post-absorptive state. The normal BMR for an adult man is 35-38 Cal/m^2 body surface/hr.*
3. *Carbohydrates are the major source of body fuel supplying about 40-70% of body calories. The non-digested carbohydrates (cellulose, pectin) are referred to as fiber. Adequate intake of fiber prevents constipation, improves glucose tolerance and reduces plasma cholesterol.*
4. *Lipids are the concentrated source of energy. They also provide essential fatty acids (linoleic and linolenic acids) and fat-soluble vitamins (A, D, E and K).*
5. *Proteins are the body building foods that supply essential amino acids, besides meeting the body energy requirement partly (10-15%).*
6. *Protein energy malnutrition (kwashiorkor and marasmus) obesity, atherosclerosis, vitamin deficiencies (e.g. beri-beri, rickets, scurvy) are important nutrition-related disorders.*

THE READER MUST REFER APPENDIX D (Page Nos : 422 - 431) FOR RELATED CLINICAL CASE STUDIES (Nos : 1, 2, 3, 4, 5, 6, 7, 17)

DNA-REPLICATION, RECOMBINATION, AND REPAIR

24

Deoxyribonucleic acid (DNA) is a macromolecule that carries genetic information from generation to generation. It is responsible to preserve the identity of the species over millions of years. ***DNA*** may be regarded as a ***reserve bank of genetic information*** or a memory bank.

The central dogma of life

The ***biological information flows from DNA to RNA, and from there to proteins***. This is the central dogma of life (***Fig.24.1***). It is ultimately the DNA that controls every function of the cell through protein synthesis.

As the carrier of genetic information, DNA in a cell must be duplicated (replicated), maintained and passed down accurately to the daughter cells. Three distinct processes are designed for this purpose. The '***three Rs' of DNA-r***eplication, ***r***ecombination, and ***r***epair, are dealt with in this chapter.

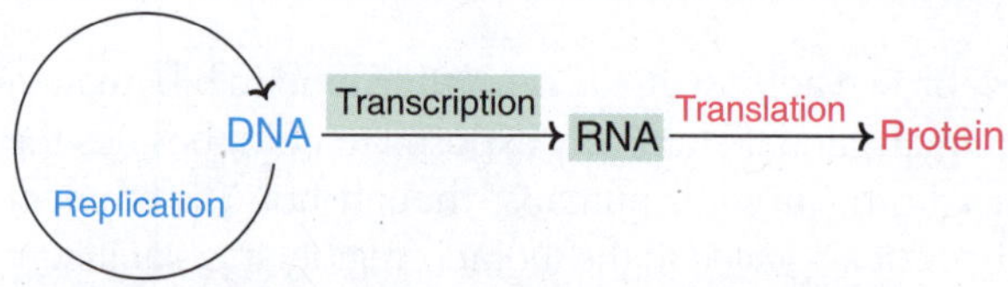

Fig.24.1 : *The central dogma of life.*

REPLICATION OF DNA

DNA is the genetic material. When the cell divides, the daughter cells receive an identical copy of genetic information from the parent cell.

Replication is a ***process in which DNA copies itself to produce identical daughter molecules of DNA***. Replication is carried out with high fidelity which is essential for the survival of the species. Synthesis of a new DNA molecule is a complex process involving a series of steps.

The salient features of replication in prokaryotes are described first. This is followed by some recent information on the eukaryotic replication.

REPLICATION IN PROKARYOTES

Replication is semiconservative

The parent DNA has two strands complementary to each other. Both the strands undergo simultaneous replication to produce two daughter molecules. Each one of the newly synthesized DNA has one-half of the parental DNA (one strand from original) and one-half of new DNA (***Fig.24.2***). This type of replication is known as semiconservative since ***half of the original DNA is conserved in the daughter DNA***. The first experimental evidence for the semiconservative DNA replication was provided by Meselson and Stahl (1958).

Initiation of replication

The initiation of DNA synthesis occurs at a site called ***origin of replication***. In case of prokaryotes, there is a single site whereas in ***eukaryotes***, there are ***multiple sites of origin***. These sites mostly consist of a short sequence of A-T base pairs. A specific protein called ***dna A*** (20-50 monomers) binds with the site of origin for replication. This causes the double-stranded DNA to separate.

Replication bubbles

The two complementary strands of DNA separate at the site of replication to form a bubble. Multiple

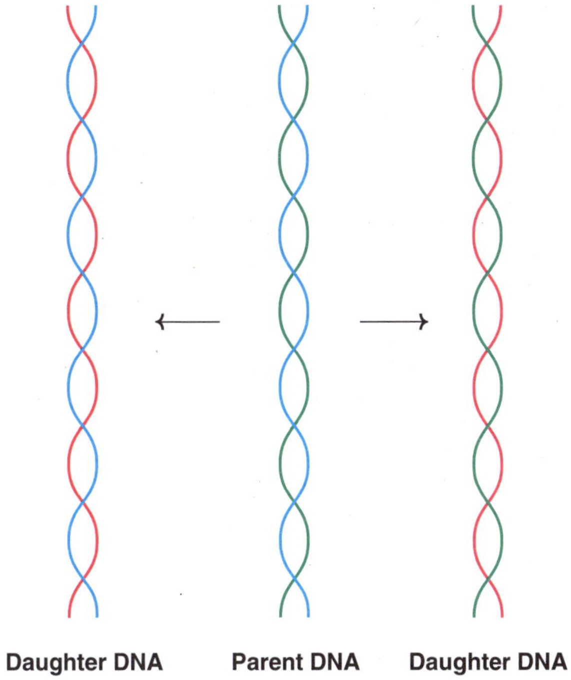

Fig.24.2 : DNA replication—semiconservative.

replication bubbles are formed in eukaryotic DNA molecules, which is essential for a rapid replication process (***Fig.24.3***).

RNA primer

For the synthesis of new DNA, a short fragment of RNA (about 5-50 nucleotides, variable with species) is required as a primer. The enzyme ***primase*** (a specific RNA polymerase) in association with single-stranded binding proteins forms a complex called ***primosome***, and produces RNA primers.

DNA synthesis is semidiscontinuous and bidirectional

The replication of DNA occurs in 5′ to 3′ direction, simultaneously, on both the strands of DNA. On one strand, the ***leading (continuous or forward) strand—the DNA synthesis is continuous***. On the other strand, the ***lagging (discontinuous or retrograde) strand—the synthesis of DNA is discontinuous***. Short pieces of DNA (15-250 nucleotides) are produced on the lagging strand.

In the replication bubble, the DNA synthesis occurs in both the directions (bidirectional) from the point of origin.

Replication fork and DNA synthesis

The separation of the two strands of parent DNA results in the formation of a replication fork. The active synthesis of DNA occurs in this region. The replication fork moves along the parent DNA as the daughter DNA molecules are synthesized.

DNA helicases : These enzymes bind to both the DNA strands at the replication fork. Helicases move along the DNA helix and separate the strands. Their function is comparable with a ***zip opener***. Helicases are dependent on ATP for energy supply.

Single-stranded DNA binding (SSB) proteins : These are also known as DNA helix-destabilizing proteins. They possess no enzyme activity. SSB proteins bind only to single-stranded DNA (separated by helicases), keep the two strands separate and provide the template for new DNA synthesis.

DNA synthesis catalysed by DNA polymerase III

The synthesis of a new DNA strand, catalysed by DNA polymerase III, occurs ***in 5′ → 3′ direction***. This is antiparallel to the parent template DNA strand. The presence of all the four deoxyribonucleoside triphosphates (dATP, dGTP, dCTP and dTTP) is an essential prerequisite for replication to take place.

The ***synthesis of two new DNA strands, simultaneously, takes place in the opposite direction***—one is in a direction (5′ → 3′) towards the replication fork which is continuous, the other in a direction (5′ → 3′) away from the replication fork which is ***discontinuous*** (***Fig.24.4***).ß

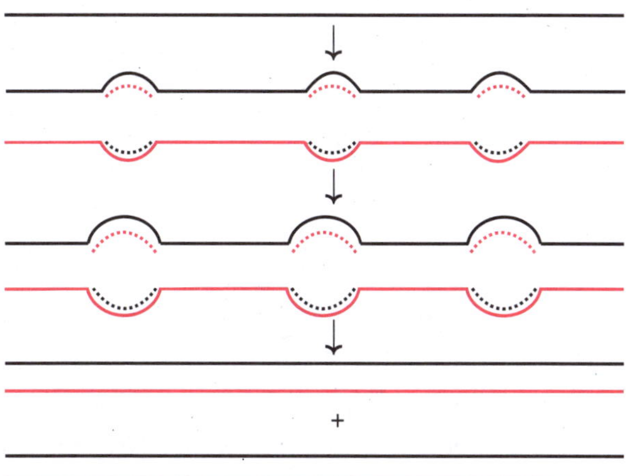

Fig.24.3 : Schematic representation of multiple replication bubbles in DNA replication.

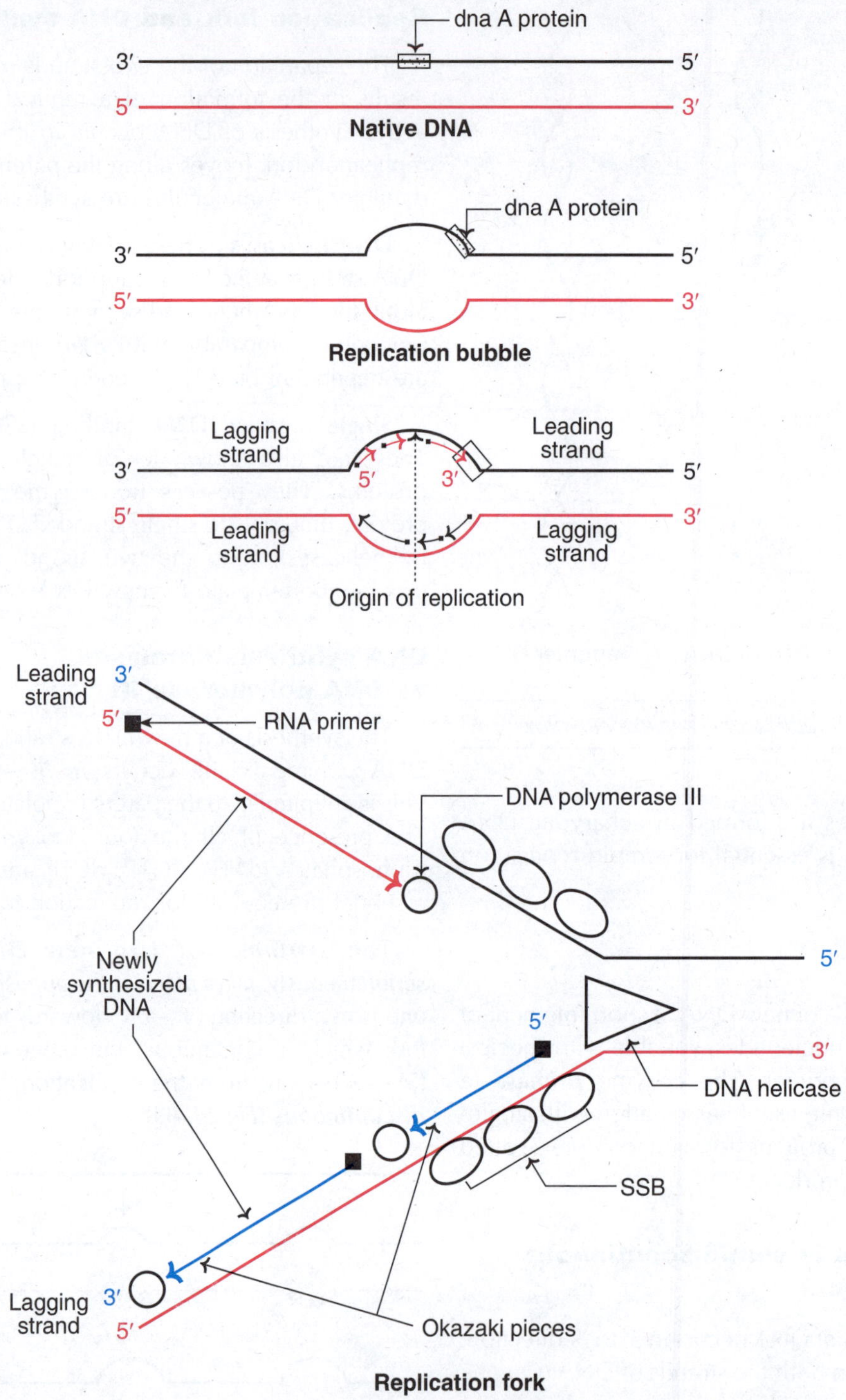

***Fig.24.4 :** Overview of DNA replication process (SSB–Single-stranded binding protein).*

The incoming ***deoxyribonucleotides*** are added one after another, to 3′ end of the growing DNA chain. A molecule of pyrophosphate (PPi) is removed with the addition of each nucleotide. The template DNA strand (the parent) determines the base sequence of the newly synthesized complementary DNA.

Polarity problem

The DNA strand (leading strand) with its 3′-end (3′-OH) oriented towards the fork can be elongated by sequential addition of new nucleotides. The other DNA strand (lagging strand) with 5′-end presents some problem, as there is no DNA

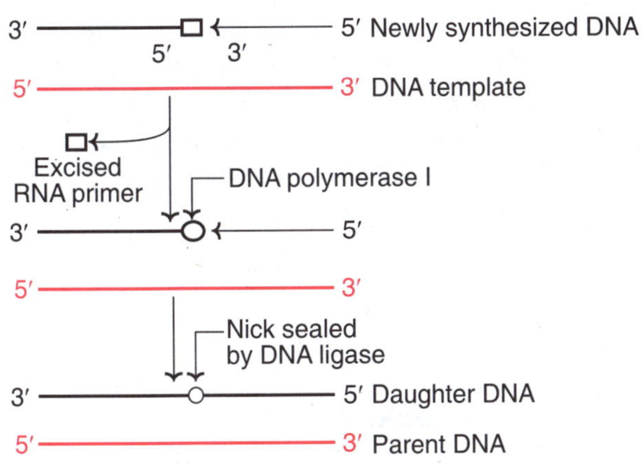

***Fig.24.5 :** Overview of the action of DNA polymerase I and DNA ligase.*

polymerase enzyme (in any organism) that can catalyse the addition of nucleotides to the 5′ end (i.e. $3' \rightarrow 5'$ direction) of the growing chain. This problem however, is solved by synthesizing this strand as a series of small fragments. These pieces are made in the normal $5' \rightarrow 3'$ direction, and later joined together.

Okazaki pieces : The small ***fragments of the discontinuously synthesized DNA*** are called Okazaki pieces. These are produced on the lagging strand of the parent DNA. Okazaki pieces are later joined to form a continuous strand of DNA. DNA polymerase I and DNA ligase are responsible for this process (details given later).

Proof-reading function of DNA polymerase III

Fidelity of replication is the most important for the very existence of an organism. Besides its $5' \rightarrow 3'$ directed catalytic function, DNA polymerase III also has a ***proof-reading activity.*** It checks the incoming nucleotides and allows only the correctly matched bases (i.e. complementary bases) to be added to the growing DNA strand. Further, DNA polymerase edits its mistakes (if any) and removes the wrongly placed nucleotide bases.

Replacement of RNA primer by DNA

The synthesis of new DNA strand continues till it is in close proximity to RNA primer. Now the DNA polymerase I comes into picture. It removes the RNA primer and takes its position. DNA polymerase I catalyses the synthesis ($5' \rightarrow 3'$ direction) of a fragment of DNA that replaces RNA primer (***Fig.24.5***).

The enzyme ***DNA ligase*** catalyses the formation of a phosphodiester linkage between the DNA synthesized by DNA polymerase III and the small fragments of DNA produced by DNA polymerase I. This process—nick sealing-requires energy, provided by the breakdown of ATP to AMP and PPi.

Supercoils and DNA topoisomerases

As the double helix of DNA separates from one side and replication proceeds, ***supercoils are formed at the other side.*** The formation of supercoils can be better understood by comparing DNA helix with two twisted ropes tied at one end. Hold the ropes at the tied end in a fixed position. And let your friend pull the ropes apart from the other side. The formation of supercoils is clearly observed.

Type I DNA topoisomerase cuts the single DNA strand (nuclease activity) ***to overcome*** the problem of ***supercoils*** and then reseals the strand (ligase activity). Type II DNA topoisomerase (also known as DNA gyrase) cuts both strands and reseals them to overcome the problem of supercoils. DNA topoisomerases are targeted by drugs (campthoterin for topoisomerase I, and amsacrime and etoposide for topoisomerase I, and for topoisomerase II) in the treatment of cancers.

REPLICATION IN EUKARYOTES

Replication of DNA in eukaryotes closely resembles that of prokaryotes. Certain differences, however, exist. ***Multiple origins of replication*** is a characteristic feature of eukaryotic cell. Further, at least ***five distinct DNA polymerases*** are known in eukaryotes. Greek letters are used to number these enzymes.

1. **DNA polymerase α** is responsible for the ***synthesis of RNA primer*** for both the leading and lagging strands of DNA.

2. **DNA polymerase β** is involved in the ***repair of DNA.*** Its function is comparable with DNA polymerase I found in prokaryotes.

3. **DNA polymerase γ** participates in the replication of ***mitochondrial DNA.***

4. **DNA polymerase δ** is responsible for the replication on the ***leading and lagging strands of DNA.*** It also possesses proof-reading activity.

5. **DNA polymerase ε** is involved in proof-reading function of DNA replication.

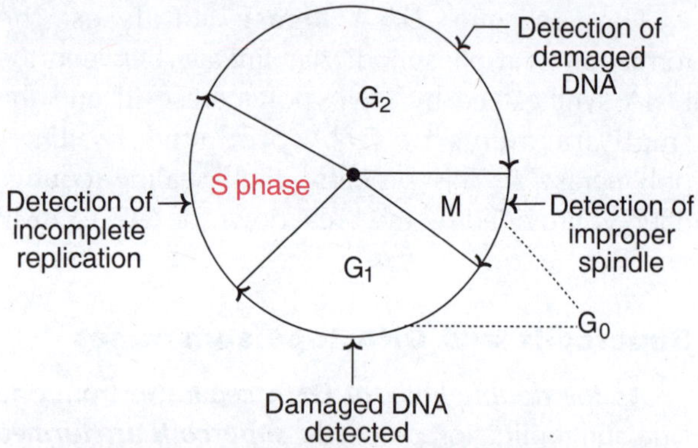

Fig.24.6 : *The cell cycle of a mammalian cell (M–Mitotic phase; G_1–Gap1 phase; G_0–Dormant phase; S phase – Period of replication; G_2–Gap 2 phase).*

The differences in the DNA replication between bacteria and human cells, attributed to the enzymes, are successfully used in antibacterial therapy to target pathogen (bacterial) replication and spare the host (human) cells.

INHIBITORS OF DNA REPLICATION

Bacteria contain a specific type II topoisomerase namely ***gyrase***. This enzyme cuts and reseals the circular DNA (of bacteria), and thus overcomes the problem of supercoils. Bacterial gyrase is ***inhibited by the antibiotics ciprofloxacin, novobiocin and nalidixic acid***. These are widely used as antibacterial agents since they can effectively block the replication of DNA and multiplication of cells. These antibacterial agents have almost ***no effect on human enzymes***.

Certain compounds that ***inhibit human topoisomerases*** are used as ***anticancer agents*** e.g. adriamycin, etoposide, doxorubicin. The nucleotide analogs that inhibit DNA replication are also used as anticancer drugs e.g. 6-mercaptopurine, 5-fluorouracil.

CELL CYCLE AND DNA REPLICATION

The cell cycle consists of four distinct phases in higher organisms—mitotic, G_1, S and G_2 phases (***Fig.24.6***). When the cell is not growing, it exists in a dormant or undividing phase (G_0). G_1 phase is characterized by active protein synthesis.

Replication of DNA occurs only once in S-phase and the chromosomes get doubled i.e. diploid genome gets converted into tetraploid. The entire process of new DNA synthesis takes place in about 8–10 hours, and a large number of DNA polymerases (500–1,000) are simultaneously involved in this process. It is believed that methylation of DNA serves as a marker to inhibit replication.

The G_2 phase is characterized by enlargement of cytoplasm and this is followed by the actual cell division that occurs in the mitotic phase.

Cell cycle check points

As depicted in ***Fig.24.6***, there occurs a continuous monitoring of the cell cycle with respect to DNA replication, chromosome segregation and integrity. If any damage to DNA is detected either in G_1 or G_2 phase of the cycle, or if there is a formation of defective spindle (i.e. incomplete chromosomal segregation), the cell cycle will not progress until appropriately corrected. If it is not possible to repair the damage done, the cells undergo ***apoptosis*** (***programmed cell death***).

Cancer and cell cycle

Cancer represents an excessive division of cells. In cancer, a large quantity of cells are in mitosis and most of them in S-phase.

Majority of the ***drugs*** used for cancer therapy are ***designed to block DNA replication*** or inhibit the enzymes that participate in replication (directly or indirectly). ***Methotrexate*** (inhibits dihydrofolate reductase) and ***5-fluorouracil*** (inhibits thymidylate synthase) block nucleotide synthesis.

In recent years, ***topoisomerase inhibitors*** are being used. They block the unwinding of parental DNA strands and prevent replication.

TELOMERES AND TELOMERASE

There are certain difficulties in the replication of linear DNAs (or chromosomes) of eukaryotic cells. The leading strand of DNA can be completely synthesized to the very end of its template. This is not possible wih the lagging strand, since the ***removal of the primer RNA leaves a small gap which cannot be filled*** (***Fig.24.7***). Consequently, the daughter chromosomes will have ***shortened DNA molecules***. This becomes significant after several cell cycles involving replication of chromosomes. The result is that over a period of time, the chromosomes may

Medical Concepts/ Clinical Correlates

- A few picograms (10^{-12} g) of DNA in a fetal cell stores the genetic information that will determine the differentiation and every function of an adult animal. This is the marvel of molecular biology.
- Topoisomerase inhibitors (e.g. adriamycin, etoposide) are useful to prevent DNA replication, and thus uncontrolled cell proliferation in cancer. These compounds block the unwinding of DNA strands.
- The progressive shortening of telomeres (DNA sequences at chromosomal ends) is prevented by telomerase. This enzyme is an attractive target for cancer therapy.

lose certain essential genes and the cell dies. This is however, avoided to a large extent.

Telomeres are the special structures that prevent the continuous loss of DNA at the end of the chromosomes during the course of replication. Thus, they protect the ends of the chromosomes, and are also responsible to prevent the chromosomes from fusing with each other. Telomeres are many repeat sequences of six nucleotides present at the ends of eukaryotic chromosomes. Human telomeres contain thousands of ***repeat TTAGGG sequences***, which can be up to a length of 1500 bp.

Telomeres are howerver, maintained to a large extent by the enzyme telomerase, also called as ***telomere terminal transferase***. Telomerase is an unusual enzyme as it is composed of both protein and RNA.

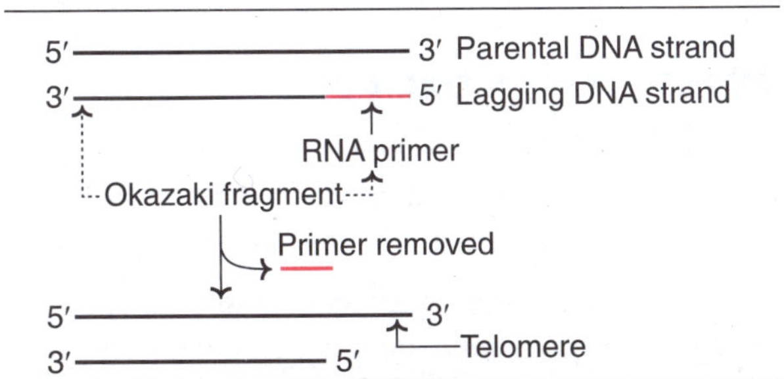

Fig.24.7 : *Formation of telomere during DNA replication.*

Telomere in senescence and cancer

Many biologists ***link the process of telomere shortening with cell senescence*** (i.e. cell death). This is mainly based on the observations made in the *in vitro* mammalian cell cultures. However, some researchers question this relation between telomere shortening and senescence.

Cancerous cells are able to divide continuously. There is a strong evidence to suggest that the ***absence of senescence in cancer cells is linked to the activation of the enzyme telomerase***. Thus, telomere length is maintained throughout multiple cell divisions.

The enzyme ***telomerase is an attractive target for cancer chemotherapy***. The drugs have been designed to inactivate telomerase, and consequently induce senescence in the cancer cells. This in turn prevents the rapid cell proliferation.

RECOMBINATION

Recombination basically involves the ***exchange of genetic information***. There are mainly two types of recombinations.

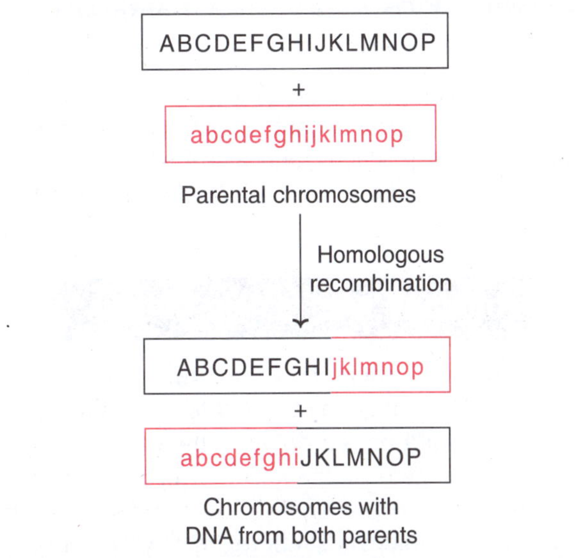

Fig.24.8 : *A diagrammatic representation of homologous recombination.*

1. **Homologous recombination :** This is also called as ***general recombination***, and occurs between identical or nearly identical chromosomes (DNA sequences). The best example is the recombination between the paternal and maternal chromosomal pairs (***Fig.24.8***).

2. **Non-homologous recombination :** This is regarded as ***illegitimate recombination*** and does not require any special homologous sequences. ***Transposition*** is a good example of non-homologous recombination. Random integration of outside genes into mammalian chromosomes is another example.

Homologous recombination

It is a known fact that the chromosomes are not passed on intact from generation to generation. Instead, they are inherited from both the parents. This is possible due to homologous recombination. Three models have been put forth to explain homologous recombinations.

- ***Holliday model***
- ***Meselson-Radding model***
- ***Double-strand break model***.

Non-homologous recombination

The recombination process without any special homologous sequences of DNA is regarded as non-homologous recombination.

Transposition primarily involves the ***movement of specific pieces of DNA in the genome***. The mobile segments of DNA are called ***transposons*** or ***transposable elements***.

Transposons are mobile and can move almost to any place in the target chromosome. There are two modes of transposition. One that involves an RNA intermediate, and the other which does not involve RNA intermediate.

DAMAGE AND REPAIR OF DNA

Being the carrier of genetic information, the cellular DNA must be replicated (duplicated), maintained, and passed down to the daughter cells accurately. In general, the accuracy of replication is extremely high. However, there do occur replication errors. It is estimated that approximately one error is introduced per billion base pairs during each cycle of replication. The cells do posses the capability to repair damages done to DNA to a large extent.

TABLE 24.1 Major types of DNA damages

Category	*Types*
Single-base alteration	Deamination (C $\rightarrow$ U; A $\rightarrow$ hypoxanthine)
	Depurination
	Base alkylation
	Insertion or deletion of nucleotides
	Incorporation of base analogue
Two-base alteration	UV light induced pyrimidine dimer alteration (T–T)
Chain breaks	Oxidative free radical formation Ionizing radiation
Cross-linkage	Between bases in the same or opposite strands
	Between the DNA and protein molecules

Consequences of DNA damage

Despite an efficient repair system for the damaged DNA, replication errors do accumulate that ultimately result in mutations.

Besides the possible errors in replication, the DNA is constantly subjected to attack by both physical and chemical agents. These include ***radiation, free radicals, chemicals*** etc., which also result in ***mutations***.

It is fortunate that a great majority of the mutations probably occur in the DNA that does not encode proteins, and consequently will not have any serious impact on the organism. This is not, however, all the time true, since mutations do occur in the coding regions of DNA also. There are situations in which the change in a single base pair in the human genome can cause a serious disease e.g. sickle-cell anemia.

TYPES OF DNA DAMAGES

The damages done to DNA by physical, chemical and environmental agents may be broadly classified into ***four categories*** with different types (***Table 24.1***).

The DNA damage may occur due to ***single-base alterations*** (e.g. depurination, deamination), ***two-base alterations*** (e.g. pyrimidine dimer) ***chain breaks*** (e.g. ionizing radiation) and ***cross-linkages*** (e.g. between bases).

Medical Concepts/ Clinical Correlates

- Mutations may sometimes result in serious diseases. e.g. sickle-cell anemia, cancer.
- Xeroderma pigmentosum is a rare disease characterized by photosensitivity and risk for skin cancer. This is due to a defect in the nucleotide excision repair of the damaged DNA (caused by UV rays).
- Hereditary nonpolyposis colon cancer is a common inherited cancer, and is due to a faulty mismatch repair of defective DNA.

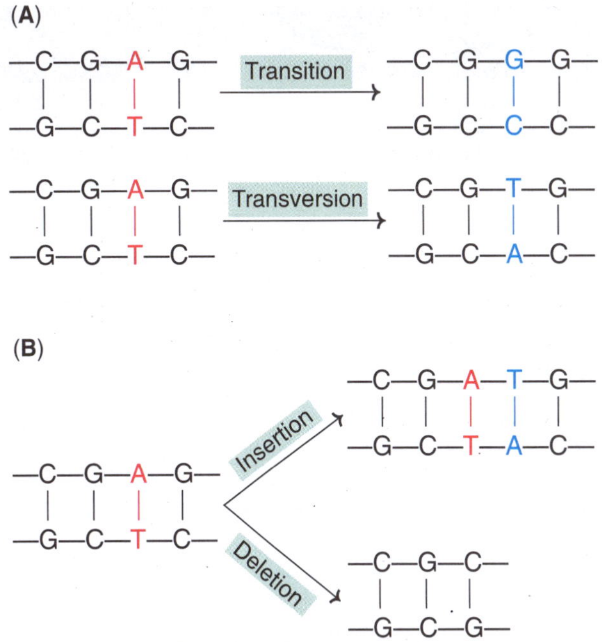

*Fig.24.9 : An illustration of mutations (**A**)-Point mutations; (**B**)-Frameshift mutations.*

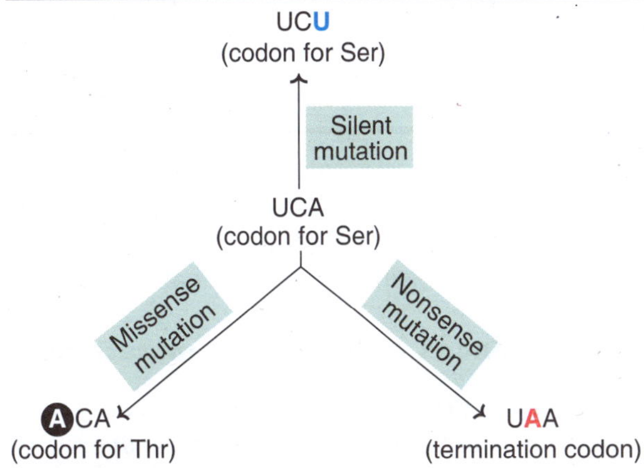

Fig.24.10 : An illustration of point mutations (represented by a codon of mRNA)

MUTATIONS

The genetic macromolecule DNA is highly stable with regard to its base composition and sequence. However, DNA is not totally exempt from gradual change.

Mutation refers to ***a change in the DNA structure of a gene.*** The substances (chemicals) which can induce mutations are collectively known as ***mutagens***.

The changes that occur in DNA on mutation are reflected in replication, transcription and translation.

Types of mutations

Mutations are mainly of two major types—point mutations, frameshift mutations (***Fig.24.9***).

1. **Point mutations :** The replacement of one base pair by another results in point mutation. They are of two sub-types.

   (a) **Transitions :** In this case, ***a purine*** (or a pyrimidine) ***is replaced by another***.

   (b) **Transversions :** These are characterized by ***replacement of a purine by a pyrimidine or vice versa***.

2. **Frameshift mutations :** These occur when one or more base pairs are inserted in or deleted from the DNA, respectively, causing ***insertion or deletion mutations.***

Consequences of point mutations

The change in a single base sequence in point mutation may cause one of the following (***Fig.24.10***).

1. **Silent mutation :** The codon (of mRNA) containing the changed base may code for the same amino acid. For instance, *UCA* codes for serine and change in the third base *(UCU)* still codes for serine. This is due to degeneracy of the genetic code. Therefore, there are ***no detectable effects*** in silent mutation.

2. **Missense mutation :** In this case, the changed base may code for a different amino acid. For

TABLE 24.2 Major mechanisms of DNA repair

Mechanism	*Damage to DNA*	*DNA repair*
Base excision-repair	Damage to a single base due to spontaneous alteration or by chemical or radiation means.	Removal of the base by N–glycosylase; abasic sugar removal, replacement.
Nucleotide excision-repair	Damage to a segment of DNA by spontaneous, chemical or radiation means.	Removal of the DNA fragment (ï 30 nt length) and replacement.
Mismatch repair	Damage due to copying errors (1-5 base unpaired loops).	Removal of the strand (by exonuclease digestion) and replacement.
Double-strand break repair	Damage caused by ionizing radiations, free radicals, chemotherapy etc.	Unwinding, alignment and ligation.

example, *UCA* codes for serine while *ACA* codes for threonine. The mistaken (or missense) amino acid may be ***acceptable***, ***partially acceptable*** or ***unacceptable*** with regard to the function of protein molecule. ***Sickle-cell anemia*** is a classical example of missense mutation.

3. **Nonsense mutation :** Sometimes, the codon with the altered base may become a ***termination*** (or nonsense) ***codon***. For instance, change in the second base of serine codon *(UCA)* may result in *UAA*. The altered codon acts as a stop signal and causes termination of protein synthesis, at that point.

Consequences of frameshift mutations

The insertion or deletion of a base in a gene results in an ***altered reading frame of the mRNA*** (hence the name frameshift). The machinery of mRNA (containing codons) does not recognize that a base was missing or a new base was added. Since there are no punctuations in the reading of codons, translation continues. The result is that the protein synthesized will have several altered amino acids and/or prematurely terminated protein.

Mutations and cancer

Mutations are permanent alterations in DNA structure, which have been implicated in the etiopathogenesis of cancer.

REPAIR OF DNA

As already stated, damage to DNA caused by replication errors or mutations may have serious consequences. The cell possesses an inbuilt system to repair the damaged DNA. This may be achieved by four distinct mechanisms (***Table 24.2***).

1. ***Base excision-repair***
2. ***Nucleotide excision-repair***
3. ***Mismatch repair***
4. ***Double-strand break repair***.

Base excision-repair

The bases cytosine, adenine and guanine can undergo spontaneous depurination to respectively form uracil, hypoxanthine and xanthine. These altered bases do not exist in the normal DNA, and therefore need to be removed. This is carried out by base excision repair (***Fig.24.11***).

A defective DNA in which cytosine is deaminated to uracil is acted upon by the enzyme uracil DNA glycosylase. This results in the removal of the defective base uracil. An endonuclease cuts the backbone of DNA strand near the defect and removes a few bases. The gap so created is filled up by the action of repair DNA polymerase and DNA ligase.

Nucleotide excision-repair

The DNA damage due to ultraviolet light, ionizing radiation and other environmental factors often results in the modification of certain bases, strand breaks, cross-linkages etc. Nucleotide excision-repair is ideally suited for such large-scale defects in DNA. After the identification of the defective piece of the DNA, the DNA double helix is unwound to expose the damaged part. An ***excision nuclease*** (exinuclease) cuts the DNA

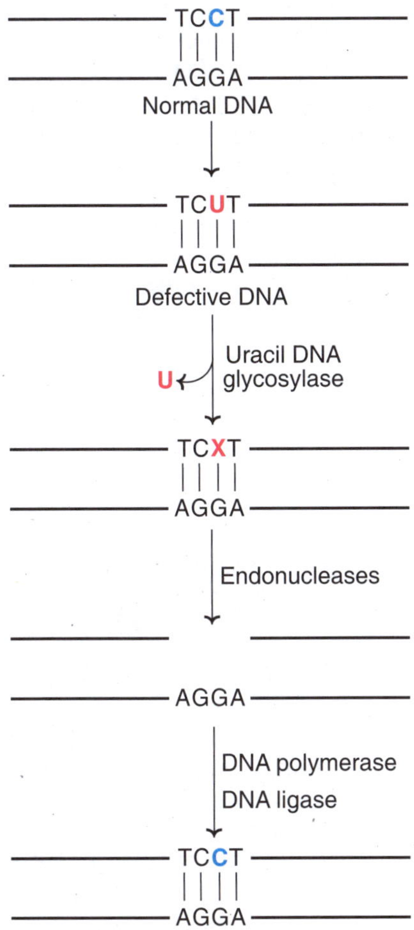

***Fig.24.11 :** A diagrammatic representation of base excision-repair of DNA.*

on either side (upstream and downstream) of the damaged DNA. This defective piece is degraded. The gap created by the nucleotide excision is filled up by DNA polymerase which gets ligated by DNA ligase (***Fig.24.12***).

Xeroderma pigmentosum (XP) is a rare autosomal recessive disease. The affected patients are photosensitive and susceptible to skin cancers. It is now recognized that XP is ***due to a defect in the nucleotide excision repair of the damaged DNA***.

Mismatch repair

Despite high accuracy in replication, defects do occur when the DNA is copied. For instance, cytosine (instead of thymine) could be incorporated opposite to adenine. Mismatch repair corrects a single mismatch base pair e.g. C to A, instead of T to A.

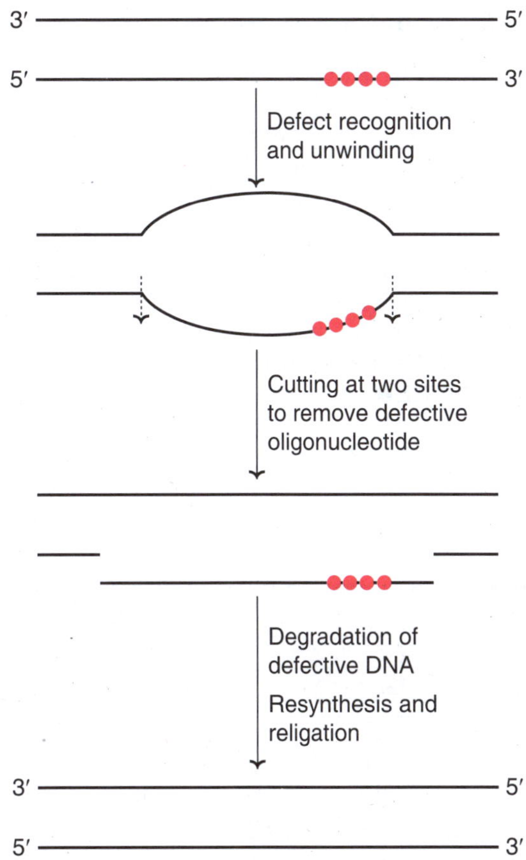

***Fig.24.12 :** A diagrammatic representation of nucleotide excision-repair of DNA.*

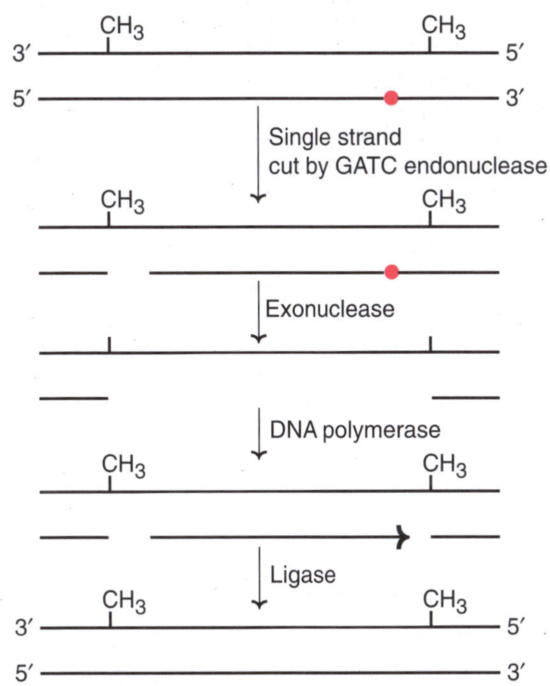

***Fig. 24.13 :** A diagrammatic representation of mismatch repair of DNA.*

Double-strand break repair

Double-strand breaks (DSBs) in DNA are dangerous. They result in genetic recombination which may lead to chromosomal translocation, broken chromosomes, and finally cell death. DSBs can be repaired by homologous recombination or non-homologous end joining. Homologous recombination occurs in yeasts while in mammals, non-homologous and joining dominates.

DEFECTS IN DNA REPAIR AND CANCER

Cancer develops when certain genes that regulate normal cell division fail or are altered. Defects in the genes encoding proteins involved in nucleotide-excision repair, mismatch repair and recombinational repair are linked to human cancers. For instance, ***hereditary nonpolyposis colon cancer (HNPCC)*** is one of the most common inherited cancers. This cancer is now linked with ***faulty mismatch repair*** of defective DNA.

The template strand of the DNA exists in a methylated form, while the newly synthesized strand is not methylated. This difference allows the recognition of the new strands. The enzyme GATC endonuclease cuts the strand at an adjacent methylated GATC sequence (***Fig.24.13***). This is followed by an exonuclease digestion of the defective strand, and thus its removal. A new DNA strand is now synthesized to replace the damaged one.

SUMMARY

1. *Replication is a process in which DNA copies itself to produce identical daughter molecules of DNA. DNA replication is semiconservative, bidirectional and occurs by the formation of bubbles and forks.*
2. *DNA syntheis is catalysed by the enzyme DNA polymerase III. This enzyme possesses proof-reading activity and edits the mistakes that might occur during nucleotide incorporation.*
3. *Replication in eukaryotes (particularly on the lagging strand) is more complex and involves several factors e.g. replication protein A, replication factor C, flap endonuclease.*
4. *Telomeres (repeat TTAGGG sequences) are the special structures that prevent the continuous loss of DNA at the end of the chromosome during the course of replication.*
5. *Recombination involves the exchange of genetic information through the exchange of DNA. Transposition refers to the movement of specific pieces of DNA (called transposons) in the genome.*
6. *Damage to DNA may be due to single base alteration, two-base alteration, chain breaks and cross linkages. The cells possess an inbuilt system to repair the damaged DNA.*

TRANSCRIPTION AND TRANSLATION

25

The conventional concept of central dogma of life which in essence is "***DNA makes RNA makes protein***" is an oversimplification of molecular biology. With the advances in cell biology and rapid developments in bioinformatics, the terms ***genome, transcriptome and proteome*** are in ***current use*** to represent the central dogma of molecular biology (***Fig.25.1***). Some information on the new concepts and terminology is given hereunder.

GENOME

The ***total DNA*** (genetic information) contained ***in an organism*** or ***a cell*** is regarded as the genome. Thus, the genome is the storehouse of biological information. It includes the chromosomes in the nucleus, and the DNA in mitochondria, and chloroplasts.

Genomics : The study of the structure and function of genome is genomics.

TRANSCRIPTOME

The ***RNA copies of*** the active protein coding ***genes*** represent transcriptome. Thus, transcriptome is the initial product of gene expression which directs the synthesis of proteins.

Transcriptomics : The study of transcriptome that involves all the RNA molecules made by a cell, tissue or an organism is transcriptomics.

PROTEOME

The ***cell's repertoire*** (repository/storehouse) of ***proteins*** with their nature ***and biological functions*** is regarded as proteome. Thus, proteome represents the entire range of proteins and their biological functions in a cell.

Proteomics : The study of the ***proteome***.

Metabolomics : The study of chemical processes involving metabolites.

Whether the central dogma of life is represented in the conventional or more recent form, replication, transcription and translation are the key or core processes that ultimately control life. Replication of DNA has been described in ***Chapter 24***, while transcription and translation are discussed in this chapter.

TRANSCRIPTION

Transcription is a process in which ribonucleic acid ***(RNA) is synthesized from DNA***. The word ***gene*** refers to the ***functional unit of the DNA*** that can be transcribed. Thus, the genetic information stored in DNA is expressed through RNA. For this purpose,

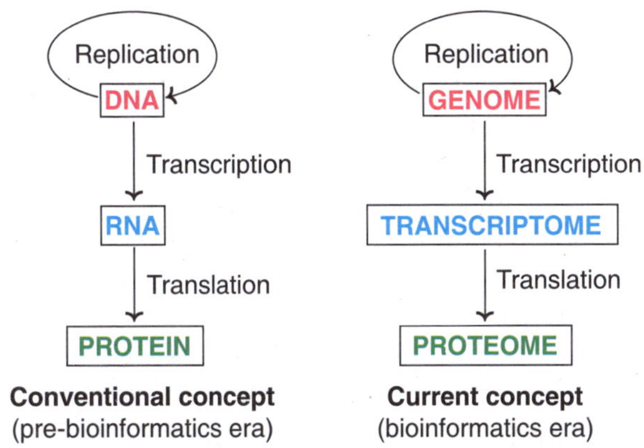

***Fig.25.1 :** The central dogma of life (or molecular biology) represented in the form of conventional and current concepts.*

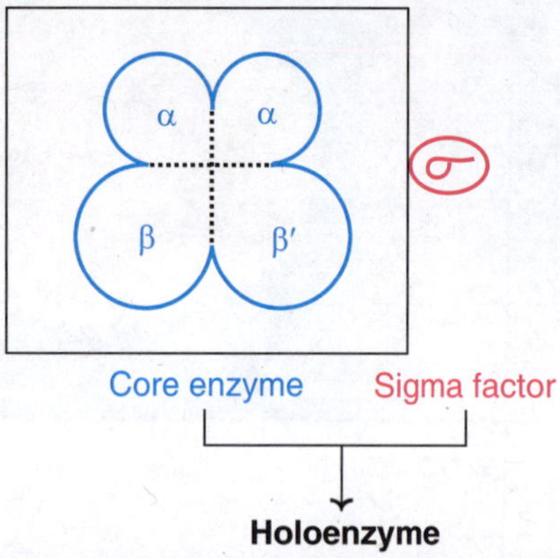

Fig.25.2 : *RNA polymerase of* E. coli.

one of the two strands of DNA serves as a ***template*** (non-coding strand or ***antisense strand***) and produces ***working copies of RNA molecules***. The ***other DNA strand*** which ***does not participate in transcription is referred to as coding strand or sense strand or non-template strand***. (Coding strand commonly used since with the exception of T for U, primary mRNA contains codons with the same base sequence).

Transcription is selective

The entire molecule of DNA is not expressed in transcription. RNAs are synthesized only for some selected regions of DNA. For certain other regions of DNA, there may not be any transcription at all. The exact reason for the selective transcription is not known. This may be ***due to some inbuilt signals in the DNA molecule***.

The product formed in transcription is referred to as ***primary transcript.*** Most often, the primary RNA transcripts are inactive. They undergo certain alterations (splicing, terminal additions, base modifications etc.) commonly known as ***post-transcriptional modifications***, to produce functionally active RNA molecules.

There exist certain differences in the transcription between prokaryotes and eukaryotes. The RNA synthesis in prokaryotes is given in some detail. This is followed by a brief discussion on eukaryotic transcription.

TRANSCRIPTION IN PROKARYOTES

A single enzyme—DNA dependent RNA polymerase or simply ***RNA polymerase***—synthesizes all the RNAs in prokaryotes. RNA polymerase of *E. coli* is a complex holoenzyme (mol wt. 465 kDa) with five polypeptide subunits—2α, 1β and 1β′ and one sigma(s) factor (***Fig.25.2***). The enzyme without sigma factor is referred to as core enzyme ($\alpha_2\beta\beta'$).

An overview of RNA synthesis is depicted in ***Fig.25.3***. Transcription involves three different stages—initiation, elongation and termination (***Fig.25.4***).

Initiation

The binding of the enzyme RNA polymerase to DNA is the prerequisite for the transcription to start. The specific region on the DNA where the enzyme binds is known as ***promoter region.*** There are two base sequences on the ***coding DNA strand*** which the sigma factor of RNA polymerase can recognize for initiation of transcription (***Fig.25.5***).

1. **Pribnow box (TATA box) :** This consists of 6 nucleotide bases (TATAAT), located on the left side about 10 bases away (upstream) from the starting point of transcription.

2. **The '–35' sequence :** This is the second recognition site in the promoter region of DNA. It contains a base sequence TTGACA, which is located about 35 bases (upstream, hence –35) away on the left side from the site of transcription start.

Elongation

As the holoenzyme, RNA polymerase recognizes the promoter region, the sigma factor is released and transcription proceeds. ***RNA is synthesized from 5′ end to 3′ end (5′ → 3′) antiparallel to the DNA template***. RNA polymerase ***utilizes*** ribonucleotide triphosphates (***ATP, GTP, CTP and UTP***) for the formation of RNA. For the addition of each nucleotide to the growing chain, a pyrophosphate moiety is released.

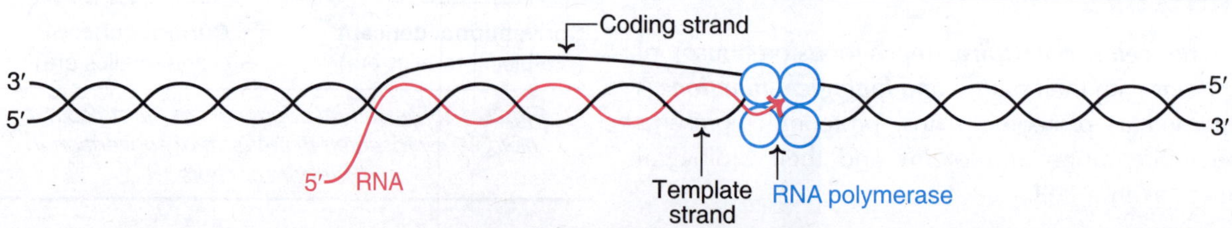

Fig.25.3 : *An overview of transcription.*

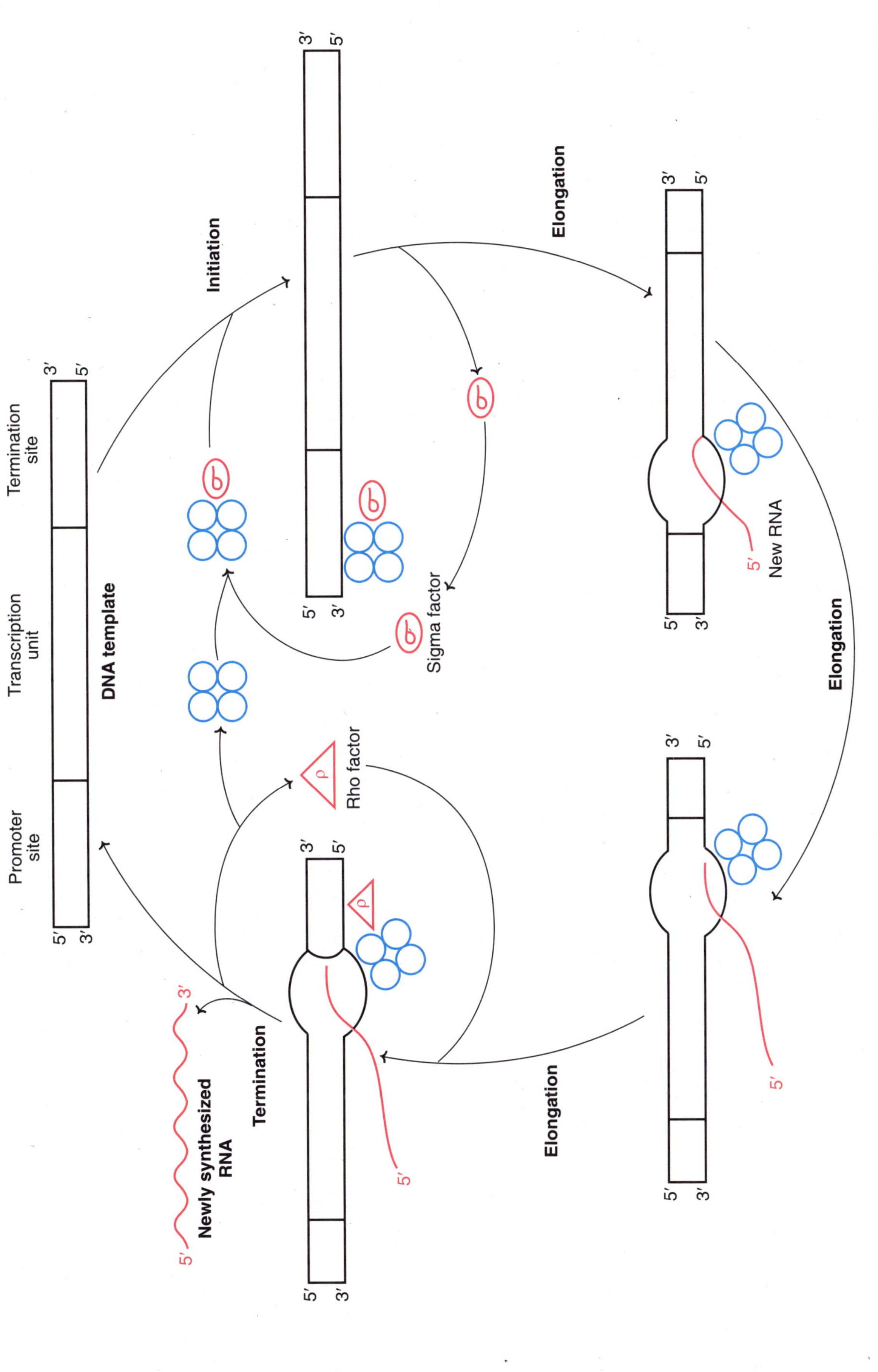

Fig.25.4 : Synthesis of RNA from DNA template (transcription).

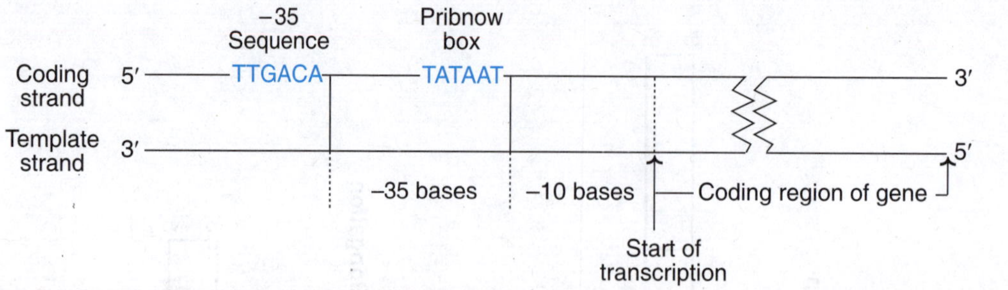

Fig.25.5 : Promoter regions of DNA in prokaryotes.

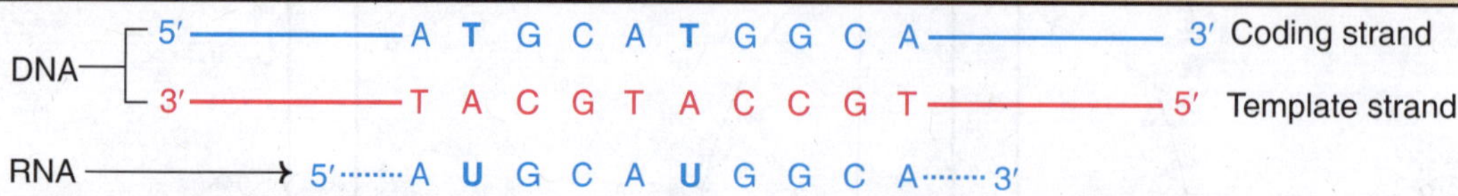

Fig.25.6 : Transcription–Complementary base pair relationship.

The sequence of nucleotide bases in the mRNA is complementary to the template DNA strand. It is however, identical to that of coding strand except that RNA contains U in place of T in DNA (***Fig.25.6***).

Termination

The process of transcription stops by termination signals. Two types of termination are identified.

1. **Rho (ρ) dependent termination :** A specific protein, named ***ρ factor***, binds to the growing RNA (and not to RNA polymerase) or weakly to DNA, and in the bound state it acts as ATPase and terminates transcription and releases RNA. The ρ factor is also responsible for the dissociation of RNA polymerase from DNA.

2. **Rho (ρ) independent termination :** The termination in this case is brought about by the formation of ***hairpins*** of newly synthesized RNA. This occurs due to the presence of ***palindromes***. A palindrome is a word that reads alike forward and backward e.g. ***madam***, ***rotor***. The presence of palindromes in the base sequence of DNA template (same when read in opposite direction) in the termination region is known. As a result of this, the newly synthesized RNA folds to form hairpins (due to complementary base pairing) that cause termination of transcription.

TRANSCRIPTION IN EUKARYOTES

RNA synthesis in eukaryotes is a much more complicated process than the transcription described above for prokaryotes. As such, all the details of eukaryotic transcription (particularly about termination) are not clearly known. The salient features of available information are given hereunder.

RNA polymerases

The nuclei of eukaryotic cells possess three distinct RNA polymerases (***Fig.25.7***).

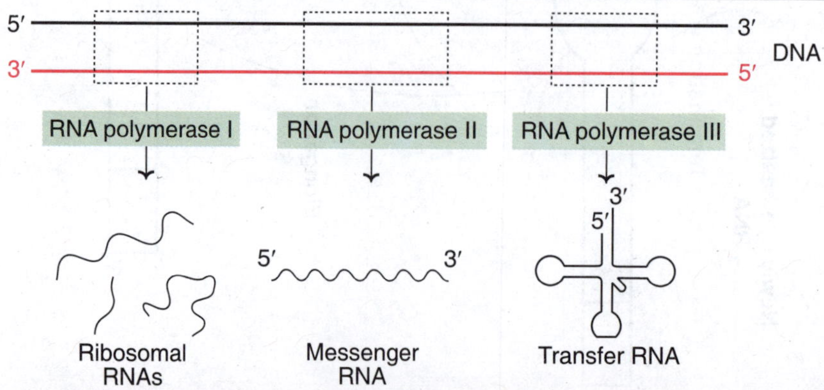

Fig.25.7 : An overview of transcription in eukaryotes.

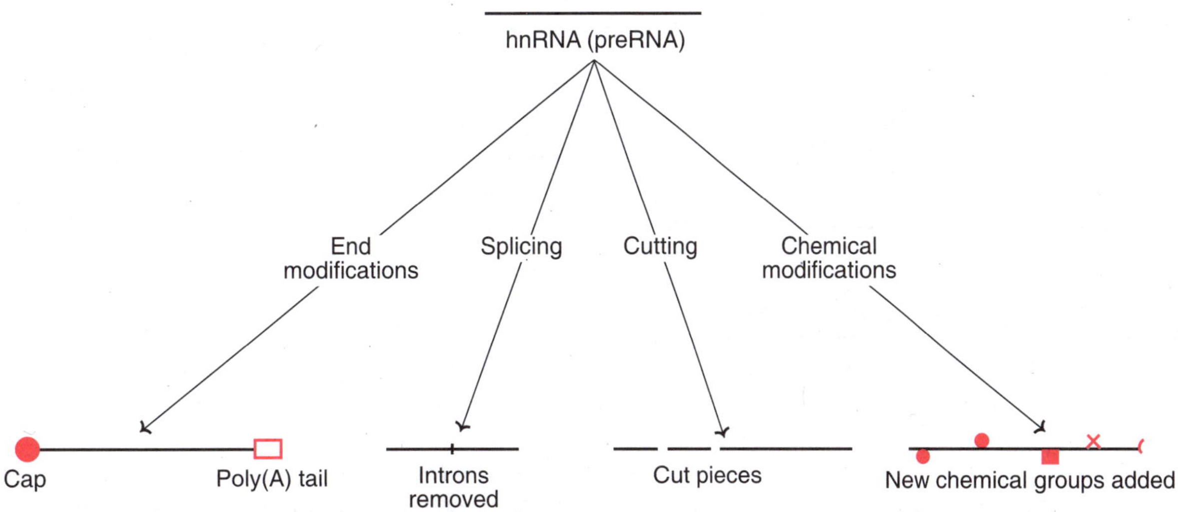

Fig.25.8 : *An outline of post-transcriptional modifications of RNA (hnRNA-Heterogeneous nuclear RNA).*

1. **RNA polymerase I** is responsible for the synthesis of precursors for the large ***ribosomal RNAs.***

2. **RNA polymerase II** synthesizes the precursors for ***mRNAs and small nuclear RNAs.***

3. **RNA polymerase III** participates in the formation of ***tRNAs and small ribosomal RNAs.***

Besides the three RNA polymerases found in the nucleus, there also exists a mitochondrial RNA polymerase in eukaryotes. The latter resembles prokaryotic RNA polymerase in structure and function.

Heterogeneous nuclear RNA (hnRNA)

The ***primary mRNA transcript*** produced by RNA polymerase II in eukaryotes is often referred to as heterogeneous nuclear RNA (hnRNA). This is then processed ***to produce mRNA*** needed for protein synthesis.

POST-TRANSCRIPTIONAL MODIFICATIONS

The RNAs produced during transcription are called primary transcripts. They undergo many alterations— ***terminal base additions, base modifications, splicing*** etc., which are collectively referred to as post-transcriptional modifications. This process is required to convert the RNAs into the active forms. A group of enzymes, namely ribonucleases, are responsible for the processing of tRNAs and rRNAs of both prokaryotes and eukaryotes.

The prokaryotic mRNA synthesized in transcription is almost similar to the functional mRNA. In contrast, eukaryotic mRNA (i.e. hnRNA) undergoes extensive post-transcriptional changes.

An outline of the post-transcriptional modifications is given in ***Fig.25.8***, and some highlights are described.

Messenger RNA

The primary transcript of mRNA is the hnRNA in eukaryotes, which is subjected to many changes before functional mRNA is produced.

1. **The 5′ capping :** The 5′ end of mRNA is capped with 7-methylguanosine by an unusual 5′→5′ triphosphate linkage. S-Adenosylmethionine is the donor of methyl group. This cap is required for translation, besides stabilizing the structure of mRNA.

2. **Poly-A tail :** A large number of eukaryotic mRNAs possess an adenine nucleotide chain at the 3′-end. This poly-A tail, as such, is not produced during transcription. It is later added to stabilize mRNA. However, poly-A chain gets reduced as the mRNA enters cytosol.

3. **Introns and their removal :** Introns are the intervening nucleotide sequences in mRNA which do not code for proteins. On the other hand, ***exons of mRNA possess genetic code, and are responsible for protein synthesis.***

Post-transcriptional modifications of mRNA occur in the nucleus. The mature RNA then enters the cytosol to perform its function (translation).

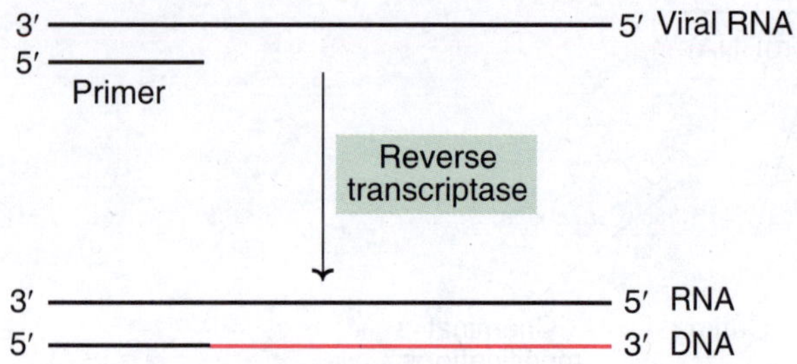

Fig.25.9 : Reverse transcription of RNA virus.

Inhibitors of transcription

The synthesis of RNA is inhibited by certain antibiotics and toxins.

Actinomycin D : Actinomycin D binds with DNA template strand and ***blocks the movement of RNA polymerase***. This was the very first antibiotic used for the treatment of tumors.

Rifampin : It is an antibiotic widely used for the treatment of tuberculosis and leprosy. Rifampin binds with the β-subunit of ***prokaryotic RNA polymerase and inhibits its activity***.

REVERSE TRANSCRIPTION

Some of the viruses—known as ***retroviruses***—possess RNA as the genetic material. These viruses cause cancers in animals, hence known as ***oncogenic***. They are actually found in the transformed cells of the tumors.

The enzyme ***RNA dependent DNA polymerase —or*** simply ***reverse transcriptase***—is responsible for the formation of ***DNA from RNA*** (***Fig.25.9***). This DNA is complementary (cDNA) to viral RNA and can be transmitted into host DNA.

Synthesis of cDNA from mRNA : As already described, the DNA expresses the genetic information in the form of RNA. And the mRNA determines the amino acid sequence in a protein. The mRNA can be utilized as a template for the synthesis of ***double-stranded complementary DNA*** (***cDNA***) by using the enzyme reverse transcriptase. This cDNA can be used as a probe to identify the sequence of DNA in genes.

TRANSLATION

The genetic information stored in DNA is passed on to RNA (through transcription), and ultimately expressed in the language of proteins. The ***biosynthesis of a protein or a polypeptide in a living cell*** is referred to as translation. The term translation is used to represent the biochemical translation of four-letter language information from nucleic acids (DNA and then RNA) to 20 letter language of proteins. The sequence of amino acids in the protein synthesized is determined by the nucleotide base sequence of mRNA.

Variability of cells in translation

There are wide variations in the cells with respect to the quality and quantity of proteins synthesized. This largely depends on the need and ability of the cells. ***Erythrocytes*** (red blood cells) lack the machinery for translation, and therefore ***cannot synthesize proteins***.

In general, the growing and dividing cells produce larger quantities of proteins. Some of the cells continuously synthesize proteins for export. For instance, liver cells produce albumin and blood clotting factors for export into the blood for circulation. The normal liver cells are very rich in the protein biosynthetic machinery, and thus the ***liver*** may be regarded as the ***protein factory in the human body***.

GENETIC CODE

The ***three nucleotide*** (triplet) ***base sequences in mRNA that act as code words for amino acids*** in protein constitute the genetic code or simply ***codons.*** The genetic code may be regarded as a ***dictionary of nucleotide bases (A, G, C and U)*** that determines the sequence of amino acids in proteins.

The codons are ***composed of*** the four nucleotide bases, namely the purines—adenine (***A***) and guanine (***G***), and the pyrimidines—cytosine (***C***) and uracil (***U***). These four bases produce 64 different combinations (4^3) of three base codons, as depicted in ***Table 25.1***. The nucleotide sequence of the codon on mRNA is written from the 5′-end to 3′ end. ***Sixty one codons code for the 20 amino acids found in protein***.

The three codons ***UAA, UAG*** and ***UGA*** do not code for amino acids. They act as ***stop signals*** in protein synthesis. These three codons are collectively known as ***termination codons*** or ***non-sense codons***. The codons UAG, UAA and UGA are often referred to, respectively, as ***amber***, ***ochre*** and ***opal*** codons.

The codons ***AUG***—and, sometimes, ***GUG***—are the chain ***initiating codons***.

Other characteristics of genetic code

The genetic code is ***universal, specific, non-overlapping and degenerate***.

1. **Universality :** The same codons are used to code for the same amino acids in all the living organisms. Thus, the genetic code has been conserved during the course of evolution. Hence genetic code is appropriately regarded as universal. There are, ***however, a few exceptions***. For instance, AUA is the codon for methionine in mitochondria. The same codon (AUA) codes for isoleucine in cytoplasm. With some exceptions noted, the genetic code is universal.

2. **Specificity :** A particular ***codon always codes for the same amino acid***, hence the genetic code is highly specific or unambiguous e.g. UGG is the codon for tryptophan.

3. **Non-overlapping :** The genetic code is read from a fixed point as a continuous base sequence. It is non-overlapping, commaless and without any punctuations. For instance, UUUCUUAGAGGG is read as UUU/CUU/AGA/GGG. Addition or deletion of one or two bases will radically change the message sequence in mRNA. And the protein synthesized from such mRNA will be totally different. This is encountered in ***frameshift mutations*** which cause an alteration in the reading frame of mRNA.

4. **Degenerate :** ***Most of the amino acids have more than one codon***. The codon is degenerate or redundant, since there are ***61 codons*** available to code ***for*** only ***20 amino acids***. For instance, glycine has four codons. The codons that designate the same amino acid are called ***synonyms***. Most of the synonyms differ only in the third (3′ end) base of the codon.

TABLE 25.1 The genetic code along with respective amino acids

First base 5′ end	*Second base (middle one)* U	C	A	G	*Third base 3′ end*
	UUU Phe	UCU	UAU Try	UGU Cys	U
U	UUC	UCC Ser	UAC	UGC	C
	UUA Leu	UCA	UAA Stop	UGA Stop	A
	UUG	UCG	UAG Stop	UGG Trp	G
	CUU	CCU	CAU His	CGU	U
C	CUC Leu	CCC Pro	CAC	CGC Arg	C
	CUA	CCA	CAA Gln	CGA	A
	CUG	CCG	CAG	CGG	G
	AUU	ACU	AAU Asn	AGU Ser	U
A	AUC Ile	ACC Thr	AAC	AGC	C
	AUA	ACA	AAA Lys	AGA Arg	A
	AUG* Met	ACG	AAG	AGG	G
	GUU	GCU	GAU Asp	GGU	U
G	GUC Val	GCC Ala	GAC	GGC Gly	C
	GUA	GCA	GAA Glu	GGA	A
	GUG	GCG	GAG	GGG	G

**AUG serves as initiating codon, besides coding for methionine residue in protein synthesis; UAA, UAG and UGA called as nonsense codons, are responsible for termination of protein synthesis.*

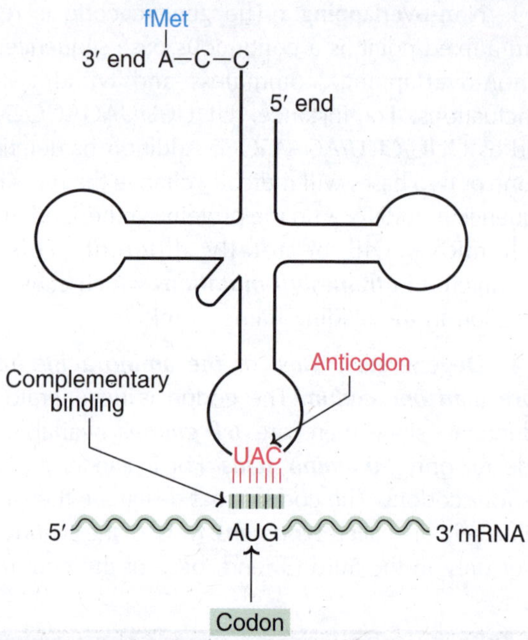

Fig.25.10 : Complementary binding of codon (of mRNA) and anticodon (of tRNA).

The Wobble hypothesis explains codon degeneracy (described later).

Codon-anticodon recognition

The codon of the mRNA is recognized by the anticodon of tRNA (***Fig.25.10***). They pair with each other in antiparallel direction (5′ → 3′ of mRNA with 3′ → 5′ of tRNA). The usual conventional complementary base pairing (A=U, C≡G) occurs between the first two bases of codon and the last two bases of anticodon. The third base of the codon is rather lenient or flexible with regard to the complementary base.

Wobble hypothesis

Wobble hypothesis, put forth by Crick, is the phenomenon in which a ***single tRNA can recognize more than one codon.*** This is due to the fact that the third base (3′-base) in the codon often fails to recognize the specific complementary base in the anticodon (5′-base). Wobbling is attributed to the difference in the spatial arrangement of the 5′-end of the anticodon. The possible pairing of 5′-end base of anticodon (of tRNA) with the 3′-end base of codon (mRNA) is given

Anticodon		Codon	
C	—	G	Conventional base pairing
A	—	U	
U	—	G or A	Non-conventional base (coloured) pairing
G	—	U or C	

Wobble hypothesis explains the degeneracy of the genetic code, i.e. existence of ***multiple codons for a single amino acid.*** Although there are 61 codons for amino acids, the number of tRNAs is far less (around 40) which is due to wobbling.

PROTEIN BIOSYNTHESIS

The ***protein synthesis*** which involves the translation of nucleotide base sequence of mRNA into the language of amino acid sequence may be divided into the ***following stages*** for the convenience of understanding.

I. Requirement of the components

II. Activation of amino acids

III. Protein synthesis proper

IV. Chaperones and protein folding

V. Post-translational modifications.

I. REQUIREMENT OF THE COMPONENTS

The ***protein synthesis may be considered as a biochemical factory operating on the ribosomes.*** As a factory is dependent on the supply of raw materials to give a final product, the protein synthesis also requires many components.

1. **Amino acids :** Proteins are polymers of amino acids. Of the ***20 amino acids*** found ***in protein structure***, half of them (10) can be synthesized by man. About ***10 essential amino acids*** have to be provided through the diet. Protein synthesis can occur only when all the amino acids needed for a particular protein are available.

As regards prokaryotes, there is no requirement of amino acids, since all the 20 are synthesized from the inorganic components.

2. **Ribosomes :** The functionally active ribosomes are the ***centres*** or ***factories for protein synthesis.*** Ribosomes may also be considered as workbenches of translation. Ribosomes are huge complex structures (70S for prokaryotes and 80S for eukaryotes) of proteins and ribosomal RNAs. Each ribosome consists of two subunits—one big and one small. The functional ribosome has two sites—A site and P site. Each site covers both the subunits. ***A site*** is for binding of aminoacyl tRNA and ***P site*** is for binding peptidyl tRNA, during the course of translation. Some authors consider A site as acceptor site, and P site as donor site. In case of eukaryotes, there is another site called

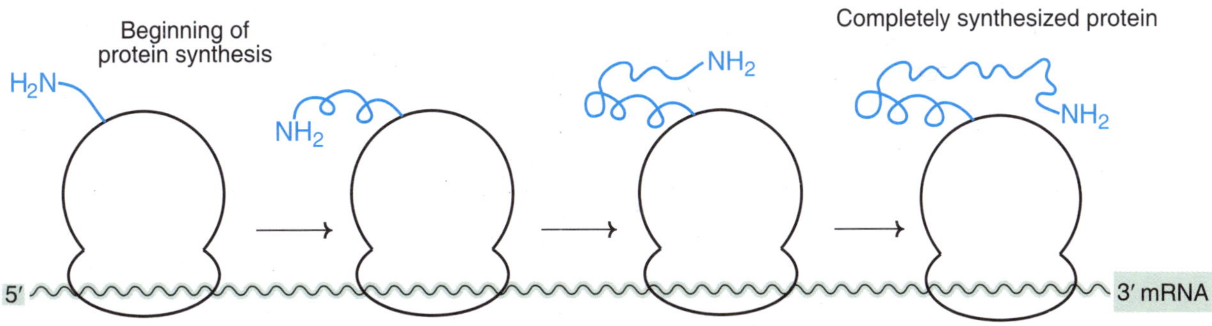

Fig.25.11 : A polyribosome in protein synthesis.

exit* site** or ***E site. Thus, eukaryotes contain three sites (A, P and E) on the ribosomes.

The ribosomes are located in the cytosomal fraction of the cell. They are found in association with rough endoplasmic reticulum (RER) to form clusters RER—ribosomes, where the protein synthesis occurs. The term ***polyribosome*** (***polysome***) is used when several ribosomes simultaneously translate on a single mRNA (***Fig.25.11***).

3. **Messenger RNA (mRNA) :** The specific information required for the synthesis of a given protein is present on the mRNA. The DNA has passed on the genetic information in the form of ***codons*** to mRNA to ***translate into a protein sequence***.

4. **Transfer RNAs (tRNAs) :** They ***carry the amino acids***, and hand them over to the growing peptide chain. The amino acid is covalently bound to tRNA at the 3′-end. Each tRNA has a three nucleotide base sequence—the ***anticodon***, which is responsible to recognize the codon (complementary bases) of mRNA for protein synthesis.

In man, there are ***about 50 different tRNAs*** whereas in bacteria around 40 tRNAs are found. Some amino acids (particularly those with multiple codons) have more than one tRNA.

5. **Energy sources :** Both ***ATP*** and ***GTP*** are required for the supply of energy in protein synthesis.

6. **Protein factors :** The process of translation involves a number of protein factors. These are needed for initiation, elongation and termination of protein synthesis. The protein factors are more complex in eukaryotes compared to prokaryotes.

II. ACTIVATION OF AMINO ACIDS

Amino acids are activated and attached to tRNAs in a two step reaction. A group of enzymes—namely aminoacyl tRNA synthetases—are required for this process. These enzymes are highly specific for the amino acid and the corresponding tRNA.

The amino acid is first attached to the enzyme utilizing ATP to form enzyme-AMP-amino acid complex. The amino acid is then transferred to the 3′ end of the tRNA to form aminoacyl tRNA (***Fig.25.12***).

III. PROTEIN SYNTHESIS PROPER

The protein or polypeptide synthesis occurs on the ribosomes (rather polyribosomes). The ***mRNA is read in the 5′→ 3′ direction and the polypeptide synthesis proceeds from N-terminal end to C-terminal end***. Translation is directional and collinear with mRNA.

The prokaryotic mRNAs are ***polycistronic,*** since a single mRNA has many coding regions that code for different polypeptides. In contrast, eukaryotic mRNA is ***monocistronic,*** since it codes for a single polypeptide.

In case of prokaryotes, translation commences before the transcription of the gene is completed. Thus, simultaneous transcription and translation are possible. This is not so in case of eukaryotic organisms since transcription occurs in the nucleus whereas translation takes place in the cytosol. Further, the primary transcript (hnRNA) formed from DNA has to undergo several modifications to generate functional mRNA.

Protein synthesis is comparatively simple in case of prokaryotes compared to eukaryotes. Further, many steps in eukaryotic translation were not understood for quite sometime. For these reasons, majority of the textbooks earlier used to describe translation in prokaryotes in detail, and give most important and

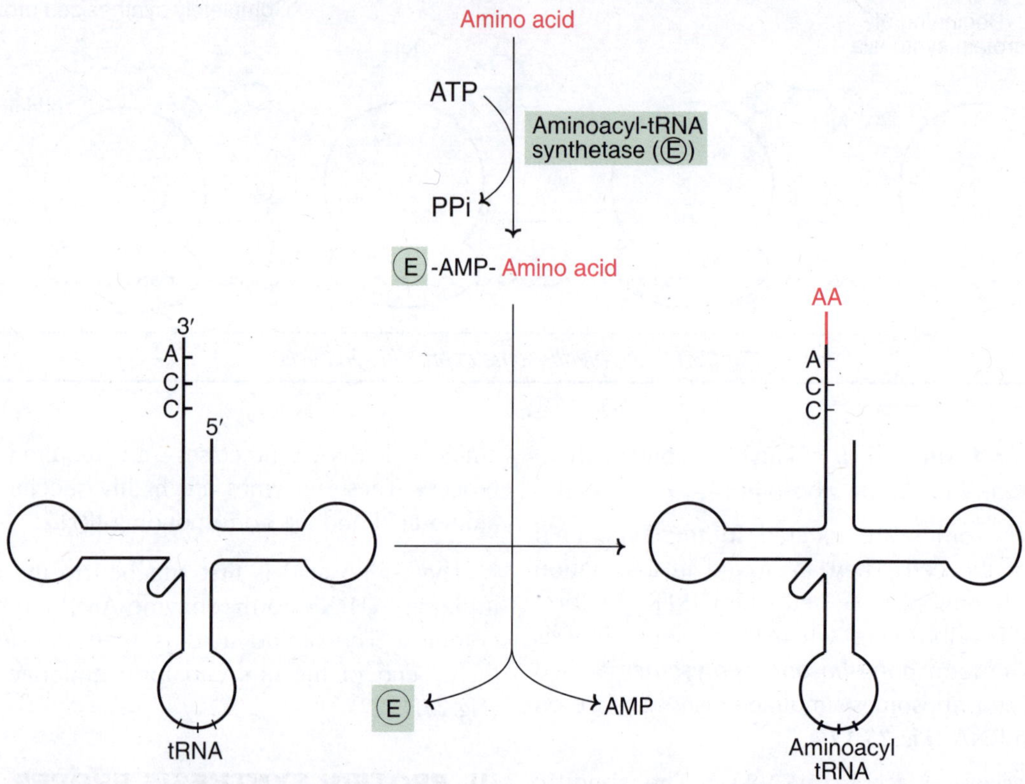

***Fig.25.12 :** Formation of aminocacyl tRNA (AA–Amino acid; E–Enzyme).*

relevant information for eukaryotic translation. With the advances in molecular biology, the process of protein biosynthesis in eukaryotes is better understood now.

Translation in eukaryotes is briefly ***described here***, along with some relevant features of prokaryotic protein biosynthesis. Translation proper is divided into three stages—***initiation, elongation and termination*** (as it is done for transcription).

INITIATION OF TRANSLATION

The initiation of translation ***in eukaryotes*** is complex, involving at least ***ten eukaryotic initiation factors (eIFs)***. Some of the eIFs contain multiple (3-8) subunits. The process of translation initiation can be divided into four steps (***Fig.25.13***).

1. Ribosomal dissociation.
2. Formation of 43S preinitiation complex.
3. Formation of 48S initiation complex.
4. Formation of 80S initiation complex.

Ribosomal dissociation

The 80S ribosome dissociates to form 40S and 60S subunits. Two initiating factors namely eIF-3 and eIF-1A bind to the newly formed 40S subunit, and thereby block its reassociation with 60S subunit. For this reason, some workers name eIF-3 as ***anti-association factor***.

Formation of 43S preinitiation complex

A ternary complex containing met-tRNA^i and eIF-2 bound to GTP attaches to 40S ribosomal subunit to form 43S preinitiation complex. The presence of eIF-3 and eIF-1A stabilizes this complex (***Note :*** Met-tRNA is specifically involved in binding to the ***i***nitiation condon AUGs; hence the superscript^i is used in met-tRNA^i).

Formation of 48S initiation complex

The binding of mRNA to 43S preinitiation complex results in the formation of 48S initiation complex through the intermediate 43S initiation complex. This, however, involves certain interactions between some of the eIFs and activation of mRNA.

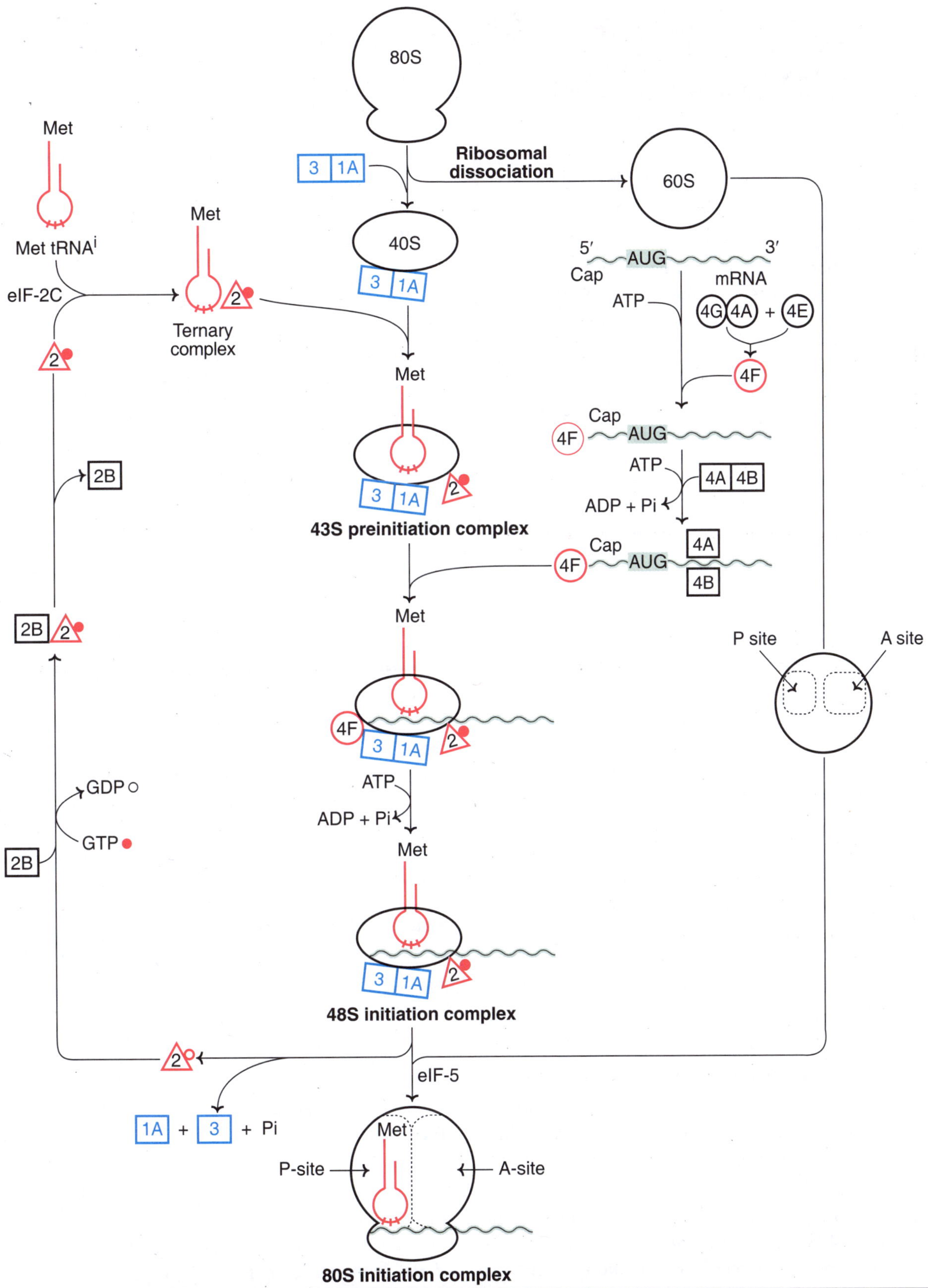

***Fig.25.13 :** A diagrammatic representation of initiation of protein biosynthesis (translation) in eukaryotic cells (The eukaryotic initiation factors are represented by symbols □, △, and ○. By prefixing with eIF, the full names of the factors are obtained e.g. [3] represents eIF–3).*

Medical Concepts/ Clinical Correlates

- **Faulty splicing of hnRNA may result in certain diseaes e.g. β-thalassemia.**
- **Inhibitors of transcription are used as therapeutic agents. Thus, actinomycin D was the first antibiotic used in the treatment of tumors. Rifampin is employed to treat tuberculosis and leprosy.**
- **Retroviruses (RNA is the genetic material) are oncogenic i.e. cause cancers.**

eIF-4F complex is formed by the association of eIF-4G, eIF-4A with eIF-4E. The so formed eIF-4F (referred to as cap binding protein) binds to the cap of mRNA. Then eIF-4A and eIF-4B bind to mRNA and reduce its complex structure. This mRNA is then transferred to 43S complex. For the appropriate association of 43S preinitiation complex with mRNA, energy has to be supplied by ATP.

Recognition of initiation codon : The ribosomal initiation complex scans the mRNA for the identification of appropriate initiation codon. 5′-AUG is the initiation codon and its recognition is facilitated by a specific sequence of nucleotides surrounding it. This marker sequence for the identification of AUG is called as ***Kozak consensus sequences***. In case of ***prokaryotes*** the recognition sequence of initiation codon is referred to as ***Shine- Dalgarno sequence***.

Formation of 80S initiation complex

48S initiation complex binds to 60S ribosomal subunit to form 80S initiation complex. The binding involves the hydrolysis of GTP (bound to eIF-2). This step is facilitated by the involvement of eIF-5.

As the 80S complex is formed, the initiation factors bound to 48S initiation complex are released, and recycled. The activation of eIF-2 requires eIF-2B (also called as guanine nucleotide exchange factor) and GTP. The activated eIF-2 (i.e. bound to GTP) requires eIF-2C to form the ternary complex.

Initiation of translation in prokaryotes

The formation of translation initiation complex in prokaryotes is less complicated compared to eukaryotes. The 30S ribosomal subunit is bound to initiation factor 3 (IF-3) and attached to ternary complex of IF-2, formyl met-tRNA and GTP. Another initiation factor namely IF-I also participates in the formation of preinitiation complex. The recognition of ***initiation codon AUG*** is done through Shine-Dalgarno sequence. A 50S ribosome unit is now bound with the 30S unit to produce 70S initiation complex in prokaryotes.

ELONGATION OF TRANSLATION

Ribosomes elongate the polypeptide chain by a sequential addition of amino acids. The amino acid sequence is determined by the order of the codons in the specific mRNA. Elongation, a cyclic process involving certain ***elongation factors (EFs)***, may be divided into three steps (***Fig.25.14***).

1. Binding of aminoacyl t-RNA to A-site.
2. Peptide bond formation.
3. Translocation.

Binding of aminoacyl—tRNA to A-site

The 80S initiation complex contains met-tRNA^i in the P-site, and the A-site is free. Another aminoacyl-tRNA is placed in the A-site. This requires proper codon recognition on the mRNA and the involvement of elongation factor 1a (EF-Ia) and supply of energy by GTP. As the aminoacyl-tRNA is placed in the A-site, EF-1α and GDP are recycled to bring another aminoacyl-tRNA.

Peptide bond formation

The enzyme ***peptidyltransferase*** catalyses the formation of peptide bond (***Fig.25.15***). The activity of this enzyme lies on 28S RNA of 60S ribosomal subunit. It is therefore the ***rRNA*** (and not protein) referred to as ***ribozyme*** that catalyses the peptide bond formation.

Translocation

As the peptide bond formation occurs, the ribosome moves to the next codon of the mRNA (towards 3′-end). This process called translocation, basically involves the ***movement of growing peptide chain from A-site to P-site***. Translocation requires EF-2 and GTP. GTP gets hydrolysed and supplies energy to move mRNA. EF-2 and GTP complex recycles for translocation.

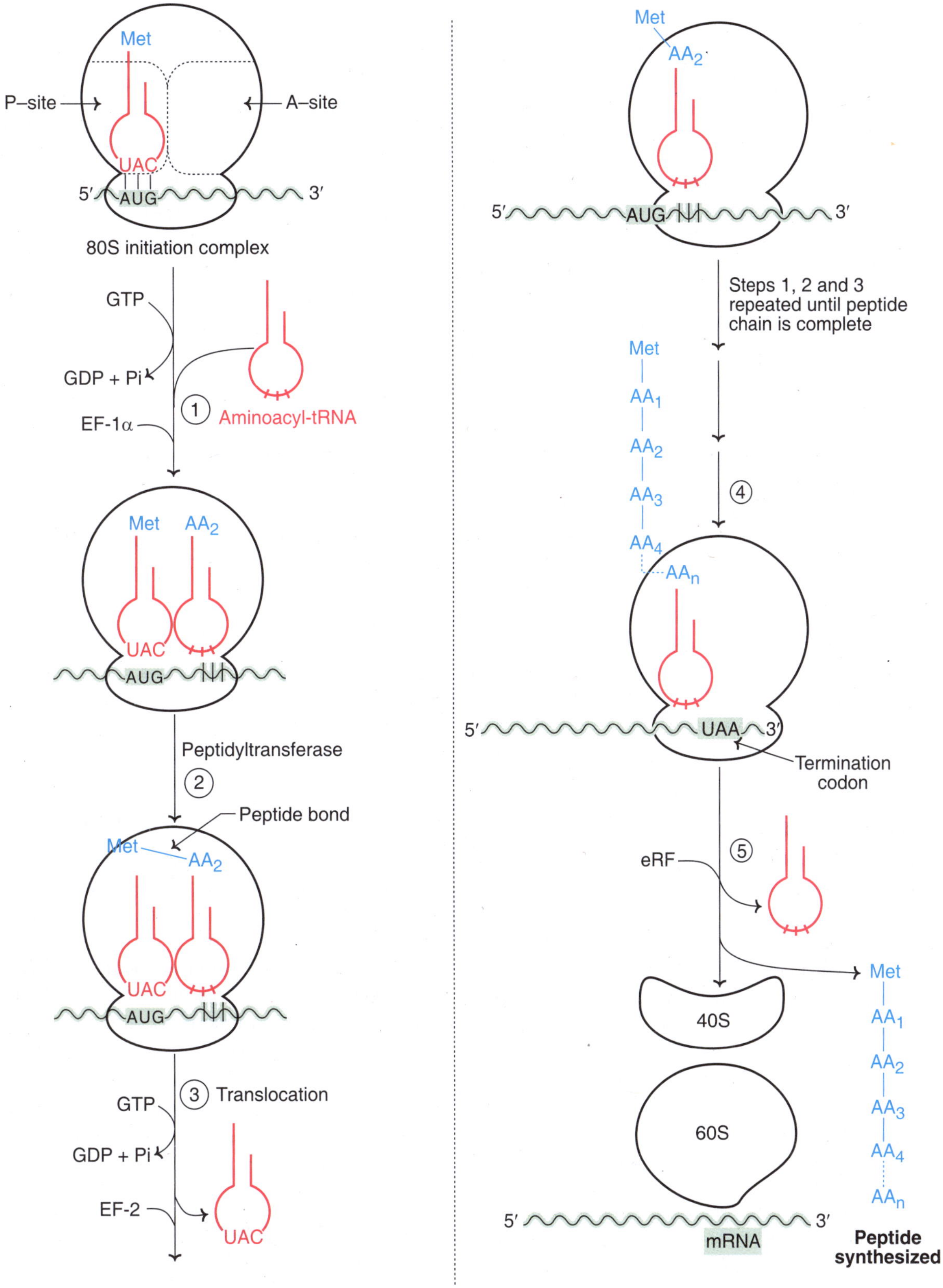

Fig. contd. next column

Fig.25.14 : *Protein biosynthesis – Elongation and termination (for initiation* ***See Fig. 25.13****). Met-Methionine; P-site – Peptidyl tRNA binding site; A-site – Aminoacyl tRNA binding site. AA-Amino acid; EF-Elongation factor; RF-Releasing factor.*

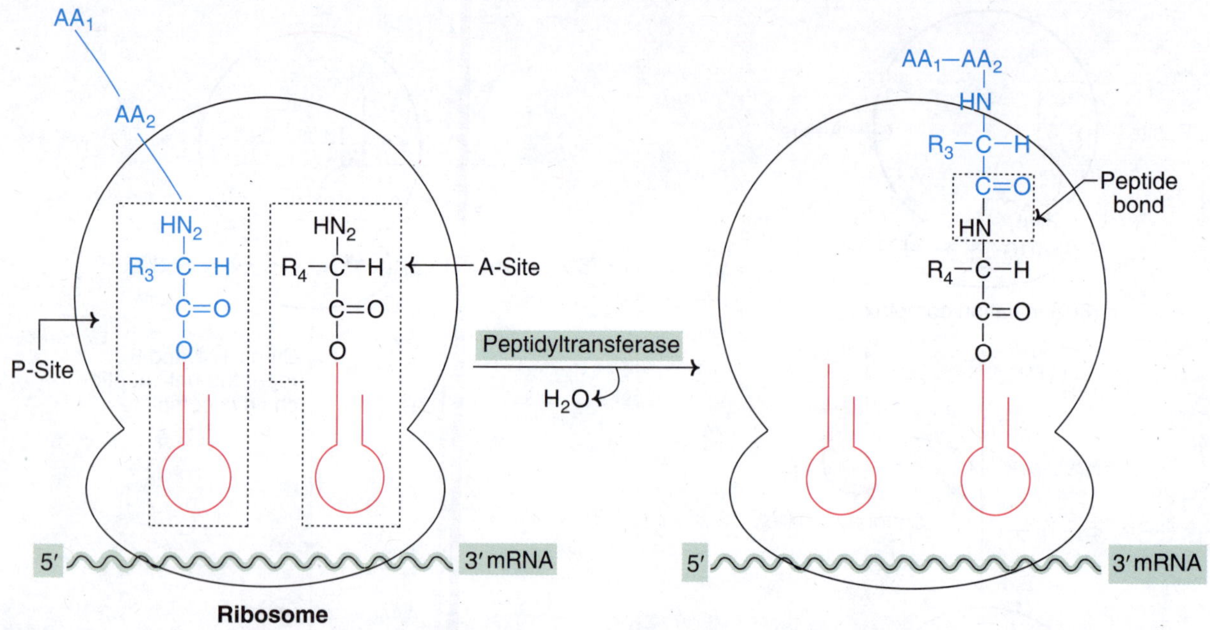

***Fig.25.15 :** Formation of peptide bond in translation (P-site—Peptidyl tRNA site; A-site—Aminoacyl tRNA site).*

In recent years, another site namely ***exit site*** (***E-site***) has been identified in eukaryotes. The deacylated tRNA moves into the E-site, from where it leaves the ribosome.

In case of prokaryotes, the elongation factors are different, and they are EF-Tu, EF-Ts (in place of of EF-1a) and EF-G (instead of EF-2).

Incorporation of amino acids

It is estimated that about ***six amino acids per second*** are incorporated during the course of elongation of translation ***in eukaryotes***. In case of prokaryotes, as many as 20 amino acids can be incorporated per second. Thus the process of protein/ polypeptide synthesis in translation occurs with ***great speed*** and ***accuracy***.

TERMINATION OF TRANSLATION

Termination is a simple process when compared to initiation and elongation. After several cycles of elongation, incorporating amino acids and the formation of the specific protein/polypeptide molecule, one of the ***stop or termination signals (UAA, UAG and UCA)*** terminates the growing polypeptide. The termination codons which act as stop signals do not have specific tRNAs to bind. As the termination codon occupies the ribosomal A-site, the ***release factor*** namely ***eRF*** recognizes the stop signal. eRF-GTP complex, in association with the enzyme peptidyltransferase, cleaves the peptide bond between the polypeptide and the tRNA occupying P-site. In this reaction, a water molecule, instead of an amino acid is added. This hydrolysis releases the protein and tRNA from the P-site. The 80S ribosome dissociates to form 40S and 60S subunits which are recycled. The mRNA is also released.

INHIBITORS OF PROTEIN SYNTHESIS

Translation is a complex process and it has become a favourite target for inhibition by antibiotics. Antibiotics are the substances produced by bacteria or fungi which inhibit the growth of other organisms. ***Majority of the antibiotics interfere with the bacterial protein synthesis and are harmless to higher organisms***. This is due to the fact that the process of translation sufficiently differs between prokaryotes and eukaryotes.

Streptomycin, ***tetracycline***, ***puromycin***, ***chloramphenicol***, ***erythromycin*** and diphtheria toxin are among the commonly used inhibitors of protein synthesis.

IV. CHAPERONES AND PROTEIN FOLDING

The three dimensional conformation of proteins is important for their biological functions. Some of the proteins can spontaneously generate the correct functionally active conformation e.g. denatured pancreatic ribonuclease. However, a vast majority

Medical Concepts/ Clinical Correlates

- **Several antibiotics selectively block bacterial translation, and thus inhibit their growth e.g. streptomycin, tetracycline, puromycin.**
- **Protein misfolding often results in the formation of prions (proteinous infectious agents) which have been implicated in many diseases e.g. mad cow disese, Alzheimer's disease.**
- **Lebers' hereditary optic neuropathy is caused by mutation in mtDNA in males. The victims become blind due to loss of central vision as a result of neuroretinal degeneration.**

of proteins can attain correct conformation, only through the assistance of certain proteins referred to as chaperones. Chaperones are ***heat shock proteins*** (originally discovered in response to heat shock). They ***facilitate*** and favour the interactions on the polypeptide surfaces to finally give the ***specific conformation of a protein***. Chaperones can reversibly bind to hydrophobic regions of unfolded proteins and folding intermediates. They can stabilize intermediates, prevent formation of incorrect intermediates, and also prevent undesirable interactions with other proteins. All these activities of chaperones help the protein to attain compact and biologically active conformation.

Protein misfolding and diseases

The failure of a protein to fold properly generally leads to its rapid degradation. ***Cystic fibrosis*** (***CF***) is a common autosomal recessive disease. Some cases of CF with mutations that result in altered protein (cystic fibrosis transmembrane conductance regulator or in short ***CFTR***) have been reported. Mutated CFTR cannot fold properly, besides not being able to get glycosylated or transported. Therefore, CFTR gets degraded.

Certain neurological diseases which are due to cellular accumulation of aggregates of misfolded proteins or their partially degraded products have been identified. The term ***prions*** (***pr***oteinous ***in***fectious agents) is used to collectively represent them (***Refer Chapter 22***).

Prions exhibit the characteristics of viral or microbial pathogens and ***have been implicated in many diseases***. e.g. ***mad cow disease***, Creutzfeldt-Jacob disease, ***Alzheimer's disease***, Huntington's disease (***Refer Chapter 22***).

It is now recognized that ***prion diseases are protein conformation diseases*** due to alteration in protein conformation resulting in altered protein properties.

V. POST-TRANSLATIONAL MODIFICATIONS OF PROTEINS

The proteins synthesized in translation are, as such, not functional. Many changes take place in the polypeptides after the initiation of their synthesis or, most frequently, ***after the protein synthesis is completed***. These ***modifications*** include protein folding (described already), trimming by proteolytic degradation, intein splicing and covalent changes which are collectively known as post-translational modifications (***Fig.25.16***).

Proteolytic degradation

Many proteins are synthesized as the precursors which are much bigger in size than the functional proteins. Some portions of precursor molecules are removed by proteolysis to liberate active proteins. This process is commonly referred to as trimming. The formation of insulin from preproinsulin, conversion of ***zymogens*** (inactive digestive enzymes e.g. trypsinogen) to the active enzymes are some examples of trimming.

Intein splicing

Inteins are ***intervening sequences in certain proteins***. These are comparable to introns in mRNAs. ***Inteins*** have to be ***removed***, *and* ***exteins ligated*** in the appropriate order for the protein to become active.

Covalent modifications

The proteins synthesized in translation are subjected to many covalent changes. By these modifications in the amino acids, the proteins may be converted to active form or inactive form. Selected examples of covalent modifications are described below.

1. **Phosphorylation :** The hydroxyl group containing amino acids of proteins, namely serine, threonine and tyrosine are subjected to

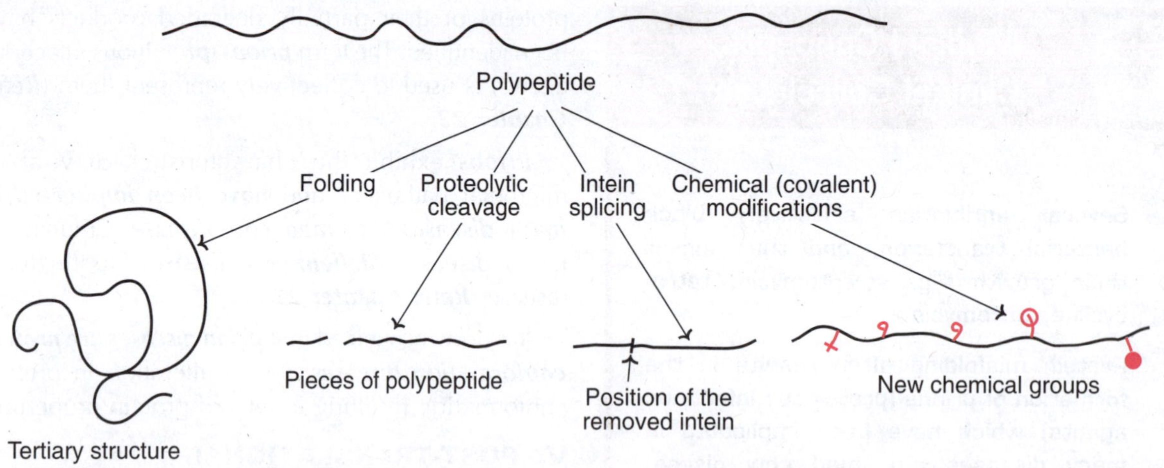

***Fig.25.16 :** An outline of post-translational modifications of proteins.*

phosphorylation. The phosphorylation may either increase or decrease the activity of the proteins. A group of enzymes called ***protein kinases*** catalyse phosphorylation while protein phosphatases are responsible for dephosphorylation (removal of phosphate group). Many enzymes that undergo phosphorylation or dephosphorylation are known in metabolisms (e.g. glycogen synthase).

2. **Hydroxylation :** During the formation of collagen, the amino acids ***proline and lysine*** are respectively converted to ***hydroxyproline and hydroxylysine***. This hydroxylation occurs in the endoplasmic reticulum and requires ***vitamin C***.

3. **Glycosylation :** The attachment of carbohydrate moiety is essential for some proteins to perform their functions. The complex carbohydrate moiety is attached to the amino acids, serine and threonine (O-linked) or to asparagine (N-linked), leading to the synthesis of glycoproteins.

Vitamin K dependent ***carboxylation*** of glutamic acid residues in certain clotting factors is also a post-translational modification.

PROTEIN TARGETING

The eukaryotic proteins (tens of thousands) are distributed between the cytosol, plasma membrane and a number of cellular organelles (nucleus, mitochondria, endoplasmic reticulum etc.). At the appropriate places, they perform their functions.

The proteins, synthesized in translation, have to ***reach*** their ***destination*** to exhibit their biological activities. This is carried out by a process called protein targeting or ***protein sorting*** or ***protein localization***. The proteins move from one compartment to another by multiple mechanisms.

The protein transport from the endoplasmic reticulum through the Golgi apparatus, and beyond uses carrier vesicles. It may be, however, noted that only the ***correctly folded proteins*** are recognized as the ***cargo for transport***. Protein targeting and post-translational modifications occur in a well coordinated manner.

Protein targeting to mitochondria

Most of the proteins of mitochondria are synthesized in the cytosol, and their transport to mitochondria is a complex process. Majority of the proteins are synthesized as larger preproteins with N-terminal presequences for the entry of these proteins into mitochondria. The ***transport of unfolded proteins*** is often ***facilitated by chaperones***.

One protein namely ***mitochondrial matrix targeting signal***, involved in protein targeting has been identified. This protein can recognize mitochondrial receptor and transport certain proteins from cytosol to mitochondria. This is an energy-dependent process.

MITOCHONDRIAL DNA, TRANSCRIPTION AND TRANSLATION

The mitochondrial DNA (***mtDNA***) has ***structural and functional resemblances with prokaryotic DNA***. This fact supports the view that mitochondria are derivatives of prokaryotes. mtDNA is circular in nature and contains about 16,000 nucleotide bases.

A vast majority of structural and functional proteins of the mitochondria are synthesized in the cytosol, under the influence of nuclear DNA. However, ***certain proteins (around 13)***, most of them being the components of electron transport chain, are ***synthesized in the mitochondria*** (e.g. cytochrome b of complex III, two subunits of ATP synthase). Transcription takes place in the mitochondria leading to the synthesis of mRNAs, tRNAs and rRNAs. Two types of rRNA and about 22 species of tRNA have been so far identified. Transcription is followed by translation resulting in protein synthesis.

The mitochondria of the sperm cell do not enter the ovum during fertilization, therefore, ***mtDNA is inherited from the mother.*** Mitochondrial DNA is subjected to high rate of mutations (about 10 times more than nuclear DNA) that causes inherited defects in oxidative phosphorylation. The best known among them are certain mitochondrial myopathies and Leber's hereditary optic neuropathy. The latter is mostly found in males and is characterized by blindness due to loss of central vision as a result of neuroretinal degeneration. ***Leber's hereditary optic neuropathy*** is a consequence of single base mutation in mtDNA. Due to this, the amino acid histidine, in place of arginine, is incorporated into the enzyme NADH coenzyme Q reductase.

SUMMARY

1. *Transcription is the process in which RNA is synthesized from DNA, which is carried out in 3 stages–initiation, elongation and termination. In case of prokaryotes, a single enzyme synthesizes all the RNAs. In eukaryotes, RNA polymerase I, II and III respectively catalyse the formation of rRNAs, mRNAs and tRNAs.*
2. *Biosynthesis of a protein or a polypeptide is known as translation. The amino acid sequence of a protein is determined by the triplet nucleoside base sequences of mRNA, arranged as codons.*
3. *The genetic code (codons)–composed of A, G, C and U–is universal, specific, non-overlapping and degenerate. Of the 64 codons, three (UAA, UAG, UGA) are termination codons while the rest code for amino acids.*
4. *Ribosomes are the factories of protein biosynthesis. Translation involves activation of amino acids, protein synthesis proper (initiation, elongation and termination), protein folding and post-translational modifications.*
5. *The post-translational modifications include proteolytic degradation, intein splicing and covalent modifications (phosphorylation, hydroxylation, glycosylation etc.). These modifications are required to make the proteins biologically active.*
6. *The mitochondria possess independent DNA with the machinery for transcription and translation. However, only a few proteins (around 13) are actually synthesized in the mitochondria.*

REGULATION OF GENE EXPRESSION

26

DNA, the chemical vehicle of heredity, is composed of ***functional units***, namely ***genes***. The term ***genome*** refers to the total genetic information contained in a cell. The bacterium *Escherichia coli* contains about 4,400 genes present on a single chromosome. The ***genome of humans*** is more complex, with 23 pairs of (diploid) chromosomes containing 6 billion (6×10^9) base pairs of DNA, with an estimated ***19,000–20,000 genes***. At any given time, only a fraction of the genome is expressed.

GENE REGULATION—GENERAL

The ***regulation of the expression of genes*** is absolutely ***essential for the growth, development, differentiation*** and the very existence of an organism. There are two types of gene regulation-positive and negative.

1. **Positive regulation :** The gene regulation is said to be positive when its ***expression is increased*** by a regulatory element (positive regulator).

2. **Negative regulation :** A ***decrease in the gene expression*** due to the presence of a regulatory element (negative regulator) is referred to as negative regulation.

It may be noted here that double negative effect on gene regulation results in a positive phenomenon.

Constitutive and inducible genes

The genes are generally considered under two categories.

1. **Constitutive genes :** The products (proteins) of these genes are ***required all the time*** in a cell. Therefore, the constitutive genes (or ***housekeeping genes***) are expressed at more or less constant rate in almost all the cells and, further, they are not subjected to regulation e.g. the ***enzymes of citric acid cycle***.

2. **Inducible genes :** The concentration of the proteins synthesized by inducible genes is ***regulated by*** various ***molecular signals***. An inducer increases the expression of these genes while a repressor decreases, e.g. tryptophan pyrrolase of liver is induced by tryptophan.

One cistron-one subunit concept

The chemical product of a gene expression is a protein which may be an enzyme. It was originally believed that each gene codes for a specific enzyme, leading to the popular concept, one gene-one enzyme. This however, is not necessarily valid due to the fact that several enzymes (or proteins) are composed of two or more nonidentical subunits (polypeptide chains).

The cistron is the smallest unit of genetic expression. It is the fragment of DNA coding for the subunit of a protein molecule. The original concept of ***one gene-one enzyme is replaced by one cistron-one subunit***.

Models for the study of gene expression

Elucidation of the regulation of gene expression in prokaryotes has largely helped to understand the principles of the flow of information from genes to mRNA to synthesize specific proteins. Some important features of ***prokaryotic gene expression*** are described first. This is followed by a brief account of eukaryotic gene expression.

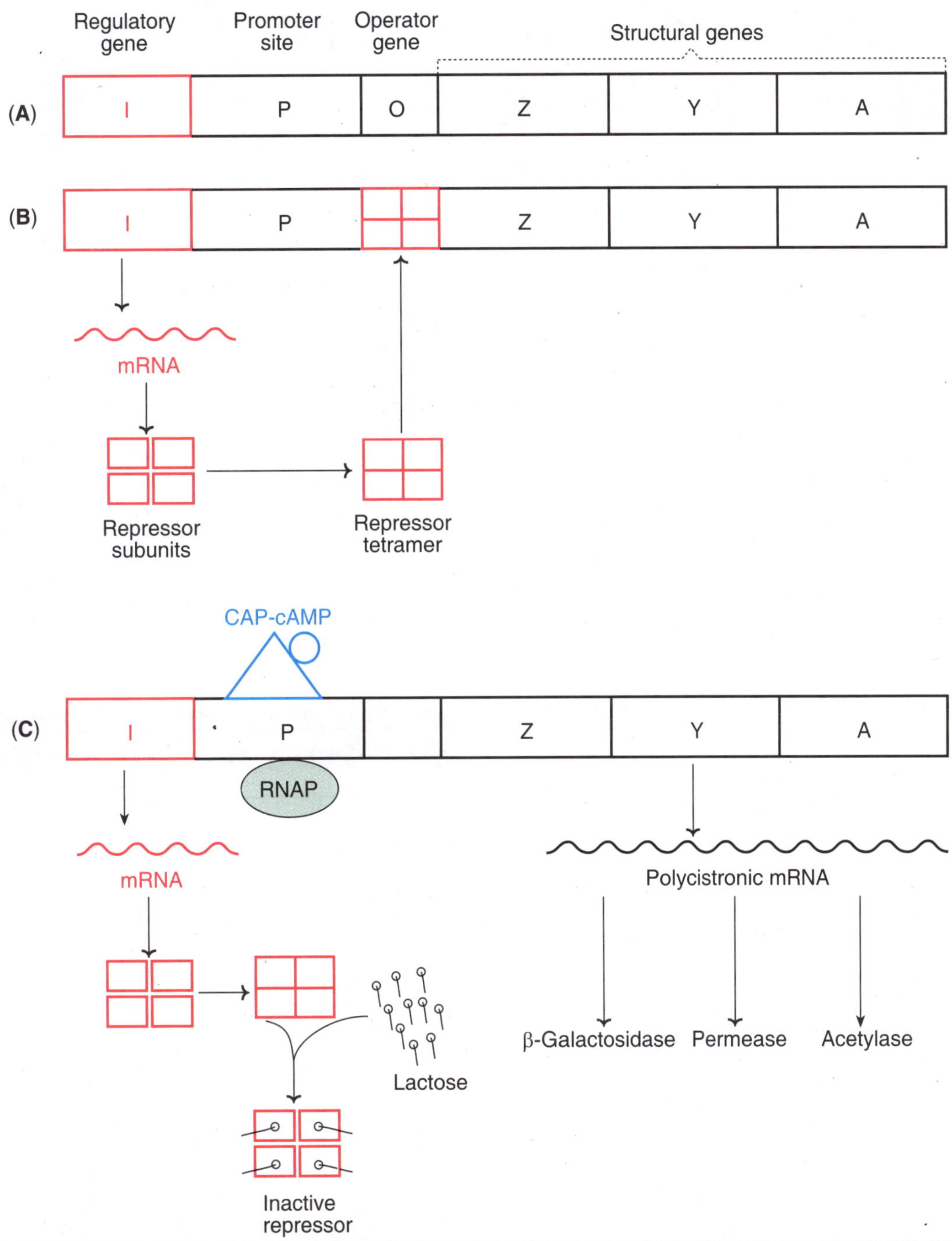

Fig.26.1 : *Model of lactose operon in* E.coli *(**A**) Structure of lac operon (**B**) Repression of lac operon (**C**) Derepression of lac operon. (CAP—cAMP–catabolite gene activator protein bound to cAMP; RNAP–RNA polymerase).*

THE OPERON CONCEPT

The operon is the ***coordinated unit of genetic expression in bacteria***. The concept of operon was introduced by ***Jacob and Monod*** in 1961 (Nobel Prize 1965), based on their observations on the regulation of lactose metabolism in *E. coli*. This is popularly known as ***lac operon***.

LACTOSE (LAC) OPERON

Structure of lac operon

The lac operon (***Fig.26.1***) consists of a regulatory gene (I; I for inhibition), operator gene (O) and three structural genes (Z, Y, A). Besides these genes, there is a promoter site (P), next to the operator gene, where the enzyme RNA polymerase binds.

Medical Concepts/ Clinical Correlates

- **Regulation of gene expression to adapt to the changes in the environment is a remarkable property of living cells e.g. synthesis of insulin by β-cells of pancreas and nowhere else.**
- **The growth, development and differentiation of an organism involves complex mechanisms which ultimately depend on gene regulation.**
- **The house-keeping genes or constitutive genes are expressed at almost a constant rate in the cells, and they are not usually subjected to regulations e.g. enzymes of Krebs cycle.**

The structural genes ***Z, Y and A respectively***, code for the enzymes ***β-galactosidase, galactoside permease and galactoside acetylase***. β-Galactosidase hydrolyses lactose (β-galactoside) to galactose and glucose while permease is responsible for the transport of lactose into the cell. The function of acetylase (coded by A gene) remains a mystery.

The structural genes Z, Y and A transcribe into a single large mRNA with 3 independent translation units for the synthesis of 3 distinct enzymes. An mRNA coding for more than one protein is known as ***polycistronic mRNA***. Prokaryotic organisms contain a large number of polycistronic mRNAs.

Repression of lac operon

The regulatory gene (I) is constitutive. It is expressed at a constant rate leading to the synthesis of lac repressor. Lac repressor is a tetrameric (4 subunits) regulatory protein (total mol. wt. 150,000) which specifically binds to the operator gene (O). This prevents the binding of the enzyme RNA polymerase to the promoter site (P), thereby blocking the transcription of structural genes (Z, Y and A). This is what happens in the absence of lactose in *E. coli*. The ***repressor molecule acts as a negative regulator of gene expression***.

Derepression of lac operon

In the presence of lactose (inducer) in the medium, a small amount of it can enter the *E. coli* cells. The repressor molecules have a high affinity for lactose. The lactose molecules bind and induce a conformational change in the repressor. The result is that the repressor gets inactivated and, therefore, cannot bind to the operator gene (O). The RNA polymerase attaches to the DNA at the promoter site and transcription proceeds, leading to the formation of polycistronic mRNA (for genes Z, Y and A) and, finally, the 3 enzymes. Thus, ***lactose induces the synthesis of the three enzymes*** β-galactosidase, galactoside permease and galactoside acetylase. ***Lactose acts by inactivating the repressor molecules***, hence this process is known as derepression of lac operon.

The attachment of RNA polymerase to the promoter site requires the presence of a ***catabolite gene activator protein (CAP)*** bound to cyclic AMP.

GENE EXPRESSION IN EUKARYOTES

Each cell of the higher organism contains the entire genome. As in prokaryotes, gene expression in eukaryotes is regulated to provide the appropriate response to biological needs. This may occur in the following ways

- Expression of certain genes (housekeeping genes) in most of the cells.
- Activation of selected genes upon demand.
- Permanent inactivation of several genes in all but a few types.

In case of prokaryotic cells, most of the DNA is organized into genes which can be transcribed. In contrast, in mammals, ***very little of the total DNA is organized into genes*** and their associated regulatory sequences. The function of the bulk of the extra DNA is not known.

Eukaryotic gene expression and its regulation are ***highly complex***. Some of the important aspects are briefly described.

CHROMATIN SRUCTURE AND GENE EXPRESSION

The DNA in higher organisms is extensively folded and packed to form ***protein-DNA complex called chromatin***. The structural organization of DNA in the form of chromatin plays an important

TABLE 26.1 A selected list of genes (represented by the products) along with respective chromosomes

Genes	Chromosome number
Alkaline phosphatase	1
Apolipoprotein B	2
Transferrin	3
Alcohol dehydrogenase	4
HMG CoA reductase	5
Steroid 21-hydroxylase	6
Arginase	7
Carbonic anhydrase	8
Interferon	9
Parathyroid hormone	11
Glyceraldehyde 3-phosphate dehydrogenase	12
Adenosine deaminase	13
α_1-Antitrypsin	14
Cytochrome P_{450}	15
Hemoglobin α-chain	16
Growth hormone	17
Prealbumin	18
Creatine phosphokinase (M chain)	19
Adenosine deaminase	20
Superoxide dismutase	21
Immunoglobulin (λ chain)	22
Glucose 6-phosphate dehydrogenase	X
Steroid sulfatase	Y

role in eukaryotic gene expression. In fact, chromatin structure provides an additional level of control of gene expression.

A selected list of genes (represented by the products) along with the respective chromosomes on which they are located is given in ***Table 26.1***.

In general, the genes that are transcribed within a particular cell are less condensed and more open in structure. This is in contrast to genes that are not transcribed which form highly condensed chromatin.

Histone acetylation and deacetylation

Eukaryotic DNA segments are wrapped around histone proteins to form nucleosome. Acetylation or deacetylation of histones is an ***important factor in determining the gene expression***. In general, acetylation of histones leads to activation of gene expression while deacetylation reverses the effect.

Acetylation predominantly occurs on the lysine residues in the amino terminal ends of histones. This modification in histones reduces the positive charges of terminal ends (tails), and decreases their binding affinity to negatively charged DNA. Consequently, nucleosome structure is disrupted to allow transcription.

Methylation of DNA and inactivation of genes

Cytosine in the sequence CG of DNA gets methylated to form 5′-methylcytosine. A major portion of CG sequences (about 20%) in human DNA exists in methylated form. In general, ***methylation leads to*** loss of transcriptional activity, and thus ***inactivation of genes***. This occurs due to binding of methylcytosine binding proteins to methylated DNA. As a result, methylated DNA is not exposed and bound to transcription factors. It is interesting to note that methylation of DNA correlates with deacetylation of histones. This provides a double means for repression of genes.

The activation and normal expression of genes, and gene inactivation by DNA methylation are depicted in ***Fig.26.2***.

ENHANCERS AND TISSUE-SPECIFIC GENE EXPRESSION

Enhancers (or activators) are DNA elements that facilitate or ***enhance gene expression***. The enhancers provide binding sites for specific proteins that regulate transcription. They facilitate binding of the transcription complex to promoter regions.

Some of the enhancers possess the ability to promote transcription in a tissue-specific manner. For instance, gene expression in lymphoid cells for the production immunoglobulins (Ig) is promoted by the enhancer associated with Ig genes between J and C regions.

Transgenic animals are frequently used for the study of tissue-specific expression. The available evidence from various studies indicates that the ***tissue-specific gene expression is largely mediated through the involvement of enhancers***.

Medical Concepts/ Clinical Correlates

- The malignant cells develop drug resistance to long term administration of methotrexate. This occurs by amplification of the genes coding for dihydrofolate reductase.
- The human body has the capability to produce around 10 billion antigen-specific immunoglobulins. This is achieved by a process called gene rearrangement.
- Knowledge on the gene expression and its regulation helps in the understanding and control of several diseases, including cancer.
- Epigenetic therapy of cancer employing certain inhibitors holds a great promise.

MOTIFS IN PROTEINS AND GENE EXPRESSION

A motif literally means a dominant element. Certain motifs in proteins mediate the binding of regulatory proteins (transcription factors) to DNA. The specific control of transcription occurs by the binding of regulatory proteins with high affinity to the correct regions of DNA.

A great majority of specific ***protein-DNA interactions*** are brought out by four unique motifs—helix-turn-helix (HTH), zinc finger, leucine zipper, helix-loop-helix (HLH).

GENE REGULATION IN EUKARYOTES

The important features of eukaryotic gene expression along with the regulatory aspects are described in the preceeding pages. Besides transcription, eukaryotic cells also employ variety of other mechanisms to regulate gene expression. The most important ones are listed below, and briefly described next.

1. Gene amplification
2. Gene rearrangement
3. Processing of RNA
4. Alternate mRNA splicing
5. Transport of mRNA from nucleus to cytoplasm
6. Degradation of mRNA.

Gene amplification

In this mechanism, the expression of a gene is increased several fold. This is commonly observed ***during the developmental stages of eukaryotic organisms***. For instance, in fruit fly (*Drosophila*), the amplification of genes coding for egg shell proteins is observed during the course of oogenesis. The amplification of the gene (DNA) can be observed under electron microscope (***Fig.26.3***).

The occurrence of gene amplification has also been reported in humans. ***Methotrexate is an anticancer drug*** which inhibits the enzyme ***dihydrofolate reductase***. The malignant cells

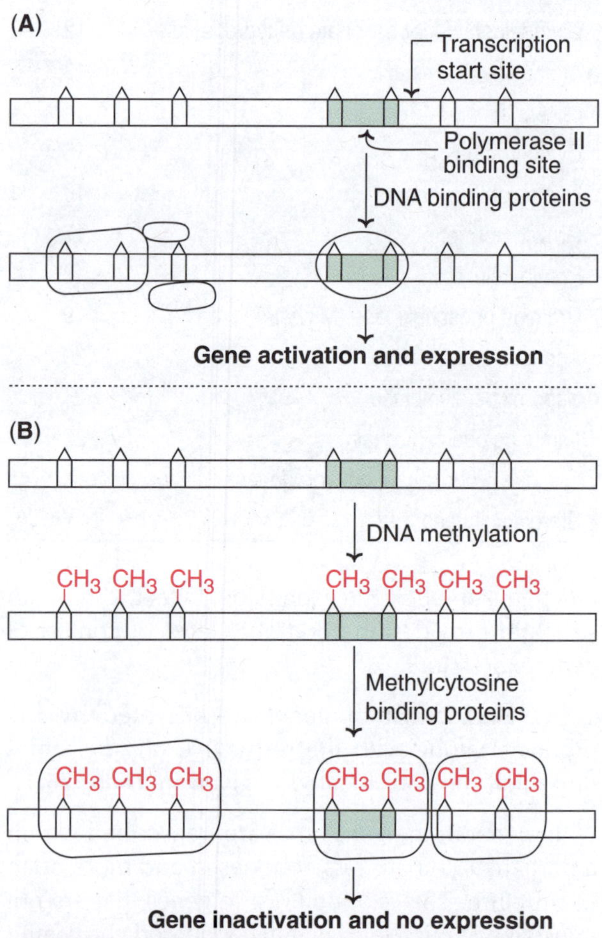

Fig.26.2 : *Methylation of DNA and inactivation of genes (**A**) Gene activation in the absence of DNA methylation (**B**) Gene inactivation due to methylation (⌂ represent CG sequences).*

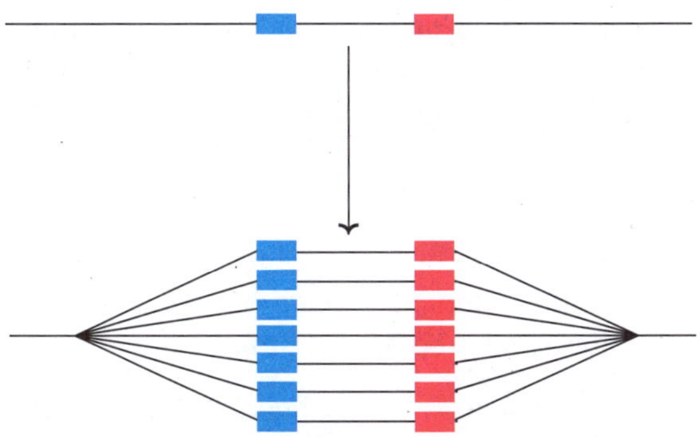

Fig.26.3 : *A diagrammatic representation of gene amplification (the genes are depicted in blue and red colours).*

develop drug resistance to long term administration of methotrexate by amplifying the genes coding for dihydrofolate reductase.

Gene rearrangment

The body possesses an enormous capacity to synthesize a wide range of antibodies. It is estimated that the human body can produce about 10 billion (10^{10}) antibodies in response to antigen stimulations. The molecular mechanism of this antibody diversity was not understood for long. It is now explained on the basis of gene rearrangement or ***transposition of genes*** or ***somatic recombination of DNA***.

The structure of a typical immunoglobulin molecule consists of two light (L) and two heavy (H) chains. Each one of these chains (L or H) contains an N-terminal variable (V) and C-terminal constant (C) regions (***Refer Fig.9.2***). The V regions of immunoglobulins are responsible for the recognition of antigens. The phenomenon of gene rearrangement can be understood from the mechanism of the synthesis of light chains of immunoglobulins (***Fig.26.4***).

Each light chain can be synthesized by three distinct DNA segments, namely the variable (V_L), the joining (J_L) and the constant (C_L). The mammalian genome contains about 500 V_L segments, 6 J_L segments and 20 C_L segments. During the course of differentiation of B-lymphocytes, one V_L segment (out of the 500) is brought closer to J_L and C_L segments. This occurs on the same chromosome. For the sake of illustration, 100th V_L, 3rd J_L and 10th C_L segments are rearranged in ***Fig.26.4***. The rearranged DNA (with V_L, J_L and C_L fragments) is then transcribed to produce a single mRNA for the synthesis of a specific light chain of the antibody. By innumerable combinations of V_L, J_L and C_L segments, the body's immune system can generate millions of antigen specific immunoglobulin molecules.

The formation of heavy (H) chains of immunoglobulins also occurs by rearrangement of 4 distinct genes—variable (V_H), diversity (D), joining (J_H) and constant (C_H).

Processing of RNA

The RNA synthesized in transcription undergoes modifications resulting in a functional RNA. The changes include intron-exon ***splicing***, polyadenylation etc. (***Chapter 25***).

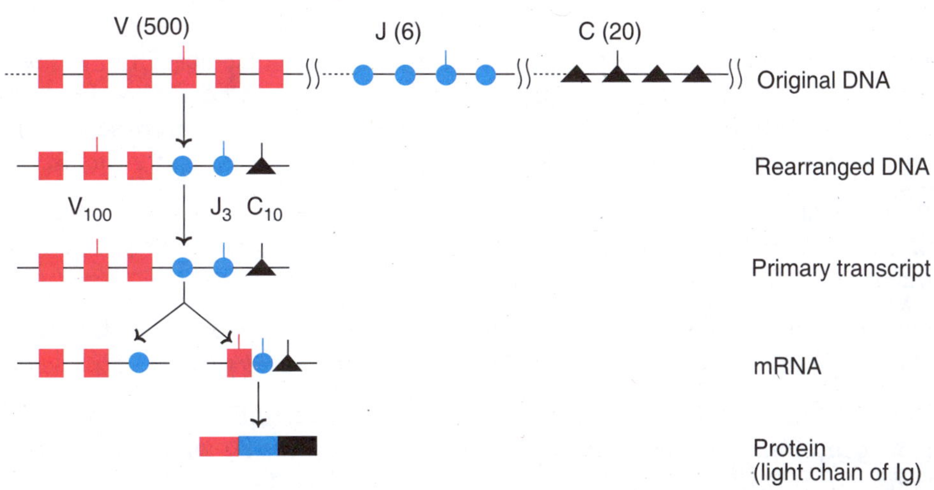

Fig.26.4 : *A diagrammatic representation of gene rearrangement for the synthesis of light chain of immunoglobulin.*

Alternate mRNA splicing

Eukaryotic cells are capable of carrying out alternate mRNA processing to control gene expression. Different mRNAs can be produced by alternate splicing which code for different proteins (for more details, ***Refer Chapter 25***).

Degradation of mRNA

The expression of genes is indirectly influenced by the stability of mRNA. Certain hormones regulate the synthesis and degradation of some mRNAs. For instance, estradiol prolongs the half-life of vitellogenin mRNA from a few hours to about 200 hours.

EPIGENETIC REGULATION OF GENE EXPRESSION

The term ***epigenetics*** (literally means above genetics) is used to describe the ***changes in the characteristics*** of a cell or an organism that are ***not due to changes in the nucleotide sequence of the DNA***.

Epigenetics regulates gene expression by modulating chromatin structure via histone modification or modification of DNA via methylation. For instance, ***acetylation of histones leads to activation of gene expression***, while DNA methylation is associated with a reduction in gene expression.

The pattern of covalent modifications in histones that provide a tunable means of regulating gene expression is referred to as ***histone code***.

Although there is no change in DNA sequence, epigenetics is heritable due to coordination between histone modification and methylation of DNA.

Epigenetic therapy of cancers

Hypermethylation of DNA in some parts of tumor suppressor genes is found in certain cancers. The enzyme DNA methyltransferase (DNMT) responsible for DNA methylation is targeted for cancer therapy. Many inhibitors of DNMT (e.g. 5-azacytidine) have been approved by FDA, and are in use for the treatment of leukemia.

Epigenetic therapy of cancer by employing various inhibitors (individually or in combination) holds a great promise.

SUMMARY

1. *DNA, the chemical vehicle of heredity, is composed of genes. The regulation of gene expression is absolutely essential for the growth, development and differentiation of an organism. A positive regulation increases gene expression while a negative regulation decreases.*

2. *The operon is the coordinated unit of gene expression. The lac operon of* E. coli *consists of regulatory genes and structural genes. The lac repressor binds to the DNA and halts the process of transcription of structural genes. However, the presence of lactose inactivates the repressor (derepression) leading to the expression of structural genes.*

4. *Eukaryotic gene expression and its regulation are highly complex. Acetylation of histones leads to gene expression while deacetylation reverses the effect. In general, methylation of DNA results in the inactivation of genes.*

5. *The protein-DNA interactions, brought out by motifs (helix-turn-helix, zinc finger, leucine zipper, helix-loop-helix), are involved in the control of gene expression.*

6. *Eukaryotic cells have developed several mechanisms to regulate gene expression. These include gene amplification, gene rearrangement, processing of RNA, transport and degradation of mRNA.*

RECOMBINANT DNA AND BIOTECHNOLOGY

27

The term biotechnology represents a fusion or an alliance between ***bio***logy and ***technology***. Frankly speaking, biotechnology is a newly discovered discipline for age-old practices e.g. preparation of wine, beer, curd, bread. These natural processes are regarded as ***old*** or ***traditional biotechnology***.

The ***new*** or ***modern biotechnology*** embraces all the ***genetic manipulations***, cell fusion techniques, and improvements made in the old biotechnological processes. The biotechnology with particular reference to recombinant DNA in human health and disease is briefly described in this chapter.

Genetic engineering primarily involves the ***manipulation of genetic material (DNA)*** to achieve the desired goal in a pre-determined way. Some ***other terms*** are also in common use to describe genetic engineering.

- ***Gene manipulation***
- ***Recombinant DNA (rDNA) technology***
- ***Gene cloning (molecular cloning)***
- ***Genetic modifications***
- ***New genetics.***

Recombinant DNA technology is a vast field. The basic principles and techniques of rDNA technology along with the most important applications are briefly described in this chapter.

BASIC PRINCIPLES OF rDNA TECHNOLOGY

There are many diverse and complex techniques involved in gene manipulation. However, the basic principles of recombinant DNA technology are reasonably simple, and broadly involve the following stages (***Fig.27.1***).

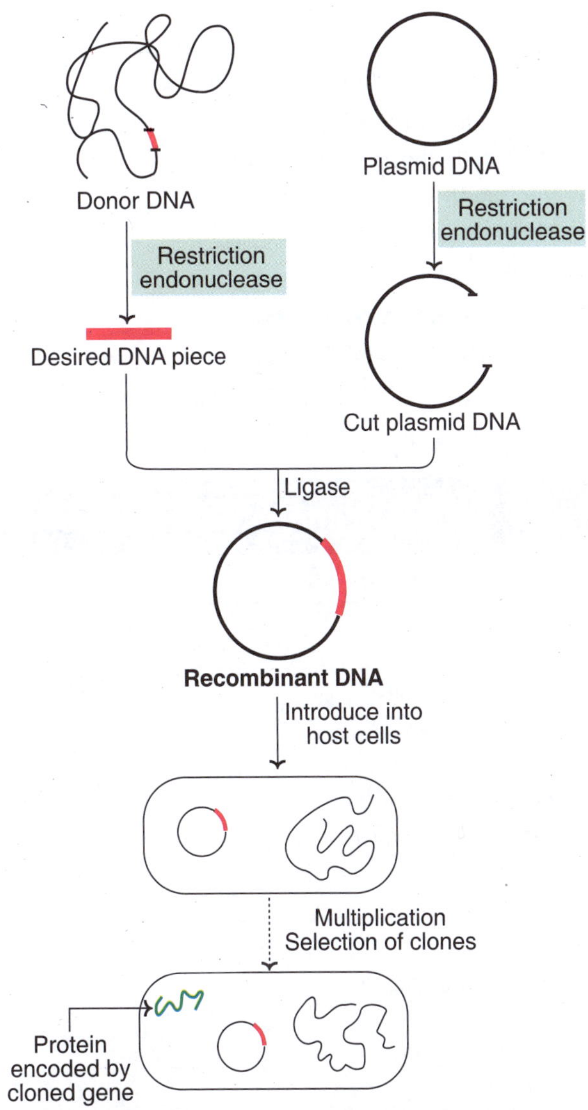

Fig.27.1 : *The basic principle of recombinant DNA technology.*

1. Generation of DNA fragments and selection of the ***desired piece of DNA*** (e.g. a human gene).

2. Insertion of the selected DNA into a cloning vector (e.g. a plasmid) to ***create*** a ***recombinant DNA*** or ***chimeric DNA*** (Chimera is a monster in Greek mythology that has a lion's head, a goat's body and a serpent's tail. This may be comparable to Narasimha in Indian mythology).

3. Introduction of the recombinant vectors into ***host cells*** (e.g. bacteria).

4. Multiplication and selection of clones containing the recombinant molecules.

5. Expression of the gene to produce the ***desired product***.

Recombinant DNA technology with special reference to the following aspects is described

- Molecular tools of genetic engineering.
- Host cells-the factories of cloning.
- Vectors-the cloning vehicles.
- Methods of gene transfer.
- Gene cloning strategies.

MOLECULAR TOOLS OF GENETIC ENGINEERING

The term genetic engineer may be appropriate for an individual who is involved in genetic manipulations. The ***genetic engineer's toolkit*** or molecular tools namely the ***enzymes***, most commonly used in recombinant DNA experiments, are briefly described.

Restriction endonucleases—DNA cutting enzymes

Restriction endonucleases are one of the most important groups of enzymes for the manipulation of DNA. These are the ***bacterial enzymes*** that can ***cut/split DNA*** (from any source) at specific sites. They were first discovered in *E.coli* restricting the replication of bacteriophages, by cutting the viral DNA (The host *E.coli* DNA is protected from cleavage by addition of methyl groups). Thus, the enzymes that restrict the viral replication are known as ***restriction enzymes*** or restriction endonucleases.

Cleavage patterns : Majority of restriction endonucleases cut DNA at defined sites within recognition sequence. A selected list of enzymes, recognition sequences, and their products formed is given in ***Table 27.1.***

The cut DNA fragments by restriction endonucleases may have mostly ***sticky ends*** (cohesive ends) or ***blunt ends***, as given in ***Table 27.1***. DNA fragments with sticky ends are particularly useful for recombinant DNA experiments. This is because the single-stranded

TABLE 27.1 Some restriction enzymes with sources, recognition sequences and the products formed

Enzyme (source)	*Recognition sequence*	*Products*
EcoRI (*Escherichia coli*)	5′······G–A–A–T–T–C······3′ 3′······C–T–T–A–A–G······5′	A–A–T–T–C······ G······ ······G C–T–T–A–A
BamHI (*Bacillus amyloliquefaciens*)	5′······G–G–A–T–C–C······3′ 3′······C–C–T–A–G–G······5′	G–A–T–C–C······ G······ ······G ······C–C–T–A–G
HaeIII (*Haemophilus aegyptius*)	5′······G–G–C–C······3′ 3′······C–C–G–G······5′	*C–C······ G–G······ ······*G–G ······ C–C
HindIII (*Haemophilus influenzae*)	5′······A–A–G–C–T–T······3′ 3′······T–T–C–G–A–A······5′	A–G–C–T–T······ A······ ······A ······T–T–C–G–A

(*Note : Scissors indicate the sites of cleavage.* * *The products are with blunt ends while for the rest, the products are with sticky ends*).

TABLE 27.2 The most commonly used enzymes in recombinant DNA technology/genetic engineering

Enzyme	*Use/reaction*
Alkaline phosphatase	Removes phosphate groups from 5′-ends of double/single-stranded DNA (or RNA).
DNA ligase	Joins DNA molecules by forming phosphodiester linkages between DNA segments.
DNA polymerase I	Synthesizes DNA complementary to a DNA template.
DNase I	Produces single-stranded nicks in DNA.
Exonuclease III	Removes nucleotides from 3′-end of DNA.
Restriction enzymes	Cut double-stranded DNA with a specific recognition site.
Reverse transcriptase	Synthesizes DNA from RNA.
Taq DNA polymerase	Used in polymerase chain reaction
SI nuclease	Degrades single-stranded DNA and RNA.
Terminal transferase	Adds nucleotides to the 3′-ends of DNA or RNA. Useful in homopolymer tailing.

sticky DNA ends can easily pair with any other DNA fragment having complementary sticky ends.

DNA ligases—DNA joining enzymes

The cut DNA fragments are covalently joined together by DNA ligases. These enzymes were originally isolated from viruses. They also occur in *E.coli* and eukaryotic cells. DNA ligases actively participate in cellular DNA repair process.

Many enzymes are used in the recombinant DNA technology/genetic engineering. A selected list of these enzymes and the reactions catalysed by them is given in ***Table 27.2***.

HOST CELLS — THE FACTORIES OF CLONING

The hosts are the ***living systems or cells*** in which the carrier of recombinant DNA molecule or ***vector can be propagated***. There are different types of host cells-prokaryotic (bacteria) and eukaryotic (fungi, animals and plants). Some examples of host cells used in genetic engineering are given in ***Table 27.3***.

Host cells, besides effectively incorporating the vector's genetic material, must be conveniently cultivated in the laboratory to collect the products. In general, ***microorganisms*** are ***preferred*** as host cells, since they ***multiply faster*** compared to cells of higher organisms (plants or animals).

VECTORS — THE CLONING VEHICLES

Vectors are the DNA molecules, which can ***carry a foreign DNA fragment to be cloned***. They are self-replicating in an appropriate host cell. The most important vectors are ***plasmids*** (e.g. pBR322), ***bacteriophages*** (e.g. phage λ, phage M_{13}), cosmids and artificial chromosome vectors.

Choice of a vector

Among the several factors, the size of the foreign DNA is very important in the choice of vectors. The efficiency of this process is often crucial for determining the success of cloning. The sizes of DNA insert that can be accepted by different vectors is shown in ***Table 27.4***.

METHODS OF GENE TRANSFER

Introducing a foreign DNA (i.e. the gene) into the cells is an important task in biotechnology. The

TABLE 27.3 Some examples of host cells used in genetic engineering

Group	*Examples*
Prokaryotic	
Bacteria	*Escherichia coli*
	Bacillus subtilis
	Streptomyces sp
Eukaryotic	
Fungi	*Saccharomyces cerevisiae*
	Aspergillus nidulans
Animals	Insect cells
	Oocytes
	Mammalian cells
	Whole organisms
Plants	Protoplasts
	Intact cells
	Whole plants

TABLE 27.4 The different cloning vectors with the corresponding hosts and the sizes of foreign insert DNAs

Vector	*Host*	*Foreign insert DNA size*
Phage λ	*E. coli*	5–25 kb
Cosmid λ	*E. coli*	35–45 kb
Plasmid artifical chromosome (PAC)	*E. coli*	100–300 kb
Bacterial artificial chromosome (BAC)	*E. coli*	100–300 kb
Yeast chromosome	*S. cerevisiae*	200–2000 kb

efficiency of this process is often crucial for determining the success of cloning. The most commonly employed gene transfer methods are ***transformation, conjugation, electroporation and lipofection, and direct transfer of DNA***.

GENE CLONING STRATEGIES

A clone refers to ***a group of*** organisms, cells, ***molecules*** or other objects, ***arising from a single individual.*** Clone and colony are almost synonymous.

Gene cloning strategies in relation to recombinant DNA technology broadly involve the following aspects (***Fig.27.2***).

- Generation of desired DNA fragments.
- Insertion of these fragments into a cloning vector.
- Introduction of the vectors into host cells.
- Selection or screening of the recipient cells for the recombinant DNA molecules.

BASIC TECHNIQUES IN GENETIC ENGINEERING

There are several techniques used in recombinant DNA technology or gene manipulation. The most frequently used methods are listed, and briefly described.

- Isolation and purification of nucleic acids.
- Nucleic acid blotting techniques.
- DNA sequencing.
- Methods of gene transfer (described already).
- Polymerase chain reaction.
- Construction of gene library.
- Production of monoclonal antibodies by hybridoma technology (***Chapter 41***).

ISOLATION AND PURIFICATION OF NUCLEIC ACIDS

Almost all the experiments dealing with gene manipulations require pure forms of either DNA or RNA, or sometimes even both. Hence there is a need for the specific isolation of nucleic acids from the cells. The purification of nucleic acids broadly involves three stages.

1. Breaking or opening of the cells to expose nucleic acids.
2. Separation of nucleic acids from other cellular components.
3. Recovery of nucleic acids in a pure form.

NUCLEIC ACID BLOTTING TECHNIQUES

Blotting techniques are very widely used analytical tools for the specific ***identification of desired DNA or RNA fragments*** from thousands of molecules. Blotting refers to the ***process of immobilization of sample nucleic acids on solid support*** (nitrocellulose or nylon membranes). The blotted nucleic acids are then used as targets in the hybridization experiments for their specific detection. An outline of the nucleic acid blotting technique is depicted in ***Fig.27.3.***

Types of blotting techniques

The most comonly used blotting techniques are listed below

- ***Southern blotting (for DNA)***
- ***Northern blotting (for RNA)***
- ***Dot blotting (DNA/RNA)***

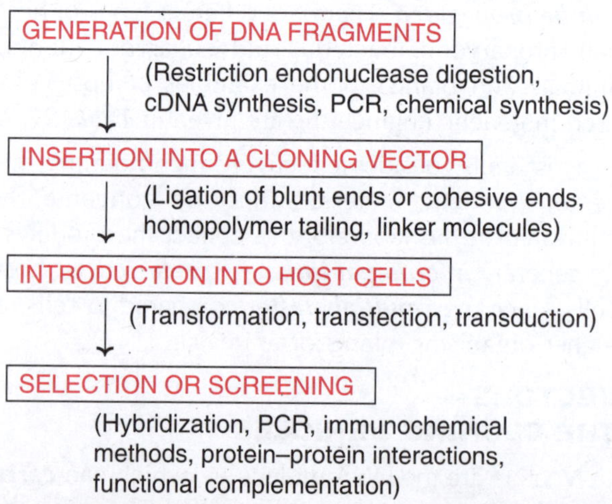

***Fig.27.2 :** An overview of cloning strategies in recombinant DNA technology.*

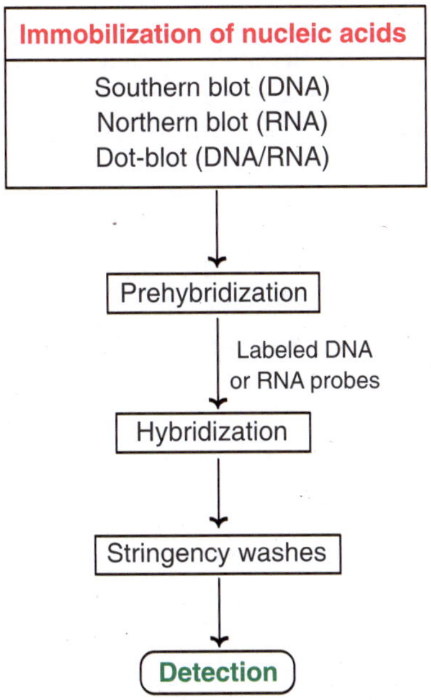

***Fig.27.3 :** An outline of the nucleic acid blotting techniques.*

The Southern blotting is named after the scientist Ed Southern (1975) who developed it. The other names Northern blotting and Western blotting are laboratory jargons which are now accepted. ***Western blotting involves the transfer of protein blots and their identification by using specific antibodies.***

SOUTHERN BLOTTING

Southern blotting technique is the first nucleic acid blotting procedure developed in 1975 by Southern. It is depicted in ***Fig.27.4***, and briefly described.

The genomic DNA isolated from cells/tissues is digested with one or more restriction enzymes. This mixture is loaded into a well in an agarose or polyacrylamide gel and then subjected to electrophoresis. DNA, being negatively charged migrates towards the anode (positively charged electrode); the smaller DNA fragments move faster.

The separated DNA molecules are denatured by exposure to a mild alkali and transferred to nitrocellulose or nylon paper. This results in an exact replica of the pattern of DNA fragments on the gel. The DNA can be annealed to the paper on exposure to heat (80°C). The nitrocellulose or nylon paper is then exposed to labeled cDNA probes. These probes hybridize with complementary DNA molecules on the paper.

The paper, after thorough washing, is exposed to X-ray film to develop autoradiograph. This reveals specific bands corresponding to the ***DNA fragments recognized by cDNA probe***.

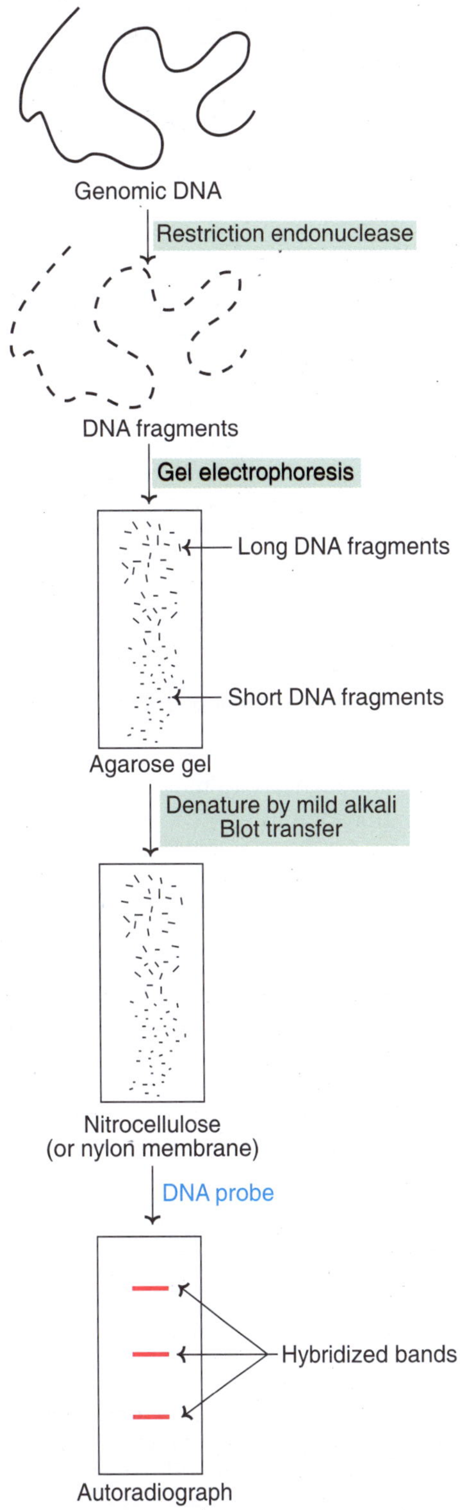

***Fig.27.4 :** Southern blotting technique.*

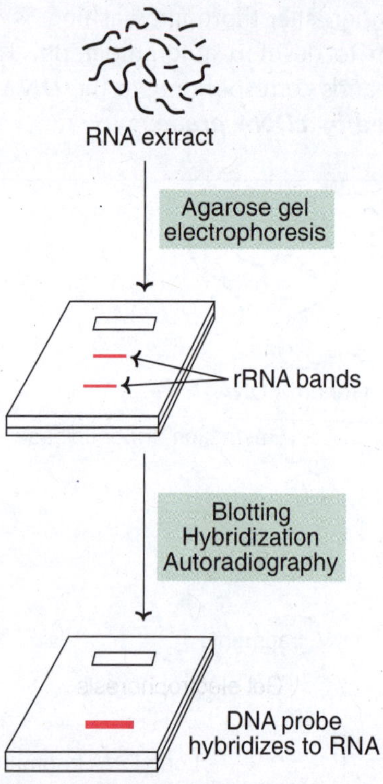

***Fig.27.5 :** An outline of Northern blotting.*

Applications of Southern blotting

Southern blotting technique is extremely specific and sensitive, although it is a simple technique. Some of the applications are listed.

- It is an invaluable method in gene analysis.
- Important for the confirmation of DNA cloning results.
- Forensically applied to detect minute quantities of DNA (***to identify parenthood, thieves, rapists etc.***).
- Highly useful for the determination of restriction fragment length polymorphism (RFLP) associated with pathological conditions.

NORTHERN BLOTTING

Northern blotting is the technique for the ***specific identification of RNA molecules***. The procedure adopted is almost similar to that described for Southern blotting, and is depicted in ***Fig.27.5***. RNA molecules are subjected to electrophoresis, followed by blot transfer, hybridization and autoradiography.

Northern blotting is theoretically, a good technique for determining the number of genes (through mRNA) present on a given DNA. But this is not really practicable since each gene may give rise to two or more RNA transcripts. Another drawback is the presence of exons and introns.

DOT-BLOTTING

Dot-blotting is a modification of Southern and Northern blotting techniques described above. In this approach, the ***nucleic acids (DNA or RNA) are directly spotted onto the filters***, and not subjected to electrophoresis. The hybridization procedure is the same as in original blotting techniques.

Dot-blotting technique is particularly useful in obtaining quantitative data for the evaluation of gene expression.

Western blotting

Western blotting involves the ***identification of proteins***. It is very useful to understand the nucleic acid functions, particularly during the course of gene manipulations.

The technique of Western blotting involves the transfer of electrophoresed protein bands from polyacrylamide gel to nylon or nitrocellulose membrane. These proteins can be detected by specific protein-ligand interactions. Antibodies or lectins are commonly used for this purpose.

Autoradiography

Autoradiography is the process of ***localization and recording of a radiolabel*** within a solid specimen, with the production of an ***image in a photographic emulsion***. These emulsions are composed of silver halide crystals suspended in gelatin.

When a β-particle or a γ-ray from a radiolabel passes through the emulsions, silver ions are converted to metallic silver atoms. This results in the development of a visible image which can be easily detected.

Applications of autoradiography

As already described, autoradiography is closely associated with blotting techniques for the detection of DNA, RNA and proteins.

DNA SEQUENCING

Determination of nucleotide sequence in a DNA molecule is the basic and fundamental requirement

Medical Concepts/ Clinical Correlates

- **Biotechnology is a newly discovered discipline for age-old practices (e.g. preparation of curd, wine, beer), with special emphasis on genetic manipulations.**
- **Southern blotting technique (that specifically detects DNA) is employed for the identification of thieves, rapists, and settlement of parenthood.**
- **Polymerase chain reaction (PCR) is useful for the diagnosis of inherited diseases, in DNA sequencing, and in forensic medicine.**
- **The analysis of genetic material DNA (gene/genes) is employed for the diagnosis of certain diseases, and in medical forensics e.g. AIDS, sickle-cell anemia, certain cancers, DNA finger-printing.**

in biotechnology. DNA sequencing is important to understand the functions of genes, and basis of inherited disorders. Further, DNA cloning and gene manipulation invariably require knowledge of accurate nucleotide sequence. The methods used for DNA sequencing are listed

- Maxam and Gilbert technique
- Dideoxynucleotide method
- Automated DNA sequencing
- DNA chips (microarrays)

POLYMERASE CHAIN REACTION (DNA AMPLIFICATION)

The polymerase chain reaction (***PCR***) is a laboratory (*in vitro*) ***technique for generating large quantities of a specified DNA***. Obviously, PCR is a ***cell-free amplification technique*** for synthesizing multiple identical copies (billions) of ***any DNA*** of interest. Developed in 1984 by Karry Mullis (Nobel Prize, 1993), PCR is now considered as a basic tool for the molecular biologist. As is a photocopier a basic requirement in an office, so is the PCR machine in a molecular biology laboratory!

Principle of PCR

The double-stranded DNA of interest is denatured to separate into two individual strands. Each strand is then allowed to hybridize with a primer (renaturation). The primer-template duplex is used for DNA synthesis (the enzyme-DNA polymerase). These three steps—***denaturation, renaturation*** and ***synthesis*** are repeated again and again to generate multiple forms of target DNA.

Technique of PCR

The essential requirements for PCR are listed below

1. A target DNA (100–35,000 bp in length).
2. Two primers (synthetic oligonucleotides of 17–30 nucleotides length) that are complementary to regions flanking the target DNA.
3. Four deoxyribonucleotides (dATP, dCTP, dGTP, dTTP).
4. A DNA polymerase that can withstand at a temperature up to 95°C (i.e., thermostable).

The actual technique of PCR involves repeated cycles for amplification of target DNA. Each cycle has three stages.

1. **Denaturation :** On raising the temperature to about 95°C for about one minute, the DNA gets denatured and the two strands separate.

2. **Renaturation or annealing :** As the temperature of the mixutre is slowly cooled to about 55°C, the primers base pair with the complementary regions flanking target DNA strands. This process is called renaturation or annealing. High concentration of primer ensures annealing between each DNA strand and the primer rather than the two strands of DNA.

3. **Synthesis :** The initiation of DNA synthesis occurs at 3′-hydroxyl end of each primer. The primers are extended by joining the bases complementary to DNA strands. The synthetic process in PCR is quite comparable to the DNA replication of the leading strand (***Refer Chapter 24***).

As is evident from the ***Fig.27.6*** (cycle I), the new DNA strand joined to each primer is beyond the sequence that is complementary to the second primer. These new strands are referred to as ***long templates***, and they will be used in the second cycle.

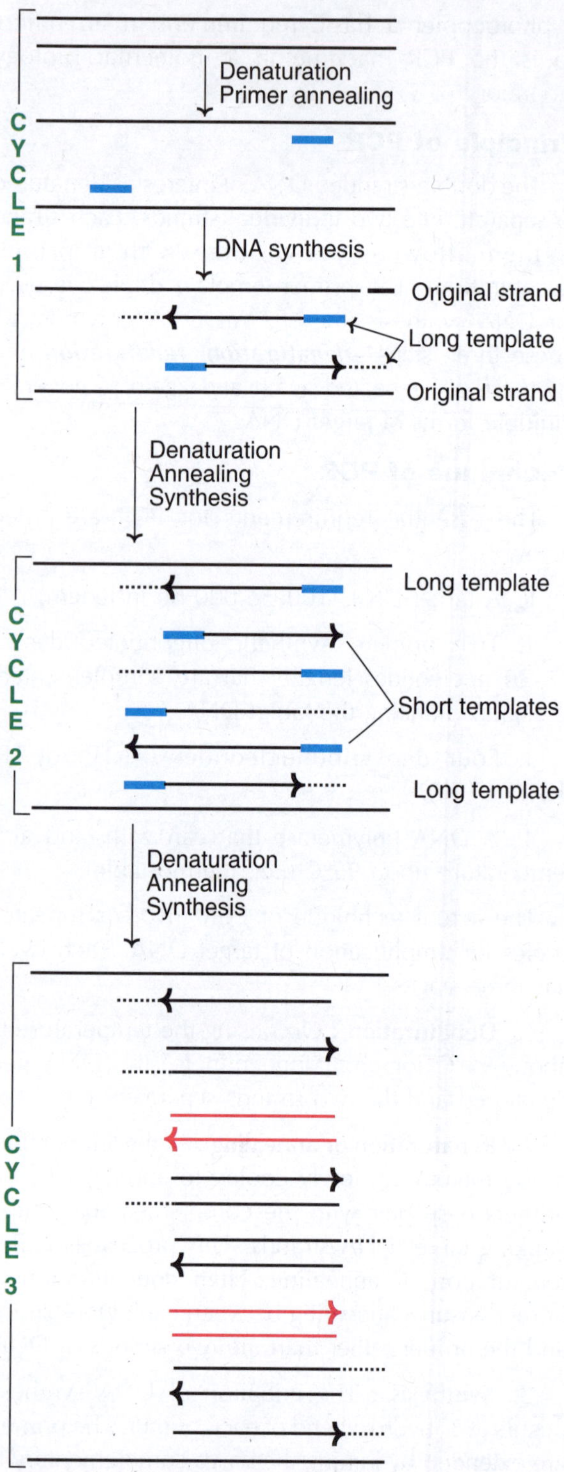

Fig.27.6 : *The polymerase chain reaction (PCR) representing the initial three cycles (▬ indicate primers).*

For the second cycle of PCR, the DNA strands (original + newly synthesized long template) are denatured, annealed with primers and subjected to DNA synthesis. At the end of second round, long templates, and short templates (DNA strands with primer sequence at one end, and sequence complementary to the other end primer) are formed.

In the third cycle of PCR, the original DNA strands along with long and short templates are the starting materials. The technique of ***denaturation, renaturation and synthesis are repeated***. This procedure is repeated ***again and again for each cycle***. It is estimated that at the end of 32nd cycle of PCR, about a million-fold target DNA is synthesized. The short templates possessing precisely the target DNA as double-stranded molecules accumulate.

Sources of DNA polymerase

In the original technique of PCR, Klenow fragment of *E. coli* DNA polymerase was used. This enzyme, gets denatured at higher temperature, therefore, fresh enzyme had to be added for each cycle. A breakthrough occurred (Lawyer 1989) with the introduction of *Taq* DNA polymerase from thermophilic bacterium, *Thermus aquaticus*. The ***Taq DNA polymerase is heat resistant, hence it is not necessary to freshly add this enzyme for each cycle of PCR***.

APPLICATIONS OF PCR

The advent of PCR had, and continues to have tremendous impact on molecular biology. The applications of PCR are too many to be listed here. Some of them are selectively and very briefly described. Other applications of PCR are discussed at appropriate places.

PCR in clinical diagnosis

The specificity and sensitivity of PCR is highly useful for the diagnosis of various diseases in humans. These include diagnosis of inherited disorders (genetic diseases), viral diseases, bacterial diseases etc.

Prenatal diagnosis of inherited diseases : PCR is employed in the prenatal diagnosis of inherited diseases by using chorionic villus samples or cells from amniocentesis. Thus, diseases like ***sickle-cell anemia, β-thalassemia and phenylketonuria*** can be detected by PCR in these samples.

Diagnosis of retroviral infections : PCR from cDNA is a valuable tool for diagnosis and monitoring of retroviral infections, e.g., ***HIV infection***.

Diagnosis of bacterial infections : PCR is used for the detection of bacterial infections e.g., ***tuberculosis*** by *Mycobacterium tuberculosis*.

Diagnosis of cancers : Several virally-induced cancers (e.g., cervical cancer caused by human papilloma virus) can be detected by PCR. Further, some cancers which occur due to chromosomal translocation (chromosome 14 and 18 in follicular lymphoma) involving known genes are identified by PCR.

PCR in sex determination of embryos : Sex of human and live stock embryos fertilized *in vitro*, can be determined by PCR, by using primers and DNA probes specific for sex chromosomes. Further, this technique is also useful to detect sex — linked disorders in fertilized embryos.

PCR in the diagnosis of COVID-19 : The diagnosis of COVID-19 pandemic done by real-time PCR against the specific genes of severe acute respiratory syndrome coronavirus 2 (SARS-CoV$_2$).

PCR in comparative studies of genomes

PCR is very important in the study of ***evolutionary biology***, more specifically referred to as ***phylogenetics***. As a technique which can amplify even minute quantities of DNA from any source (hair, mummified tissues, bone, or any fossilized material), PCR has revolutionized the studies in palaentology and archaelogy. The movie 'Jurassic Park', has created public awareness of the potential applications of PCR!

PCR in forensic medicine

A single molecule of DNA from any source (blood strains, hair, semen etc.) of an individual is adequate for amplification by PCR. Thus, PCR is very important for identification of criminals.

The reader has to refer DNA finger printing technique described later in this chapter.

GENE LIBRARIES

The ***collection of DNA fragments*** (specifically genes) from a particular species represents gene libraries. The creation or construction of gene libraries (broadly ***genomic libraries***) is accomplished by isolating the complete genome (entire DNA from a cell) which is cut into fragments, and cloned in suitable vectors. Then the specific clone carrying the desired (target) DNA can be identified, isolated and characterized. In this manner, a library of genes or clones (appropriately considered as ***gene bank***) for the entire genome of a species can be constructed.

DNA IN DISEASE DIAGNOSIS AND MEDICAL FORENSICS

DNA, being the genetic material of the living organisms, contains the information that contributes to various characteristic features of the specific organism. Thus, ***the presence of a disease-causing pathogen can be detected by identifying a gene or a set of genes of the organism*** (e.g. tuberculosis, malaria, AIDS). Likewise an ***inherited genetic defect can be diagnosed by identifying the alterations in the gene*** (e.g. sickle-cell anemia). In the modern laboratory diagnostics, DNA analysis is a very useful and a sensitive tool.

DNA FINGERPRINTING OR DNA PROFILING

DNA fingerprinting is the present day ***genetic detective*** in the practice ***of modern medical forensics***. The underlying principles of DNA fingerprinting are briefly described.

The structure of each person's genome is unique. The only exception being monozygotic identical twins (twins developed from a single fertilized ovum). The unique nature of genome structure provides a good opportunity for the specific identification of an individual.

It may be remembered here that in the traditional fingerprint technique, the individual is identified by preparing an ink impression of the skin folds at the tip of the person's finger. This is based on the fact that the nature of these skin folds is genetically determined, and thus the fingerprint is unique for an individual. In contrast, the ***DNA fingerprint is an analysis of the nitrogenous base sequence in the DNA of an individual***.

Applications of DNA fingerprinting

The amount of DNA required for DNA fingerprint is remarkably small. The ***minute quantities of DNA*** from blood strains, body fluids, hair fiber or skin fragments ***are enough. Polymerase chain reaction is used to amplify*** this DNA for use in fingerprinting. DNA profiling has wide range of applications—most of them related to medical forensics. Some important ones are listed below.

- Identification of criminals, rapists, thieves etc.
- Settlement of paternity disputes.
- Use in immigration test cases and disputes.

In general, the ***fingerprinting technique is carried out by collecting the DNA from a suspect*** (or a person in a paternity or immigration dispute) and ***matching it with that of a reference sample*** (from the victim of a crime, or a close relative in a civil case).

Table 27.5 A selected list of human proteins produced by recombinant DNA technology for treatment of human disorders

Disorder	*Recombinant protein(s)*
Anemia	Hemoglobin, erythropoietin
Delivery	Relaxin
Blood clots	Tissue plasminogen activator, urokinase
Burns	Epidermal growth factor
Cancer	Interferons, tumor necrosis factor, colony stimulating factors, interleukins
Diabetes	Insulin, insulin-like growth factor
Emphysema	α_1-Antitrypsin
Female infertility	Chorionic gonadotropin
Growth defects	Growth hormone, growth hormone-releasing factor
Heart attacks	Prourokinase
Multiple sclerosis	Interferons (α, β, γ)
Nerve damage	Nerve growth factor
Osteomalacia	Calcitonin
Pain	Endorphins and enkephalins

PHARMACEUTICAL PRODUCTS OF DNA TECHNOLOGY

The advent of recombinant DNA technology heralded a new chapter for the production of a wide range of therapeutic agents in sufficient quantities for human use. The commercial exploitation of ***recombinant DNA (rDNA) technology*** began in late 1970s by a few biotechnological companies to produce proteins. There are at least 400 different proteins being produced (by DNA technology) which may serve as therapeutic agents for humans. A selected list of some important human proteins produced by recombinant DNA technology, potential for the treatment of human disorders, is given in ***Table 27.5***. As of now, only a selected few of them (around 30) have been approved for human use, and the most important among these are given in ***Table 27.6***. Recombinant insulin (trade name ***Humulin***) was the first pharmaceutical produt of rDNA technology that was successfully used for the treatment of, the disease diabetes mellitus.

RECOMBINANT VACCINES

Recombinant DNA technology in recent years has become a boon to produce ***new generation vaccines***. By this approach, some of the limitations (low yield, high cost, side effects) of traditional vaccine production could be overcome. In addition, several new strategies, involving gene manipulation, are being tried to create novel recombinant vaccines.

Hepatitis B vaccine—the first synthetic vaccine

In 1987, the recombinant vaccine for hepatitis B (i.e. ***HBsAg***) became the first synthetic vaccine for public use. It was marketed by trade names ***Recombivax*** and ***Engerix-B***. Hepatitis B vaccine is safe to use, very effective and produces no allergic reactions. For these reasons, this recombinant vaccine has been in use since 1987. The individuals must be administered three doses over a period of six months. Immunization against hepatitis B is strongly recommended to anyone coming in contact with blood or body secretions.

Table 27.6 A selected list of rDNA-derived therapeutic agents (approved by FDA) with trade names and their applications in humans

rDNA product	*Trade name(s)*	*Applications/uses*
Insulin	Humulin	Diabetes
Growth hormone	Protropin/Humatrope	Pituitary dwarfism
α-Interferon	Intron A	Hairy cell leukemia
Hepatitis B vaccine	Recombinax HB/Engerix B	Hepatitis B
Tissue plasminogen activator	Activase	Myocardial infarction
DNase	Pulmozyme	Cystic fibrosis
Erythropoietin	Epogen/Procrit	Severe anemia with kidney damage

Medical Concepts/ Clinical Correlates

- The pharmaceutical products of rDNA technology have revolutionized the treatment of certain diseases e.g. diabetes, asthma, atherosclerosis, heart attacks, hemophilia.
- Recombinant vaccine for hepatitis B is the first synthetic vaccine. It is effective, safe and produces no allergic reactions.
- Transgenic mice that serve as animal models for human diseases have been developed. These include human mouse (model for immune system), Alzheimer's mouse, oncomouse (model for cancer), prostate mouse, knockout mice (for allergy, transplantation etc.).
- Certain pet animals (cats, dogs) are being cloned by some companies.

All the health professionals—physicians, surgeons, medical laboratory technicians, nurses, dentists, besides police officers, firefighters etc., must get vaccinated against hepatitis B.

India is the ***fourth country*** (after USA, France and Belgium) in the world to develop an indigenous hepatitis B vaccine. It was launched in 1997, and is now being used.

RECOMBINANT VACCINES FOR COVID-19

A novel coronavirus (COVID-19) was identified in December 2019 in Wuhan, China. This virus rapidly spread throughout the world and resulted in millions of deaths. To contain COVID-19 pandemic, many pharmaceutical companies have come up with recombinant vaccines.

The most important vaccines are listed below.

- ***Covaxin*** (inactivated virus), developed by India.
- ***Covishield/Astrazenca*** (viral vector-modern chimpanzee adeno), developed by U.K, Sweden and India.
- ***Moderna*** (mRNA based) developed by US.
- ***Sputnik V*** (viral vector-modified adeno), developed by Russia.
- ***Pfizer and Biotech*** (mRNA based) developed by US and Germany.

TRANSGENIC ANIMALS

With the advent of modern biotechnology, it is now possible to carry out manipulations at the genetic level to get the desired characteristics in animals. ***Transgenesis refers to the phenomenon of introduction of exogeneous DNA into the genome to create and maintain a stable heritable character***. The foreign DNA that is introduced is called ***transgene***. And the animal whose genome is altered by adding one or more transgenes is said to be transgenic. The transgenes behave like other genes present in the animals' genome, and are passed on to the offsprings. Thus, transgenic animals are ***genetically engineered*** or ***genetically modified organisms (GMOs)*** with a new heritable character.

Importance of transgenic animals—general

Transgenic animals serve as good models for understanding the human diseases. Further, several proteins produced by transgenic animals are important for medical and pharmaceutical applications. Thus, the transgenic farm animals (often referred to as animal bioreactors) are a part of the lucrative world-wide biotechnology industry, with great benefits to mankind.

BIOTECHNOLOGY AND SOCIETY

Advances in biotechnology, and their applications are most frequently associated with controversies. Based on their perception to biotechnology, the people may be grouped into three broad categories.

1. ***Strong opponents*** who oppose the new technology, as it will give rise to problems, issues and concerns humans have never faced before. They consider biotechnology as an unnatural manipulative technology.

2. ***Strong proponents*** who consider that the biotechnology will provide untold benefits to society. They argue that for centuries the society has safely used the products and processes of biotechnology.

3. A ***neutral group*** of people who have a ***balanced approach*** to biotechnology. This group believes that research on biotechnology (with regulatory systems), and extending its fruits to the society should be pursued with a cautious approach.

BENEFITS OF BIOTECHNOLOGY

The fruits of biotechnology are beneficial to the fields of ***healthcare***, ***agriculture***, ***food production***, ***manufacture of industrial enzymes*** and appropriate environmental management.

Undoubtedly, technology has largely contributed to the present day world we live in. Many pepople consider biotechnology as a technology that will improve the quality of life in every country, besides maintaining living standards at a reasonably higher level.

ELSI OF BIOTECHNOLOGY

ELSI is the short form to represent the ***ethical, legal and social implications of biotechnology***. ELSI broadly covers the relationship between biotechnology and society with particular reference to ethical and legal aspects.

Risks and ethics of biotechnology

The modern biotechnology deals with genetic manipulations of viruses, bacteria, plants, animals, fish and birds. Introduction of foreign genes into various organisms raises concerns about the safety, ethics and unforeseen consequences. Some of the popular phrases used in the media while referring to experiments on rDNA technology are listed.

- Manipulation of life
- Playing God
- Man-made evolution

The major apprehension of genetic engineering is that through recombinant DNA experiments, unique microorganisms or viruses (either inadvertently, or sometimes deliberately for the purpose of war) may be developed that would cause epidemics and environmental catastrophes. Due to these fears, the regulatory guidelines for research dealing with DNA manipulation were very stringent in the earlier years.

So far, risk assessment studies have failed to demonstrate any hazardous properties acquired by host cells/organisms due to transfer of DNA. Thus, ***the fears of genetic manipulations may be unfounded*** to a large extent.

It is now accepted that biotechnology is certainly beneficial to humans. But it should not cause problems of safety to people and environment, and create unacceptable social, moral and ethical issues.

SUMMARY

1. *Recombinant DNA (rDNA) technology is primarily concerned with the manipulation of genetic material (DNA) to achieve the desired goal in a pre-determined way.*
2. *The procedure for rDNA technology involves molecular tools (enzymes e.g. restriction endonucleases), host cells (*E. coli, S. cerevisiae*), vectors (plasmids, bacteriophages), gene transfer (transformation, electroporation) and the strategies of gene cloning.*
3. *Blotting techniques are employed for the identification of desired DNA (Southern blot), RNA (Northern blot), and protein (Western blot).*
4. *Polymerase chain reaction is an* in vitro *technique for generating large quantities of a specified DNA i.e. cell-free amplification.*
5. *Analysis of DNA (i.e. detection of gene/genes) can be used as a diagnostic system for the detection of many pathogenic and genetic diseases e.g. tuberculosis, malaria, AIDS, sickle-cell anemia, certain cancers. DNA fingerprinting or DNA profiling is the present day genetic detective in the practice of modern medical forensics.*
6. *Many pharmaceutical compounds of health importance (for disease treatment) are being produced by rDNA technology e.g. insulin, growth hormone, interferons, erythropoietin, hepatitis B vaccine.*

HUMAN GENOME PROJECT

28

The most important features of a DNA molecule are the nucleotide sequences, and the identification of genes and their activities.

THE BIRTH AND ACTIVITY OF HUMAN GENOME PROJECT

The human genome project (***HGP***) was conceived in 1984, and officially begun in earnest in October 1990. The primary objective of HGP was to ***determine the nucleotide sequence of the entire human nuclear genome***. James Watson (who elucidated DNA structure) was the first Director of HGP.

In 1997, United States established the ***National Human Genome Research Institute*** (***NHGRI***). The HGP was an international venture involving research groups from six countries — USA, UK, France, Germany, Japan and China, and several individual laboratories and a large number of scientists and technicians from various disciplines. This collaborative venture was named as ***International Human Genome Sequencing Consortium*** (***IHGSC***) and was headed by Francis Collins. A total expenditure of $3 billion, and a time period of 10–15 years for the completion of HGP was expected. A second human genome project was set up by a private company — ***Celera Genomics***, of Maryland USA in 1998. This team was led by Craig Venter.

Announcement of the draft sequence of human genome

The date ***26th June 2000*** will be remembered as one of the most important dates in the history of science or even mankind. It was on this day, Francis Collins and Craig Venter, the leaders of the two human genome projects, in the presence of the President of U.S., jointly ***announced the working drafts of human genome sequence***. The detailed results of the teams were later published in February 2001 in scientific journals Nature (IHGSC) and Science (Celera Genomics).

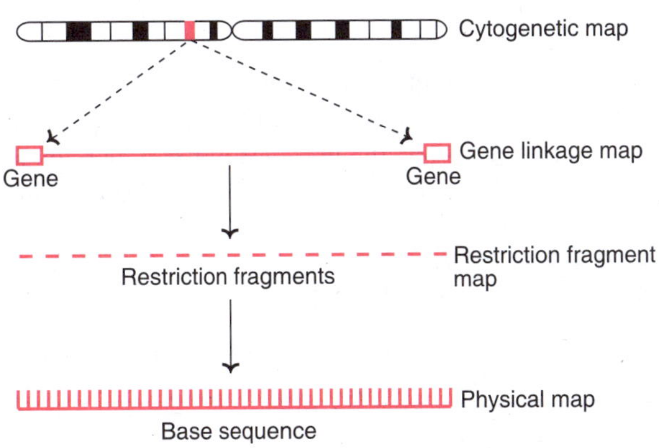

Fig.28.1 : Different types of genome maps.

The human genome project results attracted worldwide attention. This achievement was hailed with many descriptions in the media.

- The mystery of life unravelled.
- The library of life.
- The periodic table of life.

MAPPING OF THE HUMAN GENOME

The most important objective of human genome project was to construct a series of maps for each chromosome. In ***Fig.28.1***, an outline of the different types of maps is given.

1. **Cytogenetic map :** This is a map of the chromosome in which the active genes respond to a chemical dye and display themselves as bands on the chromosome.

2. **Gene linkage map :** A chromosome map in which the active genes are identified by locating closely associated marker genes. The most commonly used DNA markers are ***restriction fragment length polymorphism (RFLP), variable number tandems repeats (VNTRs) and short tandem repeats (STRs)***. VNTRs are also called as ***minisatellites*** while STRs are ***microsatellites***.

3. **Restriction fragment map :** This consists of the random DNA fragments that have been sequenced.

4. **Physical map :** This is the ultimate map of the chromosome with highest resolution base sequence. Physical map depicts the location of the active genes and the number of bases between the active genes.

APPROACHES FOR GENOME SEQUENCING

The principle methods used for mapping of genomes are listed below.

- DNA sequencing
- Use of DNA probes
- Fluorescence *in situ* hybridization (FISH)
- Cloning in vectors
- Polymerase chain reaction
- Chromosome walking
- Chromosome jumping
- Data bases utility

HUMAN GENOME SEQUENCE—RESULTS SUMMARISED

The information on the human genome projects is too vast, and only some highlights can be given (***Table 28.1***), and briefly described.

Most of the genome sequence is identified

About 90% of the human genome has been sequenced. It is composed of 3.2 billion base pairs (3200 Mb or 3.2 Gb). If written in the format of a telephone book, the base sequence of human genome would fill about 200 telephone books of 1000 pages each. Some other ***interesting analogs/ sidelights of genome*** are given in ***Table 28.2***.

TABLE 28.1 Major highlights of human genome
• The draft represents about 90% of the entire human genome. It is believed that most of the important parts have been identified.
• The remaining 10% of the genome sequences are at the very ends of chromosomes (i.e. telomeres) and around the centromeres.
• Human genome is composed of 3200 Mb (or 3.2 Gb) i.e. 3.2 billion base pairs (3,200,000,000).
• Approximately 1.1 to 1.5% of the genome codes for proteins.
• The number of protein coding genes is in the range of 30,000–40,000. (recently revised — 19,000 to 20,000)
• Genes and DNA sequences associated with many diseases such as breast cancer, muscle diseases, deafness and blindness have been identified.
• Repeated sequences constitute about 50% of the human genome.
• A vast majority of the genome (~ 97%) has no known functions.
• Between the humans, the DNA differs only by 0.2% or one in 500 bases.
• More than 3 million single nucleotide polymorphisms (SNPs) have been identified.
• Human DNA is about 98% identical to that of chimpanzees.
• About 200 genes are close to that found in bacteria.

Individual differences in genomes : It has to be remembered that every individual, except identical twins, have their own versions of genome sequences. The differences between individuals are largely due to ***single nucleotide polymorphisms (SNPs)***. SNPs represent positions in the genome where some individuals have one nucleotide (i.e. an A), and others have a different nucleotide (i.e. a G). The frequency of occurrence of SNPs is estimated to be one per 1000 base pairs. About 3 million SNPs are believed to be present and at least half of them have been identified.

Genes present in human genome

The two genome projects differ in their estimates of the total number of genes in humans. Their figures are in the range of 30,000–40,000 genes (recently revised 19,000–20,000). The main reason for this variation is that it is rather difficult to specifically recognize the DNA sequences which are genes and which are not.

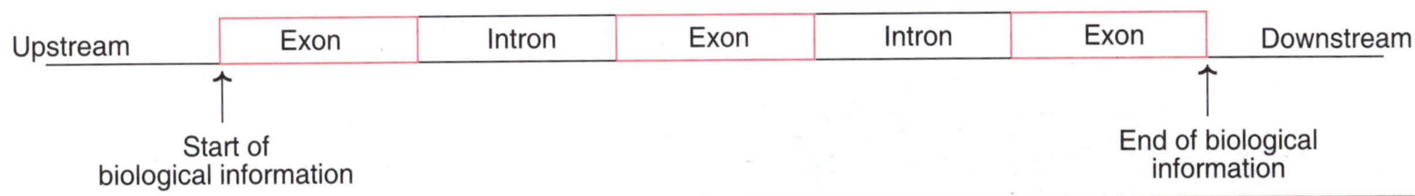

Fig.28.2 : *A diagrammatic representation of a typical structure of an average human gene.*

Before the results of the HGP were announced, the best guess of human genes was in the range of 80,000–100,000. This estimate was based on the fact that the number of proteins in human cells is 80,000–100,000, and thus so many genes expected. The fact that the number of genes is much lower than the proteins suggests that the ***RNA editing*** (RNA processing) is widespread, so that a single mRNA may code for more than one protein.

A diagrammatic representation of a typical structure of an average human gene is given in ***Fig.28.2***. It has exons and introns.

TABLE 28.2 Some interesting analogs/sidelights about human genome

- The base sequence in human genome would fill about 200 telephone books of 1000 pages each.
- If the genome is recited at the rate of one base per second for 24 hours a day, it would take a century to recite the book of life.
- If a typist types at the rate of 60 words per minute (i.e. 360 letters) for 8 hours a day, he/she would take around 50 years to type human genome.
- If the DNA sequence is typed in lines 10 cm containing 60 nucleotide bases and printed, the human genome sequence (from a single cell) would stretch a distance of 5000 km.
- If the DNA in the entire human body is put end to end, it would reach to the sun and back over 600 times (**Note :** The human body contains 100 trillion cells; the length of DNA in a cell is 6 feet; the distance between the sun and earth is 93 million miles).
- The total expenditure for human genome project was $3 billion. The magnitude of this huge amount has to be appreciated. If one starts counting at a non–stop rate of a dollar per second, it would take about 90 years to complete.

A broad categorization of human gene catalog in the form of a pie chart is depicted in ***Fig.28.3***. About 17.5% of the genes participate in the general biochemical functions of the cells, 23% in the maintenance of genome, 21% in signal transduction while the remaining 38% are involved in the production of structural proteins, transport proteins, immunoglobins etc.

Human genes encoding proteins

It is now clear that only 1.1-1.5% of the human genome codes for proteins. Thus, this figure 1.1-1.5% represents exons of genome.

As already described, a huge portion of the genome is composed of introns, and intergenic sequences (junk DNA).

The major categories of the proteins encoded by human genes are listed in ***Table 28.3***. The functions of at least 40% of these proteins are not known.

BENEFITS/APPLICATIONS OF HUMAN GENOME SEQUENCING

It is expected that the sequencing of human genome, and the genomes of other organisms will

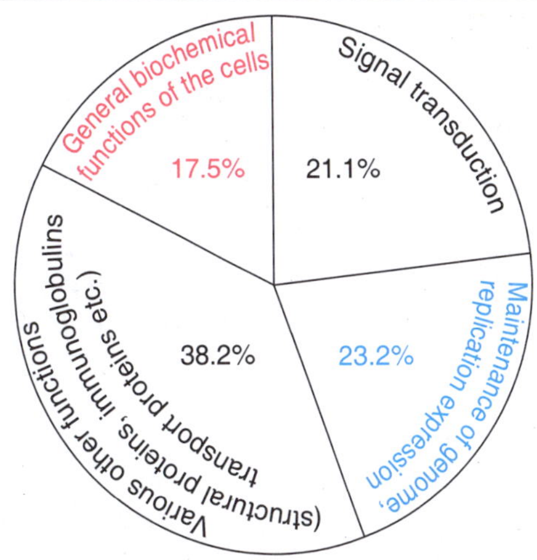

Fig.28.3 : *A pie chart showing a broad categorization of the human gene catalog (About 13000 genes whose functions are not known are not included.*

TABLE 28.3 Different categories of proteins encoded by human genes (based on the Human Genome Project report, 2001)

Category of proteins	*Percentage*	*Actual number of genes*
Unknown functions	41.0%	12,809
Nucleic acid enzymes	7.5%	2,308
Transcription factors	6.0%	1,850
Receptors	5.0%	1,543
Hydrolases	4.0%	1,227
Regulatory proteins (G-proteins, cell cycle regulators etc.)	3.2%	988
Protooncogenes	2.9%	902
Structural proteins of cytoskeleton	2.8%	876
Kinases	2.8%	868

*(**Note** : This table is based on the rough draft of human genome reported by Celera Genomics. The percentages are derived from a total of 26,383 genes)*

dramatically change our understanding and perceptions of biology and medicine. Some of the benefits of human genome project are given.

- Identification of human genes and their functions.
- Understanding of polygenic disorders e.g. cancer, hypertension, diabetes.
- Improvements in gene therapy
- Improved diagnosis of diseases
- Development of pharmacogenomics
- Genetic basis of psychiatric disorders
- Understanding of complex social trait
- Improved knowledge on mutations

ETHICS AND HUMAN GENOME

The research on human genomes will make very sensitive data available that will affect the personal and private lives of individuals. For instance, once it is known that a person carries genes for an incurable disease, what would be the strategy of an insurance company? How will the society treat him/her? There is a possibility that ***individuals with substandard genome sequences may be discriminated***. Human genome results may also promote racial discrimination categorizing the people with ***good*** and ***bad genome sequences***. Considering the gravity of ethics related to a human genome, about 3% of the HGP budget was earmarked for ethical research.

SUMMARY

1. *Human Genome Project is an international venture involving several laboratories, and a large number of scientists and technicians from various disciplines.*
2. *About 90% of the human genome has been sequenced. It is composed of 3.2 billion base pairs.*
3. *The total number of genes in the humans is in the range of 19,000–20,000.*
4. *Human genome sequencing has wide range of applications–better understanding of genetic diseases, improvements in gene therapy, development of pharmacogenomics, and advancement of biotechnology.*

GENE THERAPY

29

Advances in biochemistry and molecular biology have helped to understand the genetic basis of inherited diseases. It was a dream of the researchers to replace the defective genes with good ones, and ***cure the genetic disorders***.

Gene therapy is the process of inserting genes into cells to treat diseases. The newly introduced genes will encode proteins and correct the deficiencies that occur in genetic diseases. Thus, gene therapy primarily involves ***genetic manipulations*** in animals or humans ***to correct a disease***, and keep the organism in good health. The initial experiments on gene therapy are carried out in animals, and then in humans. Obviously, the goal of the researchers is to benefit the mankind and improve their health.

An overview of gene therapy strategies is depicted in ***Fig.29.1***. In ***gene augmentation therapy***, a DNA is inserted into the genome to replace the missing gene product. In case of ***gene inhibition therapy***, the antisense gene inhibits the expression of the dominant gene.

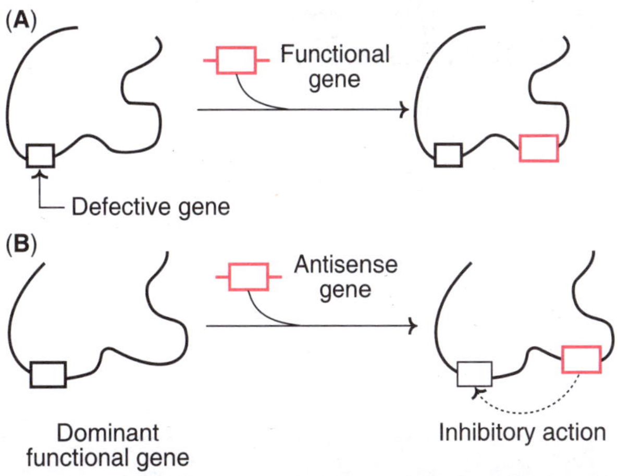

Fig.29.1 : *Overview of two major gene therapy strategies (**A**) Gene augmentation therapy (**B**) Gene inhibition therapy.*

APPROACHES FOR GENE THERAPY

There are two approaches to achieve gene therapy.

1. **Somatic cell gene therapy :** The ***non-reproductive (non-sex) cells*** of an organism are referred to as somatic cells. These are the cells of an organism other than sperm or egg cells, e.g., bone marrow cells, blood cells, skin cells, intestinal cells. At present, all the research on gene therapy is directed to correct the genetic defects in somatic cells. In essence, somatic cell gene therapy involves the ***insertion of a fully*** functional and ***expressible gene into a target somatic cell*** to correct a genetic disease permanently.

2. **Germ cell gene therapy :** The ***reproductive (sex) cells*** of an organism constitute germ cell line. Gene therapy involving the introduction of DNA into germ cells is passed on to the successive generations. For safety, ethical and technical reasons, germ cell gene therapy is ***not being attempted at present***.

The genetic alterations in somatic cells are not carried to the next generations. Therefore, ***somatic cell gene therapy is preferred*** and extensively studied with an ultimate objective of correcting human diseases.

A large number of genetic disorders and other diseases are currently at various stages of gene therapy trials. A selected list of some important ones is given in ***Table 29.1***.

There are two types of gene therapies.

I. ***Ex vivo* gene therapy :** This involves the ***transfer of genes in cultured cells*** (e.g., bone marrow cells) which are then reintroduced into the patient.

II. ***In vivo* gene therapy :** The ***direct delivery of genes into the cells*** of a particular tissue is referred to as *in vivo* gene therapy.

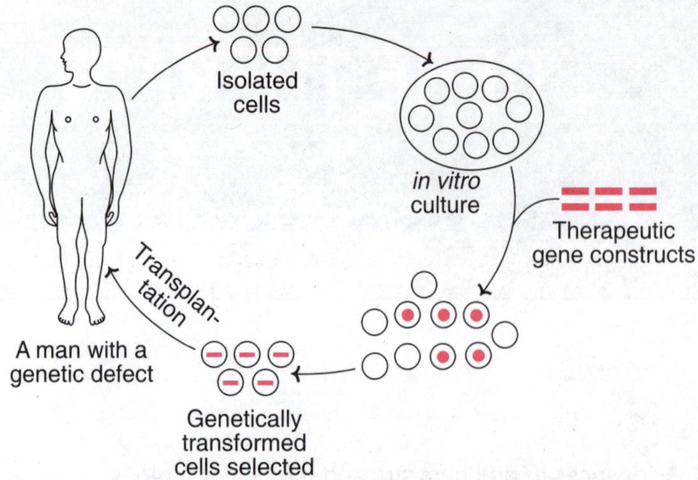

Fig.29.2 : *The procedure for* ex vivo *gene therapy.*

EX VIVO GENE THERAPY

The *ex vivo* gene therapy can be applied to only selected tissues (e.g., bone marrow) whose cells can be cultured in the laboratory.

The technique of *ex vivo* gene therapy involves the following steps (***Fig.29.2***).

1. Isolate cells with genetic defect from a patient.
2. Grow the cells in culture.
3. Introduce the therapeutic gene to correct gene defect.
4. Select the genetically corrected cells (stable transformants) and grow.
5. Transplant the modified cells to the patient.

TABLE 29.1 Human gene therapy trials

Disease	*Gene therapy*
Severe combined immunodeficiency (SCID)	Adenosine deaminase (ADA).
Cystic fibrosis	Cystic fibrosis transmembrane regulator (CFTR).
Familial hypercholesterolemia	Low density lipoprotein (LDL) receptor.
Hemophilia B	Factor IX
Thalassemia	α- or β-Globin
Sickle-cell anemia	β-Globin
Lesch-Nyhan syndrome	Hypoxanthine-guanine phosphoribosyltransferase (HGPRT).
Melanoma	Tumor necrosis factor (TNF)
Melanoma, renal cancer	Interleukin-2 (IL-2)
Glioblastoma (brain tumor), AIDS, ovarian cancer	Thymidine kinase (herpes simplex virus)
Head and neck cancer	p^{53}
Breast cancer	Multidrug resistance I
AIDS	*rev* and *env*
Duchenne muscular dystrophy	Dystrophin
Short stature*	Growth hormone
Diabetes*	Glucose transporter-2 (GLUT-2), glucokinase
Phenylketonuria*	Phenylalanine hydroxylase

* *Mostly confined to animal experiments*

Medical Concepts/ Clinical Correlates

- **Theoretically, gene therapy is the permanent solution for genetic diseases.**
- **A large number of genetic disorders and other diseases are at various stages of gene therapy trials e.g. sickle-cell anemia, cystic fibrosis, AIDS, cancer.**
- **Despite extensive research and trials, as of now, no disease has been permanently cured by gene therapy. However, a breakthrough may come at anytime.**

The procedure basically involves the ***use of the patient's own cells for culture and genetic correction, and then their return back to the patient***. This technique is therefore, not associated with adverse immunological responses after transplanting the cells. *Ex vivo* gene therapy is efficient only if the therapeutic gene (remedial gene) is stably incorporated and continuously expressed. This can be achieved by use of vectors.

VECTORS IN GENE THERAPY

The ***carrier particles or molecules used to deliver genes to somatic cells*** are referred to as vectors. The important vectors employed in *ex vivo* gene therapy are listed below

- Viruses-mostly ***retroviruses***
- ***Human artificial chromosome***
- ***Bone marrow cells***.

THERAPY FOR ADENOSINE DEAMINASE DEFICIENCY

The ***first*** and the most publicised human ***gene therapy*** was carried out to correct the deficiency of the enzyme adenosine deaminase (ADA). This was done on September 14, 1990 by a team of workers led by Blaese and Anderson at the National Institute of Health, USA (The girl's name is Ashanti, 4 years old then).

Severe combined immunodeficiency (SCID)

SCID is rare inherited immune disorder associated with T-lymphocytes, and (to a lesser extent) B-lymphocytes dysfunction. About 50% of SCID patients have a defect in the gene (located on chromosome 20, and has 32,000 base pairs and 12 exons) that encodes for adenosine deaminase. In the deficiency of ADA, deoxyadenosine and its metabolites (primarily deoxyadenosine 5′-triphosphate) accumulate and destroy T-lymphocytes. T-Lymphocytes are essential for body's immunity. Besides participating directly in body's defense, they promote the function of B-lymphocytes to produce antibodies. Thus, the ***patients of SCID*** (***lacking ADA***) ***suffer from infectious diseases and die at an young age***.

Technique of therapy for ADA deficiency

The general scheme of gene therapy adopted for introducing a defective gene in the patient has been depicted in ***Fig.29.2***. The same procedure with suitable modifications can also be applied for other gene therapies.

A plasmid vector bearing a proviral DNA is selected. A part of the proviral DNA is replaced by the ADA gene and a gene (G 418) coding for antibiotic resistance, and then cloned. The antibiotic resistance gene will help to select the desired clones with ADA gene.

A diagrammatic representation of the treatment of ADA deficient patient is depicted in ***Fig.29.3***.

Circulating lymphocytes are removed from a patient suffering from ADA deficiency. These cells are transfected with ADA gene by exposing to billions of retroviruses carrying the said gene. The genetically-modified lymphocytes are grown in cultures to confirm the expression of ADA gene and returned to the patient. These lymphocytes persist in the circulation and synthesize ADA. Consequently, the ability of the patient to produce antibodies is increased. However, there is a limitation. The lymphocytes have a short life span (just live for a few months), hence the transfusions have to be carried out frequently.

IN VIVO GENE THERAPY

The ***direct delivery of the therapeutic gene*** (DNA) ***into the target cells*** of a particular tissue of a patient constitutes *in vivo* gene therapy (***Fig.29.4***). Many tissues are the potential candidates

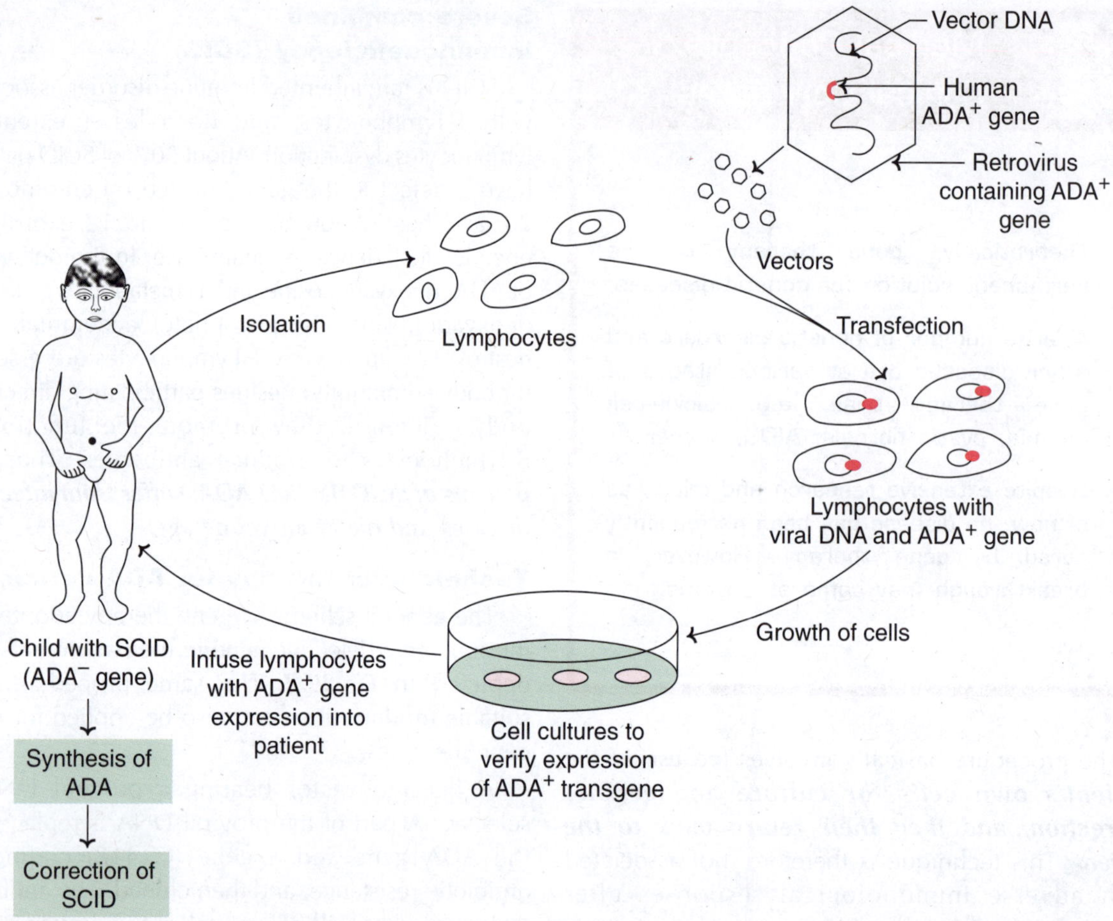

***Fig.29.3 :** Treatment of adenosine deaminase (ADA) deficient patient by somatic* ex vivo *gene therapy (SCID-Severe combined immunodeficiency).*

for this approach. These include liver, muscle, skin, spleen, lung, brain and blood cells. Gene delivery can be carried out by ***viral*** or ***non-viral vector systems***. The success of ***in vivo*** gene therapy mostly depends on the following parameters

- The efficiency of the uptake of the remedial (therapeutic) gene by the target cells.
- Intracellular degradation of the gene and its uptake by nucleus.
- The expression capability of the gene.

Gene delivery by viruses

Many viral vector systems have been developed for gene delivery. These include retroviruses, adenoviruses, adeno-associated viruses and herpes simplex virus.

Gene delivery by non-viral systems

There are certain limitations for using viral vectors in gene therapy. In addition to the prohibitive cost of

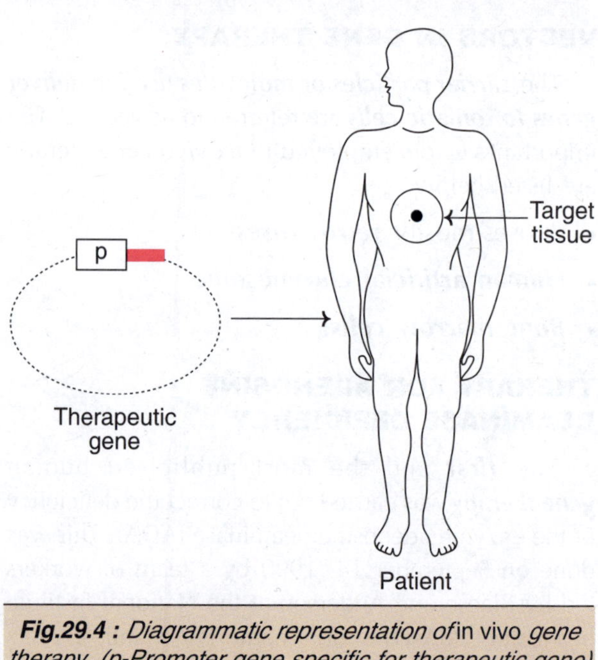

***Fig.29.4 :** Diagrammatic representation of* in vivo *gene therapy. (p-Promoter gene specific for therapeutic gene)*

maintaining the viruses, the viral proteins often induce inflammatory responses in the host. Therefore, there is a continuous search by researchers to find alternatives to viral vector systems.

The non-viral gene delivery systems are listed

- ***Pure DNA constructs*** that can be directly introduced into target tissues.
- ***Lipoplexes***, lipid-DNA complexes that have DNA surrounded by lipid layers.
- ***Human artificial chromosome*** which can carry large DNA (one or more genes).

ANTIGENE AND ANTISENSE THERAPY

In general, gene therapy is carried out by introducing a therapeutic gene to produce the defective or the lacking protein. But there are certain disorders (cancer, viral and parasitic infections, inflammatory diseases) which result in an overproduction of certain normal proteins. It is possible to treat these diseases by ***blocking transcription*** using a single-stranded nucleotide sequence (antigene oligonucleotide) that hybridizes with the specific gene, and this is called ***antigene therapy***. ***Antisense therapy*** refers to the ***inhibition of translation*** by using a single-stranded nucleotide (antisense oligonucleotide). Further, it is also possible to inhibit both transcription and translation by blocking (with oligonucleotides) the transcription factor responsbile for the specific gene expression.

Nucleic acid therapy refers to the ***use of DNA or RNA molecules for therapeutic purposes***, as stated above.

THE FUTURE OF GENE THERAPY

Theoretically, gene therapy is ***the permanent solution for genetic diseases***. But it is not as simple as it appears since gene therapy has several inbuilt complexicities.

The public, in general, have ***exaggerated expectations*** on gene therapy. The researchers, at least for the present, are ***unable to satisfy them***.

It may be true that as of now, gene therapy due to several limitations, has not progressed the way it should, despite intensive research. But ***a breakthrough may come anytime***, and of course, this is only possible with persistent research. And a day may come (it might take some years) when almost every disease will have a gene therapy, as one of the treatment modalities. And gene therapy will revolutionize the practice of medicine!

SUMMARY

1. *Gene therapy is the process of inserting genes into cells to treat diseases. Somatic cell gene therapy, involving the insertion of an expressible gene into somatic cells, is the preferred approach.*
2. Ex vivo *gene therapy involves the transfer of genes in cultured cells which are then reintroduced into the patient. The direct delivery of genes into the cells of a particular tissue is regarded as* in vivo *gene therapy.*
3. *Gene therapy was successfully carried out in a patient of severe combined immunodeficiency (caused by the deficiency of the enzyme adenosine deaminase).*
4. *Antigene therapy involves blocking of transcription (by antigene oligonucleotide) while in antisense therapy, translation is inhibited (by antisense oligonucleotide).*

BIOINFORMATICS

30

Bioinformatics is ***the combination** (**or marriage**!) **of biology and information technology***. Basically, bioinformatics is a recently developed science using information to understand biological phenomenon. It broadly involves the computational tools and methods used to manage, analyse and manipulate volumes and volumes of biological data.

Bioinformatics may also be regarded as a part of the ***computational biology***. The latter is concerned with the application of quantitative analytical techniques in modeling and solving problems in the biological systems. Bioinformatics is an inter-disciplinary approach requiring advanced knowledge of computer science, mathematics and statistical methods for the understanding of biological phenomena at the molecular level.

History and relevance of bioinformatics

The term bioinformatics was first ***introduced in 1990s***. Originally, it dealt with the management and analysis of the data pertaining to ***DNA, RNA and protein sequences***. As the biological data is being produced at an unprecedented rate, their management and interpretation invariably requires bioinformatics. Thus, bioinformatics now includes many other types of biological data. Some of the most important ones are listed below

- Gene expression profiles
- Protein structure
- Protein interactions
- Microarrays (DNA chips)
- Functional analysis of biomolecules
- Drug designing.

Bioinformatics is largely (not exclusively) a computer-based discipline. Computers are in fact very essential to handle large volumes of biological data, their storage and retrieval.

We have to accept the fact that ***there is no computer on earth*** (however advanced) ***which can store information, and perform the functions like a living cell***. Thus a highly complex information technology lies right within the cells of an organism. This primarily includes the organism's genes and their dictates for the organisms biological processes and behaviour.

BROAD COVERAGE OF BIOINFORMATICS

Bioinformatics covers many specialized and advanced areas of biology.

Functional genomics : Identification of genes and their respective functions.

Structural genomics : Predictions related to functions of proteins.

Comparative genomics : For understanding the genomes of different species of organisms.

Medical informatics : This involves the management of biomedical data with special reference to biomolecules, *in vitro* assays and clinical trials.

COMPONENTS OF BIOINFORMATICS

Bioinformatics comprises three components

1. **Creation of databases :** This involves the organizing, storage and management of the

biological data sets. The databases are accessible to researchers to know the existing information and submit new entries. e.g. protein sequence data bank for molecular structure. Databases will be of no use until analysed.

2. **Development of algorithms and statistics :** This involves the development of tools and resources to determine the relationship among the members of large data sets e.g. comparison of protein sequence data with the already existing protein sequences.

3. **Analysis of data and interpretation :** The appropriate use of components 1 and 2 (given above) to analyse the data and interpret the results in a biologically meaningful manner. This includes DNA, RNA and protein sequences, protein structure, gene expression profiles, and biochemical pathways.

BIOINFORMATICS AND THE INTERNET

The internet is an ***inter***national computer ***net***work. A computer network involves a group of computers that can communicate (usually over a telephone system) and exchange data between users.

It is the ***internet protocol*** (IP) that determines how the packets of information are addressed and routed over the network. To access the internet, a computer must have the correct hardware (modem/ network card), appropriate software and permission for access to network. For this purpose, one has to subscribe to an ***internet service provider*** (ISP).

World wide web (***www***) : www involves the exchange of information over the internet using a programme called ***browser***. The most widely used browsers are Internet explorer and Netscape navigator.

BIOLOGICAL DATABASES

The collection of the ***biological data*** on a computer which can be manipulated to appear ***in varying arrangements and subsets*** is regarded as a database. The biological information can be stored in different databases. Each database has its own website with unique navigation tools.

The biological databases are, in general, publicly accessible. Selected examples of biological databases are given in ***Table 30.1***.

APPLICATIONS OF BIOINFORMATICS

The advent of bioinformatics has revolutionized the advancements in biological science. And

TABLE 30.1 Selected examples of biological databases in bioinformatics

Database(s)	*Salient features*
Primary nucleotide sequence databases	
GenBank (www.ncbi.nih.gov/GeneBank/)	Provides nucleotide sequence databases maintained by the National Center for Biotechnology Information (NCBI), USA.
Other nucleotide sequence databases	
UniGene (www.ncbi.nih.gov/UniGene/)	The nucleotide sequences of GenBank in the form of clusters, representing genes are available.
Genome Biology (www.ncbi.nlm.nih.gov/Genomes/)	The information about the completed genomes is available.
Protein sequence database	
SWISS-PROT (www.expasy.ch/sport)	Provides the description of the structure of a protein, its domains structure, post-translational modifications, variants etc. It has high level of integration with other databases and minimal level of redundancy.
Protein sequence motif databases	
PROSITE (www.expasy.ch/prosite/)	Provides information on protein families and domains. It also has patterns and profiles for sequences and biological functions.

Medical Concepts/ Clinical Correlates

- **Bioinformatics has largely benefited biological and medical sciences, particularly related to molecular biology and biotechnology. Some applications are listed :**
  - **Sequencing of macromolecules (proteins, DNA, RNA)**
  - **Human genome sequencing**
  - **Molecular modelling of biomolecules**
  - **Handling of vast biological data**
  - **Designing of drugs for the treatment of diseases**
  - **Development of models for the functioning of cells, tissues and organs**
- **As such, there is no field of biological science that is not benefited by bioinformatics.**

biotechnology is largely benefited by bioinformatics. The best example is the sequencing of human genome in a record time which would not have been possible without bioinformatics. A selected list of applications of bioinformatics is given below.

- Sequence mapping of biomolecules (DNA, RNA, proteins).
- Identification of nucleotide sequences of functional genes.
- Finding of sites that can be cut by restriction enzymes.
- Prediction of functional gene products.
- To trace the evolutionary trees of genes.
- For the prediction of 3-dimensional structure of proteins.
- Molecular modelling of biomolecules.
- Designing of drugs for medical treatment.
- Handling of vast biological data which otherwise is not possible.
- Development of models for the functioning various cells, tissues and organs.

The above list of applications however, may be treated as incomplete, since at present there is no field in biological sciences that does not involve bioinformatics.

SUMMARY

1. *Bioinformatics (a computer-based discipline) represents an alliance between biology and information technology.*
2. *The storage, management and interpretation of vast biological data invariably require bioinformatics.*
3. *Bioinformatics comprises three components-creation of data base, development of algorithms and statistics, and analysis of data and interpretation.*
4. *Biological databases, containing the biological information, are publicly accessible e.g. GenBank (www.ncbi.nih.gov/GeneBank).*
5. *Bioinformatics has revolutionized the advancements of biological and medical sciences e.g. sequencing of human genome.*

METABOLISM OF XENOBIOTICS (DETOXIFICATION)

31

Man is continuously exposed to several foreign compounds such as drugs, pollutants, food additives, cosmetics, pesticides etc. Certain unwanted compounds are produced in the large intestine by the bacteria which enter the circulation. These include indole from tryptophan, cadaverine from lysine, tyramine from tyrosine, phenol from phenylalanine etc. In the normal metabolism of the body, certain waste compounds (e.g. bilirubin) are formed. A vast majority of the foreign compounds or the unwanted substances, produced in the body, are toxic and, therefore, they should be quickly eliminated from the body.

The term ***detoxication or detoxification*** refers to ***the series of biochemical reactions occurring in the body to convert the foreign (often toxic) compounds to non-toxic or less toxic and more easily excretable forms.***

Detoxification—a misnomer?

Detoxification is rather misleading, since sometimes a detoxified product is more toxic than the original substance. It appears that the body tries to get rid of a foreign substance by converting it into a more soluble (often polar) and easily excretable compound, and this may be sometimes associated with increased toxicity (e.g. conversion of methanol to formaldehyde).

In recent years, the term detoxification is replaced by ***biotransformation*** or metabolism of xenobiotics (*Greek* : xenos—strange, foreign) or simply ***metabolism of foreign compounds***.

Site of detoxification

The detoxification reactions are carried out ***mainly in the liver*** which is equipped with the enzyme machinery. Kidney and other organs may sometimes be involved. The products formed by detoxification are mostly excreted by the kidneys, less frequently excreted via feces or expired air.

MECHANISM OF DETOXIFICATION

The metabolism of xenobiotics may be divided into two phases which may occur together or separately (***Fig.31.1***).

Phase I : The reactions of phase I are ***oxidation, reduction and hydrolysis***.

Phase II : These are the ***conjugation reactions***, involving compounds such as glucuronic acid, amino acids (glycine), glutathione, sulfate, acetate and methyl group.

Generally, detoxification of a compound involves phase I as well as phase II reactions.

Oxidation

A large number of foreign substances are detoxified by oxidation. These include alcohols, aldehydes, amines, aromatic hydrocarbons and sulfur compounds. In general, aliphatic compounds are more easily oxidized than aromatic ones.

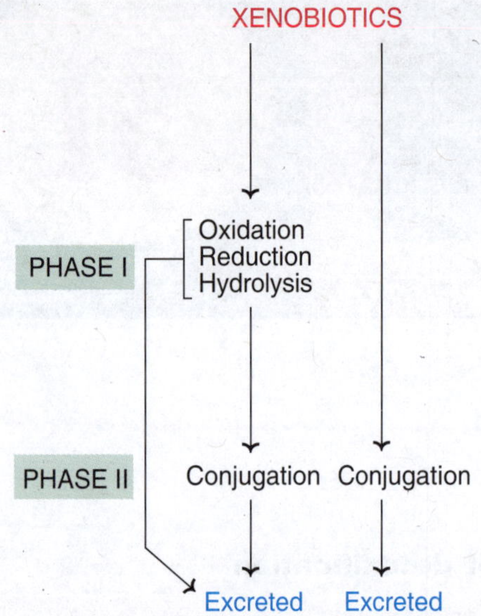

Fig.31.1 : Phase I and phase II reactions in the metabolism of xenobiotics.

Alcohols : Aliphatic and aromatic alcohols undergo oxidation to form the corresponding acids.

CH_3OH Methanol	⟶	HCOOH Formic acid
C_2H_5OH Ethanol	⟶	CH_3COOH Acetic acid

Aldehydes : Aldehydes are oxidized to produce the corresponding acids.

C_6H_5CHO Benzaldehyde	⟶	C_6H_5COOH Benzoic acid
$C.Cl_3CHO$ Chloral	⟶	CCl_3COOH Trichloroacetic acid

Amines and their derivatives : Alipahtic amines are converted to the corresponding acids, liberating urea while aromatic amino acids are oxidized to phenols.

RCH_2NH_2 Aliphatic amine	⟶	$R{-}COOH + NH_2{-}CO{-}NH_2$ Aliphatic acid Urea
$C_6H_5NH_2$ Aniline	⟶	$HO{-}C_6H_4{-}NH_2$ p-Amino phenol

Aromatic hydrocarbons : Benzene may be oxidized to mono, di- and trihydroxy phenols as shown in the next column.

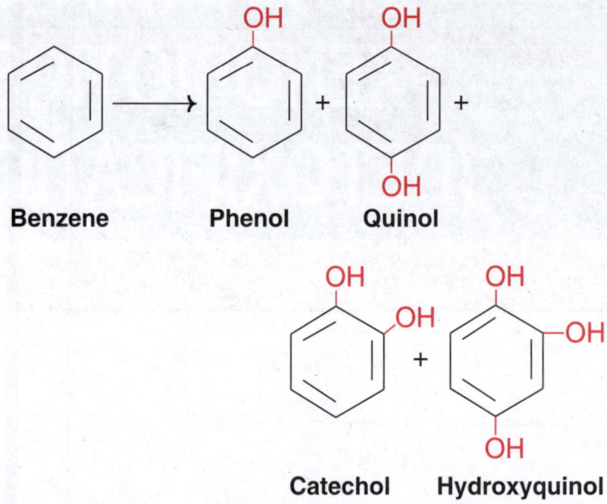

If the aromatic hydrocarbons contain alkyl group, it is oxidized to acid.

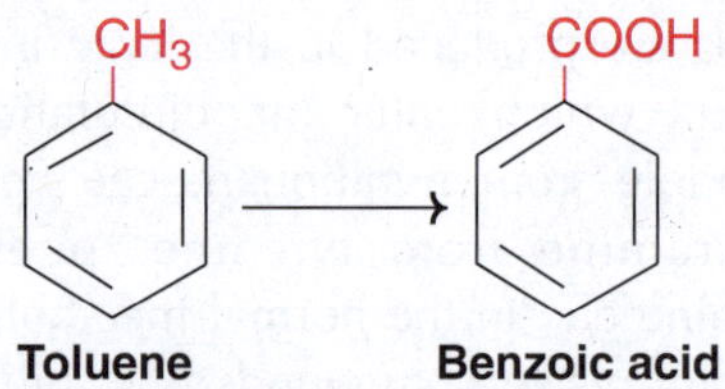

Sulfur compounds : Organic sulfur is oxidized to sulfuric acid.

Drugs : Meprobamate is a tranquilizer. It is oxidized to hydroxymeprobamate and excreted in urine.

Role of cytochrome P_{450}

Most of the oxidation reactions of detoxification are catalysed by monooxygenase or cytochrome P_{450}. This enzyme, also called ***mixed function oxidase***, is associated with the microsomes. The usage P_{450} refers to the absorption peak (at 450 nm), exhibited by the enzyme when exposed to carbon monoxide.

Salient features of cytochrome P_{450}

1. ***Multiple forms*** of cytochrome P_{450} are believed to exist, ranging from 20 to 200. At least 6 species have been isolated and worked in detail.

2. They are all ***hemoproteins***, containing heme as the prosthetic group.

3. Cytochrome P_{450} species are found in the highest concentration in the ***microsomes of liver***. In the adrenal gland, they occur in mitochondria.

4. The mechanism of action of cytochrome P_{450} is complex and is dependent on NADPH.

5. Cytochrome P_{450} is an ***inducible enzyme***. Its synthesis is increased by the administration of drugs such as phenobarbitol.

Reduction

A few examples of detoxification by reduction are given.

$$C_6H_2OH(NO_2)_3 \longrightarrow C_6H_2OH(NO_2)_2NH_2$$

Picric acid → Picramic acid

$$CCl_3.CH(OH)_2 \longrightarrow CCl_3CH_2OH$$

Chloral → Trichloroethanol

Hydrolysis

The hydrolysis of the bonds such as ester, glycoside and amide is important in the metabolism of xenobiotics. Several compounds undergo hydrolysis during the course of their detoxification. These include aspirin, diisopropylfluorophosphate. atropine and procaine.

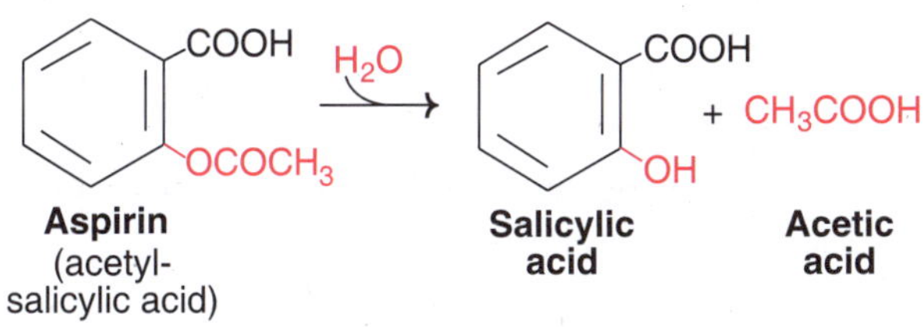

$$(C_3H_7O)_2POF \xrightarrow{H_2O} (C_3H_7O)_2PO(OH) + HF$$

Diisopropyl fluoro-phosphate → Dialkyl phosphate

$$\text{Atropine} \xrightarrow{H_2O} \text{Tropic acid} + \text{Tropine}$$

$$\text{Procaine} \xrightarrow{H_2O} \text{p-Aminobenzoic acid} + \text{Diethylaminoethanol}$$

Conjugation

Several xenobiotics undergo detoxification by conjugation to produce less toxic and/or more easily excretable compounds. Conjugation is the process in which a foreign compound combines with a substance produced in the body. The process of conjugation may occur either directly or after the phase I reactions. At least 8 different conjugating agents have been identified in the body. These are glucuronic acid, glycine, cysteine, glutamine, methyl group, sulfate, acetic acid and thiosulfate.

Glucuronic acid : Conjugation with glucuronic acid is the ***most common***. The active form of glucuronic acid is UDP-glucuronic acid produced in the uronic acid pathway (***Chapter 13***). The microsomal enzymes UDP-glucuronyl transferases participate in glucuronide formation. A general reaction of glucuronide conjugation is shown below (X-OH represents xenobiotic).

$$X{-}OH + UDP{-}\text{glucuronic acid} \xrightarrow[\text{nyltransferase}]{\text{UDP-glucuro-}} XO{-}\text{glucuronide} + UDP$$

Glucuronic acid conjugation may occur with compounds containing hydroxyl, carbonyl, sulfhydryl or amino groups. A few examples of glucuronide conjugation are given here.

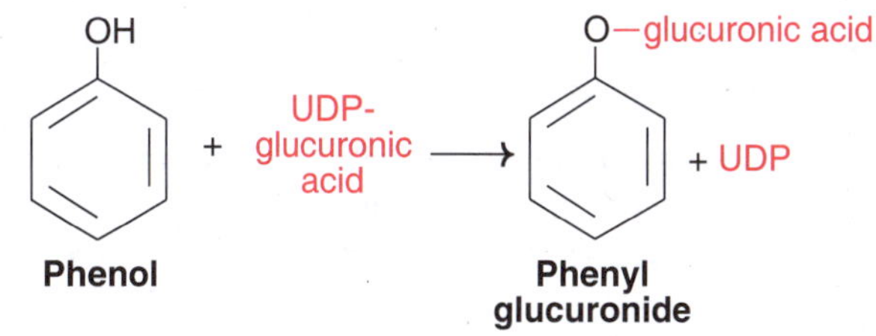

$$\text{Benzoic acid} + \text{UDP-Glucuronic acid} \longrightarrow \text{Benzoyl glucuronide} + UDP$$

$$\text{Bilirubin} + \text{UDP-Glucuronic acid} \longrightarrow \text{Bilirubin-diglucuronide}$$

Glycine : Many aromatic carboxylic acids (e.g. benzoic acid, phenylacetic acid) are conjugated with glycine. Hippuric acid is formed when glycine is conjugated with benzyl CoA.

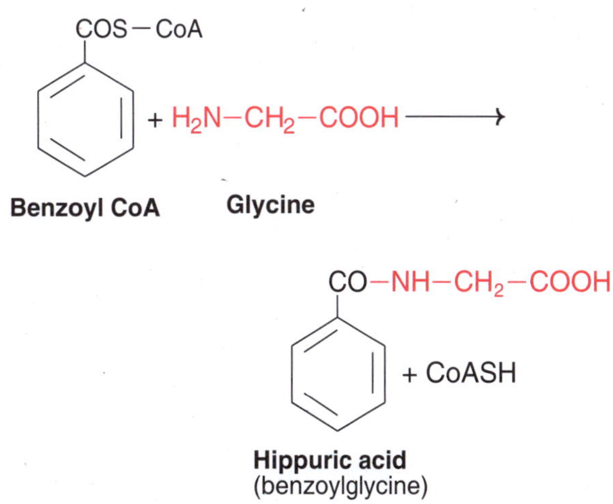

$$\text{Phenylacetic acid} + \text{Glycine} \longrightarrow \text{Phenylaceturic acid}$$

Medical Concepts/ Clinical Correlates

- The body possesses the capability to get rid of the foreign substances by converting them into more easily excretable forms.
- Detoxification is not necessarily associated with the conversion of toxic into non-toxic compounds. For instance, methanol is metabolized to a more toxic formaldehyde.
- Detoxification primarily occurs in the liver through one or more of the reactions, namely oxidation, reduction, hydrolysis and conjugation.
- British antilewisite (BAL), a compound developed during Second World War, was used to detoxify certain war poisons.

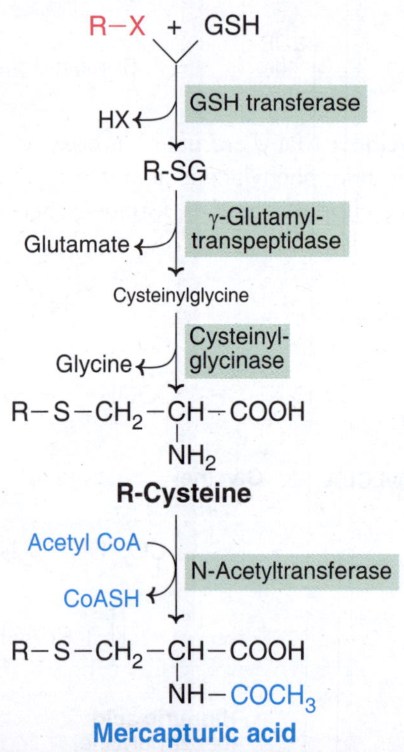

***Fig.31.2 :** Role of glutathione in conjugation to form mercapturic acid (R-X–A xenobiotic; GSH–Glutathione).*

Cysteine : Cysteine, required for conjugation process, is derived from the tripeptide ***glutathione*** (Glu-Cys-Gly), which is the active conjugating agent. A wide range of organic compounds such as alkyl or aryl halides, alkenes, nitro compounds and epoxides get conjugated with cysteine of glutathione. The formation of mercapturic acid is depicted in ***Fig.31.2***. The glutamate and glycine of glutathione are removed and an acetyl group is added to the cysteine residue.

Glutamine : Phenylacetic acid is conjugated with glutamine to form phenylacetyl glutamine. Conjugation with glutamine is, however, relatively less important.

Methyl group : The methyl group ($-CH_3$) of ***S-adenosylmethionine*** is frequently used to methylate certain xenobiotics. This is catalysed by the enzyme methyltransferase.

S-Adenosylmethionine + X–OH $\xrightarrow{\text{Methyltransferase}}$ S-Adenosylhomocysteine + XO–CH_3

Sulfate : The active form of sulfate—3′-phosphoadenosine 5-phosphosulfate (***PAPS***)—participates in conjugation reactions and the enzyme sulfotransferase is involved in this process. Several aliphatic and aromatic compounds undergo sulfation.

Phenol + PAPS $\xrightarrow{\text{Sulfotransferase}}$ Phenyl sulfate + PAP (Phosphoadenosylphosphate)

Acetic acid : ***Acetyl CoA*** is the active form of acetic acid that takes part in conjugating reactions. Drugs such as sulfanilamide are converted to acetyl derivatives.

Sulfanilamide + Acetyl CoA ⟶ Acetyl sulfanilamide + CoASH

Thiosulfate : The highly toxic cyanides are conjugated with thiosulfate to form less toxic thiocyanate.

Cyanide + Sodium thiosulfate ⟶ Thiocyanate + Sodium sulfate

Detoxification by drugs : It may be surprising to know that some drugs are administered to detoxify foreign substances. The toxic effects of certain metals such as arsenic, mercury and cadminum could be overcome by administering BAL (British antilewisite). This compound was developed during the World War II and was used as a detoxifying agent for certain war poisons.

PROSTAGLANDINS AND RELATED COMPOUNDS

32

Prostaglandins and their related compounds—***prostacyclins (PGI)***, ***thromboxanes (TXA)*** and ***leukotrienes (LT)*** are collectively known as ***eicosanoids***, since they all contain 20 carbons (*Greek* : eikosi-twenty). Eicosanoids are considered as locally acting hormones with a wide range of biochemical functions.

Structure of prostaglandins

Prostaglandins are derivatives of a hypothetical 20-carbon fatty acid namely ***prostanoic acid***, hence known as prostanoids. This has a ***cyclopentane ring*** (formed by carbon atoms 8 to 12) and two side chains, with carboxyl group on one side. Prostaglandins differ in their structure due to substituent group and double bond on cyclopentane ring.

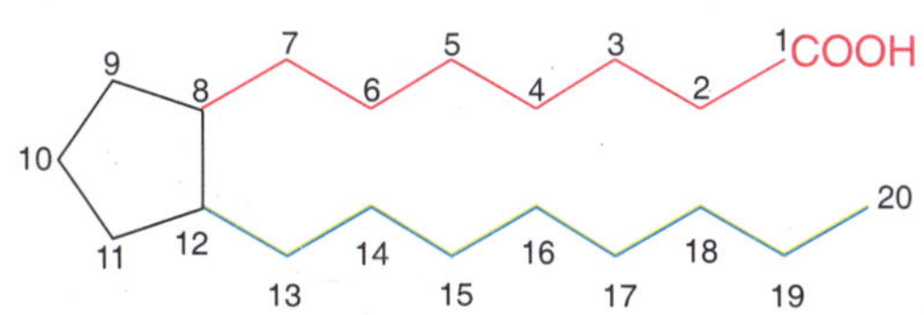

Prostanoic acid

The structure of the most important prostaglandins (PGF_2 and $PGF_2\alpha$), prostacyclins (PGI_2), thromboxanes (TXA_2) and leukotrienes (LTA_4) along with arachidonic acid are depicted in ***Fig.32.1***. A subscript numeral indicates the number of double bonds in the two side chains. A subscript α-denotes that the hydroxyl group at C_9 of the ring and the carboxyl group are on the same side of the ring.

Synthesis of prostaglandins

Arachidonic acid is the precursor for most of the prostaglandins in humans. The biosynthesis of PGs occurs in the endoplasmic reticulum in the following stages, as depicted in ***Fig.32.2***.

1. Release of ***arachidonic acid from*** membrane bound ***phospholipids*** by phospholipase A_2—this reaction occurs due to a specific stimuli by hormones such as epinephrine or bradykinin.

2. Oxidation and cyclization of arachidonic acid to PGG_2 which is then converted to PGH_2 by a reduced glutathione dependent peroxidase.

3. PGH_2 serves as the immediate precursor for the synthesis of a number of prostaglandins, including prostacyclins and thromboxanes.

The above pathway is known as ***cyclic pathway of arachidonic acid***. In the linear pathway of arachidonic acid, leukotrienes are synthesized (details given later).

Inhibition of PG synthesis : A number of structurally unrelated compounds can inhibit prostaglandin synthesis. Corticosteroids (e.g. cortisol) prevent the formation of arachidonic acid by inhibiting the enzyme phospholipase A_2.

Many non-steroidal anti-inflammatory drugs inhibit the synthesis of prostaglandins, prostacyclins and thromboxanes. They do so by blocking the action of the enzyme cyclooxygenase.

Aspirin inhibits PG synthesis : Aspirin irreversibly inhibits the enzyme ***cyclooxygenase***. Other antiinflammatory drugs, such as indomethacin and phenylbutazone act as reversible inhibitors of the enzyme cyclooxygenase.

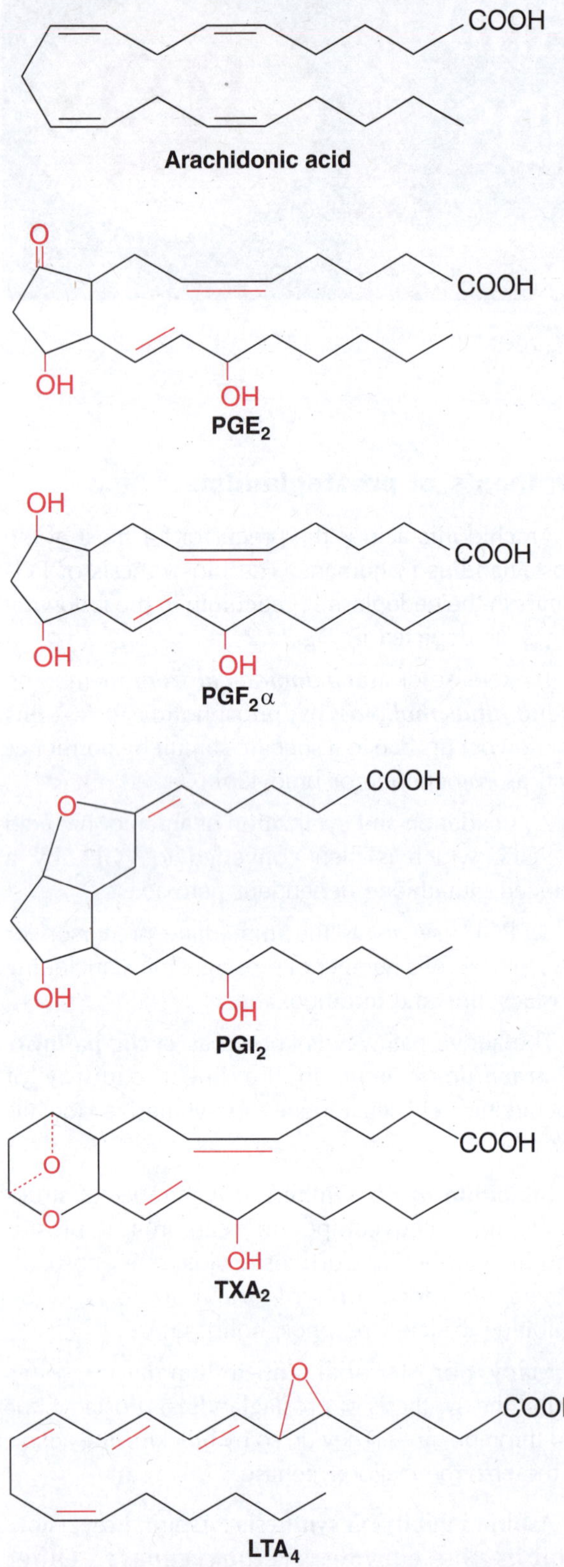

Fig.32.1 : *The structures of arachidonic acid, common prostaglandins (PGE_2 and $PGF_2\alpha$), prostacyclins (PGI_2), thromboxanes (TXA_2) and leukotrienes (LTA_4).*

Degradation of prostaglandins : Almost all the eicosanoids are metabolized rapidly. The lung and liver are the major sites of PG degradation. Two enzymes, namely 15-α-hydroxy PG dehydrogenase and 13-PG reductase, convert hydroxyl group at C_{15} to keto group and then to C_{13} and C_{14} dihydro-derivative.

Biochemical actions of prostaglandins

Prostaglandins act as ***local hormones*** in their function. They, however, differ from the true hormones in many ways. Prostaglandins are produced in almost all the tissues in contrast to hormonal synthesis which occurs in specialized glands. PGs are not stored and they are degraded to inactive products at the site of their production. Further, PGs are produced in very small amounts and have low half-lives.

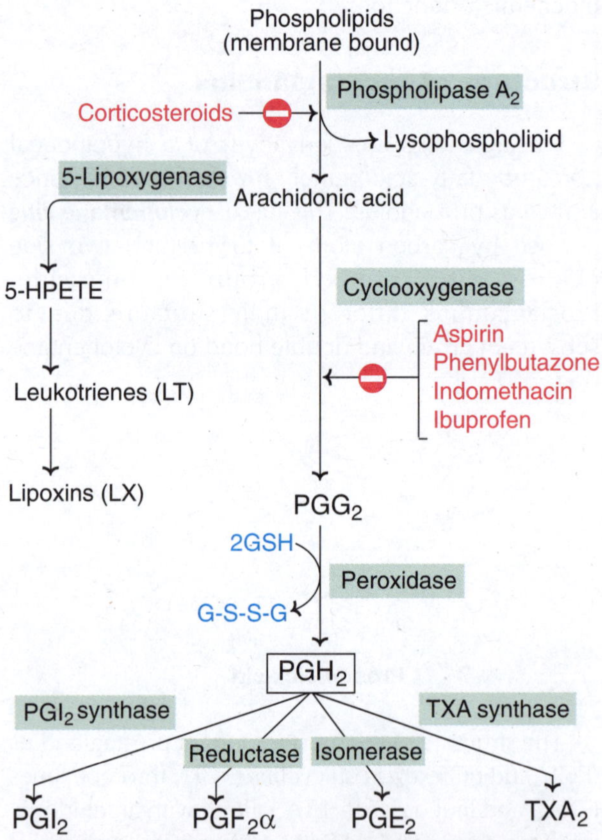

Fig.32.2 : *Overview of biosynthesis of prostaglandins and related compounds (5-HPETE–5-Hydroxy-peroxyeicosatetraenoic acid; PG–Prostaglandins; PGI_2–Prostacyclin I_2; TXA_2–Thromboxane A_2).*

Medical Concepts/ Clinical Correlates

- **PGs act as local hormones and perform diversified biochemical functions. These include lowering of blood pressure, inhibition of gastric HCl secretion, decrease in immunological response and induction of labor.**
- **Overproduction of PGs causes symptoms such as pain, fever, vomiting, nausea, inflammation etc. Aspirin/ibuprofen/ corticosteroid administration inhibits PG synthesis and relieves these symptoms.**
- **Platelet aggregation that may lead to thrombosis is promoted by thromboxanes and prostaglandins E_1 and inhibited by prostacyclins.**
- **Leukotrienes are implicated in hypersensitivity (allergy) and asthma.**
- **Consumption of fish foods containing the unsaturated fatty acid namely eicosapentaenoic acid is advocated to prevent heart attacks.**

Prostaglandins are involved in a variety of biological functions.

1. **Regulation of blood pressure :** The prostaglandins (PGE, PGA and PGI_2) are vasodilator in function. This results in increased blood flow and decreased peripheral resistance to ***lower the blood pressure***. PGs serve as agents in the treatment of hypertension.

2. **Inflammation :** The prostaglandins PGE_1 and PGE_2 ***induce*** the symptoms of ***inflammation*** (redness, swelling, edema etc.) due to arteriolar vasodilation. This led to the belief that PGs are natural mediators of inflammatory reactions of rheumatoid arthritis (involving joints), psoriasis (skin), conjunctivitis (eyes) etc. Corticosteroids are frequently used to treat these inflammatory reactions, since they inhibit prostaglandin synthesis.

3. **Reproduction :** Prostaglandins have widespread application in the field of reproduction. PGE_2 and PGF_2 are used for the medical ***termination of pregnancy*** and ***induction of labor***.

4. **Pain and fever :** It is believed that pyrogens (fever producing agents) ***promote prostaglandin*** biosynthesis leading to the formation of PGE_2 in the hypothalamus, the site of regulation of body temperature. Migraine is also due to PGE_2. Aspirin and other non-steroidal drugs inhibit PG synthesis and thus control fever and relieve pain.

5. **Regulation of gastric secretion :** In general, prostaglandins (PGE) ***inhibit gastric secretion***. PGs are used for the treatment of gastric ulcers.

6. **Influence on immune system :** Macrophages secrete PGE which ***decreases the immunological functions*** of B-and T-lymphocytes.

7. **Effects on respiratory function :** PGE is a bronchodilator whereas PGF acts as a constrictor of bronchial smooth muscles. Thus, PGE and PGF oppose the actions of each other in the lungs. PGE_1 and PGE_2 are used in the treatment of asthma.

8. **Influence on renal functions :** PGE increases glomerular filtration rate (GFR) and ***promotes urine output***. Excretion of Na^+ and K^+ is also increased by PGE.

9. **Effects on metabolism :** Prostaglandins influence certain metabolic reactions, probably through the mediation of cAMP. PGE decreases lipolysis, increases glycogen formation and promotes calcium mobilization from the bone.

10. **Platelet aggregation and thrombosis :** The prostaglandins, namely ***prostacyclins*** (PGI_2), ***inhibit platelet aggregation***. On the other hand, thromboxanes (TXA_2) and prostaglandin E_1 promote platelet aggregation and blood clotting that might lead to thrombosis. PGI_2, produced by endothelial cells lining the blood vessels, prevents the adherence of platelets to the blood vessels. TXA_2 is released by the platelets, and is responsible for their spontaneous aggregation when the platelets come in contact with foreign surface, collagen or thrombin. Thus, prostacyclins and thromboxanes are antagonists in their action. In the overall effect ***PGI_2 acts as a vasodilator, while TXA_2 is a vasoconstrictor***.

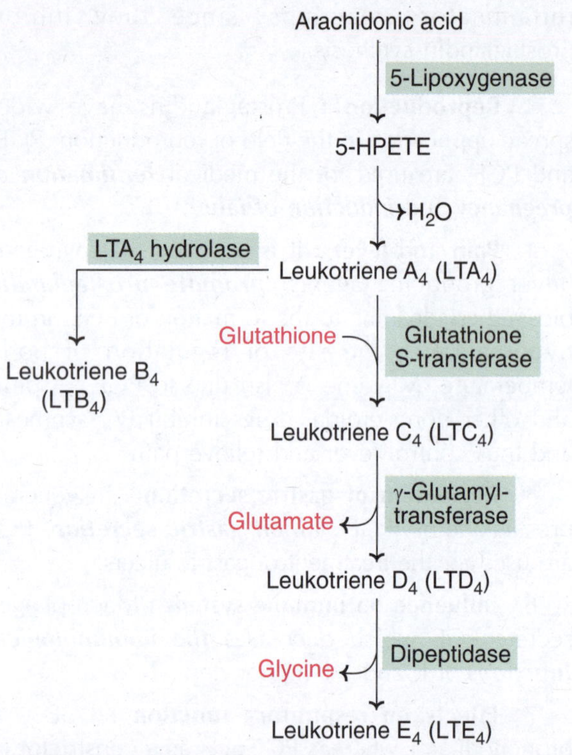

***Fig.32.3 :** Synthesis of leukotrienes from arachidonic acid (5-HPETE–5-Hydroperoxyicosatetraenoic acid).*

Mechanism of action of PGs

The mechanism of action of prostaglandins is not known for certain. They bind to the specific cellular receptors and bring about their action at the molecular level. It is believed that PGs may act through the mediation of cyclic nucleotides. PGE increases cAMP levels whereas PGF elevates cGMP.

Biomedical applications of PGs

As described above, prostaglandins perform diversified functions. And for this reason, PGs (or other derivatives) are the most exploited in therapeutic applications. They are used in the treatment of gastric ulcers, hypertension, thrombosis, asthma etc. Prostaglandins are also employed in the medical termination of pregnancy, prevention of conception, induction of labor etc.

Inhibitors of prostaglandin synthesis (e.g. aspirin, ibuprofen) are utilized in controlling fever, pain, migraine, inflammation etc.

LEUKOTRIENES

Leukotrienes are synthesized by leucocytes, mast cells, lung, heart, spleen etc., by lipoxygenase pathway of arachidonic acid. The synthesis of different leukotrienes (A_4, B_4, C_4, D_4 and E_4) through the intermediate, 5-hydroperoxyeicosatetraenoic acid (5-HPETE) is depicted in ***Fig.32.3***.

Anaphylaxis is a violent and fatal allergic reaction. It is now known that leukotrienes (C_4, D_4 and E_4) are the components of ***slow-reacting substances of anaphylaxis*** (SRS-A), released after immunological challenge. SRS–A is 100–1,000 times more potent than histamine or prostaglandins in its action as a stimulant of allergic reactions. Leukotrienes are ***implicated in asthma, inflammatory reactions, hypersensitivity (allergy) and heart attacks***.

Leukotrienes cause contraction of smooth muscles, bronchoconstriction, vasoconstriction, adhesion of white blood cells and release of lysosomal enzymes.

Dietary marine lipids in relation to PGs, LTs and heart diseases

Eskimos of Greenland have a low incidence of coronary heart diseases, despite the fact that they consume high quantities of fat and cholesterol. This is due to the high intake of marine lipids containing unsaturated fatty acids (UFA). The most predominant UFA in the fish foods consumed by Eskimos is 5,8,11,14, 17-eicosapentaenoic acid (EPA). EPA is the precursor for leukotrienes-5 series which are much lower in their activity than the leukotriene-4 series, produced from arachidonic acid. Further, eicosapentaenoic acid inhibits the formation thromboxanes (TXA_2). As already described, TXA_2 promotes platelet aggregation and thrombosis.

The ***diet rich in marine lipids (with EPA) decreases plasma cholesterol and triacylglycerols***. These factors, along with reduced synthesis of TXA_2 are believed to be responsible for the low incidence of heart attacks in Eskimos.

BIOLOGICAL MEMBRANES AND TRANSPORT

33

The plasma membrane is an envelope surrounding the cell (***Refer Fig.1.1***). It separates and protects the cell from the external hostile environment. Besides being a protective barrier, plasma membrane provides a connecting system between the cell and its environment. The subcellular organelles such as nucleus, mitochondria, lysosomes are also surrounded by membranes.

Chemical composition

The membranes are composed of ***lipids, proteins and carbohydrates***. The actual composition differs from tissue to tissue. Among the lipids, amphipathic lipids (containing hydrophobic and hydrophilic groups) namely phospholipids, glycolipids and cholesterol, are found in the animal membranes.

Structure of membranes

A ***lipid bilayer model*** originally proposed for membrane structure in 1935 by Davson and Danielle has been modified.

Fluid mosaic model, proposed by Singer and Nicolson (1972), is a more recent and acceptable model for membrane structure. The biological membranes usually have a thickness of 5-8 nm. A membrane is essentially composed of a lipid bilayer. The hydrophobic (nonpolar) regions of the lipids face each other at the core of the bilayer while the hydrophilic (polar) regions face outward. Globular proteins are irregularly embedded in the lipid bilayer (***Fig.33.1***). Membrane proteins are categorized into two groups.

1. **Extrinsic (peripheral) membrane proteins** are loosely held to the surface of the membrane and they can be easily separated e.g. cytochrome c of mitochondria.

2. **Intrinsic (integral) membrane proteins** are tightly bound to the lipid bilayer and they can be separated only by the use of detergents or organic solvents e.g. hormone receptors, cytochrome P_{450}.

The membrane is asymmetric due to the irregular distribution of proteins. The lipid and protein subunits of the membrane give an appearance of mosaic or a ceramic tile. Unlike a fixed ceramic tile, the membrane freely changes, hence the structure of the membrane is considered as fluid mosaic.

Transport across membranes

The biological membranes are relatively impermeable. The membrane, therefore, forms a barrier for the free passage of compounds across it. At least three distinct ***mechanisms*** have been identified for the ***transport of solutes*** (metabolites) through the membrane (***Fig.33.2***).

1. Passive diffusion
2. Facilitated diffusion
3. Active transport.

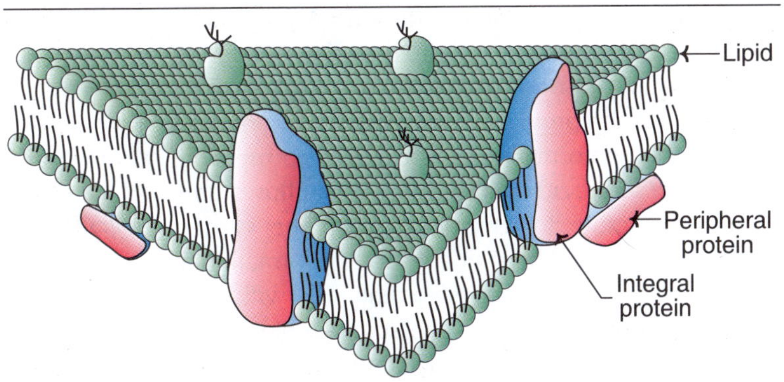

Fig.33.1 : *The fluid mosaic model of membrane structure.*

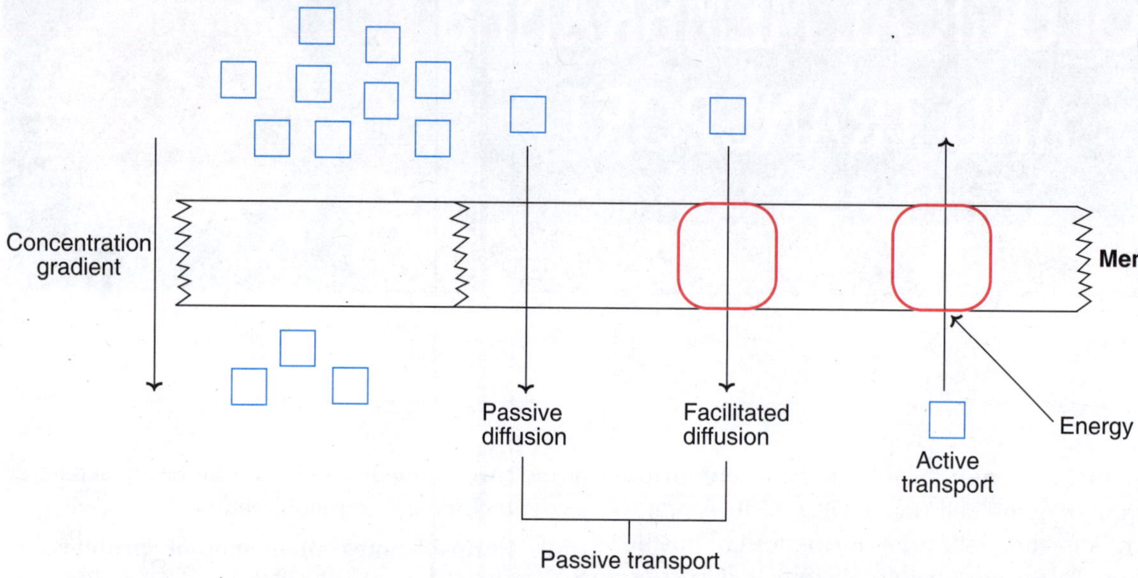

***Fig.33.2 :** Mechanism of transport across biological membrane*
*(**Note :** Transport molecules are represented in blue; the carrier proteins in red).*

1. **Passive diffusion :** This is a simple process which ***depends on the concentration gradient*** of a particular substance across the membrane. Passage of water and gases through membrane occurs by passive diffusion. This process does ***not require energy***.

2. **Facilitated diffusion :** This is somewhat comparable with diffusion since the solute moves along the concentration gradient (from higher to lower concentration) and no energy is needed. But the most important distinguishing feature is that facilitated diffusion occurs through the ***mediation of carrier or transport proteins***. Specific carrier proteins for the transport of glucose, galactose, leucine, phenylalanine etc. have been isolated and characterized.

Mechanism of facilitated diffusion : A ***ping-pong model*** is put forth to explain the occurrence of facilitated diffusion. According to this mechanism, a transport (carrier) protein exists in two conformations. In the pong conformation, it is exposed to the side with high solute concentration. This allows the binding of solute to specific sites on the carrier protein. The protein then undergoes a conformational change (ping state) to expose to the side with low solute concentration where the solute molecule is released. Hormones regulate facilitated diffusion. For instance, insulin increases glucose transport in muscle and adipose tissue; amino acid transport in liver and other tissues.

3. **Active transport :** Active transport occurs ***against a concentration gradient*** and this is dependent on the ***supply of metabolic energy (ATP)***. Active transport is also a carrier mediated process like facilitated diffusion.

Na^+-K^+ pump : The cells have a high intracellular K^+ concentration and a low Na^+ concentration. This is essentially needed for the survival of the cells. High cellular K^+ is required for the optimal glycolysis (pyruvate kinase is dependent on K^+) and for protein biosynthesis. Further, Na^+ and K^+ gradients across plasma membranes are needed for the transmission of nerve impulse.

The Na^+-K^+ pump is responsible for the maintenance of high K^+ and low Na^+ concentrations in the cells. This is brought about by an integral plasma membrane protein, namely the enzyme Na^+-K^+ ATPase. Na^+-K^+ ATPase pumps $3Na^+$ ions from inside the cell to outside and brings $2K^+$ ions from the outside to the inside with a concomitant hydrolysis of intracellular ATP. The Na^+-K^+ pump, depicted in ***Fig.33.3***, is summarized.

$$3\ Na^+\ (in) + 2K^+\ (out) + ATP \longrightarrow 3Na^+\ (out) + 2K^+\ (in) + ADP + Pi$$

A major portion of the cellular ATP (up to 70% in nerve cells) is in fact utilized by Na^+-K^+ pump to maintain the requisite cytosolic Na^+ and K^+ levels.

Medical Concepts/ Clinical Correlates

- The cells must contain high K^+ and low Na^+ concentrations for their survival. Na^+-K^+ pump, which consumes a major portion of the cellular metabolic energy (ATP), is responsible for this.
- Ouabain inhibits Na^+-K^+ ATPase (Na^+-K^+ pump). It is extracted from the seeds of an African shrub and used as poison to tip the hunting arrows by the tribals.
- Disturbances in osmosis are associated with diarrhea, edema, inflammation. Changes in membrane fluidity have been suggested to be involved in LCAT deficiency, hypertension and Alzheimer's disease.

Na^+-cotransport system : The amino acids and sugars are transported into the cells by a Na^+-cotransport system. This process essentially consists of the passage of glucose (or amino acid) into the cell with a simultaneous movement of Na^+. ATP is required to pump out the intracellular Na^+ through the mediation of ***Na^+-K^+ ATPase*** (***Refer Chapter 8***).

Transport systems

The transport systems may be divided into 3 categories (***Fig.33.4***).

1. **Uniport system :** This involves the movement of a single molecule through the membrane e.g. transport of glucose to the erythrocytes.

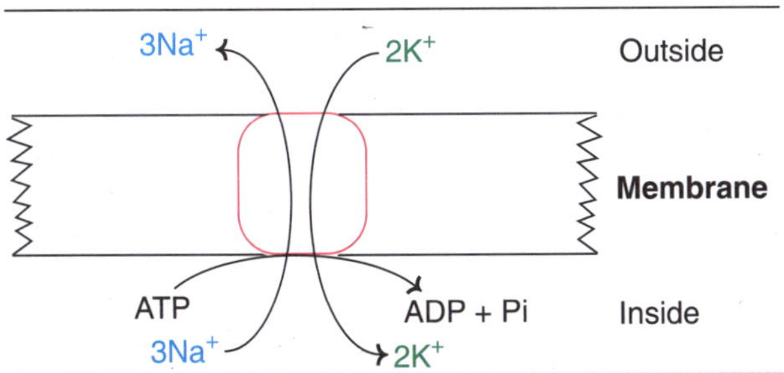

***Fig.33.3 :** Diagrammatic representation of Na^+–K^+ pump (**Note :** Red colour block represents the enzyme Na^+–K^+ ATPase).*

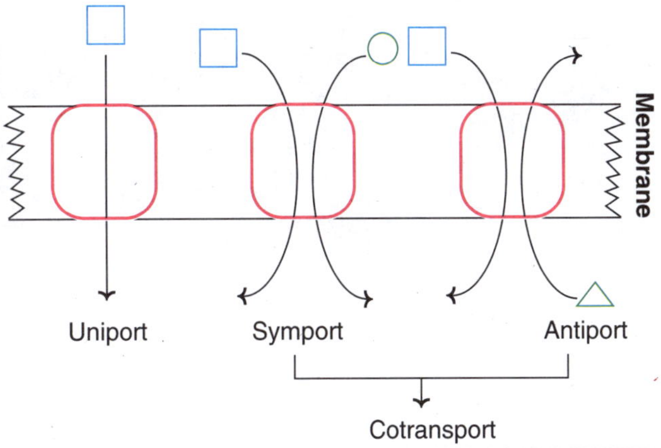

***Fig.33.4 :** Diagrammatic representation of transport systems.*

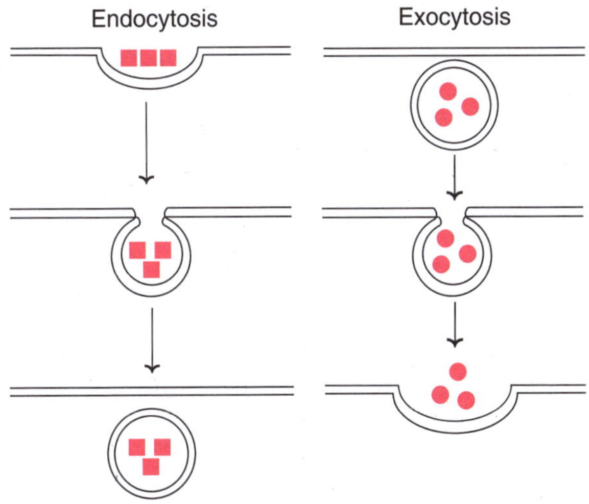

***Fig.33.5 :** Diagrammatic representation of endocytosis and exocytosis (**Note :** The red coloured particles indicate the transport material).*

2. **Symport system :** The simultaneous transport of two different molecules in the same direction e.g. transport of Na^+ and glucose to the intestinal mucosal cells from the gut.

3. **Antiport system :** The simultaneous transport of two different molecules in the opposite direction e.g exchange of Cl^- and HCO_3^- in the erythrocytes.

Cotransport system : The symport and antiport systems are collectively referred to as cotransport system.

Transport of macromolecules

The transport of macromolecules such as proteins, polysaccharides and polynucleotides across the membranes is equally important. This is brought about by two independent mechanisms namely ***endocytosis—intake of macromolecules*** (***Fig.33.5***) by the cells (e.g. uptake of LDL by cells) and ***exocytosis—release of macromolecules*** from the cells to the outside (e.g. secretion of hormones-insulin, PTH).

FREE RADICALS AND ANTIOXIDANTS

34

The supply of oxygen is absolutely essential for the existence of higher organisms. As the saying goes too much of even the best is bad. Very high concentrations of O_2 are found to be toxic, and can damage tissues. The present day concept of oxygen toxicity is due to the involvement of ***oxygen free radicals*** or ***reactive oxygen species (ROS)***. In fact, the generation of reactive metabolites of O_2 is an integral part of our daily life.

A ***free radical*** is defined as a molecule or a molecular species that ***contains one or more unpaired electrons***, and is ***capable of independent existence***.

Types of free radicals

Oxygen is required in many metabolic reactions, particularly for the release of energy. During these processes, molecular O_2 is completely reduced, and converted to water. However, if the reduction of O_2 is incomplete, a series of reactive radicals are formed, as shown in the next column.

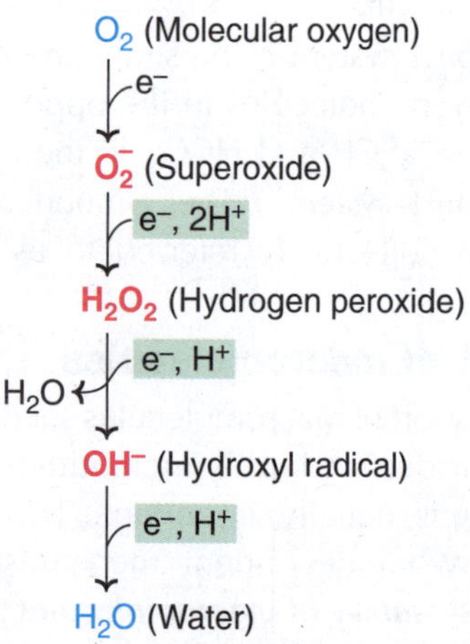

Besides the above (O_2^-, H_2O_2, OH^-), the other free radicals and reactive oxygen species of biological importance include singlet oxygen ($^1O_2$), hydroperoxy radical (HOO^-), lipid peroxide radical (ROO^-), nitric oxide (NO^-) and peroxynitrite ($ONOO^-$).

The common characteristic features of free radicals are listed

- ***Highly reactive***
- Very ***short half-life***
- Can generate new radicals by chain reaction
- Cause ***damage to biomolecules***, cells and tissues

SOURCES AND GENERATION OF FREE RADICALS

The major sources responsible for the generation of free radicals may be considered under two categories

I. Due to ***normal biological processes*** (or cellular metabolism).

II. Due to ***environmental effects***.

It is estimated that about 1-4% of the O_2 taken up by the body is converted to free radicals. A summary of the sources for generation of free radicals is given in the ***Table 34.1***, and a couple of the processes are briefly described.

Lipid peroxidation

Free radical-induced peroxidation of membrane lipids occurs in three stages-initiation, propagation and termination.

Malondialdehyde (MDA) as a marker for lipid peroxidation

Most of the products of lipid peroxidation are unstable e.g. carbonyls, esters, alkanes, alkenes,

TABLE 34.1 Sources along with some examples for generation of free radicals

I Cellular metabolism

- Leakage of electrons from the respiratory chain (ETC).
- Production of H_2O_2 or O_2^- by oxidase enzymes (e.g. xanthine oxidase, NADPH oxidase).
- Due to chain reactions of membrane lipid peroxidation.
- Peroxisomal generation of O_2 and H_2O_2.
- During the course of phagocytosis (as a part of bactericidal action).
- In the oxidation of heme to bile pigments.

II Environmental effects

- As a result of drug metabolism e.g. paracetamol, halothane, cytochrome P_{450} related reactions.
- Due to damages caused by ionizing radiations (e.g. X-rays) on tissues.
- Photoexcitation of organic molecules
- Cigarette smoke contains free radicals, and trace metals that generate OH^-.
- Alcohol, promoting lipid peroxidation.

2-alkenal, 2,4-alkadienal, MDA. Of these, malondialdehyde ($-CHO-CH_2-CHO-$) is the most extensively studied, and is used as a biochemical marker for the assessment of lipid peroxidation. MDA and other aldehydes react with thiobarbituric acid and produce red-coloured products namely ***thiobarbituric acid reactive substances (TBARS)*** which can be measured colorimetrically.

The estimation of serum MDA is often used to assess oxidative stress, and free radical damage to the body.

Generation of ROS by macrophages

During the course of ***phagocytosis, macrophages produce superoxide (O_2^-)***, by a reaction catalysed by NADPH oxidase (***Fig.34.1***). This O_2^- radical gets converted to H_2O_2, and then to hypochlorous acid (HClO). The superoxide radical along with hypochlorous ions brings about ***bactericidal action***. This truly represents the ***beneficial affects of the free radicals*** generated by the body. A large amount of O_2 is consumed by macrophages during their bactericidal function, a phenomenon referred to as ***respiratory burst***. It is estimated that about 10% of the O_2 taken up by macrophages is utilized for the generation of free radicals.

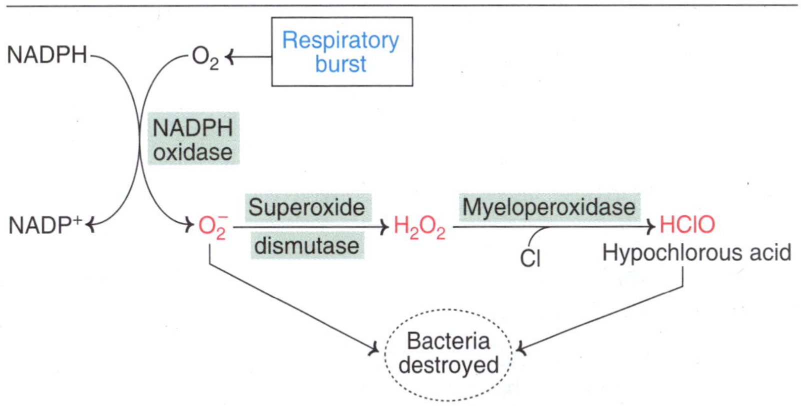

Fig.34.1 : Generation of free radicals by macrophages and respiratory burst.

HARMFUL EFFECTS OF FREE RADICALS

Free radicals and biomolecules

Free radicals are highly reactive, and are capable of damaging almost all types of biomolecules (proteins, lipids, carbohydrates, nucleic acids). The fact is that ***free radicals beget free radicals*** i.e. generate free radicals from normal compounds which continues as a chain reaction.

Free radicals and diseases

As discussed above, free radicals are harmful to biomolecules, and in turn cells and tissues. Free radicals have been implicated in the causation and progress of several diseases.

Cardiovascular diseases (CHD) : Oxidized low density lipoproteins (LDL), formed by the action of free radicals, ***promote atherosclerosis*** and CHD.

Cancer : Free radicals ***can damage DNA***, and cause mutagenicity and cytotoxicity, and thus play a key role in carcinogenesis. It is believed that ROS can induce mutations, and inhibit DNA repair process, that results in the inactivation of certain tumor suppressor genes leading to cancer.

Inflammatory diseases : *Rheumatoid arthritis* is a chronic inflammatory disease. The free radicals produced by neutrophils are the predominant causative agents. The occurrence of other inflammatory disorders—chronic glomerulonephritis and ulcerative colitis is also due to the damages caused by ROS on the extracellular components (e.g. collagen, hyaluronic acid).

Medical Concepts/ Clinical Correlates

- Free radicals have been implicated in the causation and progress of several diseases e.g. atherosclerosis and CHD, cancer, respiratory diseases.
- The respiratory burst of macrophages, accompanied by the generation of ROS (H_2O_2 and HClO), brings about bactericidal action, and is beneficial to the body.
- Dietary consumption of a variety of nutrient antioxidants (vitamins C and E, β-carotene, lycopenes, Se, β-lipoic acid) is desirable since each antioxidant targets certain types of damaging free radicals.

Respiratory diseases : ROS are also responsible for ***adult respiratory distress syndrome (ARDS)***, a disorder characterized by pulmonary edema.

Cigarette smoke, as such, contains free radicals, and further it promotes the generation of more free radicals. The damages caused to lungs in the smokers are due to ROS.

Diabetes : Destruction of islets of pancreas due to the accumulation of free radicals is one of the causes for the pathogenesis of insulin-dependent diabetes mellitus.

Cataract : Increased exposure to oxidative stress contributes to cataract formation, which is mostly related to aging.

Aging process : Free radicals are closely associated with the various biochemical and morphological changes that occur during normal aging.

Other diseases : Free radicals play a key role in ***Parkinson's disease***, ***Alzheimer's disease***, multiple sclerosis, liver cirrhosis, muscular dystrophy, toxemia of pregnancy etc.

ANTIOXIDANTS IN BIOLOGICAL SYSTEM

To mitigate the harmful/damaging effects of free radicals, the aerobic cells have developed antioxidant defense mechanisms. A biological antioxidant may be defined as a ***substance*** (present in low concentrations compared to an oxidizable substrate) that ***significantly delays*** or ***inhibits oxidation*** of a ***substrate***. Antioxidants may be considered as the ***scavengers of free radicals***. There are different ways of classifying antioxidants.

I. Antioxidants according to their location

1. ***Plasma antioxidants*** e.g. β-carotene, ascorbic acid, bilirubin, uric acid, ceruloplasmin, transferrin.

2. ***Cell membrane antioxidants*** e.g. α-tocopherol.

3. ***Intracellular antioxidants*** e.g. superoxide dismutase, catalase, glutathione peroxidase.

II. Antioxidants according to their nature and action

1. ***Enzymatic antioxidants*** e.g. superoxide dismutase, catalase, glutathione peroxidase, glutathione reductase.

2. ***Non-enzymatic antioxidants***

(a) ***Nutrient antioxidants*** e.g. carotenoids, α-tocopherol, ascorbic acid, selenium.

(b) ***Metabolic antioxidants*** e.g. glutathione, ceruloplasmin, albumin, bilirubin, transferrin, ferritin, uric acid

The antioxidant enzyme system

The antioxidant enzymes are truly the scavangers of free radicals. The major reactions of these enzymes are depicted in ***Fig.34.2***, some highlights are given below.

Superoxide dismutase : It converts superoxide (O_2^-) to hydrogen peroxide and O_2 (***Fig.34.2A***). This is the first line of defense to protect cells from the injurious effects of superoxide.

Catalase : Hydrogen peroxide, produced by superoxide dismutase, is metabolised by catalase (***Fig.34.2B***).

Glutathione peroxidase : It detoxifies H_2O_2 to H_2O, while reduced glutathione (G—SH) is converted to oxidized glutathione (GS—SG). The reduced glutathione can be regenerated by the enzyme ***glutathione reductase*** utilizing NADPH (***Fig.34.2C***).

Nutrient antioxidants

Tocopherols (vitamin E) : Vitamin E is fat soluble, and among the tocopherols, α-tocopherol is biologically the most active. It is an antioxidant present in all cellular membranes, and protects against lipid peroxidation.

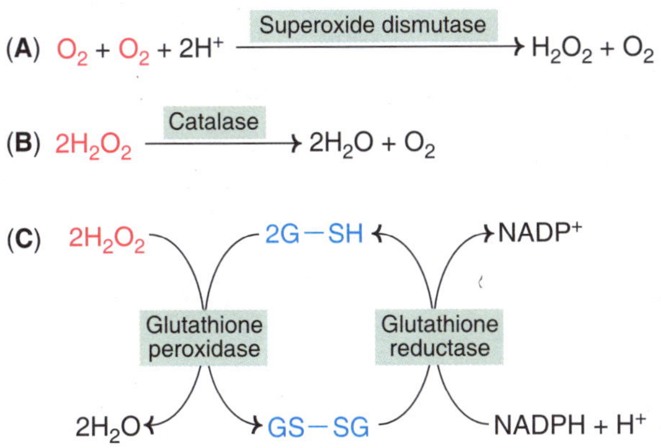

Fig.34.2 : *The antioxidant enzyme system (G—SH-reduced glutathione; GS—SG-oxidized glutathione).*

Ascorbic acid (vitamin C) : It is an important water soluble antioxidant in biological fluids. Vitamin C efficiently ***scavanges free radicals, and inhibits lipid peroxidation***.

Carotenoids : These are the natural compounds with lipophilic properties. About 500 different carotenoids have been identified, among them ***β-carotene is the most important***. It can act as an antioxidant in association with vitamins C and E. ***Lycopene***, a fat soluble pigment is a carotenoid. It is responsible for colour of certain fruits and vegetables (e.g. tomato). Lycopene possesses antioxidant property. ***Lutein*** and ***Zeaxanthin*** are also carotenoid pigments that impart yellow or green colour to fruits and vegetables. These pigments can also serve as antioxidants.

Selenium : It is an essential trace element, and is proved to be a significant antioxidant. Selenium works ***with vitamin E*** in fighting free radicals. It is also required for the function of an important antioxidant enzyme, namely ***glutathione peroxidase***.

α-Lipoic acid : It is vitamin-like compound, produced in the body, besides the supply from plant and animal sources. α-Lipoic acid plays a key role in recycling other important antioxidants such as ascorbic acid, α-tocopherol and glutathione.

Besides the above, there are many other nutrient antioxidants. In the ***Table 34.2***, some important nutrient antioxidants and their dietary sources are given. ***Consumption of a variety of nutrient antioxidants*** is important, since each antioxidant targets certain types of damaging free radicals.

Metabolic antioxidants

Glutathione : Reduced glutathione (GSH) plays a key role in the biological antioxidant enzyme system (***See Fig.34.2C***). GSH and H_2O_2 are the twin substrates for glutathione peroxidase. The reduced glutathione (GSH) gets regenerated from the oxidized glutathione (GS—SG) through the participation of glutathione reductase and NADPH.

There are many more metabolic antioxidants of biological importance. A selected few of them are listed below

- **Uric acid**, a powerful scavenger of singlet oxygen ($^1O_2$) and OH^- radicals.
- **Ceruloplasmin** inhibits iron and copper dependent lipid peroxidation.
- **Transferrin** binds to iron and prevents iron-catalysed free radical formation.
- **Bilirubin** protects the albumin bound free fatty acids from peroxidation.

TABLE 34.2 Nutrient antioxidants and their dietary sources

Antioxidant	*Dietary Source*
Vitamin E (tocopherols)	Unprocessed vegetable oils (cotton seed oil, peanut oil, sunflower oil) whole grains, leafy vegetables.
Vitamin C (ascorbic acid)	Citrus fruits (oranges, grapes) gooseberry (amla), guava, green vegetables (cabbage, spinach).
β-Carotene (provitamin A)	Carrots, green fruits and vegetables, spinach, turnip, apricots.
Lycopene	Tomatoes, papaya, pink guava.
Leutein and zeaxanthin	Egg yolk, fruits, green leafy vegetables, corn, green peas.
Selenium	Sea foods, organ meats, whole grains.
α-Lipoic acid	Red meat, liver, yeast
Coenzyme Q_{10}	Organ meats (best heart), chicken.
Proanthocyanidins	Grape seeds
Catechins	Green tea
Curcuminoids	Turmeric
Quercetin	Onions, red wine, green tea

ENVIRONMENTAL BIOCHEMISTRY

35

Environment constitutes the non-living (air, water, land, energy etc.) as well as the living (biological and social) systems surrounding man. Environmental biochemistry primarily deals with the metabolic (biochemical) responses and adaptations in man (or other organisms) due to the environmental factors.

A healthy environment is required for a healthy life which is however, not really possible or practicable. This is mainly because of the atmospheric (climatic) changes and environmental pollution.

Environmental biochemistry is a very vast subject. The basic concepts regarding the atmospheric changes and environmental pollution on humans are dealt with here.

ATMOSPHERIC (CLIMATIC) CHANGES

The climatic changes include cold, heat etc. The body makes every effort to maintain its normal temperature (despite cold and heat surroundings) for optimal physiological and biochemical functions.

EXPOSURE TO COLD

Short-term exposure to cold ***causes shivering*** (mainly due to skeletal muscle) ***to produce extra heat***. Heat is generated by the hydrolysis of ATP.

Non-shivering phase

Chronic exposure to cold results in non-shivering phase which is characterized by several metabolic adaptations.

- **Energy metabolism :** Heat generation by a process called ***chemical thermogenesis*** occurs in non-shivering phase. Elevation in BMR, and increased intake of foods are observed.
- **Lipid metabolism :** Stored fat (triacylglycerol) in the adipose tissue is mobilized to supply free fatty acids for oxidation and production of energy. Brown edipose tissue, particularly in neonatal life, significantly contributes to thermogenesis.
- **Hormonal changes :** Thyroxine, a hormone closely associated with energy metabolism, is elevated. Further, corticosteroids are increased on exposure to cold.

EXPOSURE TO HEAT

There is a continuous generation of heat by the body due to the ongoing biochemical processes, referred to as ***metabolic heat***. This heat has to be exchanged with the environment to maintain a constant body temperature. On exposure to heat in surroundings, as happens in summer, the body is subjected to an uncomfortable situation (since temperature of the surroundings is much higher than that of the body).

Heat stroke : It is characterized by the ***failure of the heat regulatory system*** (thermoregulation) ***of the body***. The manifestations of heat stroke include high body temperature, convulsions, partial (some times total) loss of consciousness. In extreme cases, heat stroke may cause irreversible damage to brain. The treatment for the heat stroke involves rapid cooling of the body.

ENVIRONMENTAL POLLUTION

Environmental pollution may be regarded as the addition of extraneous (foreign) materials to air, water or land which adversely affects the quality of life. Pollution may be caused by physical, chemical or biological processes.

The term ***pollutant*** refers to a substance which increases in quantity due to human activity and adversely affects the environment (e.g. carbon monoxide, sulfur dioxide, lead). A substance which is not present in nature but released during human activity is the ***contaminant*** (e.g. methyl isocyanate, DDT, malathion). A contaminant however, is regarded as a pollutant when it exerts detrimental effects. Environmental pollution with reference to air, water and foodstuffs is briefly discussed.

AIR POLLUTION

The major components of air include nitrogen (78.1%), oxygen (20.93%) and carbon dioxide (0.03%), along with water vapour and suspended particles. The rapid growth of industries coupled with changing lifestyles of man (urbanization, smoking, use of motor vehicles etc.) largely contribute to air pollution. The major chemical constituents of air pollution are ***sulfur dioxide***, ***oxides of carbon*** (CO_2 and CO), oxides of nitrogen, and particulates. The biochemical affects of air pollution are briefly described.

Sulfur dioxide

Sulfur dioxide (SO_2) is the ***most dangerous pollutant gas to man***. Industrial activities—burning of coal and oil emit large quantities of SO_2.

Sulfur dioxide pollution primarily ***affects respiratory system*** in man. Irritation of the respiratory tract and increasing airway resistance (breathing difficulty) are observed. Lung tissue may get damaged due to acidic pH. Continuous exposure to SO_2 (> 1 ppm) for several days causes bronchitis and in some individuals lung cancer.

Carbon monoxide

Carbon monoxide (CO) is mostly produced by incomplete combustion of fuel or carbon-containing compounds. Automobiles, aircrafts, rail engines and burning of coal in factories contribute to CO pollution.

Carbon monoxide combines with hemoglobin to form carboxyhemoglobin (***Refer Chapter 10***). This causes a drastic reduction in the supply of O_2 to tissues. At a CO concentration around 1 ppm, impairment in mental performance and visual perception take place. With a further increase in CO level, headache, dizziness and loss of consciousness occur. Death may be inevitable in persons exposed to above 750 ppm of CO.

Carbon dioxide

Carbon dioxide (CO_2), constituting only a fraction (0.03%) of the atmospheric gases, plays a significant role in controlling the climate. This is done by trapping the heat radiation from the earth's surface. Without the presence of CO_2, the earth would be as cold as moon!

Carbon dioxide is often referred to as greenhouse gas. The term ***greenhouse effect*** refers to an ***elevation in*** CO_2 near earth's surface that traps sunlight and ***increases atmospheric temperature***. Deforestation, burning of coal, oils etc., elevate atmospheric CO_2 resulting in greenhouse effect. Hence the global propaganda for increased plantation of trees!

Nitrogen dioxide

Nitrogen dioxide (NO_2) like carbon monoxide (CO), combines with hemoglobin and reduces the supply of O_2 to the tissues. NO_2 is more harmful to human health than CO. It is fortunate that the atmospheric concentration of NO_2 is relatively lower.

Nitrogen dioxide (in the form of HNO_3) along with SO_2 (as H_2SO_4) contributes to ***acid rain***.

Particulates

The ***solid dust particles*** suspended ***in the atmosphere*** constitute particulates. The sources of particulates are grinding, spraying, smoking etc.

The particulates have ill-affects on humans. These include interference in respiratory function (coughing, sneezing) and toxicity caused by the absorption particulate chemicals. Further, the dust particles carry microorganisms and other infective agents to spread diseases.

Ozone layer

Ozone is formed from atmospheric oxygen during high energy radiations of electrical discharges. This ozone forms a layer above the earth's surface (15-35 km). It ***absorbs harmful ultraviolet radiations of sun*** which would otherwise cause skin diseases and mutations, besides increasing temperature of earth.

In recent years, a decrease in the ozone layer is observed due to chemical pollution in the air. Nitrogen oxides (released from engines of aero-planes) and chlorofluoro carbons (used in refrigerators and air conditioners) deplete the ozone layer.

WATER POLLUTION

Water is the most predominant constituent of living matter. The very existence of life is unimaginable without water.

Medical Concepts/ Clinical Correlates

- The body makes every effort to maintain its normal temperature, despite cold and heat surroundings, for optimal physiological and biochemical functions.
- Sulfur dioxide (SO_2) is the most dangerous industrial pollutant gas to man. It primarily affects the respiratory system, and may result in bronchitis, and even lung cancer.
- Carbon monoxide (CO) combines with hemoglobin to form carboxyHb. This reduces O_2 supply to tissues.
- Pollution of water with pathogenic organisms causes many diseases e.g. typhoid, cholera, amoebiasis.
- Lead toxicity affects central nervous system—learning disabilities, mental retardation etc.
- Radioactive pollution may lead to gene mutations, cancer etc.

Pollution of water occurs due to waste disposal from industries, agriculture and municipalities. The pollutants may be organic, inorganic, sediments, radioactive, thermal etc., in nature.

ORGANIC POLLUTANTS

The organic pollutants include agents carrying water borne diseases, oxygen demanding wastes and organic chemicals.

Water-borne disease agents

Several pathogenic organisms find their entry into water and cause diseases. The water borne disease include typhoid, paratyphoid, cholera, amoebiasis, giardiasis and infectious hepatitis. These diseases can be prevented by disinfection techniques employed for the treatment of water.

Oxygen demanding wastes

Sewage, and wastes from industries and agriculture provide good nutrients for algae. As the algae grow utilizing the wastes, oxygen depletion occurs. This phenomenon of ***water deoxygenation*** is referred to as ***eutrophication***. As a consequence of eutrophication, fish and other acquatic animals die (due to lack of O_2), causing foul smell.

Organic chemicals

The organic chemical pollutants of water include pesticides and several synthetic compounds (detergents, paints, plastics, pharmaceuticals, food additives etc.)

Pesticides

Pesticides is a broad term used for ***insecticides, herbicides, fungicides*** and ***rodenticides***. Based on their structure, pesticides are classified as follows.

- **Chlorinated hydrocarbons** e.g. aldrin, endrin, dichlorodiphenyl trichloroethane (DDT).
- **Organophosphates** e.g. malathion, diazinon.
- **Carbamates** e.g. baygon, carbaryl (sevin).

The use of pesticides has helped in controlling certain diseases (malaria, typhus), besides boosting food production. However, pesticides pollute water and cause several health complications to humans, besides damaging acquatic life.

Dichloro-diphenyl trichloroethane (DDT) is a widely used pesticide to control cotton and peanut pests, besides malaria. However, continuous use of DDT leads to its accumulation in foods causing ill effects (hence banned in some countries like USA).

DDT, being fat soluble, ***accumulates in the adipose tissue*** and is not excreted. Thus, its concentration in the body goes on increasing. DDT causes nervous irritability, muscle twitching and convulsions.

INORGANIC POLLUTANTS

Heavy metals (lead, mercury, cadmium, etc.) are the most dangerous among the inorganic pollutants.

Lead

Lead is the ***most common inorganic pollutant*** found in water, air, foods and soils. The sources of lead pollution include petrol, gasoline, paints, cigarettes, news papers, lead pipes and xerox copies. The plasma concentration of > 25 μg/dl in adults and > 10 μg/dl in children results in toxic manifestations.

The principal target of lead toxicity is ***central nervous system***. In the growing children, Pb causes learning disabilities, behavioural changes (hyperexcitability) and mental retardation. In adults, confusion, irritability, abdominal colic and severe anemia are associated with lead toxicity.

There has been an increasing awareness worldover on the toxic manifestations of lead. This has lead to the supply of ***unleaded petrol*** in most countries.

Mercury

Mercury is a common industrial (plastic, paints, electrical apparatus, fungicides) pollutant. Acute mercuric poisoning causes gastritis, vomiting and pulmonary edema. Chronic manifestations of Hg include emotional changes, loss of memory and other neuropsychiatric disturbances. In addition, deposition of mercuric salts may cause renal failure.

Cadmium

The outbreak of cadmium toxicity was reported in Japan in the form of ***itai itai*** or ***ouch disease***. Cadmium poisoning causes fragile bones, anemia, bone marrow disorders and kidney damage. Biochemically, cadmium replaces zinc and adversely influences several metabolic reactions.

NOISE POLLUTION

The unwanted sound is noise, which is a major urban environmental pollutant. Man can tolerate noise upto 100 decibels (speaking – 60 decibels; telephone bell 70 decibels; motor cycle 110 decibels; rockets 170 decibels). A noise above 150 decibels is uncomfortable.

The affects of noise pollution include headache, increased blood pressure, irritability, neuromuscular tension, confusion, disturbed vision and digestion, depression and loss of hearing.

TOXIC COMPOUNDS IN FOODSTUFFS

The foodstuffs consumed by humans contain several toxic compounds which may be either normally present or enter foodstuffs during the course of cultivation, processing or storage.

Natural toxins in foodstuffs

Neurotoxins : Kesari dal (*Lathyrus sativus*) is a pulse grown in some parts of Madhya Pradesh, Bihar and Uttar Prodesh. Excessive consumption of kesari dal causes ***paralysis of lower limbs*** referred to as ***lathyrism***. This is due to a neurotoxin namely ***β-oxalylaminoalanine*** (***BOAA***). Cooking of kesari dal 2-3 times and removal of the supernatant water will eliminate the toxin.

Protease inhibitors : Certain legumes (soya bean, peanut) contain inhibitors of protease enzymes particularly trypsin. Normally, protease inhibitors are destroyed by cooking. However, partial cooking does not totally destroy them. In such a case, protease inhibitors ***can inhibit digestion and proteins***.

Goitrogens : These compounds ***prevent uptake and utilization of iodine by thyroid gland***. Goitrogens are found in cabbage and turnips (thioglycosides), mustard and rape seed oils (thiocyanates), ground nuts and almonds (polyphenolic glycosides).

Toxic pollutants of foodstuffs

The foodstuffs may get polluted with several toxic chemicals which might occur during cultivation, processing or storage.

Cultivation : Pesticides and other unnatural chemicals used during cultivation do find an entry into the foodstuffs. It is fortunate that most of these chemicals can be removed by peeling the outer layers of vegetables and fruits, besides repeated washings.

Processing : Defects in freezing, and packing provide a suitable environment for the growth of several organisms which release toxic products e.g. milk contamination by *Salmonella*.

Several food additives are in use for preservation and enchancing flavour. Not all of them are safe e.g. aniline dyes used as colouring agents are carcinogenic; ***sweetening agent cyclamate may cause bladder cancer***.

Storage : Contamination of stored foods occurs mostly due to fungal infections. ***Aflatoxins*** are produced by *Aspergillus favus* when ground nuts or coconuts are stored in moist conditions. Aflatoxins are ***hepatotoxic and carcinogenic***.

CARCINOGENS

The group of chemicals that cause cancer in man and animals are collectively referred to as carcinogens (***Refer Chapter 37***). Environmental pollution is undoubtedly associated with increased risk of cancer. The topic '***cancer***' may be considered as a part of environmental biochemistry for learning purpose.

INSULIN, GLUCOSE HOMEOSTASIS, AND DIABETES MELLITUS

36

Diabetes mellitus is the ***third leading cause of death*** (after heart disease and cancer) in many developed countries. It affects about ***6 to 8% of the general population***. The complications of diabetes affect the eye, kidney and nervous system. Diabetes is a ***major cause of blindness, renal failure, amputation, heart attacks and stroke***. (The term diabetes, whenever used, refers to diabetes mellitus. It should, however, be noted that ***diabetes insipidus*** is another disorder characterized by ***large volumes of urine excretion*** due to ***antidiuretic hormone deficiency***).

Diabetes mellitus is a clinical condition characterized by ***increased blood glucose level (hyperglycemia)*** due to insufficient or inefficient (incompetent) insulin. An important feature of diabetes is that the body cells are starved of glucose despite its very high concentration around i.e. scarcity in plenty. For a comprehensive understanding of diabetes, the relevant hormones, namely insulin and glucagon, homeostasis of blood glucose, besides the biochemical aspects of diabetes, are discussed in this chapter.

INSULIN

Insulin is a polypeptide hormone ***produced by the β-cells of islets of Langerhans of pancreas***. It has profound influence on the metabolism of carbohydrate, fat and protein. Insulin is considered as ***anabolic hormone***, as it promotes the synthesis of glycogen, triacylglycerols and proteins. This hormone has been implicated in the development of diabetes mellitus.

Structure of insulin

Human insulin (mol. wt. 5,734) contains 51 amino acids, arranged in two polypeptide chains. The chain A has 21 amino acids while B has 30 amino acids. Both are held together by two interchain disulfide bridges, connecting A_7 to B_7 and A_{20} to B_{19}. In addition, there is an intrachain disulfide link in chain A between the amino acids 6 and 11 (***Refer Fig.4.6***).

Metabolic effects of insulin

Insulin plays a key role in the regulation of carbohydrate, lipid and protein metabolisms (***Table 36.1***). Insulin exerts anabolic and anticatabolic influences on the body metabolism.

1. **Effects on carbohydrate metabolism :** Insulin ***lowers blood glucose level (hypoglycemic effect)*** by promoting its utilization and storage and by inhibiting its production.

- Effect on glucose uptake by tissues : Insulin is required for the uptake of glucose by muscle (skeletal, cardiac and smooth), adipose tissue, leukocytes and mammary glands. As regards liver, glucose entry into hepatocytes does not require insulin. However, insulin stimulates glucose utilization in liver and, thus, indirectly promotes its uptake.
- The effects of insulin on the important enzymes of the major pathways of glucose are given in ***Table 36.1***.

2. **Effects on lipid metabolism :** The net effect of insulin on lipid metabolism is to reduce the release of fatty acids from the stored fat and decrease the production of ketone bodies.

TABLE 36.1 Metabolic effects of insulin—a summary

Metabolism	Net effect	Effect on important enzymes
Carbohydrate metabolism		
1. Glycolysis	Increased	Glucokinase ↑ Phosphofructokinase ↑ Pyruvate kinase ↑
2. Gluconeogenesis	Decreased	Pyruvate carboxylase ↓ Phosphoenol pyruvate carboxykinase ↓ Glucose 6-phosphatase ↓
3. Glycogenesis	Increased	Glycogen synthatase ↑
4. Glycogenolysis	Decreased	Glycogen phosphorylase ↓
5. HMP shunt	Increased	Glucose 6-phosphate dehydrogenase ↑
Lipid metabolism		
6. Lipogenesis	Increased	Acetyl CoA carboxylase ↑
7. Lipolysis	Decreased	Hormone sensitive lipase ↓
8. Ketogenesis	Decreased	HMG CoA synthetase ↓
Protein metabolism		
9. Protein synthesis	Increased	RNA polymerase ↑
10. Protein degradation	Decreased	Transaminases ↓ Deaminases ↓

3. **Effects on protein metabolism :** Insulin is an anabolic hormone. It stimulates the entry of amino acids into the cells, enhances protein synthesis and reduces protein degradation.

Mechanism of action of insulin

It is now recognized that insulin binds to specific plasma membrane receptors present on the target tissues, such as muscle and adipose. This results in a series of reactions ultimately leading to the biological action. Three distinct mechanisms of insulin action are known. One concerned with the induction of transmembrane signals (signal transduction), second with the glucose transport across the membrane and the third with induction of enzyme synthesis.

GLUCAGON

Glucagon, ***secreted by α-cells*** of the pancreas, opposes the actions of insulin. It is a polypeptide hormone composed of 29 amino acids (mol. wt. 3,500) in a single chain. Glucagon is actually synthesized as proglucagon (mol. wt. 9,000) which on sequential degradation releases active glucagon. Unlike insulin, the ***amino acid sequence*** of glucagon is the ***same in all mammalian species*** (so far studied). Glucagon has a shorter half-life in plasma i.e. about 5 minutes.

Metabolic effects of glucagon

Glucagon influences carbohydrate, lipid and protein metabolisms. In general, the effects of this hormone ***oppose that of insulin.***

1. **Effects on carbohydrate metabolism :** Glucagon is the most potent hormone that enhances the blood glucose level (hyperglycemic). Primarily, glucagon acts on liver to cause increased synthesis of glucose (gluconeogenesis) and enhanced degradation of glycogen (glycogenolysis).

2. **Effects on lipid metabolism :** Glucagon promotes fatty acid oxidation resulting in energy production and ketone body synthesis (ketogenesis).

3. **Effects on protein metabolism :** Glucagon increases the amino acid uptake by liver which, in turn, promotes gluconeogenesis. Thus, glucagon lowers plasma amino acids.

REGULATION OF BLOOD GLUCOSE LEVEL (HOMEOSTASIS OF BLOOD GLUCOSE)

A healthy individual is capable of maintaining the blood glucose concentration within a narrow range. The ***fasting blood glucose*** level in a post-

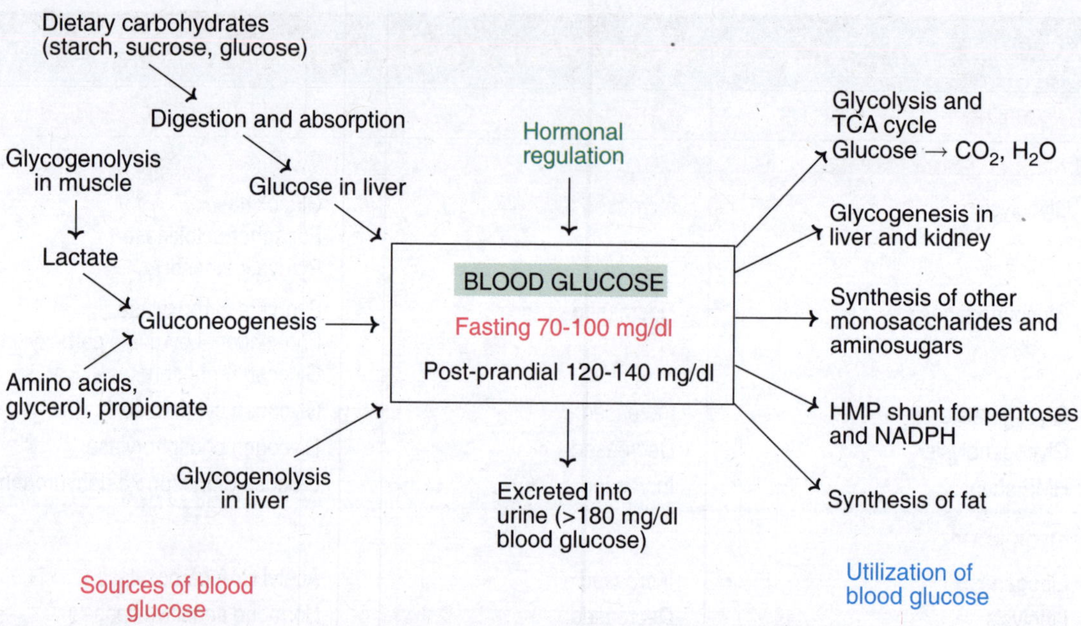

Fig.36.1 : *Overview of blood glucose homeostasis.*

absorptive state is ***70-100 mg/dl*** (plasma glucose 80-120 mg/dl). Following the ingestion of a carbohydrate meal, blood glucose may rise to 120-140 mg/dl. The fasting blood glucose value is comparatively lower in ruminant animals (sheep 30-40 mg/dl; cattle 50-60 mg/dl), while it is higher in birds (250-300 mg/dl).

The term ***hyperglycemia*** refers to an increase in the blood glucose above the normal level. ***Hypoglycemia*** represents a decreased blood glucose concentration. Excretion of glucose in urine is known as ***glycosuria.*** The concentration of blood glucose is dependent on the quantity of glucose that enters the circulation from various sources (dietary carbohydrates, glycogenolysis, gluconeogenesis etc.) and the amount that is utilized for different metabolic purposes (glycolysis, glycogenesis, fat synthesis etc.) as illustrated in ***Fig.36.1***.

Sources of blood glucose

1. **Dietary sources :** The dietary carbohydrates are digested and ***absorbed as monosaccharides*** (glucose, fructose, galactose etc.). The liver is capable of ***converting*** fructose and galactose ***into glucose***, which can readily enter blood.

2. **Gluconeogenesis :** The degradation of glycogen in muscle results in the formation of lactate. Breakdown of fat in adipose tissue will produce free glycerol and propionate. Lactate, glycerol, propionate and some amino acids are good precursors for ***glucose synthesis*** (gluconeogenesis) that actively occurs in liver and kidney. Gluconeogenesis continuously adds glucose to the blood.

3. **Glycogenolysis :** Degradation of glycogen in liver produces free glucose. This is in contrast to muscle glycogenolysis where glucose is not formed in sufficient amount due to lack of the enzyme glucose 6-phosphatase.

Utilization of blood glucose

Certain tissues like brain, erythrocytes, renal medulla and bone marrow are exclusively dependent on glucose for their energy needs. When the body is at total rest, about ***two-thirds of the blood glucose*** is utilized ***by the brain***. The remaining one-third by RBC and skeletal muscle.

The different metabolic pathways (glycolysis, glycogenesis, HMP shunt etc.) responsible for the utilization of blood glucose are already discussed (***Chapter 13***). The synthesis of fat from acetyl CoA and glycerol is described in lipid metabolism (***Chapter 14***). The organs such as liver, kidney, skeletal muscle, adipose tissue, brain etc. exert their influence on blood glucose level. This is closely related to the metabolic pathways of glucose existing in them.

Kidney plays a special role in the homeostasis of blood glucose. Glucose is continuously filtered by the glomeruli, reabsorbed and returned to the blood. If the level of glucose in blood is above

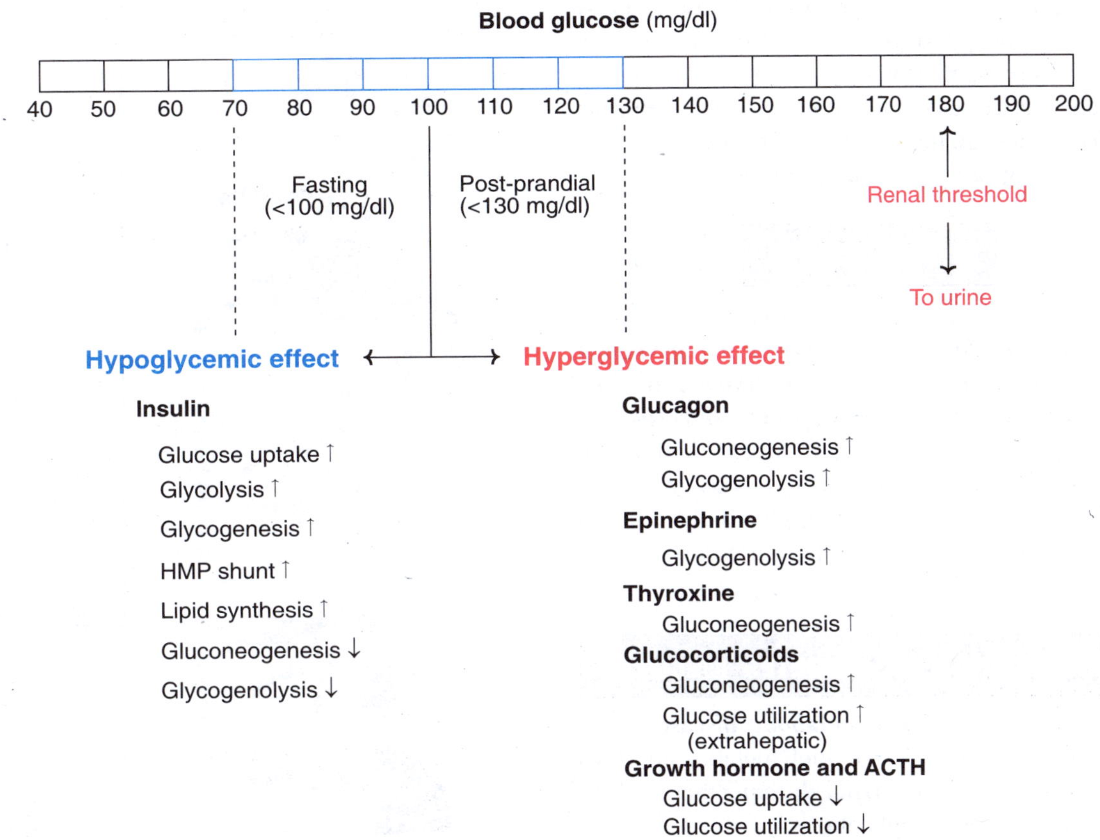

Fig.36.2 : Hormonal regulation of blood glucose.

160-180 mg/dl, glucose is excreted in urine (glycosuria). This value (160-180 mg/dl) is referred to as ***renal threshold for glucose.*** The maximum ability of the renal tubules to reabsorb glucose per minute is known as ***tubular maximum*** for glucose (TmG). The value for ***TmG is 350 mg/minute.***

Role of hormones in blood glucose homeostasis

Hormones play a significant role in the regulation of blood glucose concentration (***Figs. 36.2*** and ***36.3***). Primarily, ***insulin lowers blood glucose*** level (hypoglycemic) while the ***rest of the hormones oppose*** the actions of insulin (hyperglycemia).

Insulin : It is basically a hypoglycemic hormone that ***lowers in blood glucose level*** through various means. It is an anti-diabetogenic hormone. For details of insulin action on glucose homeostasis refer metabolic effects of insulin (carbohydrate metabolism) in this chapter.

Glucagon : It is involved in ***elevating blood glucose concentration***. It enhances gluconeogenesis and glycogenolysis. Glucagon increases liver glucose 6-phosphatase and phosphorylase activities so that more glucose is produced.

Epinephrine : This hormone is secreted by adrenal medulla. It acts both on muscle and liver to bring about glycogenolysis. The net outcome is that epinephrine ***increases blood glucose level***.

Thyroxine : It is a hormone of thyroid gland. It ***elevates blood glucose*** level by stimulating hepatic glycogenolysis and gluconeogenesis.

Glucocorticoids : These hormones stimulate protein metabolism and increase gluconeogenesis. The glucose utilization by extrahepatic tissues is inhibited by glucocorticoids. The overall effect of glucocorticoids is to ***elevate blood glucose concentration***.

Growth hormone and adrenocorticotropic hormone (ACTH) : The anterior pituitary gland secretes growth hormone and ACTH. The uptake of glucose by certain tissues (muscle, adipose tissue etc.) is decreased by growth hormone. ACTH decreases glucose utilization. The net effect of both these hormones is ***hyperglycemic***.

[In ***Fig.36.3***, regulation of blood glucose level by hormones is depicted as a game of tug of war with elephant (representing insulin) on one side and the other animals (as rest of the hormones) on the opposite side. This is just an illustration (a cartoon) for a quick understanding of glucose homeostasis].

HYPOGLYCEMIA

When the blood glucose concentration falls to ***less than 45 mg/dl***, the symptoms of hypoglycemia appear. The symptoms include ***headache***, ***anxiety***, ***confusion***, ***sweating***, ***slurred speech***, ***seizures*** and ***coma***, and, if not corrected, death. All these manifestations are directly and indirectly related to the deprivation of glucose supply to the central nervous system (particularly the brain) due to a fall in blood glucose level.

CLASSIFICATION OF DIABETES MELLITUS

Diabetes mallitus is ***a metabolic disease***, more appropriately a disorder of fuel metabolism. It is mainly characterized by ***hyperglycemia*** that leads to several long terms complications.

Diabetes mellitus is broadly divided into 2 groups, namely ***insulin-dependent diabetes mellitus*** (***IDDM***) and ***non-insulin dependent diabetes mellitus*** (***NIDDM***). This classification is mainly based on the requirement of insulin for treatment.

Insulin-dependent diabetes mellitus (IDDM)

IDDM, also known as ***type 1 diabetes*** or (less frequently) ***juvenile onset diabetes,*** mainly occurs in childhood (particularly between 12-15 yrs age). IDDM accounts for about 10 to 20% of the known diabetics. This disease is characterized by almost total deficiency of insulin due to destruction of β-cells of pancreas. The β-cell destruction may be caused by drugs, viruses or autoimmunity. The patients of IDDM require insulin therapy.

Non-insulin dependent diabetes mellitus (NIDDM)

NIDDM, also called ***type 2 diabetes*** or (less frequently) ***adult-onset diabetes,*** is the most common, accounting for 80 to 90% of the diabetic population. NIDDM occurs in adults (usually above 35 years) and is less severe than IDDM. The causative factors of NIDDM include genetic and environmental. NIDDM more commonly occurs in obese individuals.

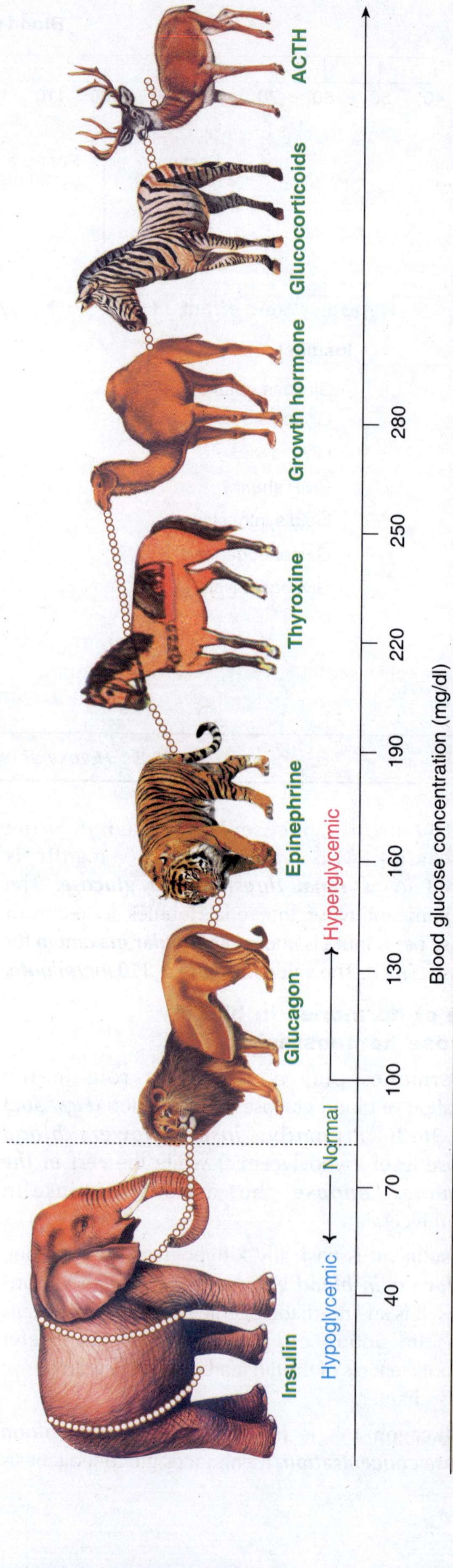

Fig.36.3 : A cartoon of tug of war illustrating hormonal action on blood glucose regulation.

Medical Concepts/ Clinical Correlates

- Diabetes affects about 6-8% of the population and is a major cause of blindness, renal failure, heart attack and stroke. The hormone insulin has been implicated in the development of diabetes.
- Hypoglycemia (blood glucose <45 mg/dl) is a serious condition that may lead to coma and death, if not corrected.
- GTT is commonly used for confirmation of diabetes mellitus.

Glucose tolerance test (GTT)

The final diagnosis of diabetes can be made on the basis of individual's response to the oral glucose load, commonly referred to as oral glucose tolerance test (OGTT).

Preparation of the subject for GTT

The person should have been taking ***carbohydrate-rich diet for at least 3 days prior to the test***. All drugs known to influence carbohydrate metabolism should be discontinued (for at least 2 days). The subject should avoid strenuous exercise on the previous day of the test. He/she should be in an ***overnight*** (at least 10 hr) ***fasting state***. During the course of GTT, the person should be comfortably seated and should refrain from smoking and exercise.

Procedure for GTT

Glucose tolerance test should be conducted ***preferably in the morning*** (ideal 9 to 11 AM). A fasting blood sample is drawn and urine collected. The subject is given 75 g glucose orally, dissolved in about 300 ml of water, to be drunk in about 5 minutes. Blood and urine samples are collected at 30 minute intervals for at least $2^1/_2$ hours. All blood samples are subjected to glucose estimation while urine samples are qualitatively tested for glucose.

Interpretation of GTT

The graphic representation of the GTT results is depicted in ***Fig.36.4***. The fasting plasma glucose level is 75–110 mg/dl in normal persons. On oral glucose load, the concentration increases and the peak value (140 mg/dl) is reached in less than an hour which returns to normal by 2 hours. Glucose is not detected in any of the urine samples.

In individuals with ***impaired glucose tolerance***, the fasting (110-125 mg/dl) as well as 2 hour (140-199 mg/dl) plasma glucose levels are elevated. These subjects slowly develop frank diabetes at an estimated rate of 2% per year. Dietary restriction and exercise are advocated for the treatment of impaired glucose tolerance.

The WHO criteria for the diagnosis of diabetes by OGTT is presented in ***Table 36.2***. A person is said to be suffering from diabetes mellitus if his/her ***fasting plasma glucose exceeds*** 7.0 mmol/l (***126 mg/dl***) and, ***at 2 hrs.*** 11.1 mmol/l (***200 mg/dl***).

Other relevant aspects of GTT

1. For conducting GTT in children, oral glucose is given on the basis of weight (1.5 to 1.75 g/kg).

2. In case of pregnant women, 100 g oral glucose is recommended. Further, the diagnostic criteria for diabetes in pregnancy should be more stringent than WHO recommendations.

3. In the ***mini GTT*** carried out in some laboratories, fasting and 2 hrs. sample (instead of $^1/_2$ hr. intervals) of blood and urine are collected.

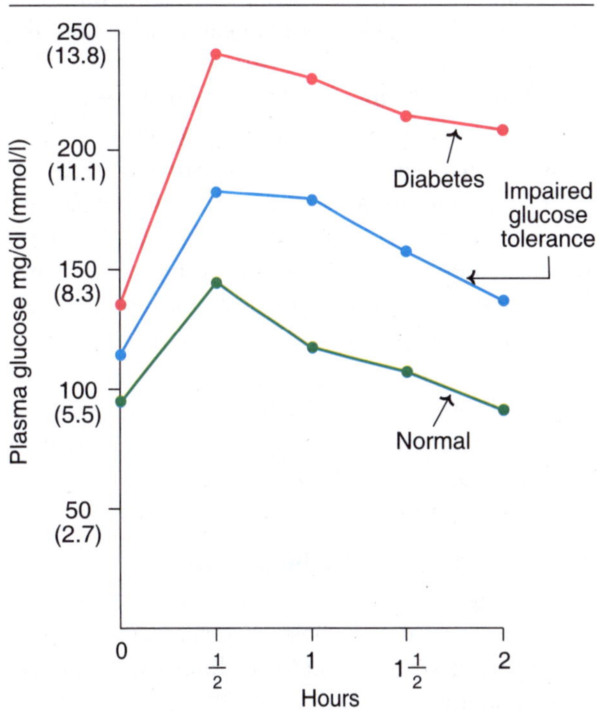

Fig.36.4 : *Oral glucose tolerance test.*

TABLE 36.2 Diagnostic criteria for oral glucose tolerance test (WHO criteria)

Condition	Plasma glucose concentration as mmol/l (mg/dl)		
	Normal	Impaired glucose tolerance	Diabetes
Fasting	< 6.1 (< 110)	< 7.0 (< 126)	≥ 7.0 (≥ 126)
2 hours after glucose	< 7.8 (< 140)	< 11.1 (< 200)	≥ 11.1 (≥ 200)

4. The GTT is rather unphysiological. To evaluate the glucose handling of the body under physiological conditions, fasting blood sample is drawn, the subject is allowed to take ***heavy breakfast***, blood samples are collected at ***1 hour and 2 hrs*** (***post-prandial***—meaning after food). Urine samples are also collected. This type of test is commonly employed in established diabetic patients for monitoring the control.

5. In individuals with suspected malabsorption, ***intravenous GTT is carried out.***

6. **Corticosteroid stressed GTT** is employed to detect latent diabetes.

Glycosuria

The commonest cause of ***glucose excretion in urine*** (glycosuria) is diabetes mellitus. Therefore, glycosuria is the first line screening test for diabetes. Normally, glucose does not appear in urine until the plasma glucose concentration exceeds ***renal threshold (180 mg/dl)***. As age advances, renal threshold for glucose increases marginally.

Renal glycosuria : Renal glycosuria is a benign condition due to a reduced renal threshold for glucose. It is unrelated to diabetes and, therefore, should not be mistaken as diabetes. Further, it is not accompanied by the classical symptoms of diabetes.

Alimentary glycosuria : In certain individuals, blood glucose level rises rapidly after meals resulting in its spill over into urine. This condition is referred to as ***alimentary (lag storage) glycosuria.*** It is observed in some normal people and in patients of hepatic diseases, hyperthyroidism and peptic ulcer.

Metabolic changes in diabetes

Diabetes mellitus is associated with several metabolic alterations. Most important among them are hyperglycemia, ketoacidosis and hypertriglyceridemia.

1. **Hyperglycemia :** Elevation of blood glucose concentration is the hallmark of uncontrolled diabetes.

Glucose toxicity : High concentrations of glucose can be harmful causing ***osmotic effects/***hypertonic effects (water drawn from cells into extracellular fluid and excreted into urine, resulting in dehydration), ***β-cell damage by free radicals*** (due to enhanced oxidative phosphorylation, oxidative stress, and increased free radicals) and ***glycation of proteins*** (associated with diabetic complications-neuropathy, nephropathy, retinopathy etc.).

2. **Ketoacidosis :** Increased mobilization of fatty acids results in overproduction of ketone bodies which often leads to ketoacidosis.

3. **Hypertriglyceridemia :** Conversion of fatty acids to triacylglycerols and the secretion of VLDL and chylomicrons is comparatively higher in diabetics. Further, the activity of the enzyme lipoprotein lipase is low in diabetic patients. Consequently, the plasma levels of ***VLDL, chylomicrons*** and triacylglycerols are ***increased. Hypercholesterolemia*** is also frequently seen in diabetics.

Long term effects of diabetes

Hyperglycemia is directly or indirectly associated with several complications. These include ***atherosclerosis, retinopathy, nephropathy*** and ***neuropathy***. ***Diabetic retinopathy*** causes loss of transparency of the eye lens, resulting in impairment of vision and blindness. ***Diabetic neuropathy*** primarily affects and damages peripheral nervous system. ***Diabetic nephropathy*** progressively damages kidney and results in renal failure.

Management of diabetes

Diet, exercise, drug and, finally, insulin are the management options in diabetics. Approximately, 50% of the new cases of diabetes can be adequately controlled by diet alone, 20-30% need oral antidiabetic drugs while the remaining 20-30% require insulin.

Medical Concepts/ Clinical Correlates

- **Diabetic ketoacidosis is frequently encountered in severe uncontrolled diabetics. The management includes administration of insulin, fluids and potassium.**
- **The hypoglycemic drugs commonly used in diabetic patients include tolbutamide, glibenclamide and acetohexamide.**
- **The long term complications of uncontrolled diabetes include atherosclerosis, retinopathy, nephropathy and neuropathy.**
- **Measurement of glycated hemoglobin (HbA_{1c}) serves as a marker for diabetic control.**

Dietary management : A diabetic patient is advised to consume ***low calories*** (i.e. low carbohydrate and fat), ***high protein and fiber rich diet***. Carbohydrates should be taken in the form of starches and complex sugars. As far as possible, ***refined sugars*** (***sucrose***, ***glucose***) ***should be avoided***. Fat intake should be drastically reduced so as to meet the nutritional requirements of unsaturated fatty acids. ***Diet control and exercise*** will help to a large extent obese NIDDM patients.

Antidiabetic drugs : The oral hypoglycemic drugs are broadly of two categories-***sulfonylureas*** and ***biguanides***. They are commonly used these days.

Sulfonylureas such as acetohexamide, ***tolbutamide and gibenclamide*** are frequently used. They promote the secretion of endogenous insulin and thus help in reducing blood glucose level.

Management with insulin : Two types of insulin preparations are commercially available—***short acting*** and ***long acting.*** The short acting insulins are unmodified and their action lasts for about 6 hours. The long acting insulins are modified ones (such as adsorption to protamine) and act for several hours, which depends on the type of preparation.

The advent of genetic engineering is a boon to diabetic patients since bulk quantities of insulin can be produced in the laboratory.

Biochemical indices of diabetic control

For a diabetic patient who is on treatment (drug or insulin therapy), ***periodical assessment*** of the efficacy of the ***treatment is essential***. Urine glucose detection and blood glucose estimations are traditionally followed in several laboratories. In recent years, more reliable and long-term biochemical indices of diabetic control are in use.

Glycated hemoglobin : Glycated or ***glycosylated hemoglobin*** refers to the glucose derived products of normal adult hemoglobin (HbA). Glycation is a post-translational, non-enzymatic addition of sugar residue to amino acids of proteins. Among the glycated hemoglobins, the most abundant form is ***HbA_{1C}***.

Diagnostic importance of HbA_{1C} : The rate of synthesis of HbA_{1C} is directly related to the exposure of RBC to glucose. Thus, the concentration of HbA_{1C} serves as an indication of the blood glucose concentration over a period, approximating to the half-life of RBC (hemoglobin) i.e. 6–8 weeks. A close correlation between the blood glucose and HbA_{1C} concentrations have been observed when simultaneously monitored for several months.

Normally, ***HbA_{1c} concentration is about 4–5.6% of the total hemoglobin***. In diabetic patients, HbA_{1c} is elevated (to as high as 15%). Determination of HbA_{1c} is used for monitoring of diabetes control. HbA_{1c} reflects the mean ***blood glucose level over 3 months period prior to its measurement***.

Microalbuminuria : Microalbuminuria is defined as the excretion of ***30-300 mg of albumin in urine per day***. It may be noted that microalbuminuria represents an intermediary stage between normal albumin excretion (2.5–30 mg/d) and macroalbuminuria (> 300 mg/d). The small increase in albumin excretion predicts impairment in renal function in diabetic patients. Microalbuminuria serves as a signal of early reversible renal damage.

THE READER MUST REFER APPENDIX D (Page Nos : 422 - 431) FOR RELATED CLINICAL CASE STUDIES (Nos : 30, 31, 32, 36, 40)

CANCER

37

In the normal circumstances, the proliferation of body cells is under strict control. The cells differentiate, divide and die in a sequential manner in a healthy organism. Cancer is characterized by ***loss of control of cellular growth*** and development leading to ***excessive proliferation and spread of cells***.

About 100 different types of human cancers have been recognized. Cancers arising from ***epithelial cells*** are referred to as ***carcinomas*** while that from ***connective tissues*** are known as ***sarcomas***. Methods for the early detection and treatment of cancers have been developed. However, little is known about the biochemical basis of cancer.

Incidence

Cancer is the ***second largest killer disease*** (the first being coronary heart disease) in the developed countries. It is estimated that cancer accounts for more than 20% of the deaths in United States. Based on the current rate of incidence, it is believed that one in every 3 persons will develop cancer at sometime during his/her life.

ETIOLOGY

In general, cancers are multifactorial in origin. The causative agents include ***physical***, ***chemical***, ***genetic*** and ***environmental factors***. A survey in USA has shown that about 90% of all cancer deaths are due to avoidable factors such as tobacco, pollution, occupation, alcohol and diet.

Most of the cancers are caused by chemical carcinogens, radiation energy and viruses. These agents may damage DNA or interfere with its replication or repair.

Chemical carcinogens

It is estimated that ***almost 80% of the human cancers are caused by chemical carcinogens*** in nature. The chemicals may be organic (e.g. dimethylbenzanthracene, benzo (a) pyrene, dimethyl nitrosamine) or inorganic (arsenic, cadmium) in nature.

In general, a chemically non-reactive pro-carcinogen is converted to an ultimate carcinogen by a series of reactions.

The carcinogens can covalently bind to purines, pyrimidines and phosphodiester bonds of DNA, often causing unrepairable damage. The chemical carcinogens frequently ***cause mutations*** (a change in the nucleotide sequence of DNA) which may finally lead to the development of cancer, hence they are regarded as mutagens.

Ames assay : This is a ***laboratory test to check the carcinogenicity of chemicals***. Ames assay employs the use of a special mutant strain of bacterium, namely *Salmonella typhimurium* (His^-).

Radiation energy

Ultraviolet rays, X-rays and γ-rays have been proved to be mutagenic in nature causing cancers. These rays ***damage DNA*** which is the ***basic mechanism to explain the carcinogenicity*** of radiation energy. For instance, exposure to UV rays results in the formation of pyrimidine dimers in DNA while X-rays cause the production of free radicals. This type of molecular damages are responsible for the carcinogenic effects of radiations.

Carcinogenic viruses

The presence of viral particles and the enzyme reverse transcriptase, besides the occurrence of base sequence in the DNA of malignant cells, complementary to tumor viruses indicate the involvement of viruses in cancer. The viruses involved in the development of cancer, commonly known as ***oncogenic viruses***, may contain either DNA or RNA.

DNA—the ultimate in carcinogenesis

DNA is the ultimate critical macromolecule in carcinogenesis. This fact is supported by several evidences.

1. Cancers are transmitted from mother to daughter cells. In other words, cancer cells beget cancer cells.

2. Chromosomal abnormalities are observed in many tumor cells.

3. Damage to DNA caused by mutations often results in carcinogenesis.

MOLECULAR BASIS OF CANCER

Cancer is caused by a genetic change in a single cell resulting in its uncontrolled multiplication. Thus, ***tumors are monoclonal***. Two types of regulatory genes—***oncogenes*** and ***antioncogenes*** are involved in the development of cancer (carcinogenesis). In recent years, a third category of genes that control the cell death or ***apoptosis*** are also believed to be involved in carcinogenesis.

Oncogenes

The ***genes capable of causing cancer*** are known as oncogenes (*Greek* : oncos—tumor or mass). Oncogenes were originally discovered in tumor causing viruses. These viral oncogenes were found to be closely similar to certain genes present in the normal host cells which are referred to as proto-oncogenes. Now, about 40 viral and cellular proto-oncogenes have been identified. Protooncogenes encode for growth-regulating proteins. The ***activation of protooncogenes to oncogenes*** is an important step in the causation of cancer.

Activation of protooncogenes to oncogenes

There are several mechanisms for converting the protooncogenes to oncogenes, some of the important ones are described hereunder.

1. **Viral insertion into chromosome :** When certain retroviruses (genetic material RNA) infect cells, a complementary DNA (***cDNA***) is made from their RNA by the enzyme ***reverse transcriptase***. The cDNA so produced gets inserted into the host genome (***Fig.37.1***). The integrated double-stranded cDNA is referred to as provirus. This pro-viral DNA takes over the control of the transcription of cellular chromosomal DNA and transforms the cells. Activation of protooncogene *myc* to oncogene by viral insertion ultimately causing carcinogenesis is well known (e.g. avian leukemia).

Some DNA viruses also get inserted into the host chromosome and activate the protooncogenes.

2. **Chromosomal translocation :** Some of the tumors exhibit chromosomal abnormalities. This is due to the rearrangement of genetic material (DNA) by chromosomal translocation i.e. splitting off a small fragment of chromosome which is joined to another chromosome. Chromosomal translocation usually results in ***overexpression of protooncogenes***.

3. **Gene amplification :** Severalfold amplifications of certain DNA sequences are observed in some cancers. Administration of anticancer drugs methotrexate (an inhibitor of the enzyme dihydrofolate reductase) is associated with gene amplification. The drug becomes inactive due to gene amplification resulting in a severalfold (about 400) increase in the activity of dihydrofolate reductase.

4. **Point mutation :** The ***ras protooncogene*** is the best ***example*** of activation by point mutation (change in a single base in the DNA). The mutated *ras* oncogene produces a protein (GTPase) which differs in structure by a single amino acid. This alteration diminishes the activity of GTPase, a key enzyme involved in the control of cell growth.

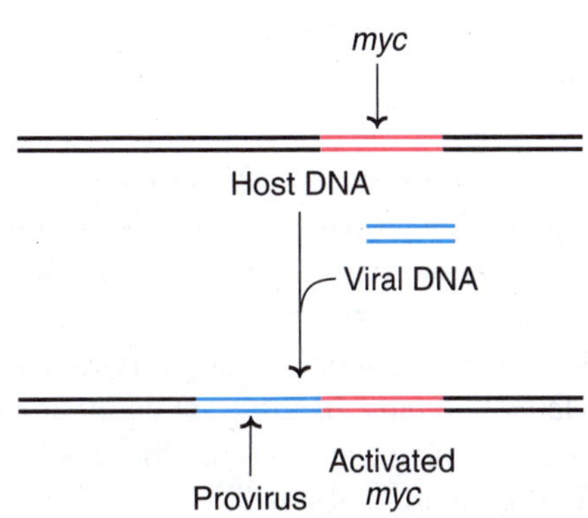

Fig.37.1 : *Integration of viral DNA into host DNA.*

TABLE 37.1 Selected polypeptide growth factors

Growth factor	Source(s)	Major function(s)
Epidermal growth factor (EGF)	Salivary gland, fibroblasts	Stimulates growth of epidermal and epithelial cells
Platelet derived growth factor (PDGF)	Platelets	Stimulates growth of mesenchymal cells, promotes wound healing
Transforming growth factor (TGF-α)	Epithelial cell	Similar to EGF
Transforming growth factor (TGF-β)	Platelets, kidney, placenta	Inhibitory (sometimes stimulatory) effect on cultured tumor cells
Erythropoietin	Kidney	Stimulates development erythropoietic cells
Nerve growth factor (NGF)	Salivary gland	Stimulates the growth of sensory and sympathetic neurons
Insulin like growth factors (IGF-I and IGF-II, respectively known as somatomedins C and A)	Serum	Stimulates incorporation of sulfates into cartilage; exerts insulinlike action on certain cells
Tumor necrosis factor (TNF-α)	Monocytes	Necrosis of tumor cells
Interleukin-1 (IL-1)	Monocytes, leukocytes	Stimulates synthesis of IL-2.
Interleukin-2 (IL-2)	Lymphocytes (mainly T-helper cells).	Stimulates growth and maturation of T-cells

Mechanism of action of oncogenes

Oncogenes encode for certain proteins, namely ***oncoproteins***. These proteins are the altered versions of their normal counterparts and are involved in the transformation and multiplication of cells. Some of the products of oncogenes are discussed below.

Growth factors : Several growth factors stimulating the proliferation of normal cells are known. They regulate cell division by transmitting the message across the plasma membrane to the interior of the cell (transmembrane signal transduction). It is believed that growth factors play a key role in carcinogenesis.

A selected list of polypeptide growth factors, their sources and major functions are given in ***Table 37.1***.

The ***cell proliferation*** is ***stimulated by growth factors***. In general, a growth factor binds to a protein receptor on the plasma membrane. This binding activates cytoplasmic protein kinases leading to the phosphorylation of intracellular target proteins. The phosphorylated proteins, in turn, act as intracellular messengers to stimulate cell division, the mechanism of which is not clearly known.

Transforming growth factor (TGF-α) is a protein synthesized and required for the growth of epithelial cells. ***TGF-α is produced in high concentration in individuals suffering from psoriasis***, a disease charcterized by excessive ***proliferation of epidermal cells***.

Antioncogenes

A special category of genes, namely ***cancer suppressor genes*** or, more commonly, antioncogenes, have been identified. The products of these genes apply breaks and regulate cell proliferation. The loss of these suppressor genes removes the growth control of cells and is believed to be a key factor in the development of several tumors, e.g. retinoblastoma, one type of breast cancer, carcinoma of lung, Wilms' kidney tumor.

Genes that regulate apoptosis

A new category of genes that regulate ***programmed cell death*** (apoptosis) have been discovered. These genes are also important in the development of tumors.

Unified hypothesis of carcinogenesis

The multifactorial origin of cancer can be suitably explained by oncogenes. The physical and chemical agents, viruses and mutations all lead to the activation of oncogenes causing carcinogenesis.

Medical Concepts/ Clinical Correlates

- About 80% of the human cancers are caused by chemical carcinogens.
- The physical and chemical agents, viruses and mutations result in the activation of oncogenes causing carcinogenesis.
- The abnormal products of tumor cells, referred to as tumor markers (CEA, AFP, PSA) are useful for the diagnosis and prognosis of cancer.
- Anticancer drugs (e.g. methotrexate, cisplatin) are commonly used in the treatment of cancer. Antioxidants (vitamins E and C, β-carotene, Se) decrease the risk of carcinogenesis and hence their increased consumption is advocated.

The antioncogenes and the genes regulating apoptosis are intimately involved in development of cancer. A simplification of a unified hypothesis of carcinogenesis is depicted in ***Fig.37.2***.

Tumor markers

The biochemical indicators employed to ***detect the presence of cancers*** are collectively referred to as tumor markers. These are the ***abnormally produced molecules of tumor cells such as surface antigens,*** cytoplasmic proteins, enzymes and hormones. Tumor markers can be measured in serum (or plasma). In theory, the tumor markers must ideally be useful for screening the population to detect cancers. In practice, however, this has not been totally true. As such, the ***tumor markers support the diagnosis of cancers, besides being useful for monitoring the response to therapy and for the early detection of recurrence***.

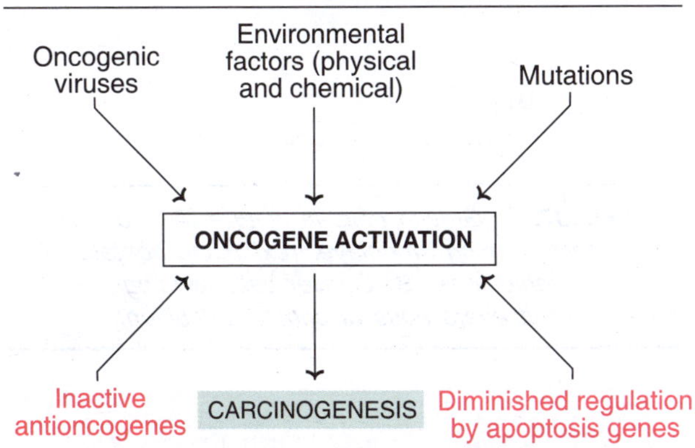

Fig.37.2 *: A simplified hypothesis for the development of cancer.*

A host of tumor markers have been described and the list is evergrowing. However, only a few of them have proved to be clinically useful. A selected list of tumor markers and the associated cancers are given in ***Table 37.3***.

A couple of the most commonly used tumor markers are discussed hereunder.

1. **Carcinoembryonic antigen** (CEA) **:** This is a complex glycoprotein, normally produced by the embryonic tissue of liver, gut and pancreas. The presence of CEA in serum is detected in several cancers (***colon, pancreas, stomach, lung***). In about 67% of the patients with colorectal cancer, CEA can be identified.

2. **Alpha-fetoprotein** (AFP) **:** It is chemically a glycoprotein, normally synthesized by yolk sac in early fetal life. Elevation in serum levels of AFP mainly ***indicates the cancers of liver and germ cells of testis*** and, to some extent, carcinomas of lung, pancreas and colon.

Characteristics of growing tumor cells

The morphological and biochemical changes in the growing tumor cells are briefly described here. These observations are mostly based on the *in vitro* culture studies.

1. **General and morphological changes**
   (a) **Shape of cells :** The tumor cells are much ***rounder*** in shape compared to normal cells.
   (b) **Loss of contact inhibition :** The normal cells are characterized by contact inhibition i.e. they form monolayers. Further, they cannot move away from each other. The cancer cells form multilayers due to ***loss of contact inhibition*** (***Fig.37.3***). As a result, the cancer cells freely move and get deposited in any part of the body, a property referred to as ***metastasis***.
2. **Biochemical changes**
   (a) **Increased replication and transcription :** The synthesis of DNA and RNA is increased in cancer cells, indicating an increase in anabolic processes.

TABLE 37.3 Selected tumor markers and associated cancers

Tumor marker	*Associated cancer(s)*
Oncofetal antigens	
Carcinoembryonic antigen (CEA)	Cancers of colon, stomach, lung, pancreas and breast
Alpha fetoprotein (AFP)	Cancer of liver and germ cells of testis
Cancer antigen-125 (CA-125)	Ovarian cancer
Hormones	
Human chorionic gonadotropin (hCG)	Choriocarcinoma
Calcitonin	Carcinoma of medullary thyroid
Catecholamines and their metabolites (mainly vanillyl mandelic acid)	Pheochromocytoma and neuroblastoma
Enzymes	
Prostatic acid phosphatase	Prostate cancer
Neuron specific enolase	Neuroblastoma
Specific proteins	
Prostate specific antigen (PSA)	Prostate cancer
Immunoglobulins	Multiple myeloma

(b) **Increased glycolysis :** The fast growing tumor cells are characterized by ***elevation in aerobic and anaerobic glycolysis***. This truly reflects the increased energy demands of multiplying cells.

(c) **Reduced requirement of growth factors :** The tumor cells require much less quantities of growth factors. Despite this fact, there is an increased production of growth factors by these cells.

(d) **Synthesis of fetal proteins :** During fetal life, certain genes are active, leading to the synthesis of specific proteins. These genes are suppressed in adult cells. However, the tumor cells synthesize the fetal proteins e.g. carcinoembryonic antigen, alpha fetoprotein.

(e) **Alterations in antigens :** A loss of regularly occurring antigens coupled with the appearance of new antigens in tumor cells is reported.

Metastasis

Metastasis refers to the ***spread of cancer cells from the primary site of origin to other tissues*** of the body where they get deposited and grow as secondary tumors. Metastasis is the major cause of cancer related morbidity and mortality. The biochemical basis of metastasis is not clearly known. It is believed that the morphological changes in tumor cells, loss of contact inhibition, loss of anchorage dependence and alterations in the structure of certain macromolecules are among the important factors responsible for metastasis.

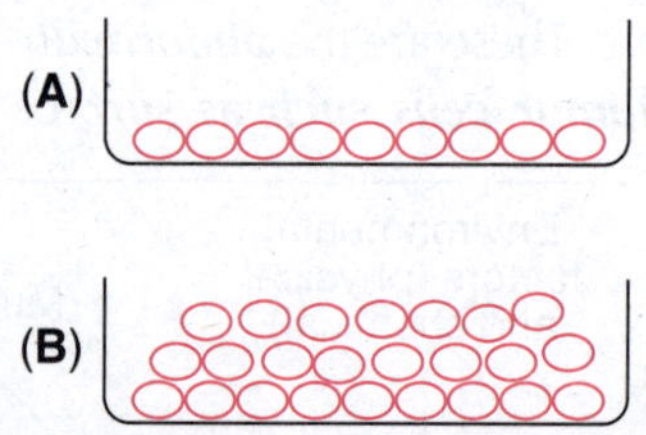

***Fig.37.3 :** Growth cells in culture (**A**) Normal cells forming monolayer (exhibiting contact inhibition); (**B**) Cancer cells forming multilayers (loss of contact inhibition).*

ACQUIRED IMMUNODEFICIENCY SYNDROME (AIDS)

38

Acquired immunodeficiency syndrome (AIDS) was first reported in 1981 in homosexual men. AIDS is a retroviral disease caused by human immunodeficiency virus (HIV). The disease is ***characterized by immunosuppression***, ***secondary neoplasma*** and ***neurological manifestations***. AIDS is invariably fatal since there is no cure. In the USA, it is the fourth leading cause of death in men between the ages 15 to 55 years.

Epidemiology

AIDS was first described in USA and this country has the majority of reported cases. The prevalence of AIDS has been reported from almost every country. The number of people living with HIV worldwide is estimated to be around ***40 million in the year 2016***. (India alone has about 5 million persons). At least 5 million deaths occurred in 2016, due to AIDS. AIDS is truely a global disease with an alarming increase in almost every country.

Transmission of HIV : Transmission of AIDS essentially requires the ***exchange of body fluids*** (semen, vaginal secretions, blood, milk) containing the virus or virus-infected cells. There are three major routes of HIV transmission—sexual contact, parenteral inoculation and from infected mothers to their new borns.

The distribution of risk factors for AIDS transmission are as follows.

Sex between men (homosexuals)	– 60%
Sex between men and women	– 15%
Intravenous drug abusers	– 15%
Transfusion of blood and blood products	– 6%
All others	– 4%

The predominant methods of HIV transmission (about 75%) are through anal or vaginal intercourse. The risk for the transmission is much higher with anal than with vaginal intercourse. The practice of 'needle sharing' is mainly responsible for the transmission of HIV in drug abusers. Pediatric AIDS is mostly caused by vertical transmission (mother to infant).

It should, however, be noted that HIV cannot be transmitted by casual personal contact in the household or work place. Further, the transmission of AIDS from an infected individual to health personnel attending on him is extremely rare.

Virology of HIV

AIDS is caused by a retrovirus, namely ***human immunodeficiency virus*** (HIV), belonging to lentivirus family. Retroviruses contain RNA as the genetic material. On entry into the host cell, they transcribe DNA which is a complementary copy of RNA. The DNA, in turn is used, as a template to produce new viral RNA copies.

Two different forms of HIV, namely HIV-1 and HIV-2 have been isolated from AIDS patients. HIV-1 is more common, being found in AIDS patients of USA, Canada, Europe and Central Africa while HIV-2 is mainly found in West Africa. Both the viruses are almost similar except that they differ in certain immunological properties.

HIV-1 is described in some detail.

Structure of HIV : The virus is spherical with a diameter of about 110 nm. It contains a core, surrounded by a lipid envelop derived from the host

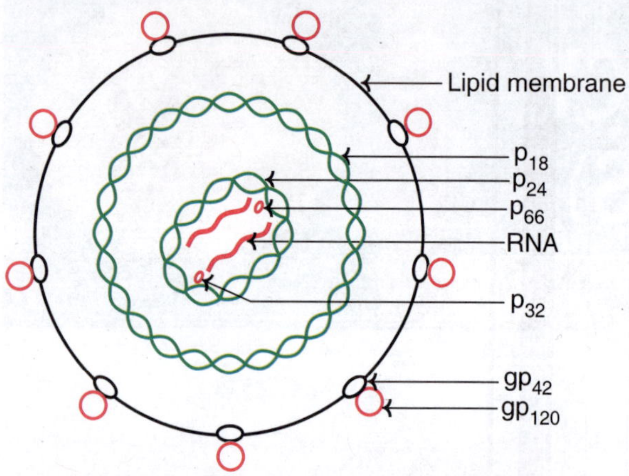

Fig.38.1 : *Diagrammatic representation of HIV (p represents protein with molecular weight e.g.* p_{18} *is with molecular weight 18,000; gp represents glycoprotein).*

plasma membrane (***Fig.38.1***). The core of the HIV has two strands of genomic RNA and four core proteins, p_{24}, p_{18}, reverse transcriptase (p_{66}/p_{51}) and endonuclease (p_{32}). Note that the naming of the proteins is based on the molecular weight. For instance, a protein with a molecular weight of 24,000 is designated as p_{24}.

The lipid membrane of the virus is studded with two glycoproteins gp_{120} and gp_{41}. The surface antigen gp_{120} is very important for the viral infection and the detection of AIDS.

Immunological abnormalities in AIDS

As is evident from the name AIDS, immunodeficiency (or immunosuppression) is the hallmark of this disease. AIDS ***primarily affects*** the ***cell-mediated immune system*** which protects the body from intracellular parasites such as viruses, protozoa and mycobacteria. This is caused by a reduction in CD_4 (cluster determinant antigen 4) cells of T-lymphocytes, besides impairment in the functions of surviving CD_4 cells.

CD_4 cells may be regarded as master cells of cell mediated immunity. They produce cytokines, macrophage chemotactic factors, hemopoietic growth factors, and others involved in the body immunity.

Entry of HIV and lysis of CD_4 cells : The virus enters the CD_4 T-lymphocytes. HIV binds to the specific receptors on CD_4 cells by using its surface membrane glycoprotein (gp_{120}). Following the entry into the host cells, RNA of HIV is transcribed into DNA by the viral enzyme reverse transcriptase. The viral DNA gets incorporated into the host genomic DNA. The virus may remain locked in the host genome for months or years and this is considered as the latent period. The viral DNA may undergo replication and translation, respectively, producing viral RNA and viral proteins. The latter two,. on assembly, result in new viruses. The newly synthesized viruses leave the host cells by forming buds on plasma membrane. Extensive viral budding is associated with lysis and death of CD_4 cells (***Fig.38.2***). The new viral particles infect other host cells and repeat the whole process, ultimately resulting in a profound loss of CD_4 cells from the blood. ***Most of the immunodeficiency symptoms of AIDS are associated with the reduction in CD_4 cells.***

Other immunological abnormalities

The viral membrane protein gp_{120} binds with normal T-helper cells and kills them. AIDS patients also display abnormalities in antibody production by B-lymphocytes (humoral immunity).

Abnormalities of central nervous system : HIV also infects the cells of central nervous system. It is believed that HIV infected monocytes enter the brain and cause damage, the mechanism of which remains obscure.

Consequences of immunodeficiency : The various clinical symptoms (fever, diarrhea, weight loss, neurological complications, multiple opportunistic infections, generalized lymphadenopathy, secondary neoplasma etc.) of AIDS are directly or indirectly related to the immunosuppression caused by HIV. Due to the deficiency in the immune system, the body of AIDS patient is freely exposed to all sorts of infections (viral, bacterial, fungal).

Natural course of AIDS

Three distinct phases of HIV interaction with the immune system of infected body have been identified. These are the early, acute phase; the intermediate, chronic phase; the final, crisis phase (***Fig.38.3***).

1. **Acute phase :** This represents the initial body response to HIV infection. It is characterized by high rate of virus production which are lodged in the lymphoid tissues and the antiviral immune response of the body. This period may last for about 8-12 weeks.

Medical Concepts/ Clinical Correlates

- AIDS is a global disease with an alarming increase in the incidence of occurrence. By the year 2014, more than 40 million AIDS deaths occurred globally.
- Homosexuality (predominantly in men) and intravenous drug abuse are the major factors in the risk of AIDS transmission.
- The patients of AIDS are destined to die (within 5–10 years after infection), since there is no cure. However, administration of certain drugs (AZT, DDI) prolongs life.
- The clinical manifestations of AIDS are directly or indirectly related to immunosuppression (mostly due to reduced CD_4 cells). AIDS patients are freely exposed to all sorts of infections (viral, bacterial, fungal).

2. **Chronic phase :** During this period that may last for 5 to 10 years or even more, the body tries to contain the virus. The immune system is largely intact. The person obviously appears normal, although he/she is the carrier of HIV which can be transmitted to others. Antibodies to HIV are found in the circulation, hence this phase is also referred to as ***seropositive period.***

3. **Crisis phase :** A failure in the defense system of the body, caused by immunosuppression by HIV, represents the crisis phase. The plasma level of virus is tremendously increased. CD_4 T-lymphocyte concentration drastically falls. A patient with lower than 200 CD_4 T-lymphocytes/μl blood is considered to have developed AIDS. Crisis phase is characterized by opportunistic infections and the related clinical manifestations. In Western countries, a cancer—Kaposi sarcoma—is associated with AIDS.

In general, AIDS patients die between 5-10 years after HIV infection. Treatment may, however, prolong the life.

Laboratory diagnosis of AIDS

The following laboratory tests are employed to diagnose the HIV infection.

1. The detection of antibodies in the circulation by ELISA (enzyme-linked immunosorbant assay).

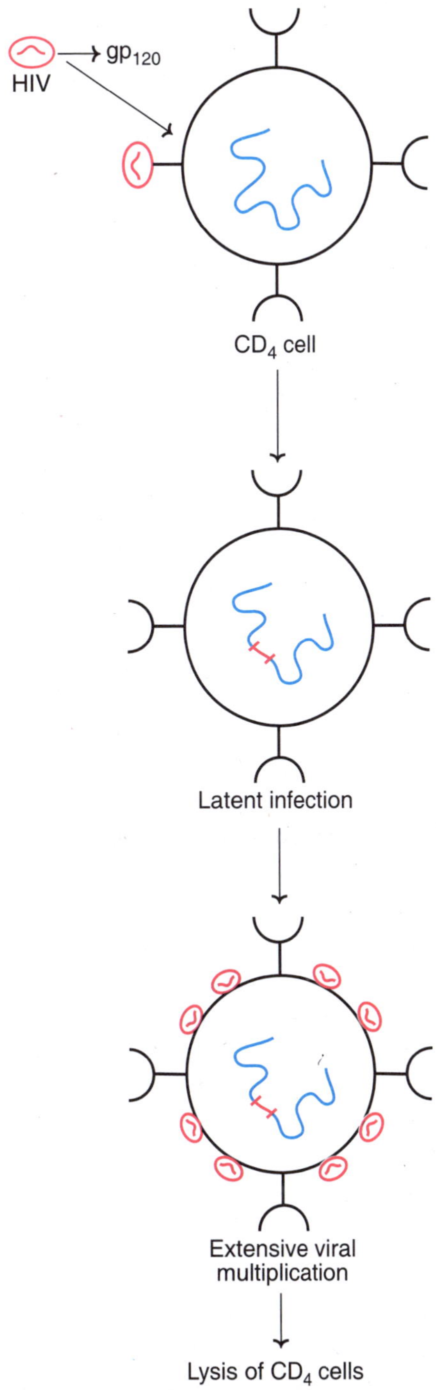

Fig.38.2 : Immunological abnormalities in CD_4 cells on HIV infection.

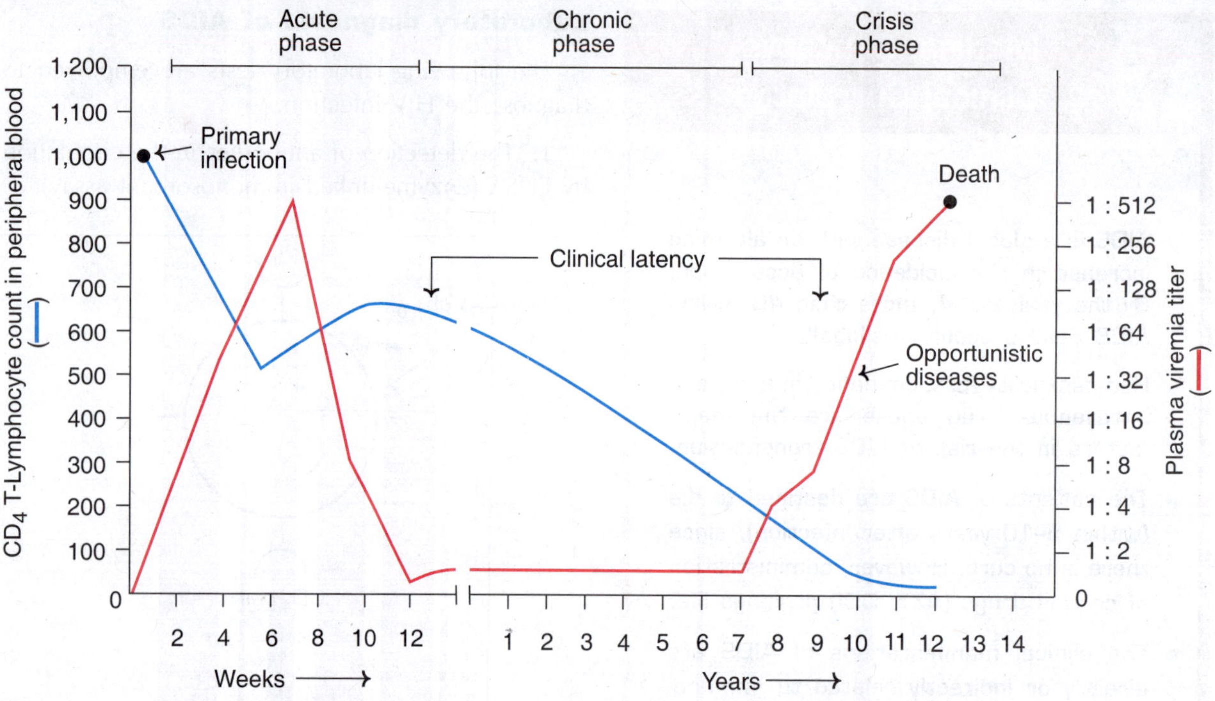

Fig.38.3 : *Graphic representation of a typical course of HIV infection.*

2. Western blot technique, a more specific test for the HIV antibodies, is employed for confirmation of ELISA positive cases.

3. A more recent and sophisticated PCR can be used to detect the presence of the HIV genome in the peripheral blood lymphocytes.

Drugs for the treatment of AIDS

Although there is no cure for AIDS, use of certain drugs can prolong the life of AIDS patients. ***Zidovudine*** or ***AZT*** (3′-azido 2′,3′-dideoxy thymidine), a structural analog of deoxythymidine was the first drug used and continues to be the drug of choice for the treatment of AIDS. ***Didanosine*** (dideoxyionosine, DDI) is another drug employed to treat AIDS.

Vaccine against AIDS—a failure so far

HIV virus exhibits genetic heterogenecity with a result that several species of virus may be found in the same AIDS patient. The principal cause for the ***genetic variation*** is the lack of proof-reading activity by the enzyme reverse transcriptase. This leads to ***very frequent alterations in the DNA base sequence*** synthesized from viral RNA which, in turn, influences the amino acid sequence of proteins. Thus, the protein products of HIV are highly variable in the amino acid composition and, therefore, the antigenic properties. For this reason, it has not been possible to develop a vaccine against AIDS. However, there have been some encouraging animal and *in vitro* experiments which rise fresh hopes for a vaccine in the near future.

INTRODUCTION TO BIOORGANIC CHEMISTRY

39

As life comes from the existing life, it was believed for a long that the carbon compounds of organisms (hence the name organic) arose from life only. This is referred to as ***vital force theory***. Friedrich Wohler (1825) first discovered that urea ($NH_2-CO-NH_2$), the organic compound, could be prepared by heating ammonium cyanate (NH_4NCO), in the laboratory. Thereafter, thousands and thousands of organic compounds have been synthesized outside the living system.

Organic chemistry broadly deals with the chemistry of carbon compounds, regardless of their origin. Biochemistry, however, is concerned with the ***carbon chemistry of life only***. The general principles of organic chemistry provide strong foundations for understanding biochemistry. However, biochemistry exclusively deals with the reactions that occur in the living system in aqueous medium.

Most common organic compounds found in living system

The organic compounds, namely carbohydrates, lipids, proteins, nucleic acids and vitamins are the most common organic compounds of life. Their chemistry has been discussed in Section I (***Chapters 1-7***).

Common functional groups in biochemistry

Most of the physical and chemical properties of organic compounds are determined by their functional groups. Biomolecules possess certain functional groups which are their reactive centres. The common functional groups of importance in biomolecules are presented in ***Table 39.1***.

Common ring structures in biochemistry

There are many homocyclic and heterocyclic rings, commonly encountered in biomolecules. A selected list of them is given in ***Fig.39.1***.

Homocyclic rings : Phenyl ring derived from benzene is found in several biomolecules (phenylalanine, tyrosine, catecholamines). Phenanthrene and cyclopentane form the backbone of steroids (cholesterol, aldosterone).

Heterocyclic rings : Furan is the ring structure found in pentoses. Pyrrole is the basic unit of porphyrins found in many biomolecules (heme). Thiophen ring is a part of the vitamin biotin. The amino acid histidine contains imidazole.

Pyran structure is found in hexoses. Pyridine nucleus is a part of the vitamins-niacin and pyridoxine. Pyrimidines (cytosine, thymine) and purines (adenine, guanine) are the constituents of nucleotides and nucleic acids. Indole ring is found in the amino acid tryptophan. Purine and indole are examples of fused heterocyclic rings.

ISOMERISM

The compounds possessing identical molecular formulae but different structures are referred to as ***isomers.*** The phenomenon of existence of isomers is called isomerism (*Greek* : isos—equal; meros—parts). Isomers differ from each other in physical and chemical properties. ***Isomerism is partly responsible for the existence of a large number of organic molecules***.

TABLE 39.1 Common functional groups of importance in biomolecules

Functional group Name	Functional group Group	General structural formula	Type of compound	Examples of biomolecule(s)
Hydroxyl	$-OH$	$R-OH$	Alcohol	Glycerol, ethanol
Aldehyde	$-C(=O)-H$	$R-C(=O)-H$	Aldehyde	Glyceraldehyde, glucose
Keto	$-C(=O)-$	$R_1-C(=O)-R_2$	Ketone	Fructose, sedoheptulose
Carboxyl	$-C(=O)-OH$	$R-C(=O)-OH$	Carboxylic acid	Acetic acid, palmitic acid
Amino	$-NH_2$	$R-NH_2$	Amino acid	Alanine, serine
Imino	$-N(H)-$	$R-N(H)-$	Imino acid	Proline, hydroxyproline
Sulfhydryl	$-SH$	$R-SH$	Thiol	Cysteine, coenzyme A
Ether	$-O-$	R_1-O-R_2	Ether	Thromboxane A_2
Ester	$-C(=O)-O-R_1$	$R_2-C(=O)-O-R_1$	Ester	Cholesterol ester
Amido	$-C(=O)-N(R_1)(R_2)$	$R_3-C(=O)-N(R_1)(R_2)$	Amide	N-Acetylglucosamine

Consider the molecular formula—C_2H_6O. There are two important isomers of this—ethyl alcohol (C_2H_5OH) and diethyl ether (CH_3OCH_3) as shown below :

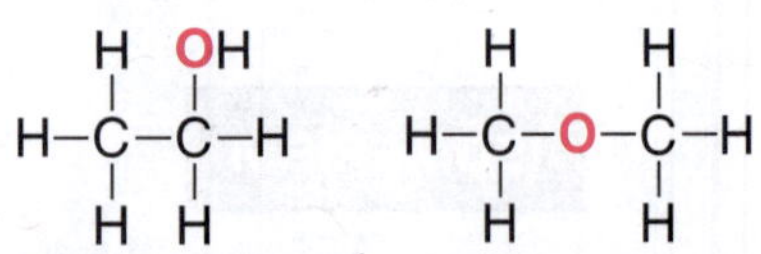

Ethyl alcohol **Diethyl ether**

Isomerism is broadly divided into two categories—structural isomerism and stereo-isomerism.

Structural isomerism

The difference in the arrangement of the atoms in the molecule (i.e. molecular framework) is responsible for structural isomerism. This may be due to variation in carbon chains *(**chain isomerism**)* or difference in the position of functional groups *(**position isomerism**)* or difference in both molecular chains and functional groups *(**functional isomerism**)*.

Structural isomerism, as such, is more common in general organic molecules. ***Tautomerism,*** a type of structural isomerism, occurs due to the migration of an atom or group from one position to the other e.g. purines and pyrimidines (***Chapter 5***).

Stereoisomerism

Stereoisomerism (*Greek* : stereos—space occupying) is, perhaps, more relevant and important to biomolecules. The differential space arrangement of atoms or groups in molecules gives rise to stereoisomerism. Thus, stereoisomers have the same structural formula but differ in their spatial arrangement.

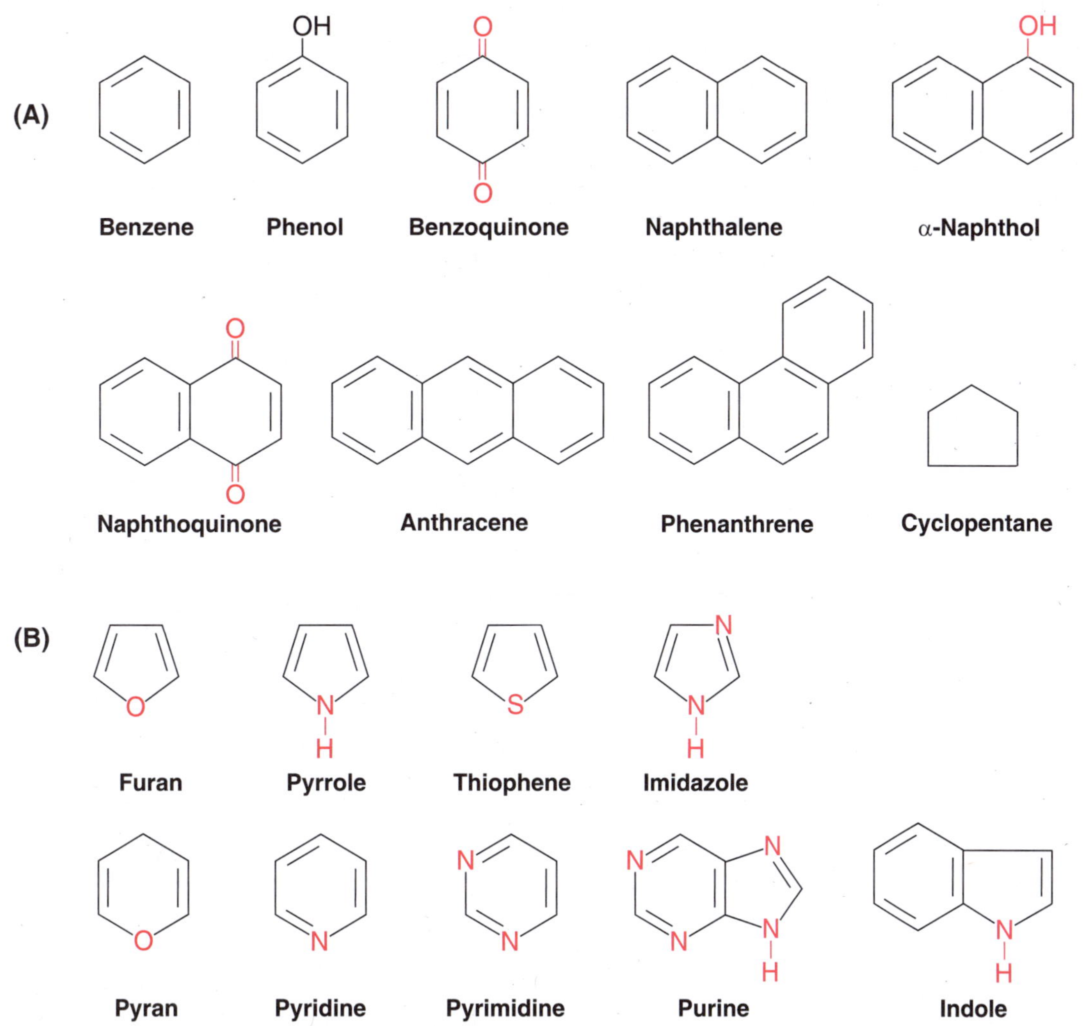

***Fig.39.1 :** Common ring structures found in biomolecules **(A)** Homocyclic rings **(B)** Heterocyclic rings.*

Stereoisomerism is of two types—***geometric isomerism*** and ***optical isomerism.***

Geometrical isomerism : This is also called ***cis-trans isomerism*** and is exhibited by certain molecules possessing double bonds. Geometrical isomerism is due to restriction of freedom of rotation of groups around a carbon-carbon double bond (C=C). Maleic acid and fumaric acid are classical examples of *cis-trans* isomerism.

H–C–COOH‖H–C–COOH
Maleic acid *(cis)*

H–C–COOH‖HOOC–C–H
Fumaric acid *(trans)*

When similar groups lie on the same side, it is called *cis* isomer (*Latin* : cis—on the same side). On the other hand, when similar groups lie on the opposite sides, it is referred to as *trans* isomer (*Latin* : trans—across). As is observed from the above structure, maleic acid is a *cis* form while fumaric acid is a *trans* form.

Geometric isomerism is also observed in sterols and porphyrins. *cis-trans* isomers differ in physical and chemical properties.

Optical isomerism : Optical isomers or enantiomers occur due to the presence of an ***asymmetric carbon*** (a chiral carbon). Optical isomers differ from each other in their optical activity to rotate the plane of polarized light.

What is an asymmetric carbon?

An object is said to be symmetrical if it can be divided into equal halves e.g. a ball. Objects which cannot be divided into equal halves are asymmetric, e.g. hand. An asymmetric object cannot coincide

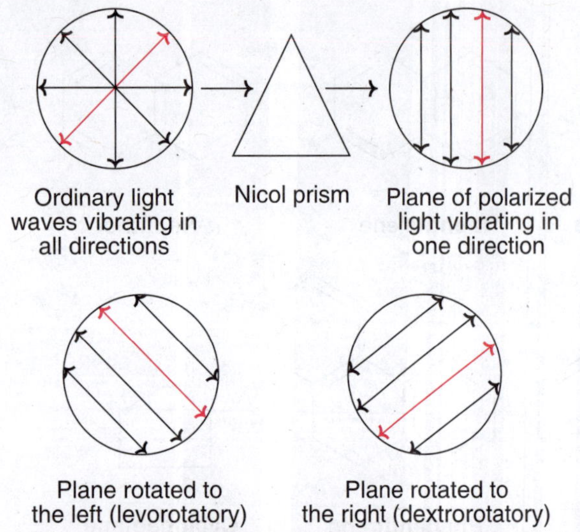

Fig.39.2 : Diagrammatic illustration of optical activity.

with its mirror image. For instance, left hand is the mirror image of right hand and these two can never be superimposed. In contrast, a symmetrical object like a ball superimposes its image.

A carbon is said to be ***chiral*** (*Greek* : hand) or asymmetric when it is attached to four different groups. Their mirror images do not superimpose with each other.

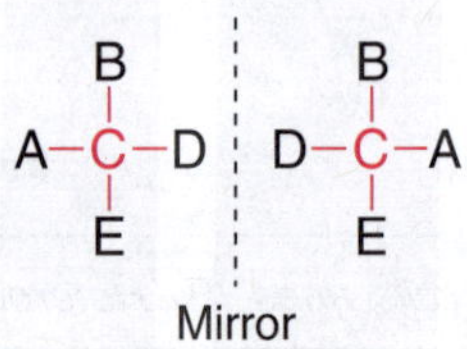

The number of possible optical isomers of a molecule depends upon the specific number of chiral carbon (n). It is given by 2^n.

What is optical activity?

The ordinary light propagates in all directions. However, on passing ordinary light through a Nicol prism, the plane of polarized light vibrates in one direction only (***Fig.39.2***).

Certain organic compounds (optical isomers) which are said to exhibit optical activity rotate the plane of polarized light either to the left or to the right.

The term ***levorotatory*** (indicated by 1 or (–) sign) is used for the substances which rotate the plane of polarized light to the left. On the other hand, the term ***dextrorotatory*** (indicated by d or (+) sign) is used for substances rotating the plane of polarized light to right (***Fig.39.2***).

The term ***racemic*** mixture represents equal concentration of d and l forms which cannot rotate the plane of polarized light.

Configuration of chiral molecules

While representing the configuration of chiral molecules, the configuration of ***glyceraldehyde*** is taken as a ***reference standard.***

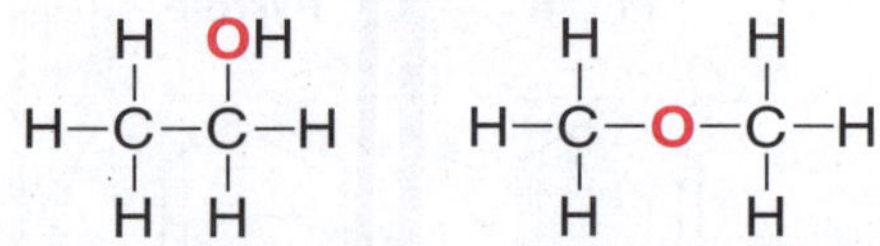

Ethyl alcohol **Diethyl ether**

It must, however, be remembered that D- and L- do not represent the direction of the rotation of plane of polarized light.

Existence of chiral biomolecules

As you know, you can never come across anybody who is your mirror image. The same is true with biomolecules. Only one type of molecules (D or L) are found in the living system. Thus, the naturally occurring amino acids are of L-type while the carbohydrates are of D-type.

OVERVIEW OF BIOPHYSICAL CHEMISTRY

40

The general laws and principles of chemistry and physics are applicable to biochemistry as well. It is, therefore, worthwhile to have a brief understanding of some of the basic chemical and physical principles that have direct relevance to life.

It must, however, be remembered that this chapter deals with quite unrelated topics to each other.

ACIDS AND BASES

According to Lowry and Bronsted, an acid is defined as a substance that gives off protons while base is a substance that accepts protons. Thus, ***an acid is a proton (H^+) donor while a base is a proton acceptor.*** A few examples of acids and their corresponding bases are given below

Acid	Base
HCl	$H^+ + Cl^-$
H_2CO_3	$H^+ + HCO_3^-$
H_2PO_4	$H^+ + HPO_4^-$
NH_4^+	$H^+ + NH_3$
H_2O	$H^+ + OH^-$
CH_3COOH	$H^+ + CH_3COO^-$
HA (general)	$H^+ + A^-$ (general)

It is evident that an acid dissociates to form proton and base. On the other hand, the base combines with proton to form acid.

Alkalies : The metallic hydroxides such as NaOH and KOH are commonly referred to as alkalies. These compounds do not directly satisfy the criteria of bases. However, they dissociate to form metallic ion and OH^- ion. The latter, being a base, accepts H^+ ions.

Ampholytes : The substances which can function both as acids and bases are referred to as ampholytes. Water is the best example for ampholytes.

Dissociation of water

Water is a weak electrolyte and dissociates as follows.

$$H_2O \rightleftharpoons H^+ + OH$$

The proton reacts with another molecule of water to form hydronium ion (H_3O^+).

$$H^+ + H_2O \rightleftharpoons H_3O^+$$

For the sake of convenience, the presence of proton as H_3O^+ is ignored.

By applying the law of mass action for the dissociation of water.

$$K = \frac{[H^+][OH^-]}{[H_2O]}$$

Here K is a constant; the concentrations are expressed in molarity. Since the degree of dissociation is very small, the concentration of undissociated $[H_2O]$ may be taken as constant.

$$K_\omega = [H^+]\ [OH^-]$$

TABLE 40.1 pH of important biological fluids

Fluid	pH
Pancreatic juice	7.5 – 8.0
Blood plasma (or whole blood)	7.35 – 7.45
Cerebrospinal fluid	7.2 – 7.4
Tears	7.2 – 7.4
Interstitial fluid	7.2 – 7.4
Human milk	7.2 – 7.4
Saliva	6.4 – 7.0
Intracellular fluid (cytosol)	6.5 – 6.9
Gastric juice	1.5 – 3.0
Urine	5.0 – 7.5

K_ω is the dissociation constant for water. Its value is 10^{-14} at 25°C.

$$[H^+]\ [OH^-] = 10^{-14}$$

In a neutral solution

$$[H^+] = [OH^-] = 10^{-7}$$

Hydrogen ion concentration (pH)

The acidic or basic nature of a solution is measured by H^+ ion concentration. The strength of H^+ ions in the biological fluids is exceedingly low. For this reason, the conventional units such as moles/l or g/l are not commonly used to express H^+ ion concentration.

Sorenson (1909) introduced the term pH to express H^+ ion concentration. ***pH is defined as the negative logarithm of H^+ ion concentration.***

$$pH = -\log\ [H^+]$$

The pH of important biological fluids are presented in ***Table 40.1***.

BUFFERS

The pH of a given solution can be easily altered by the addition of acids or bases. ***Buffers are defined as the solutions which resist change in pH by the addition of small amounts of acids or bases.*** A buffer usually consists of a weak acid and its salt (e.g. acetic acid and sodium acetate) or a weak base and its salt (e.g. ammonium hydroxide and ammonium chloride). Several buffers can be prepared in the laboratory. Nature has provided many buffers in the living system.

Mechanism of buffer action

Let us consider the buffer pair of acetic acid and sodium acetate. Acetic acid, being a weak acid, feebly ionizes. On the other hand, sodium acetate ionizes to a large extent.

$$CH_3COOH \rightleftharpoons CH_3COO^- + H^+$$

$$CH_3COONa \rightleftharpoons CH_3COO^- + Na^+$$

When an acid (say HCl) is added, the acetate ions of the buffer bind with H^+ ions (of HCl) to form acetic acid which is weakly ionizing. Therefore, the pH change due to acid is resisted by the buffer.

$$H^+ + CH_3COO^- \longrightarrow CH_3COOH$$

When a base (say NaOH) is added the H^+ ions of the buffer (acetic acid) combine with OH^- ions to form water, which is weakly dissociated. Thus, the pH change due to base addition is also prevented by the buffer.

$$OH^- + H^+ \longrightarrow H_2O$$

Buffering capacity : The efficiency of a buffer in maintaining a constant pH on the addition of acid or base is referred to as buffering capacity. It mostly depends on the concentration of the buffer components. The maximum buffering capacity is usually achieved by keeping the same concentration of the salt as well as the acid.

For a comprehensive discussion on blood buffers, refer ***Chapter 21***.

SOLUTIONS

Solutions may be regarded as mixtures of substances. In general, a solution is composed of two parts—***solute*** and ***solvent.*** The substance that is dissolved is solute and the medium that dissolves the solute is referred to as solvent. The particle size of a solute in solution is < 1 nm.

The relative concentrations of substances in a solution can be measured by several ways.

Per cent concentration : This represents parts per 100. The most frequently used is weight per volume (w/v) e.g. 9% saline (9 g/100 ml solution).

Parts per million (ppm) : This refers to the number of parts of a substance in one million parts of the solution. Thus 10 ppm chlorine means 10 μg of chlorine in 1 g of water.

Molarity (M) : It is defined as the ***number of moles of solute per liter solution***. NaCl has a molecular weight of 58.5. To get one molar (1 M) or one mole solution of NaCl, one gram molecular weight (58.5 g) of it should be dissolved in the solvent (H_2O) to make to a final total volume of 1 liter. For smaller concentrations, millimole and micromole are used.

Molality : It represents the ***number of moles of solute per 1,000 g of solvent***. One molal solution can be prepared by dissolving 1 mole of solute in 1,000 g of solvent.

Normality : Molarity is based on molecular weight while normality is ***based on equivalent weight***. One gram equivalent weight of an element or compound represents its capacity to combine or replace 1 mole of hydrogen.

COLLOIDAL STATE

Thomas Graham (1861), regarded as the 'father of colloidal chemistry', divided substances into two classes—***crystalloids*** and ***colloids.***

Crystalloids are the substances which in solution can freely pass (diffuse) through parchment membrane e.g. sugar, urea, NaCl. Colloids (*Greek* : glue-like), on other hand, are the substances that are retained by parchment membrane e.g. gum, gelatin, albumin. The above classification of Graham is no longer tenable, since any substance can be converted into a colloid by suitable means. For instance, sodium chloride in benzene forms a colloid.

Colloidal state : As such, there are no group of substances as colloids, rather, substances can exist in the form of ***colloidal state*** or ***colloidal system.*** Colloidal state is characterized by the particle size of 1 to 100 nm. When the particle size is <1 nm, it is in true solution. For the particle sizes >100 nm, the matter exists as a visible precipitate. Thus, the colloidal state is an intermediate between true solution and precipitate.

Phases of colloids : Colloidal state is heterogeneous with two phases.

1. **Dispersed phase** (internal phase) which constitutes the colloidal particles.

2. **Dispersion medium** (external phase) which refers to the medium in which the colloidal particles are suspended.

Classification of colloids

Based on the affinity of dispersion medium with dispersed phase, colloids are classified as ***lyophobic*** and ***lyophilic*** colloids.

1. **Lyophobic** (*Greek* : solvent-hating) **:** These colloids do not have any attraction towards dispersion medium. When water is used as dispersion medium, the colloids are referred to as ***hydrophobic.***

2. **Lyophilic** (*Greek* : solvent-loving) **:** These colloids have distinct affinity towards dispersion medium. The term ***hydrophilic*** is used for the colloids when water is the dispersion medium.

Properties of colloids

1. **Brownian movement :** The continuous and haphazard motion of the colloidal particles is known as Brownian movement.

2. **Optical properties :** When light is passed through a colloidal solution, it gets scattered. This phenomenon is referred to as ***Tyndal effect***.

3. **Electrical properties :** The colloidal particles carry electrical charges, either positive or negative. The separation of charged colloids can be achieved by the analytical technique—electrophoresis (***Refer Chapter 34***).

4. **Osmotic pressure :** Since the colloidal particles are larger in size, their contribution to osmotic pressure is relatively less.

5. **Donnan membrane equilibrium :** The presence of non-diffusible colloidal particles (e.g. protein) in the biological systems influences the concentration of diffusible ions across the membrane.

Biological importance of colloids

1. **Biological fluids as colloids :** These include blood, milk and cerebrospinal fluid.

2. **Biological compounds as colloidal particles :** The complex molecules of life, the high molecular weight proteins, complex lipids and polysaccharides exist in colloidal state.

3. **Fat digestion and absorption :** The formation of emulsions, facilitated by the emulsifying agents bile salts, promotes fat digestion and absorption in the intestinal tract.

4. **Formation of urine :** The filtration of urine is based on the principle of dialysis.

DIFFUSION

The molecules in liquids or gases are in continuous motion. ***Diffusion may be regarded as the movement of solute molecules from a higher concentration to a lower concentration.*** Diffusion is more rapid in gases than in liquids. The smaller particles diffuse faster than the larger ones. The greater the temperature, the higher is the rate of diffusion.

Diffusion occurs in true solutions as well as in colloidal solutions.

Applications of diffusion

1. Exchange of O_2 and CO_2 in lungs and in tissues occurs through diffusion.

2. Certain nutrients are absorbed by diffusion in the gastrointestinal tract e.g. pentoses, minerals, water soluble vitamins.

3. Passage of the waste products namely ammonia, in the renal tubules occurs due to diffusion.

OSMOSIS

Osmosis (*Greek* : push) refers to the ***movement of solvent*** (most frequently water) ***through a semipermeable membrane***.

The flow of solvent occurs from a solution of low concentration to a solution of high concentration, when both are separated by a semipermeable membrane. In a strict sense, the semipermeable membrane is expected to be permeable to the solvent and not to the solute.

Osmotic pressure

Osmotic pressure may be defined as the excess pressure that must be applied to a solution to prevent the passage of solvent into the solution, when both are separated by a semipermeable membrane.

The solutions that exert the same osmotic pressure are said to be ***isoosmotic.*** The term ***isotonic*** is used when a cell is in direct contact with an isoosmotic solution (0.9% NaCl) which does not change the cell volume and, thus, the cell tone is maintained. A solution with relatively greater osmotic pressure is referred to as ***hypertonic***. On the other hand, a solution with relatively lower pressure is ***hypotonic.***

The term ***oncotic pressure*** is commonly used to represent the osmotic pressure of colloidal substances (e.g. albumin, globulin).

Units of osmotic pressure : Osmotic pressure of biological fluids is frequently expressed as ***milliosmoles.*** The osmotic pressure of plasma is 280-300 milliosmoles/l.

Applications of osmosis

1. **Fluid balance and blood volume :** The fluid balance of the different compartments of the body is maintained due to osmosis. Further, osmosis significantly contributes to the regulation of blood volume and urine excretion.

2. **Red blood cells and fragility :** When the RBC are kept in ***hypotonic solution*** (say 0.4% NaCl), the cells bulge due to entry of water which often causes rupture of plasma membrane of RBC (***hemolysis***).

Osmotic fragility test for RBC is employed in laboratories for diagnostic purposes. For a normal human blood, RBC begin to hemolyse in 0.45% NaCl and the hemolysis is almost complete in 0.33% NaCl. Increased fragility of RBC is observed in hemolytic jaundice while it is decreased in certain anemias.

3. **Transfusion :** Isotonic solutions of NaCl (0.9%) or glucose (5%) or a suitable combination of these two are commonly used in transfusion in hospitals for the treatment of dehydration, burns etc.

4. **Action of purgatives :** The mechanism of action of purgatives is mainly due to osmotic phenomenon. For instance, epsom ($MgSO_4$ $7H_2O$) or Glauber's (Na_2SO_4 $10H_2O$) salts withdraw water from the body, besides preventing the intestinal water absorption.

5. **Edema due to hypoalbuminemia :** Disorders such as kwashiorkor and glomerulonephritis are associated with lowered plasma albumin concentration and edema. Edema is caused by reduced oncotic pressure of plasma, leading to the accumulation of excess fluid in tissue spaces.

DONNAN MEMBRANE EQUILIBRIUM

When membrane is freely permeable to ions (say Na^+, Cl^-) and if the concentration of ions on both the sides is different, the ions freely diffuse to attain equal concentration. Gibbs-Donnan observed that the ***presence of a non-diffusible ion (usually a protein) on one side of the membrane alters the diffusion of diffusible ions***.

Applications of Donnan membrane equilibrium

1. **Difference in the ionic concentrations of biological fluids :** The lymph and interstitial fluids have lower concentration of inorganic cations (Na^+, K^+) and higher concentration of anions (Cl^-) compared to plasma. This is attributed to the higher protein (Pr^-) content in the plasma.

2. **Lower pH in RBC :** The hemoglobin of RBC is negatively charged and, therefore, causes the accumulation of positively charged ions including H^+. Therefore, the pH of RBC is slightly lower (7.25) than that of plasma (7.4).

3. **Osmotic imbalance :** Donnan membrane equilibrium—which results in the differential distribution of ions in different compartments of the body—partly explains the osmotic pressure differences.

VISCOSITY

Liquid or fluid has a tendency to flow which is referred to as fluidity. The term viscosity may be defined as the ***internal resistance offered by a liquid or a gas to flow***. The property of viscosity is due to frictional forces between the layers while their movement occurs.

Viscosity of colloidal solutions, particularly lyophilic colloids, is generally higher than true solutions.

Units of viscosity : The unit of viscosity is ***poise***, after the scientist Poiseuille, who first systematically studied the flow of liquids. A poise represents dynes/cm^2.

Applications of viscosity

1. **Viscosity of blood :** Blood is about 4 times more viscous than water. The viscosity of blood is mainly attributed to suspended blood cells and colloidal plasma proteins.

2. **Viscosity change in muscle :** Excitation of the muscle is associated with increase in the viscosity of the muscle fibres. This delays the change in the tension of the contracting muscle.

3. **Vitreous body :** This is an amorphous viscous body located in the posterior chamber of the eye. It is rich in albumin and hyaluronic acid.

4. **Synovial fluid :** It contains hyaluronic acid which imparts viscosity and helps in the lubricating function of joints.

SURFACE TENSION

A molecule in the interior of a liquid is attracted by other molecules in all directions. In contrast, a molecule on the surface is attracted only downwards and sideways and not upwards. Due to this, the surface layer behaves like a stretched film. ***Surface tension is the force with which the molecules on the surface are held together.*** It is expressed as ***dynes/cm***. Surface tension decreases with increase in temperature.

Applications of surface tension

1. **Digestion and absorption of fat :** ***Bile salts*** reduce the surface tension. They act as detergents and cause emulsification of fat, thereby allowing the formation of minute particles for effective digestion and absorption.

2. **Hay's sulfur test :** This is a common laboratory test employed for the ***detection of bile salts in urine*** of jaundice patients. Sulfur powder, when sprinkled on the surface of urine possessing bile salts, sinks. This is in contrast to a normal urine where sulfur powder floats. Hay's test is based on the principle that bile salts in urine lower surface tension which is responsible for sulfur to sink.

3. **Surfactants and lung function :** The low surface tension of the alveoli keeps them apart and allows an efficient exchange of gases in lungs. In fact, certain surfactants, predominantly dipalmitoyl phosphatidyl choline (***dipalmitoyl lecithin***) are responsible for maintaining low surface tension in the alveoli. Surfactant deficiency causes respiratory distress syndrome in the infants.

4. **Lipoprotein complex membranes :** The structure of plasma membrane is composed of surface tension reducing substances, namely lipids and proteins. This facilitates absorption of these compounds.

ADSORPTION

Adsorption is a surface phenomenon. It is the process of accumulation of a substance (adsorbate) on the surface of another substance (adsorbent). Adsorption differs from absorption, as the latter involves the diffusion into the interior of the material.

Applications of adsorption

1. **Formation of enzyme-substrate complex :** For the catalysis to occur in biological system, formation of enzyme-substrate complex is a prerequisite. This happens by adsorption of substrate on the enzyme.

2. **Action of drugs and poisons :** On adsorption at the cell surface, drugs and poisons exert their action.

3. **Adsorption in analytical biochemistry :** The principle of adsorption is widely employed in the chromatography technique for the separation and purification of compounds (enzymes, immunoglobulins).

ISOTOPES

Isotopes are defined as the elements with same atomic number but different atomic weights. They possess the same number of protons but differ in the neutrons in their nuclei. Therefore, isotopes (*Greek* : iso—equal; tope—place) occupy the same place in the periodic table. The chemical properties of different isotopes of a particular element are identical.

Isotopes are of two types—***stable*** and ***unstable.*** The latter are more commonly referred to as radioactive isotopes and they are of particular interest to biochemists. Conventionally while representing isotopes, the atomic weight is written on upper left side of the element symbol.

Stable isotopes

They are naturally occurring and do not emit radiations (non-radioactive) e.g. deuterium (heavy hydrogen) ^{2}H; ^{13}C; ^{15}N; ^{18}O. Stable isotopes can be identified and quantitated by mass spectrometry or nuclear magnetic resonance (NMR). They are less frequently used in biochemical investigations.

Radioactive isotopes

The atomic nucleus of radioactive isotopes is unstable and, therefore, undergoes decay. The radioactive decay gives rise to one of the following 3 ionizing radiations.

1. **α-Rays**—an α particle possessing 2 protons i.e. helium nuclei.

2. **β-Rays**—due to the emission of electrons.

3. **γ-Rays**—due to emission of high energy photons.

The β and γ emitting radioisotopes are employed in biochemical research.

Applications of radioisotopes in biochemistry

Radioactive isotopes have become indispensable tools of biochemistry. They can be conveniently used as tracers in biochemical research since the chemical properties of different isotopes of a particular element are identical. Therefore, the living cells cannot distinguish the radioactive isotope from a normal atom.

A few important application of radioisotopes are

1. By the use of isotope tracers, the metabolic origin of complex molecules such as heme, cholesterol, purines and phospholipids can be determined. As early as 1945, it was established that nitrogen atom of heme was derived from glycine. This was done by feeding rats with (^{15}N) glycine and detecting (^{15}N) heme.

2. The precursor-product relationship in several metabolic pathways has been investigated by radioisotopes. e.g. Krebs cycle, β-oxidation of fatty acids, urea cycle, fatty acid synthesis.

3. Radioisotopes are conveniently used in the study of metabolic pools (e.g. amino acid pool) and metabolic turnovers (e.g. protein turnover).

4. Certain endocrine and immunological studies also depend on the use of radioisotopes e.g. radioimmunoassay.

5. Radioisotopes are employed in elucidating drug metabolism.

TOOLS OF BIOCHEMISTRY 41

Biochemistry is an experimental rather than a theoretical science. The understanding and development of concepts in biochemistry are a result of continuous experimentation and evidence obtained therein. It is no exaggeration to state that the foundations for the present (and the future, of course!) knowledge of biochemistry are based on the laboratory tools employed for biochemical experimentation.

A detailed discussion on the tools of biochemistry is beyond the scope of this book. The basic principles of some of the commonly employed tools are described in this chapter. The reader must, however, refer elsewhere for an understanding of the techniques of recombinant DNA technology and polymerase chain reaction (***Chapter 27***), and radioactive isotopes (***Chapter 40***).

CHROMATOGRAPHY

Chromatography is one of the most useful and popular tools of biochemistry. It is an analytical technique dealing with the ***separation of closely related compounds from a mixture***. These include proteins, peptides, amino acids, lipids, carbohydrates, vitamins and drugs.

Principles and classification

Chromatography (*Greek* : chroma—colour; graphein—to write), usually consists of a ***mobile phase*** and a ***stationary phase.*** The mobile phase refers to the mixture of substances (to be separated), dissoved in a liquid or a gas. The stationary phase is a porous solid matrix through which the sample contained in the mobile phase percolates. The interaction between the mobile and stationary phases results in the separation of the compounds from the mixture. These interactions include the physico-chemical principles such as ***adsorption, partition, ion-exchange, molecular sieving*** and ***affinity.***

The interaction between stationary phase and mobile phase is often employed in the classification chromatography e.g. partition, adsorption, ion exchange. Further, the classification of chromatography is also based either on the nature of the stationary phase (paper, thin layer, column), or on the nature of both mobile and stationary phases (gas-liquid chromatography). A summary of the different methods (classes) of chromatography is given in ***Fig.41.1***.

1. **Partition chromatography :** The molecules of a mixture get partitioned between the stationary phase and mobile phase depending on their relative affinity to each one of the phases.

(a) **Paper chromatography :** This technique is commonly used for the separation of amino acids, sugars, sugar derivatives and peptides. In paper chromatography, a few drops of solution containing a mixture of the compounds to be separated is applied (spotted) at one end, usually ~2 cm above, a strip of filter paper (Whatman No. 1 or 3). The paper is dried and dipped into a solvent mixture consisting of butanol, acetic acid and water in 4 : 1 : 5 ratio (for the separation of amino acids). When the migration of the solvent is upwards, it is referred to as ***ascending chromatography.*** In ***descending chromatography***, the solvent moves downwards (***Fig.41.2***).

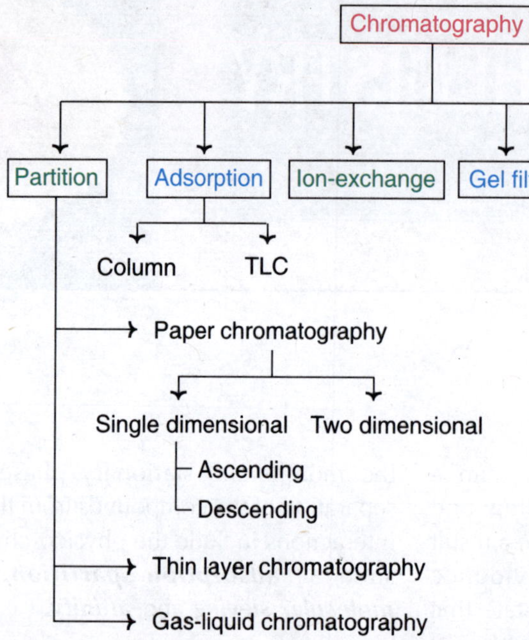

***Fig.41.1 :** Important types of chromatography (HPLC–High performance liquid chromatography; TLC–Thin layer chromatography).*

After a sufficient migration of the solvent front, the paper (chromatogram) is removed, dried and developed for the identification of the specific spots. ***Ninhydrin***, which forms purple complex with ***α-amino acids***, is frequently used as a colouring reagent. The chemical nature of the individual spots can be identified by running known standards with the unknown mixture.

The migration of a substance is frequently expressed as ***R_f value*** (**r**atio of **f**ronts)

$$R_f = \frac{\text{Distance travelled by the substance}}{\text{Distance travelled by solvent front}}$$

(b) **Thin layer chromatography (TLC) :** The principle of TLC is the same as described for paper chromatography (partition). In place of a paper, an inert substance, such as

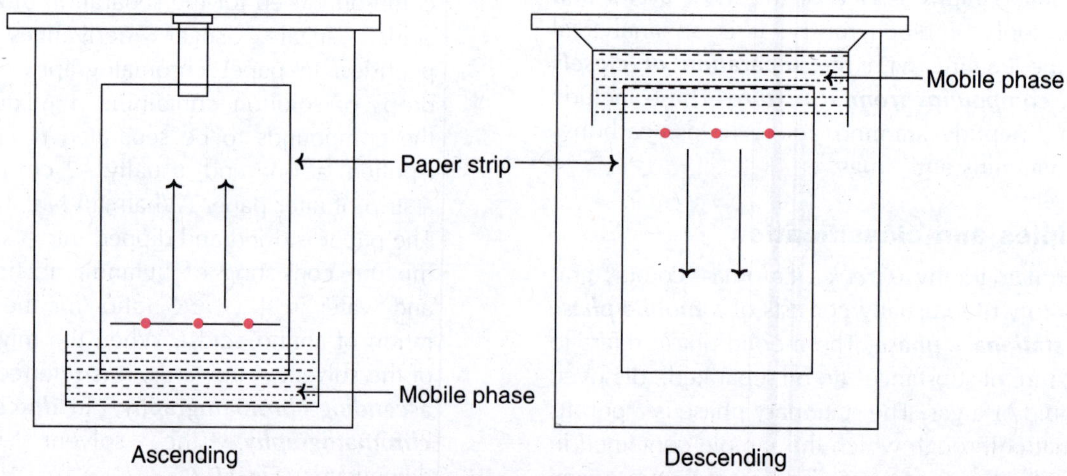

***Fig.41.2 :** Paper chromatography—ascending and descending types.*

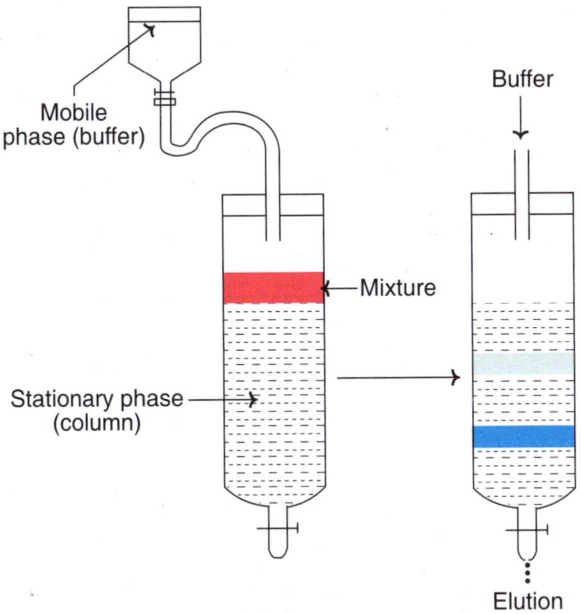

Fig.41.3 : Diagrammatic representation of adsorption column chromatography.

cellulose, is employed as supporting material. Cellulose is spread as a thin layer on glass or plastic plates. The chromatographic separation is comparatively rapid in TLC.

(c) **Gas-liquid chromatography (GLC) :** This is the method of choice for ***the separation of volatile substances*** or the volatile derivatives of certain non-volatile substances.

Gas-liquid chromatography is sensitive, rapid and reliable. It is frequently used for the quantitative estimation of biological materials such as lipids, drugs and vitamins.

2. **Adsorption column chromatography :** The adsorbents such as silica gel, alumina, charcoal powder and calcium hydroxyapatite are packed into a column in a glass tube. This serves as the stationary phase. The sample mixture in a solvent is loaded on this column. The individual components get differentially adsorbed on to the adsorbent. The elution is carried out by a buffer system (mobile phase). The individual compounds come out of the column at different rates which may be separately collected and identified (***Fig.41.3***). For instance, amino acids can be identified by ninhydrin colorimetric method. An automated column chromatography apparatus—***fraction collector***—is frequently used nowadays.

3. **Ion-exchange chromatography :** Ion-exchange chromatography involves the separation of molecules on the basis of their electrical charges. Ion-exchange resins—***cation exchangers*** and ***anion exchangers***—are used for this purpose. An anion exchanger (R^+A^-) exchanges its anion (A^-) with another anion (B^-) in solution.

$$R^+A^- + B^- \rightleftharpoons R^+B^- + A^-$$

Similarly, a cation exchanger (H^+R^-) exchanges its cation (H^+) with another cation (C^+) in solution.

$$H^+R^- + C^+ \rightleftharpoons C^+R^- + H^+$$

4. **Gel filtration chromatography :** In gel filtration chromatography, the separation of molecules is based on their size, shape and molecular weight. This technique is also referred to as ***molecular sieve or molecular exclusion chromatography.*** The apparatus consists of a column packed with spongelike gel beads (usually cross-linked polysaccharides) containing pores. The gels serve as molecular sieves for the separation of smaller and bigger molecules.

The larger molecules cannot pass through the pores of gel and, therefore, move faster. On the other hand, the smaller molecules enter the gel beads and are left behind which come out slowly (***Fig.41.4***).

The gel-filtration chromatography can be used for an approximate determination of molecular weights. This is done by using a calibrated column with substances of known molecular weight.

5. **Affinity chromatography :** The principle of affinity chromatography is based on the property of specific and non-covalent binding of proteins to other molecules, referred to as ***ligands.*** For instance, enzymes bind specifically to ligands such as substrates or cofactors.

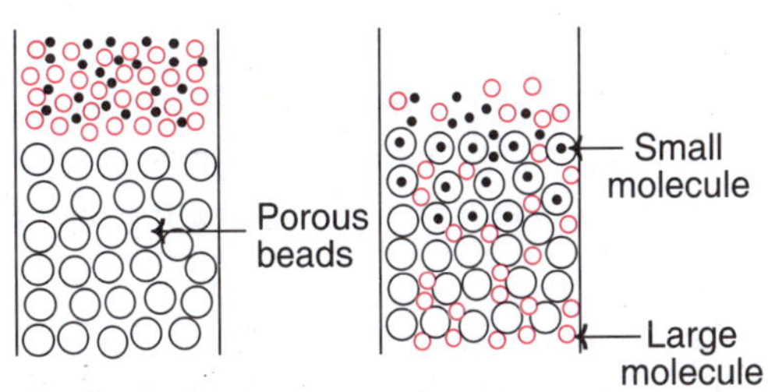

Fig.41.4 : The principle of gel-filtration chromatography.

Affinity chromatography is useful for the purification of enzymes, vitamins, nucleic acids, drugs, hormone receptors, antibodies etc.

6. **High performance liquid chromatography (HPLC) :** In general, the chromatographic techniques are slow and time consuming. The separation can be greatly improved by applying high pressure in the range of 5,000-10,000 psi (pounds per square inch), hence this technique is also referred (less frequently) to as high pressure liquid chromatography.

ELECTROPHORESIS

The ***movement of charged particles*** (ions) ***in an electric field*** resulting in their migration towards the oppositely charged electrode is known as electrophoresis. Molecules with a net positive charge (cations) move towards the negative cathode while those with net negative charge (anions) migrate towards positive anode (***Fig.41.5***). Electrophoresis is a widely used analytical technique for the separation of biological molecules such as plasma proteins, lipoproteins and immunoglobulins.

Different types of electrophoresis

1. **Zone electrophoresis :** An inert supporting material such as paper or gel are used.

(a) **Paper electrophoresis :** In this technique, the sample is applied on a strip of filter paper wetted with desired buffer solution.

For the separation of serum proteins Whatman No. 1 filter paper, veronal or tris buffer at pH 8.6 and the stains amido black or bromophenol blue are employed. The ***serum proteins are separated into five distinct bands***—albumin, α_1-, α_2-, β- and γ-globulins (***Refer Fig.9.1***). For the electrophoretic pattern of serum lipoproteins, ***refer Fig.14.21***.

(b) **Gel electrophoresis :** This technique involves the separation of molecules based on their size, in addition to the electrical charge.

2. **Isoelectric focussing :** This technique is primarily based on the ***immobilization of the molecules at isoelectric*** pH during electrophoresis. Stable pH gradients are set up (usually in a gel) covering the pH range to include the isoelectric points of the components in a mixture. Isoelectric focussing can be conveniently used for the purification of proteins.

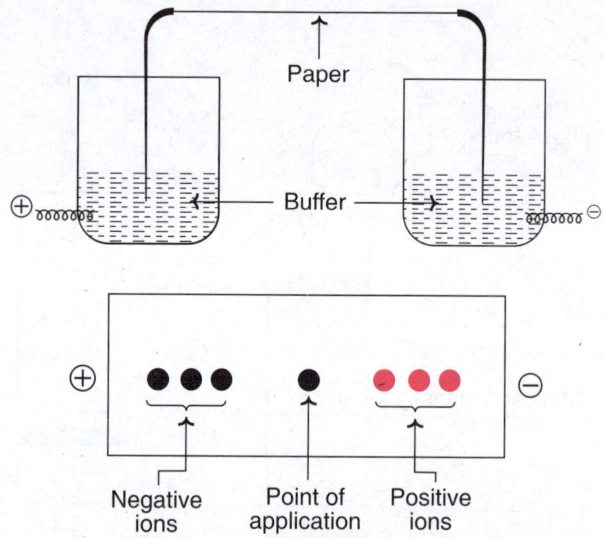

Fig.41.5 : Diagrammatic representation of paper electrophoresis.

3. **Immunoelectrophoresis :** This technique involves combination of the principles of electrophoresis and immunological reactions. Immunoelectrophoresis is useful for the analysis of complex mixtures of antigens and antibodies.

The complex proteins of biological samples (say human serum) are subjected to electrophoresis. The antibody (antihuman immune serum from rabbit or horse) is then applied in a trough parallel to the electrophoretic separation. The antibodies diffuse and, when they come in contact with antigens, precipitation occurs, resulting in the formation of ***precipitin bands*** which can be identified.

PHOTOMETRY—COLORIMETER AND SPECTROPHOTOMETER

Photometry broadly deals with the study of the phenomenon of light absorption by molecules in solution. The specificity of a compound to absorb light at a particular wavelength (monochromatic light) is exploited in the laboratory for quantitative measurements. From the biochemist's perspective, photometry forms an important laboratory tool for accurate estimation of a wide variety of compounds

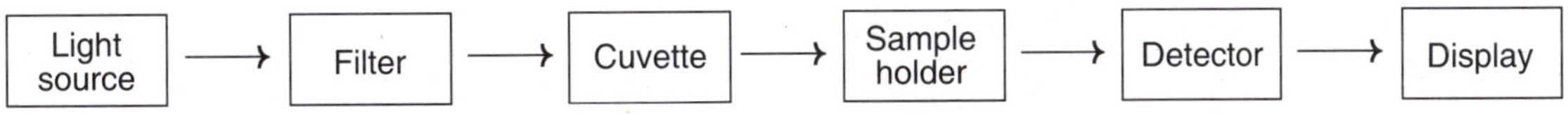

Fig.41.6 : Diagrammatic representation of the components in a colorimeter.

in biological samples. Colorimeter and spectrophotometer are the laboratory instruments used for this purpose. They work on the principles discussed below.

When a light at a particular wavelength is passed through a solution (incident light), some amount of it is absorbed and, therefore, the light that comes out (transmitted light) is diminished. The nature of light absorption in a solution is governed by ***Beer-Lambert law.***

Beer's law states that the amount of transmitted light decreases exponentially with an increase in the concentration of absorbing material (i.e. the amount of light absorbed depends on the concentration of the absorbing molecules). And according to Lambert's law, the transmitted light decreases exponentially with increase in the thickness of the absorbing molecules (i.e. the amount of light absorbed is dependent on the thickness of the medium).

The ratio of transmitted light (I) to that of incident light (I_O) is referred to as transmittance (T).

$$T = \frac{I}{I_O}$$

Absorbance (A) or optical density (OD) is very commonly used in laboratories. The relation between absorbance and transmittance is expressed by the following equation.

$$A = 2 - \log_{10} T$$
$$= 2 - \log\% \ T$$

Colorimeter

Colorimeter (or photoelectric colorimeter) is the instrument used for the measurement of coloured substances. This instrument is operative in the visible range (400-800 nm) of the electromagnetic spectrum of light. The working of colorimeter is based on the principle of Beer-Lambert law.

The colorimeter, in general consists of light source, filter sample holder and detector with display (meter or digital). A filament lamp usually serves as a light source.The filters allow the passage of a small range of wave length as incident light. The sample holder is a special glass cuvette with a fixed thickness. The photoelectric selenium cells are the most common detectors used in colorimeter. The diagrammatic representation of a colorimeter is depicted in ***Fig.41.6***.

Spectrophotometer

The spectrophotometer primarily differs from colorimeter by covering the ultraviolet region (200-400 nm) of the electromagnetic spectrum. Further, the spectrophotometer is ***more sophisticated*** with several additional devices that ultimately increase the sensitivity of its operation severalfold when ***compared to a colorimeter***. A precisely selected wavelength (say 234 nm or 610 nm) in both ultra- violet and visible range can be used for measurements. In place of glass cuvettes (in colorimeter), quartz cells are used in a spectrophotometer.

The spectrophotometer has similar basic components as described for a colorimeter (***Fig.41.6***), and its operation is also based on the Beer-Lambert law (already discussed).

FLAME PHOTOMETRY

Flame photometry primarily deals with the quantitative measurement of electrolytes such as sodium, potassium and lithium. The instrument, namely flame photometer, works on the following principle. As a solution in air is finally sprayed over a burner, it dissociates to give neutral atoms. Some of these atoms get excited and move to a higher energy state. When the excited atoms fall back to the ground state, they emit light of a characteristic wavelength which can be measured. The intensity of emission light is proportional to the concentration of the electrolyte being estimated.

ULTRACENTRIFUGATION

Ultracentrifugation is an indispensable tool for the ***isolation of subcellular organelles***, proteins and nucleic acids. In addition, this technique is also employed for the determination of molecular weights of macromolecules.

The rate at which the sedimentation occurs in ultracentrifugation primarily depends on the size and shape of the particles or macromolecules (i.e. on the molecular weight). It is expressed in terms of ***sedimentation coefficient***(s).

The sedimentation coefficient has the units of seconds. It was usually expressed in units of 10^{-13} s (since several biological macromolecules occur in this range), which is designated as one ***Svedberg unit.*** For instance, the sedimentation coefficient of hemoglobin is 4 + 10^{-13} s or 4S; ribonuclease is 2 + 10^{-13} s or 2S. Conventionally, the subcellular organelles are often referred to by their S value e.g. 70S ribosome.

Isolation of subcellular organelles by centrifugation

The cells are subjected to disruption by sonication or osmotic shock or by use of homogenizer. This is usually carried out in an isotonic (0.25 M) sucrose. The advantage with sucrose medium is that it does not cause the organelles to swell. The subcellular particles can be separated by differential centrifugation. The most commonly employed laboratory method separates subcellular organelles into 3 major fractions—nuclear, mitochondrial and microsomal (***Fig.41.7***).

When the homogenate is centrifuged at 700 g for about 10 min, the nuclear fraction (includes plasma membrane) gets sedimented. On centrifuging the supernatant (I) at 15,000 g for about 5 min mitochondrial fraction (that includes lysosomes, peroxisomes) is pelleted. Further centrifugation of the supernatant (II) at 100,000 g for about 60 min separates microsomal fraction (that includes ribosomes and endoplasmic reticulum). The supernatant (III) then obtained corresponds to the cytosol.

The purity (or contamination) of the subcellular fractionation can be checked by the use of ***marker enzymes.*** DNA polymerase is the marker enzyme for nucleus, while glutamate dehydrogenase and glucose 6-phosphatase are the markers for mitochondria and ribosomes, respectively. Hexokinase is the marker enzyme for cytosol.

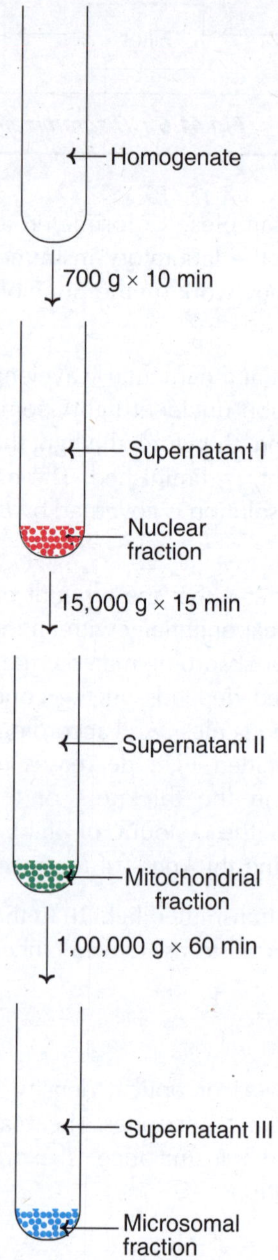

Fig.41.7 : *Separation of subcellular fractions by differential centrifugation.*

RADIOIMMUNOASSAY

Radioimmunoassay (RIA) was developed in 1959 by Solomon, Benson and Rosalyn Yalow for the estimation of insulin in human serum. This technique has revolutionized the estimation of several compounds in biological fluids that are found in exceedingly low concentrations (nanogram or picogram). RIA is a highly sensitive and specific analytical tool.

Principle

Radioimmunoassay combines the principles of ***radioactivity*** of isotopes and ***immunological*** reactions of antigen and antibody, hence the name.

The principle of RIA is primarily based on the competition between the labelled and unlabelled antigens to bind with antibody to form antigen-antibody complexes (either labelled or unlabelled). The unlabelled antigen is the substance (say insulin) to be determined. The antibody to it is produced by injecting the antigen to a goat or a rabbit. The specific antibody (Ab) is then subjected to react with unlabelled antigen in the presence of excess amounts of isotopically labelled (^{131}I) antigen (Ag^+) with known radioactivity. There occurs a competition between the antigens (Ag^+ and Ag) to bind the antibody. Certainly, the labelled Ag^+ will have an upper hand due to its excess presence.

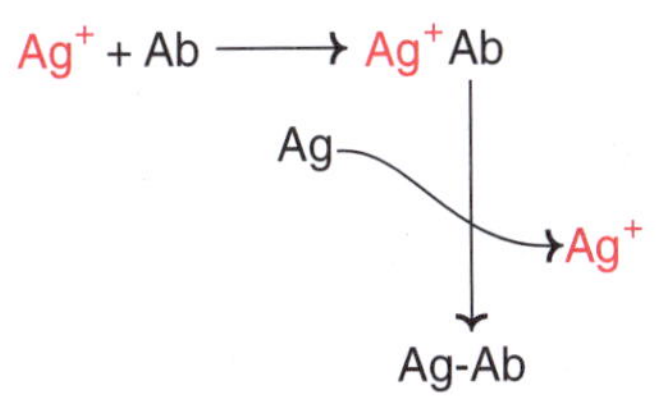

As the concentration of unlabelled antigen (Ag) increases the amount of labelled antigen-antibody complex (Ag^+-Ab) decreases. Thus, the concentration of Ag^+-Ab is inversely related to the concentration of unlabelled Ag i.e. the substance to be determined. This relation is almost linear. A standard curve can be drawn by using different concentrations of unlabelled antigen and the same quantities of antibody and labelled antigen.

The labelled antigen-antibody (Ag^+-Ab) complex is separated by precipitation. The radioactivity of ^{131}I present is Ag^+-Ab is determined.

Applications

RIA is no more limited to estimation of hormones and proteins that exhibit antigenic properties. By the use of ***haptens*** (small molecules such as dinitrophenol, which, by themselves, are not antigenic), several substances can be made antigenic to elicit specific antibody responses. In this way, a wide variety of compounds have been brought under the net of RIA estimation. These include peptides, steroid hormones, vitamins, drugs, antibiotics, nucleic acids, structural proteins and hormone receptor proteins.

Radioimmunoassay has tremendous application in the diagnosis of hormonal disorders, cancers and therapeutic monitoring of drugs, besides being useful in biomedical research.

ENZYME-LINKED IMMUNOSORBENT ASSAY

Enzyme-linked immunosorbent assay (ELISA) is a non-isotopic immunoassay. An enzyme is used as a label in ELISA in place of radioactive isotope employed in RIA. ELISA is as sensitive as or even more sensitive than RIA. In addition, there is no risk of radiation hazards (as is the case with RIA) in ELISA.

Principle

ELISA is based on the immunochemical principles of antigen-antibody reaction. The stages of ELISA, depicted in ***Fig.41.8***, are summarized.

1. The antibody against the protein to be determined is fixed on an inert solid such as polystyrene.

2. The biological sample containing the protein to be estimated is applied on the antibody coated surface.

3. The protein antibody complex is then reacted with a second protein specific antibody to which an enzyme is covalently linked. These enzymes must be easily assayable and produce preferably coloured products. Peroxidase, amylase and alkaline phosphatase are commonly used.

4. After washing the unbound antibody linked enzyme, the enzyme bound to the second antibody complex is assayed.

5. The enzyme activity is determined by its action on a substrate to form a product (usually coloured). This is related to the concentration of the protein being estimated.

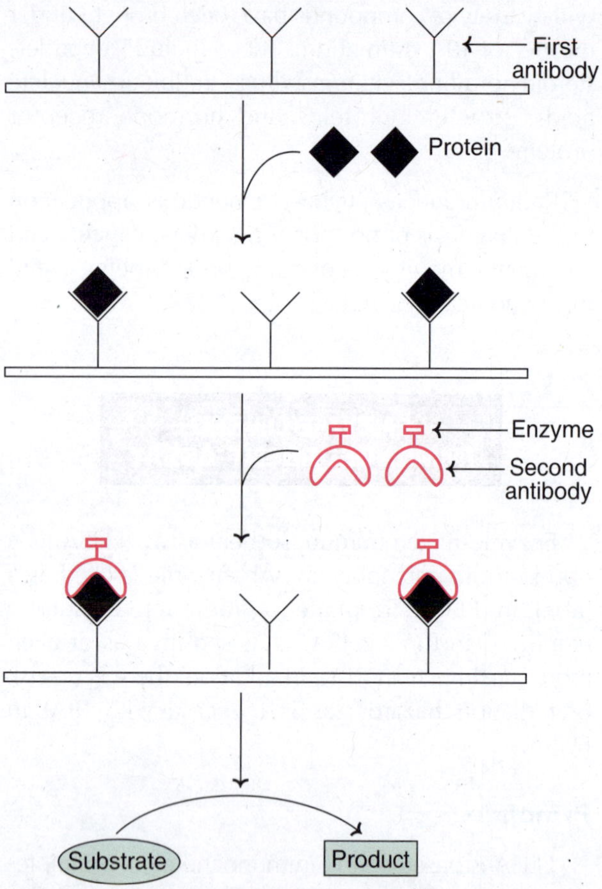

Fig.41.8 : Diagrammatic representation of enzyme-linked immunosorbant assay (ELISA).

Applications

ELISA is widely used for the determination of small quantities of proteins (hormones, antigens, antibodies) and other biological substances. The most commonly used ***pregnancy test*** for the detection of human chorionic gonadotropin (hCG) in urine is based on ELISA. By this test, pregnancy can be detected within few days after conception. ELISA is also been used for the diagnosis of AIDS.

HYBRIDOMA TECHNOLOGY

In 1975, George Kohler and Cesar Milstein (Nobel Prize 1984) created hybrid cells that will make unlimited quantities of antibodies with defined specificities, which are termed as ***monoclonal antibodies (McAb).*** This discovery, often referred to as hybridoma technology, has revolutionized methods for antibody production.

Principle

This is based on the ***fusion between myeloma cells*** (maligant plasma cells) ***and spleen cells*** from a suitably immunized animal. Spleen cells die in a short period under ordinary tissue culture conditions while myeloma cells are adopted to grow permanently in culture. Mutants of myeloma cells lack the enzyme hypoxanthine guanine phosphoribosyltransferase (azaquinine resistant) or thymidine kinase (bromodeoxyuridine resistant). These mutants cannot grow in a medium containing aminopterin, supplemented with hypoxanthine and thymidine (HAT medium). Hybrids between the mutant myeloma cells and spleen cells can be selected and cultured in HAT medium.

From the growing hybrids, individual clones can be chosen that secrete the desired antibodies (monoclonal origin). The selected clones like ordinary myeloma cells can be maintained indefinitely.

Hybridoma technology can make available highly specific antibodies in abundant amounts. The clones once developed are far cheaper than the traditionally employed animals (horses, rabbits) for producing antibodies. The clones developed from the hybrids will also ensure constancy of the quality of the product and will also avoid the batch to batch variation inherent in the conventional methods.

Applications of monoclonal antibodies

The antibodies produced by hybridoma technology have been widely used for a variety of purposes. These include the early detection of pregnancy, detection and treatment of cancer, diagnosis of leprosy and treatment of autoimmune diseases.

IMMUNOLOGY

42

Immunology deals with the study of immunity and immune systems of vertebrates. Immunity (immunis literally means exempt/free from burden) broadly involves the ***resistance shown***, and ***protection offered by the host organism against the infectious diseases***. The immune system consists of a complex network of cells and molecules, and their interactions. It is specifically designed to eliminate infectious organisms from the body. This is possible since the organism is capable of distinguishing the self from non-self, and eliminate non-self.

Immunity is broadly divided into two types — ***innate (non-specific) immunity and adaptive*** or ***acquired (specific) immunity***.

INNATE IMMUNITY

Innate immunity is non-specific, and represents the inherent capability of the organism to offer resistance against diseases. It consists of defensive barriers.

First line of defense

The ***skin*** is the largest organ in the human body, constituting about 15% of the adult body weight. The skin provides mechanical barrier to prevent the entry of microorganisms and viruses. The acidic (pH 3-5) environment on the skin surface inhibits the growth of certain microorganisms. Further, the sweat contains an enzyme lysozyme that can destroy bacterial cell wall.

Second line of defense

Despite the physical barriers, the microorganisms do enter the body. The body defends itself and ***eliminates the invading organisms by non-specific mechanisms*** such as sneezing and secretions of the mucus. In addition, the body also tries to kill the pathogens by phagocytosis (involving macrophages and complement system). The inflammatory response and fever response of the body also form a part of innate immunity.

THE IMMUNE SYSTEM

The immune system represents the third and most potent defense mechanism of the body. ***Acquired*** (adaptive or specific) ***immunity*** is capable of specifically recognizing and eliminating the invading microorganisms and foreign molecules (antigens). In contrast to innate immunity, the acquired immunity displays four distinct characteristics

- Antigen specificity
- Recognition diversity
- Immunological memory
- Discrimination between self and non-self.

The ***body*** possess tremendous capability to specifically ***identify various antigens*** (antigen is a foreign substance, usually a protein or a carbohydrate that elicits immune response). Exposure to an antigen leads to the development of immunological memory. As a result, a second encounter of the body to the same antigen results in a heightened state of immune response. The immune system recognizes and responds to foreign antigens as it is capable of distinguishing self and non-self. Autoimmune diseases are caused due to a failure to discriminate self and non-self antigens.

ORGANIZATION OF IMMUNE SYSTEM

The immune system consists of several organs distributed throughout the body (***Fig.42.1***). These lymphoid organs are categorized as primary and secondary.

Primary lymphoid organs

These organs provide appropriate micro-environment for the development and maturation of antigen-sensitive lymphocytes (a type of white blood cells). The ***thymus*** (situated above the heart) and bone marrow are the central or primary lymphoid organs. T-lymphocyte maturation occurs in the thymus while B-lymphocyte maturation takes place in the bone marrow.

Secondary lymphoid organs

These are the sites for the initiation of immune response. e.g. spleen, tonsils, lymph nodes, appendix, Peyers patches in the gut. Secondary lymphoid organs provide the microenvironment for interaction between antigens and mature lymphocytes.

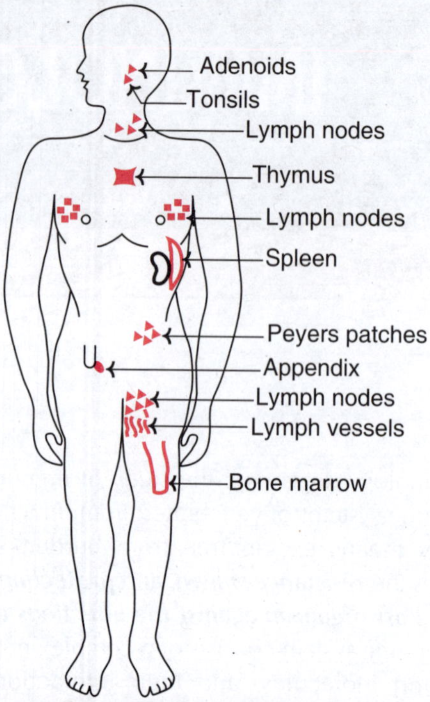

***Fig.42.1 :** A diagrammatic representation of human lymphatic system.*

CELLS OF THE IMMUNE SYSTEM

Two types of lymphocytes namely B-cells and T-cells are critical for the immune system. In addition, several accessory cells and effector cells also participate.

B-lymphocytes

The site of development and maturation of B-cells occurs in ***b***ursa fabricius in birds, and ***b***one marrow in mammals. During the course of immune response. B-cells mature into plasma cells and secrete ***antibodies (immunoglobulins)***.

The B-cells possess the capability to specifically recognize each antigen and produce antibodies (i.e. immunoglobulins) against it. B-lymphocytes are intimately associated with ***humoral immunity***. ***Immunoglobulins are described in Chapter 9***.

T-lymphocytes

The maturation of T-cells occurs in the ***t***hymus, hence the name. The ***T-cells*** can identify viruses and microorganisms from the antigens displayed on their surfaces. There are at least four different types of T-cells.

- ***Inducer T-cells*** that mediate the development of T-cells in the thymus.
- ***Cytotoxic T-cells*** (T_C), capable of recognizing and killing the infected or abnormal cells.
- ***Helper T-cells*** (T_H) that initiate immune responses.
- ***Suppressor T-cells*** mediate the suppression of immune response.

T-lymphocytes are responsible for the ***cell-mediated immunity***.

MAJOR HISTOCOMPATIBILITY COMPLEX

The major histocompatibility complex (***MHC***) represents a special group of ***proteins, present on the cell surfaces of T-lymphocytes***. MHC is involved in the recognition of antigens on T-cells. It may be noted here that the B-cell receptors (antibodies) can recognize antigens on their own, while T-cells can do so through the mediation of MHC.

In humans, the MHC proteins are encoded by a cluster of genes located on chromosome 6 (it is on chromosome 17 for mice). The major histo-compatibility complex in humans is referred to as ***human leukocyte antigen*** (***HLA***). Three classes of MHC molecules (chemically glycoproteins) are known in human. Class I molecules are found on almost all the nucleated cells of the body. Class II molecules are associated only with leukocytes involved in cell-mediated immune response. Class III molecules are the secreted proteins possessing immune functions e.g. complement components (C_2, C_4), tumor necrosis factor.

THE COMPLEMENT SYSTEM

The complement system is composed of about 20 plasma proteins that 'complement' the function of antibodies in defending the body from the invading antigens. The complementary factors are heat labile and get inactivated if heated at 56°C for about 30 minutes. The complement system helps the body immunity in 4 ways

1. **Complement fixation :** The complement system binds to the foreign invading cells and causes lysis of the cell membranes.
2. **Opsonization :** The process of promoting the phagocytosis of foreign cells is referred to as opsonization.
3. **Inflammatory reaction :** The complement system stimulates local inflammatory reaction and attracts phagocytic cells.
4. **Clearance of antigen-antibody complexes :** The complement system promotes the clearance of antigen-antibody complexes from the body.

Nomenclature of complement system : The complement proteins are designated by the letter 'C', followed by a component number—C_1, C_2, C_3 etc.

Types of reaction : The complement system brings about two sets of reactions :

1. Antibody dependent classical pathway.
2. Antibody independent alternative pathway.

Each one of the pathways consists of a series of reactions converting inactive precursors to active products by serine proteases which resembles blood coagulation.

THE IMMUNE RESPONSE

The immune response refers to the series of reactions carried out by the immune system in the body against the foreign invader. When an infection takes place or when an antigen enters the body, it is trapped by the macrophages in lymphoid organs. The phagocytic cells which are guarding the body by constant patrolling engulf and digest the foreign substance. T-helper cells (T_H) play a key role the immune response.

CYTOKINES

Cytokines are a group of ***proteins*** that ***bring*** about ***communication between different cell types*** involved in immunity. They are low molecular weight glycoproteins and are produced by lymphoid and non-lymphoid cells during the course of immune response. Cytokines may be regarded as soluble messenger molecules of immune system. They can act as short messengers between the cells or long range messengers by circulating in the blood and affecting cells at far off sites. The latter function is comparable to that of hormones.

The term ***interleukin (IL)*** is frequently used to represent cytokines. There are more than a dozen interleukins (IL-I......IL_{12}), produced by different cells with wide range of functions. The main function (directly or indirectly) of cytokines is to ***amplify immune responses*** and inflammatory responses.

Therapeutic uses of cytokines

It is now possible to produce cytokines *in vitro*. Some of the cytokines have potential applications in the practice of medicine. For instance, IL-2 is used in cancer immunotherapy, and in the treatment of immunodeficiency diseases. IL-2 induces the proliferation and differentiation of T- and B-cells, besides increasing the cytotoxic capacity of natural killer cells.

A group of cytokines namely interferons can combat viral infection by inhibiting their replication.

IMMUNITY IN HEALTH AND DISEASE

The prime function of immune system is to protect the host against the invading pathogens. The body tries its best to overcome various strategies of infectious agents (bacteria, viruses), and provides immunity.

Some of the important immunological aspects in human health and disease are briefly described.

AUTOIMMUNE DISEASES

In general, the immune system is self-tolerant i.e. not responsive to cells or proteins of self. Sometimes, for various reasons, the ***immune system fails to discriminate between self and non-self***. As a consequence, the cells or tissues of the body are attacked. This phenomenon is referred to as ***autoimmunity*** and the diseases are regarded as autoimmune diseases. The antibodies produced to self molecules are regarded as ***autoantibodies***. Some examples of autoimmune diseases are listed.

- Insulin-dependent diabetes (pancreatic β-cell autoreactive T-cells and antibodies).
- Rheumatoid arthritis (antibodies against proteins present in joints).
- Myasthenia gravis (acetylcholine receptor autoantibodies).
- Autoimmune hemolytic anemia (erythrocyte autoantibodies).

Mechanism of autoimmunity : It is widely accepted that autoimmunity generally occurs as a consequence of ***body's response against bacterial, viral or any foreign antigen***. Some of the epitopes of foreign antigens are similar (homologous) to epitopes present on certain host proteins. This results in cross reaction of antigens and antibodies which may lead to autoimmune diseases.

ORGAN TRANSPLANTATION

The phenomenon of transfer of cells, tissues or organs from one site to another (in the same organism, ***autograft*** or from another organism ***allograft***) is regarded as organ transplantation. In case of humans, majority of organ transplantations are allografts (between two individuals). The term ***xenograft*** is used if tissues/organs are transferred from one species to another e.g. from pig to man.

Organ transplantation is associated with immunological complications, and tissue rejection. This is because the host body responds to the transplanted tissue in a similar way as if it were an invading foreign organism. Major histocompatibility complex (MHC) is primarily involved in allograft rejection. This is due to the fact that MHC proteins are unique to each individual, and the immune system responds promptly to foreign MHCs.

Organ transplantation between closely related family members is preferred, since their MHCs are also likely to be closely related. And major immunological complications can be averted.

CANCERS

Growth of tumors is often associated with the formation of novel antigens. These ***tumor antigens*** (also referred to as oncofetal antigens e.g. α-fetoprotein) are recognized as non-self by the immune systems. However, tumors have developed several mechanisms to evade immune responses.

AIDS

Acquired immunodeficiency syndrome (AIDS), caused by human immunodeficiency virus, is characterized by immunosuppression, secondary neoplasma and neurological manifestations. AIDS primarily affects the cell-mediated immune system which protects the body from intracellular parasites such as viruses, and bacteria. Most of the immunodeficiency symptoms of AIDS are associated with a reduction in CD_4 (cluster determinant antigen 4) cells.

GENETICS

43

Genetics is the ***study of heredity***. It is appropriately regarded as the science that explains the similarities and differences among the related organisms.

The blood theory of inheritance in humans

For many centuries, it was customary to explain inheritance in humans through blood theory. People used to believe that the children received blood from their parents, and it was the union of blood that led to the blending of characteristics. That is how the terms '***blood relations***', '***blood will tell***', and '***blood is thicker than water***' came into existence. They are still used, despite the fact that blood is no more involved in inheritance. With the advances in genetics, the more appropriate terms should be as follows

- ***Gene relations*** in place of blood relations.
- ***Genes will tell*** instead of blood will tell.

BRIEF HISTORY AND DEVELOPMENT OF GENETICS

Genetics is relatively young, not even 150 years. The blood theory of inheritance was questioned in 1850s, based on the fact that the semen contained no blood. Thus, blood was not being transferred to the offspring. Then the big question was what was the hereditary substance.

Mendel's experiments : It was in 1866, an Austrian monk named Gregor Johann Mendel, for the first time reported the fundamental laws of inheritance. He conducted several experiments on the breeding patterns of pea plants. Mendel put forth the ***theory of transmissible factors*** which states that inheritance is controlled by ***certain factors*** passed from parents to offsprings. His results were published in 1866 in an obscure journal Proceedings of the Society of Natural Sciences.

For about 35 years, the observations made by Mendel went unnoticed, and were almost forgotten. Two European botanists (Correns and Hugo de Vries) in 1900, independently and simultaneously rediscovered the theories of Mendel. The year 1900 is important as it marks the beginning the modern era of genetics.

The origin of the word gene : In the early years of twentieth century, it was believed that the Mendel's inheritance factors are very closely related to chromosomes (literally coloured bodies) of the cells. It was in 1920s, the term ***gene*** (derived from a Greek word *gennan* meaning to produce) was introduced by Willard Johannsen. Thus, ***gene replaced the earlier terms inheritance factor or inheritance unit***.

Chemical basis of heredity : There was a controversy for quite sometime on the chemical basis of inheritance. There were two groups—the protein supporters and DNA supporters. It was in 1944, Avery and his associates presented convincing evidence that the chemical basis of heredity lies in DNA, and not in protein. Thus, ***DNA*** was finally identified as ***the genetic material***. Its structure was elucidated in 1952 by Watson and Crick.

Importance of genes in inheritance—studies on twins

Monozygotic or ***identical twins*** contain the ***same*** genetic material — ***DNA or genes***. Studies conducted on identical twins make startling revelations with regard to inheritance. One such study is described here.

Oskar Stohr and Jack Yufe were identical twins separated at birth. Oskar was taken to Germany where he was brought up by his grandmother as a Christian. Jack was raised by his father in Israel as a Jew. The two brothers were reunited at the age of 47. Despite the different environmental influences, their behavioural patterns and personalities were remarkably similar

- Both men had moustaches, wore two pocket shirts, and wire-rimmed glasses.
- Both loved spicy foods and tended to fall asleep in front of television.
- Both flushed the toilet before using.
- Both read magazines from back to front.
- Both stored rubber bands on their wrists.
- Both liked to sneeze in a room of strangers.

Besides Oskar and Jack, many other studies conducted on identical twins point out the importance of genes on the inherited characters related to personality and mannerisms.

BASIC PRINCIPLES OF HEREDITY IN HUMANS

The understanding of how genetic characteristics are passed on from one generation to the next is based on the principles developed by Mendel.

As we know now, the human genome is organized into a diploid (2n) set of 46 chromosomes. They exist as ***22 pairs of autosomes*** and ***one pair of sex chromosomes (XX/XY)***. During the course of meiosis, the chromosome number becomes haploid (n). Thus, haploid male and female gametes — sperm and oocyte respectively, are formed. On fertilization of the oocyte by the sperm, the diploid status is restored. This becomes possible as the zygote receives one member of each chromosome pair from the father, and the other from the mother. As regards the sex chromosomes, the males have X and Y, while the females have XX. The sex of the child is determined by the father.

Monogenic and polygenic traits

The genetic traits or characters are controlled by single genes or multiple genes. The changes in genes are associated with genetic diseases.

Monogenic disorders : These are the ***single gene disease traits*** due to alterations in the corresponding gene e.g. sickle-cell anemia, phenylketonuria. Inheritance of monogenic disorders usually follows the Mendelian pattern of inheritance.

Polygenic disorders : The genetic traits conferred ***by more than on gene*** (i.e multiple genes), and the disorders associated with them are very important e.g. height, weight, skin colours, academic performance, blood pressure, aggressiveness, length of life.

PATTERNS OF INHERITANCE

The heredity is transmitted from parent to offspring as individual characters controlled by genes. The genes are linearly distributed on chromosomes at fixed positions called ***loci***.

A gene may have different forms referred to as ***alleles***. Usually one allele is transferred from the father, and the other from the mother. The allele is regarded as ***dominant*** if the trait is exhibited due to its presence. On the other hand, the allele is said to be ***recessive*** if its effect is masked by a dominant allele. The individuals are said to be ***homozygous*** if both the alleles are the same. When the alleles are different they are said to be ***heterozygous***.

The pattern of inheritance of monogenic traits may occur in the following ways (***Fig.43.1***).

1. Autosomal dominant
2. Autosomal recessive
3. Sex-linked.

1. **Autosomal dominant inheritance :** A normal allele may be designated as ***a*** while an autosomal dominant disease allele as ***A*** (***Fig.43.1A***). The male with Aa genotype is an affected one while the female with aa is normal. Half of the genes from the affected male will carry the disease allele. On mating, the male and female gametes are mixed in different combinations. The result is that ***half of the children*** will be heterozygous (Aa) and ***have the disease***. Example of autosomal dominant inherited diseases are familial hypercholesterolemia, β-thalassemia, breast cancer genes.

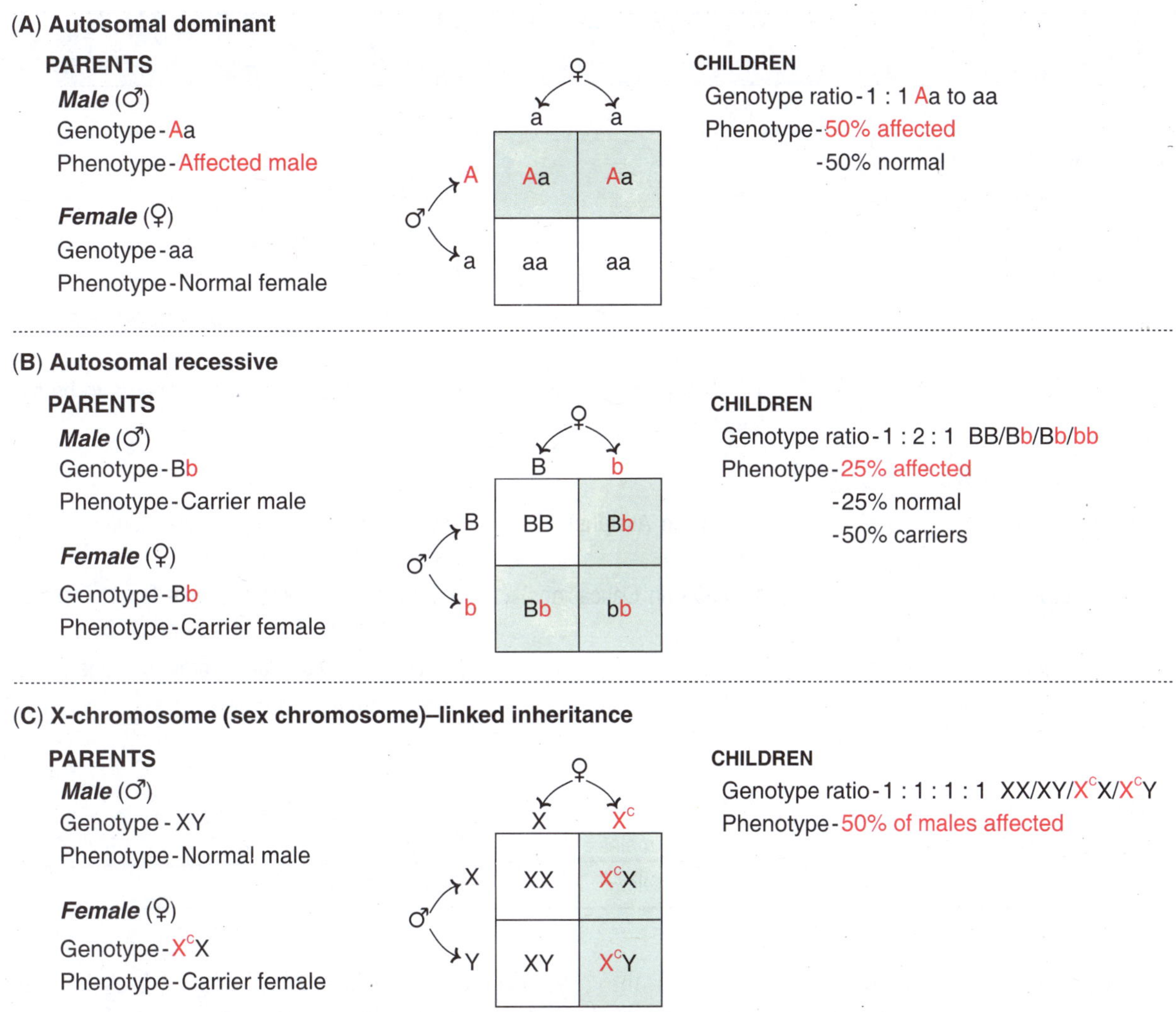

***Fig.43.1 :** Patterns of inheritance-autosomal dominant, autosomal recessive and X-linked (**Note :** Genotype refers to the description of genetic composition, while phenotype represents the observable character displayed by an organism).*

2. **Autosomal recessive inheritance :** In this case, the normal allele is designated as ***B*** while the disease-causing one is ***a*** (***Fig.43.1B***). The gametes of carrier male and carrier female (both with genotype Bb) get mixed. For these heterozygous carrier parents, there is one fourth chance of having an affected child. Cystic fibrosis, sickle-cell anemia and phenylketonuria are some good examples of autosomal recessive disorders.

3. **Sex (X)-linked inheritance :** In the ***Fig.43.1C***, sex-linked pattern of inheritance is depicted. A normal male (XY) and a carrier female (X^cY) will produce children wherein, half of the male children are affected while no female child is affected. This is due to the fact that the male children possess only one X chromosome, and there is no dominant allele to mark its effects (as is the case with females). Colour blindness and hemophilia are good examples of X-linked diseases.

A selected list of genetic disorders (monogenic traits) due to autosomal and sex-linked inheritance in humans is given in ***Table 43.1***.

GENETIC DISEASES IN HUMANS

The pattern of inheritance and monogenic traits along with some of the associated disorders are described above (***Table 43.1***). Besides ***gene mutations***, ***chromosomal abnormalities*** (aberrations) also result in genetic diseases.

Aneuploidy : The presence of ***abnormal number of chromosomes*** within the cells is referred to as aneuploidy. The most common aneuploid condition is ***trisomy*** in which three copies of a particular

Table 43.1 Selected examples of genetic disorders (monogenic traits) in humans

Inherited pattern/disease	*Estimated incidence*	*Salient features*
Autosomal dominant		
Familial hypercholesterolemia	1 : 500	High risk for heart diseases
Huntington's disease	1 : 5000	Nervous disorders, dementia
Familial retinoblastoma	1 : 12000	Tumors of retina
Breast cancer genes *(BRAC 1 and 2)*	1 : 800	High risk for breast and ovarian cancers
β-Thalassemia	1 : 2500 (in people of Mediterranean descent)	A blood disorder; the blood appears to be blue instead of red
Autosomal recessive		
Sickle-cell anemia	1 : 100 (in Africans)	Severe life threatening anemia; confers resistance to malaria
Cystic fibrosis	1 : 2500 (in Caucasians)	Defective ion transport; severe lung infections and early death (before they reach 30 years)
Phenylketonuria	1 : 2000	Mental retardation due to brain damage
α_1-Antitrypsin deficiency	1 : 5000	Damage to lungs and liver
Sex-linked		
Colour blindness	1 : 50 males	Unable to distinguish colours
Hemophilia (A/B)	1 : 10,000 males	Defective blood clotting
Duchenne muscular dystrophy	1 : 7000 males	Muscle wastage

chromosome are present in a cell instead of the normal two e.g. trisomy-21 causing ***Down's syndrome***; trisomy-18 that results in ***Edward's syndrome***. These are the examples of autosomal aneuploidy.

In case of sex-linked aneuploidy, the sex chromosomes occur as three copies. e.g. phenotypically male causing Klinefelter's syndrome has XXY; trisomy-X is phenotypically a female with XXX.

EUGENICS

Eugenics is a ***science of improving human race based on genetics***. Improving the traits of plants and animals through breeding programmes has been in practice for centuries.

Eugenics is a highly controversial subject due to social, ethical, and political reasons. The proponents of eugenics argue that people with desirable and good traits (good blood) should reproduce while those with undersirable characters (bad blood) should not. The advocates of eugenics, however, do not force any policy, but they try to convince the people to perform their duty voluntarily. The object of eugenics is to ***limit the production of people who are unfit to live in the society***.

Eugenics in Nazi Germany

Germany developed its own eugenic programme during 1930s. A law on eugenic sterilization was passed in 1933. In a span of three years, compulsory sterilization was done on about 250,000 people, who allegedly suffered from hereditary disabilities, feeble mindedness, epilepsy, schizophrenia, blindness, physical deformities, and drug or alcohol addiction.

The German Government committed many atrocities in the name of racial purity. Other countries, however do not support this kind of eugenics.

SELF-ASSESSMENT EXERCISES

ESSAY QUESTIONS

1. Describe the structure of eukaryotic cell.

2a. Classify carbohydrates with suitable examples. Write briefly the functions of carbohydrates.

2b. Describe the structure and functions of 3 disaccharides.

3a. Give an account of the structure and functions of phospholipids.

3b. Write briefly about the saturated and unsaturated fatty acids of biological importance.

4a. Classify amino acids with suitable examples.

4b. Describe the structure of proteins.

5a. Write an account of the structure and functions of nucleotides.

5b. Describe the structure of DNA.

6a. Give an account of the important serum enzymes in the diagnosis of diseases.

6b. Discuss the factors affecting enzyme activity.

7a. Describe the chemistry, biochemical functions, daily requirements, sources and deficiency manifestations of any one vitamin (fat soluble or water soluble).

7b. Classify vitamins and describe briefly their functions and deficiency disorders.

8a. Write an account of the digestion and absorption of proteins.

8b. Discuss the enzymes of gastrointestinal tract involved in the digestion of carbohydrates and fats.

9a. Describe the plasma proteins along with their characteristics and functions.

9b. Give an account of different types, structures and functions of immunoglobulins.

10a. Describe the structure and functions of hemoglobin.

10b. Give an account of the biosynthesis of heme. Add a note on the regulation.

11. Describe the components of electron transport (respiratory) chain. Add a note on oxidative phosphorylation.

11b. Give an account of the high-energy compounds in metabolism.

12. Define the terms—catabolism and anabolism. Describe the methods employed for the study of metabolism.

13a. Write briefly on the metabolism of glucose.

13b. Describe the events in cirtric acid cycle. Add a note on the energetics.

Note : *Question numbers (for essays/short notes/MCQs) correspond to the chapters in the book.*

14a. Give an account of fatty acid oxidation.

14b. Discuss cholesterol biosynthesis. Add a note on the significance of plasma cholesterol.

15a. Describe the synthesis of urea.

15b. Give an account of the metabolism of phenylalanine and tyrosine.

16. Describe the integration of major metabolic pathways with reference to energy metabolism.

17. Give an account of the degradation of purine nucleotides. Add a note on gout.

18a. Give an account of the trace elements and their metabolism in the body.

18b. Discuss the biochemical functions, dietary requirements, sources and absorption of calcium.

19. Give an account of the anterior pituitary hormones.

20a. Write briefly on the laboratory tests to assess liver function.

20b. Describe the renal function tests.

21a. Discuss the role of kidney in the regulation of acid-base balance.

21b. Classify acid-base disorders. Add a note on the compensatory mechanisms.

22. Describe the contractile proteins. Add a note on muscle contraction.

23a. Define basal metabolic rate (BMR) and discuss the factors affecting BMR.

23b. Give an account of the recommended dietary allowances (RDA) for macro-and micronutrients.

24a. Write briefly on the replication (biosynthesis) of DNA.

24b. Describe the different types of DNA recombination.

25a. Give a brief account of transcription.

25b. Describe protein biosynthesis (translation).

26. Write briefly on the regulation of gene expression with special reference to lactose (lac) operon.

27a. Describe the principles underlying the recombinant DNA technology.

27b. Give an account of the nucleic acid blotting techniques. Add a note on their importance.

28. Describe briefly the human genome project.

29. Give a brief account of the gene therapy.

30. Describe the various aspects of bioinformatics.

31. Describe briefly the metabolism of foreign compounds (xenobiotics) in the body.

32. Discuss the structure and functions of prostaglandins.

33. Write briefly on the structure of biological membranes and mechanisms of transport across membranes.

34a. Write an account of the types, sources and generation of free radicals. Add a note on their harmful effects.

34b. Describe the various antioxidants in biological systems.

35. Write briefly on the environmental biochemistry.

36. Discuss the regulation of blood glucose level (homeostasis of blood glucose).

37. Write briefly on the etiology and molecular basis of cancer.

38. Describe epidemiology of AIDS. Add a note on the immunological abnormalities in AIDS.

39. Discuss the common ring structures found in biomolecules.

40. Write briefly on the applications of diffusion, osmosis, and Donnan membrane equilibrium in biological systems.
41. Describe 5 different tools (techniques) employed in the biochemistry laboratroy.
42. Give a brief account of the immune system with special reference to B-lymphocytes and T-lymphocytes.
43. Describe the basic principles of heredity in humans with special reference to patterns of inheritance.

SHORT NOTES

1. (a) Differences between prokaryotic and eukaryotic cells,
   (b) Nucleus, (c) Mitochondria, (d) Lysosomes.
2. (a) Epimers, (b) Mutarotation, (c) Mucopolysaccharides, (d) Deoxysugars.
3. (a) Essential fatty acids, (b) Sphingomyelins, (c) Steroid nucleus, (d) Amphipathic lipids.
4. (a) Essential amino acids, (b) Denaturation, (c) Peptide bond, (d) Glutathione.
5. (a) Chargaff's rule, (b) Transfer RNA, (c) Nucleosides, (d) Complementary base pairs.
6. (a) Classification of enzymes, (b) Coenzymes, (c) Enzyme inhibition, (d) Isoenzymes.
7. (a) Thiamine pyrophosphate, (b) Calcitriol, (c) Pyridoxal phosphate, (d) Ascorbic acid.
8. (a) Intestinal absorption of glucose, (b) Disaccharidases, (c) Absorption of lipids, (d) Lactose intolerance.
9. (a) Blood clotting, (b) Bence–Jones proteins, (c) Electrophoresis of plasma proteins, (d) Albumin.
10. (a) Porphyrias, (b) Degradation of heme, (c) Sickle-cell anemia, (d) Thalassemias.
11. (a) Inhibitors of ETC, (b) Cytochromes, (c) P : O ratio, (d) ATP as energy currency.
12. (a) Intermediary metabolism, (b) Energy metabolism.
13. (a) Glycolysis, (b) Glycogen metabolism, (c) Hexose monophosphate shunt, (d) Glycogen storage diseases.
14. (a) Fatty acid biosynthesis, (b) Lipopropteins, (c) Ketone bodies, (d) Lipoptropic factors.
15. (a) Transamination, (b) Specialized products from glycine, (c) Serotonin, (d) Inborn errors of amino acid metabolism.
16. (a) Role of liver in metabolic integration, (b) Regulation of metabolic pathways.
17. (a) Salvage pathways, (b) Hyperuricemia, (c) Lesch—Nyhan syndrome, (d) PRPP.
18. (a) Regulation of plasma calcium, (b) Factors affecting iron absorption, (c) Wilson's disease, (d) Iodine.
19. (a) Second messengers, (b) Hormones of adrenal gland, (c) Sex hormones, (d) Aldosterone.
20. (a) van den Bergh reaction, (b) Types of jaundice, (c) Creatinine clearance, (d) Gastric function tests.
21. (a) Water balance in the body, (b) Regulation of electrolyte balance, (c) Blood buffers, (d) Anion gap.

22. (a) Collagen, (b) Prion diseases, (c) Composition of milk, (d) Diagnostic importance of CSF.
23. (a) Balanced diet, (b) Caloric value of foods, (c) Specific dynamic action, (d) Nitrogen balance.
24. (a) Telomeres, (b) Okazaki pieces, (c) Mutations, (d) DNA repair.
25. (a) Reverse transcription, (b) Post-translational modifications, (c) Genetic code, (d) Wobble hypothesis.
26. (a) One cistron-one subunit concept, (b) Gene inactivation by DNA methylation (c) Gene amplification, (d) Gene rearrangement.
27. (a) Restriction endonucleases, (b) Polymerase chain reaction, (c) Restriction fragment length polymorphism, (d) Recombinant vaccines.
28. (a) Methods used for mapping of genomes, (b) Applications of human genome sequencing.
29. (a) Vectors in gene therapy, (b) Antigene and antisense therapy.
30. (a) Biological databases, (b) Applications of bioinformatics.
31. (a) Cytochrome P_{450}, (b) Detoxification by conjugation.
32. (a) Synthesis of prostaglandins, (b) Leukotrienes.
33. (a) Active transport, (b) Facilitated diffusion.
34. (a) Free radicals, (b) Antioxidant enzyme system, (c) Nutrient antioxidants, (d) Metabolic antioxidants.
35. (a) Sulfur dioxide as a pollutant, (b) Toxic compounds in foodstuffs.
36. (a) Insulin, (b) Types of diabetes.
37. (a) Oncogenic viruses, (b) Tumor markers.
38. (a) Structure of HIV, (b) Drugs for the treatment of AIDS.
39. (a) Heterocyclic rings, (b) Isomerism.
40. (a) Surface tension, (b) Isotopes.
41. (a) Electrophoresis, (b) Chromatography.
42. (a) Innate immunity, (b) Major histocompatability complex, (c) Cytokines, (d) Autoimmune diseases.
43. (a) Monogenic and polygenic traits, (b) Sex-linked inheritance.

MULTIPLE CHOICE QUESTIONS

1(i). The most predominant chemical constituent of life

(a) water (b) protein (c) lipid (d) carbohydrate.

1(ii). The cellular organelles regarded as the digestive tract of the cell

(a) nucleus (b) Golgi apparatus (c) mitochondria (d) nucleus.

2(i). The naturally occurring monosaccharides in the mammalian tissues are of

(a) D-configuration (b) L-configuration (c) racemic mixture (d) DL-mixture.

2(ii). All of the following are ploymers of glucose except

(a) starch (b) inulin (c) dextrin (d) cellulose.

3(i). The unsaturated fatty acids that are essential to humans

(a) oleic acid and linoleic acid (b) linoleic acid and linolenic acid

(c) linolenic acid and arachidonic acid (d) arachidonic acid and palmitoleic acid.

3(ii). The nitrogenous base present in lecithins

(a) choline (b) ethanolamine (c) phosphatidic acid (d) plasmalogen.

4(i). Proteins are the polymers of

(a) L-α-amino acids (b) D-α-amino acids (c) L-β-amino acids (d) D-β-amino acids.

4(ii). At isoelectric pH (p^I), amino acids exist as

(a) cations (b) anions (c) zwitterions (d) acidic ions.

5(i). DNA is the polymer of

(a) ribonucleotides (b) deoxyribonucleotides (c) deoxyriboses (d) ribose sugars.

5(ii). RNA contains all of the following except

(a) adenine (b) guanine (c) uracil (d) thymine.

6(i). K_m value indicates the affinity between

(a) enzyme and coenzyme (b) enzyme and substrate

(c) coenzyme and substrate (d) enzyme and cofactor.

6(ii). Most of the coenzymes are derivatives of

(a) water soluble vitamins (b) fat soluble vitamins

(c) amino acids (d) nucleic acids.

6(iii). Serum alkaline phosphatase is elevated in

(a) obstructive jaundice (b) hepatitis

(c) pancreatitis (d) myocardial infarction.

6(iv). All of the following serum enzymes are elevated in myocardial infarction except

(a) creatine phosphokinase (b) aspartate transaminase

(c) alkaline phosphatase (d) lactate dehydrogenase.

7(i). The following two vitamins are regarded as water soluble hematopoietic

(a) E and K (b) folic acid and B_{12} (c) B_1 and B_6 (d) B_2 and biotin.

7(ii). Vitamin K acts as a coenzyme and brings about a post-translational modification namely

(a) hydroxylation (b) carboxylation (c) phosphorylation (d) methylation.

7(iii). One of the following amino acids is a component of pantothenic acid (coenzyme A)

(a) glycine (b) α-alanine (c) β-alanine (d) proline.

7(iv). The coenzyme that is regarded as the coenzyme of metabolic integration

(a) coenzyme A (b) thiamine pyrophosphate

(c) tetrahydrofolate (d) pyridoxal phosphate.

8(i). The monosaccharide that is most rapidly absorbed in the small intestine

(a) glucose (b) fructose (c) galactose (d) arabinose.

8(ii). The major vehicles for intestinal absorption of lipids
(a) chylomicrons (b) liposomes (c) emulsions (d) micelles.

9(i). The main difference between plasma and serum is the presence or absence of
(a) fibrinogen (b) albumin (c) globulins (d) plasmin.

9(ii). The following immunoglobulin is elevated in allergic reactions
(a) IgA (b) IgM (c) IgD (d) IgE.

10(i). Sickle-cell anemia is characterized by replacement of
(a) glutamate at 6th position of β-chain by valine
(b) glutamate at 6th position of α-chain by valine
(c) valine at 6th position of β-chain by glutamate
(d) valine at 6th position of α-chain by glutamate.

10(ii). The regulatory enzyme for heme biosynthesis in liver
(a) δ-aminolevulinate synthase (b) δ-aminolevulinate dehydratase
(c) heme synthase (d) ferrochelatse.

11(i). All of the following are high-energy compounds except
(a) phosphoenolpyruvate (b) carbamoyl phosphate
(c) creatine phosphate (d) fructose 6-phosphate.

11(ii). Cytochrome oxidase of electron transport chain is inhibited by
(a) carbon monoxide (b) rotenone (c) British antilewisite (d) amytal.

12. The energy liberated during the metabolic reactions is primarily
(a) ADP (b) CTP (c) ATP (d) GTP.

13(i). The enzymes of glycolysis are located in
(a) mitochondria (b) cytosol (c) nucleus (d) lactose.

13(ii). Rapoport-Leubering cycle, operative in RBC, is primarily concerned with the production of
(a) 1,3-bisphosphoglycerate (b)2,3-bisphosphoglycerate
(c) pyruvate (d) lactate.

13(iii). The number of NADH + H^+ produced in citric acid cycle
(a) 1 (b) 2 (c) 3 (d) 4.

13(iv). The enzyme defect in galactosemia
(a) galactose 1-phosphate uridyltransferase (b) UDP-hexose epimerase
(c) aldolase B (d) L-gulonolactose oxidase.

14(i). The fatty acids in the body are predominantly oxidized by
(a) α-oxidation (b) β-oxidation (c) γ-oxidation (d) ω-oxidation.

14(ii). The synthesis of ketone bodies occurs in
(a) muscle (b) liver (c) brain (d) kidney.

14(iii). The rate limiting enzyme in cholesterol biosynthesis
(a) HMG CoA synthase (b) HMG CoA reductase
(c) thiolase (d) mevalonate kinase.

14(iv). Fatty liver is primarily due to the accumulation of

(a) triacylglycerols (b) phospholipids (c) cholesterol (d) glycolipids.

15(i). Transamination is dependent on the coenzyme

(a) thiamine pyrophosphate (b)flavin adenine dinucleotide

(c) biotin (d) pyridoxal phosphate.

15(ii). The excretory end product of protein (amino acid) metabolism in humans

(a) ammonia (b) urea (c) uric acid (d) allantoin.

15(iii). Phenylketonuria is due to a defect in the enzyme

(a) tyrosine hydroxylase (b) phenylalanine hydroxylase

(c) tyrosinase (d) homogentisate oxidase.

15(iv). Transmethylation involves the transfer of methyl group from

(a) methionine (b) S-adenosylmethionine

(c) S-adenosylhomocysteine (d) S-adenosylcysteine.

16. The metabolic pathway of integration

(a) Embden-Meyerhoff pathway (b) electron transport chain

(c) Krebs cycle (d) fatty acid oxidation.

17(i). The purines are built upon a pre-existing

(a) ribose 5-phosphate (b) deoxyribose 5-phosphate

(c) phosphoribosyl pyrophosphate (d) ribulose 5-phosphate.

17(ii). The drug used for the treatment of gout

(a) sulfonamide (b) allopurinol (c) methotrexate (d) reserpine.

18(i). The regulation of plasma calcium (homeostasis of Ca) occurs by the action of

(a) calcitriol (b) parathyroid hormone

(c) calcitonin (d) all of the above.

18(ii). The element that can prevent lipid peroxidation

(a) iron (b) copper (c) selenium (d) chromium.

19(i). Cyclic AMP acts as a second messenger for a majority of

(a) steroid hormones (b) polypeptide hormones

(c) neurotransmitters (d) pheromones.

19(ii). The hormones synthesised by adrenal medulla

(a) epinephrine and norepinephrine (b) thyroxine and triiodothyronine

(c) cortisol and cortisone (d) pregnenolone and progesterone.

20(i). Hemolytic jaundice can be detected by van den Bergh reaction which gives

(a) indirect positive (b) biphasic (c) direct positive (d) none of these.

20(ii). The most sensitive test to assess renal function

(a) blood urea estimation (b) serum creatinine estimation

(c) creatinine clearance test (d) urea clearance test.

21(i). The hormone controlling water excretion by the kidney
(a) vasopressin (b) oxytocin (c) aldosterone (d) cortisone.

21(ii). The normal pH of blood
(a) 7.15-7.25 (b) 7.25-7.35 (c) 7.35-7.45 (d) 7.45-7.55.

22. The most abundant protein in mammals
(a) collagen (b) myosin (c) actin (d) elastin

23(i). The endocrine gland largely responsible for the regulation of BMR
(a) pituitary (b) pancreas (c) adrenal (d) thyroid.

23(ii). The protein with highest biological value
(a) meat protein (b) fish protein (c) milk protein (d) egg protein.

24(i). The biological information flows from DNA to RNA, and from RNA to
(a) lipids (b) carbohydrates (c) proteins (d) nucleotides

24(ii). The problem of supercoils in DNA replication is solved by a group of enzymes called
(a) DNA ligases (b) DNA polymerases (c) DNA topoisomerases (d) telomerases

25. Genetic code is located on the following RNA
(a) mRNA (b) tRNA (c) rRNA (d) any one of these

26. The structural 'Z' gene of lactose (lac) operon is responsible for the synthesis of the enzyme(s)
(a) β-galactosidase (b) permease (c) acetylase (d) all of them

27(i). The first pharmaceutical product of recombinant DNA technology, approved for human use
(a) insulin (b) growth hormone (c) interferon (d) hepatitis B vaccine

27(ii). Genetic immunization involves the administration of
(a) antigen (b) antibodies (c) DNA (d) RNA

28. The total number of genes in humans is in the range of
(a) 10,000—20,000 (b) 30,000—40,000 (c) 50,000—60,000 (d) 70,000—80,000

29. The direct delivery of genes into the cells is referred to as
(a) *ex vivo* gene therapy (b) *in vivo* gene therapy (c) antigene herapy (d) antisense therapy

30. Bioinformatics deals with the management and analysis of the data pertaining to
(a) DNA (b) RNA (c) proteins (d) all of them

31. The site of the metabolism of foreign compounds
(a) brain (b) liver (c) heart (d) kidney.

32. The precursor for the biosynthesis of prostaglandins
(a) linoleic acid (b) linolenic acid (c) arachidonic acid (d) prostanoic acid.

33. According to the fluid mosaic model, the biological membranes are composed of
(a) lipid monolayer (b) lipid bilayer (c) protein-lipid layer (d) protein bilayer.

34. Vitamin E (α-tocopherol) is a nutrient antioxidant that works in association with the following element
(a) iron (b) zinc (c) selenium (d) iodine

35. The neurotoxin present in the kesari dal (*Lathyrus sativus*) that causes paralysis of lower limbs (lathyrism)
(a) β-oxalylaminoalanine (b) goitrogens (c) biogenic amines (d) aflatoxins

36(i). Insulin lowers blood glucose level by
(a) promoting glucose uptake of tissues (b) increasing glycolysis in liver and muscle
(c) decreasing gluconeogenesis (d) all of the above.

36(ii). Normally, glucose gets excreted into urine when the plasma glucose concentration exceeds
(a) 100 mg/dl (b) 140 mg/dl (c) 180 mg/dl (d) 220 mg/dl.

37. The genes that regulate the development of cancer
(a) oncogenes (b) antioncogenes (c) apoptosis genes (d) all of the above.

38. Most of the immunodeficiency symptoms of AIDS are associated with
(a) B-lymphocytes (b) T-lymphocytes (c) CD_4-cells (d) complement system.

39(i). The compound possessing identical molecular formulae but different structures are referred to as
(a) monomers (b) oligomers (c) isomers (d) epimers.

39(ii). For the representation of chiral (asymmetric carbon containing) molecules, the following compound is taken as a reference standard
(a) glycerol (b) glyceraldehyde (c) glucose (d) amino acid.

40(i). The term ampholyte refers to a substance which can function as
(a) an acid (b) an alkali (c) a base (d) both acid and base.

40(ii). Hay's sulfur test, commonly employed in the laboratory for the detection of bile salts, is based on the principle of
(a) surface tension (b) diffusion (c) viscosity (d) adsorption.

41(i). The laboratory technique used for the separation of serum proteins
(a) chromatography (b) electrophoresis (c) photometry (d) ultracentrifugation.

41(ii). The marker enzyme to check the purity of mitochondia isolated by centrifugation
(a) DNA polymerase (b) hexokinase
(c) glutamate dehydrogenase (d) glucose 6-phosphatase.

42. The compounds that bring about communication between different cell types involved in immunity
(a) antigens (b) antibodies (c) complements (d) cytokines

43. One of the following is not a monogenic disorder
(a) sickle-cell anemia (b) skin colour (c) phenylketonuria (d) alkaptonuria

ANSWERS TO MULTIPLE CHOICE QUESTIONS

1(i). **(a)**	1(ii). **(b)**	2(i). **(a)**	2(ii). **(b)**	3(i). **(b)**	3(ii). **(a)**	4(i). **(a)**	4(ii). **(c)**	5(i). **(b)**
5(ii). **(d)**	6(i). **(b)**	6(ii). **(a)**	6(iii). **(a)**	6(iv). **(c)**	7(i). **(b)**	7(ii). **(b)**	7(iii). **(c)**	7(iv). **(a)**
8(i). **(c)**	8(ii) **(d)**	9(i). **(a)**	9(ii). **(d)**	10(i). **(a)**	10(ii). **(a)**	11(i). **(d)**	11(ii). **(a)**	12. **(c)**
13(i). **(b)**	13(ii). **(b)**	13(iii). **(c)**	13(iv). **(a)**	14(i). **(b)**	14(ii). **(b)**	14(iii). **(b)**	14(iv). **(a)**	15(i). **(d)**
15(ii). **(b)**	15(iii). **(b)**	15(iv). **(b)**	16. **(c)**	17(i). **(a)**	17(ii). **(b)**	18(i). **(d)**	18(ii). **(c)**	19(i). **(b)**
19(ii). **(a)**	20(i). **(c)**	20(ii). **(c)**	21(i) **(a)**	21(ii). **(c)**	22. **(a)**	23(i). **(d)**	23(ii). **(d)**	24(i). **(c)**
24(ii). **(c)**	25. **(a)**	26. **(a)**	27(i). **(a)**	27(ii). **(c)**	28. **(b)**	29. **(b)**	30. **(d)**	31. **(b)**
32. **(c)**	33. **(b)**	34. **(c)**	35. **(a)**	36(i). **(d)**	36(ii). **(c)**	37. **(d)**	38. **(c)**	39(i). **(c)**
39(ii). **(b)**	40(i). **(d)**	40(ii). **(a)**	41(i). **(b)**	41(ii) **(c)**	42. **(d)**	43. **(b)**.		

ABBREVIATIONS USED IN THIS BOOK

Appendix A

A	adenine, adenosine
Ab	antibody
ACP	acyl carrier protein
ACTH	adrenocorticotropic hormone
Acyl CoA	fatty acid derivative of coenzyme A
ADA	adenosine deaminase
ADH	alcohol dehydrogenase
ADH	antidiuretic hormone
ADP	adenosine diphosphate
AFP	α-fetoprotein
Ag	antigen
A/G	albumin/globulin (ratio)
AIDS	acquired immunodeficiency syndrome
ALA	δ-aminolevulinic acid
ALP	alkaline phosphatase
ALT	alanine transaminase
AMP	adenosine monophosphate
Apo-A	apoprotein-A
AP sites	apurinic sites
AST	aspartate transaminase
AT	α_1-antitrypsin
ATCase	aspartate transcarbamoylase
ATP	adenosine triphosphate
BAL	British antilewisite
BAO	basal acid output
BMR	basal metabolic rate
BOAA	β-oxalylaminoalanine
bp	base pair
BP	blood pressure
BPG	bisphosphoglycerate (2, 3-BPG, 1, 3-BPG)
BUN	blood urea nitrogen
BV	biological value
C	cytosine, cytidine
CA	carbonic anhydrase
Cal	calorie
cAMP	3′,5′-cyclic adenosine monophosphate (cyclic AMP)
CAP	catabolite activator protein
CCK	cholecystokinin
cDNA	complementary DNA
CD_4	cluster determinant antigen 4
CDP	cytidine diphosphate
CEA	carcinoembryonic antigen
CF	cystic fibrosis
CFTR	cystic fibrosis transmembrane regulator
C_H	constant heavy chain
CHD	coronary heart disease
ChE	cholinesterase
C_L	constant light chain
CMP	cytidine monophosphate
CNS	central nervous system
CoA or CoASH	coenzyme A
COHb	carboxyhemoglobin
CoQ	coenzyme Q (ubiquinone)
CPK (CK)	creatine phosphokinase (creatine kinase)
CPPP	cyclopentanoperhydrophenanthrene
CPS	carbamoyl phosphate synthase
CSF	cerebrospinal fluid
CT	calcitonin
CTP	cytidine triphosphate
dA	deoxyadenosine
dADP	deoxyadenosine diphosphate
DG	diacylglycerol
DAM	diacetyl monoxime

DCT	distal convoluted tubule
DEAE	diethyl aminoethylamine
DFP or DIFP	diisopropyl fluorophosphate
dGMP	deoxyguanosine monophosphate
DHAP	dihydroxyacetone phosphate
DHCC	dihydroxycholecalciferol (1, 25-DHCC; 24, 25-DHCC)
DHEA	dehydroepiandrosterone
DHF	dihydrofolate
DHT	dihydrotestosterone
DIT	diiodotyrosine
dl	deciliter
DMB	dimethyl benzimidazole
DNA	deoxyribonucleic acid
DNase	deoxyribonuclease
DNP	2, 4-dinitrophenol
DOPA	dihydroxy phenylalanine
DPG	diphosphoglycerate
DPP	dimethyl allyl pyrophosphate
E_O	redox potential
EC	enzyme commission
ECF	extracellular fluid
EDRF	endothelium-derived releasing factor
EDTA	ethylene diamine tetraacetate
EF	elongation factor
EFA	essential fatty acids
elFs	eukaryotic initiation factors
EGF	epidermal growth factor
ELISA	enzyme-linked immunosorbent assay
EM	Embden-Meyerhof
ER	endoplasmic reticulum
ES	enzyme-substrate complex
ES cells	embryonic stem cells
E-site	exit site
ETC	electron transport chain
FA	fatty acid
Fab	antigen binding fragment
FAD	flavin adenine dinucleotide
$FADH_2$	reduced FAD
FAS	fatty acid synthase
F 1, 6-BP	fructose 1,6-bisphosphate
F 2, 6-BP	fructose 2,6-bisphosphate
FDNB	1-fluoro 2,4-dinitrobenzene
FFA	free fatty acid
FH_4	tetrahydrofolate
FIGLU	formiminoglutamic acid
fMet	N-formylmethionine
FMN	flavin mononucleotide
$FMNH_2$	reduced FMN
F 1-P	fructose 1-phosphate
F 6-P	fructose 6-phosphate
Fp	flavoprotein
FSH	follicle stimulating hormone
FTM	fractional test meal
G	guanine, guanosine
g	gram
ΔG	free energy change
GABA	γ-aminobutyric acid
GAG	glycosaminoglycans
GDH	glutamate dehydrogenase
GDP	guanosine diphosphate
GFR	glomerular filtration rate
GGT (GT)	γ-glutamyl transpeptidase
GH	growth hormone
GHRH	growth hormone releasing hormone
GIP	gastric inhibitory peptide
GIT	gastrointestinal tract
Gla	γ-carboxy glutamate
GLC	gas liquid chromatography
Gly	glycine
GLUT	glucose transporters
GMP	guanosine monophosphate
GnRH	gonadotropin releasing hormone
GRH	growth hormone releasing hormone
G 6-P	glucose 6-phosphate
G 6-PD	glucose 6-phosphate dehydrogenase
GPP	geranyl pyrophosphate
GSH	glutathione (reduced form)
GSSG	glutathione (oxidized form)
GTP	guanosine triphosphate
GTT	glucose tolerance test
ΔH	change in enthalpy
HAC	human artificial chromosome
Hb	hemoglobin
HbA_1	adult hemoglobin
HbA_{1C}	glycosylated hemoglobin
HbF	fetal hemoglobin
HbO_2	oxyhemoglobin
HbS	sickle-cell hemoglobin
hCG	human chorionic gonadotropin
HDL	high density lipoproteins
HGPRT	hypoxanthine guanine phosphoribosyltransferase
HIAA	hydroxy indole acetic acid
HIV	human immunodeficiency virus
HLA	human leukocyte antigen

HLH	helix-loop-helix
HMG CoA	β-hydroxy β-methylglutaryl CoA
HMP	hexose monophosphate
HNPCC	hereditary nonpolyposis colon cancer
hnRNA	heterogeneous nuclear RNA
Hp	haptoglobin
HPLC	high performance liquid chromatography
HRE	hormone responsive element
Hsp	heat shock protein
5HT	5-hydroxytryptamine
HTH	helix-turn-helix
IDDM	insulin dependent diabetes mellitus
IDL	intermediate density lipoproteins
IF	initiation factor
Ig	immunoglobulin
IGF	insulin-like growth factor
IL	interleukins
IMP	inosine monophosphate
INH	isonicotinic acid hydrazide (isoniazid)
IPP	isopentenyl pyrophosphate
IR	infrared
IU	international unit
IV	intravenous
K	dissociation constant
KA	King Armstrong
K_a	dissociation constant of acid
Kbp	kilo base pair
KD	kilodalton
Keq	equilibrium constant
α-KG	α-ketoglutarate
Ki	inhibition constant
K_m	Michaelis constant
KJ	kilojoule
LCAT	lecithin cholesterol acyltransferase
LDH	lactate dehydrogenase
LDL	low density lipoproteins
LFT	liver function tests
LH	luteinizing hormone
LT	leukotrienes
Lp-a	lipoprotein-a
M	molar
MAO	maximal acid output
MAO	monoamine oxidase
Mb	myoglobin
MbO_2	oxymyoglobin
mEq	milliequivalents
mg	milligram
MHC	major histocompatibility complex
MI	myocardial infarction
MIT	monoiodotyrosine
mM	millimolar
mol. wt.	molecular weight
mRNA	messenger RNA
MSH	melanocyte stimulating hormone
mtDNA	mitochondrial DNA
MW	molecular weight
NAD^+	nicotinamide adenine dinucleotide
NADH	reduced NAD^+
$NADP^+$	nicotinamide adenine dinucleotide phosphate
NADPH	reduced $NADP^+$
NAG	N-acetylglutamate
NANA	N-acetylneuraminic acid
NE	niacin equivalents
NEAT	Non-exercise activity thermogenesis
NEPA	Non-exercise physical activity
ng	nanogram (10^{-9} g)
NIDDM	non-insulin dependent diabetes mellitus
NMR	nuclear magnetic resonance
NPN	non-protein nitrogen
NPU	net protein utilization
OAA	oxaloacetate
Ob	obese
OD	optical density
OMP	orotidine monophosphate
Osm	osmoles
PABA	para amino benzoic acid
PAF	platelet-activating factor
PAGE	polyacrylamide gel electrophoresis
PAPS	phosphoadenosine phosphosulfate
PBG	porphobilinogen
PBI	protein bound iodine
PCM	protein-calorie malnutrition
pCO_2	partial presence of CO_2
PCR	polymerase chain reaction
PCT	proximal convoluted tubule
PDGF	platelet derived growth factor
PDH	pyruvate dehydrogenase
PEG	polyethylene glycol
PEM	protein-energy malnutrition
PEP	phosphoenol pyruvate
PER	protein efficiency ratio
PFK	phosphofructokinase

PG	prostaglandins
PGA	pteroyl glutamic acid
pH	negative log of (H^+)
Pi	inorganic phosphate
p^I	isoelectric pH
p^{Ka}	negative log of Ka
PKU	phenylketonuria
PL	phospholipid
PLP	pyridoxal phosphate
pO_2	partial pressure of O_2
POMC	pro-opiomelanocortin
PPi	inorganic pyrophosphate
ppm	parts per million
PRL	prolactin
PRPP	5-phosphoribosyl 1-pyrophosphate
PT	prothrombin time
PTH	parathyroid hormone
PTH	phenyl thiohydantoin
PUFA	polyunsaturated fatty acids
QPRT	quinolinate phosphoribosyltransferase
RAPD	random amplified polymorphic DNA
ras	rat sarcoma
RBC	red blood cells
RBP	retinol binding protein
RDA	recommended dietary (daily) allowance
rDNA	recombinant DNA
RE	retinol equivalents
RER	rough endoplasmic reticulum
RF	releasing factor
Rf	ratio of fronts
RFLP	restriction fragment length polymorphism
RIA	radioimmunoassay
RMR	resting metabolic rate
RNA	ribonucleic acid
RNAP	RNA polymerase
RNase	ribonuclease
R 5-P	ribose 5-phosphate
RQ	respiratory quotient
rRNA	ribosomal RNA
RSV	rous sarcoma virus
RT	reverse transcriptase
SAM	S-adenosylmethionine
SCID	severe combined immunodeficiency
SDA	specific dynamic action
Sf	Svedberg floatation
SGOT	serum glutamate oxaloacetate transaminase
SGPT	serum glutamate pyruvate transaminase
SIDS	sudden infant death syndrome
sn	stereospecific number
SNPs	single nucleotide polymorphisms
sRNA	soluble RNA
SRS	slow reacting substance
T	thymine, thymidine
T	thymus (T-lymphocyte)
T_3	3,5,3′-triiodothyronine
T_4	3,5,3′,5′-tetraiodothyronine (thyroxine)
TBG	thyroxine binding globulin
TCA	tricarboxylic acid
TF	tissue factor
TG	triacylglycerol
Tgb	thyroglobulin
THF	tetrahydrofolate
TIBC	total iron binding capacity
TLC	thin layer chromatography
Tm	tubular maximum
TMP	thymidine monophosphate
TNF	tumor necrosis factor
TPP	thiamine pyrophosphate
TRH	thyrotropin releasing hormone
tRNA	transfer RNA
TSH	thyroid stimulating hormone
TX	thromboxane
μm	micrometer (10^{-6} m)
UCP	uncoupling protein
UDP	uridine diphosphate
UDPG	uridine diphosphate glucose
μl	microliter (10^{-6} l)
μM	micromoles (10^{-6} M)
UMP	uridine monophosphate
UTP	uridine triphosphate
UV	ultraviolet
V_H	variable heavy chain
V_L	variable light chain
VLDL	very low density lipoproteins
VMA	vanillyl mandelic acid
V_{max}	velocity maximum
WBC	white blood cells
XP	xeroderma pigmentosum
YAC	yeast artificial chromosome

PRACTICAL BIOCHEMISTRY—PRINCIPLES

Appendix B

QUALITATIVE EXPERIMENTS

Several laboratory qualitative experiments are performed to identify the compounds of biochemical importance (carbohydrates, proteins/amino acids, non-protein nitrogenous substances) and to detect the abnormal constituents of urine. The principles of the reactions pertaining to the most widely employed qualitative tests are described here.

I. REACTIONS OF CARBOHYDRATES

The carbohydrates used in the laboratory for the qualitative tests include glucose and fructose (monosaccharides), sucrose, lactose and maltose (disaccharides) and starch (polysaccharide). The principles of the reactions of carbohydrates are given below.

1. **Molisch test :** It is a general test for the detection of ***carbohydrates***. The strong H_2SO_4 hydrolyses carbohydrates (poly- and disaccharides) to liberate monosaccharides. The monosaccharides get dehydrated to form furfural (from pentoses) or hydroxy methylfurfural (from hexoses) which condense with α-naphthol to form a violet coloured complex.

2. **Iodine test :** Polysaccharides combine with iodine to form a coloured complex. Thus, ***starch*** gives blue colour while dextrins give red colour with iodine.

3. **Benedict's test :** This is a test for the identification of ***reducing sugars***, which form enediols (predominantly under alkaline conditions). The enediol forms of sugars reduce cupric ions (Cu^{2+}) of copper sulfate to cuprous ions (Cu^{+}) which form a yellow precipitate of cuprous hydroxide or a red precipitate of cuprous oxide.

4. **Barfoed's test :** The principle of this test is the same as that of Benedict's test except that the reduction is carried out in mild acidic medium. Since acidic medium is not favourable for reduction, only strong reducing sugars (monosaccharides) give this test positive. Thus, Barfoed's test serves as a key reaction to distinguish ***monosaccharides*** from disaccharides.

5. **Seliwanoff's test :** This is a specific test for ***ketohexoses***. Concentrated hydrochloric acid dehydrates ketohexoses to form furfural derivatives which condense with resorcinol to give a cherry red complex.

6. **Foulger's test :** This is also a test for ***ketohexoses***. The furfural derivatives formed from ketohexoses condense with urea in the presence of stannous chloride to give a blue colour.

7. **Rapid furfural test :** ***Ketohexoses*** are converted to furfural derivatives by HCl which form a purple colour complex with α-naphthol.

8. **Osazone test :** Phenylhydrazine in acetic acid, when boiled with reducing sugars forms osazones. The first two carbons (***C_1*** and ***C_2***) are involved in this reaction. The sugars that differ in their configuration on these two carbons give the same type of osazones, since the difference is marked by binding with phenylhydrazine. Thus, glucose, fructose and mannose give the same type (needle shaped) of osazones. However, the osazones of reducing disaccharides differ—maltose gives sunflower-shaped while lactose powder-puff shaped.

9. **Sucrose hydrolysis test :** Sucrose is a ***non-reducing sugar***, hence it does not give Benedict's and Barfoed's tests. Sucrose can be hydrolysed by concentrated HCl, to be converted to glucose and fructose (reducing monosaccharides) which answer the reducing reactions. However, after sucrose hydrolysis, the medium has to be made alkaline (by adding Na_2CO_3) for effective reduction process.

II. REACTIONS OF PROTEINS

The proteins employed in the laboratory for the qualitative tests include albumin, globulins, casein, gelatin and peptones. The principle of the most common reactions of proteins/amino acids performed in the laboratory are given hereunder.

A. PRECIPITATION REACTIONS

Proteins exist in colloidal solution due to hydration of polar groups (—COO^-, —NH_3^+, —OH). They can be precipitated by dehydration or neutralization of polar groups. Several methods are in use to achieve protein precipitation.

1. **Precipitation by neutral salts :** The process of protein precipitation by the addition of neutral salts such as ammonium sulfate or sodium sulfate is referred to as salting out. This phenomenon is explained on the basis of dehydration of protein molecules by salts. This causes increased protein-protein interaction, resulting in molecular aggregation and precipitation.

The amount of salt required for protein precipitation depends on the size (molecular weight) of the protein molecule. In general, the higher is the protein molecular weight, the lower is the salt required for precipitation. Thus, serum globulins are precipitated by half saturation with ammonium sulfate while albumin is precipitated by full saturation.

2. **Precipitation by salts of heavy metals :** Heavy metal ions like Pb^{2+}, Hg^{2+}, Fe^{2+}, Zn^{2+}, Cd^{2+} cause precipitation of proteins. These metals being positively charged, when added to protein solution (negatively charged) in alkaline medium result in precipitate formation.

3. **Precipitation by anionic or alkaloid reagents :** Proteins can be precipitated by trichloroacetic acid, sulphosalicylic acid, phosphotungstic acid, picric acid, tannic acid, phosphomolybdic acid etc. By the addition of these acids, the proteins existing as cations are precipitated by the anionic form of acids to produce protein-sulphosalicylate, protein-tungstate, protein-picrate etc.

The anionic reagents such as phosphotungstic acid and trichloroacetic acid are used to prepare protein-free filtrate of blood needed for several estimations (e.g., urea, sugar) in the laboratory.

4. **Precipitation by organic solvents :** Organic solvents such as alcohol are good protein precipitating agents. They dehydrate the protein molecule by removing that water envelope and cause precipitation.

B. COLOUR REACTIONS

The proteins give several colour reactions which are often useful to identify the nature of the amino acids present in them as shown in the table.

Colour reactions of proteins/amino acids

Reaction	*Specific group or amino acid*
1. Biuret reaction	Two peptide linkages
2. Ninhydrin reaction	α–Amino acids
3. Xanthoproteic reaction	Benzene ring of aromatic amino acids (Phe, Tyr, Trp)
4. Millons reaction	Phenolic group (Tyr)
5. Hopkins-Cole reaction	Indole ring (Trp)
6. Sakaguchi reaction	Guanidino group (Arg)
7. Nitroprusside reaction	Sulfhydryl groups (Cys)
8. Sulfur test	Sulfhydryl groups (Cys)
9. Pauly's test	Imidazole ring (His)
10. Folin–Coicalteau's test	Phenolic groups (Tyr)

1. **Biuret reactions :** Biuret is a compound formed by heating urea to 180°C. When biuret is treated with dilute copper sulfate in alkaline medium, a purple colour is obtained. This is the basis of biuret test used for identification of proteins and peptides.

Biuret test is answered by compounds containing two or more CO—NH groups i.e., ***peptide bonds***. All proteins and peptides possessing atleast two peptide linkages i.e., tripeptides (with 3 amino acids) give positive biuret test. The principle of biuret test is conveniently used to detect the presence of proteins in biological fluids. The mechanism of biuret test is not clearly known. It is believed that the colour is due to the formation of a copper co-ordinated complex.

2. **Ninhydrin reaction :** The α-amino acids react with ninhydrin to form a purple, blue or pink colour complex (Ruhemann's purple).

$$\text{Amino acid} + \text{Ninhydrin} \longrightarrow \text{Keto acid} + NH_3 + CO_2 + \text{Hydrindantin}$$

$$\text{Hydrindantin} + NH_3 + \text{Ninhydrin} \longrightarrow \text{Ruhemann's purple}$$

3. **Xanthoproteic reaction :** Xanthoproteic reaction is due to nitration of ***aromatic amino acids*** (tryptophan, tyrosine and phenylalanine) on treatment with strong nitric acid at high temperature.

4. **Millon's test :** This test is given by the amino acid ***tyrosine***, or any other compound containing hydroxyphenyl ring. A red colour or precipitate is obtained in this reaction due to the formation of mercury complex of nitrophenol derivative.

5. **Hopkins-Cole reaction :** This reaction is specific for the indole ring of ***tryptophan***. It combines with formaldehyde in the presence of the oxidizing agent (sulfuric acid with mercuric sulfate) to form a violet or purple coloured compound.

6. **Sakaguchi reaction :** Arginine, containing guanidino group, reacts with α-naphthol and alkaline hypobromite to form a red colour complex.

7. **Sulfur test :** This is a test specific for sulfur containing amino acids namely ***cysteine*** and ***cystine***, but not methionine. When cysteine and cystine are boiled with sodium hydroxide, organic sulfur is converted to inorganic sodium sulfide. This reacts with lead acetate to form a black precipitate of lead sulfide. Methionine does not give this test, since sulfur of methionine is not split by alkali.

8. **Pauly's test :** This reaction is specific for ***histidine*** (imidazole ring). Diazotised sulfanilic acid reacts with imidazole ring in alkaline medium to form a red coloured complex.

9. **Molisch test :** This is a specific test for the detection of ***carbohydrates***. The proteins containing carbohydrates (e.g., glycoproteins) give this test positive. Albumin contains carbohydrate bound to it, hence answers Molisch test.

III. REACTIONS OF NON-PROTEIN NITROGENOUS SUBSTANCES

The non-protein nitrogenous (NPN) substances of biochemical importance include urea, uric acid and creatinine.

1. **Sodium hypobromite test :** This is a test for the detection of ***urea***. Sodium hypobromite decomposes urea to liberate nitrogen. The latter can be identified by brisk effervescence.

2. **Specific urease test :** The enzyme urease (source-horse gram) specifically acts on ***urea*** to liberate ammonium carbonate (alkali). The latter can be identified by a colour change in phenophthalein indicator (pink colour in alkaline medium).

3. **Benedict's uric acid test :** ***Uric acid***, being a strong reducing agent, reduces phosphotungstate to tungsten blue in alkaline medium.

4. **Murexide test :** ***Uric acid*** is oxidized by nitric acid to give purpuric acid (reddish yellow). This in turn combines with ammonia to form purple red colour ammonium purpurate (murexide).

5. **Jaffe's test :** Creatinine reacts with picric acid in alkaline medium to form orange red colour complex.

IV. ABNORMAL CONSTITUENTS OF URINE

Urine is the most important excretory fluid from the body. Some of the diseases are associated with an excretion of abnormal constituents in urine. The identification of such compounds in urine is of great diagnostic importance.

Urine abnormal constituent	*Associated disorder*
Albumin	Kidney damage (glomerulonephritis)
Hemoglobin	Damage to kidneys or urinary tract.
Glucose	Diabetes mellitus, renal glycosuria.
Ketone bodies	Diabetes mellitus, starvation.
Bile salts	Obstructive jaundice
Bile pigments	Obstructive jaundice and hepatic jaundice.

1. **Sulfosalicylic acid test :** ***Proteins*** get precipitated by sulfosalicylic acid by forming protein-sulfosalicylate.

2. **Heat coagulation test :** This is a test for the detection of ***albumin*** and/or ***globulins*** in urine. Heat coagulation test is based on the principle of denaturation of proteins, followed by coagulation.

(**Note :** Small amounts of dilute acetic acid are added to dissolve the phosphates and sulfates that get precipitated on heating.)

3. **Benzidine test :** This test detects the presence of ***blood***. Hemoglobin (acts like peroxidase) decomposes hydrogen peroxide to liberate nascent oxygen (O–) which oxidises benzidine to a green or blue coloured complex.

(**Note :** Pus cells of urine possess peroxidase activity which interferes in benzidine test. This can be eliminated by boiling the urine prior to the test to inactivate the enzyme).

4. **Benedict's test :** This is a semiquantitative test for the detection of urine reducing sugars (primarily glucose). Benedict's test is based on the principle of reducing property of sugars (described in detail under reactions of carbohydrates). Colour of the precipitate formed indicates the approximate amount of ***glucose*** present in urine. Thus, green turbidity = traces; green precipitate = 0.5%; yellow precipitate = 1%; orange precipitate = 1.5% brick red precipitate = 2%.(***Note :*** Benedict's test is not specific to glucose, since it can be answered by any reducing substance).

5. **Glucose oxidase test :** This is a strip test for the ***specific detection of glucose***. The enzyme glucose oxidase oxidizes glucose to liberate hydrogen peroxide which in turn is converted to nascent oxygen (O–) by peroxidase enzyme. The compound O-diansidine combines with nascent oxygen to form a coloured (yellow to red) complex.

6. **Rothera's test :** Nitroprusside in alkaline medium reacts with keto group of ***ketone bodies*** (acetone and acetoacetate) to form a purple ring. This test is not given by β-hydroxybutyrate.

7. **Hay's test :** This test is based on the surface tension lowering property of ***bile salts*** (sodium glycocholate and sodium taurocholate). Sulfur powder sprinkled on the surface of urine containing bile salts sinks to the bottom.

8. **Pettenkofer's test :** This test is employed for the detection of ***bile salts***. The furfural derivatives (by reacting sugar with concentrated H_2SO_4) condense with bile salts to form a purple ring.

9. **Gmelin's test :** Nitric acid oxidizes the bile pigment bilirubin to biliverdin (green) or bilicyanin (blue). Gmelin's test gives a play of colours and is used for the identification of ***bile pigments***.

10. **Fouchet's test :** This test is also employed for the detection of ***bile pigments***. Bile pigments are adsorbed on barium sulfate. Fouchet's reagents (containing ferric chloride in trichloroacetic acid) oxidizes bilirubin to biliverdin (green) and bilicyanin (blue).

QUANTITATIVE EXPERIMENTS

Quantitative experiments, dealing with the determination of concentrations of several biologically important compounds and the assay of many enzymes, are of great significance in the laboratory practice. Very often, the ultimate diagnosis and prognosis of a large number of diseases are guided by the quantitative biochemical investigations.

The principles involved in some of the quantitative experiments, commonly employed in the biochemistry laboratory by an undergraduate student, are briefly described here.

1. Blood glucose estimation

The quantitative determination of blood (plasma/serum) glucose is of great importance in the diagnosis and monitoring of diabetes mellitus.

(i) ***Folin Wu method*** **:** Alkaline copper (cupric ions) is reduced by glucose when boiled with protein-free blood filtrate to cuprous oxide. The cuprous oxide in turn reacts with phosphomolybdic acid to form blue coloured oxides of molybdenum. The intensity of the colour can be measured in a colorimeter at a wavelength 680 nm. [Folin Wu method is rather old and is not specific for glucose determination, since other substances (e.g., fructose, lactose, glutathione) also bring about reduction. Consequently the blood glucose level when estimated by Folin Wu method is higher i.e., normal fasting is 80-120 mg/dl against true glucose 60-100 mg/dl]

(ii) ***O-Toluidine method*** **:** Glucose combines with O-toluidine when boiled in acid medium to form a green coloured complex which can be measured in a colorimeter at a wavelength 630 nm. (This method determines glucose alone).

(iii) ***Glucose oxidase-peroxidase (GOD—POD) method*** **:** This is an enzymatic determination of blood glucose. Glucose gets oxidized by glucose oxidase to gluconic acid and hydrogen peroxide. The enzyme peroxidase converts hydrogen peroxide to water and oxygen. The oxygen in turn reacts with 4-aminophenzone in the presence of phenol to form a pink coloured complex, the intensity of which can be measured at 530 nm.

2. Blood urea estimation

Determination of blood urea (reference range 10-40 mg/dl) is important for the evaluation of kidney (renal) function. Elevation of blood urea is associated with pre-renal (diabetic coma, thyrotoxicosis), renal (acute glomerulonephritis, polycystic kidney) and post-renal (obstruction in the urinary tract, due to tumors, stones) conditions.

Diacetyl monoxime (DAM) method : Urea when heated with diacetyl monoxime forms a yellow coloured complex of dioxime derivatives which can be measured at 520 nm.

3. Serum creatinine estimation

Estimation of serum creatinine (reference range 0.5-1.5 mg/dl) is used as a diagnostic test to assess kidney function. Serum creatinine is not influenced by endogenous and exogenous factors, as is the case with urea. Hence, some workers consider serum creatinine as a more reliable indicator of renal function.

Alkaline picrate method : This method is based on Jaffe's reaction. Creatinine reacts with alkaline picrate to form creatinine picrate, an orange red coloured complex, which can be measured in a colorimeter at 530 nm.

(**Note** : Urinary creatinine can also be determined by employing the same principle given above).

4. Determination of serum proteins

The normal concentration of total serum proteins is in the range 6-8 g/dl (albumin 3.5-5.0 g/dl; globulins 2.5-3.5 g/dl; A/G ratio is 1.2 to 1.5 : 1). The A/G ratio is lowered either due to a decrease in albumin or an increase in globulins.

Serum albumin concentration is decreased in liver diseases, severe protein malnutrition, and excretion of albumin in urine (due to renal damage). Serum globulin concentration is elevated in chronic infections and multiple myeloma.

Biuret method : Peptide bonds (−CO−NH) of proteins react with cupric ions in alkaline medium to form a violet colour complex which is measured at a wavelength 530 nm. This method is suitable for total serum proteins with estimation.

Bromocresol green (BCG) dye method : This technique is employed for the estimation of serum albumin. BCG dye reacts with albumin to form an intense blue-green coloured complex which can be measured at 628 nm.

5. Estimation of serum bilirubin

The total bilirubin concentration in serum is 0.2-1 mg/dl (conjugated ~ 0.6 mg/dl; unconjugated ~ 0.4 mg/dl). Elevation in serum bilirubin concentration is observed in jaundice. Unconjugated bilirubin is increased in hemolytic jaundice, conjugated bilirubin in obstructive jaundice, while both of them are increased in hepatic jaundice.

van den Bergh reaction : Serum bilirubin estimation is based on van den Bergh reaction. The principle of the reaction is that diazotised sulfanilic acid (formed by mixing equal volumes of sulfanilic acid in HCl and sodium nitrite) reacts with bilirubin to form a purple coloured azobilirubin which can be measured at 540 nm.

6. Estimation of serum cholesterol

Serum cholesterol concentration (reference range 150-225 mg/dl) is elevated in atherosclerosis, diabetes mellitus, obstructive jaundice and hypothyroidism. Decreased levels are observed in hyperthyroidism.

Acetic anhydride method : Serum cholesterol reacts with acetic anhydride in the presence of glacial acetic acid and concentrated H_2SO_4 to form a green coloured complex. Intensity of this colour is measured at 560 nm.

7. Estimation of serum uric acid

Uric acid is the end product of purine metabolism. Its concentration in serum is increased (reference range - men 4-8 mg/dl; women 3-6 mg/dl) in gout.

Henry-Caraway's method : Uric acid in the protein-free filtrate when treated with phosphotungstic acid in the presence of sodium carbonate (alkaline solution) gives a blue coloured complex which can be measured at 660 nm.

8. Estimation of serum calcium

Serum calcium level is elevated (reference range 9-11 mg/dl) in hyperparathyroidism and decreased in hypothyroidism.

O-Cresolphthalein complexone method : Calcium reacts with the dye, O-cresolphthalein complexone (CPC) in alkaline solution to form a complex which can be measured at a wavelength 660 nm.

9. Estimation of serum phosphorus (inorganic)

Serum phosphate (reference range 3-4.5 mg/dl) is increased in hypoparathyroidism, and decreased in hyperparathyroidism and renal rickets.

For the determination of serum phosphate, serum proteins are precipitated by trichloroacetic acid. The protein-free filtrate containing inorganic phosphate is reacted with molybdic acid reagent to form phosphomolybdate. The latter in turn is reduced to molybdenum blue by treatment with 1-amino 2-naphthol-4 sulfonic acid (ANSA). The intensity of the blue colour is measured at 689 nm.

10. Determination of SGPT and SGOT

Serum glutamate pyruvate transaminase (SGPT; alanine transaminase) and serum glutamate oxaloacetate transaminase (SGOT; aspartate transaminase) are two important diagnostic enzymes. SGPT activity (reference range 5-40 IU/L) is more specifically increased in liver diseases (hepatic jaundice). SGOT activity is elevated (reference range 5-45 IU/L) in heart diseases (myocardial infarction).

Principle of assay : SGPT catalyses the following reaction

$$\text{L-Alanine} + \alpha\text{-ketoglutarate} \longrightarrow \text{L-glutamate} + \text{pyruvate}$$

SGOT brings about the following reaction

$$\text{L-Aspartic acid} + \alpha\text{-ketoglutarate} \longrightarrow \text{L-glutamate} + \text{oxaloacetate}$$

The keto acid (pyruvate or oxaloacetate), formed in the above reaction, when treated with 2,4-dinitrophenyl hydrazine forms dinitrophenyl hydroazone (brown colour) in alkaline medium which can be measured at 505 nm.

11. Determination of serum alkaline phosphatase

The activity of the enzyme serum alkaline phosphatase (normal range 3-13 KA Units/dl) is elevated in rickets and obstructive jaundice.

Principle of assay : Alkaline phosphatase hydrolyses disodium phenylphosphate liberating phenol. On treatment with 4-amino antipyrine in alkaline medium, phenol gives ferricyanide (reddish colour) which can be measured at 520 nm.

12. Determination of serum amylase

Serum amylase activity is increased (reference range 80-180 Somogyi Units/dl) in acute pancreatitis.

Principle of assay : Amylase acts on starch and hydrolyses to dextrins and maltose. Starch forms blue coloured complex with iodine, a decrease in the colour (measured at 670 nm) is proportional to the activity of amylase.

13. Analysis of cerebrospinal fluid

Cerebrospinal fluid (CSF) is the aqueous medium surrounding the brain and spinal cord. From the biochemical perspective, estimation of proteins and glucose in CSF is important. Increase in protein (reference range 15-40 mg/dl) and decrease in glucose (reference range 50-75 mg/dl) in the cerebrospinal fluid are observed in tuberculosis meningitis.

CSF protein estimation : Sulfosalicylic acid (in sodium sulfate solution) precipitates CSF proteins and the turbidity is measured at 680 nm.

CSF glucose estimation : Any one of the standard methods employed for the determination of blood glucose (already described) can be used for CSF glucose estimation.

CLINICAL BIOCHEMISTRY LABORATORY

Appendix C

The ultimate application of the biochemistry subject is for the health and welfare of mankind. Clinical biochemistry (also known as ***clinical chemistry*** or ***chemical pathology***) is the laboratory service absolutely essential for medical practice. The results of the biochemical investigations carried out in a clinical chemistry laboratory will help the clinicians to determine the diseases (diagnosis) and for follow-up of the treatment/recovery from the illness (prognosis). Biochemical investigations hold the key for the ***diagnosis and prognosis of diabetes mellitus, jaundice, myocardial infarction, gout, pancreatitis, rickets, cancers, acid-base imbalance*** etc. Successful medical practice is unimaginable without the service of clinical biochemistry laboratory.

The ***biological fluids*** employed in the clinical biochemistry laboratory include ***blood***, ***urine***, ***cerebrospinal fluid*** and pleural fluid. Among these, ***blood*** (directly or in the form of plasma or serum) is ***frequently used*** for the investigations in the clinical biochemistry laboratory.

On rare occassions, tissue biopsy samples may also be used for diagnosis.

COMMON ABNORMALITIES AND BIOCHEMICAL PROFILE

A large number of conditions, particularly diseases (diabetes mellitus, genetic diseases, MI, organ failure etc.), are associated with abnormal biochemical profiles. A selected list of them is given in ***Table C***, e.g. cancer can be diagnosed by estimating tumor markers.

Table C : Common abnormalities associated with blood/plasma biochemical profile (selected examples listed)

Abnormality	Biochemical profile
Organ function failure	***Renal failure***......serum creatinine and urea elevated ***Liver failure***......serum bilirubin and ALT increased.
Myocardial infarction	Cardiac troponin I and CPK in serum elevated.
Cancer	Tumor markers elevated e.g. PSA in prostate cancer.
Diabetes mellitus	Increased blood glucose and HbA_{1C}
Genetic diseases	Serum phenylalanine elevated in phenylketonuria; NH_3 increased in urea cycle disorders.
Nutritional disorders	Serum albumin low in kwashiorkor.
Endocrine dysfunction	TSH high in hypothyroidism and low in hyperthyrodism.
Changes in blood pH	Metabolic acidosis—serum bicarbonate low; metabolic alkalosis—serum bicarbonate high.
Inflammation/infection	C-reactive protein elevated.
Use of drugs	Increase in serum uric acid in cancer chemotherapy.
Poisons	Decrease in the activity of serum butylcholinesterase in organo-phosphorus poisoning.
Stress	Increase in serum cortisol and catecholamines.
Altered physiological conditions	Elevated levels of serum hCG in pregnancy; increased blood lactate after strenuous exercise

THE REFERENCE RANGE

There exists biological variations between normal and healthy individuals for any analyte/compound measured in the laboratory. This is referred to as reference range around a mean (or average) which is considered to be normal. The most important reference biochemical values are given inside back cover.

If may benoted that the reference values of some analytes differ between sexes, different age groups, and in different ethinic groups. Further, the biochemical resuls differ from laboratory to laboratory due to variations in the methods employed.

In general, for a healthy population (depending on the gender, age and ethenicity), the range that includes 95% of target population is considered to the reference range (for a given analyte). In statistical terms, a mean ±2 SD is considered to be a reference range. The values of the analytes outside the reference ranges are considered to be abnormal, and therefore need to be further investigated.

TECHNIQUES USED IN CLINICAL BIOCHEMISTRY LABORATORY

The reader may refer tools of biochemistry (**Chapter 41**) and principles of practical biochemistry (***appendix B***) for a brief knowledge on the techniques employed in clinical biochemistry. Broadly, the following principles or techniques are used

- Absorption spectrophotometry
- Fluorescence spectrophotometry (or spectrophotofluorimetry)
- High pressure liquid chromatography
- Flame photometry
- Ion-specific electrodes
- Atomic absorption spectrometry
- Electrophoresis
- Radioimmunoassay
- Enzyme-linked immunosorbant assay

Dry chemistry dipstick

For certain assays, the reagents, enzymes or antibodies can be put together on ***specially designed plastic strip*** namely dipstick. Dipsticks are handy devices for use at the bedside in a hospital, doctors clinic or even at home by individuals. Some commonly used tests employing dry chemistry dipsticks are listed—

- Finger-prick blood glucose measurement by test strip and glucometer.
- Strips to semiquantitatively detect urinary levels of glucose, ketone bodies, proteins etc. (individual dipsticks or multidipsticks are available)
- Dipstick for home pregnancy test to detect human chorionic gonadotropin (hCG) in urine.

COLLECTION OF BLOOD

Venous blood is ***most commonly*** used for a majority of biochemical investigations. It can be drawn from any prominent vein (usually from a vein on the front of the elbow). ***Capillary blood*** (<0.2 ml) obtained from a finger or thumb, is less frequently employed. ***Arterial blood*** (usually drawn under local anesthesia) is used for ***blood gas determinations***.

Precautions for blood collection : Use of sterile (preferably disposable) needles and syringes, cleaning of patients skin, blood collection in clean and dry vials/tubes are some of the important precautions.

CHOICE OF BLOOD SPECIMENS

Biochemical investigations can be performed on 4 types of blood specimens-whole blood, plasma, serum and red blood cells. The selection of the specimen depends on the parameter to be estimated. ***Whole blood*** (usually mixed with an anticoagulant) is used for the estimation of hemoglobin, carboxyhemoglobin, pH, glucose, urea, non-protein nitrogen, pyruvate, lactate, ammonia etc. (Note : for glucose determination, plasma is prefered in recent years).

Plasma, obtained by centrifuging the whole blood collected with an anticoagulant, is employed for the parameters—fibrinogen, glucose, bicarbonate, chloride, ascorbic acid etc.

Serum is the supernatant fluid that can be ***collected after centrifuging the clotted blood.*** It is the most frequently used specimen in the clinical biochemistry laboratory. The parameters estimated in serum include proteins (albumin/globulins), creatinine, bilirubin, cholesterol, uric acid, electroylets (Na^+, K^+, Cl^-), enzymes (ALT, AST, LDH, CK, ALP, ACP, amylase, lipase) and vitamins.

It may be noted that plasma is physiologic fluid while serum is prepared in the laboratory.

Red blood cells are employed for the determination of abnormal hemoglobins, glucose 6-phosphate dehydrogenase, pyruvate kinase etc.

ANTICOAGULANTS

Certain biochemical tests require unclotted blood. Anticoagulants are employed for collecting such specimens.

Heparin : Heparin (inhibits the conversion of prothrombin to thrombin) is an ideal anticoagulant, since it does not cause any change in blood composition. However, other anticoagulants are prefered to heparin, due to the cost factor.

Potassium or sodium oxalate : These compounds precipitate calcium and inhibit blood coagulation. Being more soluble, potassium oxalate (5-10 mg per 5 ml blood) is prefered.

Potasium oxalate and sodium fluoride : These anticoagulants are employed for collecting blood to estimate glucose. Further sodium fluoride inhibits glycolysis and preserves blood glucose concentration.

Ammonium oxalate and potassium oxalate : A mixture of these two compounds in the ratio 3 : 2 is used for blood collection to carry out certain hematological tests.

Enthylene diaminetetracetic acid (EDTA) : It chelates with calcium and blocks coagulation. EDTA is employed to collect blood for hematological examinations.

HEMOLYSIS

The rupture or lysis of RBC, releasing the cellular constituents interferes with the laboratory investigations. Therefore, utmost care should be taken to avoid hemolysis when plasma or serum are used for biochemical tests. Use of dry syringes, needles and containers, allowing slow flow of blood into syringe are among the important precautions to avoid hemolysis.

PRESERVATION OF BLOOD SPECIMENS

Plasma or serum should be separated within 2 hours after blood collection. It is ideal and advisable to analyse blood, plasma or serum, immediately after the specimen collection. This however, may not be always possible. In such a case, the samples (usually plasma/serum) can be stored at 4°C until analysed. For enzyme analysis, the sample are preserved at –20°C.

TYPES OF LABORATORY TESTS

The biochemical investigations (on blood/plasma/serum) carried out in the clinical biochemistry laboratory may be grouped into different types.

1. **Discretionary or on-off tests :** Most common clinical biochemistry tests that are designed to answer specific questions. e.g., does the patient have increased blood urea/glucose concentration? Normally, these tests are useful to support the diagnosis.

2. **Biochemical profiles :** These tests are based on the fact that more useful information on the patients disease status can be obtained by analysing more constituents rather than one e.g., ***plasma electrolytes*** (Na^+, K^+, Cl^-, bicarbonate, urea); ***liven function tests*** (serum bilirubin, ALT, AST).

3. **Dynamic function tests :** These tests are designed to measure the body's response to external stimulus e.g., oral glucose tolerance test (to assess glucose homeostasis) : bromosulphthalein test (to assess liver function).

4. **Screening tests :** These tests are commonly employed to identify the inborn errors of metabolism, and to check the entry of toxic agents (pesticides, lead, mercury) into the body.

5. **Metabolic work-up tests :** The programmed intensive investigations carried out to identify the endocrinological disorders come under this category.

The term ***emergency tests*** is frequently used in the clinical laboratory. It refers to the tests to be performed immediately to help the clinician for proper treatment of the patient e.g., blood glucose, urea, serum electrolytes.

COLLECTION OF URINE

Urine, containing the metabolic waste products of the body in water is the most important excretory fluid. For biochemical investigations, urine can be collected as a single specimen or for 24 hours. Single specimens of urine, normally collected in the morning, are useful for qualitative tests e.g., sugar, proteins. Twenty four hour urine collections (done between 8 AM to 8 AM) are employed for quantitative estimation of certain urinary constituents e.g., proteins, hormones, metabolites.

Preservatives for urine : For the collection of 24 hr urine samples, preservatives have to be used or else urine undergoes changes due to bacterial action. Hydrochloric acid, toluene, light petroleum, thymol, formalin etc., are among the common preservatives used.

CEREBROSPINAL FLUID (CSF)

CSF is a fluid of the nervous system. It is formed by a process of selective dialysis of plasma by the choroid plexuses of the ventricles of the brain. The total volume CSF is 100-200 ml.

Collection of CSF : CSF is collected by puncturing the interspace between the 3rd and the 5th lumbar vertebrae, under aseptic conditions and local anesthesia.

Biochemical investigations on CSF : Protein, glucose and chloride estimations are usually performed in the clinical biochemistry laboratory.

ANALYSIS IN CLINICAL BIOCHEMISTRY LABORATORY

The details on the biochemistry of health and disease states in relation to normal and abnormal biochemical data have been described in this book. The reader must refer the following chapters for examples of laboratory tests

- Chapter 6 for enzyme alterations in disease states.
- Chapter 19 for hormonal tests.
- Chapter 20 for organ function tests.
- Chapter 21 for acid-base abnormalities.
- Chapter 37 for tumor markers.

For ready reference, the most common reference biochemical values are given on inside back cover .

QUALITY CONTROL

Quality control in clinical biochemistry laboratory refers to the reliability of investigative service. Any error in the laboratory will jeopardize the lives of patients. It is therefore utmost important that the laboratory errors are identified and rectified.

Quality control comprises of four interrelated factors namely precision, accuracy, specificity and sensitivity.

Precision refers to the reproducibility of the result when the same sample is analysed on different occasions (replicate measurements) by the same person. For instance, the precision is good, if the blood glucose level is 78, 80 and 82 mg/dl on replicates.

Accuracy means the closeness of the estimated result to the true value e.g., if true blood urea level is 50 mg/dl, the laboratory reporting 45 mg/dl is more accurate than the one reporting 35 mg/dl.

Specificity refers to the ability of the analytical method to specifically determine a particular parameter e.g., glucose can be specifically estimated by enzymatic glucose oxidase method.

Sensitivity deals with the ability of a particular method to detect small amounts of the measured constituent.

METHODS OF QUALITY CONTROL

Internal quality control refers to the analysis of the same pooled sample on different days in a laboratory, the results should vary within a narrow range.

External quality control deals with the analysis of a sample received from outside, usually from a national or regional quality control centre. The results obtained are then compared.

AUTOANALYSERS IN CLINICAL CHEMISTRY

The heavy work load in the clinical biochemistry laboratory has lead to the discovery of autoanalysers. These modern equipment are useful to analyse hundreds of samples in a short time. Single channel and multi-channel machines (autoanalysers) based on the principles of either continuous or discrete analysis are available on the market.

CASE STUDIES WITH BIOCHEMICAL CORRELATIONS

Appendix D

I. Nutritional Disorders

CASE STUDY 1

A 3-year old female child had stunted growth, edema (particularly on legs and hands) discoloration of skin and hair, apathy and moon-face. She also had frequent respiratory infections and diarrhea. On enquiring, the mother informed the physician that the child was mostly breast-fed until 2 years of age, and for the past one year she was being given dilute buffalo milk and a small quantities of rice with ghee and dhal. The following are the laboratory data of the child

Parameter	Subject	Reference range
Hemoglobin	7 g/dl	12-14 g/dl
Serum proteins (total)	4 g/dl	6-8 g/dl
Serum albumin	2 g/dl	3-4.5 g/dl
Serum potassium	2.8 mEq/1	3.5-4.5 mEq/1

Comments and diagnosis

The clinical manifestations and the nutritional history, supported by the laboratory data clearly indicate that the child was suffering from ***kwashiorkor***, a predominant nutritional disorder, in the developing countries. Edema occurred due to lack of adequate serum proteins to maintain water distribution between blood and tissues. The immunological response of the child to infections was very low. Deficiency of serum K^+ was observed due to diarrhea.

CASE STUDY 2

A one year old male baby had growth retardation, reduced physical activity, muscle wasting (emaciation), loose folds of skin wrapped over bones. The mother of the child, now in third month of gestation, informed the physician that she had been giving the child rice gruel, along with breast milk. The laboratory data of the child are given hereunder.

Parameter	Subject	Reference range
Hemoglobin	7 g/dl	12-14 g/dl
Serum proteins (total)	6 g/dl	6-8 g/dl
Serum albumin	3 g/dl	3-4.5 g/dl
Serum potassium	3.5 mEq/1	3.5-5 mEq/1

Comments and diagnosis

The child was suffering from ***marasmus***, a nutritional disorder, predominantly due to the deficiency of calories. Marasmus mostly occurs in children less than one year of age. It can be distinguished from kwashiorkor (Case 1) by lack of edema and almost unaltered serum albumin level.

CASE STUDY 3

A 12-year old boy had difficulty to see in dim light (evenings and while he enters cine theatres). His plasma retinol level was very low.

Comments and diagnosis

This is a case of ***night blindness (nyctalopia)*** due to vitamin A deficiency. Since vitamin A is closely associated with dim light vision, its deficiency causes prolonged dark adaptation time. Severe deficiency of vitamin A leads to dryness of conjunctiva and cornea (xerophthalmia) and the occurrence of triangular plaques known as Bitot's spots.

CASE STUDY 4

A 6-year old boy had bone deformities such as bow legs and pigeon chest. He had a history of delayed eruption of teeth. On enquiring, the mother informed the physician that the boy had been on a strict vegetarian diet with low intake of milk as well as fats and oils. The laboratory findings of this boy are given on next page.

Parameter	*Subject*	*Reference range*
Calcium	8.1 mg/dl	9-11 mg/dl
Phosphate	2.5 mg/dl	3-4.5 mg/dl
Alkaline phosphatase	300 U/l	50-100 U/l

Diagnosis and discussion

This case is classical example of ***rickets*** due to vitamin D deficiency. As a result of low serum calcitriol (the biochemically active form of vitamin D), calcium and phosphate levels are not maintained in the circulation, hence mineralization is impaired, ultimately causing bone deformities such as bow legs. Serum alkaline phosphatase is elevated in rickets in a vain attempt to result in bone formation.

CASE STUDY 5

A 25-year old working woman had been in the habit of eating freeze-stored vegetables and fruits for over 2 years. She developed the symptoms of spongy and sore gums, swollen joints. She used to have frequent infections (common cold and respiratory tract infections) and observed delay in the healing of her wounds. She was mildly anaemic (hemoglobin 10 g/dl against normal 12-14 g/dl). Her friend, who happened to be a nutritionist, advised her to eat fresh fruits and vegetables, besides the supplementation of vitamin C. The situation improved in about 3 months.

Comments and diagnosis

The manifestations described along with the response of the woman to vitamin C supplementation clearly point that the woman was suffering from ***scurvy***, a vitamin C deficiency nutritional disorder. This is observed in people who do not eat vitamin C rich foods and/or fresh fruits and vegetables. Vitamin C (ascorbic acid) participates in the synthesis of collagen, besides its involvement as an antioxidant. The symptoms observed in scurvy are due to impairment in these two functions.

CASE STUDY 6

A 25-year old strict vegetarian woman, with two children aged 5 years and 3 years, complained of tiredness and appeared pale. In the recent few months, the woman had heavy and prolonged menstrual flow. Enquiries revealed that her consumption of milk and milk products was reasonably good, but leafy vegetables was low. The laboratory investigations showed that the woman's hematocrit was 28% (reference range 40% to 50%), while her hemoglobin concentration was 8 g/dl (reference range 12-14 g/dl).

Diagnosis and discussion

The woman with a low hemoglobin concentration and reduced hematocrit depicts a very common nutritional disorder — ***iron deficiency anemia***. A strict vegetarian diet coupled with low consumption of leafy vegetables led her to iron deficiency.

CASE STUDY 7

A 20-year old medical student had a delicious dinner with mushroom dishes in a Chinese restaurant. The next day morning, he was admitted in a hospital with abdominal cramps, vomiting and diarrhea.

Comments and diagnosis

This appears to be a case of ***food-poisoning***, caused by a dangerous species of ***mushroom*** (most probably *Amanita phalloides*). The toxicity is due to toxin α-amanitin which is not inactivated by cooking. α-Amanitin is a cyclic octapeptide and it strongly inhibits RNA polymerase II and thus transcription (synthesis of RNA).

II. METABOLIC DISEASES/INBORN ERRORS

CASE STUDY 8

A boy, aged 12 years was given treatment for prolonged diarrhea. After improvement, he complained abdominal discomfort and diarrhea with a feeling of being bloated, after consumption of milk. He was taken to a physician who advised him to stop his intake of milk. He felt better in 3 days.

Diagnosis and discussion

The boy was suffering from ***acquired lactose intolerance***. This was precipitated by diarrhea where the intestinal mucosal cells were denuded faster. The brush border of the intestine houses the enzyme lactase which is lost due to diarrhea, and hence typical symptoms of flatulence.

CASE STUDY 9

A normal one month old baby had a history of vomiting and diarrhea that frequently occurred after breast feeding. The urine gave a positive test for reducing sugars (Benedict's test) while the test was negative by Glucostix (specific for glucose). The RBC were found to be totally deprived of activity of the enzyme galactose 1-phosphate uridyltransferase (Reference range 4-30 units/g of hemoglobin).

Comments and diagnosis

The baby was suffering from ***galactosemia***, a metabolic disorder, due to the deficiency of the enzyme galactose 1-phosphate uridyltransferase. The disease is characterized by increase in blood galactose level, and its excretion into urine. For the babies suffering from galactosemia, milk has to be removed from the diet and replaced with infant formula containing sucrose.

CASE STUDY 10

A 30-year old man was admitted in a cardiology ward after he complained chest pain. His clinical and biochemical investigations (ECG changes, isoenzymes of creatine phosphokinase etc.) indicated that he suffered a mild myocardial infarction. Timely intervention and appropriate treatment saved him from death. The lipid profile data of this man are given here.

Parameter	Subject (mg/dl)	Reference range
Serum total cholesterol	450	150-200
LDL-cholesterol	380	<100
HDL-cholesterol	40	40-60
VLDL-cholesterol	30	20-40
Serum triglycerides	150	75-150

Comments and diagnosis

This man had highly elevated total and LDL-cholesterol while other lipid profile parameters are within the normal limits. This is a case of ***familial hypercholesterolemia (hyperlipoproteinemia type IIa)***. The victims of this disorder have decreased number of LDL receptors. They have very high risk of coronary heart diseases in the 3rd and 4th decades of life. Lipid-lowering drugs, besides low cholesterol, low fat diet and regular exercise are useful for these people.

CASE STUDY 11

A 5-year old boy had delayed developments and was unable to speak or walk properly. He was found to be mentally retarded with characteristic seizures and tremors. The boy had a plasma phenylalanine level of 30 mg/dl (reference range of 1-2 mg/dl). His urine gave a positive ferric chloride test, indicating an elevated excretion of phenylpyruvate.

Diagnosis and discussion

An elevation is plasma phenylalanine concentration with an increased excretion of phenylpyruvate in urine, supported by the clinical manifestations, clearly indicate that the boy was a victim of ***phenylketonuria (PKU)***. This is an inborn error of phenylalanine metabolism due to a defect in the enzyme phenylalanine hydroxylase. As a result of this, phenylalanine cannot be converted to tyrosine, hence gets diverted to alternate pathways, producing metabolites which are directly or indirectly inolved in growth failure and mental retardation. The treatment for PKU is to provide diet low or deficient in phenylalanine, so that their plasma phenylalanine levels are kept within the normal limits.

CASE STUDY 12

A 50-year old man had a complaint of arthritis. His urine gave a positive Benedict's test for reducing substances. On careful examination, his urine was found to turn to blackish (coke-like) colour on standing. The laboratory investigations revealed that the old man had normal fasting (90 mg/dl) and post-prandial (120 mg/dl) blood glucose and normal serum uric acid (4 mg/dl) concentrations.

Comments and diagnosis

This is a case of ***alkaptonuria***, an inborn error of tyrosine metabolism, due to a defect in the enzyme homogentisate oxidase. The urine of these subjects contains large quantities of homogentisate — a strong reducing substance (hence Benedict's test positive). Homogentisate on standing gets oxidized to impart the characteristic coke colour to urine. Arthritis in alkaptonuric patients is believed to be due to the deposition of pigment alkapton (in the joints) produced from homogentisate.

CASE STUDY 13

A 55-year old man complained severe pain in the joints. He was a non-vegetarian and consumed alcohol occasionally. His laboratory findings are as follows.

Parameter	Subject	Reference range
Serum uric acid	12 mg/dl	3-7 mg/dl
Blood urea	25 mg/dl	15-40 mg/dl
Urinary uric acid	2.5 g/day	0.5-0.7 g/day
Urinary pH	4.5	5.5-7.0

The subject responded to the treatment by the drug-allopurinol.

Diagnosis and discussion

Elevation of uric acid in serum and urine along with case history, further supported by the response of the subject to the drug allopurinol, clearly indicate that the patient was suffering from ***gout.*** Over-production of uric acid, the excretory end product of purine metabolism, causes gout. Crystals of sodium urate get deposited in the joints causing gouty arthritis. Allopurinol competitively inhibits the enzyme xanthine oxidase and lowers the formation of uric acid, hence the relief by administering this drug.

CASE STUDY 14

A 4-year old boy showed signs of learning disability and aggressive behaviour, besides pain in the joints. It was observed that he had an irresistible urge to bite his fingers and lips. The laboratory investigations revealed that the boy had serum uric acid concentration of 10 mg/dl (reference range 4-6 mg/dl).

Diagnosis and discussion

The clinical manfestations along with elevated serum uric acid level support that the boy was suffering from ***Lesch-Nyhan syndrome***. This is an inborn error of purine metabolism (salvage pathway). It is a sex-linked disorder affecting only males due to the deficiency of the enzyme hypoxanthine-guanine phosphoribosyl transferase (HGPRT). Lesch Nyhan syndrome is characterized by increased production of uric acid, often causing gouty arthritis, besides the neurological abnormalities.

CASE STUDY 15

A 2-year old male child with retarded growth was found to have anemia. This was non-responsive to treatment by iron, folic acid, pyridoxine and cobalamin. The following are his laboratory investigative data

Parameter	Subject	Reference range
Hemoglobin	7 g/dl	14-16 g/dl
Hematocrit	30 %	40-50 %
Orotic acid (urine)	800 mg/day	<2 mg/day

Comments and diagnosis

The child was suffering from the metabolic disorder ***orotic aciduria***. This is due to deficiency of the enzymes orotate phosphoribosyl transferase and OMP decarboxylase of pyrimidine biosynthesis. Feeding a diet rich in uridine and/or cytidine is an effective treatment for orotic aciduria.

CASE STUDY 16

An African negro boy (aged 12 years), studying in an Indian High School, was admitted in a hospital with complaints of fever and severe pain in the arms and legs. His laboratory investigative data are given hereunder.

Parameter	Subject	Reference range
Hemoglobin	7 g/dl	14-16 g/dl
Hematocrit	20 %	40-50 %
Hemoglobin electrophoresis	A distinct HbS band with slower movement than adult hemoglobin (HbA) detected	—
Microscopic examination of blood smear	Crescent (sickle) shaped erythrocytes	—

Comments and diagnosis

The laboratory data of the body with sickle-shaped erythrocytes and altered hemoglobin electrophoretic pattern with a distinct HbS band supported by anemia and reduced hematocrit, along with the clinical manifestations support the diagnosis of ***sickle cell anemia***. This disorder primarily occurs in black population and the patients may die before they reach adulthood (< 20 years). Sickle cell anemia is caused by an abnormality in hemoglobin. It occurs due to the replacement of glutamate by valine at the 6th position of β-chain of hemoglobin.

CASE STUDY 17

A school boy aged 12 years complained of abdominal pain and was admitted in a hospital. He was weak and tired, and had a history of behavioural disturbance and epileptic form of seizures. Clinical examination revealed that the boy had an enlarged liver. His cornea showed the presence of Kayser Fleischer ring. The laboratory data of the boy are given below.

Parameter	Subject	Reference range
Serum copper	40 μg/dl	100-200 μg/dl
Serum ceruloplasmin	5 mg/dl	25-50 mg/dl
Urinary copper	200 μg/dl	< 25 μg/dl

Diagnosis and discussion

This boy had low serum level of copper and ceruloplasmin with an elevated copper excretion in urine. The clinical manifestations and the biochemical data indicate that the boy was suffering from ***Wilson's disease***. This disorder is associated with the deposition of copper in liver, kidney and brain.

III. Cases Related to Organ Function

CASE STUDY 18

An 18-year old male medical student had the complaints of loss of appetite (anorexia), nausea, headache and malaise. On clinical examination, his liver was found to be slightly enlarged. He was passing pale coloured stools but dark coloured urine. The data of his laboratory investigations are as follows.

Parameter (serum)	Subject	Reference range
Bilirubin (total)	8 mg/dl	0.2-1 mg/dl
(van den Bergh reaction	Biphasic	Negative)
Conjugated bilirubin	4.5 mg/dl	0.4 mg/dl
Unconjugated bilirubin	3.5 mg/dl	0.6 mg/dl
Alanine transaminase (ALT)	180 U/l	10-45 U/l
Aspartate transaminase (AST)	80 U/l	10-50 U/l
Alkaline phosphatase (ALP)	10 U/l	50-100 U/l

Comments and diagnosis

A biphasic van den Bergh positive reaction (elevated serum conjugated and unconjugated bilirubin), along with increased activity of ALT and AST clearly indicates that this is a case of ***hepatic jaundice***. This is due to impairment in liver cell function caused by viral infection (viral hepatitis), poisons and toxins or cirrhosis of liver. Damage to the liver cells adversely affects the bilirubin uptake and its conjugation (hence conjugated and unconjugated bilirubins are elevated). The pale and yellow coloured stools are due to the absence of stercobilinogen while the dark coloured urine is due to urobilinogen.

CASE STUDY 19

A 20-year old man was treated for malaria a month ago. For the past one week, he was excreting slightly dark coloured urine and dark brown coloured stools. His laboratory findings are given next.

Parameter	Subject	Reference range
Bilirubin (total)	7 mg/dl	0.2-1 mg/dl
(van den Bergh reaction	Indirect positive	Negative)
Conjugated bilirubin	0.4 mg/dl	<0.4 mg/dl
Unconjugated bilirubin	6.6 mg/dl	<0.6 mg/dl
Alanine transaminase (ALT)	30 U/l	10-45 U/l
Aspartate transaminase (AST)	40 U/l	10-50 U/l
Alkaline phosphatase (ALP)	10 U/l	50-100 U/l

Diagnosis and discussion

Following a malarial attack, the man developed ***hemolytic jaundice***. This was due to an excessive breakdown of erythrocytes releasing hemoglobin. The heme degraded to bilirubin cannot be effectively conjugated, as a result unconjugated bilirubin in the serum was elevated. The laboratory data — indirect van den Bergh reaction positive with no increase in the activities of serum ALT, AST and ALP support the diagnosis of hemolytic jaundice. The urine was dark coloured due to increased excretion of urobilinogen while the feces were dark brown due to sterocobilinogen.

CASE STUDY 20

A 30-year old man had fever and abdominal pain. His stools were clay in colour and contained fat (steatorrhea). The man's urine was found to be dark in colour. His laboratory findings are given in the next page.

Parameter	Subject	Reference range
Bilirubin (total)	12 mg/dl	0.2-1 mg/dl
(van den Bergh reaction	Direct positive	Negative)
Conjugated bilirubin	11.5 mg/dl	<0.4 mg/dl
Unconjugated bilirubin	0.5 mg/dl	<0.6 mg/dl
Alanine transaminase (ALT)	70 U/l	10-45 U/l
Aspartate transminase (AST)	75 U/l	10-50 U/l
Alkaline phosphatase (ALP)	450 U/l	50-100 U/l

In addition, the urine gave a positive test for bile salts (Hay's test) and bile pigments (Fouchet's test).

Diagnosis and discussion

A direct van den Bergh reaction reflecting an increase in conjugated bilirubin, along with an elevation in serum alkaline phosphatase indicates that the man was suffering from ***obstructive jaundice (cholestasis)***. A marginal increase in the activity of

ALT and AST is also observed in this disorder. Cholestasis is due to an obstruction (caused by gall stones, tumors etc.) that prevents the passage of bile into the intestine. The urine was dark in colour due to the excretion of conjugated bilirubin while the stools were clay in colour due to the deficiency of stercobilinogen, a pigment derived from bilirubin. Feces contained excess fat indicating impairment in fat digestion and absorption due to the absence of bile sats.

CASE STUDY 21

A 50-year old man, an industrialist by profession, had a routine medical check up. He was found to have slightly enlarged liver. His laboratory findings are given below.

Parameter	Subject	Reference range
Blood glucose, fasting	100 mg/dl	70-100 mg/dl
Blood urea	30 mg/dl	15-40 mg/dl
Serum bilirubin	0.8 mg/dl	< 1 mg/dl
Serum alanine transaminase	250 U/l	10-45 U/l
Serum aspartate transaminase (AST)	200 U/l	10-50 U/l
Serum γ-glutamyl transpeptidase (GGT)	700 U/l	5-50 U/l

Comments and diagnosis

The biochemical data indicate that this man had chronic liver disease. This is evident from the highly elevated activity of ALT, AST and GGT in the serum. These 3 enzymes leak into the circulation due to hepatocellular damage. The increased GGT indicates that the man was suffering from ***alcoholic hepatitis***. This can be confirmed by liver biopsy.

CASE STUDY 22

A 55-year old man was brought to the hospital with severe chest pain, breathlessness and vomiting. He could be rushed to the city hospital 5 hours after the onset of chest pain. His blood was immediately drawn, and the laboratory data are given below.

Parameter	Subject	Reference range
Creatine phosphokinase (CPK)	600 U/l	50-200 U/l
Aspartate transaminase (AST)	85 U/l	10-50 U/l
Lactate dehydrogenase (LDH)	315 U/l	50-200 U/l

Further laboratory evaluation revealed that the isoenzymes CPK_2(MB) and LDH_1 were highly elevated.

Diagnosis and discussion

The man obviously had an attack of ***myocardial infarction***. This is supported by the laboratory findings — elevation in CPK (MB), LDH (isoenzyme LDH_1) and aspartate transaminase. It may be noted that CPK starts to rise at 4-6 hours and reaches a peak value within 24-30 hours. As regards LDH and AST, they rise from the second day onwards. Thus, CPK is the earliest marker enzyme for the diagnosis of myocardial infarction.

CASE STUDY 23

A 50-year old man, an occasional alcoholic, had the complaint of severe abdominal pain following a large meal or alcohol intake. He also had the symptoms of nausea and vomiting. His laboratory data are given hereunder.

Parameter	Subject	Reference range
Serum amylase	650 U/l	30-150 U/l
Urine amylase	1350 U/l	200-800 U/l

Diagnosis and discussion

Elevation in the amylase activity is an indication of ***pancreatitis***. It may be noted that serum amylase activity is elevated on the first day of the disease and falls rapidly due to renal clearance. Very frequently, the urinary amylase activity is more important for the diagnosis of pancreatitis.

CASE STUDY 24

A 70-year old man had the complaints of frequent urination and pain in the lower abdomen. He had a sensation of obstruction in his urine flow. His rectal examination revealed the presence of hard mass. The laboratory findings of the man are as follows.

Parameter	Subject	Reference range
Serum acid phosphatase	30 U/l	0.5-5 U/l
Serum alkaline phosphatase	60 U/l	50-100 U/l
Serum prostatic specific antigen (PSA)	8 ng/ml	< 4 ng/ml

Comments and diagnosis

The old man with an elevation in serum acid phosphatase and PSA, and rectal examination with prostate hard mass was suffering from ***cancer of prostate gland***. It may be noted that estimation of serum PSA (elevated) is highly specific for prostate carcinoma.

CASE STUDY 25

A 4-year old male child was passing less quantity of urine that was red in colour. He was found to have marked edema around the eyes. Enquiries revealed that the child had the history of repeated upper respiratory tract infections during the past 3 years. Microscopic examination of urine showed erythrocytes, leucocytes, epithelial cells and granular casts. The child had almost normal concentrations of blood urea, serum creatinine, serum total proteins and albumin. His 24 hour urine contained 1000 mg protein (normal < 150 mg).

Comments and diagnosis

This child with oliguria (low urine output), hematuria and proteinuria, with characteristic erythrocyte casts in urine and marked edema around the eyes represents a case of ***acute glomerulonephritis***. This is supported by the history of respiratory tract infections. In most cases, acute glomerulonephritis is reversible.

CASE STUDY 26

A 20-year old man had generalized edema of the body with puffiness of the face in the mornings. His laboratory findings are given below.

Parameter	Subject	Reference range
Serum total proteins	4.5 g/dl	6-8 g/dl
albumin	1.5 g/dl	3.5-5 g/dl
globulins	3.0 g/dl	2.5-3.5 g/dl
Serum cholesterol	350 mg/dl	150-200 mg/dl
Blood urea	30 mg/dl	15-40 mg/dl
Serum creatinine	1.2 mg/dl	0.5-1.5 mg/dl
Urinary proteins	15 g/day	< 100 mg/day

The serum electrophoresis of the subject showed a sharp and prominent α_2-globulin band.

Diagnosis and discussion

This is a case of ***nephrotic syndrome***, characterized by generalized edema heavy proteinuria and hypoproteinemia (particularly low albumin). Increased serum cholesterol level in this disorder is quite common. The serum electrophoretic pattern (a prominent α_2-globulin band) supports the diagnosis of nephrotic syndrome. Generalized edema is due to low plasma colloidal osmotic pressure, as a result of reduced albumin concentration.

IV. Cases Related to Endocrine Function

CASE STUDY 27

A 30-year old woman, married five years ago, had no children. She complained of tiredness, weight gain, cold intolerance, neuromuscular pains and constipation. She was found to be anemic. Her laboratory data are given hereunder.

Parameter (serum)	Subject	Reference range
Triiodothyronine (T_3)	100 ng/dl	110-180 ng/dl
Thyroxine, total (T_4)	4 μg/dl	5-12 μg/dl
Thyroid stimulating hormone (TSH)	50 μU/ml	0.5-4 μU/ml
Cholesterol	250 mg/dl	150-200 mg/dl

Comments and diagnosis

The woman with the described clinical manifestations, elevated serum TSH and marginally lower T_4 and T_3 was a case of ***hypothyroidism***. Serum cholesterol level is also increased in hypothyroidism. (Note : This has no diagnositic importance, since serum cholesterol gets elevated in several other diseases e.g. diabetes, nephrotic syndrome).

CASE STUDY 28

A 35-year old woman complained of weight loss, despite normal or even increased appetite. She was found to have the symptoms of nervousness, irritability, anxiety, diarrhea, sweating, sensitive to heat and had increased pulse rate (tachycardia). Her laboratory findings are given below.

Parameter (serum)	Subject	Reference range
Triiodothyronine (T_3)	350 ng/dl	110-180 ng/dl
Thyroxine, total (T_4)	14 μg/dl	5-12 μg/dl
Thyroid stimulating hormone (TSH)	0-0.1 μU/ml	0.5-4 μU/ml

Comments and diagnosis

The clinical manifestations of the woman, along with reduced serum TSH and increased T_3 and T_4 indicate that the woman was suffering from ***hyperthyroidism***.

CASE STUDY 29

A 50-year old post-menopausal woman had the symptoms of lethargy, depression, muscle weakness, loss of memory, loss of appetite, constipation, polyuria and polydipsia. Over the past 10 years, she had 2 kidney stones that passed spontaneously. Chemical analysis of the stones revealed that calcium phosphate was the major constituent. The following are the laboratory findings of this subject.

Parameter (serum)	*Subject*	*Reference range*
Calcium	13 mg/dl	9-11 mg/dl
Phosphate	2.5 mg/dl	3-4.5 mg/dl
Chloride	110 mEq/l	95-105 mEq/l
Alkaline phosphatase	110 U/l	50-100 U/l

Comments and diagnosis

The clinical manifestations and the laboratory data — serum calcium elevation with decreased phosphate and a marginal increase in chloride and alkaline phosphatase strongly suggest that the woman was suffering from ***hyperparathyroidism***. This diagnosis was further supported from the fact that the woman had a recurrent stone formation (containing calcium and phosphate).

CASE STUDY 30

Oral glucose tolerance test (mini GTT) was performed on a 45-year old obese women with a family history of diabetes mellitus (her mother was a diabetic patient). The following are the laboratory data given.

	Fasting	*2 hr*
Blood glucose (mg/dl)	120	160
(Normal	<110	<140)
Urine glucose	–ve	–ve
Urine ketone bodies	–ve	–ve

Comments and diagnosis

The woman had marginally elevated fasting blood glucose and ***impaired oral glucose tolerance test***. Although she cannot be labelled as a diabetic patient at present, she requires careful periodical check up.

CASE STUDY 31

A 13-year old boy had the complaints of increased frequency of urination, and urinary tract infection. On routine examination, his urine was found to contain glucose and ketone bodies. He had a random blood glucose concentration 190 mg/dl. His laboratory data on the oral glucose tolerance test (mini GTT) are given below.

	Fasting	*2 hr*
Blood glucose (mg/dl)	150	240
(Normal	< 110	< 140)
Urine glucose	–ve	++
Ketone bodies	+	++

Diagnosis and discussion

The laboratory data of the boy-increase in blood glucose coupled with urinary excretion of glucose and ketone bodies, along with the results obtained in the OGTT point out that this is a good example of ***juvenile onset diabetes mellitus*** (type I diabetes or insulin dependent diabetes mellitus, IDDM). IDDM, which mainly occurs in childhood, is characterized by almost total deficiency of insulin due to the destruction of β-cells of pancreas. The boy has to be treated by insulin administration.

CASE STUDY 32

A 45-year old woman visited her physician with complaints of increased appetite and thirst with high frequency of urination. She complained of generalized weakness, fatigue and sweating. Her fasting blood glucose level was 160 mg/dl, with no presence of glucose or ketone bodies in urine. Her laboratory findings on the oral glucose tolerance test (mini GTT) are as follows.

	Fasting	*2 hr*
Blood glucose (mg/dl)	155	205
(Normal Reference	<110	<140)
Urine glucose	–ve	++
Ketone bodies	–ve	–ve

Diagnosis and discussion

The woman's fasting (elevated) blood glucose and the data obtained on OGTT indicate that she was suffering from ***non-insulin dependent diabetes mellitus*** (NIDDM or type II diabetes). NIDDM mainly occurs in adults (>30 yrs) with a strong genetic predisposition. The patients of NIDDM can be usually treated with oral hypoglycemic drugs. Before prescribing the drugs, diet control and exercise are tried.

CASE STUDY 33

A 55-year old woman, who had menopause at the age of 48, fell down in bathroom and fractured her leg. She informed the orthopaedician that she had been experiencing severe back pain for the past 2 years. She was not on hormonal replacement therapy (HRT). Radiographic studies revealed that her bone density was decreased. Her laboratory data are given below.

Parameter (serum)	Subject	Reference range
Calcium	9 mg/dl	9-11 mg/dl
Phosphate	3.5 mg/dl	3-4.5 mg/dl
Alkaline phosphatase	70 KAU/dl	3-13 KAU/dl

Comments and diagnosis

The post-menopausal woman was suffering from ***osteoporosis***, a common bone disease in the elderly (particularly woman). Osteoporosis is characterized by demineralization of bones resulting in progressive loss of bone mass (hence decreased bone density). Usually, the biochemical parameters in osteoporosis patients are in the normal range. But serum alkaline phosphatase may get elevated following a fracture. The etiology of osteoporosis is largely unknown. However, in post-menopausal woman, deficiency of sex hormones are implicated in the development of osteoporosis.

CASE STUDY 34

A 28-year old man complained constant headache and visual disturbance. He also had the symptoms of anxiety, tremors and was found to be hypertensive. His laboratory data are given below.

Parameter	Subject	Reference range
Serum epinephrine	10 ng/ml	0.1 ng/ml
Serum norepinephrine	26 μg/l	0.5 μg/l
Urine vanillylmandelic acid (VMA)	62 mg/day	<8 mg/day

Comments and diagnosis

Elevation in the serum levels of catecholamines (epinephrine and norepinephrine) along with an increased urinary excretion of their principal metabolite - VMA, supported by the clinical manifestations (anxiety, hypertension) indicates that this man was suffering from ***pheochromocytoma***, a tumor of adrenal medulla.

CASE STUDY 35

A 25-year old woman visited her gynecologist as soon as she missed the menstrual cycle. The presence of human chorionic gonadotropin (hCG) hormone could be detected in her urine.

Comments and diagnosis

Human chorionic gonadotropin is synthesized by the placenta. The levels of hCG in plasma and urine increase almost immediately after the implantation of fertilized ovum. Therefore, detection of hCG in urine is used for the ***detection of pregnancy***. Thus, this woman was pregnant, as confired by the hormonal test.

V. MISCELLANEOUS

CASE STUDY 36

A 55-year old man was brought to the hospital in a confused and semiconscious state. He had low BP and feeble pulse. His breath had fruity odour. The data of his laboratory investigations are given below.

Parameter	Subject	Reference range
Blood pH	7.1	7.35-7.45
Plasma bicarbonate (HCO_3^-)	12 mmol/l	22-26 mmol/l
Plasma carbonic acid (H_2CO_3)	1.2 mmol/l	1.2 mmol/l
Blood glucose (random)	580 mg/dl	<140 mg/dl
Blood urea	40 mg/dl	15-40 mg/dl
Serum creatinine	1.5 mg/dl	0.5-1.5 mg/dl
Urine sugar	4+	negative
Urine ketone bodies	3+	negative

Diagnosis and discussion

The blood pH was reduced due to a decrease in plasma bicarbonate (HCO_3^-) concentration. This is case of ***metabolic acidosis***. The blood glucose level was highly increased along with urinary excretion of glucose and ketone bodies. The breath of the subject was fruity due to exhalation of acetone. Obviously, this diabetic patient was not under control. The acidosis was due to the accumulation of ketone bodies. This may be appropriately regarded as ***diabetic ketoacidosis***.

CASE STUDY 37

One patient with severe vomiting and hyokalemia was admitted in an intensive care unit. His laboratory data and acid-base status are given below.

Parameter	*Subject*	*Reference range*
Blood pH	7.7	7.35-7.45
Plasma bicarbonate (HCO_3^-)	42 mmol/l	22-26 mmol/l
Plasma carbonic acid (H_2CO_3) (pCO_2 40 mm Hg)	1.2 mmol/l	1.2 mmol/l

Diagnosis and discussion

This patient with increased blood pH due to an elevation in the plasma bicarbonate concentration is a typical case of ***metabolic alkalosis***.

CASE STUDY 38

A 60-year old woman complained of back pain, non-responsive to drugs. The physical examination and enquiries revealed that a small marble-sized mass in the neck of the patient, with an existence of 15 years, had started increasing in size during the recent 4 months. However, there was no pain associated with this mass. But he had recurrent infections. Serum electrophoresis showed an abnormal protein pattern with a distinct band of monoclonal immunoglobin between β- and γ-globulins. The urine gave a positive test for Bence Jones proteins.

Comments and diagnosis

The characteristic electrophoretic pattern of serum proteins with an 'M' band along with the excretion of Bence Jones proteins in urine and the clinical manifestations led to the diagnosis that the patient was suffering from ***multiple myeloma***. This is due to the malignancy of a single clone of plasma cells in the bone marrow resulting in an abnormal production of immunoglobulins, mostly IgG (75%) and in some cases IgA (25%). The patients of multiple myeloma have depressed immunity, hence recurrent infections are common in them.

CASE STUDY 39

A 6-year old boy had high temperature with sweats during sleep, marked tachycardia and loss of weight. He also had the complaints of headache, neck stiffness and vomiting. The liver was found to be slightly enlarged. The following are the laboratory findings of this boy.

Parameter	*Subject*	*Reference range*
Cerebrospinal fluid (CSF)		
Glucose	30 mg/dl	50-75 mg/dl
Proteins	100 mg/dl	15-40 mg/dl
Chloride	105 mEq/l	120-130 mEq/l
Blood glucose	100 mg/dl	70-100 mg/dl
(simultaneously collected)		

Diagnosis and discussion

A decrease in the CSF glucose and chloride concentrations along with an increase in protein level indicates that the boy was a victim of ***tuberculosis meningitis***. This is supported by the clinical manifestations.

CASE STUDY 40

A 30-year old man was found to excrete glucose in urine in a random sample. He did not have any symptoms of diabetes mellitus. His fasting and post-prandial blood glucose respectively were 90 mg/dl and 140 mg/dl. Urine contained glucose but there were no ketone bodies. Oral glucose tolerance test (OGTT) also revealed that the man had a normal glucose tolerance except the presence of glucose in urine at 1 hr.

Comments and diagnosis

This is a case of ***renal glycosuria***. It is a benign condition due to reduced renal threshold for blood glucose (normal around 180 mg/dl). It may be noted that renal glycosuria is unrelated to diabetes mellitus and therefore not accompanied by the classical symptoms of diabetes. This disorder does not require any specific treatment.

INDEX

A

D

I

J

K

L

P

T

U

V

W

X

Z